Lecture Notes in Artificial Intelligence   2358
Subseries of Lecture Notes in Computer Science
Edited by J. G. Carbonell and J. Siekmann

Lecture Notes in Computer Science
Edited by G. Goos, J. Hartmanis, and J. van Leeuwen

Springer
*Berlin
Heidelberg
New York
Barcelona
Hong Kong
London
Milan
Paris
Tokyo*

Tim Hendtlass   Moonis Ali (Eds.)

Developments in Applied Artificial Intelligence

15th International Conference
on Industrial and Engineering Applications
of Artificial Intelligence and Expert Systems
IEA/AIE 2002
Cairns, Australia, June 17-20, 2002
Proceedings

Springer

Series Editors

Jaime G. Carbonell, Carnegie Mellon University, Pittsburgh, PA, USA
Jörg Siekmann, University of Saarland, Saarbrücken, Germany

Volume Editors

Tim Hendtlass
Swinburne University of Technology
Centre for Intelligent Systems and Complex Processes
John Street, Hawthorn, Victoria, Australia 3122
E-mail: thendtlass@swin.edu.au

Moonis Ali
Southwest Texas State University
Department of Computer Science
601 University Drive, San Marcos, TX 78666, USA
E-mail: ma04@swt.edu

Cataloging-in-Publication Data applied for

Die Deutsche Bibliothek - CIP-Einheitsaufnahme

Developments in applied artificial intelligence : proceedings / 15th
International Conference on Industrial and Engineering Applications of
Artificial Intelligence and Expert Systems, IEA/AIE 2002, Cairns, Australia,
June 17 - 20, 2002. Tim Hendtlass ; Moonis Ali (ed.). - Berlin ; Heidelberg ;
New York ; Barcelona ; Hong Kong ; London ; Milan ; Paris ; Tokyo : Springer, 2002
  (Lecture notes in computer science ; Vol. 2358 : Lecture notes in
artificial intelligence)
  ISBN 3-540-43781-9

CR Subject Classification (1998): I.2, F.1, F.2, I.5, F.4.1, D.2

ISSN 0302-9743
ISBN 3-540-43781-9 Springer-Verlag Berlin Heidelberg New York

This work is subject to copyright. All rights are reserved, whether the whole or part of the material is
concerned, specifically the rights of translation, reprinting, re-use of illustrations, recitation, broadcasting,
reproduction on microfilms or in any other way, and storage in data banks. Duplication of this publication
or parts thereof is permitted only under the provisions of the German Copyright Law of September 9, 1965,
in its current version, and permission for use must always be obtained from Springer-Verlag. Violations are
liable for prosecution under the German Copyright Law.

Springer-Verlag Berlin Heidelberg New York
a member of BertelsmannSpringer Science+Business Media GmbH

http://www.springer.de

© Springer-Verlag Berlin Heidelberg 2002
Printed in Germany

Typesetting: Camera-ready by author, data conversion by DA-TeX Gerd Blumenstein
Printed on acid-free paper   SPIN 10870180   06/3142   5 4 3 2 1 0

Preface

Artificial Intelligence is a field with a long history, which is still very much active and developing today. Developments of new and improved techniques, together with the ever-increasing levels of available computing resources, are fueling an increasing spread of AI applications. These applications, as well as providing the economic rationale for the research, also provide the impetus to further improve the performance of our techniques. This further improvement today is most likely to come from an understanding of the ways our systems work, and therefore of their limitations, rather than from ideas 'borrowed' from biology. From this understanding comes improvement; from improvement comes further application; from further application comes the opportunity to further understand the limitations, and so the cycle repeats itself indefinitely.

In this volume are papers on a wide range of topics; some describe applications that are only possible as a result of recent developments, others describe new developments only just being moved into practical application. All the papers reflect the way this field continues to drive forward. This conference is the 15th in an unbroken series of annual conferences on Industrial and Engineering Application of Artificial Intelligence and Expert Systems organized under the auspices of the International Society of Applied Intelligence. This series of conferences has played an important part in promulgating information about AI applications, and an inspection of proceedings of the conferences in this series shows just how both the magnitude and pace of development has increased over the past 15 years.

I would like to welcome delegates to this conference. I know that you will find much to learn from the papers as well as from the informal discussion that is so important a part of every conference.

Acknowledgements

People who have not organized an international conference often have little idea of all that is involved. In the case of this conference the first call for papers was issued at IEA/AIE 2001, a year ago. Papers were submitted electronically at a dedicated conference website. A technical committee of some 50 referees were recruited, lodged their details on the conference website, and were issued with passwords and special access privileges. Over 150 papers from 28 countries were each matched up with at least 2 suitable referees, who downloaded an electronic copy of the whole paper and then lodged their review electronically. The papers to be accepted were selected using these responses, and emails advising them of the acceptance of their paper were sent to the authors. These emails also gave them the reviewers' comments and advised of the format requirements for the final modified paper. The final files were collected, again at the website, before being sent to form the proceedings that you are holding.

Clearly, without the website, the logistics of handing so much correspondence from all over the world would have been much, much worse, and so it is with deep gratitude that I acknowledge the efforts of Clinton Woodward who designed and programmed the website which, together with all the software behind it, was able to either automate or substantially assist with the whole process.

While I am grateful to all the members of the organizing committee, I wish to especially acknowledge contributions to the conference organization from David Braendler, David Liley, and Graham Forsyth together with the entire program committee.

And, last but not least, I would like to thank all our authors and speakers, without whom this conference would have no reason to exist.

April 2002   Tim Hendtlass

Table of Contents

Neural Networks 1

An Error Back-Propagation Artificial Neural Networks Application in
Automatic Car License Plate Recognition 1
Demetrios Michalopoulos and Chih-Kang Hu

Use of Artificial Neural Networks for Buffet Loads Prediction 9
Oleg Levinski

Computational Cost Reduction by Selective Attention for Fast Speaker
Adaptation in Multilayer Perceptron 17
In-Cheol Kim and Sung-Il Chien

A Comparison of Neural Networks with Time Series Models
for Forecasting Returns on a Stock Market Index 25
Juliana Yim

Image/Speech 1

Automatic Detection of Film Orientation with Support Vector Machines ... 36
Dane Walsh and Christian Omlin

A Generic Approach for the Vietnamese Handwritten
and Speech Recognition Problems 47
*Vu Hai Quan, Pham Nam Trung, Nguyen Duc Hoang Ha, Lam Tri Tin,
Hoang Kiem, and An H Nguyen*

Efficient and Automatic Faces Detection Based on Skin-Tone
and Neural Network Model 57
Bae-Ho Lee, Kwang-Hee Kim, Yonggwan Won, and Jiseung Nam

Efficient Image Segmentation Based on Wavelet and Watersheds
for Video Objects Extraction 67
Jong-Bae Kim and Hang-Joon Kim

Evolutionary and Genetic Algorithms 1

FPGA-Based Implementation of Genetic Algorithm
for the Traveling Salesman Problem and Its Industrial Application 77
Iouliia Skliarova and António B. Ferrari

Minimal Addition Chain for Efficient Modular Exponentiation
Using Genetic Algorithms 88
Nadia Nedjah and Luiza de Macedo Mourelle

Course Scheduling Using Genetic Algorithm Methods 99
Raed Abu Zitar

An Evolutionary Algorithm for the Synthesis of RAM-Based FSMs 108
Valery Sklyarov

Genetic Algorithms for Design of Liquid Retaining Structure 119
K. W. Chau and F. Albermani

Autonomous Agents

Modelling Crew Assistants with Multi-Agent Systems
in Fighter Aircraft 129
Arjen Vollebregt, Daan Hannessen, Henk Hesselink, and Jelle Beetstra

Learning from Human Decision-Making Behaviors –
An Application to RoboCup Software Agents 136
Ruck Thawonmas, Junichiro Hirayama, and Fumiaki Takeda

Distributed Deadlock Detection in Mobile Agent Systems 146
Bruce Ashfield, Dwight Deugo, Franz Oppacher, and Tony White

An Agent-Based Approach for Production Control Incorporating
Environmental and Life-Cycle Issues, together with Sensitivity Analysis ... 157
Elisabeth Ilie Zudor and László Monostori

Feasibility Restoration for Iterative Meta-heuristics Search Algorithms 168
Marcus Randall

Best Paper Candidates 1

Optimization of Pulse Pattern for a Multi-robot Sonar System
Using Genetic Algorithm 179
*George Nyauma Nyakoe, Makoto Ohki, Suichiro Tabuchi,
and Masaaki Ohkita*

The Suitability of Particle Swarm Optimisation
for Training Neural Hardware 190
David Braendler and Tim Hendtlass

Evolutionary Multi-objective Integer Programming
for the Design of Adaptive Cruise Control Systems 200
Nando Laumanns, Marco Laumanns, and Hartmut Kitterer

The Macronet Element: A Substitute for the Conventional Neuron 211
Tim Hendtlass and Gerrard Murray

Genetic Algorithm Optimisation of Mathematical Models
Using Distributed Computing 220
S. Dunn and S. Peucker

Best Paper Candidates 2

Genetic Algorithm Optimisation of Part Placement
Using a Connection-Based Coding Method 232
*Alan Crispin, Paul Clay, Gaynor Taylor, Robert Hackney, Tom Bayes,
and David Reedman*

A Fast Evolutionary Algorithm for Image Compression in Hardware 241
Mehrdad Salami and Tim Hendtlass

Automatic Speech Recognition: The New Millennium 253
Khalid Daoudi

Applying Machine Learning for Ensemble Branch Predictors 264
Gabriel H. Loh and Dana S. Henry

A Customizable Configuration Tool for Design of Multi-part Products 275
Niall Murtagh

Neural Networks 2

Phase-to-Phase Wave Parameters Measurement
of Distribution Lines Based on BP Networks 284
Fengling Han, Xinghuo Yu, Yong Feng, and Huifeng Dong

Learning Capability: Classical RBF Network vs. SVM
with Gaussian Kernel 293
Rameswar Debnath and Haruhisa Takahashi

Trading off between Misclassification, Recognition and Generalization
in Data Mining with Continuous Features 303
Dianhui Wang, Tharam Dillon, and Elizabeth Chang

Interacting Neural Modules 314
Garry Briscoe

The Application of Visualization and Neural Network Techniques
in a Power Transformer Condition Monitoring System 325
Zhi-Hua Zhou, Yuan Jiang, Xu-Ri Yin, and Shi-Fu Chen

Internet Applications 1

Entrepreneurial Intervention in an Electronic Market Place 335
John Debenham

Intelligent Auto-downloading of Images 346
Vikram Natarajan and Angela Goh

Intelligent Facilitation Agent
for Online Web-Based Group Discussion System 356
*Junalux Chalidabhongse, Wirat Chinnan, Pichet Wechasaethnon,
and Arpakorn Tantisirithanakorn*

TWIMC: An Anonymous Recipient E-mail System 363
Sebon Ku, Bogju Lee, and Dongman Lee

Mental States of Autonomous Agents in Competitive
and Cooperative Settings 373
Walid S. Saba

Expert Systems

An Expert System Application for Improving Results
in a Handwritten Form Recognition System 383
Silvana Rossetto, Flávio M. Varejão, and Thomas W. Rauber

A Knowledge-Based System for Construction Site Level Facilities Layout .. 393
K.W. Chau and M. Anson

A Decision-Support System to Improve Damage Survivability
of Submarine 403
D. Lee, J. Lee, and K. H. Lee

On the Verification of an Expert System: Practical Issues 414
Jorge Santos, Zita Vale, and Carlos Ramos

DOWNSIZINGX: A Rule-Based System for Downsizing
the Corporation's Computer Systems 425
J. L. Mitrpanont and T. Plengpung

Credit Apportionment Scheme for Rule-Based Systems:
Implementation and Comparative Study 435
N. M. Hewahi and H. Ahmad

Internet Applications 2

An Adaptive Web Cache Access Predictor Using Neural Network 450
Wen Tian, Ben Choi, and Vir V. Phoha

A Designated Bid Reverse Auction
for Agent-Based Electronic Commerce 460
Tokuro Matsuo and Takayuki Ito

Design of a Fuzzy Usage Parameter Controller for Diffserv and MPLS 470
K. K. Phang, S. H. Lim, M. Hj. Yaacob, and T. C. Ling

A Tool for Extension
and Restructuring Natural Language Question Answering Domains 482
Boris Galitsky

Effective Retrieval of Information in Tables on the Internet 493
Sung-Won Jung, Kyung-Hee Sung, Tae-Won Park, and Hyuk-chul Kwon

Evolutionary and Genetic Algorithms 2

A Fitness Estimation Strategy for Genetic Algorithms 502
Mehrdad Salami and Tim Hendtlass

Derivation of L-system Models from Measurements
of Biological Branching Structures Using Genetic Algorithms 514
Bian Runqiang, Phoebe Chen, Kevin Burrage, Jim Hanan, Peter Room, and John Belward

Evolving a Schedule with Batching, Precedence Constraints, and
Sequence-Dependent Setup Times: Crossover Needs Building Blocks 525
Paul J. Darwen

The Development of the Feature Extraction Algorithms
for Thai Handwritten Character Recognition System 536
J. L. Mitrpanont and S. Kiwprasopsak

Route Planning Wizard: Basic Concept and Its Implementation 547
Teruaki Ito

AI Applications

The Design and Implementation of Color Matching System Based
on Back Propagation 557
HaiYi Zhang, JianDong Bi, and Barbro Back

Component-Oriented Programming as an AI-Planning Problem 567
Debasis Mitra and Walter P. Bond

Dynamic CSPs for Interval-Based Temporal Reasoning 575
Malek Mouhoub and Jonathan Yip

Efficient Pattern Matching of Time Series Data 586
Sangjun Lee, Dongseop Kwon, and Sukho Lee

A Multi-attribute Decision-Making Approach toward Space System Design
Automation through a Fuzzy Logic-Based Analytic Hierarchical Process .. 596
Michelle Lavagna and Amalia Ercoli Finzi

Best Paper Candidates 3

A Case Based System for Oil and Gas Well Design 607
Simon Kravis and Rosemary Irrgang

Ant Colony Optimisation Applied to a Dynamically Changing Problem ... 618
Daniel Angus and Tim Hendtlass

A GIS-Integrated Intelligent System for Optimization
of Asset Management for Maintenance of Roads and Bridges 628
M. D. Salim, T. Strauss, and M. Emch

A Unified Approach for Spatial Object Modelling
and Map Analysis Based on 2nd Order Many-Sorted Language 638
Oscar Luiz Monteiro de Farias and Sueli Bandeira Teixeira Mendes

Training and Application of Artificial Neural Networks
with Incomplete Data 649
Zs. J. Viharos, L. Monostori, and T. Vincze

Knowledge Processing

Message Analysis for the Recommendation of Contact Persons
within Defined Subject Fields 660
Frank Heeren and Wilfried Sihn

An Intelligent Knowledge Processing System on Hydrodynamics
and Water Quality Modeling 670
K. W. Chau, C. Cheng, Y. S. Li, C. W. Li, and O. Wai

Uncertainty Management and Informational Relevance 680
M. Chachoua and D. Pacholczyk

Potential Governing Relationship and a Korean Grammar Checker
Using Partial Parsing 692
Mi-young Kang, Su-ho Park, Ae-sun Yoon, and Hyuk-chul Kwon

Image/Speech 2

On-Line Handwriting Character Recognition Using Stroke Information 703
Jungpil Shin

Face Detection by Integrating Multiresolution-Based Watersheds
and a Skin-Color Model 715
Jong-Bae Kim, Su-Woong Jung, and Hang-Joon Kim

Social Interaction of Humanoid Robot Based on Audio-Visual Tracking ... 725
Hiroshi G. Okuno, Kazuhiro Nakadai, and Hiroaki Kitano

Hybrid Confidence Measure for Domain-Specific Keyword Spotting 736
Jinyoung Kim, Joohun Lee, and Seungho Choi

Model Based Reasoning

Model-Based Debugging or How to Diagnose Programs Automatically 746
Franz Wotawa, Markus Stumptner, and Wolfgang Mayer

On a Model-Based Design Verification
for Combinatorial Boolean Networks 758
Satoshi Hiratsuka and Akira Fusaoka

Optimal Adaptive Pattern Matching 768
Nadia Nedjah and Luiza de Macedo Mourelle

Analysis of Affective Characteristics and Evaluationof Harmonious Feeling
of Image Based on *1/f* Fluctuation Theory 780
Mao Xia, Chen Bin, Zhu Gang, and Muta Itsuya

Adaptive Control

Collective Intelligence and Priority Routing in Networks 790
Tony White, Bernard Pagurek, and Dwight Deugo

An Agent-Based Approach to Monitoring and Control
of District Heating Systems 801
Fredrik Wernstedt and Paul Davidsson

Using Machine Learning to Understand Operator's Skill 812
Ivan Bratko and Dorian Šuc

Reactive Load Control of Parallel Transformer Operations
Using Neural Networks 824
Fakhrul Islam, Baikunth Nath, and Joarder Kamruzzaman

Author Index 831

An Error Back-Propagation Artificial Neural Networks Application in Automatic Car License Plate Recognition

Demetrios Michalopoulos and Chih-Kang Hu

Department of Computer Science, California State University
Fullerton, California 92831

Abstract. License plate recognition involves three basics steps: 1) image preprocessing including thresholding, binarization, skew detection, noise filtering, and frame boundary detection, 2) character and number segmentations from the heading of the state area and the body of a license plate, 3) training and recognition on an Error Back-propagation Artificial Neural Networks (ANN). This report emphasizes on the implementation of modeling the recognition process. In particular, it deploys classical approaches and techniques for recognizing license plate numbers. The problems of recognizing characters and numbers from a license plate are described in details by examples. Also, the character segmentation algorithm is developed. This algorithm is then incorporated into the license plate recognition system.

1 Introduction

The purpose of this project is to implement license plate recognition application using Artificial Neural Networks. This application model performs trainings on image files in BMP format of license plates and stores all the features into a database. This system then performs the recognition process on license plate image based on the training database

The scope of the project utilizes two major computer techniques —Document Image Analysis and Artificial Neural Networks. Document Image Analysis focus on license plate image analysis, which contain global thresholding, binarization, skew detection, noise filtering, frame boundary detection, and feature extraction for generating features of an image.

After certain features are extracted from the image preprocessing processes, Error Back-propagation ANN performs training processes on these image features to formula classification neural networks.

2 Preprocessing of License Plate Image

2.1 Global Thresholding

A captured license plate image file stores color intensity in pixels. These picture elements are the raw data for document analysis. Depend on number of bits representing in a pixel and color map in a BMP file, each pixel in the BMP file is read for further image processing [1].

In figure 2.1, the example shows the histogram of a 4-bit image. The histogram is plotted based on the intensity of 16 colors. The global threshold value in the project is to evaluate two maximum values, which separate two high-density areas from the histogram. Color Histogram of a 4-bit image file In this example, the thresholding area is around color indexes 12 and 13.

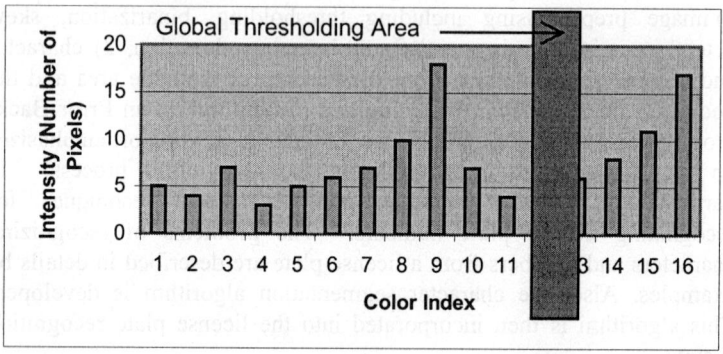

Fig. 2.1 Color Histogram of a 4-bit image file

There are some other ways for implementing thresholding to determine a better threshold value. O'Gorman[2] describes several adaptive thresholding techniques. Once the thresholding value is generated, this value then decides the foreground and background of an image data. However, an image file may not have a clear contrast. A possibility solution to a poor contrast of image is to adjust contrast level during receiving or scanning the image.

2.2 Noise Filtering

This project adopts the 3 * 3 window filtering algorithm for filtering noise of an image file. The 3 * 3 window filter can only filter out the noise such as single-pixel noise and single-pixel hole. However, for noise larger than one-pixel can apply kFill noise reduction algorithm [2].

The algorithm works as follows: the centroid walks through all the pixels from left to right and top to bottom. At any moment, if the centroid pixel value is different from all the surrounding neighborhood pixels, the centroid pixel value flips to be the same value as surrounding pixels. If one of the surrounding pixels is outside of the image pixels, set the out off boundary surrounding pixels to the opposite value of the centroid. In step 2 of figure 2.2, the upper three window pixels are set to OFF pixels, and so the centroid is flipped into an OFF pixel.

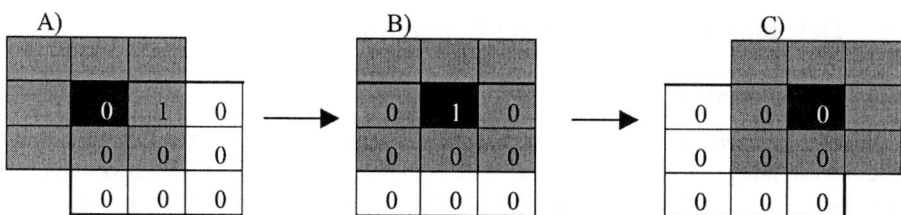

Fig. 2.2 A 3 * 3 window scanning through a 3 * 3 pixel grids where the centroid is shown in black

The filtering result applying 3 * 3 window filtering is shown in figure 2.3 in which noises in the image of a character E are filtered

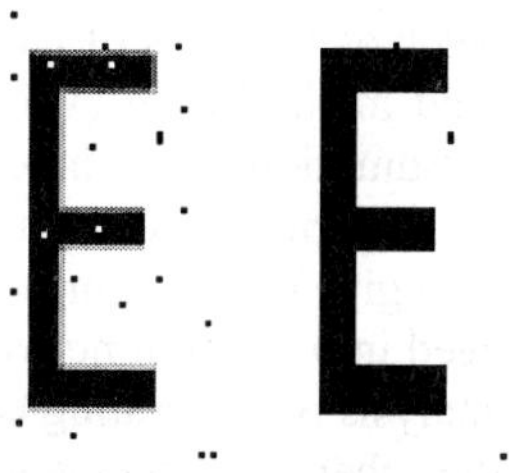

Fig. 2.3 Two image files before and after applying window filtering

2.3 Skew Detection

In many cases, the captured image files or licenses are not proper aligned. This problem may decrease the reorganization rate. It is important to perform skew detection and correction before extracting features. This project deploys the Projection Profile technique [4] to correct a skewed image.

Basically, the most straightforward use of the projection profile for skew detection is to compute it at a number of angles close to the expected orientation [2]. As shown in figure 2.4, an image file is rotated from the skew angle 0 to 15. The result in table 2.1 shows that the number of ON pixels on X-histogram and Y-histogram has the maximum peak, 3932, when the image is proper aligned (skew angle = 0).

Fig. 2.4 An image file rotated 15 degrees

Table 2.1 A Project Profile histogram on an image with the sum of X and Y

Skew Angel	0	2	4	6	8	10	12	14
Number of ON Pixels (X+Y)	3932	3920	3918	3918	3918	3920	3920	3918

3 Development of Character Segmentation Technique

3.1 Layout Analysis

After skew correction, a license image is proper aligned. The structural analysis on the license image is then performed. The layout analysis is done by performing the top-down analysis. Basically, this analysis tries to isolate structural blocks of license frame boundaries, state area, and body area.

The top-down analysis utilizes horizontal and vertical project profiles for finding the structural blocks. For some cases, license plates may have license frames which the frames' color are similar to the license body color. This problem can be solved in two ways: 1) contextual analysis, 2) projection profile.

Contextual analysis is to find the features of the rectangular frame. The features should contain coded contours such as two sets of parallel straight lines [5] and four 90-degree curves [6]. The second method is to evaluate the vertical and horizontal projection profile peaks. The assumption made here is that a frame has very high connectivity in turns of the same color pixels. Once a frame, if there is any, is found, the rest of project profile peaks can give us the information of state and body areas.

The entire state area is later feed into a neural network classifier. The body text of the license plate needs further analysis on separating each character. Separating each character can be done by the character segmentation algorithm described in section 3.2.

3.2 Character Segmentation Algorithm

The purpose of character segmentation is to find out the isolate character in a license plate by analyzing the peaks and valleys on the vertical and horizontal projection profiles. The major problem on examining the projection profiles is the determination of threshold values discriminating noise and text.

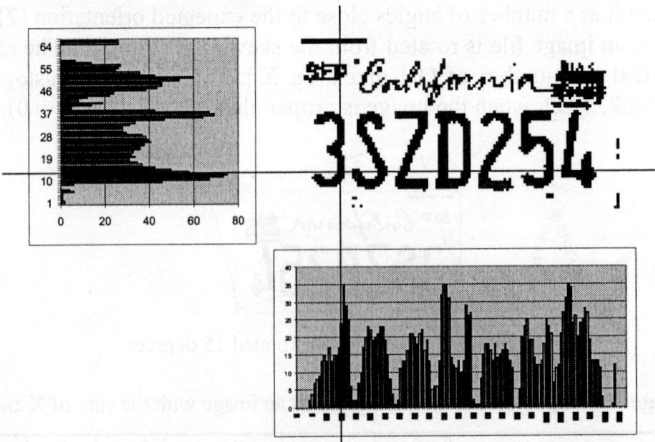

Fig. 3.1 Cross section of the highest vertical and horizontal projection profile peaks

The character segmentation algorithm is based on the assumption that 1) all the characters in a license body are capitalized, 2) each character or number has rectangular shape, 3) all the characters are isolated characters. Based on the assumption, the cross section on the highest horizontal and vertical projection profile peaks points to one of the character in the body area. By starting at this point, this algorithm expends the boundaries outward literally until the character or number is confined in the block. The algorithm walks through all the vertical projection profile peaks until all the characters and numbers are found. The algorithm is described as following:

Step 1: Set Xmin = maximun value in X-histogram
Set Ymin = maximun value in Y-histogram
Set Xman = maximun value in X-histogram
Set Yman = maximum value in Y-histogram

Step 2: Let window ω (Xmin, Xmax, Ymin, Ymax) the starting point and ω is an ON pixel.

Step 3: Starting at W and searching for the directions of left, right, top, and down for all the pixels having same value as ω

Step 4: Once all the directions reach the OFF pixels,
Set Xmin = left
Xmax = right
Ymin = top

Ymax = down

Step 5: Expend the window ω for all the ON pixels, until there is no ON pixels can be found goto step 6,
Otherwise goto step 3.

Step 6: Scan through entire image from left to right based on the range of Ymin and Ymax.

Step 7: Apply same algorithm for step 2, 3, 4, and 5 for generating ω_i where i is the number of significant features in the image

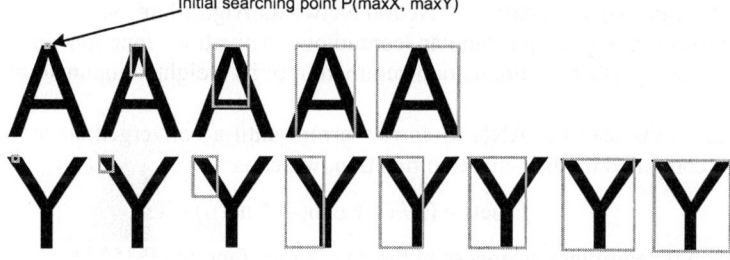

In worse case, this algorithm has the efficiency (finding a character) of $O((n^2+m^2)^{1/2}) \approx O(n)$ where m = width of a character and n = height of the character

4 Implementation of Error Back-Propagation Artificial Neural Networks

4.1 Setup the Error Back-Propagation ANN Model

Three necessary elements for constructing an ANN contain the architecture of the neural networks, activation function, and learning rule. This project adopts the nonparametric training technique with delta learning rule for constructing the multiperceptron layer neural networks. The activation function used in this project is bipolar continuous activation function

In this project, the architecture of the neural networks contains three layers—input layer, hidden layer, and output layer. The input layer contains 106 (7 * 15 pixels + 1 fixed input) input nodes mapped into 36 possible solutions (26 characters + 10 numeric values). For 36 possible solutions, the number of output nodes is 6, since $\log_2^n >= 36$ where n is an integer, and n = 6. The hidden layer contains 112 (106 input + 6 output nodes) neurons. The architecture of the neural networks is shown in figure 4.1 where $i = 106, j = 142$, and $k = 6$.

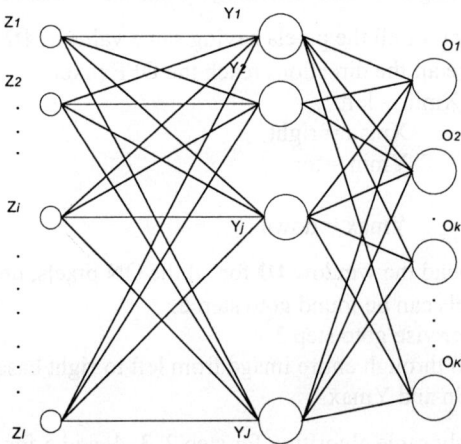

Fig. 4.1 Error Back-Propagation Neural Networks

Error Back-propagation Artificial Neural Networks Algorithm

This project deploys the bipolar continuous activation function in which a processing node performs computation summation of its weighted inputs to obtain net results.

The net results lead the ANN to correct errors until a converged stage. The net results are computed by the activation function, which is

$$f'(net) = (2 / (1 + \exp(-\lambda * net))) - 1$$

where $\lambda > 0$ determines steepness of the continuous function [3].

This project also deploys the delta learning rule for the utilization of continuous activation function. Basically, the processing nodes are trained based on the desired outputs. This supervised learning algorithm minimizes the error and converge to the stage of separating all the classes. Figure 4.2 shows the model of the classifier

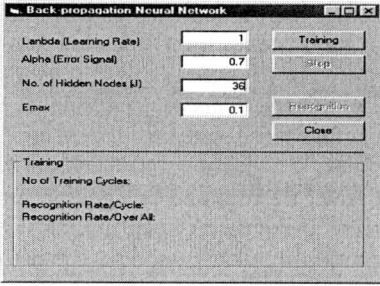

Fig. 4.2 The Error Back-propagation Neural Network Classifier

Finally, the classification algorithm for this project is described as follows:

ω_1 is the weight matrix of layer 1 and layer 2 where $\omega_1 = (J * I)$.

ω_2 is the weight matrix of layer 2 and layer 3 where $\omega_2 = (K * J)$.

O_1 is the output vector with size J of layer
O_2 is the output vector with size K of layer 3.
D is the desired output vector (known answers).

δ_1 is the error matrix of layer 1 and layer 2.

δ_2 is the error matrix of layer 2 and layer 3.
Z is the input matrix of training samples where $Z = (I * K)$
η is the learning constant and $\eta > 0$
E_{max} is the maximum tolerate error where $0 < E_{max} < 1$
$p \leftarrow 0$ where p is the counter of trained samples
$q \leftarrow 0$ where q is the total counter of trained samples
P is the total number of training samples.
$E \leftarrow 0$ where E is the total training error

```
InitializeW
p = 0 : q = 0 : E = 0
While E < Emax
    While p < P
        O₁ ← f(ω₁ * Z)
        O₂ ← f(ω₂ * O₁) 'Weighted sum of layer 2 and 3
        E = E + ((O₂ - D)² ) / 2
        δ₂= (D - O₂) * (1 - O₂²) / 2
        δ₁= (1 - O₁²) * (δ₂ * ω₂) / 2
        ω₂ = ω₂ + η (δ₂ * O₁)
        ω₁= ω₁+ η (δ₁* Z)
        p = p + 1 : q = q + 1
    Loop
    If (E > Emax)
        E = 0 : p = 0
Loop
```

5   Conclusions

The license plate recognition application can be used in human's life in many aspects—registration tracking systems by DMV, toll road services by toll road companies, parking fee collecting applications by many organizations, and car theft detection systems by police officers. This project provides practical approaches and techniques for the build-up of a license plate recognition application.

This is an on going project for gathering statistics on the comparison results of image preprocessing and ANN training and recognition processes. There are also some future works to refine the projects such as applying adaptive thresholding techniques, kFill noise reduction algorithm, and image depth detection and correction. Hopefully, finalizing the project by adopting the above adaptive methods can improve the quality and efficiency of a license plate recognition project

References

1. John Miano, 1999. *Compressed Image File Formats*. Reading, Mass.: Addison Wesley Publishing Co.
2. L. O'Gorman „Image and Document Processing Techniques for the RightPage Electronic Library System," Proc. Int'l Conf. Pattern Recognition (ICRP), IEEE CS Press, Los Alamitos, Calif., 1992, pp. 260-263.
3. Jacek Zurada, *Introduction to Artificial Neural Systems*, West Publishing Company, 1992
4. W. Postl, „Detection of Linear Oblique Structures and Skew Scan in Digitized Document," *Proc. Int'l Conf. Document Analysis and Recognition (ICPR)*, IEEE CS Press, Los Alamitos, Calif., 1993, pp. 478-483.
5. R. O. Duda and P. E. Hart, Pattern Classification and Scene Analysis, Wiley-Interscience, New York, 1973, pp. 330-336.
6. W-Y. Wu and M-J. J. Wang. „Detecting the Dominant Points by the Curvature-Based Polygonal Approximation," *CVGIP: Graphical Models and Image Processing*, Vol. 55, No. 2, Mar. 1993, pp.69-88.

Use of Artificial Neural Networks for Buffet Loads Prediction

Oleg Levinski

Aeronautical and Maritime Research Laboratory
506 Lorimer St, Fishermens Bend, Melbourne, Victoria, Australia 3207
Oleg.Levinski@dsto.defence.gov.au

Abstract. The use of Artificial Neural Networks (ANN) for predicting the empennage buffet pressures as a function of aircraft state has been investigated. The buffet loads prediction method which is developed depends on experimental data to train the ANN algorithm and is able to expand its knowledge base with additional data. The study confirmed that neural networks have a great potential as a method for modelling buffet data. The ability of neural networks to accurately predict magnitude and spectral content of unsteady buffet pressures was demonstrated. Based on the ANN methodology investigated, a buffet prediction system can be developed to characterise the F/A-18 vertical tail buffet environment at different flight conditions. It will allow better understanding and more efficient alleviation of the empennage buffeting problem.

1 Introduction

The unsteady pressures acting on the aircraft lifting surfaces, referred to as buffet, are broadband random fluctuations having predominant frequencies associated with the primary aerodynamic characteristics of the aircraft. Twin-tail fighter aircraft such as F/A-18 have proven to be especially susceptible to empennage buffet at high angles of attack. The vortices emanating from the wing root leading edge extensions (LEXs) tend to burst producing highly energetic swirling flow, which convects downstream and impinges upon the vertical tails and horizontal stabilisers. The turbulent airflow following a burst LEX vortex excites the tail surfaces and large oscillatory structural responses result at the low order resonant frequencies of the tail. After prolonged exposure to this dynamic environment, the tail structure begins to fatigue and repairs must be initiated. The maintenance costs and aircraft down-time associated with these repairs are often quite high.

It is known that buffet data are extremely difficult to model using traditional regression techniques due to the multiple number of noisy parameters that interact in a non-linear manner, see Ferman *et al.* [1]. So it was suggested that Artificial Neural Networks (ANN) are especially adept at modelling this kind of data because their inter-connected algorithms can accommodate these nonlinearities, see Jacobs *et al.*

[2]. One of the major features of neural networks is their ability to generalise, that is, to successfully classify patterns that have not been previously presented, which makes them a good candidate for sparse data modelling.

A number of sub-scale and full-scale experiments [3, 4, 5] on F/A-18 tail buffet, as well as numerical predictions [6, 7, 8] performed so far provide valuable information about buffet pressure distributions, dynamic response of the vertical tails and some details of the flowfield in the vertical tail region. These experiments and flight tests form a database of dynamic load environments, which can be utilised in order to develop an accurate and robust buffet prediction method.

The main objective of this study is to determine the feasibility of using Artificial Neural Networks (ANN) to characterise the unsteady buffet loads as a function of aircraft state using aerodynamic pressures measured on the twin tails of an F/A-18. This work has been done as part of an effort to develop a buffet prediction system, which incorporates experimental and flight test data, computational unsteady aerodynamics and artificial neural networks and can be used to further improve the service life of fleet aircraft and reduce costly post-production repairs.

2 Full-Scale Tail Buffet Test

The data were acquired from a full-scale F/A-18 tail buffet test in the 80ft X 120ft wind tunnel of the National Full-Scale Aerodynamic Complex at the NASA Ames Research Center. The main purpose of this test was to obtain further information on the unsteady excitation experienced by the vertical tails at high angle of attack, using unsteady surface pressure measurements, and on the tail response, using acceleration measurements of the tail structure. Buffet pressures and the resulting structural vibrations of the vertical tails were obtained over a range of angles of attack and sideslip. The test was conducted in an attempt to quantify the F/A-18 tail buffet loads and to provide data for use in the development of potential solutions to counter the twin tail buffet problem.

Vertical tail buffet studies were conducted on a full-scale production F/A-18 fighter aircraft. It was tested over an angle of attack range of 18 to 50 degrees, a sideslip range of -15 to 15 degrees, and at wind speeds up to 100 knots. All of the runs available were conducted at a free stream velocity of 168 ft/s. This corresponds to a dynamic pressure of approximately 33 psf and a Mach number of 0.15.

The F/A-18 tail buffet test instrumentation consisted of 32 pressure transducers, eight accelerometers, six strain gauges, and a surface temperature sensor. The pressure transducers were mounted on the surface of the port vertical tail in a four by four matrix on both the inboard and outboard surfaces. Data were sampled at a rate of 512 Hz per channel for a period of 32 seconds.

The method of data reduction was chosen due to the random nature of tail buffet pressures. As the unsteady buffet pressures measured are assumed to be zero-mean and stationary random process, it was subjected to standard analysis techniques in the time and frequency domains. The surface pressure fluctuations were used to calculate root-mean-square (RMS) values and power spectral densities (PSD) of the buffet pressures.

3 Modelling of Vertical Tail Buffet

Buffet data provided from the full-scale F/A-18 tail buffet test contained the inboard and outboard pressure histories as a function of free stream dynamic pressure, angle of attack, angle of sideslip and position of the pressure sensors. Since dynamic excitation of the vertical tail depends on the net contribution of unsteady loads, the buffet loading was described in terms of differential buffet pressures measured at each of the transducer pairs. As the unsteady buffet pressures can be normalised with the free-stream dynamic pressure, the buffet loads were reduced to differential pressure coefficients that allow for easy incorporation of other experimental and flight data into the integrated buffet database.

3.1 Modelling the RMS Values of Buffet Pressures

All available test conditions were examined and 73 representative test points with unique combinations of angles of attack and sideslip were selected to form a training set. The set of network input parameters included angle of attack, angle of sideslip, chordwise and spanwise locations of pressure transducer with RMS value of differential pressure as the network output. A total of 1168 input/output pairs were available for neural network training and validation. Note that one particular test point supposedly acquired at 50 degrees angle of attack and −10 degrees of angle of sideslip was not included in the training set as its test conditions were not positively identified. This test point was used later for validation of network generalisation abilities.

Two neural network architectures were initially investigated in order to assess their ability to model the buffet data, namely the Multi-Layer Perceptron (MLP) network trained with Back Propagation (BP) and the Radial Basis Function (RBF) network trained with an orthogonal least squares algorithm. For the case of the RBF network, an appropriate set of transfer functions to be used for buffet data modelling was determined on a sample buffet data. Several transfer functions including Gaussian, cubic, multiquadric and inverse multiquadric functions were assessed for function approximation, interpolation and extrapolation tasks. It was found that for most of the transfer functions evaluated, the outputs unrealistically diverged or approached zero, as testing inputs were placed further away from the nearest training input. Transfer functions with such a property cannot be used for extrapolation or even interpolation over long distances.

After extensive experimentation it was concluded that an RBF network with multiquadric transfer functions is better suited for buffet modelling as it provides acceptable fits to complex functions and allows for function extrapolation. This is consistent with the findings of Jacobs *et al.* [2], who also employed multiquadric transfer functions for sparse buffet data modelling.

Initially, the ability of the ANNs to predict the RMS differential pressures along the vertical tail at various test conditions was investigated on two reduced datasets where one set contained only data acquired at various angles of attack and zero sideslip and the other included data measured at 30 degrees angle of attack and variable sideslip. The datasets were broken down by angle of attack into a training set and a test set to show generalisation of the neural network to new inputs. The output

RMS pressures were generated at a fine grid of evenly spaced points along the surface of the tail and then interpolated using tension splines.

After inspection of the results it was concluded that both the MPL and RBF networks performed reasonably well in estimating the pressure distribution along the tail. As there was no restriction placed on the complexity of the networks, the RBF network was able to fit all the data points with a specified accuracy using up to 110 neurons while only 20 neurons with tan-sigmoid transfer functions were required by the MLP network for the same task.

Comparison of pressure maps generated by MPL and RBF network architectures showed that the results are quite similar for most of the test conditions. However, some inconsistency in magnitude of RMS differential pressures was observed for moderate negative values of sideslip, where pressure distributions produced by the RBF network appeared to better match the training data. It was found that the use of supervised training is required to improve accuracy of MLP network predictions. Despite this shortfall, the MLP network still appeared to be a more suitable choice for modelling the buffet data since it had the best training time and greater generalisation capability by capturing the most significant features in the training dataset.

After showing that both of the neural networks can generalise well after training on the reduced datasets, the next step was to find if reasonable pressure distribution can be predicted at any angles of attack and sideslip over the entire test matrix. In this case, both the networks were presented with all the available data and allowed to train until the convergence of performance functions. Although the MLP showed the most promise for buffet modelling during initial trials, the network training on a complete set of data encountered serious difficulties. Numerous attempts to train the MLP network using up to 50 neurons failed to produce a satisfactory solution, as the trained network showed an unacceptable fit to existing data and poor generalisation.

The use of Bayesian regularisation to optimise network parameters did not improve the training results as, in most of the cases, the search vector tended to settle at the local minimum preventing further training progress. After exhaustive testing of various algorithms for feed-forward network training [9, 10] it was concluded that existing learning methods cannot accommodate complex non-linear relations in buffet data. Here, the use of global optimisation techniques such as evolutionary and genetic algorithms may be required to determine an optimal set of network parameters. This is left for future studies.

The difficulties experienced with the MLP network training prevented its further use in the study and only an RBF network with multiquadric functions was used in following investigations.

Having established the ability of the RBF network to interpolate over the tail surface, pressure maps were generated for test conditions of 50 degrees angle of attack and −10 degrees of sideslip, as this test point was originally excluded from the training set. Comparison of computed and measured results revealed remarkable similarity of patterns and RMS values of differential pressure distribution allowing the identification of the test conditions of the suspected test point and confirming the prediction abilities of the network, see Fig. 1.

To ensure that the RBF network could accomplish distance interpolation, the data measured from the pressure transducer pair at 60% span, 45% chord location on the tail were removed from the training set. The trained network was then tested to verify its ability to predict the pressures near the centre of the tail for various angles of attack and sideslip.

It was found that the RBF network is able to predict the missing data reasonably well as for most of the test conditions, the network's predictions were fairly accurate with maximum error well below 15%. Only in limited regions of high angle of attack and negative sideslip did the prediction error reached 35% possibly due to the presence of large local pressure gradients, which affected the network prediction ability, especially if training occurred on a coarse grid of pressure sensors.

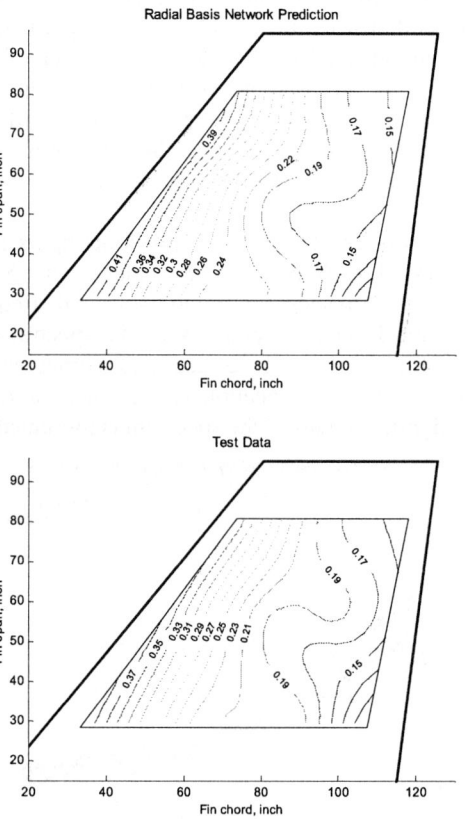

Fig. 1. Predicted and measured RMS differential pressure distribution on vertical tail at 50 degrees angle of attack and -10 degrees of sideslip

3.2 Modelling the Spectral Content of Buffet Pressures

The next step was to assess the network's ability to predict the Power Spectral Density (PSD) of differential buffet pressures on the tail. Again, the test point at 50 degrees angle of attack and −10 degrees of sideslip was selected for validation. The training set contained spectral characteristics of buffet pressure data for all the other angles of attack and sideslip in the form of PSD functions, each containing the 513 nondimensional frequencies and their corresponding outputs (nondimensional pressure spectra values). The network was expected to predict power spectral densities of buffet pressures at any point over the tail surface at new input conditions.

It should be noted that the average magnitude of the buffet pressure spectrum values can vary significantly over the frequency range, as buffet energy is concentrated in a relatively narrow frequency band. Because of the variation of dominant frequencies with test conditions and large scatter of pressure power values, it was found impractical to characterise the spectral content of buffet pressures in terms of the shape of the PSD curve alone. In order to equally emphasise both the lower and higher spectral density magnitudes, a PSD prediction system has been developed by utilising an independent RBF network for each of the spectral lines and combining output from all of the basic networks to predict the complete PSD curve. Thus, it is assumed that the magnitude of power spectral density at a particular spectral line is independent of the rest of the spectrum. The choice of RBF architecture for each of the basic networks was based on its robustness and greater interpolation ability.

The network system was developed as having four inputs (angles of attack and sideslip and coordinates of each pressure transducer) and 513 outputs (pressure spectrum magnitudes) at corresponding nondimensional frequencies. A total of 599,184 input/output pairs formed a training set. The spectral content of the buffet pressure was predicted over the tail surface and comparison of predicted and measured PSD curves for selected location is presented in Fig. 2. Note that, for clarity, only the most significant part of the spectrum is presented on the plots.

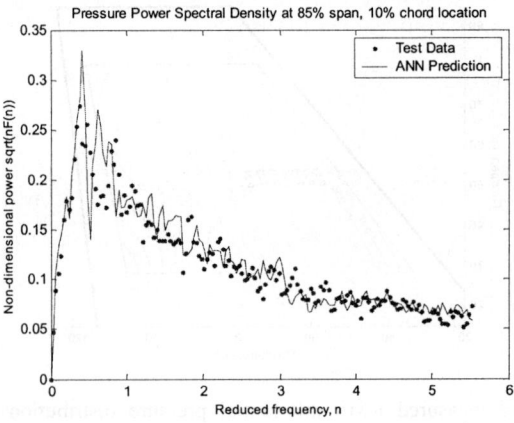

Fig. 2. ANN predictions of pressure power density at 50 degrees angle of attack and -10 degrees of sideslip

As shown by the results, the neural network system is able to predict the correct shape of the PSD curves as well as to identify dominant frequencies in the PSD spectra. Despite some discrepancy in the prediction of power pressure values, the network's ability to reproduce the overall trends of the PSD curves confirmed the validity of the adopted approach. It is anticipated that the neural network system can also be used to model other important buffet characteristics such as spatial correlation and frequency response functions between unsteady pressures on the tail, which are required for correct simulation of buffet loading.

Most of the experimental buffet data obtained so far [4, 5] are measured on a finite grid of pressure transducers limiting our knowledge of dynamic pressure distribution on the rest of the tail. Numerical simulations of vertical tail buffet loading can provide some insight into pressure distributions in the areas where the buffet data have not yet been acquired, see Levinski [8]. The results of numerical simulations can be used to supplement the sparse experimental data allowing the ANN buffet prediction system to generate pressure maps over the entire tail surface.

4 Conclusions

The feasibility of using artificial neural networks for predicting empennage buffet pressures as a function of geometric conditions has been investigated. Two network architectures have been assessed for buffet data modelling. The RBF network with multiquadric functions was selected based on its robustness and good generalisation abilities for differing input conditions. Although the MLP network showed the most promise during initial trials, its training on a large dataset resulted in an unacceptable fit to existing data and poor generalisation. The use of global optimisation techniques such as evolutionary and genetic algorithms will be investigated in the future to determine an optimal set of network parameters.

The study revealed that artificial neural networks have a great potential as a method for modelling the complex nonlinear relationships inherent in buffet data. The ability of neural networks to accurately predict RMS values and frequency content of unsteady buffet pressures was confirmed. Based on the ANN methodology investigated, a buffet prediction system can be developed to provide detailed information about the F/A-18 vertical tail buffet environment through the use of additional experimental and flight test data, as well as results of computational unsteady aerodynamics. It will allow for a better understanding of the empennage buffeting problem and can be used in fatigue usage monitoring systems for fleet aircraft.

References

1. Ferman, M. A., Patel, S. R., Zimmerman, N. H., and Gerstenkorn, G.: A Unified Approach to Buffet Response, *70th Meeting of Structures and Materials Panel*, AGARD 17, Sorrento, Italy, April (1990).
2. Jacobs, J. H., Hedgecock, C. E., Lichtenwalner, P. F., and Pado, L. E.: Use of Artificial Neural Networks for Buffet Environments, *Journal of Aircraft*, Vol. 31, No. 4, July-Aug (1994).
3. Martin, C.A., Thompson, D.H.: Scale Model Measurements of Fin Buffet Due to Vortex Bursting on F/A-18, *AGARD Manoeuvring Aerodynamics*, AGARD-CP-497, (1991).

4. Meyn, L. A., James, K. D.: Full Scale Wind Tunnel Studies of F/A-18 Tail Buffet, *AIAA Applied Aerodynamics Conference*, AIAA 93-3519, August 9-11, (1993), Monterey, CA.
5. Pettit C.L., Brown D.L., and Pendleton E.: Wind Tunnel Tests of Full-Scale F/A-18 Twin Tail Buffet: A Summary of Pressure and Response Measurements, AIAA Paper 94-3476, (1994).
6. Kandil, O.A., Sheta, E.F., Massey, S.J.: Twin Tail/Delta Wing Configuration Buffet due to Unsteady Vortex Breakdown Flow, AIAA Paper 96-2517-CP, (1996).
7. Gee, K., Murman, S. M., and Schiff, L. B.,: Computation of F/A-18 Tail Buffet, *Journal of Aircraft*, Vol. 33, No. 6, November-December (1996).
8. Levinski, O.: *Prediction of Buffet Loads on Twin Vertical Tail Using Vortex Method*, Australian Department of Defence, Aeronautical and Maritime Research Laboratory, DSTO-RR-0217, (2001).
9. Hagan, M. T., Demuth, H., Beale, M.: *Neural Network Design*, PWS Publishing Company, (1996).
10. Demuth, H., Beale, M.: *Neural Network Toolbox, For Use with MATLAB*, The Mathworks, Inc, (2000).

Computational Cost Reduction by Selective Attention for Fast Speaker Adaptation in Multilayer Perceptron

In-Cheol Kim and Sung-Il Chien

School of Electrical Engineering and Computer Science
Kyungpook National University, Taegu, 702-701, Korea
kiminc@palgong.knu.ac.kr
sichien@ee.knu.ac.kr

Abstract. Selective attention learning is proposed to improve the speed of the error backpropagation algorithm of a multilayer Perceptron. Class-selective relevance for evaluating the importance of a hidden node in an off-line stage and a node attention technique for measuring the local errors appearing at the output and hidden nodes in an on-line learning process are employed to selectively update the weights of the network. The acceleration of learning time is then achieved by lowering the computational cost required for learning. By combining this method with other types of improved learning algorithms, further improvement in learning speed is also achieved. The effectiveness of the proposed method is demonstrated by the speaker adaptation task of an isolated word recognition system. The experimental results show that the proposed selective attention technique can reduce the adaptation time more than 65% in an average sense.

1 Introduction

A slow learning speed is one of the major drawbacks of the error backpropagation (EBP) algorithm generally employed in training a multilayer Perceptron (MLP). In case of some related works employing off-line learning strategy, this problem is not a serious restriction. However, it can be a main obstacle to applying the MLP into on-line learning or relearning based applications. To accelerate the EBP algorithm, several modified methods, such as varying the learning rate during the learning process [1, 2] or using different types of error function [3, 4], have been suggested, which mainly focus on decreasing the number of iterations to reduce the learning time. In this paper, we propose a selective attention method that lowers the computational burden required for MLP learning by selectively updating the weights of the network to speed up the learning process.

To specify the weights participating in MLP learning, two attention criteria are introduced. Derived from the concept of relevance [5] for network pruning, the class-selective relevance is newly defined to determine the importance of a hidden node in minimizing the mean square error (MSE) function for a given class. Those weights connected to the hidden nodes that are irrelevant to the considered class are then fixed without updating during the learning process for the input patterns belonging to that class, thereby lowering the overall computational cost of MLP learning. As another attention criterion, a node attention technique is also introduced to further improve the learning speed by omitting the updating procedure for the weights incoming to the output and hidden nodes whose local error is marginally small. The proposed learning method based on the above selective attention criterions is quite attractive, since it can be easily coupled with other types of existing learning methods proposed to speed up the EBP algorithm.

The proposed method is particularly effective when applied to a relearning task reconsidering already trained weights [6]. Thus, its effectiveness is demonstrated using a representative example of a relearning task, speaker adaptation which is a training procedure for constructing a speaker dependent speech system by adapting a speaker independent system to a new speaker using a small amount of speaker-specific training data [7].

2 Selective Attention Learning

The EBP algorithm for MLP learning can be mainly divided into two distinct procedures of computation, MSE calculating procedure (forward pass) and weight updating procedure (backward pass). Our selective attention strategy is to reduce the computational cost related to the backward pass. This can be done by performing partial computation of that pass only for a portion of the network guided by the given attention criterions detailed below.

2.1 Class-Selective Relevance

The concept of relevance is originally used to measure the importance of a given hidden node for producing the appropriate output correctly [5]. The relevance of the mth hidden node is defined by the incremental MSE computed without that node.

$$R_m = \sum_{p=1}^{P}\sum_{j=1}^{N} \{t_j^p - \varphi(\sum_{i=0}^{M} w_{ji}\rho_{mi}v_i^p)\}^2 - \sum_{p=1}^{P}\sum_{j=1}^{N}(t_j^p - o_j^p)^2 \quad (1)$$

Here, o_j^p and t_j^p denote the actual output and corresponding target value of the jth output node, respectively, and v_i^p denotes the output of the ith hidden node, upon presentation of input pattern p. The jth output node is connected to the ith hidden node via weight w_{ij}. Plus $\varphi(\cdot)$ is a sigmoid function, and ρ_{mi} is 0 for $m = i$ and one, otherwise. From eqn. 1, it is apparent that a hidden node with large relevance plays a very important role in learning, since removing that node results in a significant increase in the MSE. This idea has been successfully applied in several classification problems related to network pruning [8].

For selective attention learning, we propose to measure the relevance of each hidden node separately according to the class. Let Ω_k be a set of training patterns belonging to the kth class, ω_k. Then, the effect of the removal of the mth hidden node on increasing the MSE for class ω_k is measured as follows:

$$R_{mk} = \sum_{p \in \Omega_k} \sum_{j=1}^{N} \{t_j^p - \varphi(\sum_{i=0}^{M} w_{ji}\rho_{mi}v_i^p)\}^2 - \sum_{p \in \Omega_k} \sum_{j=1}^{N} (t_j^p - o_j^p)^2. \qquad (2)$$

As an off-line step for speaker adaptation, this class-selective relevance, R_{mk} is calculated in advance using a baseline speech system. In an adaptation stage, the hidden nodes, depending on a class, are divided into relevant and irrelevant ones according to the threshold determined from the histogram of R_{mk} values. Then selective attention relearning is performed by updating only the incoming and outgoing weights of the relevant nodes.

Conceptually, this method has the same effect as weight pruning. However, the weights connected to the hidden nodes that are found to be irrelevant to a specific class are not actually pruned but rather frozen, and then become active again for input patterns belonging to other classes. Since the input patterns used for adaptation are characteristically similar with the initial training set whereby the relevance of each hidden node has been measured, it is expected that selective learning by the class-selective relevance can effectively remove redundant or unnecessary computation.

2.2 Node Attention

Node attention based learning is the method to update a portion of weights instead of the whole ones of a network for reducing the learning time. In MLP learning, to minimize MSE, each weight w_{ji} is updated by

$$\Delta w_{ji}(n) = -\frac{\partial E(n)}{\partial w_{ji}(n)}$$
$$= \delta_j(n) o_j(n) \qquad (3)$$

Here,

$$\delta_j(n) = -\frac{\partial E(n)}{\partial e_j(n)} \frac{\partial e_j(n)}{\partial o_j(n)} \frac{\partial o_j(n)}{\partial v_j(n)}$$
$$= e_j(n) \varphi'_j(v_j(n)) \qquad (4)$$

is the local gradient [9] in which $e_j(n)$ and $v_j(n)$ represent the local error signal and the net output in the jth output node at iteration n, respectively. Ignorance of the weights connected to the output nodes whose local error $e_j(n)$ is near to zero can reduce the computational cost related to the network learning, since those weights make no contribution to decreasing MSE. This method is also extended to estimating the importance of the hidden nodes of the MLP. Unlike the case of the output nodes, the error contribution of the hidden nodes is not directly observable, but it can be easily inferred from the errors of output nodes as follows:

$$\delta_j(n) = -\sum_k \delta_k(n) w_{kj}(n) \varphi'_j(v_j(n)) \qquad (5)$$

where $\delta_k(n)$ is local gradient of the kth output node.

Since the local errors of the output and hidden nodes are iteratively changed during the learning process, the weights to be updated are renewed every learning iteration in node attention based learning. On the other hand, in relevance based learning, the weights for updating are specified in an initial off-line stage and kept unchanged, as the relevant hidden nodes have been assigned in advance by their class-selective relevance.

3 Experiments

Now, the effectiveness of the proposed method is demonstrated in a speaker adaptation task building a speaker dependent recognition system based on a small amount of training data newly added from a new speaker when prior knowledge derived from a rich speaker-independent database is already available. Although there are many research issues related to speaker adaptation, we concentrate our focus on how fast such adapting can be made by the selective attention learning method.

3.1 Baseline System Build-up

A baseline speech system recognizing 20 Korean isolated-words was firstly constructed based on an MLP with one hidden layer. Two thousand speech data for 20 words were used for the initial training of the MLP. To extract the same dimensional feature vectors regardless of the signal length, the speech signal was partitioned into 19 frames with a length-dependent width; the longer the length of the speech signal, the wider the frame. Finally, 777-dimensional feature vectors consisting of magnitude coefficients derived from mel-scale filter banks [10] were extracted after applying a Hamming window with a half-frame shift.

The MLP consisted of 777 inputs, 35 hidden nodes, and 20 output nodes. The learning rate and momentum were assigned as 0.05 and 0.8, respectively. These learning parameters remained unchanged during the adaptation experiments to provide a fair comparison. For the initial MLP training before adaptation, the weights were initialized with random values drawn from a range [-5×10^{-3}, 5×10^{-3}] and the learning termination criterion was determined as an MSE of 0.001. All experiments were carried out on a Pentium III-500 based PC with a Linux operating system. This MLP based system is then to be adapted to a new speaker through the adaptation process.

3.2 Adaptation Experiments

First, two attention criteria described in the previous section were separately employed in selective learning for the speaker adaptation and then their adaptation speeds were evaluated. For each selective attention learning method, ten simulations were performed with different input pattern presentations, and then the results were averaged. Furthermore, such adaptation simulation was repeated for five speakers to investigate the effectiveness of our method more exactly. As an adaptation database, 10 speech data for each word obtained from the person to be newly adapted were used. At the beginning of the adaptation task, the weights are initialized to the original weights obtained from the previous initial learning.

In selective learning based on the class-selective relevance, the irrelevant hidden nodes, depending on the word class, are labeled based on their class-selective relevance computed in advance. In the experiment, a hidden node whose relevance value for a given class was lower than 0.01 was considered as irrelevant to that class. Figure 1 shows the distribution of the irrelevant nodes on the hidden layer for 20 classes. On average, 20% of the nodes were determined as irrelevant, yet their distribution was strongly class-dependent. The overall network size seems to be appropriate for our task because no nodes are commonly assigned as irrelevant across all classes.

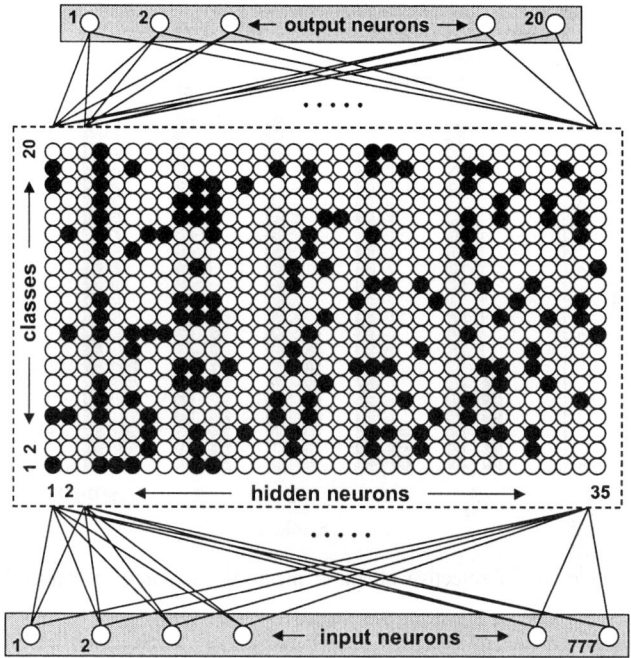

Fig. 1. Distribution of irrelevant hidden nodes (denoted by filled circle) depending on word class

In node attention based learning, the output and hidden nodes whose local errors were lower than 0.001 were sorted out every learning iteration in order to skip the updating procedure for their incoming weights. As learning goes on to minimize the MSE function, the number of skipping becomes greater. Therefore, it is estimated that the computation time per iteration will be very shorten at the end of learning.

The results in Table 1 show that each selective attention method produced faster convergence than the standard EBP, although the overall learning time somewhat varied depending on the speaker being adapted. Specially, node attention method shows a remarkable improvement in adaptation speed even when compared to learning by class-selective relevance.

Table 1. Adaptation results using standard EBP and selective attention methods

Speakers	Learning Time (sec)		
	Standard EBP	Class-selective relevance	Node attention
A speaker	11.30	8.23	4.59
B speaker	35.32	29.50	15.04
C speaker	9.72	6.68	3.54
D speaker	3.38	2.21	1.01
E speaker	20.41	14.60	7.50

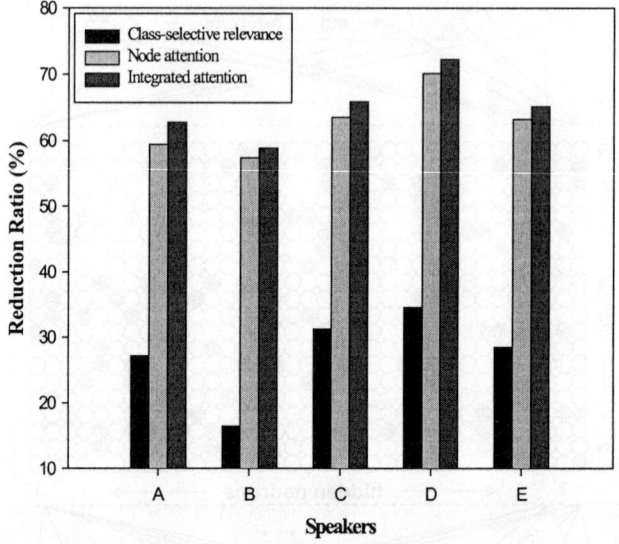

Fig. 2. Reduction ratios of selective attention methods compared to standard EBP for five speakers

Two attention criteria can be simultaneously applied to the selective learning procedure, by which further improvement in learning speed can be expected. Thus, we conducted the second part of adaptation experiments where two attention criteria are integrated for MLP learning. Figure 2 shows the reduction ratio of the learning time for the integrated attention method as well as the individual two selective attention methods against the standard EBP algorithm. As expected, the integrated method performs best with its reduction ratio of more than 65% on the average.

The next simulation was performed to show that our selective attention scheme could be successfully combined with other types of improved learning methods. We introduced Fahlman's learning method [3] that was proposed to shorten the learning iterations by solving the problem of premature saturation [4] inherent in EBP learning. Figure 3 shows that Fahlman's method achieved significant time reduction of about 29% in an average sense when compared to the standard EBP. The learning time could be further reduced with an average reduction ratio of 70% by combining it with the proposed selective attention scheme.

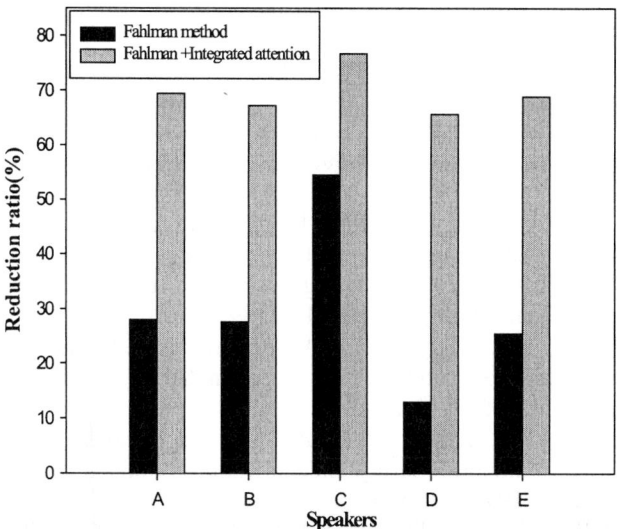

Fig. 3. Reduction ratios of relearning times compared to standard EBP for Fahlman's method and its selective attention version

4 Conclusions

MLP learning for speaker adaptation was accelerated by selectively updating the weights of the network, thereby reducing the computational cost per iteration related to the computation of the backward pass in the EBP algorithm. The weights actually participating in the learning process are specified by two attention criterions: class-selective relevance for labeling the class-dependent irrelevant hidden nodes in an initial off-line stage and node attention for iteratively sorting out the output and hidden nodes whose local errors were very close to zero during the on-line learning process.

Through a series of speaker adaptation experiments, we found that MLP learning could be considerably accelerated by the proposed selective learning method and further improvement was also achieved by integrating two attention criterions. Additionally, the results of the simulation using Fahlman's method show that our attention scheme can be successfully combined with other types of improved learning algorithms.

Acknowledgement

This research was supported by the Bain Korea 21 Project and a grant No. (1999-2-30300-001-3) from the Basic Research Program of the Korea Science and Engineering Foundation.

References

1. Jacobs, R.A.: Increased Rates of Convergence Through Learning Rate Adaptation, Neural Networks, **1** (1988) 295-307
2. Park, D.J., Jun, B.E., Kim, J.H.: Novel Fast Training Algorithm for Multilayer Feedforward Neural Network, Electronic Letters, **28(6)** (1992) 543-544
3. Fahlman, S.E.: Faster-Learning Variations on Backpropagation: An Empirical Study, Proc. Connectionist Models Summer School, Carnegie Mellon University, (1988) 38-51
4. Oh, S.H.: Improving the Error Back-Propagation Algorithm with a Modified Error Function, IEEE Trans. Neural Networks, **8(3)** (1997) 799-803
5. Mozer, M.C., Smolensky, P.: Using Relevance to Reduce Network Size Automatically, Connection Science, **1(1)** (1989) 3-16
6. Abrash, V., Franco, H., Sankar, A., Cohen, M.: Connectionist Speaker Normalization and Adaptation, Proc. European Conf. Speech Communication and Technology, (1995) 2183-2186
7. Lee, C.H., Lin, C.H., Juang, B.H.: A Study on Speaker Adaptation of the Parameters of Continuous Density Hidden Markov Models, IEEE Trans. Signal Processing, **39(4)** (1991) 806-814
8. Erdogan, S.S., Ng, G.S., Patrick, K.H.C.: Measurement Criteria for Neural Network Pruning," Proc. IEEE TENCON Digital Signal Processing Applications, **1** (1996) 83-89
9. Rumelhart, D.E., Hinton, G.E., Williams, R.J.: Learning Internal Representations by Error Propagation, in Parallel Distributed Processing, Cambridge, MA: MIT Press, (1986) 318-364
10. Davis, S.B., Mermelstein, P.: Comparison of Parametric Representations for Monosyllabic Word Recognition in Continuously Spoken Sentences, IEEE Trans. Acoustics, Speech, and Signal Processing, **28(4)** (1980) 357-366

A Comparison of Neural Networks with Time Series Models for Forecasting Returns on a Stock Market Index

Juliana Yim

School of Economics and Finance, RMIT University
255 Bourke Street, Victoria 3000 Australia
Juliana.Yim@ems.rmit.edu.au

Abstract. This paper analyses whether artificial neural networks can outperform traditional time series models for forecasting stock market returns. Specifically, neural networks were used to predict Brazilian daily index returns and their results were compared with a time series model with GARCH effects and a structural time series model (STS). Further, using output from ARMA-GARCH model as an input to a neural network is explored. Several procedures were utilized to evaluate forecasts, RMSE, MAE and the Chong and Hendry encompassing test. The results suggest that artificial neural networks are superior to ARMA-GARCH models and STS models and volatility derived from the ARMA-GARCH model is useful as an input to a neural network.

1 Introduction

The fundamental concept of an artificial neural network (ANN) is inspired by the neural architecture of the human brain. As a non-parametric model, ANN does not rely on assumptions such as normality and stationarity that are often adopted to make traditional statistical methods tractable. ANN systems learn by example and dynamically modify themselves to fit the data presented. They, also have the ability to learn from very noisy, distorted or incomplete sample data.

ANN models have outperformed the traditional statistical models in forecasting stock prices, stock returns, inflation, imports and exchange rate. Swales and Yoon [1] investigated the ability of ANN to predict stock prices compared to multivariate discriminant analysis (MDA). Their results indicated that the ANN approach gave more accurate predictions of stock prices. Boyd and Kaastra [2] examined whether neural networks can outperform traditional ARIMA models for forecasting commodity prices. Their results showed neural network forecasts were considerably more accurate. Lim and McNelis [3] investigated the predictability of daily stock price returns in Australia, and their response to shocks in foreign markets, using three empirical methods - an autoregressive linear model, a GARCH-M model and non-linear neural network. Their results indicated that the ANN presented the lowest root mean squares errors (RMSE) for some forecast horizons, however, the results from Diebold and Mariano test showed that there is no significant difference in the predictive accuracy between the ANN model and GARCH-M. Correa and Portugal [4] provided an empirical evaluation of the ANN and structural models forecasting performance of Brazilian inflation and total imports in presence of structural change. The results for one-step-ahead forecasts show that the ANN model has marginally better performance than the STS in the periods just after the structural change. Toulson [5] presented a comparison between the multi-layer perceptron neural network and structural model by investigating the predictive power of both models for one-step-ahead forecast of the USD/DEM exchange rate. The results indicated that the multi-layer perceptron's performance is superior to that of structural model.

The aim of this article is to examine whether artificial neural networks can outperform traditional time series models for forecasting stock market returns in emerging markets. More precisely, neural networks were used to predict Brazilian daily stock index returns and the results were compared with an ARMA-GARCH model and a STS model. In addition, the question of using output from ARMA-GARCH model as input to a neural network is explored.

This paper is organized as follows. Section 2 briefly describes the structure and training of the popular backpropagation algorithm, which is used in this study. Section 3 describes the GARCH models. Section 4 describes the STS model. Section 5 describes the data. Section 6 presents the results of estimated models. Section 7 collects the forecasting results from all models, and presents an evaluation and comparison of neural network models, the ARMA-GARCH model and STS model. Section 8 presents concluding comments.

2 Artificial Neural Network Model

The ANN used in this study uses a multi-layer perceptron network (MLP), trained by a gradient descent algorithm called by backpropagation [6]. It is the most common type of formulation and is used for problems that involve supervised learning. The first phase of the backpropagation algorithm consists of repeatedly presenting the network with examples of input and expected output. Suppose that the qth neuron of the hidden layer receives the activation signal, H_q, given by:

$$H_q = \sum_j v_{qj} x_j, \qquad (2.1)$$

where x_j is the signal to the input neuron j and v_{qj} is the weight of the connection between input neuron j and the hidden neuron q.

This activation signal is then transformed by a transfer function f in the hidden layer to give the output

$$h_q = f(H_q). \qquad (2.2)$$

The output neuron i now receives the activation signal, O_i, from the hidden nodes given by

$$O_i = \sum_q w_{iq} h_q, \qquad (2.3)$$

where w_{iq} is the weight of the connection between hidden neuron q and output neuron i. This is transformed again to give the output signal

$$o_i = f(O_i). \tag{2.4}$$

This is then compared with the desired, or actual value of the output neuron, and the function of squared errors for each node, which is to be minimized, is given by

$$E(w) = \frac{1}{2}\sum_i (d_i - o_i)^2, \tag{2.5}$$

In the second phase the weights are modified to reduce the squared error. The change in weights, Δw_{iq}, used by the backpropagation is given by

$$\Delta w_{iq} = -\gamma \frac{\partial E(w)}{\partial w_{iq}}. \tag{2.6}$$

where $0 < \gamma < 1$ is the learning rate.

Using the chain rule, it can easily be shown that

$$\Delta w_{iq} = -\gamma \frac{\partial E}{\partial o_i}\frac{\partial o_i}{\partial O_i}\frac{\partial O_i}{\partial w_{iq}} = \gamma(d_i - o_i)f'(O_i)h_q = \gamma \delta_{oi} h_q, \tag{2.7}$$

where δ_{oi} is the error signal of neuron i and is given by

$$\delta_{oi} = (d_i - o_i)f'(O_i). \tag{2.8}$$

To avoid oscillation at large γ, the change in the weight is made dependent on the past weight change by adding a momentum term

$$\Delta w_{iq}(t+1) = \gamma \delta_{oi} h_q + \alpha \Delta w_{iq}(t), \tag{2.9}$$

where α is a constant chosen by the operator. Similarly it can be shown that the change in the weight between the hidden neuron i and the input neuron j, Δv_{ij}, is given by

$$\Delta v_{qj} = \gamma \delta_{hq} x_j, \tag{2.10}$$

where δ_{hq} is the error signal of neuron q and is given by

$$\delta_{hq} = f'(H_q)\sum_i \delta_{oi} w_{iq}. \tag{2.11}$$

As before a momentum term can be used to prevent oscillation.

3 ARCH Models

Mandelbrot [7] suggests that for asset prices, large changes tend to be followed by large changes and small changes tend to be followed by small changes. The independent and identically distributed normal assumption is convenient for financial models, however research on stock returns typically finds the distribution is leptokurtic. In order to capture the non-linearity involved in this type of series, Bollerslev [8] introduces GARCH(p,q) model, specifying the current conditional variance as a function of the past conditional variance as well as the past squared error terms derived from the model. The GARCH(p,q) model can be represented by the following system:

$$R_t = \eta + \sum_{i=1}^{m} \phi_i R_{t-i} + \sum_{j=1}^{n} \theta_j \varepsilon_{t-j} + \varepsilon_t \quad (3.1)$$

where R_t is series of continuous stock returns, $\varepsilon_t \sim N(0, h_t)$ and the conditional variance of errors, h_t is specified as:

$$h_t = \alpha_0 + \sum_{i=1}^{p} \alpha_i \varepsilon_{t-i}^2 + \sum_{j=1}^{q} \beta_j h_{t-j} \quad (3.2)$$

The parameters in equation 3.2 should satisfy:

$$\sum_{i=1}^{q} \alpha_i + \sum_{j}^{p} \beta_j < 1 \quad \text{and} \quad \alpha_0 > 0, \alpha_i \geq 0 \text{ and } \beta_j \geq 0$$

Nelson [9] suggested that the exponential GARCH model (EGARCH) is more general than the GARCH model, because it allows innovation of different signs to have a differential impact on the volatility unlike the standard GARCH model.

The specification for the conditional variance of the EGARCH model is

$$\log(h_t) = \alpha_0 + \sum_{j=1}^{q} \beta_j \log(h_{t-j}) + \sum_{i=1}^{p} \gamma_i \frac{\varepsilon_{t-i}}{\sqrt{h_{t-i}}} + \sum_{i=1}^{p} \alpha_i \left(\frac{|\varepsilon_{t-i}|}{\sqrt{h_{t-i}}} - \sqrt{\frac{2}{\pi}} \right) \quad (3.3)$$

The left side is the log of the conditional variance. This shows that the leverage effect is exponential and that forecasts of the conditional variance are guaranteed to be nonnegative.

4 Structural Model

The structural time series model (STS), introduced by Harvey [10], was used in this study. The complete structural model suggests that a time series, y_t, is modeled by decomposition into its basic components: trend (μ_t), cyclical (ψ_t), seasonal (γ_t) and irregular (ε_t) components, and is given by:

$$y_t = \mu_t + \gamma_t + \psi_t + \varepsilon_t. \quad (4.1)$$

The structural model used in this paper consists of a random walk component to capture the underlying level μ_t, plus a random disturbance term, ε_t,

$$y_t = \mu_t + \varepsilon_t \quad (4.2)$$

$$\mu_t = \mu_{t-1} + \eta_t \tag{4.3}$$

where η_t is independently distributed terms of the form:

$$\eta_t \sim NID(0, \sigma^2_\eta) \tag{4.4}$$

The irregular component is assumed to be stationary and consists of a white noise disturbance term of the form:

$$\varepsilon_t \sim NID(0, \sigma^2_\varepsilon) \tag{4.5}$$

The hyperparameters, σ^2_ε and, σ^2_η were calculated using maximum likelihood. After estimation, the state μ_t is found by using the Kalman filter [11]. In this study it was necessary to use lagged log-price in the model. The equation for this model may be written as:

$$y_t = \mu_t + \sum_{i=1}^{K} \rho_i Lp_{t-i} + \varepsilon_t \tag{4.6}$$

where Lp_{t-i} are the lagged prices.

5 Data Description

Figure 5.1 consists of daily observations of a spot stock price and continuously compounded returns index, the IBOVESPA of the Sao Paulo Stock Exchange. The periods of high volatility indicate the Mexican crisis (December/1994) and the Asian crisis (July/1997). The data starts from July, 30 1994 through to June, 30 1998, a total of 988 observations. They were obtained from the *Macrometrica* database. Forty observations were reserved for forecasting performance evaluation using one step-ahead prediction.

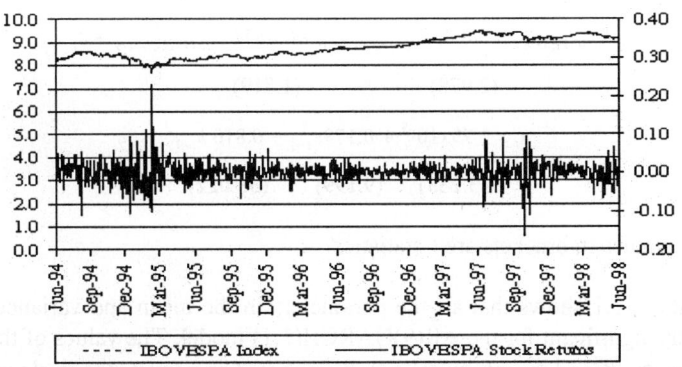

Fig. 5.1. Daily IBOVESPA Index and IBOVESPA Stock Returns-Jun-94/Jun-98

The descriptive statistics for these stock index returns are given in Table 5.1. The kurtosis, skewness and Jarque-Bera statistics indicate the data are non-normal.

Table 5.1. Descriptive statistics for the stock index returns

Mean	Std. Dev.	Skewness	Kurtosis	Jarque-Bera	Sample
0.000994	0.228182	0.113256	10.62632	2396.407	988

The Ljung-Box statistic for the returns in Table 5.2 shows the presence of correlation and the squared return series indicate a high degree of non-linearity in the data.

Table 5.2. The Ljung-Box statistics for the return and squared return

Q(12)	Q(24)	Q(36)	$Q(12)^2$	$Q(24)^2$	$Q(36)^2$
60.286*	73.545*	89.401*	419.75*	461.69*	506.62*

*Denotes statistical significance at 5% level.

6   Estimated Models

This section presents the best specifications of the ARMA-GARCH model, STS models and ANN models for daily IBOVESPA return series.

6.1   ARMA-GARCH Model

The GARCH(p,q) and EGARCH(p,q) models were tried for p =1,2,..., 9 and q= 1, 2,....,9. From all possible models estimated, the final specification was decided using likelihood-ratio tests, tests for the standardized residuals, and other statistics such as the standard deviation, the AIC and BIC.

The equation of the final ARMA-GARCH(1,1)[1] model was found to be

$$R_{IBOVESPA\,t} = 0.094\, R_{IBOVESPA\,t-1} + 0.0557\, R_{IBOVESPA\,t-9} + \varepsilon_t$$

$$(2.978) \qquad\qquad (1.819)$$

$$h_t = 1.59 \times 10^{-5} + 0.178 \varepsilon_{t-1}^2 + 0.810\, h_{t-1} \qquad (6.1.1)$$

$$(3.233) \quad (9.199) \quad (38.022)$$

where the values in brackets are t-statistics.

Equation 6.1.1 shows that all the parameters in the mean and variance model are statistically significant for the AR(9)GARCH(1,1) model. The values of the estimated parameters α_0, β and α_1 satisfy $\alpha_0 > 0$, β, $\alpha_1 \geq 0$. The persistence is close to but less than, unity. This result suggests that the model has high persistence in the conditional

---

[1] Estimated using the software package EVIEWS 3.1.

volatility. According to Bollerslev, Engle and Nelson [8, 12, 9], it can be observed in most of the empirical applications of GARCH(1,1) model.

The equation of the final ARMA-EGARCH(1,1) model may be written as

$$R_{IBOVESPAt} = 0.0105\, R_{IBOVESPA\,t-1} + 0.065\, R_{IBOVESPA\,t-9} + \varepsilon_t$$

$$\quad\quad\quad\quad (3.314) \quad\quad\quad\quad\quad\quad (2.474)$$

$$\log(h_t) = -0.747 + 0.931 \log(h_{t-1}) - 0.167\, \frac{\varepsilon_{t-1}}{\sqrt{h_{t-1}}} + 0.294 \left(\frac{|\varepsilon_{t-1}|}{\sqrt{h_{t-1}}} - \sqrt{\frac{2}{\pi}} \right) \quad (6.1.2)$$

$$(-6.649)\ (69.949) \quad\quad\quad\quad (-7.092) \quad\quad (8.213)$$

Equation 6.1.2 indicates that for the AR(9)EGARCH(1,1) model the leverage effect term (γ) is negative and statistically different from zero, indicating the existence of the leverage effect on IBOVESPA returns during the sample period. The positive value of the estimated parameter $\alpha = 0.294$ confirms the existence of volatility clustering.

Table 6.1.2 indicates that the model that presented lowest AIC and BIC was AR(9)EGARCH(1,1). In addition, the Ljung-Box (Q(K) and Q(K)^2)^{2}statistic tests for the residuals show that the most of the linear dependence (autocorrelation) in the mean and variance has been captured by AR(9)EGARCH(1,1) specification.

Table 6.1.2. Diagnostic Tests of GARCH model

Model/Statistic	AIC	BIC	Q(12)	Q(24)	Q(36)	Q(12)^2	Q(24)^2	Q(36)^2
AR(9)GARCH	-4.74	-4.72	6.57	20.2	26.8	17.4	27.3	35.4
AR(9)EGARCH	-4.79	-4.76	6.68	21.1	26.8	12.1	21.7	29.9

6.2 Structural Model

From all models that were estimated, the best specification of the structural model[3] may be written as

$$Lp_t = 6.5708 + 0.3559\, Lp_{t-1} - 0.0945\, Lp_{t-2} - 0.097\, Lp_{t-5} + 0.1206\, Lp_{t-9} \quad (6.2.1)$$

$$(12.663)\ (11.256)\quad (-3.0097)\quad\quad (-3.3053)\quad\quad (4.0605)$$

with $\hat{\sigma}_\xi^2 = 0.0002$ and $\hat{\sigma}_\eta^2 = 0.0003$

As can be observed in Equation 6.2.1, all the parameters are statistically significant at level of 5% and the log of IBOVESPA prices (Lp_t) values has a stochastic trend. In Table 6.2.1 the normality test statistic indicates that the null hypothesis of a normal

---

[2] Q(K) is Ljung-Box statistic for standardized residual and Q(K)^2 is Ljung-Box statistic for squared standardised residual.
[3] Estimated using the software package STAMP 5.0.

distribution is rejected. In addition, the heteroskedasticity test shows no evidence for heteroskedasticity. The Box & Ljung Q(K) statistic test for the residual indicates that the model captured the serial correlation in the series. From all the models considered, this model presented the lowest AIC and BIC.

Table 6.2.1. Diagnostic Test of STS model

Normality	$H(326)^4$	Q(12)	Q(24)	Q(36)	AIC	BIC
2298	0.68	10.97	20.75	30.38	-7.2117	-7.1818

6.3 Artificial Neural Networks

The multi-layer model used in this paper is a fully connected, three-layer feed forward network and is trained to map lagged returns to current returns using a backpropagation algorithm.

During the training, the RMSE and MAE were used to compare and select the networks. The networks were trained for 20,000 iterations. After considerable experimentation the learning rate and momentum term were set for each network to 0.5 and 0.7, respectively. The number of hidden neuron and hidden layers were selected experimentally based on the testing set performance of each neural network. The tanh function was the activation function specified.

A large number of lagged returns were combined in order to select the best set of input variables. But, only ten levels of the number of input nodes ranging from one to ten produced RMSE and MAE comparable to the other two econometric models used in this study. Table 6.3.1 presents the networks that produced the lowest RMSEs and MAEs. These are a network that has nine lagged returns in the input layer (lags 1 to 9), four neurons in the single hidden layer and one neuron in the output layer, one that included only four neurons in the input layer (lags 1, 2, 5 and 9), two neurons in the single hidden layer and one neuron in the output layer and another that consisted of two neurons (lags 1 and 9) in the input layer, two neurons in the single hidden layer and one neuron in the output layer. This final small network produced the best overall results. The question of using the conditional volatility estimated by ARMA-GARCH model as an input to a neural network was also explored. This hybrid model was proposed in order to reduce the overall errors. The network that combined the specification from the final small network with the volatility from the ARMA-GARCH model presented a performance superior to the simple network during the training phase. The ability of generalization from both models was tested in the next section. The networks were implemented using the software package NeuroShell2.

Table 6.3.1. Training Results

Statistic/Model	ANN (lags-1 to 9)	ANN (lags-1, 2, 5 and 9)	ANN (lags-1, 9)	ANN (lags 1, 9) volatility
RMSE	0.0317	0.0318	0.0315	0.0304
MAE	0.0220	0.0241	0.0201	0.0198

---

[4] The heteroskedasticity test statistic, H(h), is the ratio of the squares of the last h residuals to the squares of the first h residuals where h is the set to the closest integer of T/3. It is centred around unity and should be treated as having a F distribution with (h,h) degrees of freedom.

7   Forecasting Performance

In this section the ARMA-GARCH models, structural models[5] and ANN models were used to forecast one-step-ahead, for the 40 remaining days in the IBOVESPA return series. The forecasts of stock index returns were evaluated and compared, considering the RMSE and MAE forecast statistics. Table 7.1 reports that the MAE from STS model presents the highest value and there is no real difference between the remaining models. Considering the RMSE forecast statistic, both ANN models outperform the other models, with the ANN with volatility from AR(9)EGARCH(1,1) model having slightly better performance.

Table 7.1. Out-of-sample, 40 days of one step ahead, forecasting performance of stock returns

	AR(9)GARCH	AR(9)EGARCH	STS	ANN (lags 1, 9)	ANN (lags 1, 9) + volatility
RMSE	0.0243	0.0242	0.0248	0.0238	0.0236
MAE	0.0188	0.0187	0.0192	0.0188	0.0188

Among the network models, the ANN-(lags 1, 9) and ANN-(lags 1, 9) with volatility produced the best results. Although useful, these forecasting evaluation criteria cannot discriminate between forecasting models when they are very close to each other. For that reason, the Chong and Hendry encompassing test [13, 14] was used to evaluate the statistical significance of the competing forecasting models. According to the authors, a model (1) encompasses another model (2) if model 1's forecast significantly explain model 2's forecasting errors. Thus, mode 1 has more relevant information than model 2. The significance of the parameters β and δ is tested on the following regression equations:

$$(r_t - r_{2t}) = \beta_{21} r_{1t} + u_{1t} \quad (7.1)$$

$$(r_t - r_{1t}) = \delta_{1t} r_{2t} + u_{2t} \quad (7.2)$$

where (r_t-r_{1t}) and (r_t-r_{2t}) are the forecasting errors from model 1 and 2, respectively; r_{1t} and r_{2t} are the forecasts of the two models; and u_{1t} and u_{2t} are random errors.

The null hypothesis is that neither model can be considered superior to another. If the t-tests show β to be significant, but δ is not, this indicates that the model 1 encompasses model 2. On the other hand, when δ is significant but β is not, this indicates that model 2 encompasses model 1. If both β and δ are insignificant, or both β and δ are significant, then the null hypothesis is not rejected, so neither model is clearly superior.

The results of the encompassing test for forty days of one-step-ahead forecasting are presented in Table 7.2. The p-values less than 0.10 along the structural model row indicates that ANN return forecasts can explain a portion of the structural model's forecast error. Conversely, any p-value less than 0.10 were not observed in the structural model column. So, the structural model returns forecasts cannot explain

---

[5] The forecast IBOVESPA returns were calculated as the change in the logarithm of IBOVESPA prices of successive days.

errors from ARMA-GARCH or ANN models. Therefore, the ANN models encompass the structural model at the 10% level. The ARMA-GARCH row indicates that ANN models can explain ARMA-GARCH's forecast errors. Also, the ARMA-GARCH column shows that ARMA-GARCH's forecast errors cannot explain structural model's errors or the ANN model's errors. From the comparison to ARMA-GARCH, ARMA-EGARCH as well as the structural model, it can be observed that ANN-(lags 1, 9) and ANN-(lags 1, 9) with volatility forecast errors are the only errors that are not explained by any other statistical model. Also, the ANN with volatility from AR(9)EGARCH(1,1) model encompasses the ANN-(lags 1, 9) model. Therefore, the ANN with volatility model encompasses all the other models at the 10% level.

Table 7.2. Encompassing tests of forecasting performance of alternative models (p-values)

Dependent Variable:	Independent Variable: Forecast Errors from				
Forecasting Errors from	AR(9)GARCH	AR(9)EGARCH	STS	ANN (lags 1, 9)	ANN (lags 1, 9) + volatility
AR(9)GARCH	—	0.33	0.35	0.09	0.08
AR(9)EGARH	0.37	—	0.40	0.10	0.09
STS	0.62	0.62	—	0.08	0.10
ANN-AR(9)GARCH	0.66	0.67	0.67	—	0.02
ANN + volatility	0.57	0.58	0.61	0.80	—

8 Conclusions

This paper investigated whether artificial neural networks can outperform time series models for forecasting stock market returns. More precisely, neural networks were applied to forecast IBOVESPA daily stock index returns. The results were compared with a time series model with GARCH effects and STS model. Furthermore, the question of using the volatility estimated by ARMA-GARCH model as an input to a neural network was also examined. The RMSE, MAE and the Chong and Hendry encompassing test were used to evaluate and compare forecasts. Comparison of the methods suggests that artificial neural networks are superior to the ARMA-GARCH models and structural models. According to the RMSE and MAE, the performance of the ANN models and econometric models are almost the same, but the encompassing test results indicate that ANN models outperform all other models. Also, the forecasting error was slightly reduced when the volatility derived from the ARMA-GARCH model was added as an input to a neural network. Although the addition of the conditional volatility input did not dramatically improve performance, it shows some promise and should be considered as an input when high volatility exists. Therefore, the results indicate neural network approach is a valuable tool for forecasting stock index returns in emerging markets.

Acknowledgements

The author wishes to thank Heather Mitchell, Michael McKenzie and Vera Lúcia Fava for some helpful comments. Most of this work was completed while the author was Master student at Universidade de São Paulo.

References

1. Swales, G. and Yoon, Y., 'Predicting Stock Price Performance: A Neural Network Approach', Proceedings of the IEEE 24th Annual Conference of Systems Science, 1991, pp. 156-162.
2. Boyd, M., Kaastra, I., Kermanshahi, B. and Kohzadi, N., 'A Comparison of Artificial Neural Network and time series models for forecasting commodity prices', *Neurocomputing*, Vol. 10, 1996, pp. 169-181.
3. McNelis, P. and Lim, G.C., 'The Efect of the Nikkei and the S&P on the All-Ordinaries: A Comparison of Three Models', *International Journal of Finance and Economics*, Vol.3, 1998, pp.217-228.
4. Corrêa, W. R. and Portugal, M.S. 1998. Previsão de Séries de Tempo na Presença de Mudanças Estruturais: Redes Neurais Artificiais e modelos estruturais, *Economia Aplicada*, V. 2 N.3, pp. 486-514.
5. Toulson, S. 1996. Forecasting Level and Volatility of High Frequency Exchange Rate Data., *Working Paper in Financial Economics*, N.9, pp.1-5.
6. Rumelhart, D., J. McClelland and PDP Group, 'Parallel distributed processing. Exploration in the Microstructure of Cognition, Vol.1: Foundation. Cambridge, Mass.: MIT Press, 1986.
7. Mandelbrot, B., 'The Variation of Certain Speculative Prices', *Journal of Business*, Vol. 36, 1963, pp.394-419.
8. Bollerslev, T., 'Generalized Autoregressive Conditional Heteroskedasticity', *Journal of Econometrics*, Vol. 1, 1986, pp. 307-327.
9. Peter J. Brockwell and Richard A Davis, ' Time Series: Theory and Methods', 2nd Edition, 1991 Springer-Verlag New York.
10. Nelson, D. B., 'Conditional Heteroskedasticity in Asset Returns: A New Approach', *Econometrica*, V. 59, 1991, pp. 347–370.
11. Harvey, A. C., 'Forecasting Structural Time Series Models and the Kalman Filter' Cambridge: Cambridge University Press, 1991.
12. Engle, R. F., 'Autoregressive Conditional Heteroskedasticity with Estimates of the Variance of U.K. Inflation', *Econometrica*, Vol. 50, 1982, pp. 987–1008.
13. Chong, Y.Y. and Hendry, D.F., 'Econometric Evaluation of Linear Macroeconomic models', *Review of Economic Studies*, Vol.53, 1986, pp. 671-90.
14. Donaldson, R. G. and Kamstra M., 'An artificial neural network-GARCH model for international stock return volatility', *Journal of Empirical Finance*, Vol.4, 1997, pp. 17-46.

Automatic Detection of Film Orientation with Support Vector Machines

Dane Walsh and Christian Omlin

[1] Department of Computer Science, University of Stellenbosch
Stellenbosch 7600, South Africa
dwalsh@cs.sun.ac.za
[2] Department of Computer Science, University of Western Cape
Bellville 7535, South Africa
comlin@uwc.ac.za

Abstract. In this paper, we present a technique for automatic orientation detection of film rolls using Support Vector Machines (SVMs). SVMs are able to handle feature spaces of high dimension and automatically choose the most discriminative features for classification. We investigate the use of various kernels, including heavy tailed RBF kernels. Our results show that by using SVMs, an accuracy of 100% can be obtained, while execution time is kept to a mininum.

1 Introduction

Current services in the photofinishing industry include the transfer of images from film negatives into digital form. Customer satisfaction for this service requires that the orientation of images be manually corrected. Manual rotation is a time consuming process and requires human intervention. Thus a system that is able to automatically classify images and ultimately films with high speed and accuracy is essential.

Humans are able to make use of semantic objects such as trees, buildings, mountains, etc. and their relationships in order to determine the orientation of an image. Automatic extraction of this information by computers is difficult and time consuming. This makes object recognition as a basis for detecting image orientation infeasible. Instead we have to rely on low level visual features such as colour distribution, textures and contrasts as a means of orientation detection. The large number of such features may make many traditional classifiers inappropriate tools for orientation detection. Support Vector Machines (SVMs) are able to handle high dimensional feature spaces and can automatically choose the most discriminative features through training.

The remainder of the paper is organised as follows: We give a short desciption of the film orientation problem in Section 2. We give a brief introduction to SVMs in Section 3. We discuss the image database and the features we extract from the images, aswell as the classification of a film in Section 4. In Section 5 we present different kernels which can improve the classification performance over standard kernels. We present emperical results of SVM classification performace in Section 6.

2 Background

In the transfer of images from film to digital format, the developed film negative is mounted on a spool, which is later fed into the digital image scanner (Fig. 1). Before the film can be scanned, the scanner needs to gather some information about the film. The bar code reader scans a bar code that is located at the front of the film and contains information about the customer, such as batch number and film number. The DX code reader scans the DX code (Fig. 2), which is located on the top edge of the film, and identifies the type and make of the film. Because the DX code is only located on one edge of the film, the film has to be correctly orientated when being fed into the scanner so that the DX code reader can read it.

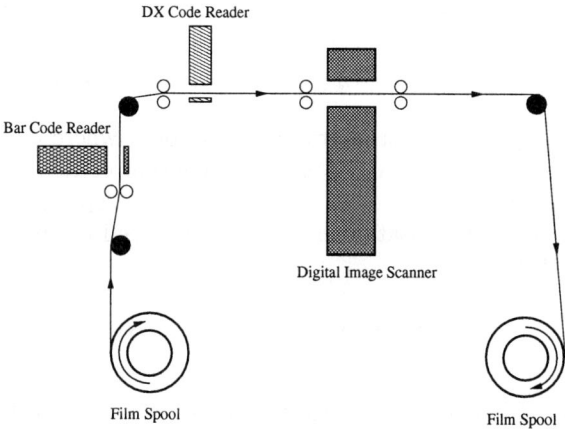

Fig. 1. Schematic representation of a scanner

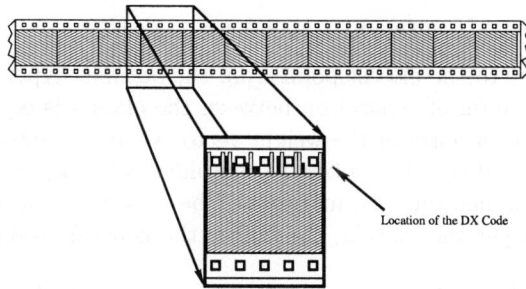

Fig. 2. The DX Code location on a film negative

There would be no need to determine the orientation of a film if all cameras fed the film in the same direction. But this is not the case. A spool of film can either be placed on the left hand side of the camera and fed towards the right, or placed on the right hand side and fed towards the left. In order to do this, the spool has to be inserted "upside-down", and this is why an entire film will appear incorrectly orientated once the images have been digitally scanned.

3 Support Vector Machines

3.1 Introduction

Unlike many other classifiers such as neural networks which are trained by minimising the output error, SVMs operate on an induction principle called structural risk minimisation, which minimises an upper bound on the generalisation error. We give a brief overview of the SVM algorithm. [2][7]

3.2 Optimal Hyperplane for Linearly Separable Patterns

Let $D = (\mathbf{x}_i, y_i)_{1 \leq i \leq N}$ be a set of training examples, where example $\mathbf{x}_i \in \Re^d$, belongs to class $y_i \in \{-1, 1\}$ and d is the dimension of the input space. The objective is to determine the *optimal hyperplane* which separates the two classes and has the smallest generalisation error. Thus, we need to find a weight vector $\mathbf{w}$ and bias b such that

$$y_i(\mathbf{w} \cdot \mathbf{x}_i + b) > 0, i = 1, ..., N \qquad (1)$$

If there exists a hyperplane satisfying Eq. (1), the set is said to be *linearly separable*. In this case it is always possible to rescale $\mathbf{w}$ and b so that

$$\min_{1 \leq i \leq N} y_i(\mathbf{w} \cdot \mathbf{x}_i + b) \geq 1, i = 1, ..., N$$

i.e. so that the algebraic distance of the closest point to the hyperplane is $1/||\mathbf{w}||$. Then Eq. (1) becomes

$$y_i(\mathbf{w} \cdot \mathbf{x}_i + b) \geq 0, i = 1, ..., N \qquad (2)$$

As the distance to the closest point from the optimal hyperplane is $1/||\mathbf{w}||$, maximising the margin of separation between the classes is equivalent to minimising the Euclidean norm of the weight vector $\mathbf{w}$ under constraints (2). This margin is $2/||\mathbf{w}||$, and thus the optimal hyperplane is the separating hyperplane that maximises the margin. The margin can be viewed as the measure of generalisation: the larger the margin, the better the generalisation is expected to be [1].

Since to Euclidean norm of $\mathbf{w}$ ($||\mathbf{w}||^2$) is convex and the constraints (2) are linear in $\mathbf{w}$, the constrained optimization problem can be solved with Lagrange multipliers $\{\alpha\}_{i=1}^N$ associated with constraints (2) by maximising

$$W(\alpha) = \sum_{i=1}^{N} \alpha_i - \frac{1}{2} \sum_{i,j=1}^{N} \alpha_i \alpha_j y_i y_j \mathbf{x}_i \cdot \mathbf{x}_j \qquad (3)$$

with $\alpha_i \geq 0$, for $i = 1, ..., N$, with the constraint $\sum_{i=1}^{N} \alpha_i y_i = 0$. This can be achieved by using of standard quadratic programing methods. Having determined the optimum Lagrange multipliers, denoted $\alpha_{o,i}$, the optimum weight vector $\mathbf{w}_o$ has the following expansion

$$\mathbf{w}_o = \sum_{i=1}^{N} \alpha_i^o y_i \mathbf{x}_i \qquad (4)$$

and the hyperplane decision function would consist of calculating the sign of

$$f(\mathbf{x}) = \sum_{i=1}^{N} \alpha_i^o y_i \mathbf{x}_i \cdot \mathbf{x} + b^o \qquad (5)$$

3.3 Linearly Nonseparable Patterns

For data that is not linearly separable, an optimal hyperplane is still required that minimises the probability of classification error, averaged over the training set. *Slack variables* $\{\xi_i\}_{i=1}^{N}$, where $0 \leq \xi_i \leq 1$, are introduced into the definition of the separating hyperplane such that

$$y_i(\mathbf{w} \cdot \mathbf{x}_i + b) \geq 1 - \xi_i, i = 1, ..., N \qquad (6)$$

with $\xi_i > 1$ to allow the for examples that violate Eq. (2). Therefore, $\sum \xi_i$ is an upper bound on the number of training errors. Minimising

$$\frac{1}{2} \mathbf{w} \cdot \mathbf{w} + C \sum_{i=1}^{N} \xi_i \qquad (7)$$

subject to constraints of (6) and $\xi_i \geq 0$ results in the optimal hyperplane. Formulation of Eq. (7) is therefore in accord with the principle of structural risk minimisation. The user defined parameter C controls the trade-off between the complexity of the machine and the number of nonseparable points.

3.4 Nonlinear Support Vector Machines

When the input data is mapped to some other, possibly infinite dimensional, Euclidean space, $\mathbf{x}$ is replaced by its mapping in the *feature space* $\Phi(\mathbf{x})$, and Eq. (3) is replaced by

$$W(\alpha) = \sum_{i=1}^{N} \alpha_i - \frac{1}{2} \sum_{i,j=1}^{N} \alpha_i \alpha_j y_i y_j \Phi(\mathbf{x})_i \cdot \Phi(\mathbf{x})_j$$

There exist *kernel functions* K such that $K(\mathbf{x}_i, \mathbf{x}_j) = \Phi(\mathbf{x})_i \cdot \Phi(\mathbf{x})_j$, i.e Φ need not be given explicitly. we do not need an explicit form for then only K would be needed in the training algorithm, and Φ would not be explicitly used. Thus, if $\mathbf{x}_i \cdot \mathbf{x}_j$ were to be replaced by $K(\mathbf{x}_i, \mathbf{x}_j)$, the training algorithm would consist of maximizing

$$W(\alpha) = \sum_{i=1}^{N} \alpha_i - \frac{1}{2} \sum_{i,j=1}^{N} \alpha_i \alpha_j y_i y_j K(\mathbf{x}_i \mathbf{x}_j) \qquad (8)$$

and classification of an input point $\mathbf{x}$ then simply consist of calculating the sign of

$$f(\mathbf{x}) = \sum_{i=1}^{N} \alpha_i y_i K(\mathbf{x}_i, \mathbf{x}) + b \qquad (9)$$

4 The System

In this section the images in the database are discussed and the classification process of the orientation of the films is explained.

Fig. 3. Grouping of images to form an inverted and correctly orientated film

4.1 Image Database

The image database used contained 5229 images, of various dimensions. The images were taken from the Corel Stock Photo Library, and included some of the following groups : *Sunrises*, *Deserts*, *Aviation*, *Glaciers*, *Lions* and *Landscapes*. The images were then randomly rotated, either 0° or 180°. The SVM was then trained on a subset of the database. The remaining images where then put into groups of 36 images, according to their orientation, to represent a complete film (Fig. 3). A few examples taken from the image database are shown in Fig. 4. We decided to limit the images in the training database to those of natural, unposed scenes. We included a few "unnatural" images such as still lifes, portraits, indoor scenes and images with unusual distribution of colours, contrasts and textures (see Fig. 7).

Fig. 4. A few of the pictures in the picture database

4.2 Film Classification

Once the SVM has been trained on the training set, the film of images is then classified. The output f of the SVM for each image derived in Eq. (5) and Eq. (9) for linearly separable and nonseparable patterns respectively, is then passed through a sigmoidal function

$$P(y=1|f) = \frac{1}{1 + e^{(Af+B)}}$$

where P(class—input) is the posterior probability (Fig. 5) and A and B are fitted using maximum likelihood estimation from the training set, as described in[5].

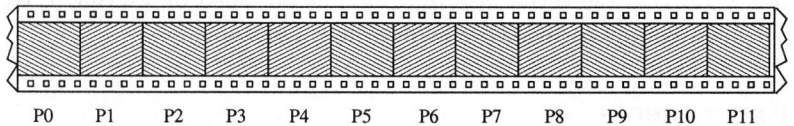

Fig. 5. Probabilities calculated for each image on the film

The further away an input vector is from the separating hyperplane, the stronger the probablity that it has been correctly classified. A negative probability means that the image is classified as inverted, and a positive probability means that the image is classified as correctly orientated.

$$P_{0°} = \log(P_{0°}1) + \cdots + \log(P_{0°}n) + \log(1 - P_{180°}1) + \cdots + \log(1 - P_{180°}n) \quad (10)$$

$$P_{180°} = \log(P_{180°}1) + \cdots + \log(P_{180°}n) + \log(1 - P_{0°}1) + \cdots + \log(1 - P_{0°}n) \quad (11)$$

All the postive probabilites for the entire film are then logarithmically added together (Eq. 10), as are all the negative probabilites (Eq. 11), and the larger of the two numbers is used to determine the orientation of the film.

5 Kernel Selection and Input Mapping

In [8] we investigated how different kernels and input remappings can improve classification accuracy. We specifically looked at the kernel of the form

$$K_{gauss}(\mathbf{x}, \mathbf{y}) = e^{-\rho \|\mathbf{x}-\mathbf{y}\|^2}$$

where ρ is a parameter specified by the user.

K_{gauss} produces a Gaussian Radial Bias Function (RBF) classifier. For K_{gauss}, the number of centers , the centers themselves , the weights α_i, and the threshold b are all determined by the SVM training algorithm; they have better results than classical RBF classifiers trained with standard methods [6].

More general forms of RBF kernels were investigated in [3] such as

$$K_{d-RBF}(\mathbf{x}, \mathbf{y}) = e^{-\rho d(\mathbf{x},\mathbf{y})}$$

where $d(\mathbf{x}, \mathbf{y})$ can be chosen to be any distance in the input space. It was shown that non-Gaussian kernels with exponential decay rates that are less than quadratic can lead to very good performance. That kernel can be further generalised to $K(\mathbf{x}, \mathbf{y}) = e^{-\rho d_{a,b}(\mathbf{x},\mathbf{y})}$ with

$$d_{a,b}(\mathbf{x}, \mathbf{y}) = \sum_i |x_i^a - y_i^a|^b$$

Special forms of the kernel K_{d-RBF} could be obtained by varying b. We limited our investigation to the three kernels shown in Table 1.

6 Experiments

In this section, we present results from various experiments using different input sets and kernels in the classification of film orientation. We used Thorsten Joachims SVM^{light} [4] implementation of SVMs.

Table 1. Kernels obtained for the different choices of the parameter b

b	Kernel
2	$K_{gauss}(\mathbf{x},\mathbf{y}) = e^{-\rho(\mathbf{x}^a - \mathbf{y}^a)^2}$
1	$K_{lap}(\mathbf{x},\mathbf{y}) = e^{-\rho|\mathbf{x}^a - \mathbf{y}^a|}$
0.5	$K_{sublinear}(\mathbf{x},\mathbf{y}) = e^{-\rho\sqrt{(\mathbf{x}^a - \mathbf{y}^a)}}$

6.1 Feature Extraction

Global image features are not invariant to image rotation; thus, we selected local regions within the images, and extracted low-level features. Each image was represented by $N \times N$ blocks; from each block, we extracted the mean and variance of RGB colours. An example of this feature extraction is shown in Fig. 6. Thus for each image we extracted $6N^2$ features.

We randomly assigned 4706 of the 5299 images in the data base to the training set; we used the remaining 523 images for testing purposes. We trained linear SVMs on the training set and investigated the impact of the number of regions on their classification performance. We show in Table 2 the classification accuracies of the training and test sets, respectively. In [8] we observe that the classification performance increases with increasing number of regions in an image for the test set, but this is not the case when a larger database is used; the performance on the training set is dependent on the number of regions.

(a) Orignal image (b) Average of local regions for N=10

Fig. 6. The representation of the image in terms of the average of each local region

Table 2. Classification performance of SVMs of individual images with linear kernel on training and test set as a function of the number of local regions

N^2	Test Set (%)	Training Set (%)
6	77.44	80.10
8	77.86	81.59
10	78.24	82.24
12	77.97	83.70
15	78.28	84.27
17	77.82	85.46
20	78.09	86.53

6.2 Number of Local Regions

As mentioned in the Section 6.1, the number of local regions in an image influences both the classification performance and the processing time. An increase in the number of local regions is particularly noticeable for SVMs that use RBF kernels (see Table 3).

6.3 Kernel Selection

In [8] we discussed the use of modified versions of the RBF kernel to improve the classification of individual images. While very similar results were obtained on the larger database, the use of these RBF kernels in the classification of an entire film showed no advantages when compared to the linear kernel. Although the number of individual images in the film that are correctly classified increases slightly, the incurred computational delay does not make these methods feasible when classifying entire films.

6.4 Input Remapping

In [8] we made use of input remapping when using the RBF kernels to improve the classification accuracy. On the larger database, the use of input remapping showed very little improvement in classification performance when using individual images, and so it was decided not to make use of input remapping when it came to the classification of entire films.

6.5 Results

Using ten fold cross vaildation on the database, all films that were presented to the classifier were correctly classified, for all the kernels used. In Table 3the classification time[1] of each film is compared for the various types of kernels. As is evident, the linear kernel has the fastest classification time, and is considerably

---
[1] On a Pentium III 450Mhz

faster than any of the RBF kernels. The Gaussian kernel function suffers from over generalisation as the number of local regions per image is increased. As a result of this an accuracy of only 80% was achieved on classification of films, while an accuracy of 100% was achieved using the other kernels.

Table 3. Classification time (secs) of an entire film using different kernels as a function of the number of local regions

N^2	Classifier			
	Linear	Gaussian	Laplacian	Sublinear
6	¡1	19	23	17
8	¡1	37	44	30
10	¡1	59	54	48
12	¡1	85	106	71
15	¡1	135	137	114
17	¡1	170	171	147
20	1	235	241	208

Fig. 7. A selection of images incorrectly classified

7 Conclusions and Future Work

Automatic detection of film orientation is a difficult problem due to the high variability of image content. We were able to achieve 100% correct detection of test films by training SVMs on low level features; these features consisted of the mean and variance of RGB colours of local regions of images. SVMs train quickly and show excellent generalisation performance. Techniques such as kernel modification and input remapping can improve classification.

Future work includes detection of four orientations of images with multiclass SVMs, automatic detection and correction of exposure levels of images, and extraction of rules in symbolic form from trained SVMs.

References

1. P. Bartlett and J. Shawe-Taylor. Generalization performance of support vector machines and other pattern classifiers. *Advances in Kernel Methods - Support Vector Learning*, 1998.
2. C. Burges. A Tutorial on Support Vector Machines for Pattern Recognition. *Data Mining and Knowledge Discovery*, 2(2):121–167, 1998.
3. O. Chapelle, P. Haffner, and V. Vapnik. SVMs for Histogram-Based Image Classification. *IEEE Trans. on Neural Networks*, 9, 1999.
4. T. Joachims. Making Large Scale SVM learning practical. In B. Scholkopf, C. Burges, and A. J. Smola, editors, *Advances in Kernel Methods - Support Vector Learning*, pages 169–184. MIT Press, 1999.
5. J. Platt. Probabilistic outputs for support vector machines and comparison to regularized likelihood methods. In A.J. Smola, P. Bartlett, B. Scholkopf, and D. Schuurmans, editors, *Advances in Large Margin Classiers*. MIT Press, 1999.
6. B. Scholkopf, K. Sung, C. Burges, F. Girosi, P. Niyogi, T. Poggio, and V. Vapnik. Comparing support vector machines with Gaussian kernels to radial basis function classifiers. *IEEE Trans. Signal Processing*, 45(11):2758–2765, 1997.
7. Vladimir N. Vapnik. *The Nature of Statistical Learning Theory*. Springer Verlag, Heidelberg, DE, 1995.
8. D. Walsh and C.W. Omlin. Automatic Detection of Image Orientation with Support Vector Machines. Technical Report TR-UWC-CS-01-01, University of the Western Cape, 2001.

A Generic Approach for the Vietnamese Handwritten and Speech Recognition Problems

Vu Hai Quan[1], Pham Nam Trung[1], Nguyen Duc Hoang Ha[1],
Lam Tri Tin[1], Hoang Kiem[1], and An H Nguyen[2]

[1] Faculty of Information Technology, University of HoChiMinh, Vietnam
vhquan@fit.hcmuns.edu.vn
[2] Digital Vision Lab, Northwestern Polytechnic University, CA, USA
ahnguyen@npu.edu

Abstract. In this paper, we propose an approach based on HMM and linguistics for the Vietnamese recognition problem, including handwritten and speech recognition. The main contribution is that our method could be used to model all Vietnamese isolated words by a small number of HMMs. The method is not only used for handwritten recognition but also for speech recognition. Furthermore, it could be integrated with language models to improve the accuracy. Experimental results show that our approach is robust and considerable.

1 Introduction

Vietnamese speech has been created for 4000 years, closely related with Indo-European languages. Today there are more than 70.000.000 people using this language. The main feature which makes it differs from Western languages is that it belongs among the group of mono - syllable languages. That means it never changes its morphology. In order to express grammatical sense we usually use means of the outside word as grammatical words, order words...

Vietnamese script is the unique Orient language derived from Latin. Written literature first appeared around the 10th century. Until 19th, literature works were written in Han (classical Chinese) and Chu Nom. In the 20th century, the country's literature has been written in Quoc Ngu (Romanized national language which was found by West Clergymen).

Vietnamese has about 8000 isolated words [1]. Each of them consists of two main components: consonant and syllable for isolated speech word (see Fig 1) or consonant and syllable coordinating with accent in the accent set for isolated handwritten word (see Fig 2).

There are 15 accents (Fig. 3), 26 consonants (Table 1), and 434 syllables.
In the recent publications, we have reported a series of papers for the recognition of Vietnamese document images, including both handwritten images and printed document images based on HMM and linguistics [2][3].

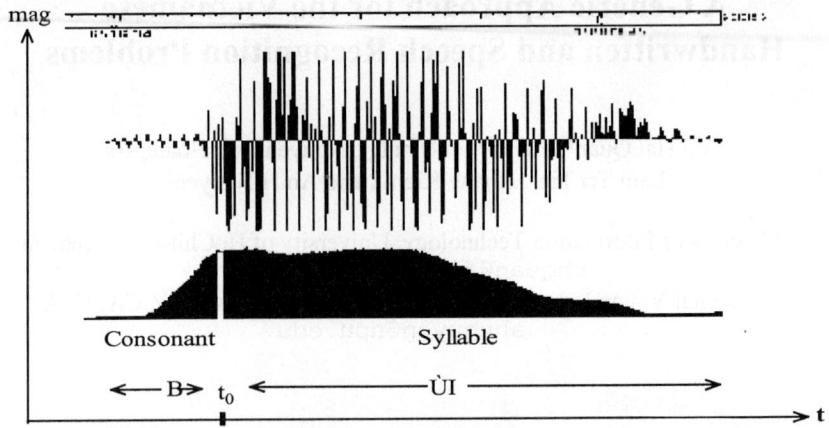

Fig. 1. Structure of Vietnamese isolated word speech

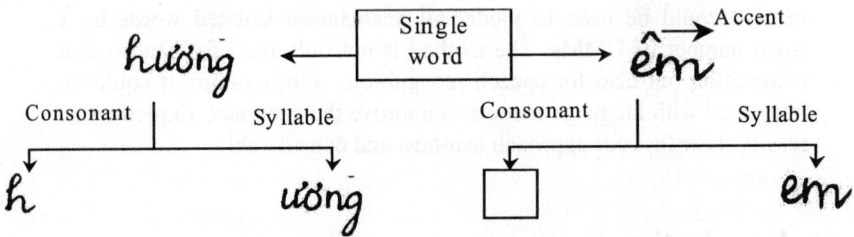

Fig. 2. Structure of Vietnamese isolated handwritten word

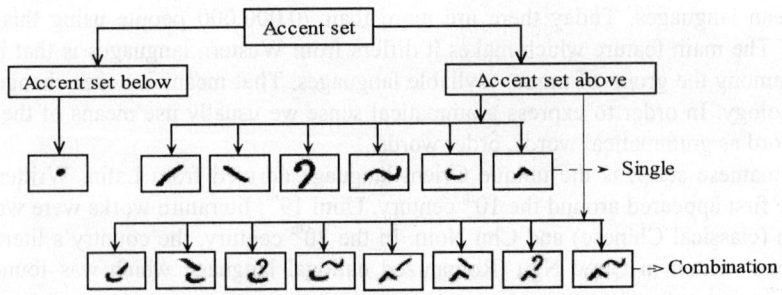

Fig. 3. Accent set

Table 1. Consonants

b	c	ch	d	đ	gh	h	k	kh	l	m	
n	ng	nh	p	ph	q	r	s	t	tr	v	x

Table 2. Syllabes without accents

```
a  oa  ac  oac  ach  oach  ai  oai  am  oam  an  oan  ang
oang  anh  oanh  ao  oao  ap  oap  at  oat  au  oau  ay  oay
uac  uan  uang  uap  uat  uay
e  oe  ec  em  en  oen  eng  oeng  eo  oeo  ep  oep  et  oet
ue  ech  uech  uen  enh  uenh  uet  eu
i  uy  ia  uya  ich  uych  iec  iem  ien  uyen  ieng  iep  iet  uyet
ieu  yeu  im  in  uynh  ip  uyp  it  uyt  iu  uyu
o  oc  oi  om  on  ong  ooc  oong  op  ot
uơ  ơi  ơm  ơn  ơp  ơt
ua  uc  ui  um  un  ung  uoc  uoi  uom  uon  uong  uot  up  ut
ưa  ưc  ưi  ưm  ưn  ưng  ươc  ươi  ươm  ươn  ương  ươp  ươt  ưu  ưt  ưu
```

In this paper, we will present a new approach, which could be used to model both handwritten and speech words in the unique system. The main contribution is based on the structure of a Vietnamese word (see Fig. 1, 2), we could model all Vietnamese isolated words by a small number of HMMs. Each isolated word will be modeled by two HMMs. The first one is used for consonant and the second one is used for syllable. There are total of 460 HMMs used in the system, 26 HMMs for consonants and 434 for syllables.

Furthermore, our method also allowed integrating with language models to improve the accuracy. An overview of the system is illustrated in Figure 4.

In the next section, we will describe the system in detail.

2 Vietnamese Recognition System

2.1 Hidden Markov Model

A hidden Markov model (HMM) has the same structure as a Markov chain, with states and transition probabilities among the states, but with one important difference: associated with each state in a Markov chain is a single "output" symbol, while in a HMM, associated with each state is a probability distribution on all symbols. Thus, given a sequence of symbols produced by a model, we cannot unambiguously determine which state sequence produced that sequence of symbols; we say that the sequence of state is hidden. However, we can compute the sequence of states with the highest probability of having produced the observed symbol sequence. If we associate symbols with feature vectors, then the recognition problem can be formulated as finding the sequence of states that could have produced the sequence of feature vectors with the highest probability. Because of the Markov property of HMMs, the search for the most likely word sequence can be computed very efficiently using the Viterbi algorithm. HMMs have some properties that make them desirable:

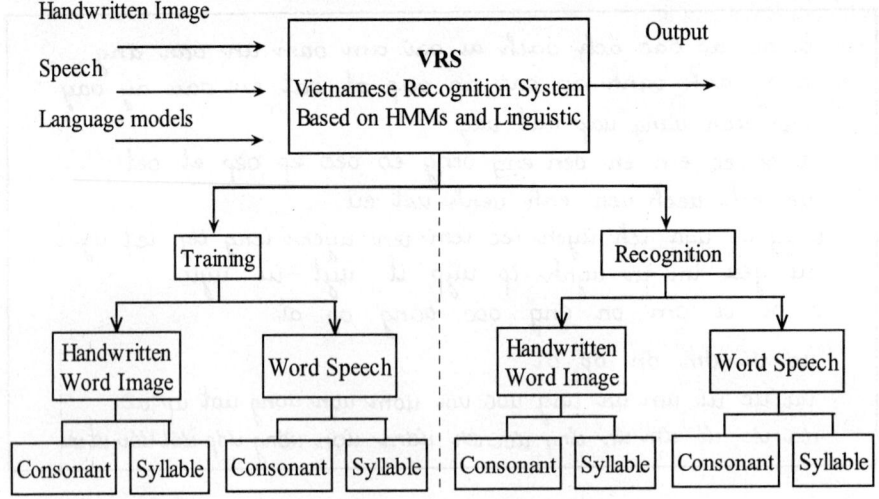

Fig. 4. Vietnamese Recognition System

HMMs provide a rigorous and flexible mathematical model of variability in feature space as a function of an independent variable (time for speech and position for text).

Segmentation and recognition are performed jointly using efficient training and search techniques.

Training is performed automatically without any manual segmentation if data.

Higher-level constraints (in the form of language models, for example) can be applied as part of the recognition process, if desired, instead of applying them as a post-processing.

The techniques are language-independent in principle, requiring only sufficient data and a lexicon from the new language in order to recognize that language.

Perhaps the most important of these properties is that the HMM parameters can be estimated automatically from the training data, without the need to either pre-segment the data or align it with the text. The training algorithm requires:
- A set of data to be used for training.
- The transcription of the data into a sequence of words
- A lexicon of the allowable a set of consonants, syllables and words.

HMM training algorithm automatically estimates the parameters of the models and performs the segmentation and recognition simultaneously, using an iterative scheme that is guaranteed to converge to a local optimum.

HMM is an effective method for speech recognition system [4]. The good performances are also reported in the on-line [5], off-line [6-9] and printed character recognition system [10-11]. For recognition of the off-line handwriting, the most difficult problem is not only because of the great variety in the shape of characters but also because of the overlapping and the interconnection of the neighboring characters.

Until now, to the best of our knowledge, there are two ways to resolve this problem. The first one is to model each character as a HMM and using Level Building [11] in recognizing process. But by this way, when observed in isolation, characters are often ambiguous, leading to fault recognition result. So it required context to minimize the classification errors. The second one is to model whose word as a HMM [9]. So, the large amount of samples to be recognized would be the most obstacles of this approach.

In the next section, we will describe our system in detail and show how we solve the above problem.

2.2 The System in Detail

2.2.1 Feature Extraction

Handwritten Image Features

The hidden Markov models is used for recognition expects as input a sequence of feature vectors for each unknown word to be recognized. To extract such a sequence of feature vector from a word image, a sliding window is used. A window of one column width and the word's height is moved from left to right over the word to be recognized. At each position of the window nine geometrical features are determined.

The first three features are the number of black pixels in the window, the center of gravity, and the second order momentum. This set characterizes the window from the global point of view. It describes how many pixels in which region of the window are, and how they are distributed.

For a more detailed description of the window, features fourth to ninth give the position of the upper- and the lower most contour pixel in the window, the orientation of the upper and lower contour pixel, the number of black-white transitions in vertical direction and the number of black pixels between the upper and lower contour. To compute the orientation of the upper and lower contour, the contour pixels of neighboring windows to the left and to the right are used. Noticed that all these features can be easily computed from the word images.

Speech Features

Mel-scale frequency cepstral (MFCC) has been successfully used in speech detection and recognition. To characterize the spectral property of cheering sound, MFCC and its derivatives are used in this system. The digitized waveform is converted into a spectral-domain representation; one of two sets of features may be used, depending on the recognizer. For the current general-purpose recognizer, we use twelve Mel-frequency cepstral coefficients (MFCC coefficients), twelve MFCC delta features that indicate the degree of spectral change, one energy feature, and one delta-energy feature (for a total of 26 features per frame).

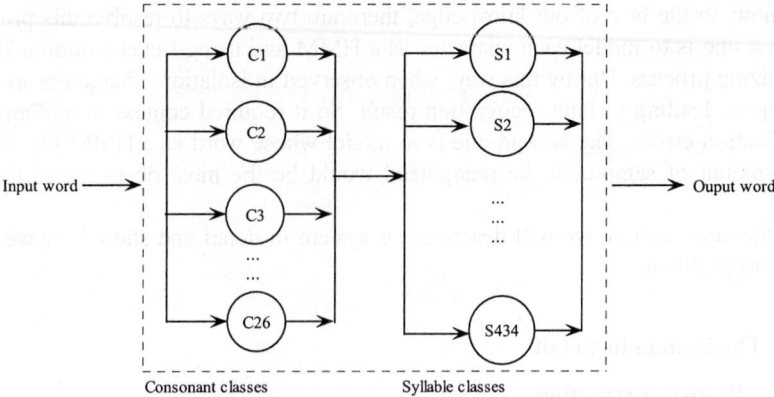

Fig. 5. Vietnamese isolated word models

2.2.2 Consonant and Syllable Model

For the purpose of word recognition, it is useful to consider left-to-right models. In a left-to-right model transition from state i to state j is only allowed if $j>=i$, resulting in a smaller number of transition probabilities to be learned. The left-to-right HMM has the desired properties that it can readily model signals whose properties change over time. It is useful for building word models from consonant and syllable models because fewer between character transition probabilities are required for constructing the word models from the corresponding character models (Fig. 5).

The first problem one faces is deciding what the states in each consonant or syllable class correspond to, and then deciding how many states should be in each model. We optimized the structure for the consonant and syllable hidden Markov models by searching for the best number of states to be fixed among all classes. We allow skipping only one state such that if a consonant or syllable class needs fewer states some of the skipping probabilities could be relatively high.

2.2.3 Training

A continuous density HMM is trained for each consonant or syllable class using the transition and gradient features computed inside word images. We varied the number of states (from N = 5 to N =18) for each model and the number of Gaussian mixtures per state (from M = 1 to M = 3). Initially, the mean and covariance matrix for each state mixture were estimated after dividing the re-sampled training transition vectors equally between the states.

The standard re-estimation formula was used inside the segmental K-mean algorithm assuming diagonal covariance matrices for the Gaussian mixtures. The Viterbi algorithm was used to find the optimal state sequence and assign each of the observation vectors to one of the states. The probability of the observation vectors given the model was used to check whether the new model is better than the old model inside the training procedure.

2.2.4 Recognition

The input to the word recognition algorithm is a data word (word image or word speech) and a lexicon. After preprocessing the data word, the resultant data is subjected to a feature extraction process. The output of this process is a sequence of observation vectors, and each of them corresponds to a data column. Using the consonant and syllable models, a word model is constructed for each string inside the lexicon. The string matching process computes a matching score between the sequence of observation vectors and each word model using the Viterbi algorithm. After postprocessing, a lexicon sorted by the matching score is the final output of the word recognition system (as shown in Fig. 6).

3 Experiments and Results

To show the efficiency of our proposed approach, we chose to work on a real application: recognizing the text line of Vietnamese handwritten name in the passport. An illustration of the application is in Fig 7.

Input to the system is a handwritten image word. This image was firstly recognized by module named "Handwritten Words Recognition". The result is words recognized and they will be corrected by module "Speech Words Recognition" if there is any mistake.

We applied our above model to the recognition of Vietnamese names including both handwritten and speech from the database of the Department of Immigration, Ministry of Public Security. The data were collected by the immigration control in the International Airport in Ho Chi Minh city from Vietnamese passengers. They were divided into three sets: first name, middle names and last name. Each of them is usually an isolated word. These data were divided into training and testing set (approximately 10% for testing and the remainder for training, as shown in table 3).

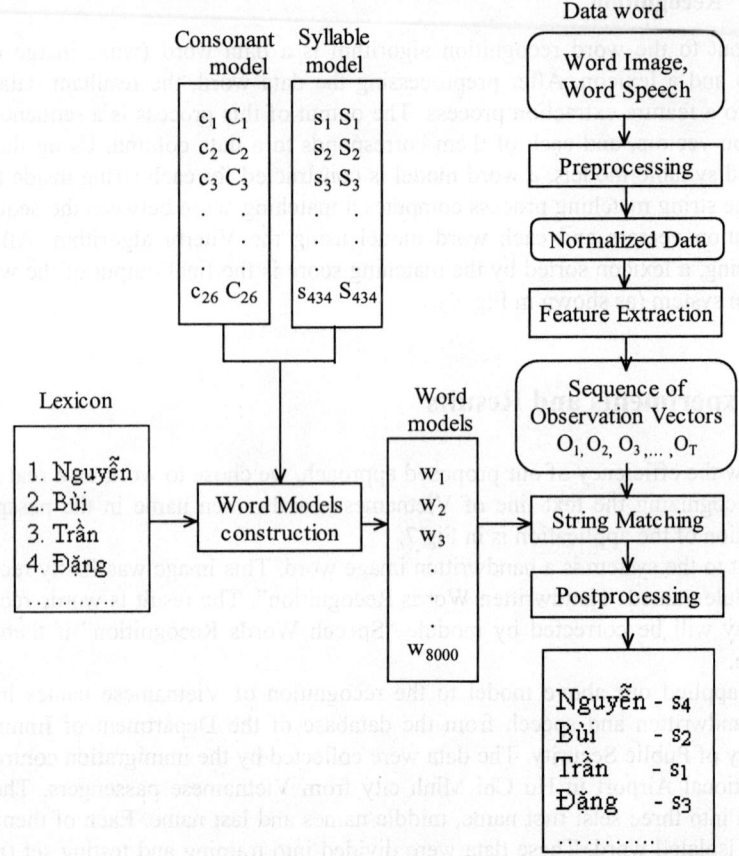

Fig. 6. Word recognition systems

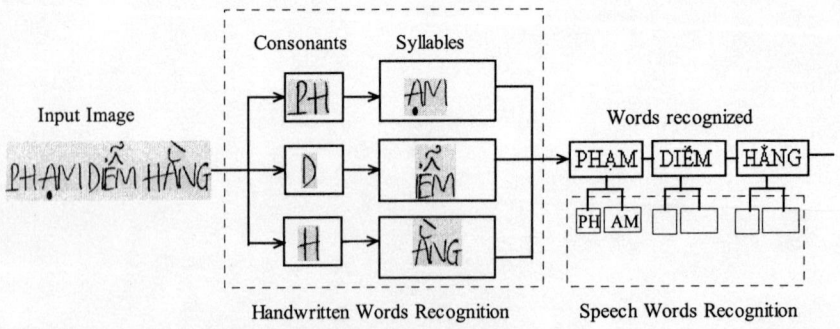

Fig. 7. Vietnamese name recognition system

Table 3. Training and testing data sets

Number of handwritten images and speech in the training and testing sets				
Subset	Training set		Testing set	
	Images	Speech	Images	Speech
First name	8480	6590	720	610
Middle name 1	6260	7230	960	760
Middle name 2	7590	8230	210	810
Middle name 3	6310	6430	200	620
Last name	6630	6720	820	640
Full name	8420	8480	370	810
Total	43690	43680	3280	4250

Table 4. Word error rates

Recognition module	Word error rates
Handwritten Words Recognition	3.8
Speech Words Recognition	2.6
Combined module	1.4

4 Conclusion

The article proposed a robust way to model and recognize Vietnamese isolated word, including both handwritten and speech based on HMM in the unique system. With this approach, we could model all Vietnamese isolated words with a small number of HMMs, which could be applied in a very large vocabulary recognition context. A real application was also introduced. Experiment results showed that this proposed approach was robust and flexible.

References

1. Hoang Phe, *Syllable Dictionary*, Danang publisher, Vietnam 1996. (In Vietnamese)
2. Vu Hai Quan et al, "A System for Recognizing Vietnamese Document Image Based on HMM and Linguistics", *Proc. Int., Conf. Document Analysis* (ICDAR'01), pp 627-630 Seattle, USA, 2001.
3. Vu Hai Quan et al, "Models for Vietnamese Document Image Recognition", ICISP2001, pp 484 – 491, Agadir, Marocco, 2001.
4. L. R Rabiner, "A Tutorial on Hidden Markov Models and Selected Applications in Speech Recognition", *Proc. IEEE,* 77(2), pp. 257-286, 1989.
5. J.-I. Lee, J.-H. Kim and M. Nakajima, "A Hierarchical HMM Network-Based Approach for On-Line Recognition of Multilingual Cursive Handwrittings ", *IEICE Trans*, E81-D(8), pp. 881-888, 1988.

6. H.-S. Park and S. -W. Lee, "A Truly 2-D Hidden Markov Model for Off-Line Handwritten Character Recognition ", *Pattern Recognition*, 31(12), pp. 1849-1864, 1998.
7. Magdi A. Mohamed, Paul Gader, "Generalized Hidden Markov Models – Part II: Application to Handwritten Word Recognition", *"IEEE Transactions on Fuzzy System"*, vol. 8(1), pp. 82-94, 2000.
8. U. Marti and H. Bunke "Text Line Segmentation Based Word Recognition with Lexicon", *Proc. Int., Conf. Document Analysis* (ICDAR'01), p Seattle, USA, 2001.
9. Didier Guillevic and Ching Y. Sen, "HMM Word Recognition Engine", http://cenparmi.concordia.ca
10. John Makhoul, Richard Schwartz, Christopher Lapre and Issam Bazzi, "A Script-Independent Methodology for Optical Character Recognition", *Pattern Recognition*, 31(9), pp.1285-1294, 1998.
11. A. J. Elms and J. Illingworth "Modelling Polyfont Printed Characters with HMMs and a Shift Invariant Hamming Distance ", *Proc. Int., Conf. Document Analysis*.

Efficient and Automatic Faces Detection Based on Skin-Tone and Neural Network Model

Bae-Ho Lee[1], Kwang-Hee Kim[1], Yonggwan Won[1], and Jiseung Nam[1]

[1] Dept. of Computer Engineering, Chonnam National University
300 Yongbong-Dong, Buk-Gu, Kwangju, 500-757, Korea
{bhlee,kshift,ygwon,jsnam}@chonnam.chonnam.ac.kr

Abstract. In this paper, we consider the problem of detecting the faces without constrained input conditions such as backgrounds, luminance and different image quality. We have developed an efficient and automatic faces detection algorithm in color images. Both the skin-tone model and elliptical shape of faces are used to reduce the influence of environments. A pre-built skin color model is based on 2D Gaussian distribution and sample faces for the skin-tone model. Our face detection algorithm consists of three stages: skin-tone segmentation, candidate region extraction and face region decision. First, we scan entire input images to extract facial color-range pixels by pre-built skin-tone model from YCbCr color space. Second, we extract candidate face regions by using elliptical feature characteristic of the face. We apply the best-fit ellipse algorithm for each skin-tone region and extract candidate regions by applying required ellipse parameters. Finally, we use the neural network on each candidate region in order to decide real face regions. The proposed algorithm utilizes the momentum back-propagation model to train it for 20*20 pixel patterns.
The performance of the proposed algorithm can be shown by examples. Experimental results show that the proposed algorithm efficiently detects the faces without constrained input conditions in color images.

1 Introduction

Automatic detection and identification are important components in a pattern recognition system. In this sense, this face detection is an essential technique in face identificnation, facial expression recognition, personal identification for security system and etc. Previous face identification methods have been focused on the recognition with the assumption that the location of the face is known *a priori*. It is not easy works to extract from general scene images without *a priori* information.

In a non-constrained scene, the face has two main features – the inner and outer feature. The inner features such as specific color range, symmetrical shapes and facial components are very important clues for the face detection. But, outer features that range various sizes, positions to the rotations of the face are one of the reasons that complicate the face detection.

There are a great number of face detection and identification methods. Some use template matching [2,3], KL(Karhunen-Loeve) expansion [4,5], Garbor wavelet transform [1] with gray-level images. These methods have made good results under constrained conditions. Most of these methods use well-framed faces or frontal-upright view of faces. However, some of recent proposed methods use skin color information [7]. Others use color and motion information [6] and use YCbCr skin-tone for MPEG video streams [10] and quantized skin color and wavelet packet analysis [8]. Neural networks are also used for more flexible results [9].

In this paper, we propose efficient and automatic faces detection algorithm in color images under non-constrained input conditions, such as complex backgrounds, various luminance and image quality. We use both a skin-tone model and an elliptical shape of face. The skin-tone model is a pre-built skin color model based on 2D Gaussian distribution and sample faces. The skin-tone can be used as a good pre-processing filter to separate facial color-range pixel from the background. Our face detection algorithm consists of three stage: skin-tone segmentation, candidate region extraction and face region decision. First, we scan entire input images to extract facial color-range pixels by skin-tone. Second, we extract candidate face regions by using the elliptical feature of face. We apply the best-fit ellipse algorithm for each region and extract candidate regions. Finally, we apply the neural network on each candidate regions to decide real face regions. We use a momentum back-propagation model to train the neural network and use 20*20 face and non-face patterns. It is shown by 310 face images and 360 random non-face images to train neural networks.

2 Face Detection Algorithm

2.1 Skin-Tone Segmentation

The skin-tone segmentation is the most important step of an entire detection process. The result of this stage affects the performance of detection. The skin-tone has specific color-ranges that distinguish from the colors of the most natural objects. It can be used as a good pre-processing filter to separate facial color regions from a complex background. Although human skin colors differ from person to person and race to race, they are mapped into a small color range on the chrominance plane. This means that skin colors are relatively consistent in hue and saturation. The only different thing is intensity [11].

Skin-Tone Generation. There are many color space models to represent the colors, such as HSI, YIQ, YCbCr, and etc. We use YCbCr color space model for skin-tone. Human skin colors are only mapped into one of quadric plane on Cb-Cr chrominance space.

The conversion relationship between RGB and YCbCr can be given the following:

$$\begin{bmatrix} Y \\ Cb \\ Cr \end{bmatrix} = \begin{bmatrix} 0.299 & 0.587 & 0.114 \\ -0.169 & -0.331 & 0.500 \\ 0.500 & -0.419 & -0.081 \end{bmatrix} \begin{bmatrix} R \\ G \\ B \end{bmatrix} \qquad (1)$$

We use 90 samples of face images from different races to generate skin-tone color statistics on the Cr-Cb plane. Table 1 shows the average Cb, Cr and each skin color category. We categorize the face image samples according to its intensity. For example, the categories can be divided into dark, light and medium. The skin-tone has a center point at (-24, 30) as shown in Table 1.

Table 1. Sample face set

Skin color	Counts	Cb	Cr
Light	30	-25.8	33.2
Medium	30	-23.5	25.7
Dark	30	-23.2	30.4
Total	90	-24.2	29.8

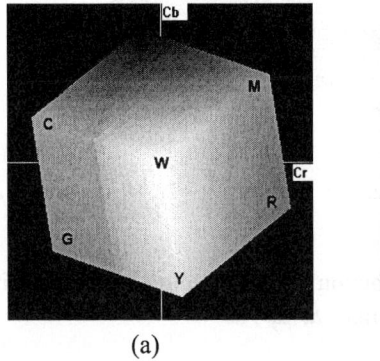
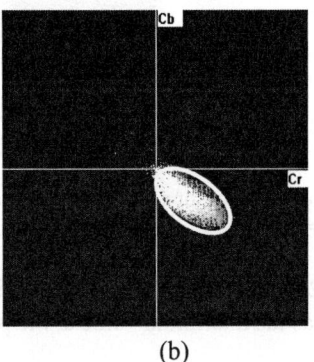

(a)   (b)

Fig.1. Skin-tone color space: (a) Cb-Cr plane, (b) Sample skin-tone

Fig. 1 shows the skin-tone color space and sample skin-tone distribution. One point in the hexagon actually corresponds to all the colors that have the same chrominance value but different intensity levels. For example, all gray color levels that range from black to white are mapped to the origin(W) in Fig. 1(a). Compared with Fig. 1(a), it is clear that skin-tone colors are distributed over a small range in the Cb-Cr plane.

Skin-Tone Modeling. Fig. 1(b) shows the sample facial color distribution sample and the entire shape of skin-tone sample is an ellipse on Cb-Cr plane. Thus, we need to make a skin-tone model to apply to input images.

Christophe *et al.*[11] use YCbCr and HSV color space model to extract skin color pixels. They use Y factor and planar approximations to measure the skin color subspace. On the other hand, Wang and Chang[16] only use chrominance place(Cb-Cr) without regarding the intensity Y factor.

In this paper, we use chrominance plane without Y factor and 2D Gaussian function to make a skin-tone model and it is shown as

$$G(x,y) = \frac{e^{-\frac{1}{2(1-\rho_{xy}^2)}\left[\left(\frac{x-\mu_x}{\sigma_x}\right)^2 - \frac{2\rho_{xy}(x-\mu_x)(y-\mu_y)}{\sigma_x\sigma_y} + \left(\frac{y-\mu_y}{\sigma_y}\right)^2\right]}}{2\pi\sigma_x\sigma_y\sqrt{(1-\rho_{xy}^2)}} \qquad (2)$$

Fig. 2 shows the sample skin-tone model based on 2D Gaussian distribution

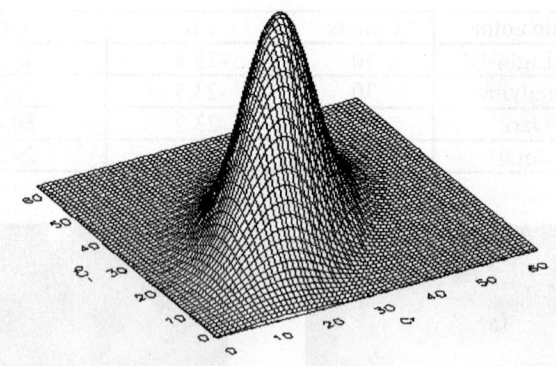

Fig. 2. Sample skin-tone model by 2D Gaussian function: σ_x =10.94, σ_y =10.98, ρ_{xy} = 0.497, μ_x=28, μ_y=-32

Fig. 2 shows the 2D Gaussian distribution $G(x,y)$ based on sample face images and $G(x,y)$ has the maximum value at the point (μ_x, μ_y) as

$$G(\mu_x, \mu_y) = \frac{1}{2\pi\sigma_x\sigma_y\sqrt{(1-\rho_{xy}^2)}} \qquad (3)$$

Skin-Tone Decision. As shown in (6), the maximum value and the other values of $G(x,y)$ are too small. Thus, we multiply the weights to $G(x,y)$ to normalize the maximum output as 1. We use it as a classification threshold(Th) for the skin-tone.

The threshold(Th) is described by

$$Th = p(x,y)\left(2\pi\sigma_x\sigma_y\sqrt{(1-\rho_{xy}^2)}\right)G(x,y) \qquad (4)$$

In (4), $p(x,y)$ is a probability at point (x,y) that evaluates from the sample skin-tone. We select a decision threshold as $Th = 0.01$ by experiments.

Fig. 3 shows the result of each threshold value. Fig. 3(a) and (b) show the original image and the skin-tone image at $Th = 0.05$, respectively. The result obtained by Th=0.05 may lead to a poorer decision result compared to the smaller Th. However, Fig. 3(c-d) show good results and the threshold values are 0.01 and 0.005.

2.2 Candidate Region Extraction

In this stage, we decide how to locate candidate face regions in the skin-tone filtered image. Skin-tone regions have to be segmented in order to form potential face regions which will be classified in the final stage of the proposed algorithm. So we compose this stage into two step: as region segmentation and shape constraints. In the first step, we segment the skin-tone image into regions. In the next step, we extract the candidate regions by shape constraint conditions.

Region Segmentation. The skin-tone extraction is performed with pixel by pixel.

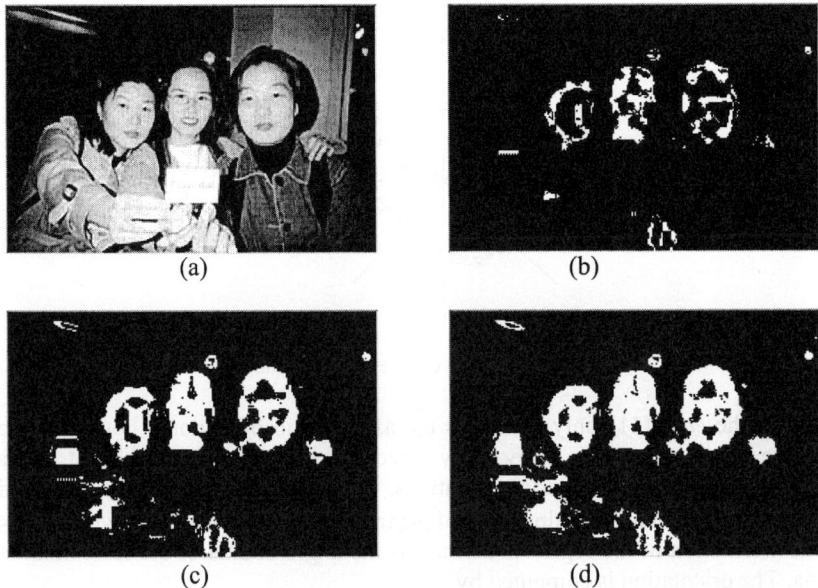

Fig.3. Skin-tone decision results: (a) original image, (b) $Th = 0.05$, (c) $Th = 0.01$, (d) $Th = 0.005$

The result images obtained by skin-tone filter are not complete region forms they show the various shapes by the background, especially when parts of the surrounding background have the same skin. Therefore we need to make regions, which are composed by neighbor pixels. This process leads to a better precision in locating the candidate regions and helps to segment the faces from the background.

This step consists of two level stages. First, we perform the filtering to skin-toned image to remove noise pixel and to make pixel groups. We apply 5*5 median filter to skin-tone image. Next, we segment pixel groups to each region for next steps. In order to segment the regions, we use the simple region segmentation algorithm that is called region-growing. And we performed the test to each region whether the size of regions is larger than pre-defined condition or not. If it is smaller than the condition, we assume that it is not a face region and remove it. We limit the use of area size to 10*10 pixels in order to obtain better performance.

Shape Constraints. The main purpose of this step is to classify the filtered regions, which are compatible with the constraint of shape. The classification procedure focuses t removing false alarms caused by the objects with colors similar to skin color and other exposed parts of the body or parts of background. The elliptical shape is one of the main features of face. We use this feature to extract the candidate face regions from filtered images

In general scene images, faces have various sizes, shapes, and angles. Therefore, the algorithm requires the techniques to compute elliptical parameters of face region. We use a algorithm called a best-fit ellipse to get parameters of each segmented region. Fig. 4 represents the parameters applying to the proposed best-fit ellipse algorithm.

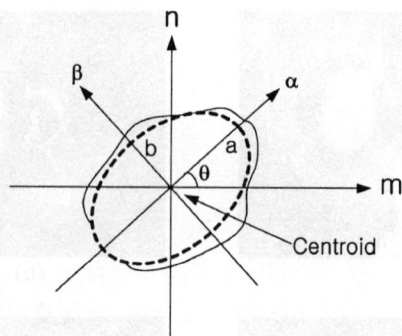

Fig.4. Best-fit ellipse representation

Christophe et al.[11] only considers the aspect ratio of region's bounding box to select the candidate regions. They allow aspect ratio range to [0.9, 2.1]. But, the faces have elliptical shapes and various rotations. So there is a need to use the rotation-invariant method. We use main elliptical parameters like θ, a and b shown in Fig. 4.

The orientation of the region is defined as the angle of axis of the least moment of inertia. The orientation is computed by

$$\theta = \frac{1}{2}\tan^{-1}\left[\frac{2\mu_{11}}{\mu_{20} - \mu_{02}}\right] \quad (5)$$

μ_{pq} is (p,q) order central moments as follows:

$$\mu_{pq} = \sum\sum_{(m,n)\in R}(m-\bar{m})^p(n-\bar{n})^q \quad (6)$$

Let a and b denote the lengths of submajor and subminor axes. The least and the greatest moments of inertia for an ellipse are described by

$$I_{min} = \frac{\pi}{4}ab^3 \quad \text{and} \quad I_{max} = \frac{\pi}{4}a^3b \quad (7)$$

For orientation θ, the above moments can be calculated as

$$I'_{min} = \sum\sum_{(m,n)\in R}\left[(n-\bar{n})\cos\theta - (m-\bar{m})\sin\theta\right]^2$$

and

$$I'_{max} = \sum\sum_{(m,n)\in R}[(n-\bar{n})\sin\theta - (m-\bar{m})\cos\theta]^2 \quad (8)$$

Finally, we calculate a and b when $I'_{min} = I_{min}$ and $I'_{max} = I_{max}$, which give

$$a = \left(\frac{4}{\pi}\right)^{1/4}\left[\frac{(I'_{max})^3}{I'_{min}}\right]^{1/8} \quad \text{and} \quad b = \left(\frac{4}{\pi}\right)^{1/4}\left[\frac{(I'_{min})^3}{I'_{max}}\right]^{1/8} \quad (9)$$

The general anatomical face ratio has a range set [1:1.4 – 1:1.6]. So, we define the ellipse ratio condition for candidate region extraction as follows:

$$1.4 \leq \frac{a}{b} \leq 2 \quad (10)$$

2.3 Face Region Decision

From this stage, we select the real face or face-contain regions from candidate regions. We use a neural network model for improved performance and use momentum back-propagation algorithm to train it.

Pattern Generation. We use a 20*20 pixel block as input subtracted from oval face mask in Fig. 5(e). The candidate regions contain various sizes, orientations, positions, and intensities of faces. To use these regions as neural network inputs we need to normalize the sizes, orientations, and intensities of regions.

The pattern normalization is as follows:

Step 1. Derotate the regions by its orientation.
Step 2. Resize the regions by an average shrinking algorithm to 20*20 patterns.
Step 3. Perform the histogram equalization to Step 2.
Step 4. Subtract the oval face mask from Step 3.
Step 5. Input Step 4 to neural network model training.

Fig. 5 shows the pattern generation procedure.

Fig. 5(b) shows the candidate region from original image, and Fig. 5(c) shows the derotated image by it's orientation, and Fig. 5(f) shows the final input pattern fo training. We don't input the pattern which has gray-level values to the neural network directly. Instead of gray-level values, we use the analog input pattern as follows:

$$p(x) = \frac{(x-127)}{M} \quad \text{and} \quad M = \begin{cases} 128 & x > 127 \\ 127 & x < 127 \end{cases} \quad (11)$$

By (11), the input patterns have a value from –1.0 to 1.0.

Neural Network Training and Decision. A large number of face and non-face images are needed to train the neural network. Initially, we use about 310 face sample images that gathered from CMU face database, WWW(World Wide Web) and scanning of photographs without constrained conditions. We process input pattern with two steps. First, we use proposed candidate extraction algorithm into color images. Second, we extract the face region without derotation from gray-level images and normalize the pattern. To gather non-face sample images, we use the random face non-contained images such as scenery images. We just extract random rectangle from target images and make a pattern using pattern normalization. Also, we collect the non-face samples during training[12,13]. For example, if the neural network incorrectly identify non-face patterns (output values > 0), we add them into training set as negative samples. We use about 360 non-face images for training. The neural network has 340 input neuron, 80 hidden neuron and 1 output neuron. And we use bipolar sigmoid function as neuron activation function. The outputs of neural network are analog values which have range from −1.0 to 1.0. We use the zero value as face decision threshold.

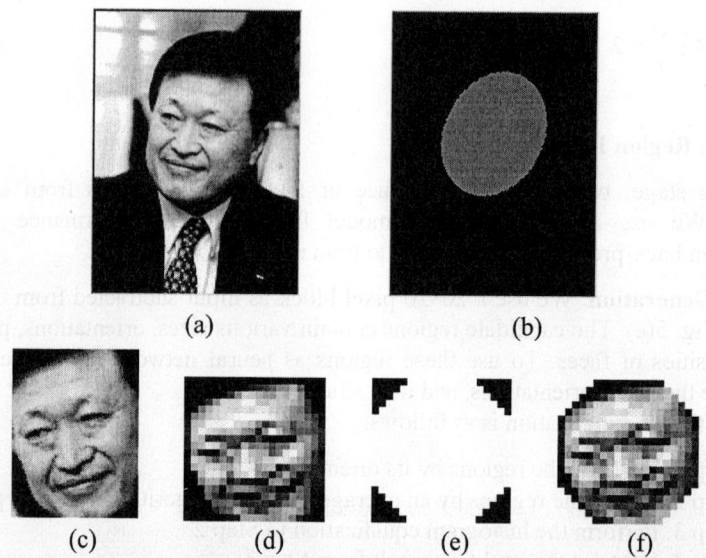

Fig.5. Pattern sample procedure: (a) original image, (b) best-fit ellipse, (c) derotate image, (d) 20*20 shrinking pattern, (e) oval mask, (f) input pattern for training

3 Experiment Results

We have implemented the system based on proposed algorithm and have tested it on two sets of images.

Test set 1 consists of a total of 150 images from the World Wide Web and photograph scanning. The images contain a total of 217 faces that have various sizes, rotations and positions. After the candidate extraction stages, the system extracts a total of 261 candidate regions. Test set 2 consists of a total 100 images from the television streams and contains a total of 148 faces. The system extracts 171 candidate regions as well.

Table 2 shows the performance of the system on Test sets. Test set 2 have low image qualities and noises. Therefore, Test set 2 has more failure to extract the face candidates from images. As shown in Table 2, the system extract 202 and 126 face candidates on each test sets. The fourth and fifth column show correct face decision and rejection for neural network. The fifth column shows the false detection rates from non-face candidates.

Table 3 shows the causes of detection failure on test sets. Most of the errors are caused by 3 categories such as skin-tone extraction failure, shape constraints error and neural network rejection.

Table 2. Detection and error rates for Test sets

	Candidate extraction		Neural network decision		False detection (%)
	Extracted faces(%)	Missed faces(%)	Face decision(%)	Face rejection(%)	
Test set1	92.2	7.8	97.0	3.0	11.9
Test set2	85.1	14.9	94.4	5.6	13.0
Total	88.9	11.1	96.6	3.4	12.3

Table 3. Detection failure causes

Detection failure cause		Rate(%)
Candidate extraction Failture	Skin-ton error	42.3
	Shape constraints error	19.2
Neural network rejection		38.5

Fig.6 and Fig.7 show the sample results of each test sets.

Fig.6. Sample results of Test set 1 for photograph scanning

Fig.7. Sample results of Test set 2 for TV scenes

4 Conclusion and Future Works

In this paper, we have proposed efficient and automatic face detection algorithms in color images under non-constrained input conditions, such as complex backgrounds, various luminance and image quality. We use skin color information, elliptical shape of faces and neural network model. 2D Gaussian distribution and the sample face set construct the skin-tone model. It is used to a good pre-processing filter to separate facial region from complex background.

We apply a momentum back-propagation algorithm for the face decision stage and use 20*20 pixel blocks for input images. We use 310 faces and over 360 non-face images to train it and 150 images to test the performance of algorithm. The training image set and test image set are gathered from CMU face database, WWW and scanning photograph scanning.

Our proposed algorithm is capable of extracting 88.9% of face candidates and detecting 96.6% of faces in a given image set with various sizes, rotations and light conditions.

In the future, we shall develop an image-tone equalization technique to improve the performance of skin-tone. We shall also search for a more accurate candidate region extraction that will not be affected by other non-face parts.

References

1. B. S. Manjunath, R. Chellappa and C. von der Malsburg: A feature based approach to face recognition, Proc. CVPR, pp. 373-378, 1992
2. L. Yullie: Deformable templates for face detection, Journal of Cognitive Neuroscience, Vol. 3, No. 1, pp. 59-70, 1991
3. R. Brunelli and T. Poggio: Face recognition: features versus templates, IEEE Trans. on PAMI, Vol. 15, No. 10, pp. 1042-1052, 1993
4. M. Kirby and L. Sirovich: Application of the Karhunen-Loeve procedure for the characterization of human faces, IEEE Trans. on PAMI, pp. 103-108, 1990
5. M. Turk and A. Pentland: Eigenfaces for recognition, Journal of Cognitive Neuroscience, Vol. 3, No. 1, pp. 71-86, 1991
6. H. Lee, J. S. Kim, K. H. Park: Automatic human face location in a complex background using motion and color information, Pattern Recognition, Vol. 29, pp. 1877-1889, 1996
7. K. Sobottka and I. Pitas: Extraction of facial region and features using color and shape information, In Proc. 10th Int'l Conf. Pattern Recognition, pp. 421-425, 1996
8. Garcia and G. Tziritas: Face detection using quantized skin color regions merging and wavelet packet analysis, IEEE Trans. on Multimedia, Vol. 1, No. 3, pp. 264-277, 1999
9. H. A. Rowley, S. Baluja, and T. Kanade: Rotation invariant neural network-based face detection, Computer Vision and Pattern Recognition, pp. 38-44, 1998
10. H. Wang and S. Chang: A highly efficient system for automatic face region detection in MPEG video, IEEE Trans. TCSVT, Vol. 7, pp. 615-628, 1997
11. Zhu Liu and Yao Wang: Face detection and tracking in video using dynamic programming, IEEE Proc. of the 2000 ICIP, Vol. 1, pp. 53-56, 2000

Efficient Image Segmentation Based on Wavelet and Watersheds for Video Objects Extraction

Jong-Bae Kim and Hang-Joon Kim

Dept. of Computer Engineering, Kyungpook National University
1370, Sangyuk-dong, Pook-gu, Dea-gu, 702-701, Korea
{kjblove,hjkim}@ailab.knu.ac.kr

Abstract. The MPEG-4 and MPEG-7 visual standard to support each frame of a video sequence should be segmented in terms of video object planes (VOPs). This paper presents an image segmentation method for extracting video objects from image sequences. The method is based on a multiresolution application of wavelet and watershed transformations, followed by a wavelet coefficient–based region merging procedure. The procedure toward complete segmentation consists of four steps: pyramid representation, image segmentation, region merging and region projection. First, pyramid representation creates multiresolution images using a wavelet transformation. Second, image segmentation is used to segment the lowest-resolution image of the created pyramid by watershed transformation. Third, region merging involves merging segmented regions using the third-order moment values of the wavelet coefficients. Finally, the region projection is used to recover a full-resolution image using an inverse wavelet transformation. Experimental results of the presented method can be applied to the segmentation of noise or degraded images as well as the reduction of over-segmentation.

1 Introduction

Image segmentation, which is defined as the partitioning of an image into homogeneous regions, is regarded as an analytic image representation method. It has been very important in many computer-vision and image-processing applications. In particular, there has been a growing interest in image segmentation mainly due to the development of MPEG-4 and MPEG-7, which enable content-based manipulation of multimedia data [1, 2, 3]. In order to obtain a content-based representation, an input video sequence must first be segmented into an appropriate set of arbitrary semantic shaped objects. Here, a semantic object represents a meaningful entity in an image. Video objects can be very useful in many different fields. For instance, in the broadcasting and telecommunication domain, the compression ratio is of great concern where the introduction of a video object could possibly improve the coding efficiency for both the storage and transmission tasks. However, a lot of computational time and power are required, since segmentation is a complex highly combinational problem [2, 3].

Although, the human visual system can easily distinguish semantic video objects, automatic image segmentation is one of the most challenging issues in the field of video processing.

Conventional segmentation approaches range from the split-and-merge method to morphological segmentation. Among them, morphological segmentation techniques are of particular interest because they rely on morphological tools which are very useful to deal with object-oriented criteria such as size and contrast. The morphological segmentation is the watershed transformation. *Beucher* and *Lantuejoul* were the first to apply the concept of watershed by dividing lines in to segmentation problems [4]. Watershed transformation is known to be a very fast algorithm for spatial segmentation of images [4, 5, 6]. It starts by creating a gradient of the image to be segmented. Each minimum of the gradient leads to a region in the resulting segmentation. Conventional gradient operators generally produce many local minima that are caused by noise or quantization error. Hence, a major problem with watershed transformation is that it produces severe over-segmentation due to the great number of minima and various noise contributions within an image or its gradient. There are two main drawbacks when applying the watershed transformation to segmentation images: sensitivity to strong noise and high computational requirements to merge the over-segmented regions. Watersheds have also been used in multiresolution methods to produce resolution hierarchies of image ridges and valleys [6]. Although these methods were successful in segmenting certain classes of images, they require significant interactive user guidance or accurate prior knowledge on the image structure.

To alleviate these problems, we developed an efficient image segmentation method which is based on the combination of multiresolution images by wavelet transformation and image segmentation by watershed transformation. In the presented approach, over-segmentation and noise problems are reduced as the watershed operation is carried out on low-resolution images from the wavelet transform. The wavelet transformation is created in a multiresolution image because the resolutions of the sub-images are reduced. In turn, the computational complexity will be simplified and reduced dramatically by operating on a lower resolution image and a full-resolution image will be projected due to the inverse wavelet transform. Experimental results show that the presented method is effective in the segmentation of an image sequence.

2   Multiresolution-Based Watershed Algorithm

A general outline of the multiresolution-based image segmentation procedure is presented in Fig. 1. First, pyramid representation creates multiresolution images by wavelet transformation. The image at different layers of the pyramid represents various image resolutions. Images are first segmented into a number of regions at the lowest layer of the pyramid by watershed transformation. The third-order moment values of the wavelet coefficient are then applied to merge the segmented regions. To recover a full-resolution image (original image), an inverse wavelet transformation is applied. The following sections provide a summary of the segmentation procedure. Pyramid representation and image segmentation will be described first, followed by the wavelet and watershed transformations. The results of this segmentation will present a bias toward over-segmentation of some homogeneous regions that will be described in further region-merging strategies. To recover a full-resolution image using inverse wavelet transformation, the region projection will be described.

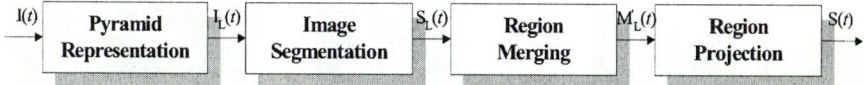

Fig. 1. Block diagram of the presented approach. $I(t)$ is the sequence image at time t to be segmented, $I_L(t)$ is the corresponding lowest resolution image of the multiresolution image, $S_L(t)$ is the segmented image, $M'_L(t)$ is the merged image, and $S(t)$ is the recovered full-resolution image

2.1 Pyramid Representation

The presented method is based on the multiresolution application of a wavelet transformation for image segmentation. A great advantage of using a multiresolution method is the possibility of determining the dimension of regions to be segmented. Thus, over-segmentation of the watershed transform and possibly noise in the image capturing process can be reduced. The application of Haar decomposition to the frames of a sequence will provide images of smaller resolution.

One of wavelet transformation special characteristics is its power in reducing resolution by using the average of consecutive function values, which in the case of images are pixel (intensity) values. The wavelet decomposition of a signal $f(x)$ is performed by convolution of the signal with a family of basis functions, $\psi_{2^s,t}(x)$:

$$\left\langle f(x), \psi_{2^s,t}(x) \right\rangle = \int_{-\infty}^{\infty} f(x)\psi_{2^s,t}(x)dx \tag{1}$$

The basis functions $\psi_{2^s,t}(x)$ are obtained through translation and dilation of a kernel function $\Psi(x)$, called the mother wavelet, and defined as follows:

$$\psi_{2^s,t}(x) = \psi_{2^s,t}(x - 2^{-s}t) = 2^s \Psi(2^s x - t) \tag{2}$$

where s and t are referred to as the dilation and translation parameters, respectively. The images can be decomposed into its wavelet coefficients using Mallat's pyramid algorithm [7]. By using Haar wavelets, the original image is first passed through low-pass filters to generate low-low (LL), low-high (LH), high-low (HL), and high-high (HH) sub-images. The decompositions are repeated on the LL sub-image to obtain the next four sub-images. Figs. 2 (a) and (b) show the 1-scale and 2-scale wavelet decompositions, respectively. The second scale decomposition on the LL part of Fig. 2 (a) is shown at the top left of Fig. 2 (b). Let W^{2^i}, $W_H^{2^i}$, $W_V^{2^i}$ and $W_D^{2^i}$ represent the decomposition results LL, LH, HL and HH at scale 2^J, respectively. For J-scale decompositions, the original signal can be represented by

$$(W^{2^J}, (W_H^{2^i}, W_V^{2^i}, \text{ and } W_D^{2^i})_{1 \leq i \leq J}) \tag{3}$$

where the size of the wavelet representation is the same as that of the original signal. This representation is composed of a coarse signal at resolution 2^J and a set of detail signals at resolution $2^1 \sim 2^J$.

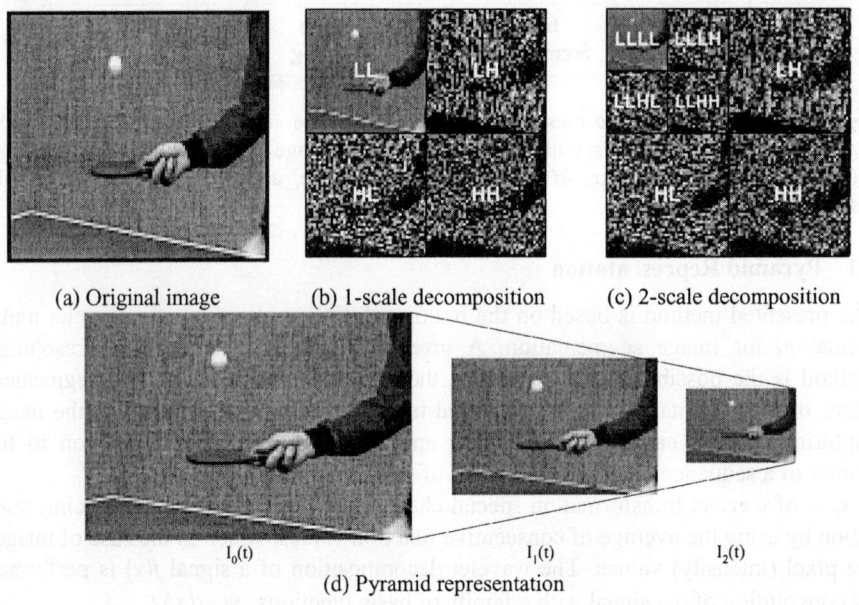

(a) Original image   (b) 1-scale decomposition   (c) 2-scale decomposition

$I_0(t)$ $I_1(t)$ $I_2(t)$
(d) Pyramid representation

Fig. 2. Wavelet decomposition at each scale

The original image to be segmented is identified as $I(t)$, where t indicates its time reference within the sequence. Stemming from this original image, also identified as $I_0(t)$, where 0 means that it is a full-resolution image ($I(t) = I_0(t)$), a set of low frequency sub-sampled images of the wavelet transformation $\{I_1(t), I_2(t), ..., I_L(t)\}$ at time t is generated. Fig. 2(d) represents the hierarchical structure, where $I_L(t)$ corresponds to the sub-sampled image at level (l) of the pyramid and $I_L(t)$, the lowest available resolution image.

2.2 Image Segmentation

After creating the pyramid representation by wavelet transform, the lowest resolution image $I_L(t)$ is segmented through the application of a watershed transformation based on the gradient image. It is basically a region-growing algorithm, starting from marks, then successively joining pixels from an uncertainty area to the nearest similar region. Watersheds are traditionally defined in terms of the drainage patterns of rainfall. Regions of terrain that drain to the same points are defined as part of the same watershed. The same analysis can be applied to images by viewing intensity as height. In this case, the image gradient is used to predict the direction of drainage in an image. By following the image gradient downhill from each point in the image, the set of points which drains to each local intensity minimum can be identified. These disjointed regions are called the watersheds of the image. Similarly, the gradients can be followed uphill to the local intensity maximum in the image, the defining the inverse watersheds of the image [4, 5, 6]. In the image segmentation phase, the ambiguity of the image segmentation is reduced and a partition $S_L(t)$ from the image $I_L(t)$ is generated. The result of this partition $S_L(t)$ will present a bias toward over-segmentation of homogeneous regions that will make further region-merging strategies necessary.

2.3 Region Merging and Projection

Generally, the above-mentioned algorithms followed by watershed transformation produce meaningful image segmentations. However, when an image is degraded by noise, it will be over-segmented. Therefore, over-segmented images may require further merging of some regions. The region merging process is formulated as a graph-based clustering problem. A graph, G, is used to represent the information upon which the region merging process is based. Fig. 3 shows the region adjacency graph. In Fig. 3, each node in G represents one of the segmented regions in the set $I = \{R_1, R_2, ..., R_k\}$. Each edge of G corresponds to a sum of the moment values (mv), which can be used to compare the mv_i of adjacent regions. Our decision on which regions to merge is determined through homogeneity and similarity criteria based on wavelet coefficients. Each of the segmented regions will have mean, second and third-order central moment values of the wavelet coefficients calculated. All the features are computed on a LL decomposed subband of a wavelet transform. For each region R_i of the segmented image $S_L(k)$ of the image segmentation phase, we calculate the mean (M), second (μ_2)- and third-order (μ_3) central moments of the region as [8].

$$M = \frac{1}{num(R_i)} \sum\sum R_i(i,j) \qquad \forall i,j \in R_i(i,j)$$

$$\mu_2 = \frac{1}{num(R_i)} \sum\sum (R_i(i,j) - M)^2 \qquad (4)$$

$$\mu_3 = \frac{1}{num(R_i)} \sum\sum (R_i(i,j) - M)^3$$

where $num(R_i)$ is the number of pixels of segmented region i. To merge the segmented regions using similarity criteria (d), we can use the following equation:

$$mv_i = \frac{1}{N}(R(M_i) + R(\mu_{2i}) + R(\mu_{3i})) \qquad i = 1, ..., N$$

(5)

$$d(R_i, R_j) = (mv_i - mv_j)^2 \qquad \forall i,j \in \{1, ..., N, \text{ for } i \neq j\}$$

where mv_i is the similarity value of segmented region i and N is the number of segmented regions. $R(M)$, $R(\mu_2)$ and $R(\mu_3)$ are the mean, second- and third-order moment values of the segmented region i. If the mv values of the adjacent regions satisfy a specified value, two adjacent regions will be merged. The specified values found in experimental.

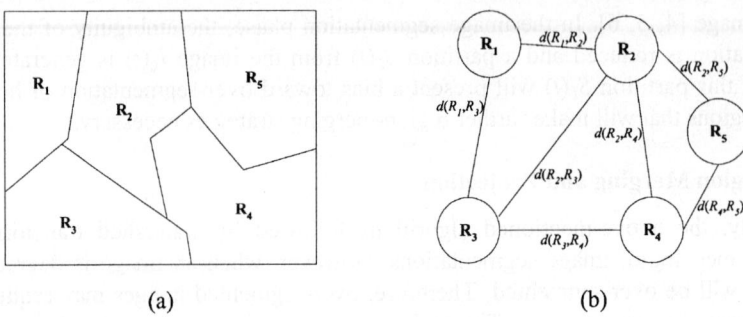

Fig. 3. Region adjacency graph (RAG) (a) Segmented regions have similarity value; (b) Segmented region adjacency graph

The results of the merging processor reduce number of segmented regions. Once the merged image $M`_L(t)$ has been generated at the image partition $S_L(t)$, it must be projected down in order to recover the full-resolution image $S(t)$. Direct projection of the segmentation, based only on pixel duplication in both vertical and horizontal directions, offers very poor results. To overcome this problem, we use an inverse wavelet transform. Doing so provides an easily recovered and efficiently projected full-resolution image.

3 Experimental Results

To evaluate the performance of the presented method, simulation was carried out on the frames of the sequences 'Table Tennis', 'Claire', 'Akiyo' and 'HM' (Hall monitor) and the frame size is 352 x 288. The pyramid image generated by the 2-scale Haar wavelet transformation and the makers for the watershed transformation were extracted from a LL subband of 2-scale wavelet transformation. Here, flat regions larger than 100 pixels were extracted as marks. The experiments were performed on a Pentium IV 1.4Ghz PC with an algorithm that was implemented using a MS Visual C++ development tool. The processing time of each frame took 0.28 sec on average. Based on both real and synthetic images, two key segmentation aspects were studied.

3.1 Evaluation of Objective Segmentation Performance

We evaluated the object segmentation results of the presented method using four common measurements: the number of segmented regions, PSNR, Goodness and computation time. As previously mentioned, the definition of image segmentation is to partition an image into flat or homogeneous regions. The flatness of a region can be represented by the mean square error. If an image is segmented into N regions, $\{Ri, i= 1, .., N\}$, the MSE is defined by

$$\text{MSE} = \sum_{i=1}^{N} \frac{\sum_{p \in R_i}(I(p)-\mu(R_i))^2}{\sum_{p \in R_i} 1} \tag{6}$$

where $I(p)$ and $\mu(R_i)$ represent the intensity value of pixel p and the average intensity value within R, respectively. Accordingly, the peak signal to noise ratio is defined by

$$PSNR = 10\log_{10}\frac{I_{peak}^2}{MSE} \qquad (7)$$

where I_{peak} is 255 that is the intensity value. The goodness function F is defined by [9]

$$F(I) = \sqrt{M} \times \sum_{i=1}^{N}\frac{e_i^2}{\sqrt{A_i}} \qquad (8)$$

where I is the image to be segmented, N is the number of regions in the segmented image, A_i is the area or number of pixels of the ith region and e_i is the sum of the Euclidean distance of the color vectors between the original image and the segmented image of each pixel in the region. Eq. (8) is composed of two terms: $\sqrt{M}$, which penalizes segmentation that forms too many regions and a second one that penalizes segmentations containing non-homogeneous regions. A larger value of e_i means that the feature of the region is not well segmented during the image segmentation process. Table 1 shows the segmentation results of the number of segmented regions, Goodness (F), PSNR and computation times using full-resolution (I_0) and low-resolution images (I_1, I_2).

Table 1. Segmentation results of the presented method: I_0 is the full-resolution image (original image), I_1 and I_2 are the 1 and 2-scale wavelet decomposed LL subband image, and M is the merged image of the segmented I_2 image

Test Images	Scale Levels	Number of Regions	Goodness	PSNR [dB]	Time (sec)
Table Tennis #2	I_0	790	439.5	28.87	0.49
	I_1	63	201.0	30.33	0.26
	I_2	23	180.4	29.55	0.19
	M	8	156.8	28.42	0.02
Claire #136	I_0	86	283.2	31.58	0.38
	I_1	36	111.7	30.79	0.24
	I_2	30	109.3	30.40	0.20
	M	7	110.5	29.98	0.02
Akiyo #2	I_0	542	302.6	31.96	0.47
	I_1	258	254.0	29.91	0.25
	I_2	65	185.2	29.23	0.22
	M	10	186.2	28.81	0.02
HA #58	I_0	479	213.5	32.61	0.42
	I_1	172	161.2	30.12	0.29
	I_2	90	140.7	29.47	0.23
	M	21	118.0	29.01	0.03

The highest PSNR was observed when using the direct application of the watershed algorithm for the full-resolution image, but with the largest number of regions and high computation times. For each set of segmentation results shown in Table 1, the full-resolution image (I_0) contains more details, but also has a larger number of regions and value of F. The low-resolution images are the opposite, where many details are lost. Therefore, the total number of regions is less and the value of PSNR is smaller. Although there are smaller PSNR values, the value of F is smaller. According to the segmentation evaluation criteria, these results are the trade–off between suppressing noise and preserving details. In each set of results, the low-resolution image always has the smallest F, which means that it is the best.

In the presented method, the use of a low-resolution image is highly preferable since it gives the best objective quality when the number of regions and F are considered at the same time. Over-segmentation and computation time after image segmentation can be reduced when operating on a low-resolution image.

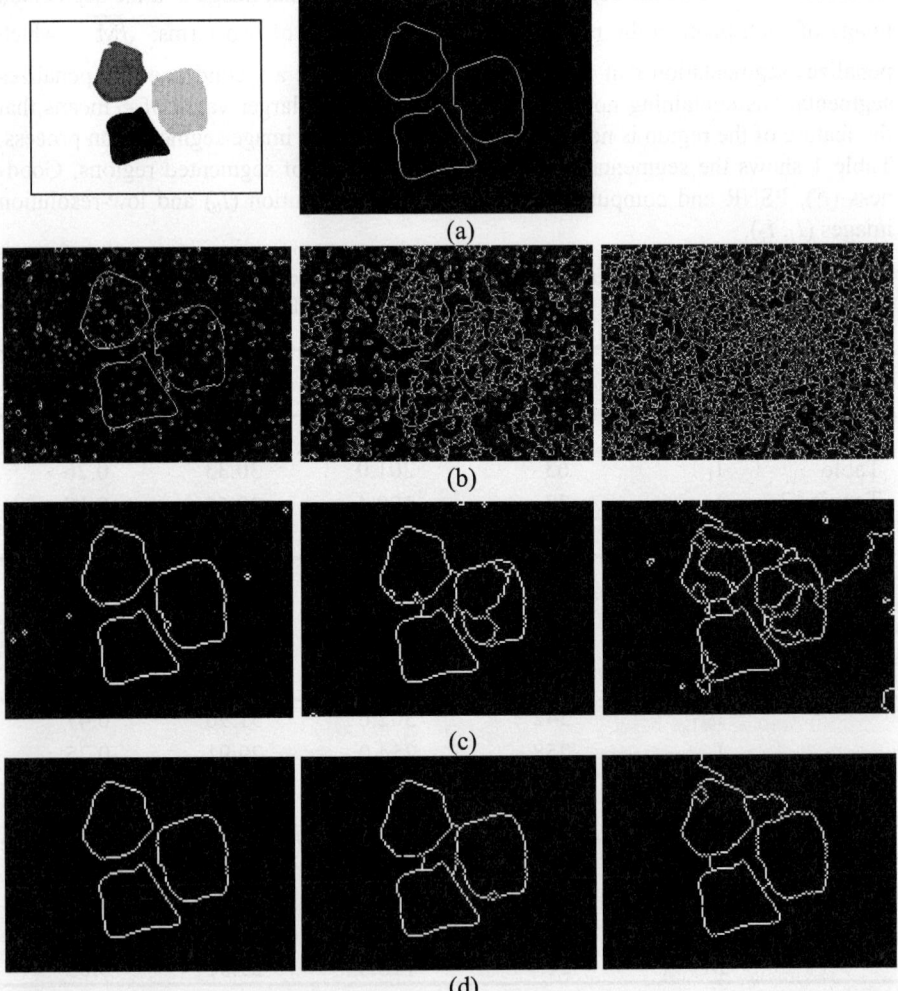

Fig. 4. Segmentation results of a synthetic image corrupted by random Gaussian noise. The left, center and right columns show the results of the segmentation of an image with random Gaussian noises of 10%, 20% and 30%. The first row shows a synthetic image and the results of segmentation. The second, third and last rows show the results of segmentation of a full-resolution (I_0), 1- (I_1), 2-scale wavelet decomposition image (I_2) of the pyramid with different Gaussian noises.

3.2 Segmentation Robustness in the Presence of Noise

The error probability of segmentation in a synthetic image (the probability if wrongly classifying pixels) when adding random Gaussian noise was analyzed. Table 2 shows the error probability evolution for different values of Gaussian noise. It was observed that the error probability decreases by approximately 70% when operating on low-resolution images, thus increasing the robustness of the segmentation while also reducing the number of required joins in the merging stage. Fig. 4 shows the segmentation results of a synthetic image corrupted by adding Gaussian noise with different SNR values.

Table 2. The error probability of the segmentation (%)

Gaussian noise (%)	0	10	20	30
I_0	6.7	34.1	43.6	58.2
I_1	2.7	8.9	17.6	23.7
I_2	2.0	3.6	10.2	12.3

4 Conclusions

An efficient image segmentation method for video objects extraction using a multiresolution application of wavelet and watershed transformation is described in this paper. The approach is based on the combination of a multiresolution image by wavelet transformation and image segmentation by watershed transformation. As shown in the simulation results, the algorithm gives visually meaningful segmentation results. In addition, our method significantly reduces the computational cost of a watershed-based image segmentation method while simultaneously improving segmentation accuracy. Moreover, this approach provides robustness in dealing with noise and prevents over-segmentation.

References

1. Sikora, T.: The MPEG-4 video standard verification model, IEEE Trans. Circuits Syst. Video Tech., Vol. 7, No. 1, (1997) 19-31
2. Kim, E. Y., Hwang, S. W., Park, S. H. and Kim, H. J.: Spatiotemporal Segmentation Using Genetic Algorithms, Pattern Recognition, Vol. 34, (2001) 2063-2066
3. Gu, C. and Lee, M. C.: Semiautomatic Segmentation and Tracking of Semantic Video Objects, IEEE Trans. Circuit Syst. Video Tech., Vol. 8, No. 5, (1998) 572-584
4. Beucher, S. and Lantuejoul, C.: Use of watershed in contour detection, int. Workshop on Image processing, Real-Time edge and motion detection, (1979) 12-21.
5. Vincent, L. and Soille, P.: Watershed in digital space: An efficient algorithm based on immersion simulation, IEEE Trans. Pattern Anal. Mach. Intell., Vol. 13, No. 6, (1991) 583-598

6   Gaush, J. M. and Pizer, S. M.: Multiresolution analysis of ridges and valleys in gray-scale images, IEEE Trans. Pattern Anal. Mach. Intell., Vol. 15, (1993) 635-646
7   Mallat, S. G.: A theory for multiresolution signal decomposition: The wavelet representation, IEEE Trans. Pattern Anal. Mach. Intell., Vol. 11, No. 7, (1989) 674-693
8   Kim, J. B., Lee, C. W., Yun, Y. S. Lee, K. M. and Kim, H. J.: Wavelet-based vehicle tracking for Automatic Traffic Surveillance, in Pro. IEEE Tencon, Singapore, (2001) 313-316
9   Liu, J. and Yang, Y. H.: Multiresolution color Image Segmentation, IEEE Trans. Pattern Anal. Mach. Intell., Vol. 16, No. 7, (1994) 689-700

FPGA-Based Implementation of Genetic Algorithm for the Traveling Salesman Problem and Its Industrial Application

Iouliia Skliarova and António B.Ferrari

Department of Electronics and Telecommunications
University of Aveiro, IEETA, 3810-193 Aveiro, Portugal
`iouliia@ua.pt, ferrari@ieeta.pt`

Abstract. In this paper an adaptive distribution system for manufacturing applications is considered and examined. The system receives a set of various components at a source point and supplies these components to destination points. The objective is to minimize the total distance that has to be traveled. At each destination point some control algorithms have to be activated and each segment of motion between destination points has also to be controlled. The paper suggests a model for such a distribution system based on autonomous sub-algorithms that can further be linked hierarchically. The links are set up during execution time (during motion) with the aid of the results obtained from solving the respective traveling salesman problem (TSP) that gives a proper tour of minimal length. The paper proposes an FPGA-based solution, which integrates a specialized virtual controller implementing hierarchical control algorithms and a hardware realization of genetic algorithm for the TSP.

1 Introduction

Many practical applications require solving the TSP. Fig. 1 shows one of them. The objective is to distribute some components from the source S to destinations $d_1,...,d_n$. The distribution can be done with the aid of an automatically controlled car and the optimization task is to minimize the total distance that the car has to travel. All allowed routes between destinations are shown in Fig. 1 together with the corresponding distances $l_1,...,l_m$. Thus we have to solve a typical TSP. Note that the task can be more complicated, where more than just one car is used for example. Indeed, some routes and some destinations can be occupied, which requires the TSP to be solved during run-time. Let us assume that the car has an automatic control system that is responsible for providing any allowed motion between destinations and for performing the sequence of steps needed to unload the car in any destination point (see Fig. 1). Thus the distribution system considered implements a sequence of control sub-algorithms and in the general case this sequence is unknown before execution time. So we have an example of adaptive control for which the optimum sequence of the required sub-algorithms has to be recognized and established during run-time.

From the definition we can see that the problem can be solved by applying the following steps:

1. Solving the TSP for a distributed system such as that is shown in Fig. 1.
2. Assembling the control algorithm from the independent sub-algorithms taking into account the results of point 1.
3. Synthesizing the control system from the control algorithm constructed in point 2.
4. Implementing the control system.
5. Repeating steps 1-4 during execution time if some initial conditions have been changed (for example, the selected route is blocked or the required destination point is occupied).

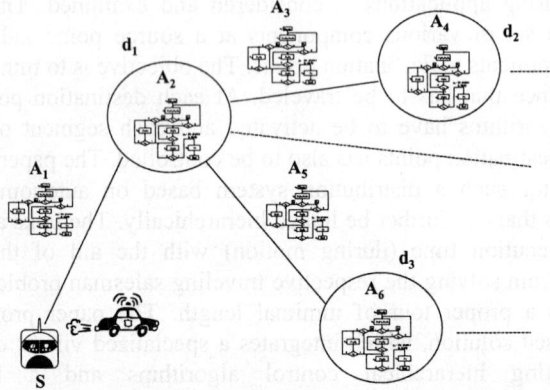

Fig. 1. Distributing system that supplies components from the source S to destinations $d_1,...,d_n$. The algorithms A_1, A_3, A_5 affect motions between the respective destinations and the algorithms A_2, A_4, A_6 describe unloading operations

We suggest realizing the steps considered above in an FPGA, integrating in this way the synthesis system with the implementation system. In order to solve the TSP, a genetic algorithm (GA) described in [1] has been used. Currently, only a part of the GA is implemented in FPGA. That is why we employ a reconfigurable hardware/software (RHS) model of the computations [2]. The task of point 2 can be handled with the aid of a hierarchical specification of the virtual control algorithms [3]. There are many known methods that can be used for synthesis of control systems (see point 3), for example [4]. The implementation of the control system (see point 4) was done based on the XCV812E FPGA.

The remainder of this paper is organized as follows. Section 2 provides a description of the GA employed. A hardware implementation of a part of the algorithm is considered in section 3. Section 4 gives some details about the specification of the control algorithms. The process of synthesis and implementation of virtual control circuits is presented in section 5. Section 6 contains the results of experiments. Finally, the conclusion is in section 7.

2   Genetic Algorithm for Solving the TSP

The TSP is the problem of a salesman who starts out from his hometown and wants to visit a specified set of cities, returning home at the end [5]. Each city has to be visited exactly once and, obviously, the salesman is interested in finding the shortest possible way. More formally, the problem can be presented as a complete weighted graph $G=(V, E)$, where $V=\{1, 2, ..., n\}$ is the set of vertices that correspond to cities, and E is the set of edges representing roads between the cities. Each edge $(i, j) \in E$ is assigned a weight l_{ij}, which is the distance between cities i and j. Thus the TSP problem consists in finding a shortest Hamiltonian cycle in a complete graph G. In this paper we consider the symmetric TSP, for which $l_{ij}=l_{ji}$ for every pair of cities.

The TSP is one of the best-known combinatorial optimization problems having many practical applications. Besides the distribution system described in the previous section, the problem is of great importance in such areas as X-ray crystallography [6], job scheduling [5], circuit board drilling [7], DNA mapping [8], etc. Although, the problem is quite easy to state, it is extremely difficult to solve (the TSP belongs to the class of NP-hard problems [9]). That is why many research efforts have been aimed at finding sub-optimal solutions that are often sufficient for many practical applications. Because of being NP-hard, having a large solution space and an easily calculated fitness function, the TSP is well suited to genetic algorithms.

GAs are optimization algorithms that work with a population of individuals and they are based on the Darwinian theory of natural selection and evolution. Firstly, an initial population of individuals is created, which is often accomplished by a random sampling of possible solutions. Then, each solution is evaluated to measure its fitness. After that, variation operators (such as mutation and crossover) are used in order to generate a new set of individuals. A mutation operator creates new individuals by performing some changes in a single individual, while the crossover operator creates new individuals (offspring) by combining parts of two or more other individuals (parents) [1]. And finally, a selection is performed, where the most fit individuals survive and form the next generation. This process is repeated until some termination condition is reached, such as obtaining a good enough solution, or exceeding the maximum number of generations allowed.

In our case a tour is represented as a path, in which a city at the position i is visited after the city at the position $i-1$ and before the city at the position $i+1$. For the TSP the evaluation part of the algorithm is very straightforward, i.e. the fitness function of a tour corresponds to its length. The mutation operator randomly picks two cities in a tour and reverses the order of the cities between them (i.e. the mutation operator tries to repair a tour that crosses its own path). We used a partially-mapped (PMX) crossover proposed in [10], which produces an offspring by choosing a subsequence of a tour from one parent and preserving the order and position of as many cities as possible from the other parent. A subsequence of a tour that passes from a parent to a child is selected by picking two random cut points. So, firstly the segments between the cut points are copied from the parent 1 to the offspring 2 and from the parent 2 to the offspring 1. These segments also define a series of mappings. Then all the cities before the first cut point and after the second cut point are copied from the parent *1* to the offspring *1* and from the parent *2* to the offspring *2*. However, this operation might result in an invalid tour, for example, an offspring can get duplicate cities. In order to overcome this situation, a previously defined series of mappings is utilized, that indicate how to swap conflicting cities.

For example, given two parents with cut points marked by vertical lines:

$$p_1 = (3 \mid 0\ 1\ 2 \mid 4)$$

$$p_2 = (1 \mid 0\ 2\ 4 \mid 3)$$

the PMX operator will define the series of mappings ($0 \leftrightarrow 0$, $1 \leftrightarrow 2$, $2 \leftrightarrow 4$) and produce the following offspring:

$$o_1 = (3 \mid 0\ 2\ 4 \mid 1)$$

$$o_2 = (4 \mid 0\ 1\ 2 \mid 3)$$

In order to choose parents for producing the offspring, a fitness proportional selection is employed. For this we use a roulette wheel approach, in which each individual is assigned a slot whose width is proportional to the fitness of that individual, and the wheel is spun each time a parent is needed. We apply also an elitist selection, which guarantees the survival of the best solution found so far.

The algorithm described was implemented in a software application developed in C++ language. After that, a number of experiments have been conducted with benchmarks from the TSPLIB [11]. The experiments were performed with different crossover rates (10%, 25% and 50%), and they have shown that a significant percentage of the total execution time is spent performing the crossover operation (it ranges from 15% to 60%). That is why we implement firstly the crossover operation in FPGA in order to estimate an efficiency of such an approach. The results of some experiments are presented in Table 1.

Table 1. The results of experiments in software

Name	Crossover rate – 25%			Crossover rate – 50%		
	t_{total} (s)	t_{cros} (s)	%$_{cros}$	t_{total} (s)	t_{cros} (s)	%$_{cros}$
a280	5.88	1.49	25.3	8.39	3.66	43.6
Berlin52	1.27	0.32	25.2	1.79	0.75	41.9
Bier127	2.75	0.68	24.7	3.92	1.68	42.9
d657	15.18	4.62	30.4	23.33	11.92	51.1
eil51	1.22	0.31	25.4	1.76	0.74	39.2
fl417	9.30	2.62	28.2	13.56	6.36	46.9
rat575	12.97	3.78	29.1	19.46	9.53	48.9
u724	17.02	5.36	31.5	25.6	13.19	51.5
Vm1084	28.09	9.69	34.5	43.63	24.13	55.3

The first column contains the problem name (the number included as part of the name shows the corresponding number of cities). The columns t_{total} store the total execution time in seconds on a PentiumIII/800MHz/256MB running Windows2000. For all the instances the population size was of 20 individuals, and 1000 generations were performed. The columns t_{cros} record the time spent performing the crossover operation. And finally, the columns %$_{cros}$ indicate the ratio of the crossover operation comparing to the total execution time (in %).

3 Hardware Implementation of the Crossover Operation

The suggested architecture of the respective circuit is depicted in Fig. 2. It includes a central control unit, which activates in the required sequence all the steps of the algorithm needed to be performed. The current version of the architecture supports tours composed of at most 1024 cities. Thus there are four memories of size $2^{10} \times 10$ that are used in order to keep two parent tours and two resulting offspring. A city at the memory address i is visited after the city at the address $i-1$ and before the city at the address $i+1$. Actually, such a large number of cities (i.e. destinations) is not necessary for the distribution system considered. The required maximum number can be much less (we think 64 is enough).

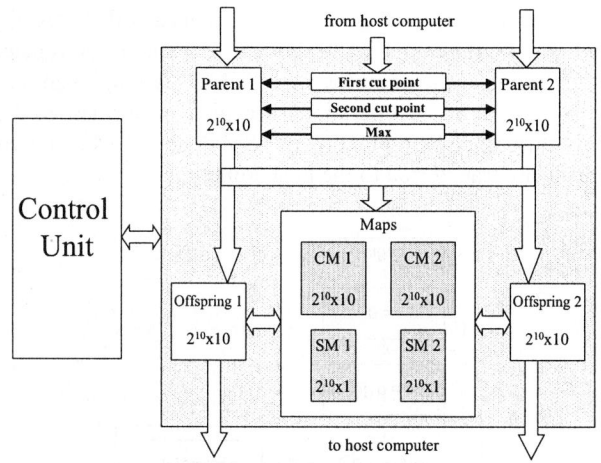

Fig. 2. The proposed architecture for realizing the crossover operation

The cut points used in a crossover are randomly chosen by the software application and stored in two special 10-bit registers ("First cut point" and "Second cut point" in Fig. 2). The third 10-bit register ("Max" in Fig. 2) stores the actual length of a tour. Taking into account this value, the control unit will only force processing of the required area of parents' and offspring memories. It allows accelerating the crossover for tours of small dimensions.

Four additional memories are utilized to help in performing the crossover operation. These memories are "CM1" and "CM2" of size $2^{10} \times 10$ (we will refer to them as complex maps) and "SM1" and "SM2" of size $2^{10} \times 1$ (we will refer to them as simple maps).

In order to perform the PMX crossover, the following sequence of operations has to be realized. Firstly, the values of cut points and the length of a tour for a given problem instance are downloaded to the FPGA. Then, the parent tours are transferred from the host computer to the memories "Parent 1" and "Parent 2". Each time a city is written to parent memories, the value "0" is also written to the same address in the respective simple map (actually, it allows the simple maps to be reset). After that the segments between the cut points are swapped from the "Parent 1" to the "Offspring 2" and from the "Parent 2" to the "Offspring 1". Each time we transfer a city $c1$ from the

"Parent 1" to the "Offspring 2", a value "1" is written to the simple map "SM2" at the address $c1$. The same thing occurs with the second parent, i.e. when we transfer a city $c2$ from the "Parent 2" to the "Offspring 1", a value "1" is written to the simple map "SM1" at the address $c2$ as well. At the same time, the value $c1$ is stored at the address $c2$ in the complex map "CM1", and the value $c2$ is stored at the address $c1$ in the complex map "CM2". At the next step, all the cities before the first cut point, and after the second cut point should be copied from the "Parent 1" to the "Offspring 1" and from the "Parent 2" to the "Offspring 2", and any conflicting situations must be resolved.

For this the following strategy is utilized (see Fig. 3). Firstly, a city c is read from the "Parent 1". If a value "0" is stored in the simple map "SM1" at the address c, then this city can be safely copied to the "Offspring 1". In the opposite case, if value "1" is stored in the simple map "SM1" at the address c, it means that this city has already been included into the "Offspring 1". Consequently, it should be replaced with some other city. For this purpose the complex map "CM1" is employed as shown in the flow-chart in Fig. 3. The same operations are also performed to fill the second offspring, the only difference being that maps "SM2" and "CM2" are employed. And finally, the offspring are transferred to the host computer.

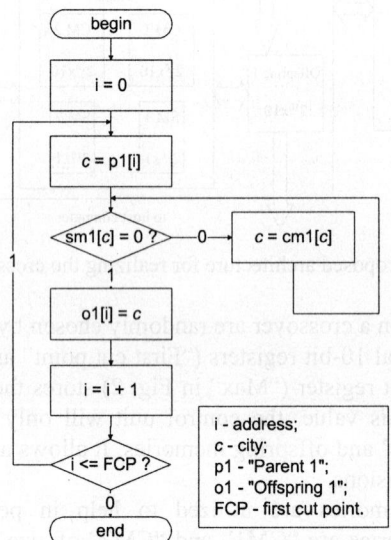

Fig. 3. Filling of the "Offspring 1" before the first cut point

We used an ADM-XRC PCI board [12] containing one XCV812E Virtex Extended Memory FPGA [13] as the hardware platform. This type of FPGA is very well suited to the proposed architecture because it incorporates large amounts of distributed RAM-blocks, which can be used to store parent and offspring tours and maps. The remaining three major parts of the GA, which are evaluation, mutation and selection, are not implemented yet. That is why currently they are performed in a software application communicating with the FPGA with the aid of the ADM-XRC API library, which provides support for initialization, loading configuration bit streams, data transfers, interrupts processing, clock management and error handling.

Table 2 contains information about the area occupied by the circuit implementation and its clock frequency. The area is shown in the number of Virtex slices (each configurable logic block (CLB) is composed of two slices). In the parentheses it is shown how much resources (in %) of XCV812E the circuit consumes.

Table 2. Parameters of the circuit implemented

Area (slices)	Number of RAM-blocks	Maximum clock frequency (MHz)
149 (1%)	20 (7%)	102.659

4 Hierarchical Specification of Control Algorithms

There are two methods that are most appropriate for the specification of hierarchical control algorithms. The first method is called Statecharts [14] and the second is called hierarchical graph-schemes (HGS) [3]. We employ the second method because it permits a synthesizable specification to be built and supports virtualization of control algorithms.

It is obvious that the flow of sub-algorithms can be obtained directly from the result of solving the respective TSP problem. For the considered technique this flow includes only unconditional transitions (see Fig. 4).

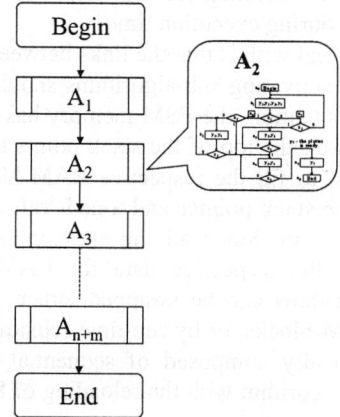

Fig. 4. The required sequence of sub-algorithms and the resulting specification in form of HGS

We assume that all the sub-algorithms are known, and the respective specifications are stored in a library. However, initially all these library sub-algorithms are unlinked. Thus the objective is to construct the main algorithm that invokes the respective sub-algorithms in the required order (see Fig. 4). In fact this task is trivial. The problem might appear when we want to synthesize a run-time reprogrammable control circuit that implements the required sequence of sub-algorithms. This is because the number of sub-algorithms and the contents of each rectangular node (see Fig. 4) are unknown before execution time. Besides, during run-time the main algorithm might need to be reconstructed. As a result the respective control circuit has to be virtual.

More complicated practical problems might also require conditional transitions between sub-algorithms. For example, such transitions will appear if there is more than one optimum result for the step 1 (for example, there are two shortest ways of the same length). Thus, in general case the main algorithm can include not only unconditional but also conditional transitions.

5 Synthesis and Implementation of Virtual Control Circuits

The property of virtualization states that the desired control algorithm has to be implemented by modifying the functionality of the existing hardware dynamically. This can be achieved with the aid of so-called hardware templates (HT). A HT includes modifiable components (such as RAM or partially dynamically reconfigurable areas of FPGA) connected to each other in a certain way. One possible architecture of HT, modeled by a RAM-based finite state machine (FSM), was considered in [4]. The HT has predefined constraints, such as the maximum number of inputs (they affect conditional transitions), the maximum number of activated sub-algorithms, etc. However, all these constraints can easily be estimated from the initial specification of the problem. In [4] it is shown that reloading the respective RAM-blocks can provide all changes in the functionality of the control circuit. This technique gives a complete solution for the virtual control circuit, making the main algorithm synthesizable during execution time.

The next problem to deal with is that the links between the rectangular nodes of the main algorithm and the activating sub-algorithms should be dynamically modifiable. In order to resolve this situation, the FSM memory has to be based on a stack. From the beginning the value on the top of the stack points to the main algorithm and the method [4] can be used to fill the respective RAM-blocks. Invocation of any sub-algorithm increments the stack pointer and a new value at the stack output indicates the respective sub-algorithm. Since all the autonomous sub-algorithms are known before execution time, the respective data for RAM-blocks can be obtained in advance. The sub-algorithms can be swapped either by changing entry points for preliminary loaded RAM-blocks, or by run-time reloading of RAM-blocks. Since the main algorithm is basically composed of sequential chains, we can combine an execution of some sub-algorithm with the reloading of RAM-blocks for the next sub-algorithm.

Finally, the process of synthesis and implementation includes the following steps.

1. Translate the main algorithm into a set of bit streams that have to be loaded into modifiable RAM-blocks of HT. This permits a skeletal frame of the main algorithm to be constructed.
2. Establish links between the rectangular nodes of the skeletal frame and real sub-algorithms that have to be implemented.
3. Load the RAM-blocks of the FPGA.

6 Experiments

A comparison between software and hardware versions of the crossover operation is presented in Table 3. The first column in Table 3 stores the number of cities in a tour. The column t_{soft} contains the time of performing the crossover in software. The software version was based on the C++ language and executed on a PentiumIII/800MHz/256MB running Windows2000. The column t_{hard} records the time for performing the crossover in hardware. The hardware part was executed on an XCV812E FPGA with a clock frequency of 40 MHz. The cut points chosen in software and hardware versions were the same in order to facilitate the comparison. The speedup resulting from our approach is given by t_{soft}/t_{hard}. The results of experiments show that the PMX crossover operator is executed in FPGA 10-50 times faster than in software.

Table 3. Comparison of software and hardware implementations of the PMX crossover operator

Number of cities	t_{soft} (ms)	t_{hard} (ms)	Speedup
280	1.0896	0.08443	12.9
52	0.4293	0.01872	22.9
127	0.6203	0.04040	15.4
657	3.838	0.19486	19.7
417	2.0605	0.12524	16.5
442	3.8394	0.13045	29.4
575	2.7703	0.17199	16.1
724	10.2145	0.20443	49.9

It should be noted that the time t_{hard} (and, consequently, the speedup achieved) strongly depends on the cut points chosen. This is because the first part of the PMX crossover (swapping the segments between the cut points) is executed in parallel for both parent-offspring pairs. Moreover, this part involves just simple reading from parent memory and sequential writing to offspring memory without performing any other operation. As a result, it concludes very quickly. On the other hand the second part of the crossover (filling offspring before the first and after the second cut points) is performed sequentially for each parent-offspring pair. Besides of executing memory reading/writing operations, some checks are also performed, that ensure only valid tours are constructed. As a result, the greater the distance between the cut points, the faster the PMX crossover will be completed in FPGA.

The acceleration achieved in FPGA compared to a software implementation can be explained by the following primary reasons. Firstly, the technique of parallel processing is employed, i.e. a part of the crossover operation is executed in parallel in FPGA. Second, the memory organization in the proposed architecture is tailored to the required data sizes. We think it is possible to provide further acceleration in the FPGA by implementing the second part of the PMX crossover also in parallel. In order to do this just simple changes in the control unit are required.

7 Conclusion

The paper presents a solution for synthesis and implementation of a reconfigurable adaptive control system, which can be used for manufacturing applications. The problem considered has been informally specified and converted to the TSP. The latter was solved with the aid of a GA. At the current implementation version, the most computationally intensive task (the crossover) is assigned to hardware, while the other tasks are performed in software. The result of solving the TSP permits the construction of a skeletal frame for future specification of the control circuit. Since the circuit is virtual, it is constructed without any knowledge of its functionality (in fact only some predefined constraints have been taken into account). It is structurally organized as a hardware template. A particular functionality can be implemented during execution time by reloading/configuring dynamically modifiable blocks. We believe that the results of the paper have shown one of the first engineering examples that demonstrates the practical implementation of virtual control circuits by combining methods and algorithms from different scientific areas, such as a modern methodology in logic synthesis and an evolutionary approach in artificial intelligence.

References

1. Michalewicz, Z., Fogel, D.B.: How to Solve It: Modern Heuristics. Springer (2000)
2. Sklyarov, V., Skliarova, I., Ferrari, A.B.: Hierarchical Specification and Implementation of Combinatorial Algorithms Based on RHS Model. Proc. of the XVI Conference on Design of Circuits and Integrated Systems – DCIS'2001, Portugal (2001) 486-491
3. Sklyarov, V.: Hierarchical Finite-State Machines and Their Use for Digital Control. IEEE Trans. on VLSI Systems, Vol. 7, No 2 (1999) 222-228
4. Skliarova, I., Ferrari, A.B.: Synthesis of reprogrammable control unit for combinatorial processor. Proc. of the 4th IEEE Int. Workshop on Design and Diagnostics of Electronic Circuits and Systems – IEEE DDECS'2001, Hungary (2001) 179-186
5. Golden, B.L., Kaku, B.K.: Difficult Routing and Assignment Problems. In: Rosen, K.H., Michaels, J.G., Gross, J.L., Grossman, J.W., Shier, D.R. (eds.): Handbook of Discrete and Combinatorial Mathematics. CRC Press, 692-705 (2000)
6. Junger, M., Reinelt, G., Rinaldi, G.: The traveling salesman problem. In: Ball, M., Magnanti, T., Monma, C., Nemhauser, G. (eds.): Network Models, 225-330 (1995)
7. Litke, J.D.: An Improved Solution to the Traveling Salesman Problem with Thousands of Nodes. Communications of the ACM, vol. 27, No. 12 (1984) 1227-1236
8. Ahuja, R.K., Magnanti, T.L., Orlin, J.B., Reddy, M.R.: Applications of network optimization. In: Ball, M., Magnanti, T., Monma, C., Nemhauser, G. (eds.): Network Models, 1-83 (1995)

9. Garey, M., Johnson, D.: Computers and Intractability: A Guide to the Theory of NP-completeness. Freeman (1979)
10. Goldberg, D.E., Lingle, R.: Alleles, Loci, and the Traveling Salesman Problem. Proc. of the 1st Int. Conf. on Genetic Algorithms, 154-159 (1985)
11. http://www.informatik.uni-heidelberg.de/groups/comopt/software/TSPLIB95/index.html
12. http://www.alphadata.co.uk
13. Xilinx: The programmable logic data book. Xilinx, San Jose (2000)
14. Harel, D.: Statecharts: A Visual Formalism for Complex Systems. Science of Computer Programming, N8 (1987) 231-271

Minimal Addition Chain for Efficient Modular Exponentiation Using Genetic Algorithms

Nadia Nedjah and Luiza de Macedo Mourelle

Department of Systems Engineering and Computation, Faculty of Engineering
State University of Rio de Janeiro, Rio de Janeiro, Brazil
{nadia,ldmm}@eng.uerj.br
http://www.eng.uerj.br/~ldmm

Abstract. Modular exponentiation is fundamental to several public-key cryptography systems such as the RSA encryption system. It is performed using successive modular multiplication. The latter operation is time consuming for large operands. Accelerating public-key cryptography software or hardware needs either optimising the time consumed by a single modular multiplication or reducing the total number of modular multiplication performed or both of them. This paper introduces a novel idea based on genetic algorithms for computing an optimal addition chain that allows us to minimise the number of modular multiplication and hence implementing efficiently the modular exponentiation.

1 Introduction

The modular exponentiation is a common operation for scrambling and is used by several public-key cryptosystems, such as the RSA encryption scheme [8]. It consists of a repetition of modular multiplications: $C = T^E \bmod M$, where T is the plain text such that $0 \leq T < M$ and C is the cipher text or vice-versa, e is either the public or the private key depending on whether T is the plain or the cipher text, and M is called the modulus. The decryption and encryption operations are performed using the same procedure, i.e. using the modular exponentiation.

The performance of such cryptosystems is primarily determined by the implementation efficiency of the modular multiplication and exponentiation. As the operands (the plain text of a message or the cipher (possibly a partially ciphered) text are usually large (i.e. 1024 bits or more), and in order to improve time requirements of the encryption/decryption operations, it is essential to attempt to minimise the number of modular multiplications performed.

It is clear that one should not compute T^E then reduce the result modulo M as the space requirements to store T^E is $E \times \log_2 T$, which is huge.

A simple procedure to compute $C = T^E \bmod M$ based on the paper-and-pencil method is described in Algorithm 1. This method requires $E-1$ modular multiplications. It computes all powers of T: $T \to T^2 \to T^3 \to ... \to T^{E-1} \to T^E$.

Algorithm 1. *simpleMethod*(T, M, E)
```
        C = T;
        for i = 1 to E-1 do
                C = (C × T) mod M;
        return C
end algorithm.
```

The paper-and-pencil method computes more multiplications than necessary. For instance, to compute T^8, it needs 7 multiplications, i.e. $T \to T^2 \to T^3 \to T^4 \to T^5 \to T^6 \to T^7 \to T^8$. However, T^8 can be computed using only 3 multiplications $T \to T^2 \to T^4 \to T^8$. The basic question is: what is the fewest number of multiplications to compute T^e, given that the only operation allowed is multiplying two already computed powers of T? Answering the above question is *NP*-hard, but there are several efficient algorithms that can find a near optimal ones.

Evolutionary algorithms are computer-based solving systems, which use evolutionary computational models as key element in their design and implementation. A variety of evolutionary algorithms have been proposed. The major ones are *genetic algorithms* [2]. They have a conceptual base of simulating the evolution of individual structures via the Darwinian natural selection process. The process depends on the performance of the individual structures as defined by its environment. Genetic algorithms are well suited to provide an efficient solution of *NP*-hard problems [2].

In the rest of this paper, we will present a novel method based on the addition chain method that attempts to minimise the number of modular multiplications, necessary to compute modular exponentiation. It does so using genetic algorithms.

This paper will be structured as follows: Section 2 presents the addition chain based methods; Section 3 gives an overview on genetic algorithms concepts; Section 4 explains how these concepts can be used to compute a minimal addition chain. Section 5 presents some useful results.

2   The Addition Chain Based Methods

The addition chain based methods attempt to find a chain of numbers such that the first number of the chain is 1 and the last is the exponent E, and in which each member of the chain is the sum of two previous members. For instance, the addition chain used by Algorithm 1 is (1, 2, 3, ... $E-2$, $E-1$, E). A formal definition of an addition chain is as in Definition 1:

Definition 1. An *addition chain* of length l for an integer n is a sequence of integers $(a_0, a_1, a_2, ..., a_l)$ such that $a_0 = 1$, $a_l = n$ and $a_k = a_i + a_j$, $0 \le i \le j < k \le l$.

A *redundant* addition chain is an addition chain in which some of its members are repeated otherwise, it is said *non-redundant*. Let *nrac* (*non-redundant addition chain*) be a function whose application to a redundant addition chain yields the corresponding non-redundant addition chain. For instance, $nrac((1, 2, 3, 5, 5, 7, 10))$ = $(1, 2, 3, 5, 7, 10)$.

The algorithm used to compute the modular exponentiation $C = T^E \mod M$, is specified by Algorithm 2.

Algorithm 2. `additionChainBasedMethod(T, M, E)`
```
    Let (a₀=1 a₁ a₂ ... aᵢ a₁=E] be the addition chain;
    powerOfT[0] = T mod M;
    for k = 1 to l
        let aₖ = aᵢ + aⱼ | i<k and j<k;
        powerOfT[k] = powerOfT[i] × powerOfT[j] mod M;
    return powerOfT[l];
end algorithm.
```

Finding a minimal addition chain for a given number is *NP*-hard. Therefore, heuristics are used to attempt to approach such a chain. The most used heuristic consists of scanning the digits of E form the less significant to the most significant digit and grouping them in partitions P_i. The size of the partitions can be constant or variable [4], [5], [6]. Modular exponentiation methods based on constant-size partitioning of the exponent are usually called *m-ary methods*, where m is a power of two and $\log_2 m$ is the size of a partition. On the other hand, modular exponentiation methods based on variable-size are usually called *sliding window methods*. There exist several strategies to partition the exponent. The generic computation of such methods is formalised in Algorithm 3, wherein V_i and L_i denote the value and length of partition P_i respectively.

Algorithm 3. `HeuristicsBasedModularExpo(T, M, E)`
```
    Partition E using the given strategy;
    for each Pᵢ in ℘(E)
        Compute TVi mod M;
    C = TVb-1 mod M;
    for i = b-2 downto 0
        C = T2Li mod M;
        if Vᵢ ≠ 0 then
            C = C×TVi mod M;
    return C;
end algorithm.
```

Assuming that there are b partitions in E, then the *m*-ary and sliding window methods computes $C = T^E \mod M$ using the following addition chain (for details see [6]):

$$nrac \left(\begin{array}{c} V_{b-1}, (V_{b-1})^{2^{L_{b-2}}}, (V_{b-1})^{2^{L_{b-2}}} + V_{b-2} \times \delta_{b-2}, \quad \cdots, \\ \left(\left((V_{b-1})^{2^{L_{b-2}}} + V_{b-2} \times \delta_{b-2} \right)^{2^{L_{b-3}}} + V_{b-3} \times \delta_{b-3} + \ldots + V_1 \times \delta_1 \right)^{2^{L_0}} + V_0 \times \delta_0 \end{array} \right) \quad (1)$$

where δ_i is 0 whenever P_i is a zero partition (i.e. 00...0) and 1 otherwise (i.e. contains at least one digit different from 0) and L_i denotes the length of partition P_i. Note that we apply function *nrac* because when δ_i is 0 the current member is the same as the previous one of the chain. With the *m*-ary method, partition length is constant and so the addition chain is as follows:

$$nrac\left(V_{b-1}, (V_{b-1})^{2^L}, (V_{b-1})^{2^L} + V_{b-2} \times \delta_{b-2}, ..., \left(\left((V_{b-1})^{2^L} + V_{b-2} \times \delta_{b-2}\right)^{2^L} + V_{b-3} \times \delta_{b-3} + ... + V_1 \times \delta_1\right)^{2^L} + V_0 \times \delta_0\right) \quad (2)$$

In order to guarantee that given lists above are addition chains, we need to show how a term V^{2^L} is decomposed: $(V, V^2, V^4, ..., V^{2^L})$.

The addition chains given above are still incomplete because they do not include the initial pre-computing addition chain. The *m*-ary pre-computing addition chain is $(1, 2, 3, ..., m-1)$ while for that of the sliding windows is $(1, 2, 3, 5, ..., 2^d-1)$ wherein *d* is the length of the longest partition of the exponent.

Particularly, in the 2-ary method or simply the binary method, each digit represents a partition. So we can safely write $V_i = e_i$. Hence the addition chain used by the binary method is as follows:

$$nrac\left(e_{b-1}, (e_{b-1})^2, (e_{b-1})^2 + e_{b-2}, ..., \left(\left((e_{b-1})^2 + e_{b-2}\right)^2 + e_{b-3} + ... + e_1\right)^2 + e_0\right) \quad (3)$$

wherein *b* is the number of bits in the binary representation of E and e_i are the digits of E. Here, we do not need to us δ_i as it plays the same role of e_i.

A variation of the sliding window method can be found in [7]. For instance, if $E = \underline{1}10\underline{111}$, i.e. $E = 55$, the non-redundant addition chain used by the binary, quaternary, octal and sliding window methods would be as follows, wherein the underlined powers are pre-computed.

Table 1. The addition chains yield by the exposed methods for exponent $E = 55$.

Method	Addition-subtraction chain	# of Mult
Binary	(1, 2, 3, 6, 12, 13, 26, 27, 54, 55)	9
Quaternary	(<u>1, 2, 3</u>, 6, 12, 13, 26, 52, 55)	8
Octal	(1, 2, 3, 4, 5, 6, 7, 12, 24, 48, 55)	10
Sliding window	(<u>1, 2, 3, 4, 5, 6, 7</u>, 12, 24, 48, 55)	10

3 Principles of Genetic Algorithms

Genetic algorithms maintain a *population* of *individuals* that evolve according to *selection* rules and other *genetic operators*, such as *mutation* and *recombination*. Each individual receives a measure of *fitness*. *Selection* focuses on high fitness individuals. Mutation and recombination or *crossover* provide general heuristics that simulate the *reproduction* process. Those operators attempt to perturb the characteristics of the parent individuals as to generate *distinct* offspring individuals.

Genetic algorithms are implemented through the following algorithm described by Algorithm 4, wherein parameters *popSize*, *fit* and *genNum* are the population maximum size, the expected fitness of the returned individual and the maximum number of generation allowed respectively.

Algorithm 4.
```
individual GA(popSize,fit,genNum)
     Generation       = 0;
     Population       = initialPopulation();
     Fitness          = evaluate(population);
     do
          parents        = select(population);
          population     = reproduce(parents);
          fitness        = evaluate(population);
          generation     = generation + 1;
     while(fitness[i]<fit, ∀ i∈population)
          and (generation < genNum);
     return fittestIndividual(population);
end algorithm.
```

In Algorithm 4, Function *intialPopulation* returns a valid random set of individuals that would compose the population of first generation, function *evaluate* returns the fitness of a given population storing the result into *fitness*. Function *select* chooses according to some random criterion that privilege fitter individuals, the individuals that should be used to generate the population of the next generation and function *reproduction* implements the crossover and the mutation process to actually yield the new population.

4 Application to Addition Chain Minimisation Problem

It is perfectly clear that the shorter the addition chain is, the faster Algorithm 2. We propose a novel idea based on genetic algorithm to solve this minimisation problem.

The addition chain minimisation problem consists of finding a sequence of numbers that constitutes an addition chain for a given exponent. The sequence of numbers should be of a minimal length.

4.1 Individual Encoding

Encoding of individuals is one of the implementation decisions one has to take in order to use genetic algorithms. It very depends on the nature of the problem to solve. There are several representations that have been used with success: *binary encoding* which is the most common mainly because it was used in the first works on genetic algorithms, represents an individual as a string of bits; *permutation encoding* mainly used in ordering problem, encodes an individual as a sequence of integer; *value encoding* represents an individual as a sequence of values that are some evaluation of some aspect of the problem [8], [9].

In our implementation, an individual represents an addition chain. We use the binary encoding wherein 1 implies that the entry number is a member of the addition chain and 0 otherwise. Let $n = 6$ be the exponent, the encoding of Fig. 1 represents the addition chain (1, 2, 4, 6):

1	2	3	4	5	6
1	1	0	1	0	1

Fig. 1. Addition chain encoding

4.2 The Genetic Algorithm

Consider Algorithm 4. Besides the parameters *popSize*, *fit* and *genNum* which represent the population maximum size, the fitness of the expected result and the maximum number of generation allowed, the genetic algorithm has several other parameters, which can be adjust by the user so that the result is up to his or her expectation. The selection is performed using some *selection probabilities* and the reproduction, as it is subdivided into crossover and mutation processes, depends on the kind of crossover and the mutation rate and degree to be used.

Selection. The selection function as described in Algorithm 5, returns two populations: one represents the population of first parents, which is *parents*[1][] and the other consists of the population of second parents, which is *parents*[2][].

Algorithm 5.
```
population[] select(population pop)
    population[] parents[2];
    for i = 1 to popultionSize
        n1 = random(0,1); n2 = random(0,1);
        for j = 1 to popSize do
            parents[1][i]=parents[2][i]=parents[popSize];
            if SelectionProbabilities[j] ≥ n1 then
                parents[1][i] = pop[j];
            else if SelectProbabilities[j] ≥ n2 then
                parents[2][i]=pop[j];
    return parents;
end algorithm.
```

The selection proceeds like this: whenever no individual that attends to the selection criteria is encountered, one of the last individuals of the population is then chosen, i.e. one of the fitter individuals of population. Note that the population from which the parents are selected is sorted in decreasing order with respect to the fitness of individuals, which will be described later on. The array *selectionProbabilities* is set up at initialisation step and privileges fitter individuals.

Reproduction. Given the parents populations, the reproduction proceeds using replacement as a reproduction scheme, i.e. offspring replace their parents in the next generation. Obtaining offspring that share some traits with their corresponding parents is performed by the *crossOver* function. There are several *types* of crossover schemes. These will be presented shortly. The newly obtained population can then suffer some mutation, i.e. some of the individuals (addition chains) of some of the genes (power numbers). The crossover type, the number of individuals that should mutated and how far these individuals should be altered are set up during the initialisation process of the genetic algorithm.

Algorithm 6. `population reproduce(population p1, p2)`
    `return mutate(crossOver(p1,p2,type),degree,rate)`
`end algorithm.`

Crossover. There are many ways how to perform crossover and these may depend on the individual encoding used [8]. We present crossover techniques used with permutation representation.

Single-point crossover consists of choosing randomly one *crossover point*, then, the part of the integer sequence from beginning of offspring till the crossover point is copied from one parent, the rest is copied from the second parent as depicted in Fig. 2(a). *Double-points crossover* consists of selecting randomly two *crossover points*, the part of the integer sequence from beginning of offspring to the first crossover point is copied from one parent, the part from the first to the second crossover point is copied from the second parent and the rest is copied from the first parent as depicted in Fig. 2(b). *Uniform crossover* copies integers randomly from the first or from the second parent. Finally, *arithmetic crossover* consists of applying some arithmetic operation to yield a new offspring.

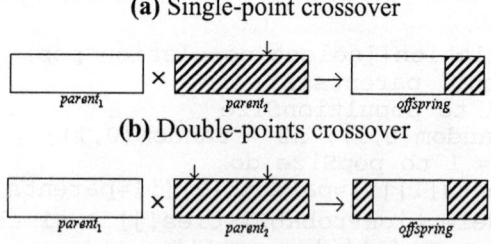

Fig. 2. Single-point vs. double-points crossover

The single point and two points crossover use randomly selected crossover points to allow variation in the generated offspring and to avoid premature convergence on a local optimum [1], [8]. In our implementation, we tested all four-crossover strategies.

Mutation. Mutation consists of changing some genes of some individuals of the current population. The number of individuals that should be mutated is given by the parameter *mutationRate* while the parameter *mutationDegree* states how many genes of a selected individual should be altered.

The mutation parameters have to be chosen carefully as if mutation occurs very often then the genetic algorithm would in fact change to *random search* [1]. Algorithm 7 describes the mutation procedure used in our genetic algorithm.

When either of *mutationRate* or *mutationDegree* is null, the population is then kept unchanged, i.e. the population obtained from the crossover procedure represents actually the next generation population.

Algorithm 7. population mutate(population pop,
                    int mutationDegree,int mutationRate)
        if(mutationRate≠0)and(mutationDegree≠0)then
            for a = 1 to PopSize do
                n = random(0,1);
                if n ≤ mutationRate then
                    for i=1 to mutationDegree do
                        gene = random(1, exponent);
                        pop[a][gene]=(pop[a][gene]+1)mod2;
        return pop;
end algorithm.

When mutation takes place, a number of genes are randomised and mutated: when the gene is 1 then it becomes 0 and vice-versa. The parameter *mutationDegree* indicates the number of gene to be mutated.

Fitness. This step of the genetic algorithm allows us to classify the population so that fitter individuals are selected more often to contribute in the formation of a new population.

The fitness evaluation of addition chain is done with respect to two aspects: *(i)* how much a given addition chain adheres to the Definition 1, i.e. how many members of the addition chain cannot be obtained summing up two previous members of the chain; *(ii)* how far the addition chain is reduced, i.e. what is the length of the addition chain. Algorithm 8 describes the evaluation of fitness used in our genetic algorithm.

For a valid addition chain, the fitness function returns its length, which is smaller than exponent E. The evolutionary process attempts to minimise the number of ones in a valid addition chain and so minimise the corresponding length. Individuals with fitness larger or equal to exponent are invalid addition chains. The constant *largePenalty* should be larger than E. With well-chosen parameters, the genetic algorithm deals only with valid addition chains.

Algorithm 8. int evaluate(individual a)
    int fitness = 0;
        for i = 2 to exponent do
            if a[i] == 1 then fitness = fitness + 1;
            if ∃j,k | 1≤j≤i & 1≤k≤i & i==j+k & a[i]==a[k]==1
                then fitness = fitness + largePenalty;
            return f;
    end algorithm.

5   Implementation Results

In applications of genetic algorithms to a practical problem, it is difficult to predict a priori what combination of settings will produce the best result for the problem. The settings consist of the population size, the crossover type, the mutation rate, when mutation does take place, the mutation degree. We investigated the impact of different values of these parameters in order to choose the more adequate ones to use. We found out that the ideal parameters are: a population of 100 individuals; the double-points crossover; a mutation rate between 0.3 and 0.5 and a mutation degree of about 1% of the value of the exponent.

The curve of Fig. 3 shows the progress made in the first 200 generations of an execution to obtain the addition chain for 250. The settings used are: a 100 individual per population, double-points crossover, a mutation rate of 0.1 and a degree of 3.

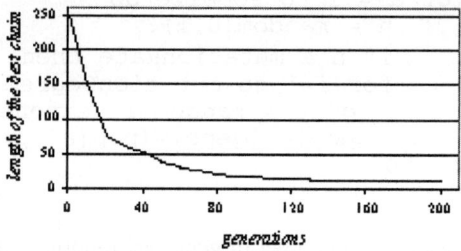

Fig. 3. The genetic algorithm result curve for the parameters given above

Let *length(E)* be the length of the shortest addition chain for exponent E, i.e. the number of multiplication necessary using the addition chain). The exact value of *length(E)* is known only for relatively small values of E. Nevertheless, it is known that for large exponents E,

$$length(E) = \log E + (1 + o(1)) \frac{\log E}{\log \log E} \qquad (4)$$

The lower bound was shown by Erdös [3] using a counting argument and the upper bound is given the *m*-ary method. For exponent 23, 55 and 250, the optimal addition chain yields more than 5, 6 and 8 multiplications.

Table 2. The addition chains yield by the exposed methods vs. the genetic algorithm for exponent $E = 23$, $E = 55$ and $E = 250$ together with the number of necessary modular multiplications

E	Method	Addition-subtraction chain	#Mult
23	Binary	(1, 2, 4, 5, 10, 11, 22, 23)	7
	Quaternary	(1, 2, 3, 4, 5, 10, 20, 23)	7
	Octal	(1, 2, 3, 4, 5, 6, 7, 8, 16, 23)	9
	Sliding window	(1, 2, 3, 5, 7, 8, 16, 23)	7
	Genetic algorithm	(1, 2, 3, 5, 10, 20, 23)	6
		(1, 2, 4, 5, 9, 18, 23)	6
55	Binary	(1, 2, 4, 6, 12, 13, 26, 27, 54, 55)	9
	Quaternary	(1, 2, 3, 6, 12, 13, 26, 52, 55)	8
	Octal	(1, 2, 3, 4, 5, 6, 7, 12, 24, 48, 55)	9
	Sliding window	(1, 2, 3, 5, 7, 12, 24, 48, 55)	8
	Genetic algorithm	(1, 2, 4, 5, 9, 18, 23, 32, 55)	8
		(1, 2, 4, 8, 9, 18, 27, 28, 55)	8
250	Binary	(1, 2, 3, 6, 7, 14, 15, 30, 31, 62, 124, 125, 250)	12
	Quaternary	(1, 2, 3, 6, 12, 15, 30, 60, 62, 124, 248, 250)	11
	Octal	(1, 2, 3, 4, 5, 6, 7, 12, 24, 31, 62, 124, 248, 250)	13
	Sliding window	(1, 2, 3, 5, 7, 8, 15, 30, 60, 120, 125, 250)	11
	Genetic algorithm	(1, 2, 3, 5, 10, 20, 30, 50, 100, 150, 250)	10
		(1, 2, 4, 8, 10, 20, 40, 80, 120, 240, 250)	10

Finding the best addition chain is impractical. However, we can find near-optimal ones. Our genetic algorithm always finds addition chains as short as the shortest addition chain used by the m-ary method independently of the value of m and by the sliding windows independently of the partition strategy used. Table 2 shows some examples for exponents 23, 55 and 250. For the sliding window method, the exponent are partitioned as follows: $E = (23)_{10} = (1\underline{0}\underline{111})_2$; $E = (55)_{10} = (1\underline{10}\underline{111})_2$ and $E = (250)_{10} = (\underline{1}\underline{111}\underline{1010})_2$.

6 Conclusions

In this paper, we presented an application of genetic algorithms to minimisation of addition chain. We first explained how individuals are encoded. Then we described the necessary algorithmic solution. Then we presented some empirical observations about the performance of the genetic algorithm implementation.

This application of genetic algorithms to the minimisation problem proved to be very useful and effective technique. Shorter addition chains compared with those obtained by the m-ary methods as well as those obtained for the sliding window methods (see Table 2 and 3 of the previous section) can be obtained with a little computational effort. A comparison of the performance of the m-ary, sliding window vs the genetic algorithm is shown in Fig. 4. A satisfactory addition chain can be obtained in a 4 to 6 second using a Pentium III with a 128 MB of RAM.

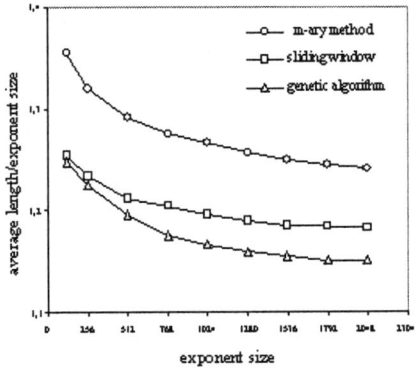

Fig. 4. Ratio for the addition chains yield by the GA vs. m-ary and sliding window methods

References

1. DeJong, K. and Spears, W.M., *An analysis of the interacting roles of the population size and crossover type in genetic algorithms*, In Parallel problem solving from nature, pp. 38-47, Springer-Verlag, 1990.
2. DeJong, K. and Spears, W.M., *Using genetic algorithms to solve NP-complete problems*, Proceedings of the Third International Conference on Genetic Algorithms, pp. 124-132, Morgan Kaufmann, 1989.

3. Erdös, P., *Remarks on number theory III: On addition chain*, Acta Arithmetica, pp 77-81, 1960.
4. Haupt, R.L. and Haupt, S.E., *Practical genetic algorithms*, John Wiley and Sons, New York, 1998.
5. Knuth, D.E., *The Art of Programming: Seminumerical Algorithms*, vol. 2. Reading, MA: Addison_Wesley, Second edition, 1981.
6. Koç, Ç.K., *High-speed RSA Implementation*, Technical report, RSA Laboratories, Redwood City, califirnia, USA, November 1994.
7. Kunihiro, N. and Yamamoto, H., *New methods for generating short addition chain*, IEICE Transactions, vol. E83-A, no. 1, pp. 60-67, January 2000.
8. Michalewics, Z., *Genetic algorithms + data structures = evolution program*, Springer-Verlag, USA, third edition, 1996.
9. Neves, J., Rocha, M., Rodrigues, Biscaia, M. and Alves, J., *Adaptive strategies and the design evolutionary applications*, Proceedings of the Genetic and the Design of Evolutionary Computation Conference, Orlando, Florida, USA, 1999.
10. Rivest, R.L., Shamir, A. and Adleman, L., *A method for obtaining digital signature and public-key cryptosystems*, Communication of ACM, vol. 21, no.2, pp. 120-126, 1978.

Course Scheduling Using Genetic Algorithm Methods

Raed Abu Zitar

Department of Computer Science, Al-Isra Private University
Amman, Jordan
rzitar@isra.edu.jo

Abstract. The timetabling problem is a high-dimensional, non-Euclidean, multi-constraint combinational optimization problem. Given data sets of classes, their days and instructors; rooms and their capacities, types; students and their preferences and requirements; the problem is to construct a feasible class schedule satisfying all the hard constraints and minimizing the medium and soft constraints. We apply Genetic Algorithms (GAs) to solve the scheduling problem mapped from the Faculty of Science at University of Jordan. This technique has achieved very good results compared to the hand made schedule result produced by the registration department.

1 Introduction

In this work we discuss the time-tabling problem [1],[2], specifically the Course Time-tabling Problem (CTTP) at the Faculty of Sciences of the University of Jordan. It is desirable to construct an automatic course schedule given the sections, teachers, students' desires, halls, time periods, and many other constraints. Some constraints must be satisfied and others are desired to be satisfied. In this work we study the problem, analyze it, suggest solutions, and finally test the applicability of these theoretical solutions. We test a technique that is based on the Genetic Algorithms methodology [3],[4]. Time-tabling is a particular form of a more general scheduling problem. In the most general case, there is a set of objects (resources) and a set of processes that need to be carried out for each of those objects.

The university course time-tabling problem occurs when scheduling a set of lectures for each course within a given number of rooms and time periods. The main differences with the high school time-tabling problem is that university courses can have common students, whereas school classes are disjoint sets of students. If two courses have common students then they conflict, and they can not be scheduled at the same period. Moreover, school teachers always teach more than one class, whereas in universities, an instructor may teach only one course. In addition, in the university problem, availability of rooms and their size plays an important role, whereas in the high school problem they are often neglected because, in most cases, we can assume that each class has its own room.

1.1 Optimization Problem

Schaert (1995) [5] covers the problem of CTTP with the following objective function:

$$\max \sum_{i=1}^{q} \sum_{k=1}^{p} d_{ik} y_{ik} \quad (1)$$

where d_{ik} is the desirability of having a lecture of C_i course at period k. This desirability may be binary, i.e. having 0 or 1 values. Also it can be a value in the interval [0, 1].

We can consider a conflict matrix M_{qxq} with integer values, such that m_{ij} represents the number of students taking both courses C_i and C_j. In this way m_{ij} represents also a measure of dissatisfaction in case a lecture of C_i and a lecture of C_j are scheduled at the same time. We have to minimize the global dissatisfaction obtained as the sum of all the dissatisfactions of the above type.

So our objective function can be written as:

Min {hp *number of hard constrains +
mp *number of medium constraints + sp *number of soft constraints},

where hp is the hard constraints weight, mp is the medium constraints weight, and sp is soft constraints weight.

2 Constraints

We divided our problem constraints into three categories hard, medium, and soft.

2.1 Hard Constraints

The hard constraints can be summarized with the following:

The hall size must be larger than or equal to the section capacity.

1. No hall is assigned to more than one lecture at the same time.
2. Each section is assigned a suitable hall with the type according to its requirements.
3. Each section must occupy the same hall for all lectures.
4. No instructor gives more than one lecture at the same time.
5. Break times, such as department meetings hours, and the families meeting hours must be free.
6. Fixed time sections must be taking into consideration.
7. Undergraduate theoretical courses are preferable to end before 2 clock in scientific faculties and before 5 clock in other faculties.
8. Graduate courses start after 3 clock and end before 5 clock in scientific faculties and starts after 5 clock in others.
9. The sections must be distributed according to the two schemes in the university. The first one with one hour period per three days a week, and the second for one and half an hour per two days a week.

10. No two or more optional or obligatory courses have their lectures. (for the same department for the same academic level).
11. No two related courses clash. (a course and its laboratory)

2.2 Medium Constraints

Distributing each department courses on all available periods in the morning and afternoon. This can be achieved by:

1. Distributing courses (sections) taught by the same teacher on all weekdays.
2. Distributing courses (sections) taught by the same teacher on all the day hours, i.e. on the morning and afternoon and having at least one free hour between every two consecutive lectures.
3. Undesired times for instructor must be taken into consideration.

2.3 Soft Constraints

No two or more optional or obligatory courses of two consecutive academic levels of the same department have their lectures at the same time.

3 Applying Genetic Algorithms to the CTTP

Genetic Algorithms (GA's) have many applications in scientific, and engineering areas including pattern recognition, function optimization, scheduling, machine learning, clustering, engineering design, expert system design, process control and many others [6],[7],[8]. In this work, we present only two experiments for the scheduling problem (DS1 and DS2) . Each experiment is based on four similar sets of data; DS1, DS2, DS3, and DS4. Data sets descriptions are shown below in Table1.

Table 1. The four data sets used in the scheduling problem

	DS2	DS3	DS4
Halls	25	26	25
Instructors	93	93	93
Sections	264	264	264
U-Sections	211	211	211
G-Sections	53	53	53
Multi-sections	34	34	34
Fixed Periods for sections	25	25	25

The two experiments, however, differ in the modifications made to the GA in the selection strategy. DS3 has one more hall available. DS4 has two more periods to use for lectures.

3.1 Experiment 1

Now we experiment with our new operators:

Enhanced Crossover:

Crossover in its standard form is not helpful at all specially when basic structure of the chromosome is a schedule. The process of recombining selected schedules in a blindly manner will result in destructive effects. With enhanced crossover, we restrict the copying operation on only those sections that make no conflicts with others in the same period. We may gradually reach better patterns for each period and then for the whole child (schedule). First, we copy the fixed time sections in their locations in both children. Then and after selecting two randomly parents, we start mating. The crossover point is a number in the interval [1, halls_num] and each child will have two opposite halves from each parent.

Medium-Mutation:

In the crossover operation we took care of hard and slightly soft constraints. However, medium constraints where not manipulated. So we decided to use the mutation for swapping periods contents between different schemes or in the same scheme, hopefully to minimize daily and weekly load constraints. We call this procedure medium-mutation since swapping periods contents affect only medium constrains.

The Obligatory-Mutation:

This operator tries to find a better location (time slot and hall) for each section having hard crashes with other sections lay in its period. This process scans the child schedule to find a new location for the section such that it makes no conflicts in that new location and without violating time and hall constraints. If one is found, we move the section to it. But if the new location does not have an empty suitable hall we are forced to swap that section with a section laying on the perfect slot.

Conflict Matrix (CM) Initialization Procedure (see Fig. 1 for Pseudo Code):

The conflict Matrix (CM) reflects the amount of conflict a section has with the other sections depending on time, space, and availability of instructors. It is used to evaluate how much constraints are violated and, consequently, calculating the total cost of every suggested solution. (The solution here is a possible schedule or a string in GA population). Fig .2 (pseudo code) shows how a timetable may be initialized using the conflict matrix of Fig. 1.

Next Generation Selection Mechanism:

This mechanism takes the best parents and children as parents for the next generation. Table.2 shows the results. They are very good in comparison with the hand made schedule, since they achieve much fewer hard violations in all runs, also medium violations are fewer in most cases. Note that adding an extra hall in DS3 does not affect the solution quality to be better as we might think. But adding two new periods one for each scheme, as in DS4, decreases the costs in general. This clarifies the problem complexity. Also we note that the best solution usually appears in early generations of the evolution.

3.2 Experiment 2

Experiment 1 gives good solutions. However, we note that the best solution is found in early generation of evolution and the parents' cost become very close, which may indicate that they hold similar patterns (contents of periods), so no new patterns may be produced and no fitter child will be created. The next generation selection strategy may have caused this or the initialization method. Accordingly, in this new experiment (experiment 2) we test for a new selection strategy. The new selection strategy adds only the best child in place of the worse parent in parents' pool. Of course this fit child must be fitter than the fittest parent existing so far in the population. This would „stir" the population pool and enhance the quality of its members away from undesired saturation. Looking at the results for the four data sets in Table. 3, experiment 2 showed improvement in overall performance measures; average or best and for both populations sizes of 50 and 100.

4 Conclusions

We conclude form the experiments that experiment 2 gives the lowest hard costs. That is the selection mechanism proposed for that experiment is better than the ordinary selection mechanism. This is because keeping few high fit children with the less fit parents from the initial population can cause different seeding patterns during the mating. While the ordinary selection method turns the parents' pool to twins, which have very close costs, which means similar patterns. The GA in general provided good solutions for the scheduling problem, however, continuous modification on the heuristics can lead to better results, of course, on the expense of reducing the generality of the GA used.

```
Procedure build_conflict_matrix()
For   section1=1 to n
         For section2=1 to n
            If section1 <> section2 then
                  test_conflict(section1,section2)
            endif
Next section2
Next section1
End
Procedure test_conflict( i,j)
If instructor clash then increment C_{ij}.inst
If  same level clash then increment C_{ij}.same-level
If related courses then increment C_{ij}.related
If successive level clash then increment C_{ij}.soft
evaluate C_{ij}.hard
evaluate C_{ij}.total
End
```

Fig. 1. Conflict Matrix Procedures

Table 2. : Experiment 1 Results, (enhanced-crossover, medium-mutation, obligatory mutation, ordinary next generation selection, CM initialization)

Data Set	Size	Solution	Total	Hard	Medium	Soft	Generation Number
DS1	50	Average	165.75	5.25	7	22.75	111
		Best	134	5	2	24	139
	100	Average	147.50	5	4.75	23.75	145.25
		Best	128	5	1	23	180
DS2	50	Average	409.75	4.25	60	24.75	146.5
		Best	399	4	59	24	184
	100	Average	440.25	3.75	68.25	24	158.5
		Best	476	3	78	26	171
DS3	50	Average	484.25	6.25	66.25	29.50	121
		Best	460	6	63	31	146
	100	Average	514.50	5.75	73.25	33.25	133.25
		Best	572	5	88	32	115
DS4	50	Average	521.75	6	74.50	29.25	115.75
		Best	448	6	60	28	188
	100	Average	482.50	6.00	66.75	28.75	148
		Best	493	5	73	28	153

Table 3. Experiment 2 Results, (enhanced-crossover, medium-mutation, obligatory mutation, best child next generation selection, CM initialization)

Data Set	size	Solution	Total	Hard	Medium	Soft	Generation Number
DS1	50	Average	141.25	4.75	4.5	23.75	86
		Best	172	4	14	22	10
	100	Average	157.50	4.25	10	22.50	28.5
		Best	154	4	10	24	6
DS2	50	Average	358	4.75	47.75	24.25	104.5
		Best	411	4	61	26	21
	100	Average	370.25	5	49	25.25	51.75
		Best	302	5	35	27	114
DS3	50	Average	484	7.25	61.00	34.00	77.25
		Best	439	7	53	34	71
	100	Average	412.75	7.50	45.25	36.50	79.5
		Best	392	7	43	37	60
DS4	50	Average	458.50	7.5	55	38.50	93
		Best	449	7	54	39	194
	100	Average	455	7	55.50	37.50	65.75
		Best	399	7	44	39	116

```
Procedure load_section(scheme, period)
1.  select randomly an unscheduled section which can
    be assigned to the specified time.
2.  choose an empty suitable hall and assign the
    section to it(if possible).
3.  for each unscheduled section which can be assigned
    to the specified time, and does not make any
    conflict with other sections in the same period
    and scheme; choose an empty suitable hall and
    assign the section to it (if possible).
End
```

```
Procedure load_unscheduled_sections()
If there is unscheduled sections, search for empty
suitable hall in a legal period with hard
conflicts=0.
If there is unscheduled sections, search for empty
suitable hall in a legal period.
If there is unscheduled sections, search for empty
suitable hall.
End

Procedure load()
for each scheme
        for each period
           load_section(scheme,period)
        next period
next scheme
load_unscheduled_sections
End
```

Fig. 2. Initializing a Timetable using The Conflict Matrix (CM)

References

1. Abdennadher,S. and Marte, M.: University Timetabling using Constraint Handling Rules.) 1998).
   http://www.pms.informatik.uni-muenchen.de/publikationen
2. Abdennadher, S. and Marte, M.: Constraint-Based Heuristics for School Timetabling. Appeared in Workshop on Integration of AI and OR techniques in Constraint Programming for Combinatorial Optimization Problems, CP-AI-OR'99, Ferrara, Italy (1999).
   http://www.pms.informatik.uni-muenchen.de/publikationen.
3. Goldberg, D. E..: Genetic algorithms in Search, Optimization, and Machin Learning. Addison-Wesely Publishing Company, Inc, USA (1989).
4. Colorni, A., Dorgio, M. and Maniezzo, V. : Genetic Algorithms a New Approach to The Timetable Problem. Published in Lecture Notes in Computer Science – NATO ASI Series, Combinatorial. Oprimization, (Ed M. Akgul and others), Springer-Verlag, Berlin Heidelberg New York 32: 235-239. (1990).

5. Schaerf, A. :A survery of automated timetabling. Centrum voor Wiskunden Infirmatica (CWI) report CS-R9567, Amsterdam, The Netherlands. (A revisted version will appear in "Artificial Intelligence Review".) (1995).
http://www.asap.cs.nott.ac.uk/ASAP/papers/index.htmlJournalPapers
6. Buckles, Bill P. and Petry, Frederick E. *Genetic Algorithms, 1st edition.* IEEE Computer Society Press, USA, (1992).
7. Burke, E.K., Elliman, D.G. and Wear, R.F. : Automated Scheduling of University Exams. Proceedings of IEE Colloguium on Resource Scheduling for large scale planning systems, Digest No. 1 *993/144.* (1993).
http://www.asap.cs.nott.ac.uk/ASAP/papers/index.htmlJournalPapers
8. Corne, D, Fang, H. and Mellish, C.: Solving the Module Exam Scheduling Problem with Genetic Algorithms. In *Proceedings of the Sixth Internatonal Conference on Industrial and Engineering). Applications of Artificial Intelligence and Expert Systems, Gordan and Breach Science Publishers, Chung, Lovegrove,* Ali (eds), pp 370-373 (1993).
http://www.asap.cs.nott.ac.uk/ASAP/ttg/resources.html

An Evolutionary Algorithm for the Synthesis of RAM-Based FSMs

Valery Sklyarov

University of Aveiro, Department of Electronics and Telecommunications, IEETA
3810-193 Aveiro Portugal
skl@ieeta.pt

Abstract. It is known that RAM-based finite state machines (FSMs) can be very effectively implemented in reconfigurable hardware with distributed memory blocks. One possible approach to the engineering design of such circuits is based on a fuzzy-state encoding technique. This approach allows rapid construction of high-speed devices based on RAM-blocks of a modest size. The key point in the engineering design process is the development of a state encoding and this problem has a high combinatorial complexity, especially for FSMs with a large number of states. This paper proposes a novel evolutionary algorithm for the solution of this problem.

1 Introduction

FSMs are probably the most widely used components in digital systems. Today almost all the available automatic design tools that are included in industrial CAD systems allow FSMs to be synthesized from their formal specifications. The rapidly expanding market for programmable logic devices (PLDs) such as FPGAs has encouraged the development of PLD-targeted CAD systems, which also include a variety of tools for FSM synthesis.

The main purpose of a general PLD is to enable the same microchip to be used of for building a variety of circuits for different engineering applications. The design process relies on programming components that implement logic functions and interconnections between the components. For many practical applications it is desirable to provide a circuit with virtual capabilities in general, and modifiability in particular. The objectives of these are very different. For one kind of application we might want to supply similar capabilities to those provided for general-purpose microprocessors, which support the virtual memory mechanism. In particular, this allows a device to be constructed on a microchip that does not have sufficient hardware resources to accommodate all the functionality of the device. For another kind of applications it may be desirable to be able to change the behavior depending on external events that cannot be predicted in advance. In some cases we need to provide sufficient flexibility to allow changes during the debugging stage, etc. These requirements force us to search for such models of digital circuits that permit modifications in the functionality of the circuit, which has been already designed and implemented in PLD. The paper [1] suggests such a model for FSMs based on RAM-blocks. The paper considers the similar technique, which differs in two ways:

1. The most complicated synthesis step (that is the fuzzy state encoding) was performed with the aid of a novel evolutionary algorithm (EA). Note that applying evolutionary strategies to FSM synthesis is not new (see for example, [2,3]). However the EA considered in this paper is used for a specific step of synthesis, which is very important for the novel model of FSM with dynamically modifiable behavior.
2. The results of the paper can be used for two architectures of RAM-based FSMs. The first is the same as in paper [1]. The second architecture, which is introduced in the paper is slightly different, but allows the elimination of significant constraints that have to be taken into account for the method [1].

2   Practical Example

The left-hand part of fig. 1 depicts a simplified state transition graph that describes a sequence of steps to control technological operations shown on the right-hand part of fig. 1. The operation has to be performed by a cutting tool **T**, which permits cuts A and B to be made in a material (such as wood, metal, etc.) with different widths. The motion of T and its operations can be controlled by signals $y_1,...y_5$ as follows: y_1 - causes the motion from left to right; y_2 - causes the motion from right to left; y_3 - reduces the width of cutting; y_4 - increases the width of cutting; $y_5=1$ sets the tool **T** in working state (**TW**) and $y_5=0$ sets the tool **T** in non-working state (**TW̵**).

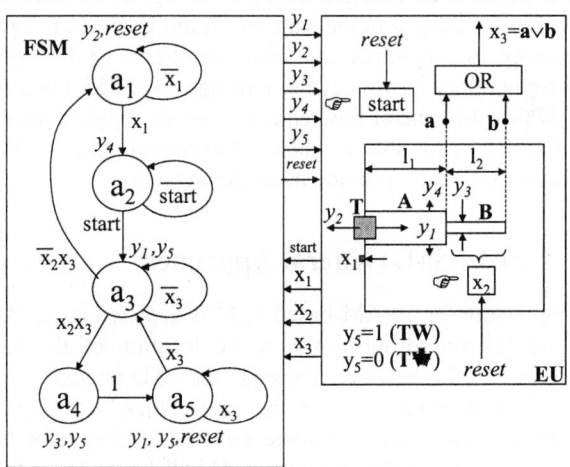

Fig. 1. Composition of FSM and execution unit (EU)

The active value of each signal (except the last one) is 1. It is obvious that some combinations of the signals, such as y_1 and y_2 or y_3 and y_4 are prohibited. Since we want to present just the concepts of a real engineering application, we will not concern ourselves with such details.

The proper sequence of signals depends on values of logic conditions $x_1,...,x_3$:

- $x_1=1$ indicates that **T** is positioned at the point from which it can start operations;
- $x_2=1$ requires the operation **B** to be carried out. In the opposite case (i.e. $x_1=0$) the operation **B** has to be omitted;
- $x_3=1$ states that the sensors either **a** or **b** are active. These sensors can be positioned manually (or with the aid of some automatic tools) before an operation, and they permit different lengths for the segments l_1 and l_2 to be specified (see fig. 1).

There is an additional button, "start", which forces the initiation of operations. The values of conditions "start" and x_2 can be reset by the signal "reset" from the FSM. Thus in total the FSM has 6 outputs, $y_1,...,y_5$, and "reset" (shown in fig. 1 in *italics*), and 4 inputs, $x_1,...,x_3$, and "start". The outputs force operations in the execution unit (EU) and the inputs are tested in the FSM in order to generate the proper sequence of outputs. The state transition graph on the left-hand part of fig. 1 describes this sequence and is considered to be a specification for the synthesis of the logic circuit that has to be implemented in reconfigurable hardware in our case.

Note that as a rule, a FSM is considered to be a small part of the engineering system. The hardware resources provided by an FPGA chip permit many other components to be implemented, so the synthesis of the FSM has to be integrated into the design of the overall system. Let us assume that the design has been finished and circuits for the total system have been implemented in FPGA. Thus we can control the operations depicted in fig. 1 and carry out some other actions implemented in the remaining hardware of the same FPGA chip. But what to do if at some later time the shape of the segments A and B needs to be changed as shown in fig. 2, for example. In fact fig. 2 provides just two trivial changes in the state transition graph (see signals y_3 and y_4 in **bold** font). However this requires the FSM synthesis to be repeated from the beginning. Besides, not only do we have to redesign the FSM, but also the remaining hardware of the FPGA. In general this requires a lot of supplementary work, such as testing, etc. Obviously, being able to change the functionality of the FSM without redesigning the hardware would provide many advantages.

3 RAM-Based FSM. General Approach

The proposed architecture for a RAM-based FSM is depicted in fig. 3. It is based on a known architecture [1] and differs only in the location of the RAM-based state transition decoder (STD). In the architecture [1] the STD is positioned after the MIX block and in fig. 3 it is situated before the MIX block. Although this is a small change, it permits the elimination of some constraints that have to be taken into account in [1]. For example, in the architecture [1] all fuzzy codes for different FSM states must be orthogonal. For many practical applications it is necessary to increase the size of the FSM memory in order to satisfy this requirement and it becomes larger than for traditional binary encoding. There are also some problems with the generation of output variables. On the other hand all the basic synthesis steps of the method [1] can be used without modifications.

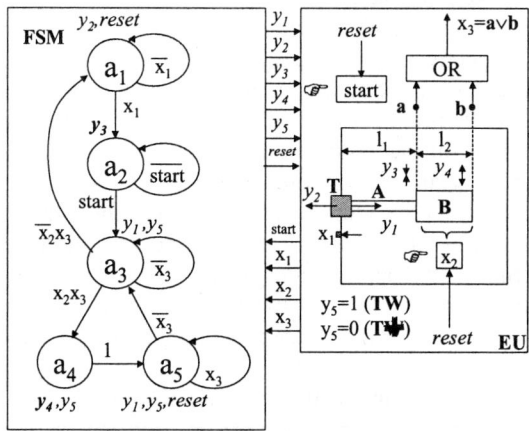

Fig. 2. Possible modifications in FSM behavior

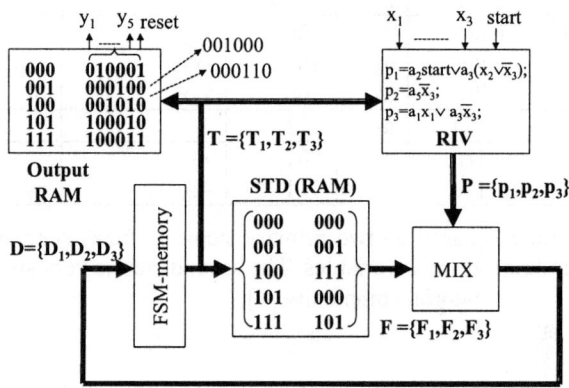

Fig. 3. Basic architecture of a RAM-based FSM

Suppose it is necessary to synthesize the FSM depicted in fig. 1 based on the architecture considered in fig. 3. The state transition graph (see fig. 1) can also be presented in the form of a table, such as that shown in columns a_{from} $Y(a)$, a_{to}, $X(a_{from}, a_{to})$ of table 1, where a_{from} and a_{to} are the initial and next states in state transitions, $X(a_{from}, a_{to})$ is a product of the input variables from the set $X=\{x_1,...,x_L\}$ that cause the transition from a_{from} to a_{to}, L is the number of FSM inputs (-x_1 denotes not x_1), $a_{from}, a_{to} \in A=\{a_1,...,a_M\}$, M is the number of FSM states, A is the set of FSM states. The FSM generates the outputs shown in column $Y(a)$ from the set $Y=\{y_1,...,y_N\}$ where N is the number of FSM outputs. For our example, L=4, N=6 and M=5. Columns a_{form} $Y(a)$, a_{to}, $X(a_{from}, a_{to})$ hold the initial data for future synthesis.

Suppose that the circuit of the FSM has already been synthesized and is described by the remaining columns of table 1, where K_{from} and K_{to} are the state codes (for the states a_{from} and a_{to} respectively), F - a fuzzy code that includes some bits marked with symbol "**i**" and these bits will be further mixed (in the block MIX) with the respective variables from the set $P=\{p_1,...,p_3\}$ (see column P), $D_1,...,D_3$ are variables that change states of FSM-memory on the assumption that this memory was built from D flip-flops. Now we can program all the components of fig. 3. The left column of each RAM-block in fig. 3 shows the address codes and the right column depicts the output vectors that have to be written to RAM at the respective addresses. The RIV block implements a system of Boolean functions p_1, p_2, p_3. It is very easy to check that the circuit functions in accordance with the given specification. The paper [1] presents all the details necessary for understanding the functionality of the circuit.

Table 1. Structural table of FSM

a_{from} Y(a)	K(a_{from})	F	a_{to}	K(a_{to})	X(a_{from},a_{to})	P	D
a_1,y_2	000	00i	a_1	000	-x_1		
reset		00i	a_2	001	x_1	p_3	D_3
a_2,	001	i01	a_2	001	-start		D_3
y_4		i01	a_3	101	start	p_1	D_1D_3
a_3,	101	i0i	a_1	000	-x_2x_3		
y_1 y_5		i0i	a_3	101	-x_3	p_1p_3	D_1D_3
		i0i	a_4	100	x_2x_3	p_1	D_1
a_4, y_3 y_5	100	111	a_5	111	1		$D_1D_2D_3$
a_5,y_1y_5	111	1i1	a_3	101	x_3		D_1D_3
reset		1i1	a_5	111	-x_3	p_2	$D_1D_2D_3$

The synthesis flow is based on two primary steps: 1) fuzzy state encoding and 2) replacement (encoding) of input variables. The second step is very simple [1] and the first step has high combinatorial complexity [1].

4 Evolutionary Algorithm for Fuzzy State Encoding

Let us designate A(a_{from}) as the subset of states that follow the state a_{from} and suppose that $k_{from} = |A(a_{from})|$ is the number of elements in the set A(a_{from}). For our example in table 1 we have A(a_1) = $\{a_1,a_2\}$, k_1=2, A(a_2) = $\{a_2,a_3\}$, k_2=2, A(a_3) = $\{a_1,a_3,a_4\}$, k_3=3, A(a_4) = $\{a_5\}$, k_4=1, A(a_5) = $\{a_3,a_5\}$, k_5=2. If k_{max}=max($k_1,k_2,k_3,...$), then G_{min} = intlog$_2k_{max}$, where G_{min} is the minimum number of variables from the set P that affect each individual transition. For our example in table 1 G_{min} = 2.

The target requirement for state encoding is the following [1]. States $a_{m1},a_{m2},...$ in each individual subset A(a_m), k_m>1, m=1,...,M, must be encoded in such a way that a maximum number of their bits with coincident indexes have constant values that are the same for all variables $a_{m1},a_{m2},...$ from A(a_m). Column K_{to} of table 1 shows such bits in bold font. The character "**i**" in fuzzy state codes indicates that the bit can be used for additional purposes (see [1] for details).

The EA for fuzzy state encoding includes the following steps:

1. Production of an initial population that is composed of individuals that represent a set of randomly generated codes for all FSM states.
2. Evaluation of the population and measuring its fitness.
3. Variation of the population by applying such operations as reproduction mutation and crossover. A reproduction operator makes a copy of the individual (with a probability based on its fitness) for inclusion to the next generation of the population. A mutation operator creates new individuals by performing some changes in a single individual, while the crossover operator creates new individuals (offspring) by combining parts of two or more other individuals (parents) [4].
4. Performing a selection process, where the fittest individuals survive and form the next generation.

Points 2-4 are repeated until a predefined termination condition is satisfied.

Let us assume that a population π including ν individuals $\pi_1,...,\pi_\nu$ has been randomly generated. In order to evaluate each individual π_t it is necessary to specify a fitness function. For our problem it is very easy. The requirements for the best solution were considered in the previous section. Indeed, if the states in each individual subset $A(a_m)$, $k_m>1$, $m=1,...,M$ have been encoded in such a way that w_m ($w_m=R-intlog_2k_m$) of their bits with coincident indexes have constant values it gives an optimal result. Here $R=intlog_2M$ is a minimum number of bits in the codes assuming binary state assignment. Thus any solution for which the fitness function W is equal to $\Sigma w_m(k_m>1, m=1,...,M)$ gives an optimal result. In fact we can discover several optimal results and for each of them the function W has the same maximum possible value. Any of these results provides the best solution to the problem so we just have to find the first of them. For example, for table 1 we have R=3, $w_1=w_2=w_5=2$, $w_3=1$ and the function $W=w_1+w_2+w_3+w_5=7$ shows the weight for the best result. This indicates that to obtain such a result the codes $F(a_1)$, $F(a_2)$, $F(a_3)$, $F(a_5)$ must be composed of 2, 2, 1 and 2 bits respectively with coincident indexes that have constant values. Since such codes have been obtained (see the bits selected with bold font in column F) we receive an optimal result of encoding.

Now the fitness can be estimated very easily. For randomly generated codes π_t we have to calculate the function W_i and compare the result with the value W. The less the difference $W-W_i$, the better the fitness that the individual π_t has. Consider an example. Let us assume that for individual π_t we have generated the following codes: $K(a_1)=000$, $K(a_2)=001$, $K(a_3)=010$, $K(a_4)=011$, $K(a_5)=100$. In this case $w^i_1=2$, $w^i_2=1$, $w^i_3=1$, $w^i_5=1$, $W_i=5$ and $W-W_i=2$, i.e. the difference is 2, the evaluation of π_i is 5 and the optimal evaluation is 7. Here w^i_j is the value w_j for individual i.

The next step produces a variation of the population. Two kinds of reproduction have been examined and compared. The first one is based on the elitist rule [4] where the best solutions in the population are certain to survive to the next generation. This rule has been implemented as follows. For reproduction purposes 10% of individuals with the best fitness have been copied to the next generation of the population. The second kind of reproduction uses the same percentage of individuals, but it is based on proportional selection [4].

The mutation operation acts on one parental individual selected with a probability based on fitness and creates one new offspring individual to be inserted into the new population at the next generation. In order to choose which parents will produce offspring, a fitness proportional selection is employed. Each parent π_i is assigned a weight W_i, calculated at the previous step. The probability of selection for each parent is proportional to its weight. The main idea of the mutation operation will be illustrated by an example of state encoding for FSM that has specification presented in [5]. The FSM has 10 states and the following transitions $a_{from} \Rightarrow A(a_{from})$: $a_1 \Rightarrow \{a_2,a_3,a_4\}$, $a_2 \Rightarrow \{a_2,a_4,a_5\}$, $a_3 \Rightarrow \{a_6,a_7,a_8,a_9\}$, $a_4 \Rightarrow \{a_5\}$, $a_5 \Rightarrow \{a_3\}$, $a_6 \Rightarrow \{a_5,a_7\}$, $a_7 \Rightarrow \{a_3,a_9\}$, $a_8 \Rightarrow \{a_2,a_{10}\}$, $a_9 \Rightarrow \{a_{10}\}$, $a_{10} \Rightarrow \{a_1\}$. Since M=10 and R=4 for each individual we can chose any 10 from 2^4=16 possible codes. Suppose that at some step of EA we found the codes for an individual I, shown in table 2, and this individual has to be mutated.

Table 2. An individual I that has been selected for mutation operation

Codes	0000	0001	0010	0011	0100	0101	0110	0111
I	0	0	4	8	0	0	0	3
Codes	1000	1001	1010	1011	1100	1101	1110	1111
I	5	6	2	9	7	0	10	1

For the individual I, all the state codes have been examined and all the weights $w^I_1,...,w^I_M$ that exceed the value 1 have been calculated. Some of these weights correspond to an optimal result. For all weights w^I_m that have an optimal value the respective states a_m have been selected (see in table 2 **bold** underlined numbers m of states a_m). For example, we have the following state transitions $a_1 \rightarrow \{K(a_2)=$**1010**, $K(a_3)=$0111, $K(a_4)=$**0010**$\}$, $a_2 \rightarrow \{K(a_2)=$**1010**, $K(a_4)=$**0010**, $K(a_5)=$**1000**$\}_{opt}$, $a_3 \rightarrow \{K(a_6)=$1001, $K(a_7)=$1100, $K(a_8)=$0011, $K(a_9)=$1011$\}$, $a_6 \rightarrow \{K(a_5)=$**1000**, $K(a_7)=$**1100**$\}_{opt}$, etc. Optimal solutions have been indicated by subscript "opt" and the respective bits (i.e. bits with coincident indexes that have constant values) of the codes were marked with bold font. The mutation operation permits a new child individual to be created and includes the following steps.

Step 1. All the elements that correspond to an optimal solution (see **bold** underlined numbers in table 2) are included into the new individual (offspring).

Step 2. The codes for the remaining elements will be randomly regenerated in such a way that just free codes (i.e. such codes that have not been chosen at step 1) can be selected.

The main ideas of crossover operation will also be illustrated by the same example [5]. Suppose that at some step of the EA we found the codes for two individuals I1 and I2 shown in table 3 and these individuals were chosen to be parents for creating a new individual that is an offspring. The codes C are shown as decimal numbers for the respective binary values. For all weights $w^{I1}_m (w^{I2}_s)$ that have an optimal value the states a_m (a_s) have been selected (see **bold** underlined numbers m of states a_m for the first individual I1 and *italic* underlined numbers s of states a_s for the second individual I2). The parents have been chosen on the basis of proportional selection [4].

The crossover operation permits a new child to be created and includes the following steps.

Step 1. The selected elements of the first individual I1 (see **bold** underlined numbers in table 3) are copied to the offspring O (see table 4).

Step 2. Permitted selected elements from the second individual I2 are added to the offspring O. The element is permitted for step 2 if: a) it was not included into the child during the first step; b) it does not have a code that has already been used during the first step. Table 5 shows the result of step 2 for our example.

Table 3. The results of encoding for two individuals

C	0	1	2	3	4	5	6	7	8	9	10	11	12	13	14	15
I1	0	0	4	8	0	0	0	3	5	6	2	9	7	0	10	1
I2	8	0	5	4	10	1	0	2	7	0	9	3	0	0	0	6

Table 4. The result of step 1

C	0	1	2	3	4	5	6	7	8	9	10	11	12	13	14	15
O			4						5		2		7		10	

Table 5. The result of step 2

C	0	1	2	3	4	5	6	7	8	9	10	11	12	13	14	15
O			4						5		2	3	7		10	

Step 3. All other permitted elements from the first and the second individuals are added to the child. The element is permitted for step 3 if it has not yet been included into the child and: a) it has the same code for both individuals I1 and I2; b) it is included in the second individual I2 and the respective code of the first individual I1 was not used for the states; c) it is included in the first individual I1 and the respective code of the second individual I2 was not used for the states.
Table 6 shows the result of step 3 for our example.

Table 6. The result of step 3

C	0	1	2	3	4	5	6	7	8	9	10	11	12	13	14	15
O	8		4			1			5	6	2	3	7		10	

Step 4. All the remaining states that have not been assigned yet are recorded in free boxes for codes from left to right. Table 7 presents the final result.

Table 7. The result of step 4 that gives the final decision

C	0	1	2	3	4	5	6	7	8	9	10	11	12	13	14	15
O	8	9	4			1			5	6	2	3	7		10	

Thus the child individual has the following codes $K(a_1)=0101$, $K(a_2)=1010$, $K(a_3)=1011$, $K(a_4)=0010$, $K(a_5)=1000$, $K(a_6)=1001$, $K(a_7)=1100$, $K(a_8)=0000$, $K(a_9)=0001$, $K(a_{10})=1110$. In order to find out fuzzy codes $F(a_{from})$ we have to replace the proper bits [1] in codes $K(a_{to})$, $a_{to} \in A(a_{from})$, $a_{from} \Rightarrow a_{to}$ with symbol "i". For example, we have $a_1 \Rightarrow \{a_2, a_3, a_4\}$ and $K(a_2)=1010$, $K(a_3)=1011$, $K(a_4)=0010$. Thus $F(a_1)= i01i$. Individuals I1, I2 and the child can be evaluated as follows: $W_{I1}=W_{I2}=11$, $W_{child}=13$ (i.e. the child is better than any of the parents I1 and I2) and an optimal weight $W=15$. There are two termination conditions for the considered EA, obtaining an optimal solution or exceeding a specified time.

5 Experiments

The proposed technique has been analyzed in several contexts. Firstly, we examined the EA for two groups of FSMs. The first group (see table 8) includes FSMs taken from practical designs and some were implemented in previously fabricated products (such devices are marked with an asterisk in table 8). The second group (see table 9) is composed of arbitrary generated FSMs. Various columns of tables 8, 9 contain the following data: "Example" - a short information about considered FSM, NG - a number of generations for EA, W - an optimal value of fitness function (see section 4), W_e - the obtained value of fitness function for the considered example, ET - an execution time in seconds for C++ program that implements the considered EA, M - number of FSM states, G_e - maximum number of variables from the set P for any individual state transition. In order to choose the value v for each population $\pi=\{\pi_1,...,\pi_v\}$ we examined the FSM of the paper [5] with various numbers v of individuals (see table 10). We found that the results do not depend much on the value v and good results were obtained with a small number of individuals. That is why for all the examples in tables 8 and 9 $v=15$. Considering a small number of individuals permits a significant reduction in the repetition of various cycles in the C++ program. A predefined termination condition is either $W_e=W$ or $ET > 1000$ s.

Table 8. Experiments with FSMs taken from practical designs

Example	NG	W	W_e	ET (s)	M	G_e
FSM for Plotter*	83	29	29	1.2	24	3
FSM for Interface block*	2	18	18	≈ 0	11	1
FSM for assembly line*	7	32	32	0.2	20	2
See fig . 1	2	7	7	≈ 0	5	2
Example 1	63	36	36	0.261	29	2
Example 2	99	7	7	0.202	12	4
Example 3	111	30	30	0.255	15	2

Table 9. Experiments with arbitrary generated FSMs

Example	NG	W	W_e	ET (s)	M	G_e
Ex1	15	21	21	0.052	21	2
Ex2	4	25	25	0.017	28	2
Ex3	146	15	15	0.216	10	2
Ex4	8396	8	8	48.1	12	2
Ex5	7932	154	154	810.7	85	2
Ex6	5534	88	87	1000	52	2
Ex7	1024	38	37	1000	49	3
Ex8	490	31	31	3.897	47	3
Ex9*	54115	20	20	575.2	16	2
Ex10*	50846	26	26	812.7	16	3

It is very important that for all the examples shown in table 8 we were able to get an optimal solution (i.e. W_e=W). It should be noted that for some state transition graphs an optimal solution that is only based on binary state codes cannot be obtained. However if don't care bits ("-") are allowed in the codes, we can find an optimal solution. The experiments were performed on a PentiumIII/800MHz/256MB running Windows2000.

Secondly we implemented in hardware and tested various FSMs. The experiments with hardware were performed with the XS40 board that contains the Xilinx XC4010XL FPGA and the Xilinx Development System (series 3.1).

Table 10. Experiments with FSM from paper [5]

ν	3	8	15	20	30	50	100	200	300	500	1000
NG	8595	951	146	177	201	212	44	7	7	6	2
ET	4.7	2.8	1.0	1.4	4.5	11.6	8.5	5.9	12	27.3	36.8

6 Conclusion

The paper considers an architecture and a method for the synthesis of fast RAM-based FSMs. The method includes two primary steps, fuzzy state encoding and replacement of input variables. The problem of the first step is the most complicated and it has a high combinatorial complexity. The proposed evolutionary algorithm enables us to solve this problem with the aid of a C++ program that gives very good results that are presented in the set of experiments.

Acknowledgements

This work was partially supported by the Portuguese Foundation of Science and Technology under grant No. POSI/43140/CHS/2001.

References

1. V.Sklyarov: Synthesis and Implementation of RAM-based Finite State Machines in FPGAs. Proceedings of FPL'2000, Villach, Austria, August, (2000), pp. 718-728.
2. L.Fogel: Autonomous Automata. Industrial Research 4, (1962).
3. P.Angeline, D.Fogel, L.Fogel: A comparison of self-adaptation methods for finite state machine in dynamic environment. Evolution Programming V. MIT Press, (1996), pp. 441-449.
4. Z. Michalewicz and D. B. Fogel: How to Solve It: Modern Heuristics, Springer-Verlag, (2000).

5. V.Sklyarov: Synthesis of Control Circuits with Dynamically Modifiable Behavior on the Basis of Statically Reconfigurable FPGAs. – 13th Symposium on Integrated Circuits and Systems Design: SBCCI2000, Manaus, Brazil, 18-24 September (2000), pp. 353-358.

Genetic Algorithms for Design of Liquid Retaining Structure

K.W. Chau[1] and F. Albermani[2]

[1] Department of Civil & Structural Engineering, Hong Kong Polytechnic University, Hunghom, Kowloon, Hong Kong
cekwchau@polyu.edu.hk
[2] Department of Civil Engineering, University of Queensland, QLD 4072, Australia

Abstract. In this paper, genetic algorithm (GA) is applied to the optimum design of reinforced concrete liquid retaining structures, which comprise three discrete design variables, including slab thickness, reinforcement diameter and reinforcement spacing. GA, being a search technique based on the mechanics of natural genetics, couples a Darwinian survival-of-the-fittest principle with a random yet structured information exchange amongst a population of artificial chromosomes. As a first step, a penalty-based strategy is entailed to transform the constrained design problem into an unconstrained problem, which is appropriate for GA application. A numerical example is then used to demonstrate strength and capability of the GA in this domain problem. It is shown that, only after the exploration of a minute portion of the search space, near-optimal solutions are obtained at an extremely converging speed. The method can be extended to application of even more complex optimization problems in other domains.

1 Introduction

In solving practical problems of design optimization, owing to the availability of standard practical sizes and their restrictions for construction and manufacturing purposes, the design variables are always discrete. In fact, it is more rational to use discrete variables during the evaluation process of optimization since every candidate design is a practically feasible solution. This may not be the case when the design variables are continuous, since some of the designs evaluated in the optimization procedures are solely mathematically feasible yet not practically feasible. However, most of the programming algorithms developed for the optimum design of structural systems during the last few decades assume continuous design variables and simple constraints, which are not always correct. Only very few algorithms have dealt with the optimization of structures under the actual design constraints of code specifications. Many of them require the approximation of derivative information and yet may only attain local optima.

Recently, there has been a widespread interest in the use of genetic algorithms (GAs), which are applications of biological principles into computational algorithm, to accomplish the optimum design solutions [1]. They are specifically appropriate to solve discrete optimum design problems and apply the principle of survival of the fittest into the optimization of structures. Although GAs require only objective function value to direct the search and do not need any derivatives of functions often necessary in other optimization programming methods, they are able to search through large spaces in a short duration. In particular, GAs have a much more global perspective than many other methods. Yet, literature review shows that only steel structures are often considered in structural optimization field [2-5].

This paper delineates a genetic algorithm (GA) for the optimum design of reinforced concrete liquid retaining structures using discrete design variables. Whereas either minimum weight or minimum cost can represent the objective function equally in the optimization of steel structures, their counterparts of reinforced concrete structures must be minimum material cost. Reinforced concrete structures involve more design variables as it involves both concrete and steel reinforcement, which have rather different unit costs.

2 Genetic Algorithms

GAs, being search techniques based on the mechanism of natural genetics and natural selections [6], can be employed as an optimization method so as to minimize or maximize an objective function. They apply the concept on the artificial survival of the fittest coupled with a structured information exchange using randomized genetic operators taken from the nature to compose an efficient search mechanism. GAs work in an iterative fashion successively to generate and test a population of strings. This process mimics a natural population of biological creatures where successive generations of creatures are conceived, born, and raised until they are ready to reproduce.

2.1 Comparisons with Conventional Algorithms

GAs differ from traditional optimization algorithms in many aspects. The following are four distinct properties of GAs, namely, population processing, working on coded design variables, separation of domain knowledge from search, and randomized operators. Whilst most common engineering search schemes are deterministic in nature, GAs use probabilistic operators to guide their search. Whereas other optimization methods often need derivative information or even complete knowledge of the structure and parameters, GAs solely entail objective function value information for each potential solution they generate and test. The population-by-population approach of GA search climbs many peaks in parallel simultaneously whilst the more common point-by-point engineering optimization search techniques often locates false local peaks especially in multi-modal search spaces.

2.2 Features of GAs

GAs are not limited by assumptions about search space, such as continuity or existence of derivatives. Through a variety of operations to generate an enhanced population of strings from an old population, GAs exploit useful information subsumed in a population of solutions. Various genetic operators that have been identified and used in GAs include, namely, crossover, deletion, dominance, intra-chromosomal duplication, inversion, migration, mutation, selection, segregation, sharing, and translocation.

2.3 Coding Representation

A design variable has a sequence number in a given discrete set of variables in GAs, which require that alternative solutions be coded as strings. Successive design entity values can be concatenated to form the length of strings. Different coding schemes have been used successfully in various types of problems. If binary codes are used for these numbers, individuals in a population are finite length strings formed from either 1 or 0 characters. Individuals and the characters are termed chromosomes and artificial genes, respectively. A string may comprise some substrings so that each substring represents a design variable.

2.4 Selection Operator

The purpose of selection operator is to apply the principle of survival of the fittest in the population. An old string is copied into the new population according to the fitness of that string, which is defined as the non-negative objective function value that is being maximized. As such, under the selection operator, strings with better objective function values, representing more highly fit, receive more offspring in the mating pool. There exist a variety of ways to implement the selection operator and any one that biases selection toward fitness can be applicable.

2.5 Crossover Operator

The crossover operator leads to the recombination of individual genetic information from the mating pool and the generation of new solutions to the problem. Several crossover operators exist in the literature, namely, uniform, single point, two points and arithmetic crossover.

2.6 Mutation Operator

The mutation operator aims to preserve the diversification among the population in the search. A mutation operation is applied so as to avoid being trapped in local optima. This operator is applied to each offspring in the population with a predetermined probability. This probability, termed the mutation probability, controls the rate of mutation in the process of selection. Common mutation operation is simple, uniform, boundary, non-uniform and Gaussian mutation.

3 Problem Formulation

The following depicts the optimization design of a reinforced concrete liquid retaining structure, subjected to the actual crack width and stress constraints in conformance to the British Standard on design of concrete structures for retaining aqueous liquid, BS 8007 [7]. The set of design variables is determined so that the total material cost of the structure comprising n groups of member,

$$\min C(x) = \sum_{i}^{n} U_i * V_i + R_i * W_i \qquad (1)$$

is minimized subject to the constraints. In eq. (1), U_i and V_i represent the unit cost and the concrete volume of member i respectively. R_i and W_i are the unit cost and the weight of steel reinforcement of member i respectively.

The serviceability limit state or crack width constraint is

$$W_a - W_{max} \le 0 \qquad (2)$$

where W_a is the actual crack width and W_{max} is the prescribed maximum crack width, which will be 0.1mm or 0.2 mm depending on the exposure environment. W_a, is determined using the following formula:-

$$W_a = \frac{3 a_{cr} \varepsilon_m}{1 + 2(\frac{a_{cr} - c}{h - x})} \qquad (3)$$

where a_{cr} is the distance from the point considered to the surface of the nearest longitudinal bar, ε_m is the average strain for calculation of crack width allowing for concrete stiffening effect, c is the minimum cover to the tension reinforcement, h is the overall depth of the member and x is the depth of the neutral axis.

The stress constraints, representing the ultimate limit states of flexure and shear resistance, are expressed in terms of the following equations for members subject to bending and shear force [8]:

$$M_{au} - M_{ult} \le 0 \qquad (4)$$

$$V_a - V_{ult} \le 0 \qquad (5)$$

where M_{au} is the actual ultimate bending moment, M_{ult} is the nominal ultimate moment capacity of the reinforced concrete section, V_a is the actual ultimate shear force and V_{ult} is the nominal ultimate shear capacity of the section. The ultimate moment capacity is determined by the following equations, depending on whether concrete or steel stresses is more critical.

$$M_{ult} = \frac{F_y}{1.15} A_s Z \text{ or } M_{ult} = 0.157 F_{cu} b d^2 \text{ whichever is the lesser} \qquad (6)$$

where F_y is the yield strength of reinforcement, A_s is area of tension steel, Z is the lever arm, F_{cu} is the characteristic concrete strength, b is the width of section and d is the effective depth of section. Ultimate shear capacity of the section ($V_{ult} = v_c b_v d$) is represented by shear strengths v_c for sections without shear reinforcement, which depend upon the percentage of longitudinal tension reinforcement [$100A_s/(b_v d)$] and the concrete grade:-

$$v_c = 0.79[100 A_s/(b_v d)]^{1/3}(400/d)^{1/4}/\gamma_m \tag{7}$$

where b_v is breadth of section, γ_m is a safety factor equal to 1.25, with limitations that [$100A_s/(b_v d)$] should not be greater than three and that (400/d) should not be less than one. For characteristic concrete strengths greater than 25 N/mm², the values given by the above expression is multiplied by $(F_{cu}/25)^{1/3}$.

Prior to applying GA, a transformation on the basis of the violations of normalized constraints [9] is employed to change the constrained problem to become unconstrained. The normalized form of constraints is expressed as:

$$\frac{W_a}{W_{max}} - 1 \leq 0 \quad i=1,\ldots n \tag{8}$$

$$\frac{M_{au}}{M_{ult}} - 1 \leq 0 \quad i=1,\ldots n \tag{9}$$

$$\frac{V_a}{V_{ult}} - 1 \leq 0 \quad i=1,\ldots n \tag{10}$$

The unconstrained objective function $\varphi(x)$ is then written as

$$\varphi(x) = C(x)[1 + K \sum_i^n \{(\frac{W_a}{W_{max}} - 1)^+ + (\frac{M_{au}}{M_{ult}} - 1)^+ + (\frac{V_a}{V_{ult}} - 1)^+\}] \tag{11}$$

where K is a penalty constant and

$$(\frac{W_a}{W_{max}} - 1)^+ = \max(\frac{W_a}{W_{max}} - 1, 0) \tag{12}$$

$$(\frac{M_{au}}{M_{ult}} - 1)^+ = \max(\frac{M_{au}}{M_{ult}} - 1, 0) \tag{13}$$

$$(\frac{V_a}{V_{ult}} - 1)^+ = \max(\frac{V_a}{V_{ult}} - 1, 0) \tag{14}$$

The penalty parameter largely depends upon the degree of constraint violation, which is found to be amenable to a parallel search employing GAs. Values of 10 and 100 have been attempted here and it is found that the results are not quite sensitive.

In order to ensure that the best individual has the maximum fitness and that all the fitness values are non-negative, the objective function is subtracted from a large constant for minimization problems. The expression for fitness here is

$$F_j = [\varphi(x)_{max} + \varphi(x)_{min}] - \varphi_j(x) \tag{15}$$

where F_j is the fitness of the j-th individual in the population, $\varphi(x)_{max}$ and $\varphi(x)_{min}$ are the maximum and minimum values of $\varphi(x)$ among the current population respectively and $\varphi_j(x)$ is the objective function value computed for the j-th individual. The calculation of the fitness of an individual entails the values of crack width and stresses, which are obtained from the finite element structural analysis.

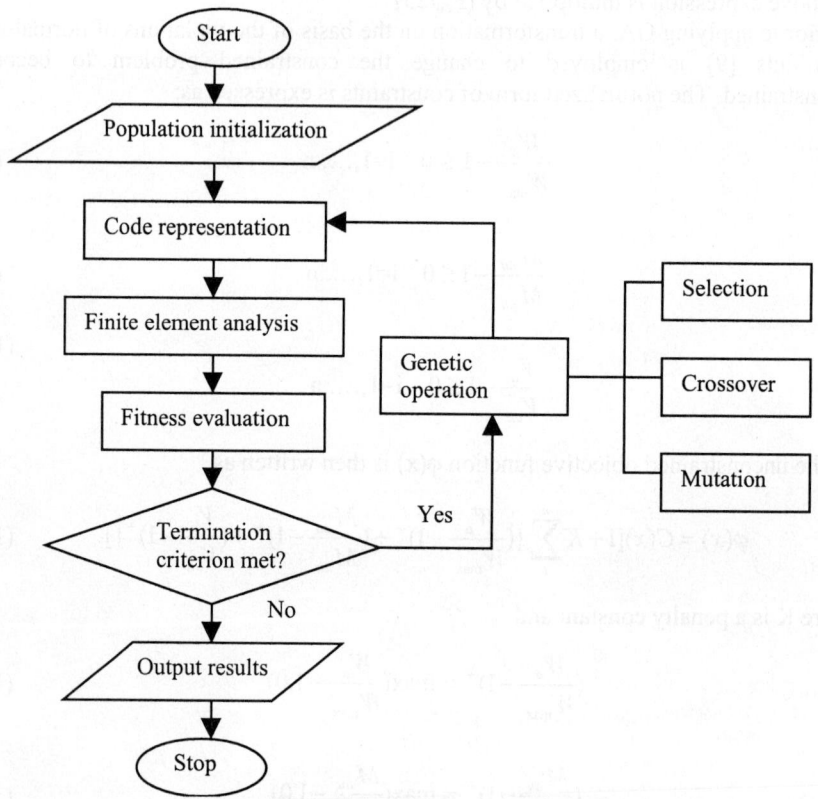

Fig. 1. Flow chart of the GA for design of liquid retaining structures

4   GA Optimization Procedure

In this study, the GA has been implemented under Microsoft Visual Basic programming environment. Figure 1 shows the flow chart of the GA for design of liquid retaining structures. The initial population containing binary digits is first randomly generated. The binary codes for the design variables of each individual are decoded and their sequence numbers are determined. The finite element analysis of the structure is performed and the unconstrained function $\varphi(x)$ for each individual is computed. From the maximum and minimum values of this function in the population, the fitness value for each individual is then found.

The population for the next generation is then reproduced by applying the selection operator. On the basis of their fitness, that is, the best strings make more copies for mating than the worst, the individuals are copied into the mating pool. In order to emulate the survival of the fittest mechanism, a rank-based scheme is employed to facilitate the mating process [10]. The chromosomes are ranked in descending order of their fitness and those with higher fitness values have a higher probability of being selected in the mating process. As the number of individuals in the next generation remains the same, the individuals with small fitness die off. Besides, the concept of elitism is also incorporated into the selection process [11]. This strategy keeps the chromosome with the highest fitness value for comparison against the fitness values of chromosomes computed from the next generation. If the ensuing generation fails to enhance the fitness value, the elite is reinserted again into the population.

A two-site crossover is employed to couple individuals together to generate offspring. It involves the random generation of a set of crossover parameters, comprising a match and two cross sites. A match is first allocated for each individual. A fixed crossover probability ($p_{crossover}$) is established beforehand so that the genetic operation of crossover is performed on each mated pair with this probability. It is necessary to find the crossover sites and to perform the crossover in the following manner. Suppose that two strings X and Y of length 11 are the mating pair with the following genes

$$X = x_1, x_2, x_3, x_4, x_5, x_6, x_7, x_8, x_9, x_{10}, x_{11} \tag{16}$$

$$Y = y_1, y_2, y_3, y_4, y_5, y_6, y_7, y_8, y_9, y_{10}, y_{11} \tag{17}$$

For each mated pairs subjected to crossover operation, two cross sites cs_1 and cs_2 are randomly generated. Two new strings are created by swapping all characters between positions cs_1 and cs_2 inclusively from one individual in the pair to the other. The cross sites for the matching pairs have to be the same. For instance, if the cross sites generated are 2 and 7, the resulting crossover yields two new strings X' and Y' following the partial exchange.

$$X' = x_1, x_2, | y_3, y_4, y_5, y_6, y_7, | x_8, x_9, x_{10}, x_{11} \tag{18}$$

$$Y' = y_1, y_2, | x_3, x_4, x_5, x_6, x_7, | y_8, y_9, y_{10}, y_{11} \tag{19}$$

The genetic operation of mutation is performed on the chromosomes with a preset mutation probability. It is applied to each offspring in the newly generated population. It operates by flipping the gene of an offspring from 1 to 0 or vice versa at random position. After the operation, the initial population is replaced by the new population.

An iteration cycle with the above steps is then continued until the termination criterion is reached. It may occur when the distance between the maximum and the average fitness values of the current population becomes less than a certain threshold, or a preset maximum number of generation is attained. At that moment, the optimum design solution is represented by the individual with the maximum fitness value in the current population.

Table 1. Composition of the individual string

Variable number	Design variable	Substring length
1	Slab thickness	4
2	Reinforcement diameter	3
3	Reinforcement spacing	4

5 Numerical Example

The test case is concerned with the optimum design of an underground circular shaped liquid retaining structure. The volume and height are 100 m^3 and 5 m, respectively. The exposure condition is very severe so that the designed crack width equals to 0.1 mm. The grades of concrete and reinforcement are 40 and high yield deformed bar respectively. The concrete cover is 40 mm and the aggregate type is granite or basalt with a temperature variation of 65 °C in determining the coefficient of expansion for shrinkage crack computation. The unit costs of concrete and reinforcement are \$500 per m^3 and \$3000 per tonne, respectively. A surcharge load of 10 kN/m^2 is specified. The level of liquid inside the tank, the ground level and the level of water table above the bottom of the tank are 5 m, 5 m and 4 m, respectively. The specific weight of soil is 20 kN/m^2 with active soil pressure coefficient of 0.3. Since it is an underground structure and the wind load is not applicable. Load combinations for both serviceability and ultimate limit states are in compliance with BS 8110.

As a demonstration, the wall and slab sections are classified under the same member group. The practically available values of the design variables are given in the lists T, D and S, representing slab thickness, bar diameter and bar spacing respectively.

$$T = (200, 225, 250, 300, 350, 400, 450, 500, 600, 700, 800, 900, 1000) \quad (20)$$

$$D = (10, 12, 16, 20, 25, 32, 40). \quad (21)$$

$$S = (100, 125, 150, 175, 200, 225, 250, 275, 300) \quad (22)$$

Since GAs work on coded design variables, now it is necessary to code the design variables into a string. Owing to the simplicity of manipulation, a binary coding is adopted here. Table 1 shows the composition of an individual string with a total length of eleven. The population size (P_{size}), the crossover probability and the mutation probability are selected as 10, 0.95 and 0.01, respectively. The population size is chosen to provide sufficient sampling of the decision space yet to limit the computational burden simultaneously. It is found that GAs are not highly sensitive to these parameters. These values are also consistent with other empirical studies.

Figure 2 shows the relationship between the minimum cost versus the number of generations for both this GA (modified GA) and the original GA used in [1] (simple GA), representing an average of 20 runs of the algorithms. It can be seen that this modified GA is slightly better than the simple GA for this numerical example. The minimum cost of $37522, representing a reinforced concrete section of member thickness 250 mm with reinforcement diameter 25 mm at spacing 200 mm, is found after 6 generations. As the generations progress, the population gets filled by more fit individuals, with only slight deviation from the fitness of the best individual so far found, and the average fitness comes very close to the fitness of the best individual. It is recalled that each generation represents the creation of $P_{size} = 10$ strings where $P_{size} * p_{crossover} = 10*0.95 = 9.5$ of them are new. It is noted that only a simple demonstration is shown here with the solution of this problem in just 6 generations. However, if the wall and slab sections are classified under different member groups, the application will become much more complicated.

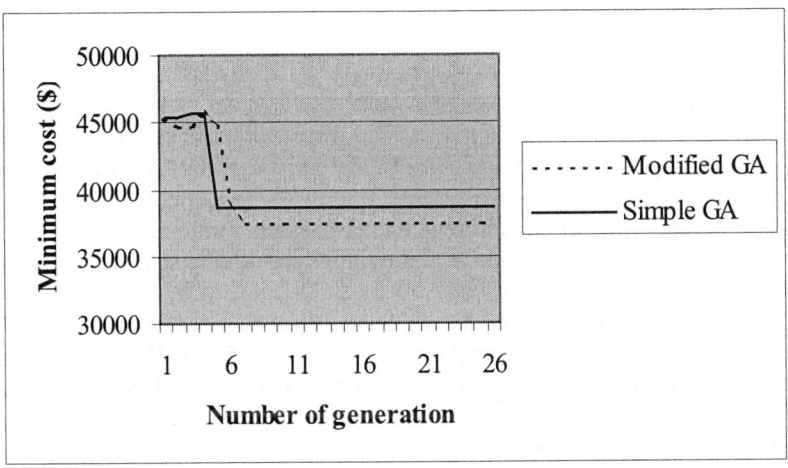

Fig. 2. Minimum cost versus number of generation for different GAs

6   Conclusions

In this paper, a GA has been successfully implemented for the optimum design of reinforced concrete liquid retaining structures involving discrete design variables. Only after examining a minute portion of the design alternatives, it is able to locate the optimal solution quickly. GAs acquire real discrete sections in a given set of standard sections. It should be noted that design variables are discrete in nature for most practical structural design problems. In most other mathematical programming techniques and in the optimality criteria approach, an approximation is often made by assigning the acquired optimum continuous design variables to the nearest standard sections. Approximate relationships may also be used between the cross-sectional properties of real standard sections in these usual methods. However, in this application, GAs remove these approximations and do not depend upon underlying continuity of the search space.

References

1. Goldberg, D.E., Kuo, C.H.: Genetic Algorithms in Pipeline Optimization. Journal of Computing in Civil Engineering ASCE **1(2)** (1987) 128-141
2. Adeli, H. and Cheng, N.T.: Integrated Genetic Algorithm for Optimization of Space Structures. Journal of Aerospace Engineering ASCE **6(4)** (1993) 315-328
3. Gutkowski, W., Iwanow, Z., Bauer, J.: Controlled Mutation in Evolutionary Structural Optimization. Structural and Multidisciplinary Optimization **21(5)** (2001) 355-360
4. Rajeev, S., Krishnamoorthy, C.S.: Genetic Algorithms-Based Methodologies for Design Optimization of Trusses. Journal of Structural Engineering ASCE **123(3)** (1997) 350-358
5. Yang, J., Soh, C.K.: Structural Optimization by Genetic Algorithms with Tournament Selection. Journal of Computing in Civil Engineering ASCE **11(3)** (1997) 195-200
6. Holland, J.H.: Adaptation in Natural and Artificial Systems. University of Michigan Press, Ann Arbor (1975)
7. British Standards Institution: BS 8007: Design of Concrete Structures for Retaining Aqueous Liquids. British Standards Institution, London (1987)
8. British Standards Institution: BS 8110: Structural Use of Concrete. British Standards Institution, London (1985)
9. Hayalioglu, M.S.: Optimum Load and Resistance Factor Design of Steel Space Frames using Genetic Algorithm. Structural and Multidisciplinary Optimization **21(4)** (2001) 292-299
10. Wu, S.J., Chow, P.T.: Steady-State Genetic Algorithms for Discrete Optimization of Trusses. Computers and Structures **56(6)** (1995) 979-991.
11. Koumousis, V.K., Georgiou, P.G.: Genetic Algorithms in Discrete Optimization of Steel Truss Roofs. Journal of Computing in Civil Engineering ASCE **8(3)** (1994) 309-325

Modelling Crew Assistants with Multi-Agent Systems in Fighter Aircraft

Arjen Vollebregt, Daan Hannessen, Henk Hesselink, and Jelle Beetstra

National Aerospace Laboratory NLR
P.O. box 90502, 1006 BM Amsterdam, The Netherlands
{vollebre,hannessn,hessel,beetstra}@nlr.nl

Abstract. As advanced crew support technologies will be available more and more in future military aircraft, it is necessary to have a good understanding of the possibilities in this area, taking into account operational demands and technical possibilities. A Crew Assistant (CA) is a decision support system for air crew, designed to improve mission effectiveness and redistribute crew workload in such a way that the crew can concentrate on its prime tasks. Designing a complex system with multiple crew assistants can be tackled by using a multi-agent system design. In this paper we will propose a multi-agent system architecture for crew assistants.

1 Introduction

Software is a major part of current and future on-board avionics. Current developments in software engineering and advanced information processing enable complex crew assistant applications. Support of the aircraft crew in carrying out primary and secondary tasks is more and more provided by electronic systems. Applying crew assistants in fighter aircraft is a challenging task, both with respect to the research and development of the software, and with respect to the human factors aspects concerning its optimal use and trust of the pilot in the system. We propose a multi-agent system approach to design complex crew assistants. We first give an introduction to crew assistants, followed by a description of multi-agent systems and then propose a multi-agent architecture for crew assistants.

2 Crew Assistants

Modern military operations take place in a complex operational environment to accomplish a spectrum of missions. The workload of fighter aircraft crews increase rapidly because of:

- the increase of the complexity of the military environment in general (e.g. combined joint task forces, peace keeping/peace enforcement).
- the increase of the complexity of the types of missions to be flown. the increase in the number of different (kinds of) threats.

Another factor is the technological development of fighter aircraft. The increase in aircraft speed, aircraft types and on-board systems causes the aircraft itself to become much more difficult to manage, putting more pressure on the crew. During high-stress situations, the crew can get overloaded with information, while it has to perform a multitude of actions. A crew assistant is an on-board

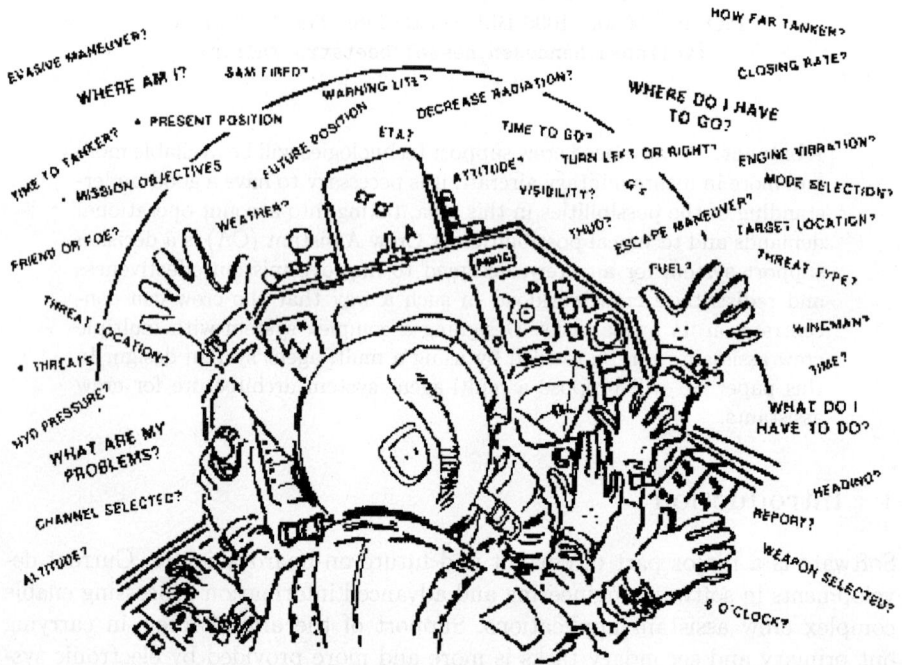

Fig. 1. The information requirements of the crew [11]

decision support system that supports the crew in performing its mission. It aims at improving mission effectiveness, flight safety and/or survivability by providing the crew with concise and relevant information, depending on the mission phase. This enables the crew to concentrate on mission decisions and make more effective decisions. [9]. Ideally, a crew assistant assists a pilot, or other crew members, by providing the following kind of functions:

- Acquire the necessary information and merge the input from different sensor and information systems into one timely and consistent view of the current situation (the status of different on-board systems, the situation outside, etc.).
- Process the merged information to give advice (weapon selection, route planning, tactics evaluation, fuel management, etc.).
- Perform tasks autonomously when allowed by the pilot or another crew member (autopilot, target tracking, systems monitoring, etc.).

3 Multi-Agent System Overview

In this chapter we will first give a definition of what we mean by an agent and then propose a multi agent architecture for crew assistants.

Definition of an Agent

The notion agent has become popular in recent years. It is used for several different applications ranging for simple batch jobs to intelligent assistants. It's good to get clear what we mean by agent. The weak notion of an agent as described in [5] (the characteristics to call the system an agent):
Autonomous behaviour: The system should be able to act without the direct intervention of humans (or other agents) and should have control over its own actions and internal state.
Reactive behaviour: Agents should perceive their environment (which may be the physical world, a user, a collection of agents, etc.) and respond in a timely fashion to changes that occur in it.
Pro-active behaviour: Agents should not simply act in response to their environment, they should also be able to exhibit opportunistic, goal-directed behaviour and take the initiative where appropriate.
Social behaviour: Agents should be able to interact, when they deem appropriate, with other artificial agents and humans in order to complete their own problem solving and to help other with their activities.

The weak agent definition describe above can be extended with intentional notions to make it a strong agent definition. These notions are for example beliefs, desires, intentions, commitments, goals, plans, preference, choice and awareness.

4 Agent Design

The agents are designed according to the DESIRE method from [1]. We chose the DESIRE method for its well-defined generic agent structure and its compositional architecture, which enables us to design the model on different levels of detail. We present in this paper the top-level design of the architecture.

4.1 Overview of the Architecture

In this section an overview of the multi-agent architecture for the PoWeR II project is presented. The proposed architecture has been based on the results of earlier projects, like the EUCLID (European Co-operation for the Long Term in Defence) Research and Technology project 6.5 [12]. The project on crew assistants for military aircraft started with extensive user interviews to establish an inventory of operational user problems and needs for pilots flying F-16, Tornado

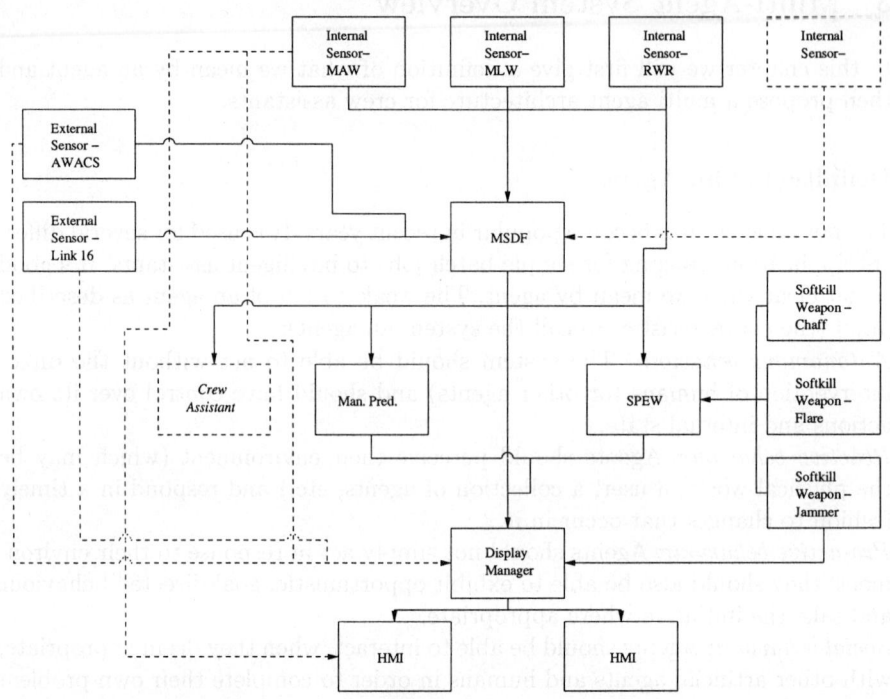

Fig. 2. Multi-Agent System Overview

and AM-X. The project came up with a generic on-board crew assistant architecture and indicated a challenge in the application of multi-agent systems and knowledge based systems. The architecture that has been set up distinguishes four groups of functional agents. The groups are (1) data and information input agents, like sensors and the multi-sensor data fusion agent, (2) data processing agents which form the actual crew assistant functions, (3) information output agents which communicate mainly with the pilot, and finally, (4) the weapon agents. Apart from these, other agents perform functions for controlling and monitoring the overall system's status and health. In this paper, we will focus on the functional part of crew assistants (figure 2).

Internal sensor agents are system components that transform the raw input data from the sensor hardware to an understandable format for the Multi-Sensor Data Fusion (MSDF) component. In our example, we included sensors to detect and distinguish SAMs and to detect incoming missiles. A Radar Warning Receiver (RWR) provides the position of ground threats (SAMs), including an indication whether the SAM is in search, track, or guidance. The Missile Launch Warning (MLW) is a passive infrared plume detector that provides missile information by detecting its engine. The Missile Approach Warning (MAW) is an active short-range radar that detects a missile body, usually in a two to three miles range.

External sensor agents are components that obtain their information from sensors or information systems that are physically located outside the aircraft, for example an AWACS or a Link-16. These sensor agents transform data and information into an understandable format for the MSDF agent or for the CA agents.

The Multi-Sensor Data Fusion agent combines the sensor information from all internal and external sensors into a combined sensor data picture. This agent may perform complex situation assessment tasks. In the current implementation, this is a fusion process that only provides the information to the crew assistants that is really necessary for the assistants to perform their task. Different projects have already shown the complexity of a multi-sensor data fusion process and have proposed architectures. [4] proposes an agent based architecture for multi-sensor data fusion, which shows the flexibility of agent systems, where agents can delegate tasks to (sub-)agents.

Crew Assistant agents are the intelligent pilot support functions. The ones mentioned in figure 1 are developed at the NLR (based on [12]), however, the range of pilot support functions is not limited to these. Crew assistants can be further classified into functions as information finding in the fused sensor picture, pilot advice (like manoeuvre prediction), pilot monitoring, mission monitoring, etc. Other classifications are possible ([2], [6]).

Weapon agents control the weapon delivery. In this example, a number of softkill weapons to countermeasure ground threats are displayed. Their intelligence for example consists of providing the correct jamming against a recognised threat or dispensing a complex pattern of chaff and flare.

The Display agent is responsible for assembling an integrated picture of crew assistant information and for prioritising information provision to the pilot. If necessary, it can hold information that is considered less important at a certain moment or less time critical, if the pilot is assumed to get overloaded with information. Once the situation is more relaxed, it can decide to provide this information. An even more intelligent display agent can decide what information should be provided on which cockpit display, or what information should be provided on which part of the cockpit display and automatically adapt the contents of the available displays if at (a certain part of) one of the displays an information overload is eminent. This technology, however, should be introduced with care [10].

The Human Machine Interface agent is the actual cockpit display that provides the information on the user interface. It may take input from the user.

4.2 Flow of Control

The internal sensor components receive raw sensor information from the sensor hardware. This sensor data is transformed to an understandable format for the Multi Sensor Data Fusion (MSDF) component. The external sensors receive information from external sources like an AWACS. This data in transformed to an understandable format for the MSDF components. The MSDF component collects all the sensor data and combines this data to derive the kind of threat. For example the MSDF component could derive a SAM-site threat. This data serves as input for the crew assistants, e.g. SPEW (derives counter measures) and Manoeuvre Prediction (predicts manoeuvres to avoid a threat). The crew assistants also get input from the Softkill Weapon components. This is data about the status of the softkill weapon components, e.g. the amount of chaff and flares available. The output of the crew assistants is passed to the Display Manager, which makes sure that the output data is put on the right HMI.

4.3 Advantages of Using Agents

We have chosen the multi-agent approach for the crew assistant application specifically because of the following characteristics of agents:

- Autonomity: The different agents in the crew assistant application should be able to operate independantly of each other, i.e. if an internal sensor agent fails the MSDF agent still derives usefull information, or when the MSDF agent fails the crew assitant agents should still be able to assist the pilot. Of course less data reduces the accuracy of the agents.
- Pro-active/Re-active behaviour : The agents have to operate in a pro- and re-active manner, for example the sensor agents have to independantly observe the environment and the crew assistant agent pro-actively warns the pilot of incoming threats. This pro-active behaviour is modelled as a goal for the agent. These agent characteristics make the design and implementation of the crew assistant application more natural.

5 Conclusions

An architecture, based on multi-agent technology has been set up, where different examples of crew assistants have been integrated. The architecture of the multi-agent system is only at top-level for simplicity, the crew assistant components are composed of other components. The multi-agent technology used to design this system allows for components to fail while the overall system still produces sensible output data. The dotted lines in figure 2 represents alternative information flow when components fail. Of course when more components fail the output will be less accurate. The architecture is now partially specified with the DESIRE tools [1]. Future work will be directed to the maintenance of the architecture provided and to the further integration of agents, both functional and for system control and monitoring.

References

1. Brazier, F. M. T., Jonker, C. M., and Treur, J.: Compositional Design and Reuse of a Generic Agent Model. Applied Artificial Intelligence Journal, vol. 14, 2000, 491-538.
2. Barrouil, Cl. et.al.: TANDEM: An Agent-oriented Approach for Mixed System Management in Air Operations. NATO Research and Technology Organisation symposium on Advanced System Management and System Integration Technologies for Improved Tactical Operations, Florence, Italy, September 1999.
3. Bossé, E., P. Valin, and A. Jouan: A Multi-Agent Data Fusion Architecture for an Airborne mission Management System, NATO Research and Technology Organisation symposium on Advanced System Management and System Integration Technologies for Improved Tactical Operations, Florence, Italy, September 1999.
4. Dee, C., Millington, P., Walls, B., and Ward, T.: CABLE, a multi-agent architecture to support military command and control.
5. Jennings, N. R. and Woolridge, M.: Agent Technology: Foundations, Applications, and Markets. Springer Verlag, 1998.
6. Taylor, R. M.: Cognitive Cockpit Engineering: Coupling Functional State Assessment, Task Knowledge Management and Decision Support for Context Sensitive Aiding, Taylor, R. M. et.al., TTCP Technical Panel 7, Human Factors in Aircraft Environments, Workshop on Cognitive Engineering, Dayton, OH, May 2000.
7. Tempelman, F., Veldman, H. E.: PoWeR-II: Crew Assistance Applications, Tasks and Knowledge, NLR-TR-2000-650
8. Tempelman, F., Van Gerwen, M. J. A. M., Rondema, E. Seljée, R., Veldman, H. E.: PoWeR-II: Knowledge and Reasoning in Crew Assistant Applications, NLR-TR-2000-659
9. Urlings, P. J. M., JJ. Brants, R. G. Zuidgeest, and B. J. Eertink: Crew Assistant Architecture, RTP 6.5 Crew Assistant, EUCLID CEPA 6, D-CAA-WP1.3-NL/03A, NLR Amsterdam, 1995.
10. Verhoeven, R. P. M., et.al.: PoWeR-II: Crew Assistant Systems: Human Factors Aspects and HMI Considerations, NLR, Amsterdam, NLR-TR-2000-649, March 2001.
11. Yannone, R. M.: The Role of Expert Systems in the Advanced Tactical Fighter of the 1990's, Proceedings of the 1985 National Aerospace and Electronics Conference, Dayton OH, IEEE, New York, 1985.
12. Zuidgeest, R.: Advanced Information Processing for On-Board Crew Assistant, NLR News, number 21, July 1995.

Learning from Human Decision-Making Behaviors – An Application to RoboCup Software Agents

Ruck Thawonmas[1], Junichiro Hirayama[2], and Fumiaki Takeda[2]

[1] Department of Computer Science, Ritsumeikan University
1-1-1 Noji Higashi Kusatsu City, Shiga 525-8577, Japan
[2] Course of Information Systems Engineering, Kochi University of Technology
185 Miyanokuchi, Tosayamada-cho, Kami-gun, Kochi 782-8502, Japan
takeda.fumiaki@kochi-tech.ac.jp

Abstract. Programming of software agents is a difficult task. As a result, online learning techniques have been used in order to make software agents automatically learn to decide proper condition-action rules from their experiences. However, for complicated problems this approach requires a large amount of time and might not guarantee the optimality of rules. In this paper, we discuss our study to apply decision-making behaviors of humans to software agents, when both of them are present in the same environment. We aim at implementation of human instincts or sophisticated actions that can not be easily achieved by conventional multiagent learning techniques. We use RoboCup simulation as an experimenting environment and validate the effectiveness of our approach under this environment.

1 Introduction

It's a not-so-easy task to build software agents or multiagents (henceforth simply called agents) so that they could perform desired actions. To define proper condition-action rules, one may choose to hand code them using if-then rules or to use online learning techniques such as reinforcement learning [1]. Hand coding requires a set of rules that must cope with various kinds of conditions. This approach requires special skills and much effort to write complete agent programs. In addition, when conditions and desired actions are complicated, rules will also become intricate and the number of rules large. On the other hand, reinforcement learning can relatively easily make agents learn from their experiences a moderate number of condition-action rules. Compared to hand coding, this approach imposes the lesser burden on the user. However, in this approach it might be difficult to learn complex rules or take long learning time to reach them. Rather, this approach is useful for adding refinement to other methods.

The main objective of our study is to solve the aforementioned problems. In our approach, we derive condition-action rules from decision-making behaviors of humans. Namely, logs are taken that consist of conditions and the corresponding actions while the latter are decided by a human who is present in the same environment as the agent. Condition-action rules extracted from these logs using C4.5 [2] are then applied to the agent. Thereby, the agent can be promptly trained to have human decision-making behaviors. Moreover, this approach has high potential to allow implementation of complex condition-action rules such as cooperative behaviors among agents or human instincts that can not be readily achieved by conventional agent learning techniques.

In this paper, we use RoboCup soccer simulation [3] as our agent platform. A system called OZ-RP [4] was recently proposed and developed by Nishino, et al., that enables human players to participate soccer games with agents. However, this system highly depends on the model for describing the environment and internal state of their agents. It thus is not compatible with our agents previously developed for both education [5] and competitions[1]. As a result, for our agents we originally develop a system called KUT-RP that has similar functions to the OZ-RP. Logs from the KUT-RP system are used for extracting human decision-making behaviors.

2 RoboCup Simulation

RoboCup[2] is an international initiative to foster artificial intelligence and intelligent robotics by providing a standard test bed (soccer) where several technologies can be integrated and examined. The grand challenge of the RoboCup initiative is to develop a team of fully autonomous humanoid robots that can win against the human world soccer champions by the year 2050. Research outcomes are to be applied to a wide range of fields such as rescuing robots, intelligent transportation systems, and other social infrastructures. At present, RoboCup consists of five leagues, namely, simulation, small size, middle size, Sony 4 legged, and humanoid leagues.

Fig. 1 shows the architecture of the simulation league system adopted in this study. They are three main components in the system, i.e., soccer clients (agents), the soccer server, and the soccer monitor. The soccer server and clients communicate with each others during a game based on a client/server model under the UDP/IP protocol. All soccer objects such as the soccer ball, the players, or the field are visualized on the soccer monitor.

In simulation, to maximally mimic real-world soccer games, a number of constraints have been introduced and applied to the system. For example, noises are intentionally imposed to sensory information (aural sensor, vision sensor, and body sensor) to be sent from the soccer server to each agent. On the other hand, three basic commands (kick, dash, and turn) to be sent from each agent to the soccer server always contain a certain degree of inaccuracy. The vision scope and stamina of each agent is also limited.

---

[1] Our soccer simulation teams NoHoHoN and NoHoHoN G2 participated in the Japan Open 2000 and 2001, respectively.
[2] The official web site is www.robocup.org.

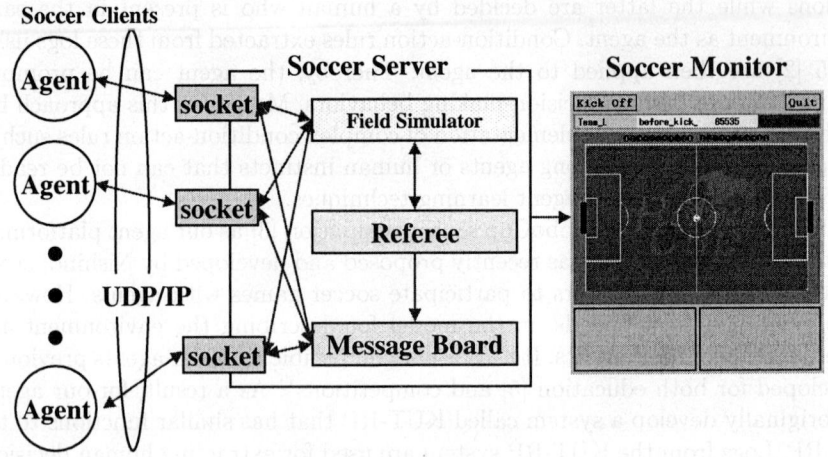

Fig. 1. Architecture of the simulation league system

3 Architecture of KUT-RP

In this section, we describe the architecture of our originally developed KUT-RP (Kochi University of Technology - Ritsumeikan University's Player) system using Java. The KUT-RP system is developed based on our original agents. The objective of the KUT-RP system is to enable human players to participate soccer games by operating agents called KUT-RP agents. Fig. 2 shows the conceptual diagram of the KUT-RP system.

The KUT-RP system is composed mainly of three components, i.e., the agent core, the real-player monitor, and the standard input devices. The agent core has the following modules, namely,

Communication module that copes with communication between the soccer server and the KUT-RP agent,

Memory module that processes and stores the information sent from the soccer server as well as the real-player monitor, and

Composite command execution module that performs composite commands composed of series of at least one of the three basic commands.

In practice, it is not feasible for a human player to keep on issuing the three basic commands. To cope with this problem, three composite commands are developed. Each of them is described as follows:

Kick to specified position that makes the KUT-RP agent kick the ball toward the specified position,

Dash to specified position that makes the KUT-RP agent dash to the specified position, and

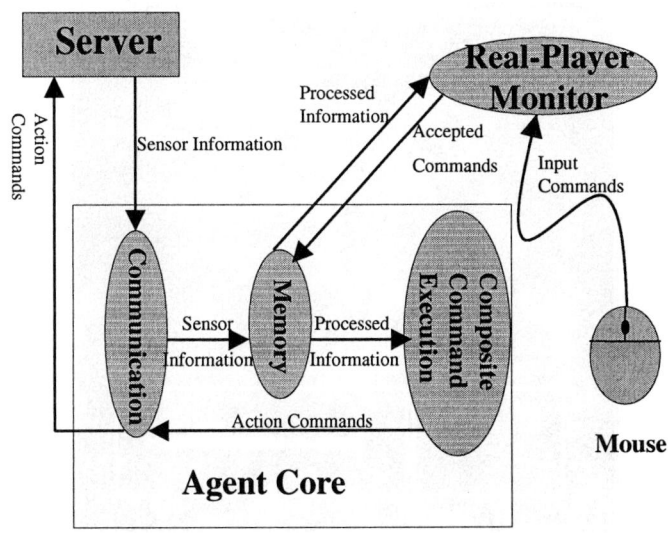

Fig. 2. Conceptual diagram of the proposed KUT-RP system

Dribble to specified position that makes the KUT-RP agent dribble the ball to the specified position.

In addition to these composite commands, the KUT-RP agent can be executed in one of the following three operation modes, namely,

Full-auto mode that will play automatically and not take into account any composite commands issued by the human player,
Semi-auto mode that will automatically trace the ball if any of the composite commands is not issued, and
Manual mode that will wait until one of the composite commands is issued.

The real-player monitor is different from the RoboCup soccer monitor where all available information and objects are visualized. In the real-player monitor, only the visual information available to the KUT-RP agent is displayed. The displayed information includes both the information sent from the soccer server and the information processed and enriched by the memory module. One example of the latter type of information is the predicted ball position when the ball can not be actually seen by the KUT-RP agent due to restricted vision scope. Fig. 3 shows a display of the real-player monitor before the ball is kicked off. In this figure, a KUT-RP agent is shown on the left side of the center circle, the ball in the middle. Two opponent players are shown in the right side of the soccer field. Our current version of the real-player monitor is also equipped with a bar showing the remaining stamina of the KUT-RP agent, a display box showing the most recently clicked button of the mouse (to be described soon below), and another display box showing the agent dashing speed.

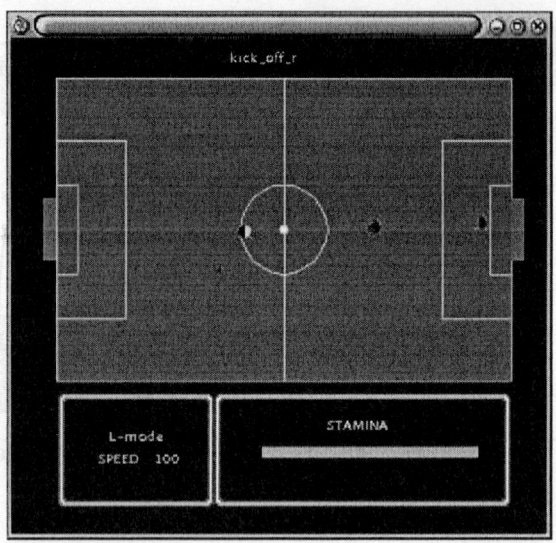

Fig. 3. Display of the real-player monitor

In this study, a mouse is used as the input device of the KUT-RP system[3]. The left, middle, and right buttons correspond respectively to **dash to specified position**, **kick to specified position**, and **dribble to specified position** composite commands. The specified position of these commands is defined by pointing the mouse cursor to the desired position in the real-player monitor while clicking one of the mouse buttons.

In addition to the components and modules described above, the KUT-RP system is also equipped with a function to record logs of the agent's commands and the sensor information when these commands are issued by the human player.

4 Experiments of Learning from Human Decision-Making Behaviors

We extensively conducted experiments in order to verify whether it is possible to learn the decision-making behaviors of the human player who operates a KUT-RP agent. In these experiments, a KUT-RP agent under the semi-auto mode played against two opponents (one defender and one goalkeeper), as shown in Fig. 3. Logs were then taken from each simulation game taking in total of 6000 simulation cycles (10 minutes in the default mode). Condition-action rules were extracted by C4.5 from the logs recorded with the KUT-RP system explained

---
[3] We are now developing a new user interface that can incorporate inputs from the keyboard

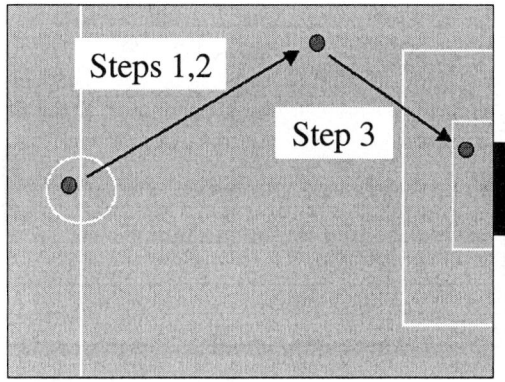

Fig. 4. Basic strategy from step 1 to 3

above. The generated decision tree was then applied to an agent henceforth called the virtual human agent. The virtual human agent was designed by two very simple if-then rules:

*If the virtual human agent holds the ball,
then consult the decision tree whether to kick or shoot,*

and

If the virtual human agent does not hold the ball, then trace the ball.

We examined the scores obtained by the virtual human agent when it was applied with different decision trees generated by accumulated logs of different numbers of training games.

The opponent defender agent was programmed in such a way that it moves at the highest speed to and forwardly kick the ball at the highest power when the distance to the ball is relatively near (below 7 meters). Otherwise, it moves back to and waits at its initial position located in front of the goalkeeper but with some distance. This situation makes it difficult for an offending agent like our KUT-RP agent to dribble or to shoot the ball directly through this defender agent toward the center of the opponent goal. A very simple but effective strategy that the human player eventually adopted for playing the KUT-RP agent can be described below as follows:

1. kick the ball to either side (top or bottom),
2. trace the ball,
3. shoot the ball while keeping it away from the goalkeeper,
4. if the ball is blocked and kicked back by the goalkeeper, then trace the ball and repeat from step 1.

Figs. 4 and 5 visually show this strategy.

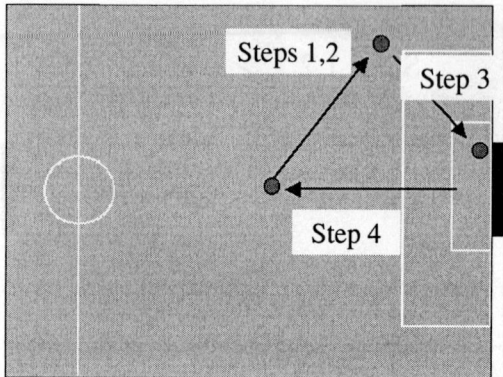

Fig. 5. Action sequences when the ball is blocked and kicked back by the goalkeeper

Table 1. Attribute names and types of the input data

Attribute Name	Attribute Type
x	continuous
y	continuous
body_angle	continuous
ball_direction	continuous
ball_distance	continuous
opponent_direction	continuous
opponent_distance	continuous

In order to generate decision trees, we have to define classes and input attributes as well as their types. In this study, they are summarized in Tables 1 and 2. The input attributes in Table 1 are selected from the information sent to the agent by the soccer server. The attributes x and y altogether represent the absolute coordinate of the agent in the soccer field. The attribute body_angle indicates the relative direction of the body of the agent toward the center of the opponent goal. The attributes ball_direction, ball_distance, opponent_direction, and opponent_distance are the relative direction and distance to the ball and to the nearest opponent, respectively. The classes in Table 2 represent agent actions. The class kick is divided into 19 subclasses according to different kicking directions relative to the agent's body direction. The class shoot is not explicitly provided by the KUT-RP system, but can be derived from the **kick to specified position** composite command by checking whether the position inside the opponent goal area is specified or not.

For comparisons, we conducted an experiment in which a programmer who has good understanding of Java but is relatively new to RoboCup simulation

Table 2. List of classes and their subclasses

Class Name	Subclass Name
kick	-90,-80,-70,-60,-50,-40,-30,-20,-10,0 ,10,20,30,40,50,60,70,80,90
shoot	Top,Bot,Mid

domain was asked to hand code an agent for playing against the two opponents described above. In addition, we also conducted an experiment using an agent trained with reinforcement learning for the same competition. For reinforcement learning, we used the Q learning with 3 states, each having 4 possible actions, namely,

State 1: Far from opponent goal in which the distance between the agent and the opponent goal is between 20 and 40 meters,

State 2: Near to opponent goal in which the distance between the agent and the opponent goal is below 20 meters,

State 3: Near to opponent player in which the distance between the agent and the nearest opponent player is below 7 meters,

Action 1: Shoot to the near-side goal post by which the ball is shot to the near-side goal post,

Action 2: Shoot to the far-side goal post by which the ball is shot to the far-side goal post,

Action 3: Dribble to the near-side goal post by which the ball is dribbled to the near-side goal post,

Action 4: Dribble to the far-side goal post by which the ball is dribbled to the far-side goal post.

In addition, when the reinforcement learning agent is not in one of the three states above, it will trace the ball and dribble the ball toward the center of the opponent goal.

5 Experimental Results

Fig. 6 shows the scores obtained by the KUT-RP agent for five games. At the first game, the human player was still not familiar with the system. As a result, the obtained scores were relatively low. More stable scoring could be seen from the second game.

Fig. 7 summarizes the averaged and highest scores from five games obtained by the hand-coded agent, the KUT-RP agent, the reinforcement learning agent and the virtual human agent, when the last two were trained by various numbers of games from one to five. The number of games used for training the virtual human agent indicates the number of games that the human player plays the KUT-RP agent; more games means more logs or training samples for extracting condition-action rules.

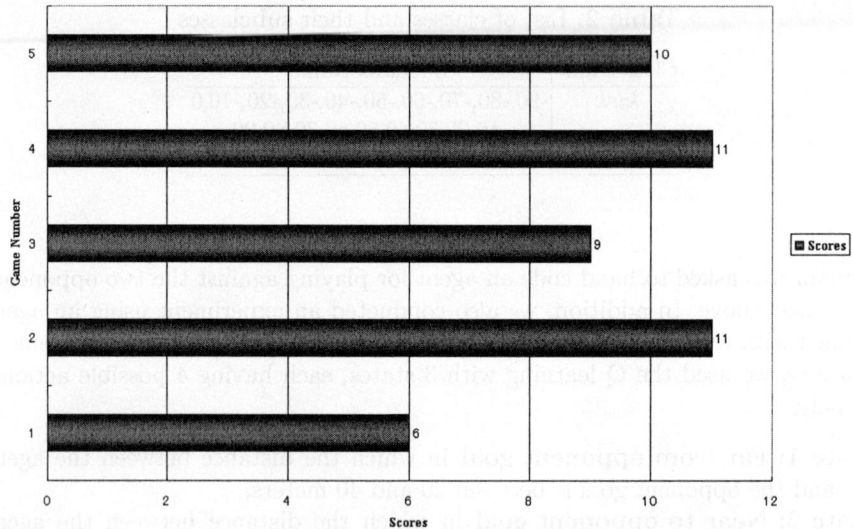

Fig. 6. Scores obtained by the KUT-RP agent

From the results in Fig. 7, it can been seen that the performance of the virtual human agent improves as the number of training games increases. When five games were used for training, its performance become comparable to that of the KUT-RP agent. Visual comparisons of the movements of both agents also confirmed that they had similar behaviors.

In addition, the performance of the virtual human agent is superior to that of the reinforcement learning agent for all numbers of training games. Though one can argue that the design of states and actions of the reinforcement learning agent were not optimal, optimizing such designs is not an easy task and might even take longer training time (more training games) due to higher degree of complexity.

Compared with the hand-coded agent, the virtual human agent has subtly lower performance. It turned out that the programmer adopted a similar strategy to the one used for the KUT-RP agent, but the hand-coded agent had more precise shooting ability. However, it took at least five hours of coding time including the time for laying out the strategy. To train the virtual human agent, it took approximately one hour for taking the logs of five games and extracting the rules, noting that 10 minutes are required for playing one game.

6 Conclusions

Applying human decision-making behaviors to software agents is an effective approach for implementing complicated condition-action rules. The experimental results given in this paper confirm this conclusion. In addition, because humans

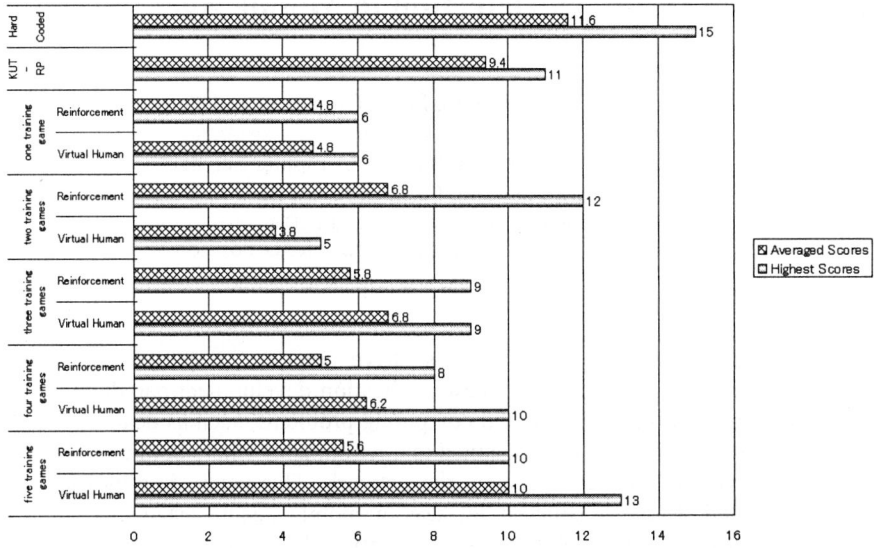

Fig. 7. Comparisons of the highest and averaged scores obtained by various kinds of agents used in the experiments

tend to incorporate prediction using their instincts into decision-making process, it can be expected that agents with such ability can be readily built using the presented approach. Validation of this conjecture, optimization of the rule extraction methodology, and applications of the approach to industrial uses are left as our future studies.

References

1. Sutton, R. S., Barto, A. G.: Reinforcement Learning (Adaptive Computation and Machine Learning). MIT Press (1998)
2. Quinlan, J. R.: C4.5 Programs for Machine Learning. San Mateo, Morgan Kaufmann (1993)
3. Kitano, H., Kuniyoshi, Y., Noda Y., Asada M., Matsubara H., Osawa, E.: RoboCup: A Challenge Problem for AI. AI Magazine, Vol. 18, No. 1 (1997), 73-85
4. Nishino, J, et al.: Team OZ-RP: OZ by Real Players for RoboCup 2001, a system to beat replicants. (2001) (submitted for publication)
5. Thawonmas, R.: Problem Based Learning Education using RoboCup: a Case Study of the Effectiveness of Creative Sheets. International Symposium on IT and Education (InSITE 2002), KochiüCJan. (2002)

Distributed Deadlock Detection in Mobile Agent Systems

Bruce Ashfield, Dwight Deugo, Franz Oppacher, and Tony White

Carleton University, School of Computer Science
Ottawa, Ontario, Canada, K1S 5B6
Bruce_Ashfield@Mitel.COM
{deugo,oppacher,white}@scs.carleton.ca

Abstract. Mobile agent systems have unique properties and characteristics and represent a new application development paradigm. Existing solutions to distributed computing problems, such as deadlock avoidance algorithms, are not suited to environments where both clients and servers move freely through the network. This paper describes a distributed deadlock solution for use in mobile agent systems. The properties of this solution are locality of reference, topology independence, fault tolerance, asynchronous operation and freedom of movement. The presented technique, called the shadow agent solution proposes dedicated agents for deadlock initiation, detection and resolution. These agents are fully adapted to the properties of a mobile agent environment.

1 Introduction

The increasing use of mobile agents for application development often stresses or breaks traditional distributed computing techniques [1]. Many existing solutions to problems [2], such as deadlock avoidance or leader election, are not suited for environments where both clients and servers move freely throughout the network. The mobility of these entities creates interactions and new conditions for distributed algorithms and applications that are traditionally not issues for them. The fundamental building blocks of many traditional distributed algorithms rely on assumptions, such as data location, message passing and static network topology. The movement of both clients and servers in mobile agent systems breaks these fundamental assumptions. Simply forcing traditional solutions into mobile agent systems changes the dynamic nature of their environments by imposing restrictions, such as limiting agent movement. As a result, new effective and efficient techniques for solving distributed problems are required that target mobile agent systems.

1.1 Problem Statement

In many distributed applications there are complex relationships between services and data. Mobile agents commonly encapsulate services and data (as do objects in object-oriented (OO) programming) [3], extending and enhancing the data-service relationship by adding movement to both services and data. Generally speaking, mobile agents, such as those involving consensus, data transfer and distributed database processing must coordinate with one another to provide services and access data. The advanced synchronization required in these mobile agent based applications can lead to complex, multi-node deadlock conditions that must be detected and resolved. Traditional distributed deadlock schemes [4] fail when agent mobility and faults are added to deadlock resolution requirements. Moreover, due to their assumptions, traditional techniques such as edge chasing on the global wait-for graph, are inadequate solutions in a mobile agent system.

In this paper, we develop a solution to the traditional problem of deadlock detection and resolution targeted for mobile agent systems. The presented solution is adapted from wait-for graph algorithms [5] to work in mobile agent environments requiring fault tolerance, asynchronous execution and absence of assumptions concerning the location or availability of data. Our solution is not bound to a particular mobile agent environment or toolkit. Therefore, properties specific to a particular mobile agent system do not influence the solution. Efficiency and speed are secondary concerns in this model, but are considered and discussed when applicable.

Section two provides a general overview of the problem of distributed deadlocks in mobile agent systems. Section three presents our approach to distributed deadlock detection in a mobile agent environment. Section four discusses the simulation experiments that were done to validate the proposed solution. Finally, in section five we draw our conclusions.

2 Deadlock

Deadlock literature formally defines a deadlock as, "A set of processes is deadlocked if each process in the set is waiting for an event that only another process in the set can cause" [4, 5]. A more informal description is that deadlocks can occur whenever two or more processes are competing for limited resources and the processes are allowed to acquire and hold a resource (obtain a lock). If a process waits for resources, any resources it holds are unavailable to other processes. If a process is waiting on a resource that is held by another process, which is in turn waiting on one of its held resources, we have a deadlock. When a system attains this state, it is effectively dead and must resolve the problem to continue operating.

There are four conditions that are required for a deadlock [4]:

1. **Mutual exclusion:** Each resource can only be assigned to exactly one resource.
2. **Hold and wait:** Processes can hold a resource and request more.
3. **No preemption:** Resources cannot be forcibly removed from a process.
4. **Circular wait:** There must be a circular chain of processes, each waiting for a resource held by the next member of the chain.

There are four techniques commonly employed to deal with deadlocks: ignore the problem, deadlock detection, deadlock prevention and deadlock avoidance [4]. Ignoring deadlocks is the easiest scheme to implement. Deadlock detection attempts to locate and resolve deadlocks. Deadlock avoidance describes techniques that attempt to determine if a deadlock will occur at the time a resource is requested and react to the request in a manner that avoids the deadlock. Deadlock prevention is the structuring of a system in such a manner that one of the four necessary conditions for deadlock cannot occur [5]. Each solution category is suited to a specific type of environment and has advantages and disadvantages. In this paper, we focus on deadlock detection and resolution, which is the most commonly implemented deadlock solution.

2.1 Mobile Agent Issues with Traditional Deadlock Detection

Agent movement and failure are two properties of a mobile agent system that traditional distributed deadlock detection approaches don't consider. For example, resources and their consumers in traditional distributed deadlock detection do not move through the network and each server has knowledge about the location of other nodes that form the network.

In a mobile agent system, agents perform transactions by moving towards the source of data and executing locally to leverage the benefits of locality of reference [6]. The mobile agent and the host server can continue to interact with additional network resources [7] and as a result, transactions can be spread across multiple hosts without informing the node that initiated the transaction. Clearly, agent movement causes problems for algorithms that depend on location information.

In distributed deadlock detection techniques, such as central server or edge chasing [8], location assumptions cannot be avoided, since information is either gathered centrally or built through a series of probes. In order to detect and resolve distributed deadlocks, these algorithms must be able to locate the nodes involved in the transaction. In a mobile agent system, an agent's movement and activities cannot be easily tracked. Therefore, the agent that initiated a transaction can be difficult to locate, as are the secondary agents that are indirectly involved. Assumptions concerning location must be adapted if an algorithm is to function properly and efficiently in a mobile agent system.

Often, distributed problem solving techniques have not dealt well with failures or crashes. For example, the distributed deadlock detection algorithm described by [5] states "The above algorithm should find any deadlock that occurs, provided that waiting transactions do no abort and there are no failures such as lost messages or servers crashing." The nature of mobile agents and their operating environment is commonly failure prone. Therefore, algorithms with strict fault assumptions will have difficulty functioning in many mobile agent systems.

3 Deadlock Shadow Agent Approach

The following sections describe our approach to distributed deadlock detection in mobile agent environments. We assume the following:

- All forms of mobile agents are separated from the topology of the network, which means they can move through the network without knowledge of how the nodes are connected.

- The topology of the network is static once the algorithm begins.
- We assume that standard deadlock avoidance techniques, such as two-phase commit or priority transactions, are being used. These mechanisms allow the detection and resolution process to be assured that an agent will not spontaneously unlock a resource (and unblock) during the detection process. This property is a significant factor in preventing phantom deadlock detection.
- Resources can only be locked and unlocked by a consumer agent when it is physically at the same environment as the resource it is manipulating. This property allows host environments to communicate the details of a consumer agent's resource requests to its associated deadlock detection counterparts.
- There is a level of cooperation between agents or shared resources. As the agents perform their tasks, resources can be "locked", which means they are made exclusive to an individual consumer agent (i.e., single writer and many readers problem).
- During the locking process consumer agents must communicate with the host environment. The host is the ultimate authority and can allow or deny access to a resource. Since the host environment can deny the lock request of an agent, a reaction is required. It is assumed that depending on the agent's task, it may block or wait on the resource (and will not move) or it may continue processing and moving through the network. It is assumed that the authority does not automatically block the agent, since this would limit flexibility and limit the dynamic nature of the mobile agent environment.
- Agents must notify the host environment if they block on the resource. This allows the environment to communicate the state of an agent to its deadlock detection counterparts and deny any additional requests made by the blocked agent. Blocked agents cannot unblock until the host environment approves their request.
- Agents must be uniquely identifiable once they hold a resource. This means that they can be found in the agent system during the deadlock detection process. The assignment of identifiers may be done before a consumer agent blocks (i.e., during creation) or only once they lock a resource (using a globally unique naming scheme).

3.1 Algorithm Overview

Our mobile agent algorithm uses agent-adapted mechanisms that are based on those found in traditional edge-pushing global wait-for graph algorithms. In particular, the division of the global wait-for graph into locally maintained segments and the initiation of deadlock detection "probes" are inspired by traditional solutions.

There are three types of agents populating the mobile agent system:

- Consumer Agent: The only agent in the system that actively performs tasks and consumes or locks resources. It represents the entity that implements algorithms. It has no active role in deadlock detection and resolution.
- Shadow Agent: This agent is created by host environments and is responsible for maintaining the resources locked by a particular consumer agent,

following it through the network and for initiating the deadlock detection phase. In addition, it analyses the data gathered by detection agents to initiate deadlock resolution and detects and recovers from faults during the deadlock detection process. It represents a portion of the global wait-for graph.
- Detection Agent: Shadow agents create this agent when informed by the host environment that their target agent has blocked. They are small, lightweight mobile agents that are responsible for visiting hosts and building the global wait-for graph and for breaking the deadlock situation.

The Host Environment is responsible for hosting the mobile agents and provides APIs to access, move, block, and unblock resources. It coordinates the consumer, shadow and detector agents as they move through the network. Coordination is performed through a "token" that is carried by all consumer agents.

In our approach, shadow agents maintain a part of the wait-for graph on behalf of their target consumer agent. During the deadlock detection process shadow agents continually check for deadlocks in the global wait-for graph and create deadlock detection agents to build the global graph.

3.2 Deadlock Initiation

As consumer agents complete tasks, they may exclusively lock resources throughout the mobile agent system. When consumer agents are initially created, they are not actively monitored by the host environments for deadlock detection reasons. This means that new agents can move freely through the network and consume resources. Consumer agent monitoring is performed through "environment tokens"; therefore, each time an agent arrives at a host environment it must present its token. This token has no meaning to the agent, and is only used by the host environments to coordinate the deadlock detection process.

Consumer agent monitoring activities begin when an agent requests an exclusive resource lock. As part of the request granting process, the host environment checks for a shadow agent that is associated with the requesting agent. If no shadow exists, one is created and associated with the consumer agent. Finally, the consumer agent's environment token is updated to reflect the existence of the newly created shadow agent. Once a shadow agent is created for a consumer agent, the host environments are able to control the deadlock detection process.

Shadow agents are notified of new exclusive locks by host environments through a defined message. This message includes deadlock detection information, such as the identifier and priority of the locked resource. Resource locking is performed through a defined API; therefore, the host environment always has a chance to update shadow agent information before granting the request. Each new lock is noted by the shadow agent and maintained as part of its deadlock detection information.

The mobile agent system's wait-for graph can be represented by the notation: C_x $[R_1, R_2, .., R_n] \rightarrow R_n$. In this notation, C_x represents a consumer agent that has exclusively locked resources. C_x can be replaced by S_x to more accurately represent the actual mobile agent system, since shadow agents are created to monitor consumer agents. The values [$R_1, R_2, .. , R_n$] are a form of distributed transaction and represent the resources that are held exclusively by a particular consumer agent. Finally, the R_b notation following the arrow (->) represents the resource on which an agent is blocked. If a consumer agent is not blocked, the -> R_b portion of the notation is not present. Combining all of the local transactions (represented by this notation), creates the global wait-for graph for the mobile agent system.

Once a shadow agent is created and associated with a consumer agent, they move through the network together. This synchronous movement is coordinated by automatically routing a consumer's shadow each time the consumer sends a migration request to the host environment. It should be noted that this pairing of agents places restrictions on consumer agents. A consumer agent cannot perform the following actions if its associated shadow agent is not present: move, lock, unlock. The consumer is notified of the failure and the request must be resubmitted. This restriction ensures that the shadow agents will contain the exact state of the wait-for graph, even if they are delayed during transmission.

When a consumer agent requests an exclusive lock that is denied by a host environment, it may decide to block and wait for the resource to be freed. If the decision to block is made, the consumer agent must inform the host environment. Host environments react to blocking notifications by informing the consumer agent's shadow to allow deadlock information to be checked. It should be noted that if the consumer holds no locks, a shadow agent will not be present and cannot be informed. This non-notification is acceptable, because if the consumer agent holds no other locks, it cannot be part of a distributed deadlock.

The host environment informs shadow agents that their target has block or unblocked through a defined message. These block and unblock events begin the deadlock initiation process. When informed of a block event, shadow agents query the host environment to determine who holds the lock on the target resource. Once the host environment provides the agent identifier who holds the lock, a second query is performed to determine if the agent is local or remote. If the locking agent is off-site, the shadow agent initiates the distributed deadlock detection sequence, otherwise, no special processing occurs.

3.3 Deadlock Detection

Shadow agents initiate the deadlock detection cycle by creating detection agents. Upon creation, detection agents are initialized with their parent shadow agent's list of locked resources and the environments where they are located. This creation of a dedicated detection agent allows a shadow to simultaneously search for deadlocks and respond to other shadow's detectors.

Once initialized, detector agents visit the resources that are locked by their target consumer agent. By recording the network location of each resource as it is locked, routing of detector agents is expedited. At each resource the detector agents visit, they query the host environment to determine if other agents are blocked on that resource. If blocked agents are found, the detector locates their associated shadow agent and queries for their deadlock detection information. This processing occurs simultaneously for every agent that is blocked on a resource held by a remote agent. The returned deadlock detection information is a list of deadlock detection entries, summarized as follows:

- **Agent Name:** The unique identifier of the consumer agent that this information belongs to
- **Blocked Resource:** The resource that the agent is blocked on, which includes the resource's unique name, the consumer agent that has this resource locked, the environment's name that contains this resource, and the priority of this resource.
- **Primary Locks:** The list of primary locks (resources) held by this agent.

Each deadlock detection entry encapsulates the relevant information about a consumer agent that is blocked on a resource. At each resource, this information is added to the detector's original detection table, increasing the amount of information it carries as it moves through the network. The deadlock information of each blocked agent is added to the detector's deadlock detection table because the agent is blocked on a resource held by the detector's target. Since these agents are blocked on a resource held by another agent, their entire detection table (and hence any agents blocked on them) is indirectly being held by that agent.

This secondary deadlock information is valid, since blocked agents cannot move to free resources while waiting for the resource locked by a detector's target. When a detector agent visits all of the resources that were placed in its initial set of locks, it returns to its initial host environment. Upon arrival, the detector agent informs its shadow that it has returned and communicates its constructed deadlock table. The shadow agent analyses this table, which represents the global wait-for graph, to determine if a deadlock is present.

Shadow agents use their target consumer agent as a deadlock detection key. If their target agent appears in the deadlock information table returned by the detector, the target is waiting on a resource held by itself. Clearly we have a deadlock (a cycle in the global wait-for graph), since the target is blocked and that resource will never be freed.

Shadow agents are responsible for recovering from failures during the deadlock detection phase. Failure detection is implemented through a running round trip delay calculation. Depending on the type of network and its properties, each shadow agent is initialised with a default round trip delay time. Shadow agents expect that their detector agents will be able to check all of their required locks in less than four times the maximum round trip delay. If a detector agent does not return in the maximum allowable time, the shadow agent assumes that it has failed and creates a new detector agent to continue the failed agent's work.

Each time a detector agent returns from checking locks its parent shadow agent updates its expected round trip delay time. This allows the shadow agents to continually tune their expectations to current network conditions. The detector agents created by a shadow are versioned to prevent errors if a shadow incorrectly assumes that its detector agent is dead. Each time a detector agent is created, the shadow agent notes the version number and will only accept deadlock information from that shadow agent. If old detector agents eventually return, they are killed and their information ignored.

The failure of a detector agent has a direct impact on the efficiency of the algorithm, since information is lost and must be recovered. The information summaries maintained by shadow agents minimize the impact of failures. When a detector agent fails, only its current information is lost and the data from previous trips is safe; therefore, only a single visit must be repeated.

If a shadow agent does not find a deadlock in the global wait-for graph, the detector agent is re-initialized and the process repeats. Depending on the requirements of the implementation and the properties of the mobile agent system, a delay may be desired between detector visits. This allows the algorithm to be tuned between implementations and to react to network conditions. Possible delay techniques include: no delay, delay for a configured duration (i.e. one second), delay for a random duration or delay for a duration based on the round trip time. Delaying based on the round trip is an example of using network conditions to provide continual feedback and adjust the algorithm accordingly.

If a shadow agent finds a deadlock in the global wait-for graph, the deadlock resolution sequence is initiated.

3.4 Deadlock Resolution

When a shadow agent detects a deadlock, there is a cycle in the global wait-for graph that must be broken. To successfully resolve the deadlock the shadow agent must identify the cycle and determine how to break the deadlock.

The information gathered by the detection agents is sufficient to build the deadlock cycle and determine which resource should be unlocked. The following figure illustrates how the deadlock detection table represents a graph.

The location of the deadlock cycle is facilitated by the entrance criteria to the resolution phase. These criteria dictate that there is a cycle in the graph, and the repeated resource is the resource on which the shadow is blocked. The detection table creates the detection graph as follows:

The resource on which the detecting agent is blocked is the root node in the detection graph. This is test.db in Figure 1.

The primary locks held in the root detection table entry form the next nodes in the graph. This is account.db in Figure 1.

As each primary lock entry is added to the graph, the entries of agents that are blocked on that resource are added as children of that node. These are user.db and foo.db in Figure 1.

Step 3 repeats for each child node as it is added to the graph, until there are no additional entries to be added or the root node's resource is located.

Note that the graph illustrated in Figure 1 is not normally constructed and it is sufficient to search a table-based data structure to determine the deadlock cycle. To determine the deadlock cycle, the shadow agent must first locate the leaf node that represents the resource on which its target consumer is blocked. The leaf nodes in Figure 1 are representations of the primary locks held by each blocked agent; therefore, the primary lock lists held in the detection tables are searched to locate the leaf node. Once that node has been located, the shadow agent must simply walk the graph to the root node (which is also the blocked resource) and record each node on the way. The information gathered for each node in the graph is sufficient to locate the deadlock and make a resolution decision.

Agent	Blocked Resource	Primary Locks
A1	account.db	[user.db]
A2	user.db	[test.db, payroll.db]
A3	test.db	[account.db]
A6	accound.db	[foo.db]

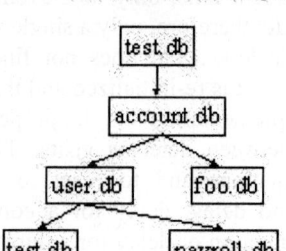

A3's Detection Graph

Fig. 1. Deadlock Detection Information

When a shadow agent detects a deadlock, it has the authority to break the deadlock. One of the easiest deadlock resolution techniques to implement is based on transaction priorities. An example of this technique is when each node is added to the mobile agent system is assigned a priority identifier. The priority of a resource, and hence any locks on that resource, have the same value as their host environment. Another option is, in a similar manner to the protocols that must be present to establish routing tables among agent environments, a protocol must be present to negotiate environment priorities. This can be implemented as a continually running and updating protocol, or it can be a central authority, which must be contacted as part of an agent environment's initialization. Regardless of the implementation, the priority protocol assigns each agent environment or resource a priority that can be used when deciding which resource should be unlocked to resolve a deadlock.

If a deadlock exists, all agents that are part of the deadlock will detect the situation and try to initiate resolution. As each shadow agent detects a deadlock and builds the deadlock cycle, it checks the priorities of all the resources that are part of the cycle. It selects the resource that has the lowest priority as the one to be unlocked. Once the selection has been made (each agent has selected the same resource), the shadow agents check to see if they are the owners of that resource, i.e., if they have it locked.

The shadow agent that owns the resource that must be unlocked is responsible for breaking the deadlock. Shadow agents break deadlocks by instructing their detector agents to travel to the deadlock resource and unlock it. Additionally, the shadow provides the name of the consumer agent that should be notified when the resource is unlocked. This ensures that in situations where multiple agents are blocked on a single resource the consumer agent that is part of the deadlock is notified and unblocked.

Once initialized, the detector agent travels to the correct host environment and requests that the resource be unlocked. If the deadlock resource is successfully unlocked, the detector agent returns to its shadow while the host environment notifies the specified consumer agent. When the shadow returns to its parent agent, the shadow removes the resource from its primary lock set and destroys the detector.

The host environment carries out the final step of the deadlock resolution. When the host environment notices that a resource that has blocked agents is unlocked, it notifies the consumer agent specified by the detector agent that the resource has been freed. The unblocked consumer agent can then complete its lock request and the deadlock is broken.

4 Experiments

We tested and validated the distributed deadlock detection approach using a mobile agent modelling system simulator we developed. The simulator had four properties that could be changed: network topology, failure rate, population and the type of deadlock. Varying these values provided a mechanism to ensure that the suggested approach works with any topology, is not dependent on synchronous execution, is fault tolerant and can resolve difficult to find deadlocks.

Mesh and ring network topologies were selected due to their significant differences and common deployment. Comparing the measurements between these configurations demonstrated that the suggested solution is topology independent. This property is significant when the context of mobile agent applications is considered. Additional topologies were used during validation, but are only variations on the ring and mesh configurations.

Measurements were gathered using the ring and mesh topologies in both fault-free and failure-prone configuration. They indicated the survivability of the approach and measured the performance impact that faults have on the deadlock detection process. In addition, four configurations of agents and agent environments were used to measure the impact of increasing the number of elements in the deadlock detection process. It should be noted that the same type of deadlock situation was measured while varying the topology and fault properties of the simulator.

Most real world IP networks are similar to the mesh configuration simulation, meaning a detector agent can often migrate directly to a desired host environment. Moreover, most deadlocks occur between two transactions [5]. A locked resource by a remote agent roughly maps to a transaction in traditional distributed deadlock detection. Therefore, the measurements for systems with two agents are of particular relevance. The complete set of experiments and results are found in [9].

5 Conclusions

The assumptions of traditional distributed deadlock algorithms prevent them from completing successfully in a mobile agent environment. To successfully detect and resolve mobile agent distributed deadlocks, our approach leverages the advantages of the mobile agent paradigm. In particular, the principle properties of the developed technique that differentiate it from traditional solutions are: locality of reference, topology independence, fault tolerance, asynchronous operation and freedom of movement. These attributes are realized through a platform independent, mobile agent distributed deadlock detection solution. The agents that consume resources in the mobile agent system are separated from the deadlock detection mechanism. This separation produces dedicated agents for deadlock initiation, detection and resolution.

These agents are fully adapted to the properties of the mobile agent environment and collaborate to implement a complete distributed deadlock detection solution.

Mobile agent environments demand topology flexibility and fault tolerance. Incorporating these properties into mobile agent solutions impacts performance. The measurements gathered during simulations of our technique confirm that these properties require additional processing and messages. Due to the parallel nature of mobile agent environments, there is no conclusive evidence that these additional messages significantly impact deadlock detection performance or efficiency. Additionally, the lack of similar mobile agent solutions makes comparison and evaluation difficult and inconclusive.

References

1. Tel, G.: Introduction to Distributed Algorithms, Cambridge University Press (1994)
2. Lynch, N.: Distributed Algorithms, Morgan Kaufmann Publishers Inc, San Francisco, California, (1996)
3. Silva, A., Romão, A., Deugo, D., and Da Silva, M.: Towards a Reference Model for Surveying Mobile Agent Systems. Autonomous Agents and Multi-Agent Systems. Kluwer Online, (2000)
4. Tanenbaum, A: Modern Operating Systems. Prentice Hall Inc., Englewood Cliffs, (1992)
5. Coulouris, G., Dollimore, J. and Kindberg, T.: Distributed Systems Concepts and Designs. 2nd ed. Addison-Wesley, Don Mills, Ontario, (1994)
6. Milojicic, D.S., Chauhan, D. and laForge, W: Mobile Objects and Agents (MOA), Design Implementation and Lessons Learned." Proceedings of the 4th USENIX Conference on Object-Oriented Technologies (COOTS), (1998), 179-194
7. White, J.E.: Telescript Technology: Mobile Agents, Mobility: Processes, Computer and Agents, Addison-Wesley, Reading, (1999)
8. Krivokapic, N. and Kemper, A.: Deadlock Detection Agents: A Distributed Deadlock Detection Scheme, Technical Report MIP-9617, Universität Passau, (1996)
9. Ashfield, B.: Distributed Deadlock Detection in Mobile Agent Systems, M.C.S. Thesis, Carleton University, (2000)

An Agent-Based Approach for Production Control Incorporating Environmental and Life-Cycle Issues, together with Sensitivity Analysis

Elisabeth Ilie Zudor and László Monostori

Computer and Automation Research Institute, Hungarian Academy of Sciences
Kende u. 13-17, Budapest, POB 63, H-1518, Hungary
Phone: (36 1) 2096-990, Fax: (36 1) 4667-503
{ilie,laszlo.monostori}@sztaki.hu

Abstract. Management of changes and uncertainties belong to the most important issues in today's production. In the approach proposed in the paper, not only these problems, but also the environmental impacts of processes are considered during a resource allocation process relying on market principles. A hierarchical rule structure called Priority Rules System is introduced for agent-based production control incorporating waste management holons. A decision support tool for selecting appropriate values of the utility factors in the rules is also presented, together with the related sensitivity analysis.

1 Introduction

Over the past few decades market demands forced production companies towards customer-specific manufacturing and the competition between these companies has drastically increased. They have become customer-order-driven, which means that a company must be able to manufacture a high variety of customer-specific products, with short delivery times against low costs.

The change in the market demands implies the need to manufacture with increased flexibility and greater efficiency. There are also two other major problems appearing in manufacturing: inaccuracies in the manufacturing information and uncertainties with regard to the progress in production.

A solution to these problems appeared to be the development of manufacturing systems from centralized, to distributed agent-based systems. Such a system is a collection of agents that can be viewed as an organization, i.e. a composition of a structure and a control regime. The system is distributed due to being physically broken up into separate components. New paradigms related to today's manufacturing, are fractal manufacturing [1], bionic manufacturing [2], random manufacturing [3], and holonic manufacturing [4].

In the concurrent environment, companies have to face basically two matters: *customer driven production* and *the imposition of new rules by the society*, such as consciousness towards the environment in which we live, in our best interest and in the interest of the generations to come.

The approach to be addressed here aims at the handling of these issues by using one of the above new manufacturing paradigms, namely the holonic one.

2 Holonic Systems, Our View on Product Holon

Holons in holonic manufacturing systems are autonomous and cooperative elements, consisting of an information processing part and often a physical processing part for transforming, transporting, storing and/or validating information and physical objects.

The holonic system regards the elements of manufacturing systems, such as machines, operations, human operators, and even the manufactured product and its components themselves, as holons.

Our approach respects the principles of the reference architecture for HMSs developed at the University of Leuven under the name Product Resource Order Staff Architecture (PROSA) [5]. PROSA comprises 4 types of holons: product, resource, order - considered basic holons, developed based on the 3 relatively independent manufacturing concerns, and staff holons.

The data maintained by holons, the functionality of holons and the knowledge exchange between them are illustrated in Figure 1.

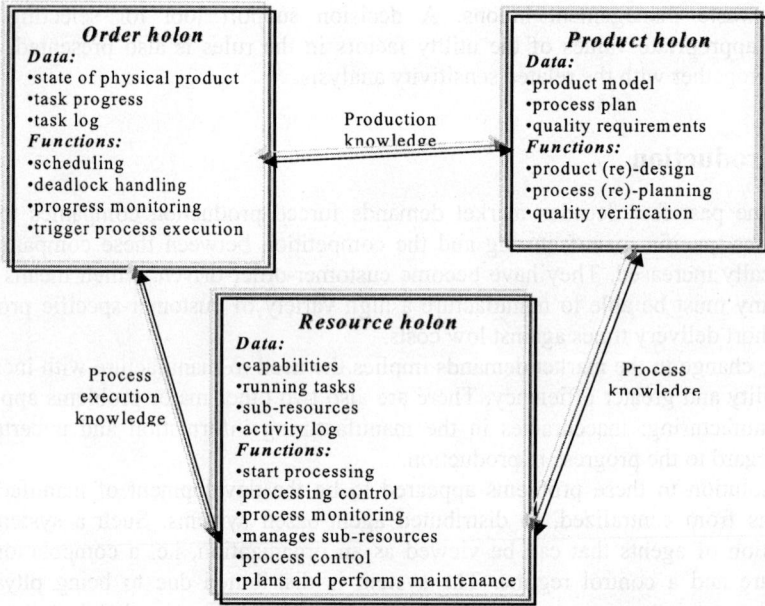

Fig. 1. Data maintained by holons, the functionality of holons and the knowledge exchange between them

In our view on holonic systems, the above representation of a *product holon* should be extended with considerations on the extension of the life of the product holon, i.e. its life does not end with accomplishing the manufacturing process, but further exists until the product disposal. According to this idea, the product holon, besides the data concerning the information on the product life cycle, user requirements, design, process plans, bill of materials, quality assurance procedures, and the process and product knowledge, contains also the information related to its end-of-life:

- the possible steps of disassembly,
- the data that would help establishing the point of maximum financial profit (from this point onward the disassembly is not worth anymore; the data might change in time, and at the end of life of product, those should be actualized) [6],
- which parts can be sent to: recycle, reuse, incinerate or landfill,
- type of waste that subparts from disassembly represent (solid, hazardous, liquid),
- the data about the environmental impact of subparts as the disassembly proceeds.

A part of the necessary information can be common to a set of products centrally (e.g. the product type specific information), the other part can be intrinsic data of the given product (e.g. product specific knowledge related to the production and usage).

3 An Approach to Solve Problems Caused by the New Customer-Driven Market and the New Society Rules

In order to keep up with the competition - among other things - an appropriate resource allocation mechanism is necessary. The one to be described here relays on the market principles for a holonic environment.

A hierarchical rule structure called Priority Rules System (PRS) is in the center of the approach proposed for finding the most appropriate resource for performing the task. This system contains the rules that the shop floor manager considers relevant when assigning a task. A more detailed presentation of the mechanism can be found in [7].

The applicability of the model is represented by the results of simulations made by the SIMPLE++ production and logistics simulation package. At this stage six rules are incorporated in the system. As in a customer driven, competition-based world market, delivering the product in time is one of the most important issues to be dealt with, 3 rules considering the time problem were enclosed, namely the: Due date rule, the Bidding time rule, the Breaks rule.

The Due date rule concerns the available interval of time for performing the task. The Bidding time rule refers to the time interval within the bids can be sent The Breaks rule concerns the trustability of the resource, meaning how many times the agent did not perform the task in the parameters established, in a certain time interval.

The second issue that the companies have to face, the environmental rules nowadays imposed by the society as mentioned above, is also considered in our model. The environmental problem comprises the matters of environmental friendly products and production and the waste management. In order to address these concerns, environment-related issues were included into the Priority Rules System, via the Environmental Effect rule and by building up the Waste management holon (WMH).

In the coming sections the *attributes of the WMH* and a *decision support tool* for shop floor managers in choosing the utility factors of the rules in the resource allocation mechanism are presented.

3.1 Waste Management Holon

In our days the responsibility for the management of the products at the end of their lives is already passed to the manufacturers, which are supposed to take-back their products and deal with the disposal problem.

The management of the product at its end-of-life can be ordered to the responsibility of waste management holons (WMHs), which, naturally, can have their jobs during the production, as well.

We consider that the responsibility over the waste management is assigned to the waste master agent (WMA). Its authority comprises the handling of

- wastes from manufacturing operations, and
- wastes from disassembly of products that reached their end-of-life.

The objectives of waste management are to *control, collect, process, utilize, minimize* and *dispose* wastes in the most economical way consistent with the protection of public health and the natural environment.

The responsibilities of WMA include the storage of wastes; inventory preparation; emergency plan preparation; waste minimization; existing waste management [8].

According to the types of wastes, the task of existing waste management can be categorized as: 1. *solid waste management*, 2. *hazardous waste management*, and 3. *wastewater management*.

After the WMA separates solid, hazardous and liquid wastes, it sends each job order to the Task Master (TM), the module responsible for job orders management. In the case of solid waste, the possible solutions (tasks) could be: solid waste recycling, solid waste combustion, solid waste landfilling. As to hazardous waste, the following techniques come into consideration: recycling, waste exchange, waste incineration, or waste secure land disposal, as to wastewater, wastewater disposal.

The schematic representation of WMA's responsibilities is presented in Figure 2.

The inventory preparation consists on compiling a detailed list of all hazardous waste sources, the characteristics of the wastes and the amounts generated of each.

Waste minimization is defined as a reduction in the toxicity or volume of wastes, which must be treated and/or disposed of.

Solid waste management:

The solid waste includes all the discarded materials not transported by water, and rejected for further use.

Hazardous waste management:

Hazardous wastes are the wastes that can be harmful to human health, other organisms, or the environment [8]. These wastes require special precautions in their storage, collection, transportation, treatment or disposal.

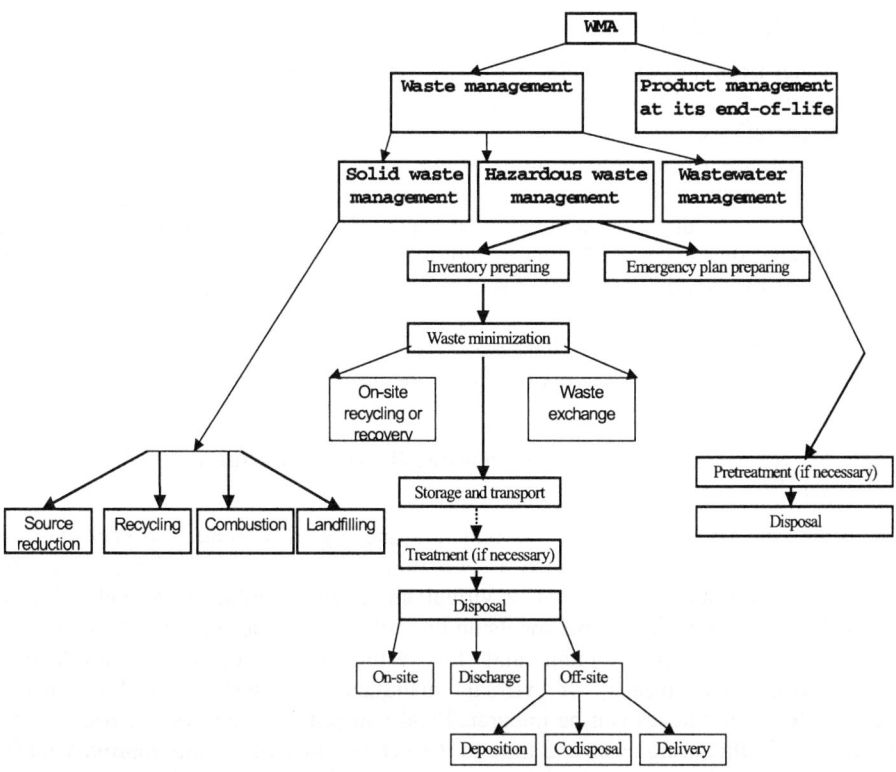

Fig. 2. Responsibilities of a waste management agent

Examples of wastes exempted from the hazardous lists are: fly ash, bottom ash, drilling fluids, wastes from crude oil, natural gas etc. The simplest way to identifying hazardous wastes is to consider them under general categories, such as: radioactive, flammable, toxic.

Many countries (as the United States, France, Germany, the United Kingdom, the Netherlands etc.) supplement a general classification system with detailed lists of substances, processes, industries, wastes considered hazardous or even classifications setting maximum concentrations for specific contaminants.

One way of reducing the quantity of hazardous materials requiring disposal is on-site recycling or recovery of waste. Another way is waste exchange. The objective of this exchange is to match the waste generated with those who could use this waste as raw material. Waste transfers are generally from continuous processes in larger companies to smaller firms able to reuse acids, alkalis, solvents, catalysts, oil of low purity, or recover valuable metals and other materials from concentrated wastes.

The storage of hazardous wastes needs special on-site tanks or basins or chemically resistant drums for holding smaller amounts of corrosive materials until these wastes can be removed. The stored wastes must be regularly collected by licensed haulers and transported by tanker truck or rail car or by flatbed truck to disposal.

Wastewater management:

Wastewater from industries includes process wastes from manufacturing, employees' sanitary wastes, wash waters, and relatively uncontaminated water from heating and cooling operations.

The wastewater from processing is the major concern. In some cases, pretreatment to remove certain contaminants or equalization to reduce hydraulic load may be mandatory before the wastewater can be accepted into the municipal system. The contaminants that cause problems include: disease-causing organisms (pathogens), organic matter, solids, nutrients, toxic substances, color, foam, heat, radioactive materials.

The primary objective of wastewater treatment is to remove or modify the contaminants detrimental to human health or the water, land, and air environment.

3.2 A Decision Support Tool for Choosing the Utility Factors; Sensitivity Analysis

With the help of the SIMPLE++ production and logistics simulation package, the communication between agents on a shop floor was simulated. The motivation of using a simulation tool is obvious, as virtual, simulated manufacturing models create a test field for conducting experiments on the influences of design on production, for supporting operations planning and for testing new methods of production management. By means of virtual manufacturing systems (VMS) many manufacturing processes can be integrated and realized into one system, resulting in reducing manufacturing costs and time-to market, and in a very good improvement of productivity. A virtual manufacturing system (VMS) is an integrated computer based model that represents the precise and whole structure of a real manufacturing system and simulates its physical and logical behavior in operation. A VMS doesn't produce output like materials and products, but information about them [9], [10].

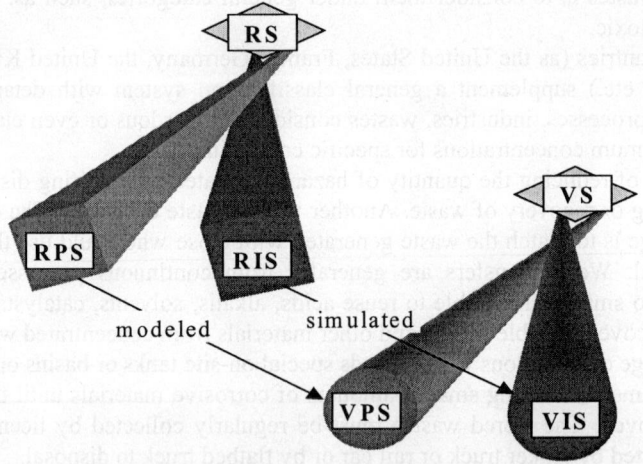

Fig. 3. Real and virtual manufacturing systems

A real manufacturing system (RMS) is composed of a physical system (RPS) and an informational system (RIS). At his turn, the VMS consists of a physical system (VPS) and an informational system (VIS). A RPS consists of entities such as: materials, products, parts, tools, robots, sensors, machines, and so on. A RIS consists of computers and the staff of the RMS, and involves the activities of information processing and decision making. A VPS models physical entities on the factory floor and simulates physical interactions between these entities. So, a VPS is a computer system that simulates the responses of the RPS. A VIS simulates manufacturing activities as design, management, planing, and so on. So, a VIS is a computer system that simulates a RIS. A schematic representation of the real and the virtual manufacturing systems can be found in Figure 3.

A shop floor manager has to establish not only the rules that he/she needs to incorporate in the task assigning mechanism, but also the importance factor for the company of these rules, which we called utility factor of the rule. In the Priority Rule System, these factors are taken into consideration in the second level of bid analysis, when calculating the bid weight (see [7]).

In order to help shop floor managers in their decisions, the model presented above was extended with a mechanism that offers them the possibility to see the impact of their decisions when establishing different values for utilities. They can rely their decisions on the records we provide them, or run the simulation with their own data, using a very simple user-interface we developed using the SimTalk programming language.

In order to illustrate the functioning of the model, 5 resources have been incorporated in the simulated system, each of them able to perform multiple operations. In the following exemplification, all the resources are able to perform the particular task T chosen to be analyzed. Therefore, all 5 resources will send bids for receiving the task (Figure 4).

For task T	Bid nr. 1	Bid nr. 2	Bid nr. 3	Bid nr. 4	Bid nr. 5
Processing cost	9	4	12	5	2
Env. effect	7	3	1	5	2
Unplanned breaks	1	2	1	2	3

Fig. 4. Bids sent by resources

The 6 rules incorporated in the PRS have been divided on 3 levels, according to their priority. The rules considered eliminatory, have priority 1, and the bids need to fulfill the requirements of those rules in order to be further analyzed (see [7]). The utility factors are to be considered for the rules with priority 2, i.e. the cost rule, the environmental effect rule and the trustability rule. Those rules are not analyzed separately, but as a sum of their impact, according to the formula:

$$BW = C*U_c + EnvEf*C_e + B*U_b \qquad (1),$$

where C is the cost variable, U_c the utility assigned to the cost rule, EnvEf is the environmental effect variable and C_e the utility assigned to the environmental effect rule, B represents the unplanned breaks (trustability) variable and U_b the utility assigned to the breaks rule. The result of the corresponding computation is going to be the weight of the bid (BW) according to the rule. The objective function is to minimize this weight. The utility of each rule shows its importance for the production factory. The sum of them is always 1 (100%).

In the model, values for utilities for 11 situations were incorporated (Figure 5). In case the study of other values is needed, those can be freely introduced. A screenshot of the simulated model is presented in Figure 6.

Nr.	UC	UE	UB	Comments
1	1	0	0	only cost is considered
2	0	1	0	only the env. ef. is considered
3	0	0	1	only the resource trustability is considered
4	1/3	1/3	1/3	all 3 parameters are taken into consideration, with same importance
5	0,4	0,4	0,2	the cost and environmental effects are more important than the trustability, but of the same importance comparing to each other
6	0,4	0,2	0,4	cost and trust. are more important than the env. ef., but of the same importance comparing to each other
7	0,2	0,4	0,4	env. ef. and trust. are more important than the cost, but of the same importance comparing to each other
8	0,5	0,5	0	only the cost and env. ef. are considered, with the same importance
9	0,5	0	0,5	only the cost and trustability are considered, with same importance
10	0	0,5	0,5	only trust. and env. ef. are considered, with the same importance
11	0.6	0.3	0.1	all 3 parameters are taken into consideration, with decreasing importance

Fig. 5. Utility values

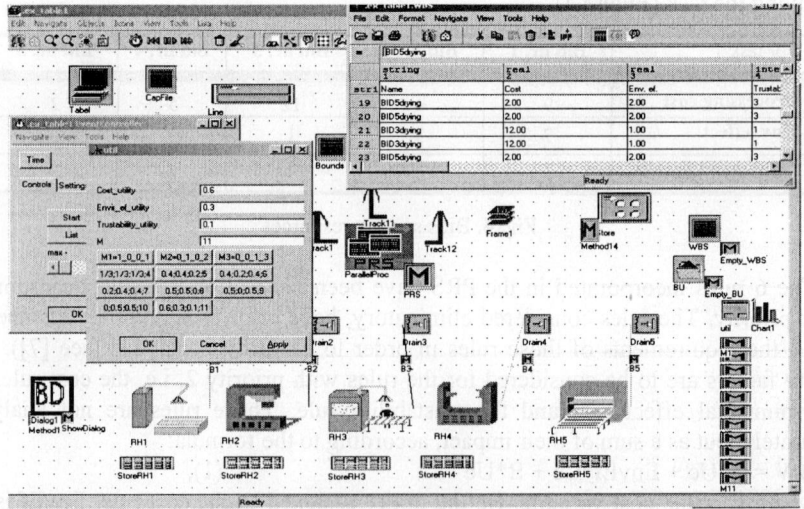

Fig. 6. Screenshot of the simulated model

The simulation has been run for all 11 situations presented above and the results (the winning bids, together with the winning resource) are gathered in Figure 7. To help decision making, the results have been represented also graphically (Figure 8), showing the cost, environmental effects and the trustability of the winning resources for all the 11 different cases. The way the change of utilities influences the weight of a certain bid can be seen in Figure 9.

Nr.	Winning bid	Cost	Environ..ef.	Trustability	Bid weight
1	BID5drying	2	2	3	2
2	BID3drying	12	1	1	1
3	BID1drying	9	7	1	1
4	BID5drying	2	2	3	7
5	BID5drying	2	2	3	2.2
6	BID5drying	2	2	3	2.4
7	BID5drying	2	2	3	2.4
8	BID5drying	2	2	3	2
9	BID5drying	2	2	3	2.5
10	BID3drying	12	1	1	1
11	BID5drying	2	2	3	2.1

Fig. 7. Winning bids, upon changing the utility's values

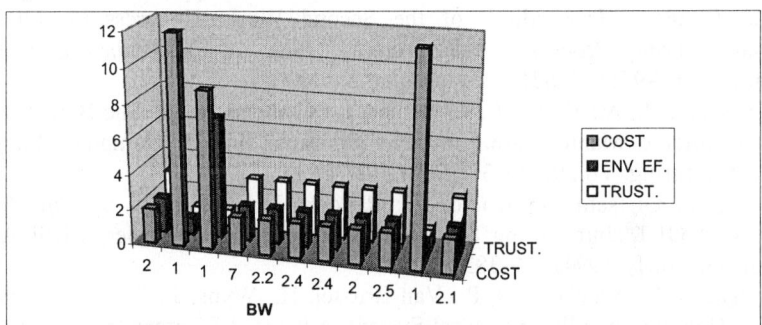

Fig. 8. Winning bids weight versus cost, env. effect., trustab., upon the change of utilities

Bid name/BW	1	2	3	4	5	6	7	8	9	10	11
BID1drying	9	7	1	17	6.6	6.6	5	8	5	4	7.6
BID2drying	4	3	2	9	3.2	3.2	2.8	3.5	3	2.5	3.5
BID3drying	12	1	1	14	5.4	5.4	3.2	6.5	6.5	1	7.6
BID4drying	5	5	2	12	4.4	4.4	3.8	5	3.5	3.5	4.7
BID5drying	2	2	3	7	2.2	2.2	2.4	2	2.5	2.5	2.1

Fig. 9. Influence of the bid weight upon the change of utilities

4 Conclusions

In the paper an approach was illustrated to deal with two important issues of today's production: *customer driven production* and *the imposition of new rules by the society,* 5such as consciousness towards the environment. Environmental aspects were introduced in agent-based dynamic production control. A decision support tool for selecting appropriate values of the utility factors in the rules was also presented, together with the related sensitivity analysis.

A more detailed elaboration of the proposed ideas is addressed in future research, and a more thorough series of experiments is foreseen.

Acknowledgements

This work was partially supported by the National Research Foundation, Hungary, Grant No. T034632.

References

1. Sihn, W.: The Fractal Company: A Practical Approach To Agility in Manufacturing, Proceedings of the Second World Congress on Intelligent Manufacturing Processes and Systems, June 10-13, Budapest, Hungary, Springer, (1997) 617-621
2. Tharumarajah, A.; Wells, A. J.; Nemes, L.: Comparison Of The Bionic, Fractal And Holonic Manufacturing System Concepts, Int. J. Computer Integrated Manufacturing, Vol. 9, No. 3, (1996) 217-226
3. Iwata, K., Onosato M., Koike, M.: Random Manufacturing System: A New Concept Of Manufacturing Systems For Production To Order, CIRP Annals, Vol. 43, No. 1, (1994) 379-383
4. Bongaerts, L., Valckenaers, P., Van Brussel, H., Wyns, J.: Schedule Execution for a Holonic Shop Floor Control System, Advanced Summer Institute 95 of the N.O.E. on Intelligent Control of Integrated Manufacturing Systems, June 24-28, Lisboa (Portugal) (1995)
5. Van Brussel, H., Wyns, J., Valckenaers, P., Bongaerts, L., Peeters, P.: Reference Architecture For Holonic Manufacturing Systems, Computers in Industry, Special Issue on Intelligent Manufacturing Systems, Vol. 37 (3), (1998) 255-274.
6. Boothroyd & Dewhurst Inc. & TNO's: Design for Environment, Wakefield, U.S.A.
7. Ilie-Zudor, E., Monostori, L., 2001, Agent-Based Support for Handling Environmental and Life-Cycle Issues, The Fourteenth International Conference on Industrial & Engineering Applications of Artificial Intelligence & Expert Systems (IEA/AIE), Lecture notes, Budapest, Hungary, 4-7 June, (1996) 812-820.
8. Henry, J., G., Heinke, G.W.: Environmental Science and Engineering, Prentice-Hall, Inc., U.S.A, (1996)

9. Onosato, M., Iwata, K.: Development Of A Virtual Manufacturing System By Integrating Product Models And Factory Models, CIRP Annals, Vol. 42, No. 1, (1993) 475-478.
10. Iwata, K., Onosato, M., Teramoto, M. K., Osaki, S.: A Modelling And Simulation Architecture For Virtual Manufacturing Systems, CIRP Annals, Vol. 44, No. 1, (1995) 399-402.

Feasibility Restoration for Iterative Meta-heuristics Search Algorithms

Marcus Randall

School of Information Technology, Bond University, QLD 4229

Abstract. Many combinatorial optimisation problems have constraints that are difficult for meta-heuristic search algorithms to process. One approach is that of feasibility restoration. This technique allows the feasibility of the constraints of a problem to be broken and then brought back to a feasible state. The advantage of this is that the search can proceed over infeasible regions, thus potentially exploring difficult to reach parts of the state space. In this paper, a generic feasibility restoration scheme is proposed for use with the neighbourhood search algorithm simulated annealing. Some improved solutions to standard test problems are recorded.

Keywords: Meta-heuristic, Combinatorial optimisation problems, Feasibility restoration, simulated annealing

1 Introduction

Iterative meta-heuristic search algorithms such as simulated annealing (SA) and tabu search (TS) are often implemented to solve combinatorial optimisation problems (COPs) using simple neighbourhood operators such as 2-opt, move and inversion [12]. In many cases, infeasible space is ignored, yet it could lead to other regions of the space in which higher quality solutions reside. One way to traverse this space is by the process of *feasibility restoration*. This technique allows the feasibility of the constraints of a problem to be broken and then brought back to a feasible state, thus traversing infeasible regions.

Feasibility restoration algorithms have typically been designed for specific problems [1,4,5,6,8,9]. A few examples will be detailed here. Chu and Beasley [5] use a heuristic to repair solutions generated by a genetic algorithm to solve generalised assignment problems (GAPs). It simply reassigns jobs from overcapacitated agents to agents that have some spare capacity. Kämpke [9] uses a similar approach in solving bin packing problems. After two items from different bins have been swapped, the capacity restriction of either of the bins may be violated. Therefore, another function is used to assign the largest item in the overfilled bin to the bin with the most spare capacity. Hertz, Laporte and Mittaz [8] use augmented local search operators on a variant of the vehicle routing problem (VRP) known as the capacitated arc routing problem. Hertz et al. [8] employ the PASTE operator which merges various vehicle routes. Often the feasibility (in the form of vehicle capacity) of the solution is violated so as such a new heuristic operator, known as SHORTEN, replaces portions of the route with shorter paths while still maintaining the required edge set.

Abramson, Dang and Krishnamoorthy [1] use an entirely different approach to any of the above. They use the general 0-1 integer linear programme (ILP feasibility restoration method of Connolly [7] (explained in Section 2), but tailor it to suit the set partitioning problem.

The characterisation of the aforementioned feasibility restoration schemes is that they are all tailored to solve specific optimisation problems. As such, if a problem definition changes or a new problem is required to be solved, new feasibility restoration algorithms would need to be developed. This paper presents the opposite approach by investigating a new generic feasibility restoration algorithm. This paper is organised as follows. Section 2 describes how feasibility restoration can be achieved in a generic manner. This is given in the context of a general modelling system for combinatorial optimisation problems (COPs) based on linked lists. Section 3 details the experiments and results of feasibility restoration using SA. Section 4 gives the conclusions.

2 Generic Feasibility Restoration

One of the first attempts at generic feasibility restoration was by Connolly [7] in a programme called GPSIMAN. GPSIMAN is a general purpose SA solver that accepts 0-1 ILPs. It incorporates a general mechanism for restoring feasibility of the system after each transition.

The feasibility restoration technique flips variables (other than the original variable changed by the SA process, called the *primary transition*) in order to obtain a new feasible solution. The scheme employed by Connolly [7] is a heuristic technique whereby a score is computed for each of the variables based on how helpful a change in the variable value would be for feasibility restoration. The most helpful variable (the one with the highest score) is flipped and the resulting feasibility/infeasibility is recalculated. If feasibility has been restored, the procedure is terminated. However, in many instances, particularly for 0-1 problems which have many related constraints, this is not the case. The algorithm proceeds to calculate the next most helpful variable. This progresses as a depth wise tree search, in which the algorithm can backtrack, should it find that the current sequence of flips cannot restore feasibility. This procedure is only useful if feasibility is relatively easy to restore, else the search for feasibility can quickly degenerate. If the process cannot restore feasibility after a fixed number of searches, then the primary transition is rejected.

According to Abramson and Randall [2], this algorithm is very slow and only effective for very small problems (such as a 5 city travelling salesman problem). In fact, they report that typically the restoration part of the algorithm takes 30% of the runtime. As such, in order for feasibility restoration to be usable, a new approach needs to be developed. The rest of this section outlines possible algorithms based on the linked list modelling system (explained next) and local neighbourhood search operators.

2.1 List Representation System

In order to develop a generic feasibility restoration scheme, it is helpful to express COPs in a uniform representation language. A language based on dynamic lists has been used to represent COPs for iterative meta-heuristics by Randall [11] and Randall and Abramson [12]. The technique can best be illustrated by an example based on the GAP, as shown in Figure 1. In a list representation, solutions can be represented using a double nested list of agents, each of which contains a list of associated jobs. Thus, in the example, jobs 1, 3, 5 and 2 are assigned to agent 1.

```
X' = X;
Status = success;
Perform the primary transition(X');
C_0 = Calculate infeasibility(X');
if (C_0 > 0)
        Status = Restore_feasibility(X,C_0,0);
End if;
if (Status == success) X = X';
End.

Function Restore_feasibility(X', C_0, level)
Perform a transition that does not disturb the elements affected by
the primary transition(X');
C_1 = Calculate infeasibility(X');
If (C_1 > C_0)
        Repeal the transition(X');
        Disallow this transition at this level;
        If (there are no more transitions at this level)
                level = level - 1;
                If (level == 0) return failed;
        End if;
        Return Restore_feasibility(X', C_1, level);
Else
        If (C_1 == 0)
                Return success;
        Else
                Return Restore_feasibility(X', C_1, level + 1);
        End if;
End if;
End Resore_feasibility;

Where:
        X is the solution.
```

Fig. 1. A 5 job, 4 agent GAP represented as a linked list

In linked list notation, the objective of the GAP can be expressed according to Equation 1.

$$Minimise \sum_{i=1}^{M} \sum_{j=1}^{|x(i)|} C(i, x(i,j)) \qquad (1)$$

Where:

x is the solution list. $x(i,j)$ is the jth job performed by agent i.
C is the cost matrix. $C(i,j)$ is the cost of assigning job j to agent i.
M is the number of agents.

Whilst this list notation appears similar to conventional vector form used in standard linear programmes, each sub-list contains a variable number of elements[1]. Thus, the second summation sign in 1 requires a bound which varies depending on the length of each sub-list (i.e. $|x(i)|$) and changes according to the current solution state. Similarly, the constraints concerning the capacity of each agent are formed across list space. In this example, Equations 3 - 6 are the *list constraints* while Equation 2 represents the *problem constraints*.

$$\sum_{j=1}^{|x(i)|} a(i, x(i,j)) \leq b(i) \qquad \forall i \quad 1 \leq i \leq M \qquad (2)$$

$$|x| = M \qquad (3)$$

$$1 \leq x(i,j) \leq N \qquad \forall i,j \quad 1 \leq i \leq M \quad 1 \leq j \leq |x(i)| \qquad (4)$$

$$min\_count(x) = 1 \qquad (5)$$

$$max\_count(x) = 1 \qquad (6)$$

Where:

a is the resource matrix. $a(i,j)$ is the resource required by agent i to perform job j.
b is the capacity vector. $b(i)$ is the capacity of agent i.
N is the number of jobs.

Equation 3 ensures that the number of agents remains static throughout the search process while Constraint 4 defines the range of element values between 1 and N. The functions $min\_count()$ and $max\_count()$ ensure that each value (as specified in Constraint 4) appears at least and at most once in the list structure (i.e. every job is represented once).

Due to the list representation, neighbourhood search operators can be applied directly to the solution. Table 1 describes each transition operator.

---

[1] An element is defined as a value occurring at a particular sub-list and position. As such, each element is unique.

Table 1. Neighbourhood search operators suitable for list representations

Operator	Description
Move	An element is moved from one list to the end of another list. This is similar to Osman's shift process [10].
Swap	The positions of two elements, from the same or different lists, are swapped. This is equivalent to 2-opt (which is generalisable the n-opt).
Inversion	The sequence between two elements on the same sub-list is reversed.
Reposition	The position of an element in a list is changed.
Add	An element is added to a list.
Drop	An element is removed from a list.
Change	The value of a list element is changed to another value.

2.2 Generic Feasibility Restoration Algorithm

As discussed in Section 1, the aim of feasibility restoration is to traverse infeasible space in order to find another feasible pocket of solutions that may be inaccessible otherwise. It is possible to apply feasibility restoration techniques to both the list constraints and problem constraints. In the implementation described in this paper, only the operation regarding the latter type is considered.

The process of feasibility restoration consists of first breaking and then repairing feasibility. A transition is made that potentially violates some or all of the problem constraints. A series of feasibility maintaining transitions is subsequently undertaken in order to return to a feasible state.

As noted in Section 2, Connolly's [7] algorithm uses a feasibility restoration scheme based on backtracking. The central idea in backtracking approaches is that the primary transition must remain after the restoration process has been completed. If feasibility cannot be restored, then the primary transition is undone. The generic backtracking feasibility restoration algorithm that is applicable for local search transition operators is given in Figure 2. Note, the amount of feasibility is calculated by comparing the leht-hand side to the right-hand side of each constraint and considering the relational operator [12].

Consider the example in Figure 3 that uses the swap operator. The example does not refer to a particular problem, though it could represent a single truck tour in a VRP under a tour length constraint.

Modified Algorithm The algorithm in Figure 2, like its Connolly [7] counterpart, will require much computational effort. Hence, a more efficient modified algorithm will be presented here. The move operator will be used to demonstrate this approach (Figure 4). In the context of this operator, the algorithm is useful for problems such as the GAP and VRP in which elements can be moved between lists.

```
X' = X;
Status = success;
Perform the primary transition(X');
C_0 = Calculate infeasibility(X');
if (C_0 > 0)
        Status = Restore_feasibility(X,C_0,0);
End if;
if(Status == success) X = X';
End.

Function Restore_feasibility(X', C_0, level)
Perform a transition that does not disturb the elements affected by
the primary transition(X');
C_1 = Calculate infeasibility(X');
If (C_1 > C_0)
        Repeal transition(X');
        Disallow this transition at this level;
        If (there are no more transitions at this level)
                level = level - 1;
                If (level == 0) return failed;
        End if;
        Return Restore_feasibility(X', C_1, level);
Else
        If (C_1 == 0)
                Return success;
        Else
                Return Restore_feasibility(X', C_1, level + 1);
        End if;
End if;
End Resore_feasibility;

Where:
        X is the solution.
```

Fig. 2. A generic feasibility restoration algorithm

3 Computational Experience

The solver software used here is the COSULOM (Combinatorial Optimisation Solver Using List Oriented Meta-heuristics) package developed by Randall and Abramson [12]. It is a general purpose system based on the linked list system and has both SA and TS search engines. The computing platform is an SUN SPARCstation 4 running at 110 MHz.

The problems that are tested are instances of the GAP and VRP (as these can utilise the move operator). Table 2 gives the name of each instance, its size and the best-known/Optimal cost. The GAP and VRP problem instances are

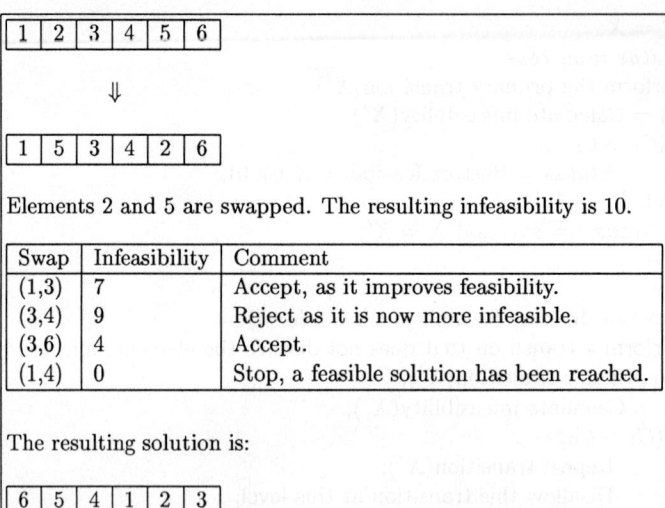

Fig. 3. Example of a backtracking feasibility restoration algorithm using the swap operator

```
Move an element to another sub-list;
Check the constraints associated with the old sub-list;
If (these constraints are violated)
        Attempt to add a combination of elements on this sub-list;
        If (this fails) try to do the same thing except remove the
        elements off the old sub-list;
        If (the old sub-list is now feasible) perform
        feasibility maintenance with the displaced elements;
        If (this fails) abort the restoration, reinstate the
        original solution and exit;
End if;
Check the constraints associated with the new sub-list;
If (these constraints are violated)
        Attempt to add a combination of elements from this list;
        if (the new sub-list is now feasible) perform
        feasibility maintenance with the displaced elements;
        If (this fails) try to add to this sub-list a combination of
        elements;
        If (this fails) abort the restoration, reinstate the
        original solution;
End If;
End.
```

Fig. 4. A modified feasibility restoration algorithm for the move operator

from Beasley [3] and Reinelt [13] respectively. In this paper, we only utilise the SA search engine.

Table 2. Problem instances used in this study

Problem	Instance	Size	Optimal/Best-Known Cost
GAP	gap1-1	15 jobs,5 agents	336
	gap2-1	20 jobs,5 agents	434
	gap3-1	25 jobs,5 agents	580
	gap4-1	30 jobs,5 agents	656
	gapA5-100	100 jobs,5 agents	1698
	gapA5-200	200 jobs,5 agents	3235
	gapA10-100	100 jobs,10 agents	1360
	gapA10-200	200 jobs,10 agents	2623
	gapA20-100	100 jobs,20 agents	1158
	gapA20-200	200 jobs,20 agents	2339
	gapC5-100	100 jobs,5 agents	1931
	gapC5-200	200 jobs,5 agents	3458
	gapC10-100	100 jobs,10 agents	1403
	gapC10-200	200 jobs,10 agents	2814
	gapC20-100	100 jobs,20 agents	1244
	gapC20-200	200 jobs,20 agents	2397
VRP	eil30	29 customers,3 trucks	545
	eil51	50 customers,5 trucks	521
	eild76	75 customers,7 trucks	692
	eilb101	100 customers,14 trucks	1114
	gil262	261 customers,25 trucks	?

Each problem uses a different set of transition operators. The operators and their probabilities have been found to give good performance in Randall [11]. The VRP uses the move, swap, inversion and reposition transition operators (with equal probabilities) whereas GAP uses only move and swap (with probabilities of 0.2 and 0.8 respectively). For more information regarding transition probabilities, see Randall and Abramson [12]. Note, only move will perform feasibility restoration transitions.

In order to describe the range of costs gained by these experiments, the minimum (denoted **Min**), median (denoted **Med**), maximum (denoted **Max**) and Inter Quartile Range (denoted **IQR**) are given (as each problem instance is run with 10 random seeds). Non-parametric descriptive statistics are used as the data are highly non-normally distributed. Table 3 shows the results of the feasibility maintenance while Table 4 shows the results for feasibility restoration.

Table 3. Feasibility Maintenance Results with SA

Problem Instance	Optimal Cost	Cost Min	Med	Max	IQR
gap1-1	336	363	363	363	0
gap2-1	434	434	434	434	0
gap3-1	580	580	580	580	0
gap4-1	656	656	656	656	0
gapA5-100	1698	1698	1700	1702	1
gapA5-200	3235	3235	3235.5	3237	1
gapA10-100	1360	1360	1361	1363	1.5
gapA10-200	2623	2623	2627	2629	4.25
gapA20-100	1158	1158	1158	1158	0
gapA20-200	2339	2339	2340	2344	2.75
gapC5-100	1931	1931	1934.5	1941	2.75
gapC5-200	3458	3432	3435.5	3442	3.75
gapC10-100	1403	1371	1382.5	1385	6.25
gapC10-200	2814	2795	2798	2806	5.5
gapC20-100	1244	1243	1246.5	1250	3.75
gapC20-200	2397	2394	2397	2400	2
eil30	545	534	535	538	3.75
eil51	524	529	545	563	18.75
eild76	692	703	740	779	35
eilb101	1114	1129	1152	1174	15.75

Table 4. Feasibility Restoration Results with SA

Problem Instance	Optimal Cost	Cost Min	Med	Max	IQR
gap1-1	336	336	336	336	0
gap2-1	434	434	434	434	0
gap3-1	580	580	580	580	0
gap4-1	656	656	656	656	0
gapA5-100	1698	1698	1700	1703	2.5
gapA5-200	3235	3235	3235.5	3236	1
gapA10-100	1360	1360	1360	1361	1
gapA10-200	2623	2623	2626	2629	4.75
gapA20-100	1158	1158	1158	1158	0
gapA20-200	2339	2339	2339	2339	0
gapC5-100	1931	1931	1936	1942	6.25
gapC5-200	345	3432	3437.5	3442	4
gapC10-100	1403	1382	1384.5	1389	2.25
gapC10-200	22814	796	2800	2807	4.75
gapC20-100	1244	1244	1247.5	1249	3.75

4 Conclusions

This paper described the use of a generic feasibility restoration techniques for meta-heuristic search engines such as SA. At the present time, code has been developed within the COSULOM framework to perform generic feasibility restoration with the move local search operator. The next step in this development is to produce restoration techniques for all operators that follow (more or less) the algorithm in Figure 2.

The results show that optimal and near optimal solutions are gained using the feasibility restoration technique. However, a cursory comparison on runtimes with Randall and Abramson [12] reveals that this algorithm is slower. There is no question though that it is an improvement on Connolly's GPSIMAN. Work needs to be done in order to increase the efficiency of the restoration algorithm. One way would be to employ feasibility restoration operations less frequently. For instance, it could be used in order to escape a local optimum.

The algorithms described in this paper each utilise one transition operator at a time. It may be possible to use a variety of operators and calculate which one is likely to restore feasibility the fastest. In addition, more research needs to be undertaken to establish a cost/benefit comparison against feasibility maintaining operations, those which are only made if feasibility is preserved. This needs to be carried out across a range of problems and use different meta-heuristics (such as TS).

Another way to approach feasibility restoration is too modify the generic initial feasible solution technique developed by Randall and Abramson [12]. In this technique, the objective function was the sum of the constraint violations. When this becomes 0, a feasible solution is generated. Therefore, a new generic algorithm (as given in Figure 5) could be trialed.

```
Function solution_feasible
      While (solution is feasible)
            Apply feasibility restoration transitions (using normal
            objective function);
      End while;
End Function.

Function solution_infeasible
      While (solution is not feasible)
            Apply transitions using the modified (sum of constraints)
            objective function;
      End while;
End Function.
```

Fig. 5. A potential feasibility restoration algorithm

References

1. Abramson, D., Dang, H. and Krishnamoorthy, M. (1996) "A Comparison of Two Methods for Solving 0-1 Integer Programs Using a General Purpose Simulated Annealing", Annals of Operations Research, 63, pp. 129-150.
2. Abramson, D. and Randall, M. (1999) "A Simulated Annealing Code for General Integer Linear Programs", Annals of Operations Research, 86, pp. 3-21.
3. Beasley, J (1990) "OR-library: Distributing Test problems by Electronic Mail", Journal of the Operational Research Society, 41, pp. 1069-1072.
4. Beasley, J. and Chu, P. (1996) "A Genetic Algorithm for the Set Covering Problem", European Journal of Operational Research, 94, pp. 392-404.
5. Chu, P. and Beasley, J. (1997) "A Genetic Algorithm for the Generalised Assignment Problem", Computers and Operations Research, 24, pp. 17-23.
6. Chu, P. and Beasley, J. (1998) "A Genetic Algorithm for the Multidimensional Knapsack Problem", Journal of Heuristics, 4, pp. 63-86.
7. Connolly, D. (1992) "General Purpose Simulated Annealing", Journal of the Operational Research Society, 43, pp. 495-505.
8. Hertz, A., Laporte, G. and Mittaz, M. (2000) "A Tabu Search Heuristic for the Capacitated Arc Routing Problem", Operations Research, 48, pp. 129-135.
9. Kämpke, T. (1988) "Simulated Annealing: A New Tool in Bin Packing", Annals of Operations Research, 16 pp. 327-332.
10. Osman, I. (1995) "Heuristics for the Generalised Assignment Problem: Simulated annealing and Tabu Search Approaches", OR Spektrum, 17, pp. 211-225
11. Randall, M. (1999) "A General Modelling Systems and Meta-heuristic based Solver for Combinatorial Optimisation Problems," PhD thesis, Faculty of Environmental Science, Griffith University. Available online at: http://www.it.bond.edu.au/randall/general.pdf
12. Randall, M. and Abramson, D. (2000) "A General Meta-heuristic Based Solver for Combinatorial Optimisation Problems", Journal of Computational Optimization and Applications, 20 (to appear).
13. Reinelt, G. (1991) "TSPLIB - a Traveling Salesman Problem Library", ORSA Journal on Computing, 3, pp. 376-384. Available on-line at http://softlib.rice.edu/softlib/tsplib.

Optimization of Pulse Pattern for a Multi-robot Sonar System Using Genetic Algorithm

George Nyauma Nyakoe, Makoto Ohki,
Suichiro Tabuchi, and Masaaki Ohkita

Dept. of Electrical and Electronic Engineering, Faculty of Engineering
Tottori University, 4-101 Koyama Minami Tottori 680-8552, Japan
{m98t3032,tabuchi}@mawell.ele.tottori-u.ac.jp
{mohki,mohkita}@ele.tottori-u.ac.jp

Abstract. Sonar sensors are widely used in mobile robotics research for local environment perception and mapping. Mobile robot platforms equipped with multiple sonars have been build and used by many researchers. A significant problem with the use of multiple sonars is that, when the sonars are operated concurrently, signal interference occurs, making it difficult to determine which received signal is an echo of the signal transmitted by a given sensor. In this paper, a technique for acquiring suitable modulation pulses for the signals emitted in a multi-sonar system is presented. We propose a technique to reduce the probability of erroneous operation due to interference by satisfying conditions for minimizing the signal length and the variation in the signal length of each sonar, using the Niched Pareto genetic algorithm (NPGA). The basic technique is illustrated for the case where two or more robots operate in the same environment.

1 Introduction

Sonar sensors are widely used for mobile robot environment recognition and mapping due to their low cost, high speed and simple output. Usually, a mobile robot is equipped with two or more sonars for accurate and precise range measurement [1],[2]. When sonars are fired concurrently, although the measurement time is shortened, the probability of obtaining incorrect measurements is high as a result of signal interference among signals transmitted from different sensors.

A desirable feature in performing accurate range measurement in the presence of signal interference is to make the variation in the signal length between sonars as small as possible [3]. Generally, acquisition of a pulse pattern with only a small variation in signal length between the sensors is considered to be difficult. The problem is further complicated in the case where two or more robots exist in the same environment. This problem is treated as a mult-objective combinatorial optimization problem, which can be solved with the Niched Pareto Genetic Agorithm (NPGA) [5]. An efficient method of implementing tournament selection, and a simplified sharing technique for obtaining the set of optimal solutions for the pulse pattern in a short time, are described.

2 Range Measurement Using Multiple Sonars

Sonars measure distance to objects by computing the *time of flight* (TOF) of the transmitted signal echoed back from an object and received by the sonar. The use of multiple concurrently operated sonars results in mutual interference of echo signals. Considering one sonar, the echo received is composed of the reflected wave transmitted by that sonar and a sum of the reflected waves transmitted from other sonars. The undesirable effect of signal interference can be avoided by applying a pulse-modulated signal to each sonar module in Fig. 1, such as that shown in Fig. 2.

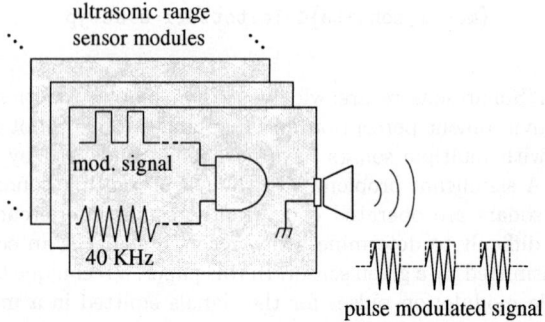

Fig. 1. Ultrasonic (sonar) sensor modules

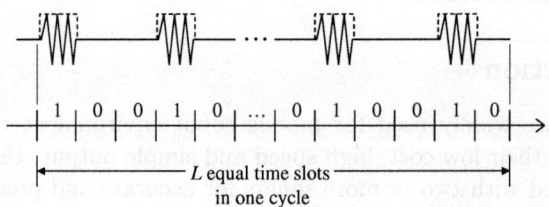

Fig. 2. Composition of a pusle modulation sequence

The pulse-width-modulated carrier signal is divided into L intervals or time slots whose lengths equal the length of one pulse. The modulation pulses of each sonar are formed by assigning 0 or 1 to the time slots, as shown in Fig. 2.

We define S as the number of sonars, T_s as a source signal transmitted from the s-th sonar, R_s as a signal received at the s-th sonar and r_{qs} as a signal transmitted from the q-th sonar, propagated on an acoustic path and received at the s-th sonar. Assuming that the S sonars operate concurrently, the signal $R_s(i)$ received by the s-th sonar is given by the summation $\sum_q r_{qs}$, $q = 1, 2, ..., S$, of the received signals and is expressed by the following equation:

$$R_s(i) = \begin{cases} 1 & \left(\sum_{j=1}^{S} r_{qs}(i) \geq 1\right) \text{ or } \left(\sum_{q=1}^{S} r_{qs}(i) = 0\right) \\ 0 & \end{cases} \quad (1)$$

where i denotes a time delay given by the time slot number. A valid distance measurement for each sonar is obtained by considering the correlation between the transmitted signal T_s and the received signal R_s, which is computed using the expression given by Eq. (2).

$$C(T_s, R_s, i) = \sum_{j=0}^{L-1} R_s(j)T_a(j+i) \quad (i = 0, 1, 2, \cdots). \quad (2)$$

The distance is determined according to the time delay i resulting in the highest value of the correlation C for all i or a value larger than a threshold α. To completely determine the time delay between the source signal and the received signal on each sonar, the following conditions must be satisfied:

$$\sum_{i=0}^{L-1} T_s(i)T_s(i+j) \begin{cases} = P \ (j = 0), \\ \leq 1 \ (1 \leq j \leq L-1), \end{cases} \quad (3)$$

$$\sum_{i=0}^{L-1} T_s(i)T_q(i+j) \leq 1 \ (1 \leq j \leq L-1) \quad (4)$$

where P denotes the number of pulses and T_q denotes a source signal transmitted by the q-th sonar. In the cases where signals satisfying Eqs. (3) and (4) are used, the value of the correlation C between different signals is less than or equal to one. If the number of sonars, S, is less than the number of pulses, P, signals with the same form of T_s cannot be composed. However, if the condition $P \geq S+1$ is satisfied, erroneous measurement can be completely avoided. Therefore, the first step to optimization of the pulse pattern is to acquire modulation pulses satisfying the above conditions.

3 Acquisition of Modulation Pulses

The technique of acquiring modulation pulses satisfying Eqs. (3) and (4) is illustrated with the example shown in Table 1 where $S = 3$ and $P = 4$. This is referred to as the primitive pulse interval sequence. In the Table, a row represents the sensor number, s, and a column represents the pulse interval number, p, which corresponds to the selection number on the checklist shown in Table 2. A similar treatment can be extended to an arbitrary number of sensors and pulses.

So as to satisfy the conditions given in Eqs. (3) and (4), the primitive pulse interval sequence undergoes the transformations explained in the following procedures. The transformation is performed on the basis of selection numbers in

Table 1. Example of a primitive pulse interval sequence

s \ p	1	2	3
1	3	13	8
2	4	5	17
3	2	10	15

Table 2. Checklist, Q, for the transformation under the initial condition

n	1	2	3	4	5	6	⋯
$Q(n)$	0	0	0	0	0	0	⋯

the checklist $Q(n)(n \in N)$ shown in Table 2. In the given procedures, h_{sp} and g_{sp} denote the selection number and the interval length, respectively, where s and p denote the sonar number and the interval number, respectively. The operation for a given sonar s is performed in the order $h_{s1}, h_{s2}, \cdots, h_{sp}$.

(**A1**) Initialize $s = 1$, $p = 1$, $\forall n \in \mathbf{N}.Q(n) = 0$.

(**A2**) Refer to h_{sp}. Identify the h_{sp}-th vacant position of the checklist, $Q(v)$, and assume the value of g_{sp} as v.

(**A3**) Examine the values $Q(m)$ at the m-th position in the checklist, where m is given by the following expression:

$$m = \sum_{k=0}^{u} g_{s(p-k)}, \quad \forall u \in \mathbf{U}_p = \{0, 1, \cdots, p-1\}. \tag{5}$$

(**A4**) In the case that (**A3**) yields $Q(m) = 0$ for all m obtained from Eq. (5), that is,

$$\forall u \in \mathbf{U}_p.Q(m) = 0, \tag{6}$$

then let $Q(m) = 1$ for all m, determine g_{sp} as v and go to (**A6**). Otherwise, proceed to (**A5**).

(**A5**) If one or more positions on the checklist $Q(m)$, where m is given by Eq. (5), are already 1, that is,

$$\exists u \in \mathbf{U}_p.Q(m) = 1, \tag{7}$$

increase h_{sp} by 1 and return to (**A2**).

(**A6**) If $p < P$, increase p by 1 and return to (**A2**). Otherwise, proceed to (**A7**).

(**A7**) If $p = P$ and $s \leq S$, increase s by 1, let $p = 1$, and return to (**A2**). Otherwise, end.

By applying procedures (**A1**)–(**A7**), appropriate pulse intervals that meet the requirements of Eqs. (3) and (4) can be uniquely acquired from arbitrary selection numbers that form the genotype code.

4 Multi-Objective Optimization of Pulse Pattern for Multiple Robots

When two or more robots equipped with multiple sonars exist in the same environment, the likelihood of performing erroneous measurement is very high. An example of such an arrangement is shown in Fig. 3.

In this paper, the idea of Pareto optimality used in solving multi-objective optimization problem [5],[6],[7] is applied in acquiring a pulse pattern that simultaneously minimizes the objective functions, i.e., the signal length L and the probability of erroneous operation E. When actually using the pulse sequence decoded from the genotype code proposed in [4], since the values of a gene occupy predetermined positions in the checklist, the variation in the signal length of each sensor results in an undesirably big difference in the length of a signal appears. The variation in the length of the signal, D, can be expressed by the following equations.

$$D = \frac{1}{L}\left(\frac{1}{S}\sum_{i=1}^{S}(\bar{l}-l_i)^2\right)^{\frac{1}{2}} \quad (8)$$

$$\bar{l} = \frac{1}{S}\sum_{i=1}^{S}l_i \quad (9)$$

where, l_i denotes the signal length of the i-th sensor and $\bar{l}$ denotes the average value of signal length.

To compensate for the difference in signal length, the variation, D, in the signal of each sensor is considered as an additional objective function. A non-dominated set of solutions providing a trade-off in the three objective functions of the optimization problem is required. We apply the Niched Pareto Genetic

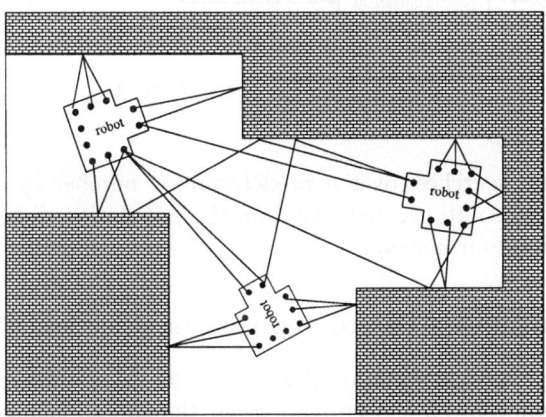

Fig. 3. Illustration of signal interference during range measurement for two or more robots existing in the same environment

Algorithm (NPGA) [5], which combines tournament selection and a simplified sharing mechanism, to determine the set of non-dominating trade-off solutions to the multi-objective optimization. We introduce an alternative selection technique called Point Assignment Pareto Tournament aimed at reducing the computation time of the Pareto domination tournament used by Horn et al.

5 Acquisition of Pulse Patterns Using Niched Pareto Genetic Algorithm

5.1 Outline of the Niched Pareto Genetic Algorithm (NPGA)

The NPGA implements a technique known as Pareto domination tournament where two competing individuals are selected and compared against a comparison set comprising of a specific number, (t_{dom}), of individuals randomly picked from the population. The non-dominated candidate individual is selected for reproduction. If both candidates are either non-dominated or dominated, fitness sharing is performed to determine a winner. The outline of the NPGA is shown in Fig. 4.

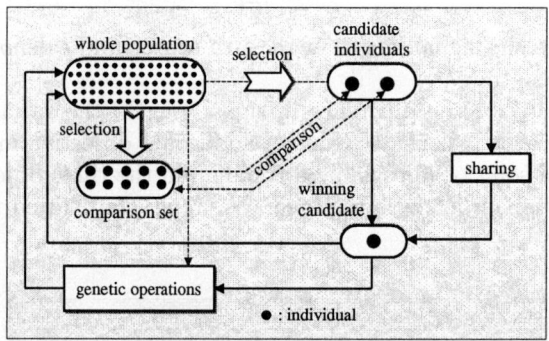

Fig. 4. Outline of the NPGA

The algorithm searches, over a predetermined number of generations, for the non-dominated solutions that maintain the population diversity during the Pareto domination tournaments.

5.2 Point Assignment Pareto Tournament Selection

When using the Pareto domination tournament [5], the domination relation between a candidate and all the individuals in the comparison set must be determined. When dealing with a large comparison set, or when there are many objective functions, a lot of computation is needed for tournament selection. To

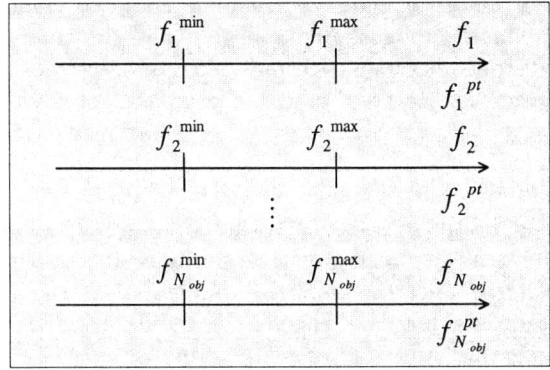

Fig. 5. The Point Assignment Pareto Tournament selection

deal with this problem, we propose the Point Assignment Tournament Selection technique, which is illustrated in Fig. 5.

The technique is clearly represented by the following procedures:
(1) Considering each objective function $f_i, (i = 1, 2, \cdots, N_{object})$, for each individual in a comparison set, a rank is attached to each value in the objective function. For every rank of each objective function, the maximum and minimum values, f_i^{max} and f_i^{min}, of the individuals in the comparison set are obtained.
(2) Comparison of the maximum and minimum values f_i^{max} and f_i^{min} obtained in (1), and the value of f_i of a candidate individual. For each objective function, if the value f_i of a candidate individual is larger than the maximum value f_i^{max}, the individual is given 0 points. If the value f_i lies between f_i^{max} and f_i^{min}, the individual gets 1 point. If f_i is smaller than f_i^{min}, it is assigned 2 points. The point value assigned to a candidate is denoted f_i^{point}.
(3) Determination of the winner. The candidate with the highest value of points, $F_i = \sum_{i=1}^{N_{object}} f_i^{point}$, is selected as the winner.

5.3 Fitness Sharing

In the fitness sharing used by Horn et al. [5], the individual with the least niche count is determined as the winner. The niche count m_i is calculated over all individuals in the population as given by the following expression:

$$m_i = \sum_{j=1}^{N} sh(d_{ij}), \qquad (10)$$

The fitness sharing function $sh(d_{ij})$ is expressed as:

$$sh(d_{ij}) = \begin{cases} 1 - d_{ij}/\sigma_{share} & (d_{ij} \leq \sigma_{share}), \\ 0 & (d_{ij} > \sigma_{share}) \end{cases} \qquad (11)$$

where, d_{ij} denotes the distance between individuals i and j and σ_{share} denotes the niche radius, which decays exponentially.

Because the distance between each individual and every other individual in the whole population has to be obtained, the amount of computation is quite enormous. In this paper, we consider a fitness sharing mechanism where the number of individuals in a niche domain is counted without the need to calculate the distance between individuals. As shown in Fig. 6, only the individuals in the domain given by the niche size of a candidate individual are counted. The candidate individual in the direction with fewer individuals within the domain is considered the winner. When the number of individuals in the domain centering on each candidate individual is the same, a candidate individual is randomly picked for reproduction. An example of this kind of sharing for candidate individuals β and γ is shown in Fig. 6. For each candidate individual, the niche domain determined by σ_{f_1} and σ_{f_2} is considered, and the candidate individual β with fewer and closer individuals in each niche domain is chosen as the winner.

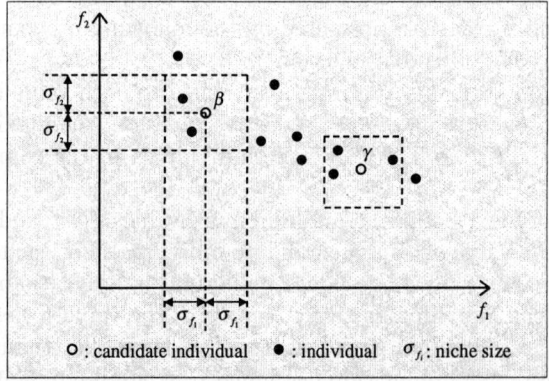

Fig. 6. Fitness sharing for candidate individuals β and γ

6 Optimization Result of the Pulse Patterns Using NPGA

6.1 Objective Functions for the Pareto Tournament Selection

In this optimization problem, the objective functions considered are: the signal length of each sensor L, the probability of erroneous operation E and the variation in the signal length D. The corresponding values of *nichesize* are: $\sigma_L = 20$, $\sigma_E = 0.05$ and $\sigma_D = 0.005$, respectively. We consider a multirobot system comprising three mobile robots each equipped with six sonars, i.e., $S = 18$. The number of pulses P is assumed to be 7. The number of individuals is considered as 100 and the size of the comparison set 10. The search is performed for 10000

generations. Two different crossover operators C_s and C_p determined according to probabilities $P_{cs} = 0.1$ and $P_{cp} = 1/(P-1)$, repectively, where P denotes the number of pulses, and a mutation operator M_u selected according to the mutation probability $P_M = 0.01$, are used as the genetic operators [4]. To ensure that individuals with short signal lengths are obtained from the entire population, two types of searches are implemented as follows :
(1) **Search of the first half of a generation:** All three objective functions are considered in the Pareto tournament selection (L_E_D).
(2) **Search of the second half of a generation:** Only two objective functions are considered in the Pareto tournament selection (L_E).

6.2 Result of the Optimization

The pulse pattern which provides the solution that mutually satisfies each of the objective functions was acquired using NPGA. From the results shown in Figs. 7 and 8, since the signal length L, the probability of erroneous operation E and the variation of the signal of each sensor D, are considered as the objective functions, we obtain a solution inclined towards the direction where E and D are minimized. Because there is a tendency for L to decrease when D increases, and since E is directly dependent on the latter, the Pareto tournament selection seldom gives the individual with short signal length as the winner. This arises because individuals with short signal lengths diminish as optimization advances as can be noted from the results . However, as shown in Figs. 9 and 10, since the Pareto tournament selection performed in the second half of the generation considers L and E as the objective functions, the solution is also distributed over the direction where L is shortened. Based on these observations, by applying the modified NPGA proposed in this paper, it is possible to acquire pulse patterns having a small variation in the signal of each sensor with a trade-off between the signal length and the probability of erroneous operation.

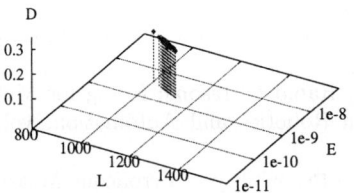

Fig. 7. Distribution of the solution (L_E_D) in the initial generation

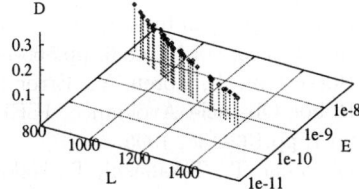

Fig. 8. Distribution of the solution (L_E_D) in the 9999-th generation

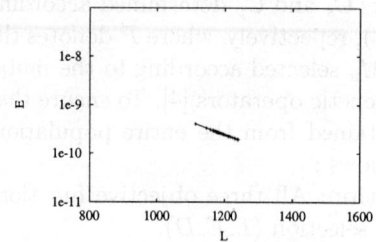
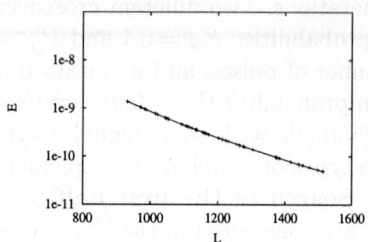

Fig. 9. Distribution of the solution (L_E) in the initial generation

Fig. 10. Distribution of the solution (L_E) in the 9999-th generation

7 Conclusion

In this paper, a technique of acquiring a pulse pattern for multiple robots, each of which is equipped with multiple concurrently operated sonars, was proposed. In the proposed technique, a pulse pattern which simultaneously reduces the probability of erroneous operation due to signal interference and provides short signal lengths, is realized. In acquiring the pulse pattern, a trade-off solution is obtained satisfying three objective functions that constitute the optimization problem, i.e., the signal length, the probability of erroneous operation and the variation in the signal length of each sensor. The Niched Pareto Genetic Algorithm based on point scores was applied in computing the set of Pareto optimal solutions of the pulse pattern. The Point Assignment Pareto Tournament selection technique can be thought to be fast in computation since it avoids calculation of distances between all individuals in the population performed in the Pareto domination tournament of Horn et al.

References

1. Barshan, B., Kuc, R.: Differentiating Sonar Reflections from Corners and Planes by Employing an Intelligent Sensor, IEEE Trans. on Pattern Analysis and Machine Intelligence, vol.12, no.6, pp.560-569, 1990.
2. Borenstein, J., Koren, Y.: Error Eliminating Rapid Ultrasonic Firing for Mobile Robot Obstacle Avoidance, IEEE Trans. on Robotics and Automation, vol.11, no.1, pp.132-138, 1995.
3. Ishiyama, T., Takahashi, T., Nakano, E.: The Probability of Erroneous Measurement Caused by Cross-talk in Multiple Sonars, Journal of Robotics Soc. of Japan, vol.17, no.4, pp.526-533, 1999 (in Japanese).
4. Nyauma, G. N., Ohki, M., Tabuchi, S., Ohkita, M.: Generation and Optimization of Pulse Pattern for Multiple Concurrently Operated Sonars Using Genetic Algorithm, IEICE Trans. Fundmentals, vol.E84-A, No.7 July 2001.
5. Horn, J., Nafpliotis, N., Goldberg, D.: A Niched Pareto Genetic Algorithm for Multiobjective Optimaization, IEEE Conference on Evolutionary Computation pp.82-97, 1994.

6. Fonseca, C. M., Fleming, P. J.: Genetic Algorithms for Multiobjective Optimization, Proc. 5th International Conference on Genetic Algorithms , pp.416-423, 1993.
7. Goldberg, D. E., Richardson, J. J.: Genetic Algorithms with Sharing for Multimodal Function Optimisation, Proc. 2nd InternationalConference on Genetic Algorithms, Cambridge, MA, July, 1987 (Lawrence Erlbaum).
8. Louis, S. J., Rawlins, G. J. E.: Pareto Optimality, GA-Easiness and Deception , Proc. of the Fifth International Conference on Genetic Algorithms , pp.118-123, 1993.

The Suitability of Particle Swarm Optimisation for Training Neural Hardware

David Braendler and Tim Hendtlass

Centre for Intelligent Systems and Complex Processes
Swinburne University of Technology, Australia

Abstract. Learning algorithms implemented for neural networks have generally being conceived for networks implemented in software. However algorithms which have been developed for software implementations typically require far greater accuracy for efficiently training the networks than can be easily implemented in hardware neural networks. Although some learning algorithms designed for software implementation can be successfully implemented in hardware it has become apparent that in hardware these algorithms are generally ill suited, failing to converge well (or at all). Particle Swarm Optimisation (PSO) is known to have a number of features that make it well suited to the training of neural hardware. In this paper the suitability of PSO to train limited precision neural hardware is investigated. Results show that the performance achieved with this algorithm does not degrade until the accuracy of the networks is reduced to a very small number of bits.

1 Desired Characteristics of Learning Algorithms

Artificial Neural Networks (ANNs) encode the knowledge needed to perform a task in terms of weights on connections between units in a network. For tasks that involve fairly simple constraints between inputs and outputs, it is sometimes possible to analytically derive a set of weights that is guaranteed to cause the network to settle into good solutions. However for tasks involving more complex relationships between inputs and outputs, correct behaviour requires such highly-complex interactions among weights that it becomes infeasible to hand-specify them. In this case, it is necessary to rely on a learning procedure that takes these interactions into account in deriving an appropriate set of weights.

A primary challenge that faces the designer of hardware neural networks is finding a learning algorithm compatible with the limited precision inherent in hardware [5].

2 Learning in Neural Networks

The purpose of machine learning algorithms is to use observations (experiences, data, patterns etc.) to improve a performance element, which determines how the network reacts when it is given particular inputs. The performance element may be a simple classifier trying to classify an input instance into a set of categories or it may be a complete agent acting in an unknown environment. By receiving feedback on the performance, the learning algorithm adapts the performance element to enhance its capabilities.

Learning in neural networks adjusts weights between connections in response to the input of training patterns, and the algorithms can be classified as supervised, reinforcement, or unsupervised.

It was decided to limit investigation to *supervised* learning, as this is simple to evaluate This type of learning can be considered as an unconstrained non-linear minimisation problem in which the objective function is defined by the error function, and the search space is defined by the weight space. The terrain modelled by the error function in its weight space (herein referred to as the 'error function') can be extremely rugged and have many local minima. A study of error functions in the supervised learning of neural networks by Shang [14] revealed the following features that may trap or mislead training algorithms.

- Flat regions.
- Numerous local minima.
- Deep but sub-optimal valleys.
- Gradients that differ by many orders of magnitude.

The learning algorithm employed should be able to escape the difficult features of the error function evidenced by neural networks. Additionally for hardware neural networks (HNNs) it is desirable that the algorithm has the following characteristics:

- Require no more computations to deal with limited precision as compared to the unlimited precision case.
- Be relatively simple so that it can be implemented in hardware.
- Use model free learning to take into account the non-idealities of the hardware.[1]

Learning algorithms used for such optimisation problems can be classified as local, global or hybrid. Local optimisation algorithms find local minima efficiently and work best on unimodal problems. They execute fast, but converge to local minima that could be much worse than the global minimum. Global optimisation algorithms, in contrast, employ heuristic strategies to look for global minima and do not stop after finding a local minimum. Hybrid algorithms have a global search phase and a local search phase. The global search locates promising regions where local search can find the local minima.

The type of algorithm most commonly employed for training neural networks are local-optimisation algorithms. Local optimisation algorithms generally require a large number of accuracy bits to perform well. This is because small variations in the error function are quantised to zero [20]. This leads to the formation of a large number of

---

[1] Model-free learning does not make any assumptions about the system or the environment. Thus, it is particularly suitable for situations in which the internal structure of the system varies, or is difficult to model (as in a complex recurrent network). Such algorithms rely on approximate methods that include the response of the network in calculating the performance, thus taking into consideration all imperfections present on the network to be trained.

plateaus in the error surface in addition to those that already exist. These plateaus can be absolutely flat due to the quantisation, which tends to trap the training procedure.

Trapping occurs more frequently if the training algorithm only makes use of the information in the vicinity of the current point in the weight space and especially if it is based on the differentiable property of the error function. Xie and Jabri have concluded that

„if as few bits as possible are to be used for the quantisation, while ensuring successful training and performance, training could not solely rely on the local information such as the gradient of the error function. Methods performing non-local search in the weight space have to be incorporated." [19]

3 Particle Swarm Optimisation

Particle Swarm Optimisation (PSO) was originally proposed and developed by Eberhart and Kennedy [6, 9]. Although PSO is population based it differs from most evolutionary algorithms in that there is no creation of new individuals. Instead the algorithm is based upon the dynamics of human social behaviour whereby information is *shared* amongst individuals in order to arrive at increasingly fitter points [9].

In a PSO system, particles fly around in a multi-dimensional search space. During flight, each particle adjusts its position according to its own experience, and according to the experience of a neighbouring particle, thus making use of the best position encountered by itself and its neighbour. PSO *combines local search methods with global search methods*, attempting to balance exploration and exploitation.

One of the first uses of PSO by Eberhart et al reported encouraging results using the algorithm to replace back-propagation in the training of a neural network. Subsequent publications on the use of PSO to train neural networks confirmed that this algorithm is well suited to this task, typically finding solutions to complex problems orders of magnitudes faster than other population based algorithms. PSO has subsequently being used to establish the network architecture in addition to finding the weight values [10].

In PSO a particle is defined as a moving point in hyperspace. For each particle, at the current time step, a record is kept of the position, velocity, and the best position found in the search space so far. The original PSO formula, as described in [9] are given by equations 1 and 2:

$$x_{id} = x_{id} + v_{id} \tag{1}$$

$$v_{id} = v_{id} + P*\{rand()*[p_{id} - x_{id}]\} + G*\{rand()*[p_{id} - x_{id}]\}$$

$$v_{id} = \begin{cases} -v_{max} &, v_{id} < -v_{max} \\ v_{max} &, v_{id} > v_{max} \end{cases} \tag{2}$$

In these equations d is the number of channels (variables), i is an individual particle in the population, g is the particle in the neighbourhood with the best fitness, v is the velocity vector, x is the location vector and p is the location vector for the individual particles best fitness yet encountered. *rand()* returns a random number uniformly drawn from the range [0,1] and *P* and *G* are weighting factors for the respective components of personal best position and neighbourhood best position. These parameters control the relative influence of the memory of the neighbourhood to the memory of the individual.

This version of PSO typically moves quickly towards the best general area in the solution space for a problem, but has difficulty in making the fine grain search required to find the absolute best point. This is a result of a conflict between the search conditions required for an aggressive broad search and those required for the final narrow search.

Two notable improvements on the initial PSO have been introduced which attempt to strike a balance between these two conditions. The first, introduced by Eberhart and Shi uses an extra 'inertia weight' term which is used to scale down the velocity of each particle and this term is typically decreased linearly throughout a run [15]. The second version introduced by Clerc involves a 'constriction factor' in which the entire right side of the formula is weighted by a coefficient [4]. Clerc's generalised particle swarm model allows an infinite number of ways in which the balance between exploration and convergence can be controlled. For the simplest of these (Type 1" PSO - T1PSO), Eberhart and Shi have shown that placing bounds on the maximum velocity of a particle in T1PSO „*provides performance on the benchmark functions superior to any other published results known by the authors*" [6]. The equations governing a T1PSO algorithm with bounds placed on the velocity is described by the equations 3 and 4.

$$x_{id} = x_{id} + v_{id} \quad (3)$$

$$v_{id} = \chi[v_{id} + P*\{rand()*[p_{id} - x_{id}]\} + G*\{rand()*[p_{id} - x_{id}]\}]$$

$$\text{where } v_{id} = \begin{cases} -v_{max} &, v_{id} < -v_{max} \\ v_{max} &, v_{id} > v_{max} \end{cases} \quad (4)$$

$$\chi = \frac{2\kappa}{\left|2 - \varphi - \sqrt{\varphi^2 - 4\varphi}\right|} \quad \text{and } \varphi = (P+G) > 4, \text{ and } \kappa \in [0,1]$$

4 Classification Problems Used to Evaluate Performance

To investigate the performance of T1PSO on the training of neural networks of limited accuracy, a number of classification problems with different levels of difficulty were used. Three real-world data-sets were used, cancer [18], diabetes [16], and glass [3][2].

The first data set (cancer) is relatively easy, with simple class boundaries, little overlap between classes and sufficient number of data points. The second data-set (diabetes) increases the degree of difficulty by having overlapping classes and

---

[2] The difficulty level of these data-sets was the result of a study of MLP by Prechelt [12].

complex boundaries. The third data-set (glass) besides complex boundaries and overlapping of classes, suffers from an insufficient number of data points.

For each of these problems, the network architecture adopted had a single hidden layer. The number of inputs and outputs are fixed by the number of attributes and number of classes for each problems, and the number of neurons in the hidden layer was determined by trial and error. The characteristics of each of these problems, and the network adopted is shown in Table 1.

The 'number of generations' shown in Table 1 is the number of generations that the swarm algorithm was allowed to run on the network. This was determined by trial and error, with the values used being conservative – they are long enough to ensure that the swarm has found the best solution that it is going to find.

Table 1 Characteristics of the three data sets

Data Set	# Cases	# Attributes	# Classes	# Generations	Network Architecture[3]
Breast Cancer	699	9	2	100	9-5-1
Diabetes	768	8	2	200	8-6-1
Glass	214	11	6	500	8-15-6

5 Equations Used to Limit the Accuracy of the Networks

Equation 5 determines the output of neuron i in the unlimited case

$$y_i = \varphi\left(\sum_{j=1}^{N} w_{ij} y_j + b_i\right) \quad (5)$$

Here y_i is the output of neuron i, w_{ij} is the weight between neuron i and j, b_i is the weight on the bias input to neuron i, and φ is a transformation function (*tanh* in this case). Limiting the accuracy of the network to 2^A bits was achieved by limiting the accuracy of each neuron according to equations 6 and 7.

$$y_i = v\left(\varphi\left(\sum_{j=1}^{N} v(v(w_{ij}, A) y_j, A) + v(b_i, A)\right), A\right) \quad (6)$$

$$\text{where} \quad v(x, A) = \frac{floor(x * 2^A)}{2^A} \quad (7)$$

That is the weight values are quantised, each weighted input to the neuron is quantised, and the output of the neuron is also quantised. This reflects the situation that would be found in neural hardware, where each step in the calculation neuron output will be subject to limitations.

---

[3] The architecture is shown as X-Y-Z where X = #inputs, Y = # hidden nodes, and Z = #outputs.

6 Results

For each of the three problems discussed above, results of 10-fold cross-validation were obtained for networks with the parameter A varied from 1 to 8 bits[4].

Table 2. 10 fold cross-validation results (testing error) for different levels of accuracy

Value of A	Breast Cancer		Diabetes		Glass	
	Mean (%)	Std Dev (%)	Mean (%)	Std Dev (%)	Mean (%)	Std Dev (%)
1	4.43	1.42	31.43	5.76	38.51	8.52
2	4.43	2.07	31.33	6.53	36.42	6.45
3	3.72	1.81	27.33	6.13	35.91	7.15
4	3.58	1.22	26.17	4.81	30.84	5.61
5	4.58	3.00	24.35	4.84	29.14	7.95
6	4.72	2.86	26.95	5.89	29.43	7.19
7	4.29	1.78	24.74	5.56	31.05	8.13
8	4.43	2.18	26.13	5.42	30.15	7.02
Unlimited	4.43	2.18	26.94	6.02	29.90	6.03

Additionally results were obtained for networks using equation 5 (the accuracy of the networks will be limited by the accuracy of the processor, however for the training of neural networks, this case is equivalent to unlimited accuracy). The results reported are the mean and standard deviation of the *testing* error. No validation set was employed to determine if training of the networks should be halted early – it is possible that better results could be achieved if such validation set was employed. The results obtained are presented in Table 2.

As can be seen from the tabular results in Table 2, and the graphical results in Figures 1 through to 4 below, the particle swarm algorithm is very robust on networks of low accuracy. The work presented here shows that although the results obtained depend upon the complexity of the problem being attempted, even for complex problems, an accuracy as low as 2^4 to 2^5 bits was sufficient for good performance.

For the simple Wisconsin breast cancer data set, results obtained with an accuracy level of only one bit were as good as the results obtained with unlimited accuracy. For the diabetes and glass data-sets, reasonable results are obtained with an accuracy of just one bit, however the results do improve with more accuracy. With $A=4$ or greater, the results achieved were as good as could be achieved with unlimited accuracy.

The best known results (to the knowledge of the authors) for each of these problems is shown below in Table 3. Also shown in this table is the results obtained using T1PSO to train networks with $A=5$ bits of accuracy. The results indicate that even at these very low levels T1PSO can produce results on a par with the best reported results for each of the three problems.

---

[4] Results are not given for A=0, as the floor function employed in equation 7 will result in the output of the network always being 0.

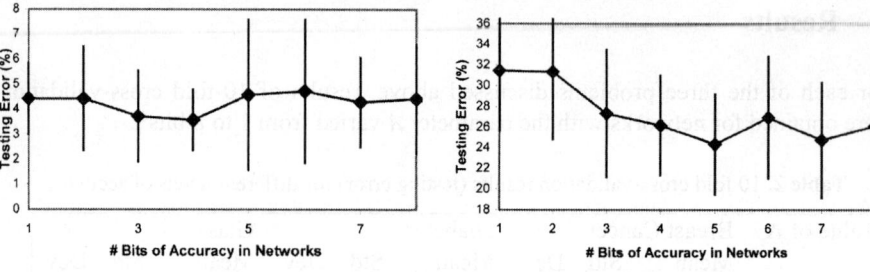

Fig. 1. Breast Cancer   **Fig. 2.** Pima Indian Diabetes

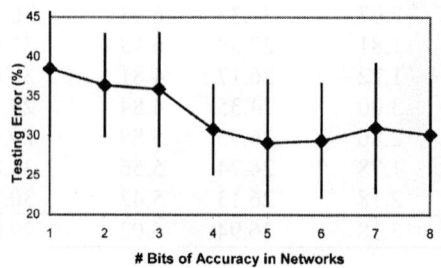

Fig. 3. Glass

Table 3. Comparison of results obtained with T1PSO for networks with 5 bits of accuracy, versus best known results

Method	PSO Accuracy Mean	Std Dev	Best Reported Accuracy (%)	Reference
Wisconsin breast cancer	3.58	1.22	3.8	[2]
Pima Indian Diabetes	26.17	4.81	22.3	[11]
Glass	30.84	5.61	29.4	[13]

7 Two Dimensional Problems

The results obtained for the three data-sets were surprisingly good even when A was as low as 1. In order to try and get a better feel for the decision boundaries that the networks were producing an additional two 2-dimensional problems ('square' and 'two circle') were investigated. Using 2-dimensional problems allows the decision boundary made by the networks to be easily examined. The two problems used are both non-standard data-sets, and were both generated from points randomly chosen (uniformly distributed) from the region $[-1,1] \times [-1,1]$. The 'square' data-set consisted of 255 , such points, and if a point was within the square bounded by $[\frac{-7}{10}, \frac{7}{10}] \times [\frac{-7}{10}, \frac{7}{10}]$ then the desired output value was set 1, otherwise it was set to -1. The 'two circles' consisted of 500 random points, with the desired output value being 1 if the point was within one of two circles and -1 otherwise. The two circles were both of radius 0.2, with one centred at (0.3, 0.3) and the other centred at $(-0.5, -0.4)$.

Table 4. Network Architecture used to train the problems

Data Set	# Cases	# Attributes	# Classes	Network Architecture
Square	255	2	2	2-6-1
Two Circle	500	2	2	2-8-1

The attributes of the problems are tabulated below, along with the network architecture used on the problems (again the network architecture was determined from trial and error).

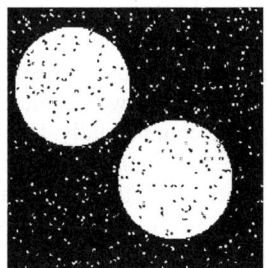
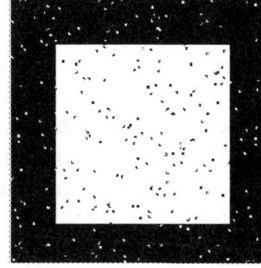
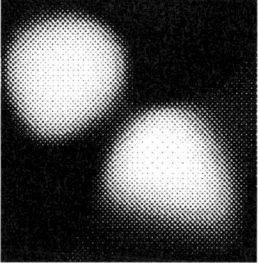

Fig. 4. Two Circle data set. **Fig. 5.** Square data set **Fig. 6.** Network approximation, Two Circle data set unlimited accuracy

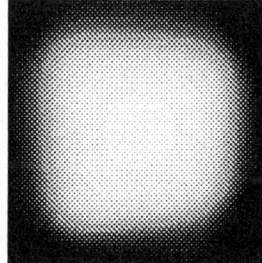
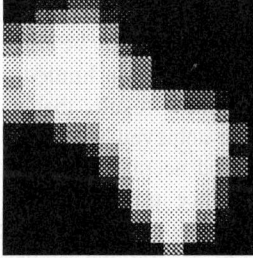
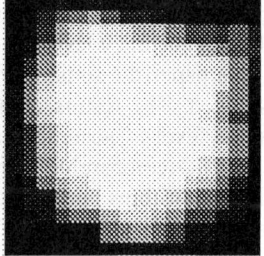

Fig. 7. Network approximation, Square data set with unlimited accuracy **Fig. 8.** Network approximation, Two Circle data set with $A=3$ **Fig. 9.** Network approximation, Square data set with $A=3$

The resulting plot of the data-sets can be seen in figures 4 and 5. In these figures the points within the data sets are shown by the dots, and the boundaries are indicated by the white background (desired output value of 1) and the black background (desired output value of –1)

These images show that for unlimited accuracy, the performance of the networks trained by T1PSO degrades due to the decision boundaries of the network being slightly misaligned with the actual boundaries. In this case the failure is in the *algorithm* to specifying weights that allow the networks to accurately map the data set. Note that the swarm algorithm has produced quite 'fuzzy' decision. This 'fuzziness' allows the decision boundaries of the networks to be shifted with small weights changes.

As the accuracy decreases, performance of the testing set begins to degrades. The reason for this is that the increased quantisation of the weights within the network results in the decision boundaries of the neural network been quantised. These quantised decision boundaries mean not only is it more difficult to place the decision boundary exactly where it should be, but also that the networks are less likely to interpolate well between data points. What is clear from figures 8 and 9 is that although the swarm algorithm has still produced reasonable results, for accuracy levels below this, the performance will be limited by the quantisation of the network.

8 Summary

A considerable effort towards hardware implementations of neural networks began in the late 80's and followed closely the wave of renewed interest in the general field of neural networks. Major research centres became interested in the neural network technology and started to design dedicated chips based on neural networks ideas. Hundreds of designs have already been built, with some available commercially. [8].

The importance of finding a training algorithm that limits the precision required of such neural hardware cannot be overstated. For neural hardware, the bits that represent the *weights* and *thresholds* have to be stored, and the *amount of hardware* required to realise them scales with the cumulative storage of weights and thresholds. Using the *minimum* accuracy levels required for neural computation means that the amount of hardware required can also be minimised.

Research has shown that if specialised training algorithms are used, then the required level of accuracy that would be adequate to allow learning (when using an appropriate learning algorithm) is reduced to around 2^4 to 2^5 bits for difficult problems – accuracy levels that would normally be considered inadequate [1].

In this paper the suitability of PSO to train limited precision neural networks was investigated. The results obtained show results that match those obtained using an unlimited level of accuracy can be achieved with an accuracy of only 2^4 to 2^5 bits, making PSO comparable in required accuracy levels to the best known specialised training algorithms.Furthermore, since training speed is typically very critical in neural hardware, PSO offers the additional advantage of being very fast. As can be seen from the results the number of generations required to achieve good results was very short in comparison to many other training algorithms.

The low accuracy and fast training of PSO combined with the facts that PSO boasts fast convergence, a small number of control parameters, and very simple computations, makes the use of this algorithm very well suited to the training of neural hardware.

References

1. R. Battiti and G. Tecchiolli, „Training Neural Nets with the Reactive Tabu Search", IEEE Transactions on Neural Networks, 6(5): 1185-1200, 1995.
2. K. Bennett and J. Blue, „A Support Vector Machine Approach to Decision Trees", R.P.I Math Report No. 97-100, Rensselaer Polytechnic Institute, Troy, NY, 1997
3. C. Blake, E. Keogh, and C. Merz, „UCI Repository of Machine Learning Databases", [http://www.ics.uci.edu/~mlearn/MLRepository.html], Irvine, CA, University of California, Department of Information and Computer Science, 1998

4. M. Clerc, „The swarm and the queen: towards a deterministic and adaptive particle swarm optimization". Proc. 1999 Congress on Evolutionary Computation, pp 1951-1957, Washington, DC, USA, 1999.
5. T. Duong, „Learning in Neural Networks: VLSI Implementation Strategies", Fuzzy Logic and Neural Network Handbook, 1995.
6. R. Eberhart and J. Kennedy, „A New Optimizer Using Particles Swarm Theory", Proc. Sixth International Symposium on Micro Machine and Human Science Nagoya, Japan, pp. 39-43, 1995.
7. R. Eberhart and Y. Shi. „Comparing inertia weights and constriction factors in particle swarm optimisation". Proceedings of the 2000 congress on evolutionary computation, pp. 84-88, 2000.
8. M. Glesner and W. Pöchmüller, „An Overview of Neural Networks in VLSI", Chapman & Hall, London, 1994.
9. J. Kennedy, „The Particle Swarm: Social Adaptation of Knowledge", Proc. IEEE International Conference on Evolutionary Computation Indianapolis, Indiana, USA, pp. 303-308, 1997.
10. J. Kennedy and R. Eberhart with Y. Shi. „Swarm Intelligence", San Francisco: Morgan Kaufmann/ Academic Press, 2001.
11. R. King, C. Feng, and A. Sutherland, „STATLOG: comparison of classification algorithms on large real-world problems. Applied Artificial Intelligence, 9(3):289-334, 1995.
12. L. Prechelt, „Proben1 - A Set of Neural Network Benchmarking Problems and Benchmarking Rules". Technical Report Nr. 21/ 94. Faculty of Computer Science, University of Karlsruhe, Germany, 1994.
13. N. Shang and L. Breiman, „Distribution Based Trees Are More Accurate International Conference on Neural Information Processing". ICONIP'96, p.133, 1996.
14. Y. Shang, „Global Search Methods for Solving Nonlinear Optimization Problems", Ph.D. Thesis, Dept. of Computer Science, Univ. of Illinois, Urbana, Illinois,August 1997.
15. Y. Shi and R.C. Eberhart, „A Modified Particle Swarm Optimizer", IEEE International Conference on Evolutionary Computation, Anchorage, Alaska, USA, 1998.
16. J. Smith, J. Everhart, W. Dickson, W. Knowlerand and R. Johannes, „Using the ADAP learning algorithm to forecast the onset of diabetes mellitus", . In Proceedings of the Symposium on Computer Applications and Medical Care, pp. 261-265, 1988.
17. W. Wolberg and O. Mangasarian, „Multisurface method of pattern separation for medical diagnosis applied to breast cytology", Proceedings of the National Academy of Sciences, U.S.A., vol. 87, pp 9193-9196, 1990.
18. Y. Xie and M. Jabri. Training algorithms for limited precision feedforward neural nets. SEDAL technical report 1991-8-3, Department of Electrical Engineering, University of Sydney, NSW 2006, Australia, 1991.
19. Y. Xie and M. Jabri, „Analysis of effects of quantisation in multilayer neural networks using statistical models", IEEE Trans. Neural Networks, 3(2):334-338, 1992.

Evolutionary Multi-objective Integer Programming for the Design of Adaptive Cruise Control Systems

Nando Laumanns[1], Marco Laumanns[2], and Hartmut Kitterer[3]

[1] RWTH Aachen, Institut für Kraftfahrwesen (ika)
D-52704 Aachen, Germany
laumanns@ika.rwth-aachen.de

[2] ETH Zürich, Computer Engineering and Networks Laboratory (TIK)
CH-8092 Zürich, Switzerland
laumanns@tik.ee.ethz.ch
http://www.tik.ee.ethz.ch/aroma

[3] WABCO Vehicle Control Systems
D-30453 Hannover, Germany
hartmut.kitterer@wabco-auto.com

Abstract. Adaptive Cruise Control (ACC) systems represent an active research area in the automobile industry. The design of such systems typically involves several, possibly conflicting criteria such as driving safety, comfort and fuel consumption. When the different design objectives cannot be met simultaneously, a number of non-dominated solutions exists, where no single solution is better than another in every aspect. The knowledge of this set is important for any design decision as it contains valuable information about the design problem at hand.

In this paper we approximate the non-dominated set of a given ACC-controller design problem for trucks using multi-objective evolutionary algorithms (MOEAs). Two different search strategies based on a continuous relaxation and on a direct representation of the integer design variables are applied and compared to a grid search method.

1 Introduction

Crowded motorways and a higher average vehicle speed create increasing difficulties for drivers. The automobile industry tries to compensate these additional demands by inventing driver assistance systems such as antilock braking system (ABS), cruise control (CC) and electronic stability control (ESC).

In contrast to the systems mentioned above, adaptive cruise control (ACC) has not been thoroughly established yet. The ACC-system is an enhanced cruise control, not only designed to keep the vehicle's speed constant, but also to analyze the traffic situation in front of the vehicle and regulate its longitudinal dynamics accordingly. Thus it especially suits the demands of truck drivers, who frequently have to follow a leading vehicle.

Used effectively, ACC-systems can increase driving safety, make driving more comfortable and reduce fuel consumption. However it is rather difficult to develop a controller that meets the drivers' requirements concerning its speed regulating behavior as well as safety criteria and fuel efficiency.

Since experimental testing of each modified controller variant would enormously raise development costs and time, a simulation tool is used to analyze the ACC-system's behavior. This offers the possibility to improve the development process further by applying numerical optimization techniques such as evolutionary algorithms to optimize the ACC-controller.

This paper now examines the design of an ACC-controller for trucks under multiple, possibly conflicting criteria. In the next section the ACC-system is explained along with its simulation environment and the corresponding optimization problem that results from the system's degrees of freedom and the different design criteria. Section 3 addresses multi-objective optimization and describes the algorithms that are used to approximate the set of non-dominated or Pareto-optimal solutions. In section 4 the results of the different optimization runs are presented and discussed.

2 Adaptive Cruise Control (ACC)

The adaptive cruise control system is intended to support the driver with controlling the vehicle's longitudinal dynamics. A radar sensor detects moving objects in front of the car. A tracking algorithm along with a yaw rate sensor determines a relevant target, including its distance and its relative velocity to the ACC-vehicle.

The ACC-controller uses this information to calculate a desired speed and a corresponding acceleration. This is transferred via CAN-bus to the appropriate parts of the car (see Fig. 1). Concerning trucks, ACC-systems usually have influence on the various elements of the braking system and on the engine control. In addition to that, the gear-box can be controlled. The interaction between the system and the driver is accomplished through an intelligent dashboard with switches, displays and a buzzer.

2.1 Design Variables

The longitudinal controller is responsible for the translation of the incoming data about the traffic situation in front of the vehicle and its own driving condition into a desired acceleration. The data produced by the sensor contains some deviation. This requires four different filters in order to create a smooth acceleration signal. The influence of these filters can be regulated by four integer parameters $x_1, \ldots, x_4$. Strong filters result in very smooth signals. However, they somewhat delay the vehicle's reaction to incoming data and that weakens its driving performance.

Two further design variables x_5, x_6 are used to define the longitudinal controller's reaction to the vehicle's distance from the leading vehicle and their relative velocity.

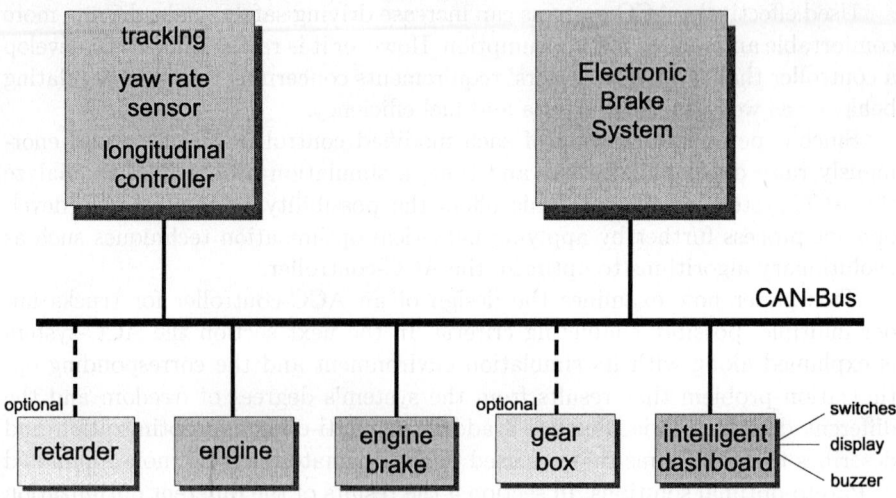

Fig. 1. Structure of the ACC-system

2.2 Simulation

Each setting of the design variables represents a decision alternative, and the resulting controller performance is determined through simulation. For this we apply the simulation tool PELOPS, which has been developed by the ika in cooperation with BMW [3]. It analyses interactions between vehicle, driver and the environment. Its three main elements are the stretch-module, the decision-module, and the handling-module (see Fig. 2). The cause-effect-principle is used to realize the relation between vehicle performance, environmental impacts and the driver's decision. In this case the ACC-system replaces the driver in terms of regulating the speed.

2.3 Design Objectives

Four objectives are defined to give a sufficient characterization of the ACC-system's longitudinal controlling behavior considering driving comfort, fuel efficiency and safety. All these objective functions are computed within the simulation. Thus, the resulting multi-objective optimization problem can be stated as follows (where the function values of f and g calculated by the simulator).

$$\text{Minimize} \quad f(x) = \begin{pmatrix} f_1(x) \\ f_2(x) \\ f_3(x) \\ f_4(x) \end{pmatrix} \quad \begin{array}{l} \text{(average fuel consumption)} \\ \text{(acceleration / deceleration time)} \\ \text{(velocity deviation)} \\ \text{(acceleration deviation)} \end{array} \quad (1)$$
$$\text{subject to} \quad g(x) \geq d_{min} \quad \text{(minimum follow-up distance)}$$
$$x \in X = \{1, 2, \ldots, 99\}^2 \times \{1, 2, \ldots, 16\} \times \{1, 2, \ldots, 8\}^3 \subseteq \mathbb{Z}^6$$

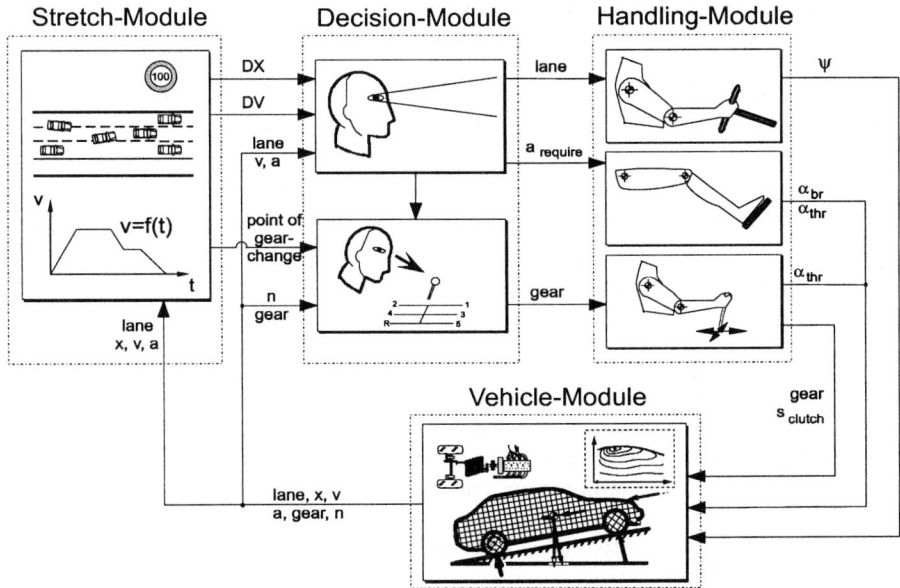

Fig. 2. Elements of simulation environment PELOPS

For safety reasons there is a constraint on the minimum distance $g(x)$ between the ACC-vehicle and the leading vehicle.

3 Multi-objective Optimization

The presence of multiple objectives in a design problem induces a partial order on the set of alternatives. This partial order is represented by the dominance relation $\prec$, where a decision alternative is said to dominate another $(a \prec b)$ if and only if it is at least as good in all objectives and strictly better in at least one objective. The solution of the constrained multi-objective integer programming problem (1) is defined as the set of feasible non-dominated decision alternatives (Pareto set) $X^* = \{a \in X : (\nexists b \in X \text{ with } b \prec a) \land g(a) \geq d_{min}\}$.

In order to approximate the non-dominated set for the constrained multi-objective integer programming problem (1), three methods are applied and compared: a grid search and two evolutionary algorithms. The computation time of the simulator makes exhaustive search or complete enumeration of all alternatives impractical. Thus, a grid search with 2^{15} representative solutions regularly distributed in the decision variable space is performed. In comparing all these alternatives to each other, the dominated ones are eliminated and the remaining represent a first approximation to the non-dominated set as a baseline.

3.1 Multi-objective Evolutionary Algorithms

Evolutionary algorithms (EA) have shown to be a useful tool for approximating the non-dominated set of multi-objective optimization problems [1,6,5]. In many engineering design problems the evaluation of decision alternatives is based on (computationally expensive) simulations which limits the use of traditional techniques like gradient-based methods. When *a priori* incorporation of the decision maker's preferences is difficult or not desired (so that aggregation of the different objectives to a scalar surrogate function is not possible), there are hardly any alternative techniques available.

Algorithm 1 $(\mu + \lambda)$-SPEA2

1: Generate an initial population P of size $\mu + \lambda$.
2: Calculate objective values of individuals in P.
3: **while** Termination criteria are not met **do**
4:     Calculate fitness values of individuals in P.
5:     Copy all non-dominated individuals in P to P'.
6:     **if** $|P'| > \mu$ **then**
7:         Reduce P' by means of the truncation procedure.
8:     **else if** $|P'| < \mu$ **then**
9:         Fill P' with dominated individuals from P in increasing order of their fitness.
10:    **end if**
11:    Create P'' of size λ by iteratively selecting parents from P', applying recombination and mutation.
12:    Calculate objective values of individuals in P''.
13:    $P \leftarrow P' + P''$
14: **end while**

Here, two algorithms based on SPEA2 ([7], an improved version of the Strength Pareto Evolutionary Algorithm, [8]) are applied. SPEA2 can be seen as a $(\mu + \lambda)$-EA with special fitness assignment and selection techniques, a pseudo-code description is given in Alg. 1. Each individual in the combined parent and offspring population P is assigned a strength value S which equals the number of solutions it dominates. On the basis of the S values, the raw fitness $R(i)$ of an individual i is calculated $R(i) = \sum_{j \in P, j \prec i} S(j)$. In addition, the local density is estimated for each individual and added to R to discriminate between individuals with equal raw fitness (line 4).

Selection is performed in two steps: environmental selection and mating selection. The best μ solutions out of P are selected to constitute the set of parents for the next generation. First, all non-dominated individuals, i.e., those which have a fitness lower than one, are selected (line 5). If there are more than μ such solutions, a truncation procedure is invoked which iteratively removes that individual which is closest to the others (line 7). If less than μ individuals are non-dominated, the space is filled with the dominated individuals in ascending order of their fitness values (line 9). In the second step the recombination partners for the next generation are selected by binary tournament selection (with replacement) based on the fitness values (line 11).

The representation of the individuals and the variation operators (recombination and mutation) are different and explained in the following.

Real-Valued Individuals Many standard search operators are based on a floating-point representation of (real-valued) decision variables. Therefore a continuous relaxation of the search space to $[0, 99]^2 \times [0, 16] \times [0, 8]^3$ is used, and the variables are rounded to their integer part (plus 1) before each run of the simulation tool. For the recombination we use the SBX-operator [1] with distribution index $\eta = 5$. The offspring individuals are then mutated by adding normal distributed random numbers, where the standard deviation σ is set to 5 per cent of the interval length.

Integer-Valued Individuals As the relaxation produces an artificial blow-up of the search space a direct representation of the decision variables as integer numbers might be more appropriate. It also eliminates the potential problem of mapping several different individuals to the same decision alternative by the rounding procedure. Search operators working directly on integer variables are not so common in evolutionary computation. Here, we adopt the techniques from Rudolph [4] who developed an EA for integer programming with maximum entropy mutation distributions that enables self-adaptive mutation control similar to real-valued evolution strategies. A successful application to a mixed integer design problem for chemical plants is reported in [2]. Here, the initial mutation step size was set to $s = 2$ for all variables.

For both version of SPEA2 the population size was set to $\mu = \lambda = 10$, and the runs were terminated after 300 generations (3000 objective function evaluations). During the run, an archive of all non-dominated solutions was maintained and output as the approximation to the non-dominated set at the end of the run.

4 Results

In order to evaluate the performance of the evolutionary algorithm a grid search over the whole parameter area is performed, along with a manual optimization of the ACC-controller. The grid search contains 16384 elements, requiring a computation time of almost 137 hours. As both instances of the evolutionary algorithm only used 3000 function evaluations each, and since their internal operations and data processing can be neglected compared to the simulation, they have a clear advantage in terms of computation time.

As a first interesting observation from the output of the different algorithms no trade-off is visible for the second objective f_2 (acceleration / deceleration time): All algorithms have found values close to the minimal value of 66.6 for

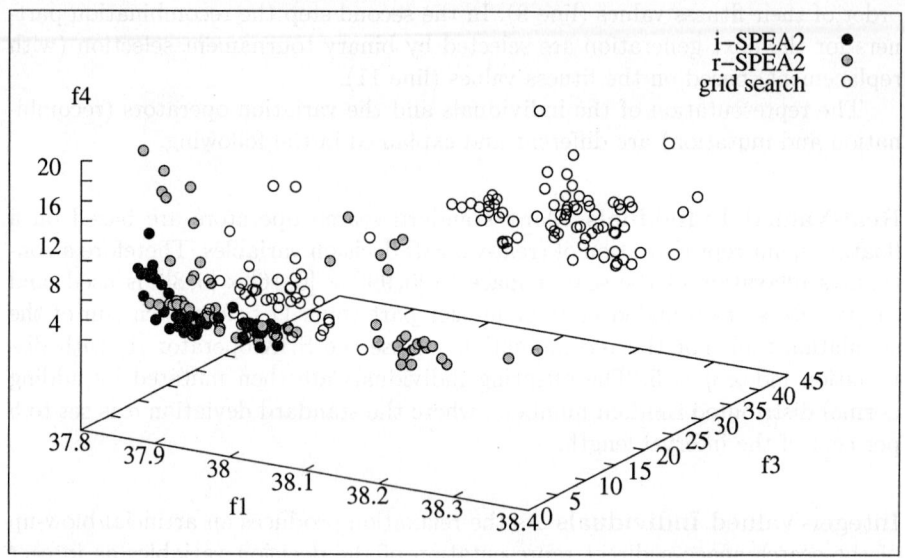

Fig. 3. Scatter plot of the non-dominated solutions produced by the grid search and the evolutionary algorithm with continuous relaxation (r-SPEA2) and direct integer coding (i-SPEA) for the objective function values f_1, f_3, f_4

almost all non-dominated alternatives. The remaining objective values of the different non-dominated sets are displayed in Fig. 3. Here, the trade-off characteristic is visible from the three-dimensional scatter plot.

Measuring the performance of different multi-objective optimizers or the quality of different non-dominated sets is not straightforward. Many performance metrics have been suggested [1] and can be applied, but there is no clear order unless one set completely dominates another.

To ensure comparability of the results one has to look at different aspects. One possibility is to define an evaluation formula based on a weighted distance to an ideal point f^* which is given by the minimal objective values in each dimension. The difference between each decision parameter and the optimum value in the referring category is multiplied with a factor, which represents the importance of the category. Thus the interpretation of the results reflects an adaptation to the user's goals. In this case the objectives f_1 and f_3 are considered most important, f_2 is least important. Representing the distance to the optimal solution the sum of those values gives the overall quality of the individual

$$D(x) = 150(f_1(x) - f_1^*) + (f_2(x) - f_2^*) + 6(f_3(x) - f_3^*) + 4(f_4(x) - f_4^*) \quad (2)$$

with $f^* = (37.8339, 66.6, 2.06935, 3.03196)$. Accordingly, a ranking of the individuals developed by the different optimization strategies can be produced. The best 100 solutions are displayed in Fig. 4. The two evolutionary algorithms create the best solutions, with a slight advantage of the integer-version in terms

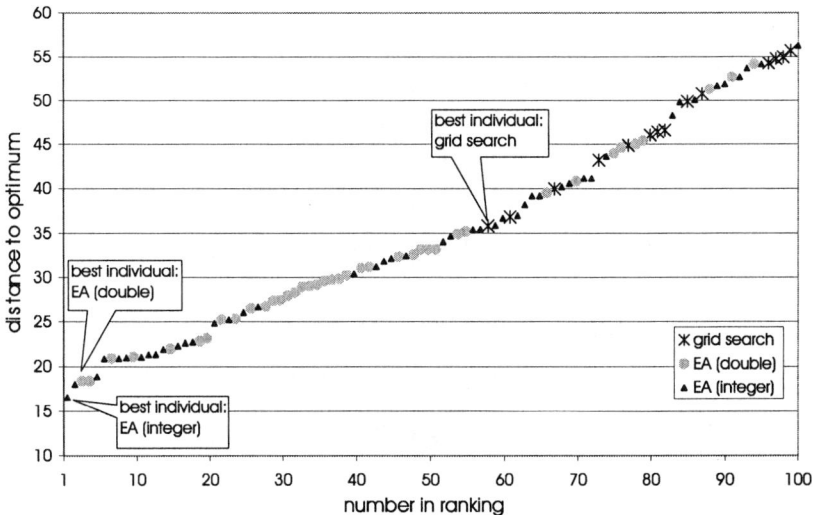

Fig. 4. Ranking of solutions according to aggregated quality measure

of density close to the optimum solution. Out of the top 100 solutions, 46 are were created by this integer-version, 40 by the double-version and only 14 by the grid-search.

Another aspect which has to be taken into consideration in order to compare the different optimization approaches is the total objective space that their solutions extend to. The minimum overall objective function value divided by the minimum value of a single approach determines the quality of the optimization in direction of the corresponding objective. Fig. 5 visualizes the performance of the different optimization approaches in terms of objective space exploration. While all three approaches reach the optimum in the objectives f_1 and f_2, the grid search shows a performance almost 10% below the optimum in f_3 and 5% in f_4. The integer-version of the EA proves to be the best optimization method with a good performance in all four objectives and an average value of 99.62%.

Table 1. Results of the coverage measure $\mathcal{C}(\cdot, \cdot)$ applied to all pairs of algorithms

$\mathcal{C}(\mathcal{A}, \mathcal{B})$	i-SPEA2	r-SPEA2	grid search
i-SPEA2		0.423567	0.991597
r-SPEA2	0.070588		0.991597
grid search	0	0	

For a pairwise comparison of different output sets, the *coverage* measure [8] gives an indication of how much of a one algorithm's output has also been reached by the other one. Specifically, $\mathcal{C}(\mathcal{A}, \mathcal{B})$ calculates the relative number of points

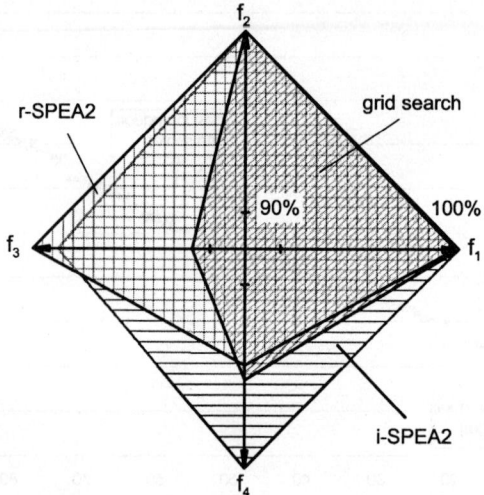

Fig. 5. This graph shows how well each algorithm could approach the minimal value in each objective dimension separately. The values on each axis is calculated by dividing the overall minimal value by the minimal value produced by each algorithm

of set B that are dominated by at least one point in set A. Table 1 shows that none of the points found by the grid search is better than any point in the non-dominated sets of the evolutionary algorithms. Also, the SPEA2 working with the floating point representation does not cover much (less than 10%) of the solutions produced by the integer version, which in turn is able to dominate nearly half of the solutions of its competitor.

Table 2. Results of the hypervolume difference measure $\mathcal{S}(\cdot,\cdot)$ applied to all pairs of algorithms and the absolute hypervolume measure $\mathcal{S}(\cdot)$ (last column)

$\mathcal{S}(\mathcal{A},\mathcal{B})$	i-SPEA2	r-SPEA2	grid search	$\mathcal{S}(\mathcal{A})$
i-SPEA2		0.0038	0.223	0.949
r-SPEA2	0.002		0.188	0.913
grid search	0.0003	0.0018		0.726

In contrast to a relative comparison by the coverage measure, the normalized volume of the dominated space is often used to evaluate a single non-dominated set alone [8]. Here, a reference cuboid between the ideal point f^* and the nadir point defined by the maximal objective function values of the union of all three non-dominated sets is chosen. $\mathcal{S}(\mathcal{A})$ gives the fraction of this reference volume that is dominated by $\mathcal{A}$. It is clear that the algorithms ideally should maximize the dominated space. The results are given in the last column of Table 2. In addition, the volume differences are depicted, where $\mathcal{S}(\mathcal{A},\mathcal{B})$ evaluates to the volume dominated by the set A, but not dominated by the set B.

5 Conclusion

The combination of PELOPS as simulation tool and an evolutionary algorithm is an efficient method to optimize the longitudinal controller of an ACC-system. Even the non-adapted version of the EA shows a better performance than a grid search over the design variable area, in spite of its lower computation time.

Nevertheless it is sensible to adapt the EA to the simulation tool. Matching parameter intervals and types increase the algorithm's performance, because they ensure that every change in the design variables, created by the EA, also changes the ACC-system's longitudinal controlling behavior within the simulation tool. This correlation is necessary for an efficient optimization process, and less redundancy is produced compared to the artificial relaxation to real variables. The superiority of results produced by the direct integer encoding shows that efficient search operators exist for this domain, and that they are applicable as well in a multi-objective environment.

Analyzing the trade-off-graph (Fig.3) the development engineer can learn about performance boundaries of the ACC-system. It helps to understand the system's behavior and shows different solutions, which can suit the company's or customer's preferences. Thus the EA can enhance the traditional developing process by quickly providing broad and detailed information about a complex optimization problem.

Acknowledgment

The second author acknowledges the support by the Swiss National Science Foundation (SNF) under the ArOMA project 2100-057156.99/1.

References

1. K. Deb. *Multi-objective optimization using evolutionary algorithms*. Wiley, Chichester, UK, 2001.
2. M. Emmerich, M. Grötzner, B. Groß, and M. Schütz. Mixed-integer evolution strategy for chemical plant optimization with simulators. In I. C. Parmee, editor, *Evolutionary Design and Manufacutre — Selected papers from ACDM'00*, pages 55–67, Plymouth, UK, April 26–28, 2000. Springer, London.
3. J. Ludmann. *Beeinflussung des Verkehrsablaufs auf Strassen: Analyse mit dem fahrzeugorientierten Verkehrssimulationsprogramm PELOPS*. Schriftenreihe Automobiltechnik. Forschungsgesellschaft Kraftfahrwesen mbH Aachen, 1998. (in German).

4. G. Rudolph. An evolutionary algorithm for integer programming. In Y. Davidor, H.-P. Schwefel, and R. Männer, editors, *Parallel Problem Solving from Nature – PPSN III, Int'l Conf. Evolutionary Computation*, pages 139–148, Jerusalem, October 9–14, 1994. Springer, Berlin.
5. P. Sen and J.-B. Yang. *Multipe Criteria Decision Support in Engineering Design*. Springer, 1998.
6. E. Zitzler, K. Deb, L. Thiele, C. A. C. Coello, and D. Corne, editors. *Proceedings of the First International Conference on Evolutionary Multi-Criterion Optimization (EMO 2001)*, volume 1993 of *Lecture Notes in Computer Science*, Berlin, Germany, March 2001. Springer.
7. E. Zitzler, M. Laumanns, and L. Thiele. SPEA2: Improving the strength pareto evolutionary algorithm for multiobjective optimization. In K. Giannakoglou et al., editors, *Evolutionary Methods for Design, Optimisation, and Control*, 2002. To appear.
8. E. Zitzler and L. Thiele. Multiobjective evolutionary algorithms: A comparative case study and the strength pareto approach. *IEEE Transactions on Evolutionary Computation*, 3(4):257–271, 1999.

The Macronet Element: A Substitute for the Conventional Neuron

Tim Hendtlass and Gerrard Murray

Centre for Intelligent Systems and Complex Processes
School of Biophysical Sciences and Electrical Engineering
Swinburne University VIC 3122 Australia
{thendtlass,gmurray}@swin.edu.au

Abstract. A potential replacement for the conventional neuron is introduced. This is called a Macronet element and uses multiple channels per signal path with each channel containing two trainable non-linear structures in addition to a conventional weight. The authors show that such an architecture provides a rich spectrum of higher order powers and cross products of the inputs using less weights than earlier higher order networks. This lower number of weights does not compromise the ability of the Macronet element to generalise . Results from training a Macronet element to develop a relationship from a sparse map of Europe are given.

1 Introduction

Artificial neural networks can at best be described as biologically inspired. In real life, biology interposes a synapse between the axon of one neuron and the dendrite of another. The synapse is the chemical conduit for signals travelling between the neurons [4] and the site of parallel modes of non-linear neurotransmission.

The artificial neural network analogue of the synapse is the weight. This configuration has worked well in multi-layered architectures as the signal conduit. However, back propagation type networks have definite performance limits and the authors contend that this is due to the simplistic properties of the artificial neurons that are used. It is suggested that more complex connections between artificial neurons are desirable and that this increased architectural complexity may lead to neurons and therefore networks capable of more complex tasks.

This idea is not new, and higher order processing units (HPU), a generic term that includes sigma-pi neurons, have been suggested for many years [3,7,2,8]. The internal activation of an HPU is a weighted sum of the inputs and combinations of various powers of the inputs. The significance of each of the combinations is defined by a separate weight. An arbitrary n-input logic function can be learned by an HPU with 2^n weights [6]. However, for a general n input function a traditional 2^n weight HPU would be limited to learn but not to generalise [5]. Reducing the number of weights restores the ability to generalise: the difficulty lies in establishing how to reduce weight numbers while preserving the ability to learn.

Numerous network architectures that incorporate HPUs have been proposed of which pi-sigma networks [9] are well known. However, work has shown [10] that the pi-sigma network is not a universal approximator. The exponential increase in the number of weights required as the number of inputs to the HPU increases has limited practical HPUs to only second order combinations of inputs, although higher orders may be needed for practical problems. *"Somewhere between multiple layers of ordinary neurons and a single higher-order n input neuron with 2^n weights there is an optimum network. As yet, no method exists for finding the optimal network."* [6].

The Macronet element is a new alternative to the classical HPU that solves this dilemma. Instead of the higher order functions and cross function of the inputs being separately generated they arise naturally from the use of a multi-channel architecture. The inspiration for this architecture is once more from biology.

Figure 1 shows one model of a synapse, derived from biological neural systems, that suggests that a synapse contain multiple pathways of neurotransmission existing in parallel. It is commonly held that populations of neurotransmitters, with different chemical structures, are released from an axon terminal and diffuse across the synaptic cleft to bind to dendritic receptors on a neighbouring neuron.

Channel architecture is a conceptualisation of the connectivity of parallel pathways and not an attempt to model synaptic form or function. Because much of the cellular interaction at the molecular level remains as yet unresolved, biological plausibility of artificial neurons is, as a consequence, impossible.

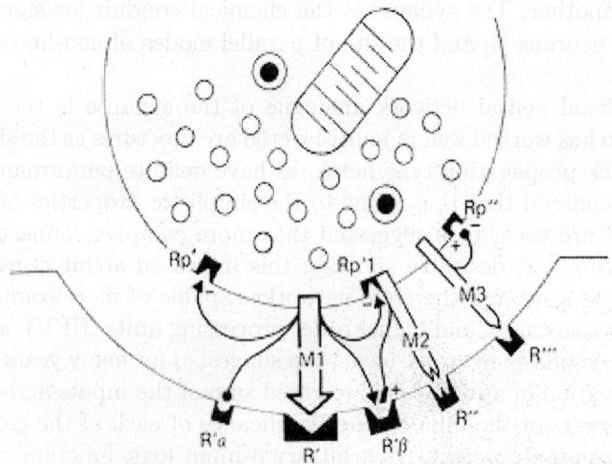

Fig. 1. A model of a synapse showing multiple pathways. Reproduced from [4]

However, analogies may be drawn that may serve to elucidate the idea of parallel channels as connections in Macronet elements. These analogies rely on the simplicity of the McCulloch type concept of neurons and the way these artificial neurons are connected.

Employing a similar reductionist reasoning, a synapse can be considered to consist of three components. A *'generator'*, that releases structurally different neurotransmitter at concentrations proportional to the electrical signal being propagated across the synaptic cleft. The synaptic cleft, an extra cellular space, is filled with extra cellular fluid, a medium the neurotransmitter can diffuse. Once across the synaptic cleft, a receptor binds the neurotransmitter and facilitates the change of the nature of the signal transfer from a chemical to an electrical form.

The authors postulate that this multiplicity of parallel pathways, each with its own non-linear transmission characteristics, when implemented in artificial networks will enhance performance. Unlike the conductivity of the extra cellular fluid, the *'generator'* and *'receptor'* processes definitely depend on the magnitude of the signal being propagated. The Macronet element has been designed to provide parallel channels, metaphors for the parallel pathways in the biological congener, with trainable non-linear transmission characteristics.

The conventional artificial neuron shown in figure 2A has a path from each input to the neuron body and each of these paths consist of only a single channel with the conductivity of that channel being set by a weight. Once training is complete the value of this weight, and therefore the conductivity of the channel,

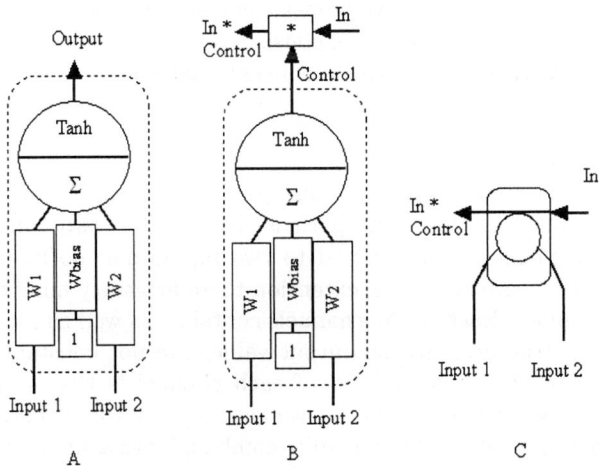

Fig. 2. A two input conventional neuron (A), a two input conventional neuron used to control the flow of an external signal (B) and the symbol used later in this paper for such a structure (C)

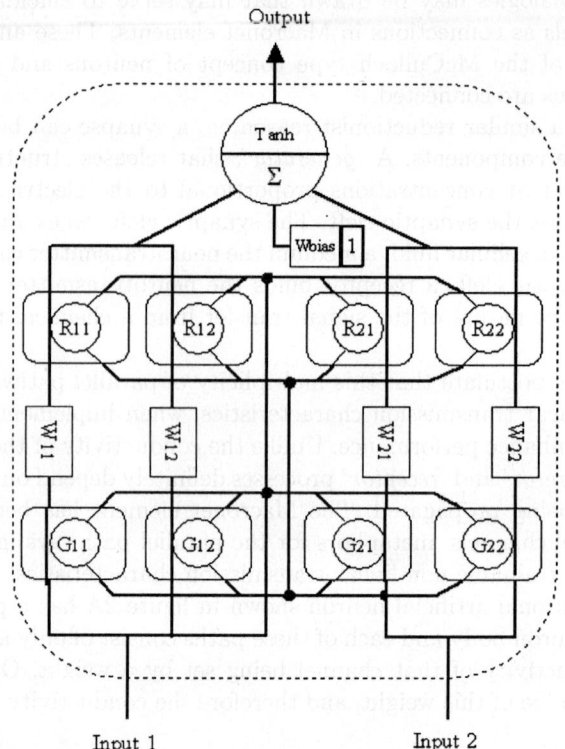

Fig. 3. A two input, two channel Macronet element. The path from each input divides into two channels, each channel consists of two control structures of the type shown in figure 2B together with a conventional weight

is fixed. The Macronet element shown in figure 3 also has a path from each input to the neuron body but that path is divided into two channels. (The figures and this explanation are limited to two inputs and no more than two channels per path to conserve space; extension to an arbitrary number of inputs and channels is simple). Each of the channels contains, as well as a conventional weight, two other structures, whose output values are not constant even after training is complete. The conductivity of each channel is the product of the outputs from these two structures and the weight value. Without these additional structures the two channels could be readily combined into a single channel with a conductivity equal to the sum of the individual channel conductivities. With the additional structures the combined effect of the channels is more complex and will be discussed in detail in the next section. However, the net effect is to generate a wide range of combinations of powers of the inputs to the neuron.

The additional structures in each channel are of the form shown in figure 2B. In the simplest form they consist of a single conventional neuron with the same inputs as the Macronet element, but could be a network of arbitrary complexity. The output of the structure (*output*) moderates the transfer function of a signal (*in*) to give an output that is *in * output*. Figure 2C shows the symbol used for the channel structures in the full diagram of the Macronet element in figure 3 for the structure shown in full in figure 2B. For convenience one of the structures in each channel in figure 3 will be referred to as a 'generator' while the other will be referred to as a 'receptor', although the terms are really interchangeable. The structures perform two functions: generating the higher orders and defining the current importance of the channel. Since the structure outputs can change sign, a channel may dynamically change from excitatory to inhibitory or visa versa.

2 A Comparison of the Properties of a Conventional Neuron and a Macronet Element

Consider a conventional neuron with an internal activation that is the sum of products of each input and its weight plus the bias weight, as shown in figure 2A.

For this two input neuron as shown above, the internal activation is given by:

$$IA = w_1 i_1 + w_2 i_2 + w_{bias} \tag{1}$$

The neuron output O is Tanh(IA). Using the Pade approximation to Tanh the output

$$O = \frac{IA + (IA)^3/15}{1 + 2(IA)^2/5} \tag{2}$$

Substituting 1 into 2 yields:

$$O = \frac{A_1 i_1^3 + B_1 i_2^3 + C_1 i_1^2 i_2 + D_1 i_1 i_2^2 + E_1 i_1^2 + F_1 i_2^2 + G_1 i_1 i_2 + H_1 i_1 + I_1 i_2 + J_1}{A_2 i_1^2 + B_2 i_2^2 + C_2 i_1 i_2 + D_2 i_1 + E_2 i_2 + F_2} \tag{3}$$

where A, B etc represent constants uniquely derived from the weight values w_1, w_2 and w_{bias}. Since the number of these constants far exceeds the weight values from which they are derived, the values of the constants are not independent.

Equation 3 is a ratio of two sums of weighted products, the upper sum containing terms involving weighted combinations of powers of each input from the third to the zeroth power and the lower involving weighted combinations of powers of each input from the second to the zeroth power.

For a convenient shorthand, let equation 3 be rewritten as:

$$O = \Re_{2,2}^{3,3}(i_1, i_2) \tag{4}$$

where the upper pair of indices show the highest power of each input on the top line of the ratio and the lower pair show the highest power of each input on the bottom line and the constants are implied.

It then follows that:

$$\Re_{c,d}^{a,b}(i_1, i_2) * \Re_{g,h}^{e,f}(i_1, i_2) = \Re_{c+g,d+h}^{a+e,b+f}(i_1, i_2) \tag{5}$$

$$\Re_{c,d}^{a,b}(i_1, i_2) + \Re_{c,d}^{a,b}(i_1, i_2) = \Re_{2c,2d}^{a+c,b+d}(i_1, i_2) \tag{6}$$

Using one control element per channel will provide some additional powers and cross products of the neuron inputs at the channel output. For the simplest control element, a single conventional neuron, the channel output is given by equation 4 multiplied by the channel input with the channel weight absorbed into the other constants. When two control elements are introduced to each channel a far wider range of input powers and cross products become available. Since the values of these control elements are multiplied together, Equation 5 shows the effect of introducing a second control element. If each control element is a single conventional neuron, the output of the channel will be $\Re_{4,4}^{6,6}(i_1 + i_2)$ multiplied by the channel input, again with the channel weight absorbed into the other constants.

Equation 6 shows that by adding a second channel to the path the available powers and cross products of the neuron inputs are further extended. Again if each control element is a single conventional neuron, the output of the two channel path will be of the form $\Re_{8,8}^{12,12}(i_1 + i_2)$ multiplied by the path input.

This complex ratio contains 144 different terms, but unlike the other approaches used to generate higher order networks that typically need a weight per product required, these are generated by only 12 weights. This augurs well for the ability of the network to generalize. However, an individual weight per product allows any combination of constants, while the multi channel per path approach is only able to provide certain combinations of constants.

3 Results

A single Macronet element has been tested on a wide range of standard test functions, both binary and analogue, and has performed no worse than a standard back propagation network. However, it is on complex problems that the Macronet element can shine. Consider, for example, the map of Europe shown at the left of figure 4 which consists of 121,446 points. For a combination of analogue input values (within the range of the axes) there is an output that has one value if that point corresponds to land and another if it corresponds to water. To test the generalization capability, a subset of points (every twentieth point) in the full map was used to train both a conventional back propagation network and a Macronet element. These 6021 training points are shown at the right of figure 4 superimposed on the outline of Europe.

A simple closest neighbour regeneration of the original map from this training set is shown at the left of figure 5. To generate this map, every map point required the calculation of the distance between it and every point in the training set with the map point taking the output associated with the training point closest to it.

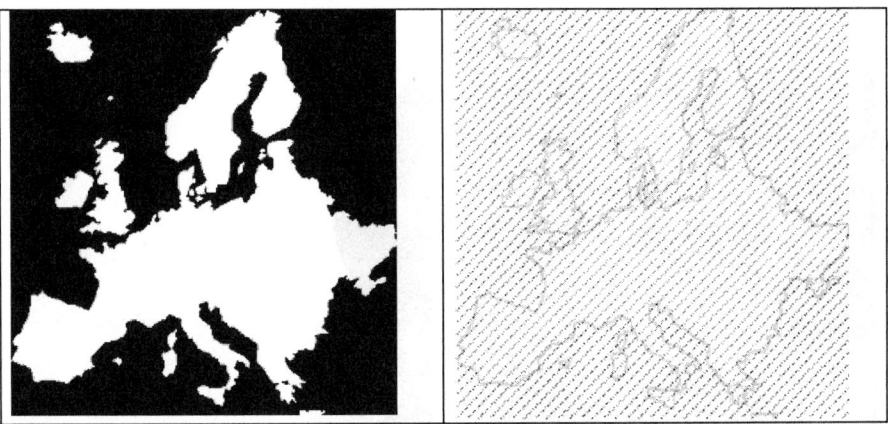

Fig. 4. A 121,444 point map of the full data set (left) and a map of the reduced data set of 6021 points (right). In the reduced data set map each data point is shown by a point: the outline of Europe has been added for convenience and is not contained in the data set in any way

This closest neighbour map represents the map available if the training set has just been learned. The map at the right of figure 5 is typical of what was achieved with a back propagation network of conventional neurons. Figure 6 shows what was achieved by the single Macronet element. The map at the left was for 8000 passes through the training set using the normal output error function, the map to the right also has a further 8000 passes through the data set using a cubic output error function. (The choice of 8000 passes was not critical in either case). As both the conventional back propagation network and the Macronet element had similar numbers of weights and conventional neurons, it is reasonable to conclude that the improved performance of the Macronet element compared to a back propagation network is a result of the architectural differences.

A comparison between figure 5 left and figure 6 shows that the Macronet element has not learned the training set but has learned a (very complex) functional relationship between the two inputs and the required output. The output maps in this paper were produced by thresholding the Macronet element output at zero (the two target levels being 0.5 and −0.5), an unthresholded map shows that in most places the transition from land to sea is very abrupt.

Some points, such as Iceland, do not appear in the training set and therefore cannot be expected in the resulting map. Some of the differences between the original map and the Macronet element generated map occur in spaces undefined in the training set. The number of training points misclassified by the Macronet element for the best map is just 163 out of 6021 (2.7%).

The nearest neighbour network required 6021 triads of numbers (X, Y and output) to be stored as the look up table, the Macronet element only requires

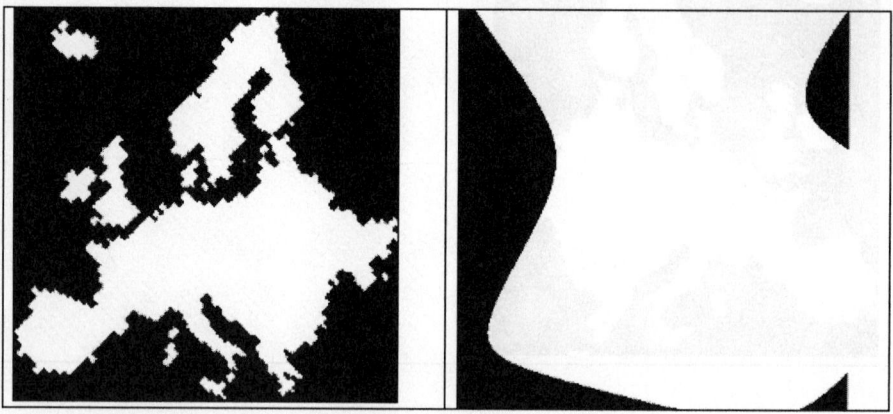

Fig. 5. A map derived from the reduced data set using a nearest neighbour algorithm (left) and a map of the relationship learned from the reduced data set by a network that uses conventional neurons (right)

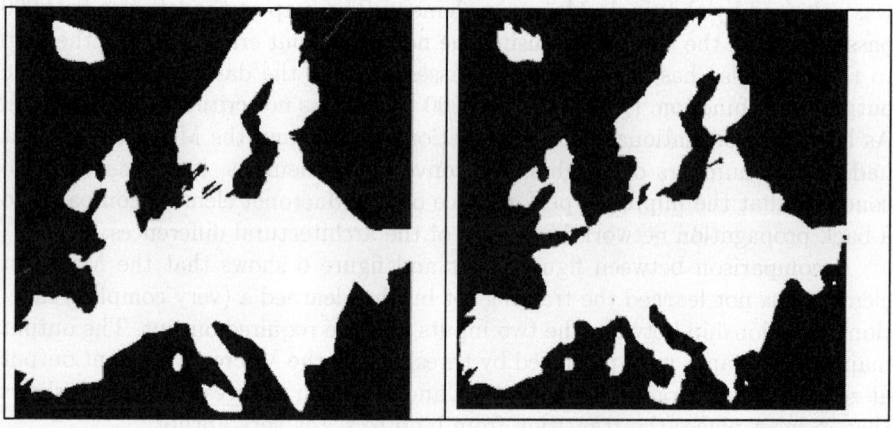

Fig. 6. Two maps of the relationships learned from the reduced data set by a single Macronet element. The map on the left is after training for 8000 passes through the data set using a nornmal output error function, the map on the right has also been trained for a further 8000 passes through the data set using a cubic error function

1021 weights values (simple numbers). Generating an output for a given pair of input values using the Macronet element took slightly longer than is required to generate the equivalent point using the closest neighbour map (40% more with these inplementations), a small factor considering the complexity of the three dimensional relationship between the inputs and output that the Macronet element has built.

4 Conclusion

It has been shown that a single Macronet element can learn an extremely complex function but that it is still able to generalise owing to the relatively small number of weights used. This believed to be as a result of having combinations of higher orders of the inputs available. It is not yet clear how to optimally distribute resources (weights), in particular how to balance the complexity of the channel structures against the number of channels per path. Preliminary work on assembling networks made of Macronet elements has begun with encouraging results. These, however, increase the problems of finding the optimum resource allocation by adding an extra decision: wether to have a small network with complex Macronet elements or a large network of simple Macronet elements.

Not withstanding these remaining questions, it is clear that the Macronet element can learn far more complex relationships than conventional neurons while retaining the all important ability to generalise. This should allow networks built from Macronet elements to be used in a range of applications that have defeated conventional artificial neural networks.

References

1. Ghosh, J. et. al. "Proceedings of the IEEE Conference on Neural Networks for Ocean Engineering" (1991).
2. Giles, C. L. and Maxwell. T. "Applied Optics" **26**, 4972-4978 (1987).
3. Klir, J. a. Valach, M. "Cybernetic Modelling" Iliffe Books, London. (1965).
4. Lundberg, J. M. and Hokfelt, T. "Neurotransmitters in Action" Elsevier Biomedical Press New York. (1985).
5. Picton, P. D. "2nd. International Conference on Artificial Neural Networks" 290-294 (1991).
6. Picton, P. D. "Neural Networks" Palgrave, New York. (2000).
7. Poggio, T. "Biological Cybernetics" **19**, 201-209 (1975).
8. Rumelhart, D. E. and McClelland, J. L. "Parallel Distributed Processing" MIT Press, Cambridge, MA. (1986).
9. Shin, Y. and Ghosh, J. "Proceedings of the Conference on Artificial Neural Networks in Engineering" (1991).
10. Shin, Y. and Ghosh, J. "Proceedings of the International Joint Conference on Neural Networks" (1992).
11. Shin, Y. and Ghosh, J. "International Journal of Neural Systems" **3** (1992).

Genetic Algorithm Optimisation of Mathematical Models Using Distributed Computing

S. Dunn[1] and S. Peucker[2]

[1] Airframes & Engines Division
Shane.Dunn@dsto.defence.gov.au
[2] Science Corporate Information Systems
Defence Science & Technology Organisation, Fishermens Bend, Victoria, Australia
Scott.Peucker@dsto.defence.gov.au

Abstract. In this paper, a process by which experimental, or historical, data are used to create physically meaningful mathematical models is demonstrated. The procedure involves optimising the correlation between this 'real world' data and the mathematical models using a genetic algorithm which is constrained to operate within the physics of the system. This concept is demonstrated here by creating a structural dynamic finite element model for a complete F/A-18 aircraft based on experimental data collected by shaking the aircraft when it is on the ground. The processes used for this problem are easily broken up and solved on a large number of PCs. A technique is described here by which such distributed computing can be carried out using desktop PCs within the secure computing environment of the Defence Science & Technology Organisation without compromising PC or the network security.

1 Introduction

There are many circumstances where mathematical models are checked against real, or experimental, data in order to gain an appreciation for the fidelity of the modelling. A circumstance where such checks are compulsory concerns the mathematical modelling that is performed to assess the aeroelastic stability of aircraft structures. If an aeroelastic instability, or flutter, were to arise on an aircraft in service, it can arise with little, or no, warning and the results can be catastrophic, resulting in the loss of the aircraft. Mathematical models are typically used to estimate the airspeed at which such an instability will arise. Given the dire consequences that can result if the mathematical model is flawed, it is a requirement of the airworthiness standards that, where such modelling is used, the structural aspects of the model must be validated against experimental data from a ground vibration test (GVT). The standards, however, are silent on how the model should be improved if it is found to be deficient. A process which has been developed by one of the authors [1], uses a genetic algorithm (GA) to create a mathematical model by determining an optimal set of mass, stiffness and damping properties such that the best agreement between the experimental data and the model predictions is achieved.

Optimisation of mathematical models can raise many questions with regards to the suitability of the nature of the model and the experimental data being used to assess the model. The best way of dealing with these issues is to create an entirely new model within the context of the information content of the experimental data available. The processes to do this - described in [2] and outlined here – involve solving many GAs in parallel, and comparing the results for the solutions with the best cost functions.

A process involving the solution of many GAs in parallel is very amenable to being broken up and solved over many computers, where each is given a complete GA run to solve. With Local Area Networks (LANs) now being commonplace, access to computers over the LAN is relatively easy. Over many LANs, there is a vast untapped computing resource that would be of great value when solving such numerically intensive problems. However, in a secure computing environment, such as that which operates within the Department of Defence, there may be security implications in allowing another user access to a wide range of users' hard disks and with leaving networked computers unattended or unprotected. A process which will be detailed here, addresses these concerns by setting up a system by which desktop PCs can be used at times when they can be dedicated to the task, such as during out-of-office hours, in such a way that the hard disk on the PC is never accessed and the network security is not compromised.

2 Structural Dynamics Modelling

The mathematical model used to estimate the structural dynamic properties of an F/A-18 is a NASTRAN finite element model and can be represented as shown in Fig. 1 where it can be seen that the 'stick model', is a vast simplification of the actual structure (as shown on the left wing tip). Even though this model is greatly simplified, there are still many thousands of separate pieces of information required to fully describe it: this information is comprised of items relating to geometry, beam and spring stiffnesses and inertial and damping properties. When a GVT is carried out, a limited amount of the experimental data is extracted - typically modal frequencies, and a qualitative description as to the mode shape. Where these results are found not to agree well with the model predictions, the question arises as to how should these many thousands of properties be updated based on limited experimental data. A typical practice is that a few parameters are chosen as variables, and these are varied until reasonable correlation is found between the limited amount of experimental data that is extracted and the model results. This is unsatisfactory from a number of perspectives:

1. only a very small amount of the available experimental data are used;
2. only a very small proportion of the model properties are available for updating; and

3. even with so few model properties available, the question remains as to whether or not there are other values of these properties that will give the same, or better correlation.

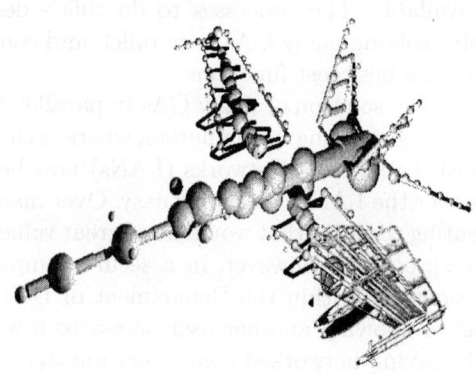

Fig. 1. Finite element model for structural dynamic analysis of an F/A -18

The challenge is: is there a technique that can make use of all of the available data; allows all the model properties to be variables; and has a greater likelihood of finding a global optimum? Genetic Algorithms have the potential to address the first two points raised above by being able to deal effectively with difficult optimisations of many dimensions (see Sect. 4). The third point has been addressed in [3] where the information content of the experimental data dictates the number of parameters in the mathematical model (see Sect. 4.1).

3 Creating an Optimal Model

From a GVT, frequency response functions (FRFs) are often collected – an experimental FRF might be as shown in Fig. 2. Such FRFs will be collected at many locations on the structure while the aircraft is being shaken. A typical model updating exercise, however, will use data only at the peaks of the FRF curves. The aim of the exercise is to use more of the FRF data from the GVT and create an optimal simple beam/mass model that gives a satisfactory representation of the data.

The object is to find the vector of unknown properties, $\{\eta\}$, such that the error, ε, is minimised by comparing the model predictions, $\chi(\eta)$, and measurements, χ', for the n freedoms measured at the N selected frequencies, as shown in (1):

$$\min(\varepsilon(\eta)) = \sum_{j=1}^{N}\sum_{i=1}^{n}||\chi_{i,j}(\eta)| - |\chi'_{i,j}|| = \sum_{j=1}^{N}\sum_{i=1}^{n}\varepsilon_{i,j} \quad (1)$$

This cost function is demonstrated graphically in Fig. 2.

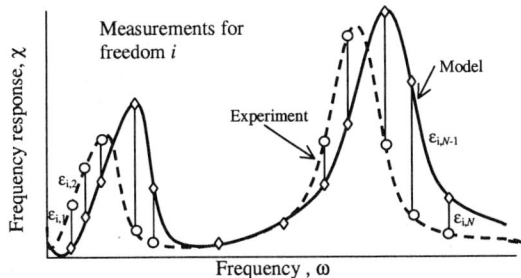

Fig. 2. Graphical depiction of the cost function for the ith freedom of (1)

The model predictions are given by:

$$\chi_{i,j}(\mu,\kappa) = \left(\phi\left([K] + i\omega[\Delta] - \omega^2[M]\right)^{-1}\right)_{i,j} \quad (2)$$

where ϕ is a vector of applied sinusoidal load amplitudes, K is the stiffness matrix, Δ is the damping matrix, M is the mass matrix, and ω is the frequency of loading. This forms the physical framework within which the optimisation is performed. The structure of the above matrices is constrained by the physics of a mass/spring/damping system, and within this optimisation procedure, this structure is maintained. If it were not maintained, better agreement between the model and the data could be achieved, but in loosing the physics behind the model it would have no predictive capabilities (it is not unknown that following other model updating procedures which have less regard for the laws of physics, the final optimised model has negative masses).

3.1 Genetic Algorithms

The inspiration for GAs arose from the realisation that the result of the principles of Darwinian evolution in nature is the attempted solution of a vast optimisation problem. Those individuals that are best suited to their environment are more likely to breed and therefore pass their genetic material onto subsequent generations. Genetic algorithms are a computer generated analogue of this process where the better individuals are chosen based on how they test against a specified cost function. For the case of using a GA to optimise mathematical models, the cost function is based on a measure of how well the model predicts actual measured data. The GA is started with a given population size of randomly generated individuals - where each individual is a potential complete solution to the problem. Based on (1), a value of *fitness*, $(1/\varepsilon)$, is then assigned to each individual. Breeding is done by selecting pairs from the original population in a weighted random process (where the weights are determined by the fitness such that the better solutions have a higher probability of breeding), and then swapping the properties of these pairs in a random manner to give rise to a new individual; this is done until a new population has been created and the processes of cost function evaluation and breeding etc. are repeated, over and over, until some stopping criterion is reached. There are many different ways of applying GAs; details of how they are applied for the case studied here can be found in [3], though a brief summary will be given as follows:

```
for j = 1 : n
   for i = 1 : N₁                              Easily parallelised through
      select random population                 distributed computing
      run GA for g generations                 environment - Beehive
      append results to Rⱼ                     (see Secn. 3.2)
   end for, i
   for i = 1 : N₂                              Serial process, best
      select random population                 handled by processors
      include previous results in Rⱼ           dedicated to the task
      run GA for g generations
      append results to Rⱼ
   end for, i
   append final result to Rf
end for, j
Examine and compare all results in R₁ to n and Rf
```

The properties the fundamental GA used here are: binary encoding; population size 30; tournament selection; effective bitwise mutation rate 0.05%; uniform crossover; and, elitism.

This process of seeding the initial population with earlier solutions has been found to more efficient than a standard 'linear' GA for these problems. Nevertheless, the computational requirements are still significant. The above process, in that it requires many independent solutions, the results of which are used as seeds for subsequent solutions, means that it is very amenable to parallel computation.

3.2 Distributed Computation – Beehive

For the solution of a simplified F/A-18 mathematical model (as shown in section 4), the system matrices in (2) are of dimension 111×111, and the most time consuming aspect of the cost function evaluation involves inverting a matrix of this size. Such an operation can easily be handled by the processor and RAM on desktop PCs without having to resort to using virtual memory on a hard disk. Therefore, it was obvious that this process could be run efficiently using desktop

PCs. With the processing power available on PCs within the Defence Science & Technology Organisation (DSTO), it was obvious that using the LAN to connect some of these computers would be a good way to process this problem.

Initially the project was set up to use PCs in the Science Corporate Information Systems Training Room when they were not in use (13 × 750MHz PIII). A dedicated Windows 98 setup was created which connected to a Windows NT system and automatically ran the GA program. This setup was distributed to the PCs via automated imaging software when the training room was not in use. This solution, while effective in this environment, had a number of shortomings: i) this solution was not easily scaleable - re-imaging client PCs every night was not an option; and, ii) if we used an "inside-Windows" environment, we left ourselves open to security concerns - users would need to be logged in to the network all the time, which would result in more user resistance to utilise *their* PC.

It was necessary to run the project in an environment that would meet with Defence security guidelines with particular respect to unattended, after-hours use of PCs. A system was needed that would work independently of the computer's usual operating system, either by a system of "dual-booting" the PC, or by using a floppy disk to boot the system. Creating a dual boot (where a separate partition is created for the Beehive operation) would be acceptable, but would require a lot more effort to set up, and would require more user intervention on bootup. With a floppy disk setup, the user can simply restart the PC with the disk in the drive and walk away. The disadvantage of the floppy disk setup was that the temporary files created by the GA program (up to 4 MB) would not fit on the disk.

It was decided to use a dedicated Novell Netware 4.11 server (Beehive), with each PC (drone) booting from a floppy disk to access this server. The server is configured with a central area from where data to be processed is retrieved and where results are written, and a separate home directory for each drone account. Intermediate results are written to temporary files in each home directory. On completion of the required number of generations, the files in the central area are updated.

Each drone PC has a customised boot disk containing the correct network card drivers, and the name of the drone account to use to login to the server. Each drone account is associated with the hardware address of the network card, so it can only be accessed by that particular PC. The operation of the drones is described as follows:

- The user starts the PC with the floppy disk in the drive.
- Network drivers are loaded, and the drone logs in to the server
- Inside a loop:
    - GA executable and data files are copied to the drone's home area
    - The program executes
- If an error occurs, the loop is restarted

The above process is controlled by files on the server. This allows new executable or data files to be placed on the server, and the drones can be remotely restarted without user intervention.

In order to allow broad expansion of the concept to as many PCs as possible, network traffic has been kept to a minimum to prevent network bottlenecks at the Beehive server. Depending on the processor speed of each Drone, each PC interrogates a file on the server approximately every 1-5 minutes to check for a 'restart' instruction, and approximately every 4-20 hours, a set of results are typically returned to the server. With PCs at DSTO laboratories in Melbourne, Adelaide and Canberra being used for this project, in order to minimise Wide Area Network traffic, another server has been installed in Adelaide, in addition to the main server in Melbourne.

There are currently around 160 PCs configured to run Beehive in five different geographic locations. The fact that we only peak at around 110 users indicates the need to 'remind' users to restart their PCs with their beehive disk in the drive.

Figure 3 shows the number of connections to the beehive server over a one month period. The sharp peaks indicate overnight use, while the flatter peaks show weekend usage. Interestingly, the troughs show that even during the day there can be some 40 PCs still processing data. This can be partly attributed to the targeting of PCs which are often not in use (training room PCs, Conference Room PCs etc.)

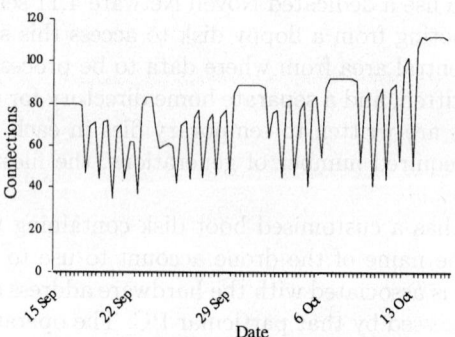

Fig. 3. Beehive usage over 1 month

The PC processors operating on Beehive range from 200MHz PII to 1.4GHz PIVs and with approximately 110 PCs connected, based on the numbers of final solutions returned, the total out-of-hours processing power is estimated at 45-50 GFLOPS using a Whetstone benchmark.

4 Optimised Structural Dynamic Models

The difficulties of solving the inverse problem for the estimation of the physical properties of a structure, based on its dynamic response, can easily be shown by simulation.

The FRFs for all 10 freedoms of a 10 degree of freedom undamped mass/spring system were simulated using (2). Assuming one of the spring stiffnesses is unknown, the cost function (1) variation with values for this stiffness is shown in Fig. 4a. Then, the process was repeated, now with two spring stiffnesses taken to be unknown; the search space now becomes a search in a two dimensional plane with the variation in cost function forming a three dimensional surface as shown in Fig. 4b. A traditional optimisation procedure will start with an initial 'guess' at the best value of the unknown spring stiffnesses, k_i, estimate the local derivative of ε with respect to k_i, and head 'downhill' – an analogy for such a process is to place a marble on the line in Fig 4a or the surface in Fig 4b and let it roll down to the valley, which is where the predicted solution will lie. It is clear from Fig. 4 that the problem is multi-modal and deceptive, in that there will be many apparent solutions, and the number of these apparent solutions grows rapidly with the dimensionality of the search space.

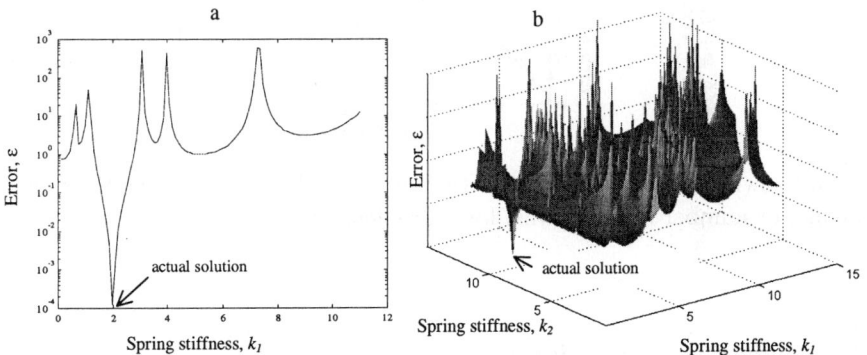

Fig. 4. a & b, The search space for the simulated 10 dof mass/spring system with one and two unknown spring stiffnesses respectively

This simulated case shows how a search in up to 10 dimensions can be searched effectively by a GA. For the case we are interested in here – optimizing the finite element model for a complete aircraft – the problems involves a search in >100 dimensional space.

The problem for the unknown spring stiffnesses was then solved for increasing numbers of unknown spring stiffnesses using a technique that relies on local derivative information (in this case a simplex algorithm was used) and a genetic algorithm. The simplex algorithm started at an initial guess within the search space and was re-started when the previous solution had converged (ie. the cost function could be improved no further within a specified tolerance). The average number of cost function evaluations (proportional to the solution time) required for a correct solution could then be taken as: (the number of correct solutions) / (the total number of cost function evaluations). In a similar manner, the average number of cost function evaluations required for a correct solution from a GA can be estimated when it is searching within the same search space as the simplex algorithm. Doing this for increasing numbers of unknowns (or, increased dimensionality of the search space) yields the results as shown in Fig. 5. In Fig. 5, it can be seen that the GA becomes relatively very efficient as the number of unknowns increases beyond 4. The dimensionality of the search space for the F/A-18 model addressed in Section 4.2 is of order 150.

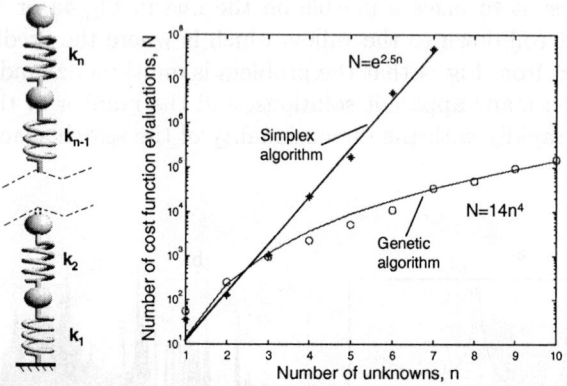

Fig. 5. Showing the advantage of a GA over a simplex algorithm as the dimensionality, or number of unknowns, for a problem grows

4.1 Model Complexity Issues

A fundamental feature of this modelling is that, *a priori*, there is very little information as to how complex the model should be to provide satisfactory agreement with the experimental data; a measure of model complexity can be taken as the number of unknown properties required. A method of determining models of minimal complexity is described in detail in [3] and will be briefly described here:

- Run the optimisation procedure - in this case a GA - a number of times such that there are a number of results where the better cost functions (1) are very similar and the model predictions give a satisfactory representation of the experimental data;
- compare the properties found for these results;

- where this comparison shows little variation, assume the property is being determined uniquely;
- where the comparison shows a great deal of variation for similar cost function, assume that either the property is not required, or that property and one or more of its neighbours can be combined into one;
- repeat this process until all parameters appear to be defined uniquely *and* the model still gives a satisfactory representation of the data.

This process for dealing with the model complexity issues is clearly very numerically intensive in that many GAs must be solved and the results compared. For such a problem, the computational power of *Beehive* is particularly valuable.

4.2 F/A-18 Model Optimisation

Bombardier, under contract to the Canadian Forces, carried out a GVT on an F/A-18, instrumented as shown in Fig. 6a. After carrying out the optimisation (where the unknowns involved the inertial properties, beam stiffnesses and locations, connecting and undercarriage spring stiffnesses and damping elements), the resulting model is represented in Fig. 6b.

Fig. 6. a. Diagram showing the instrumentation and force locations from the test described in Dickinson[4]. The accelerometers were placed symmetrically on the aircraft, although only those on one side are shown. And, b, the Optimised model for an F/A-18

The resulting correlation between experimental and model data for a selection of degrees of freedom and excitations are shown in Fig. 7.

5 Conclusion

In the mathematical modelling of complex systems, many approximations and simplifications must always be made. The implication of this for the model's fidelity is often not clear. Such models are often validated by correlating their

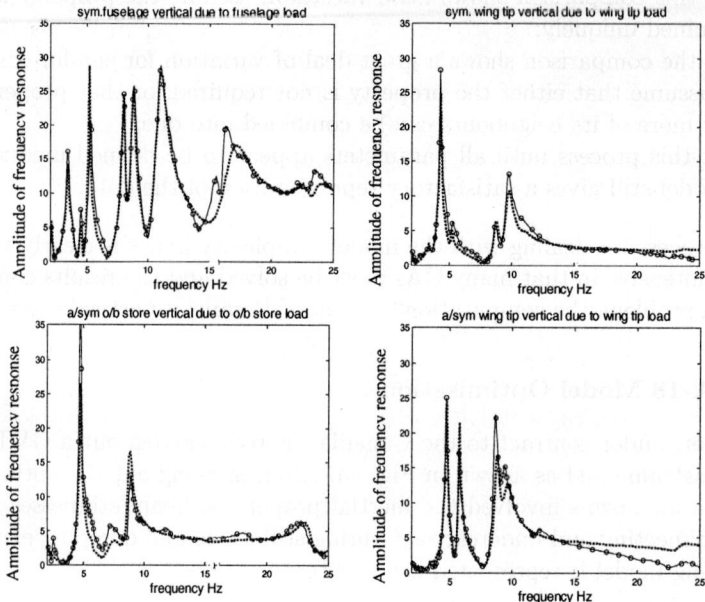

Fig. 7. Ground vibration test measurements (solid line) compared with optimised model predictions (dashed line)

predictions against experimental or historical data. A technique has been demonstrated here by which such experimental data can be used to create a simplified, physically meaningful, mathematical model using a genetic algorithm optimisation tool. This been demonstrated for structural dynamic models of aircraft structures, but there is no reason why such procedures could not be used for any other model where there is good experimental, or historical, data available, and there is a well-understood physical framework within which the model building can operate.

The optimisation procedures discussed here are highly numerically intensive, but the cost function can be determined efficiently within the confines of modern desktop PCs. A system has been described whereby a group of PCs, connected by a LAN, can be used for such a task without compromising the security of each computer or the network on which they operate.

Additional work that will be carried out, using the processes described here, will be an optimisation of the mathematical model for the flying aircraft using flight test data collected during recent Royal Australian Air Force flight trials.

Acknowledgements

The authors would like to gratefully acknowledge the assistance of David Smith and Justine Perry of DSTO's Science Corporate Information Systems whose constant support and encouragement for the Beehive project have been invaluable. Also, Chris Bee, Assistant Secretary, Science Corporate Management, Department of Defence, for her support for the distributed computing concept which led to the creation of Beehive. Lastly, many thanks are owed to the over 100 people throughout DSTO who regularly log their computer into the Beehive for use at night and over weekends making this work possible.

References

1. Dunn, S. A.: Optimised Structural Dynamic Aircraft Finite Element Models Involving Mass, Stiffness and Damping Elements. International Forum on Aeroelasticity and Structural Dynamics, Madrid, Spain, 5–7 June, (2001), 387–396
2. Dunn, S. A.: Modified Genetic Algorithm for the Identification of Aircraft Structures. J. Aircraft, **34**, 2, (1997), 251–253
3. Dunn, S. A.: Technique for Unique Optimization of Dynamic Finite Element Models. J. Aircraft, **36**, 6, (1999), 919–925
4. Dickinson, M.: CF-18/GBU24 Ground Vibration Test Report. Bombardier Inc. Canadair, Montreal, Canada, (1995).

Genetic Algorithm Optimisation of Part Placement Using a Connection-Based Coding Method

Alan Crispin[1], Paul Clay[1], Gaynor Taylor[1], Robert Hackney[2], Tom Bayes[2], and David Reedman[3]

[1] School of Engineering, Leeds Metropolitan University
Calverley Street, Leeds LS1 3HE, England.
{a.crispin,p.clay,g.e.taylor}@lmu.ac.uk
[2] SATRA Ltd.
Rockingham Road, Kettering, Northamptonshire, N16 9JH, England.
{bobh,tomb}@satra.co.uk
[3] R&T Mechatronics Ltd. The Cottage
Main Street, Wartnaby, Melton Mowbray, Leicestershire, LE14 3HY, England.
rtmechatronics@clara.net

Abstract. The problem of placing a number of specific shapes in order to minimise material waste is commonly encountered in the sheet metal, clothing and shoe-making industries. It is driven by the demand to find a layout of non-overlapping parts in a set area in order to maximise material utilisation. A corresponding problem is one of compaction, which is to minimise the area that a set number of shapes can be placed without overlapping. This paper presents a novel connectivity based approach to leather part compaction using the no-fit polygon (NFP). The NFP is computed using an image processing method as the boundary of the Minkowski sum, which is the convolution between two shapes at given orientations. These orientations along with shape order and placement selection constitute the chromosome structure.

1 Introduction

The general cutting stock problem can be defined as follows. Given a set of N shapes each of which can be duplicated k times, find the optimal non-overlapping configuration of all shapes on a sheet. The textile industry refers to the stock cutting problem as marker making and sheets are rectangular taken from rolls of material [1]. NC leather sample cutting requires that a given group of shapes be compacted so that they can be cut from small sections of graded hide using an x-y cutting table.

Many packing-driven solutions to the cutting stock problem have been reported in the literature [2]. Anand et. al. [3] and Bounsaythip et. al. [1] have explored genetic algorithm approaches. One of the main difficulties reported was finding an appropriate encoding strategy for mapping shapes evolved by the genetic algorithm to non-overlapping configurations. Anand [3] used an unconstrained approach whereby the genetic algorithm was allowed to evolve overlapping configurations and post processing was subsequently used to check and correct for shape intersection. Bounsaythip et. al. [1] have used an order based tree encoding method to pack components in strips.

Kahng [4] discusses the preoccupation with packing-driven as opposed to connectivity-driven problem formulations in the context of floor planning (i.e. automated cell placement for VLSI circuits) and suggests that approaches based on a connectivity centric should be used. This paper presents a new genetic encoding method based on shape connectivity and investigates shape compaction using this approach.

2 Shape Connectivity

To apply the idea of connectivity a description is required for the relative placement of shapes so that any two touch without overlapping. This can be achieved using the no-fit polygon [5], [6], which describes the path traced by a point on one shape as it circumnavigates around the other shape. Rules and graphs can be defined to describe which pairs of shapes are required to calculate the NFP.

A number of different methods have been devised to calculate the NFP between two shapes involving the addition of ordered vectors describing the polygonal edges of the individual shapes [7]. The basic algorithm only works correctly for convex polygons and a number of adjustments have been made to accommodate the relationship between both simple non-convex [8] and more complicated non-convex shapes [9].

Clay and Crispin [10] have used an alternative approach to obtain the NFP based on image processing techniques. The NFP is found from the Minkowski sum, which is the convolution between two shape images at a set orientation [11]. The NFP is the boundary between the area covered by the Minkowski sum and those areas not covered by the sum. Four-connected shape boundary images can be used to calculate the Minkowski sum with the NFP extracted using morphological operators.

This image processing technique for NFP calculation allows direct Boolean manipulation for fast identification and removal of crossing points of NFPs. The crossing points of two individual NFPs are required to find the points at which a shape can touch two other shapes at the same time. Possible crossing points of the NFPs may lie within the Minkowski sum between the shape to be placed and shapes other than the ones it is required to touch. These Minkowski sums are used as masks to remove these crossing points from the list of possible solutions and so prevent overlapping placements being chosen.

Figure 1 shows three shapes with fixed angles and their associated NFPs. In this example, shape1 is placed at a set position and all other shapes connect to it and each other. Shape 2 can be moved around shape1 along NFP 1-2 and one point on this boundary is selected. Valid positions for shape 3 to touch both shapes 1 and 2 are at any crossing points between NFP 1-3 and NFP 2-3. Therefore the three shapes are connected via the NFP. This connectivity approach can be scaled to any number of shapes by computing NFP intersection points. If a fourth shape is required to touch shapes 1 and 3, possible solutions may lie such that shape 2 is overlapped by shape 4.

To prevent overlapping, all points within the area of the Minkowski sum between shapes 2 and 4 are removed using a mask.

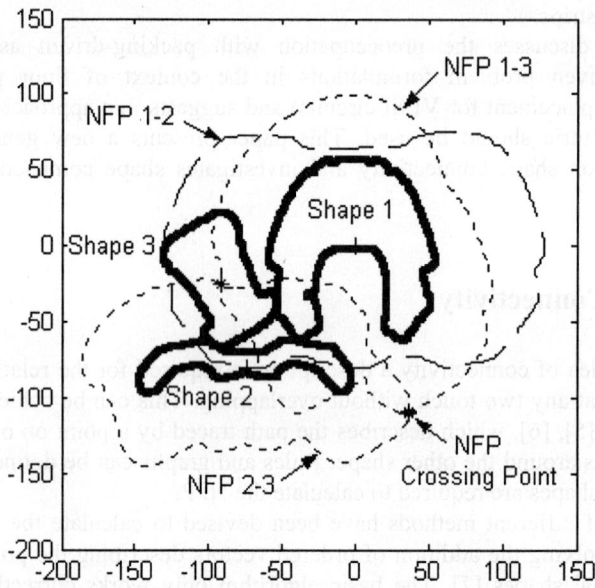

Fig. 1. No-fit polygons for three shapes. Shape 1 is placed at the origin and Shape 2 is placed on a chosen point on NFP 1-2, shown as a long dashed line. Shape 3, must be placed on a crossing point of NFP 1-3 and NFP 2-3, shown as short dashed lines. The crossing points are shown as asterisks

This approach can be described using graphs that specify the connectivity between shapes. Each of the graph nodes represents a separate shape with its corresponding properties of shape type and angle of orientation. Graph edges between nodes indicate the connectivity of those shapes that touch. For each individual connectivity graph there might be a number of placement solutions, which are dependent on shape topology and orientation. However some combinations of shapes can render a graph invalid since it would require shapes to overlap others to satisfy the connectivity criteria. To calculate the possible points of connection a shape must be placed on the intersection of the NFPs between that shape and those other shapes it touches as determined by the connectivity graph.

3 Graph Topologies

It is envisaged that a number of graph topologies can be used to describe the connectivity between shapes. Fixed topology graphs set the connectivity, while the shape type and orientations are adjusted to create a valid placement within the graphs constraints. With this type of topology the complete graph may not always be realised with fixed node properties, since the existence of NFP crossing points are dependent on the properties and the placement of those nodes placed previously. Alternatively, rules may be defined to create the graph at time of placement. Here the shape properties are fixed and the graph edges varied to fit the lay that develops due to the rules. An example of the specific case of this variable topology is shown in Figure 2. In this case each shape node is connected using the rule that it touches the previous node and an anchor node.

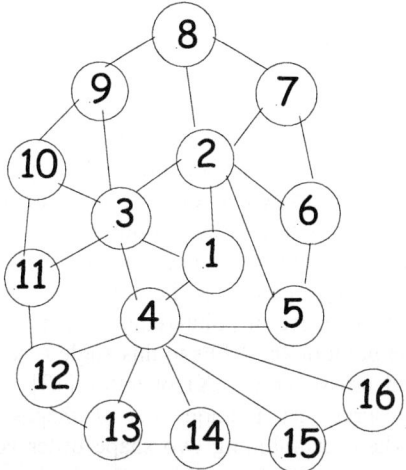

Fig. 2. An example of a variable shape connectivity graph for 16 shapes. In this specific case shape 1 is the anchor for shapes 2 to 4, while shape 5 is unable to touch both shapes 4 and 1 without overlapping another shape. Shape 2 becomes the new anchor for shapes 5 to 8 until shape 9 is unable to touch both shapes 2 and 8. Similarly shape 3 is the anchor for shapes 9 to 11 and shape 4 is the anchor for all other shapes shown

The shapes build up in a ring around a central anchor shape until no more shapes can touch it and the preceding shape without overlapping another shape. The shape placed directly following the anchor node shape now becomes the new anchor node and the overlapping shape and those after are placed around this new anchor shape until overlapping occurs again. Additional rules are required to specify the action required when the general algorithm fails to create a valid placement.

4 Optimising Compaction

A measure of compaction of a set of connected shapes for optimisation can be achieved by minimising the convex hull area as shape order, shape angle and NFP intersection point are varied. To exhaustively search all possibilities would require significant search time. It is known that evolutionary search methods such as genetic algorithms are well suited to the optimisation of combinatorial problems with large search spaces [12]. Given an adequate coding method it would be prudent to apply one to the compaction problem.

The key features of genetic algorithms are:

- They start with an initial random population of potential solutions.
- The population is indexed in terms of fitness using an objective function.
- High fitness members are randomly selected and recombined to produce an offspring.
- A small proportion of the offspring mutate to extend the search.
- Fitness values for the offspring are calculated.
- The offspring and their corresponding fitness replace poorly performing members of the original population.

Iteration is used to generate more and more offspring solutions, which it is hoped will yield better approximations to the final optimal solution [13] [14].

In this study, from an initial random selection of 20 shapes from 6 different shape types, the genetic algorithm has to evolve a combination of shape order, shape angle and NFP intersection point position in order to optimise the compaction measure. For a set of N connected shapes, vectors can be used to store shape order (represented as a list of N unique numbers) and shape angles (i.e. a list of angles associated with each shape). To select an NFP intersection point requires generating a placement angle. The closest valid crossing point on the NFPs to this angle is selected for placement.

The partially matched crossover (PMX) operator [12] is used for shape order recombination as this ensures that a unique list of shapes (i.e. without repeating elements) is produced when recombining two shape order vectors. The real valued line recombination operator [15] can be used for recombining shape angle vectors and placement angle values.

The compaction measure used for fitness evaluation has been taken as convex hull area multiplied by the square of the ratio of the number of shapes placed to the number of shapes required. If all the required shapes are placed the value is equivalent to the absolute area of the convex hull, any less would increase the value sufficiently to effectively eliminate sub-standard placements from the solutions. This ensures that the GA evolves the smallest convex hull area for the maximum number of shapes.

5   Chromosome Structure

Initially the chromosome is separated into three separate sections:

Shape order   -   set of 20 unique non-repeating integers.
Angle of shape   -   20 modulo 2π real values.
Angle for next placement -   19 modulo 2π real values.

Each chromosome section has it's own population, all with the same number of members. The members in all three sections are indexed by a single position in the population so that a single member will consist of three separate sections from the chromosome lists. On placement the radian value at a chosen locus in the shape angle chromosome section rotates the shape indicated in the corresponding locus in the shape order chromosome section.

Each of the angles in the shape angle chromosome section is associated with the shape at the corresponding locus in the shape order chromosome section. When the shapes are reordered, the associated angles are also reordered to the same positions as the corresponding shapes. Consequently, a shape angle is a property of a shape since it is linked to the shape order.

To associate the properties of shape types it was decided to only allow the shape angles from corresponding shapes in the list to recombine. This requires the angles to be reordered so that the corresponding shapes from all chromosomes are in the same order and returning their offspring to their parent's individual orders after creation. The result would be that the shape angle properties of a shape would only recombine with a like shape irrespective of shape order. The angles are then reordered to follow their respective loci in the shape order chromosome section as described before.

The angles for next placement describe a method of positioning a shape relative to the last shape placed, when a choice is given, and remain independent of the shape properties. Observation and analysis of the effect of these values suggest that the angles in the first two loci show the greatest importance to the final lay, while the rest have more subtle consequences.

6 Results

The GA was allowed to optimise for 30 generations after the calculation of the initial population. Population sizes of 5, 15 and 25 were tested and averaged over a number of runs. Each run began with a random set of twenty from a selection of the same six shapes. Compaction is measured as a ratio of the area of the convex hull to the area actually covered by the selection of shapes made. The theoretical maximum compaction of unity can only be achieved by the placement of shapes that tessellate leaving no gaps and such that the extreme boundary of all the shapes placed would also be the convex hull of the tessellation.

Figure 3 shows the average compaction ratios for the tested population sizes. It can be seen that the initial compactions lie within the 60% to 75% wastage range. Larger random populations are more likely to result in good initial compactions and although the wastage begins at a reasonably small value, the GA has been able to produce compaction improvements in the order of 10% to 15%.

Figure 4 shows the compaction of 20 leather shoe parts using the method described in this paper. The connection between shape centres is drawn to illustrate the spiral nature of the placement, which results from the graph topology as shown in Figure 2.

7 Conclusions

This paper has presented a novel method for using a genetic algorithm to compact a set of connected shapes. The method is novel being based on using rules to connect shapes at NFP crossing points so that non-overlapping placements are always made. A GA is used to evolve shape order, shape angle and placement angle to minimise convex hull area of a set of twenty connected shapes. It has been observed that the connection-based coding method results in compact placements even for the initial random populations. Experiments have shown that, even with a good initial lay, iterative improvement of compaction still results from this approach.

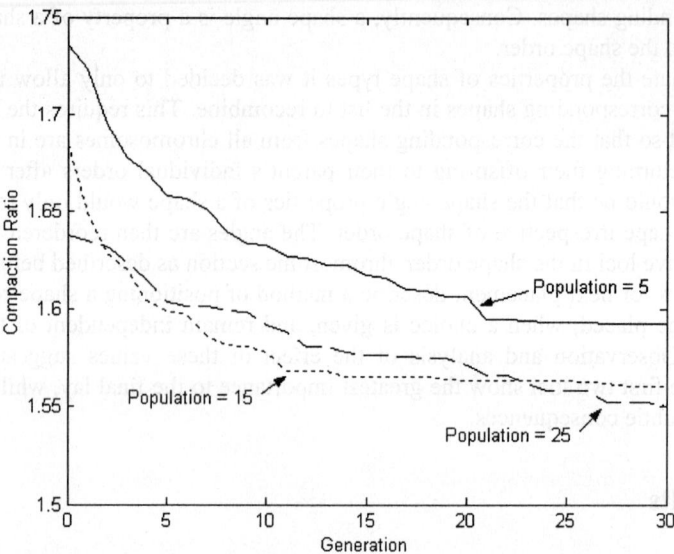

Fig. 3. Compaction area improvement over 30 generations as a ratio of area of convex hull to total area of shapes placed. Population sizes of 5 (solid line), 15 (dotted line) and 25 (dashed line) are shown

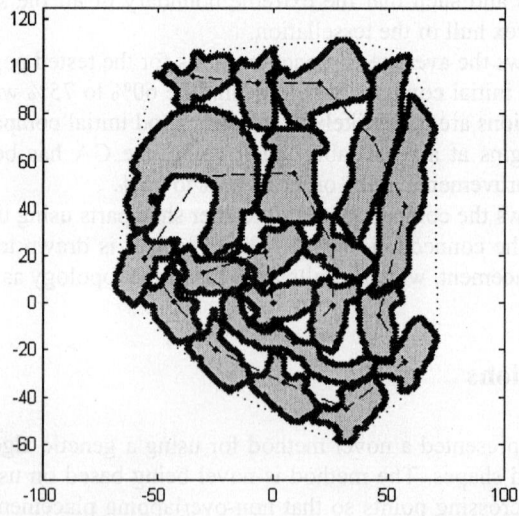

Fig. 4. Placement of shapes after compaction using a GA for 30 generations. The order of placement begins with the shape at the origin and follows the dashed line in a spiral. The ratio of compaction between the area within the convex hull (shown as a dotted line) and the area actually covered by shapes is 1.413, giving 41.3% wasted area

Acknowledgements

The authors gratefully acknowledge the support of this research by the U.K. E.P.S.R.C. grant on "A Genetic Algorithm Approach to Leather Nesting Problems", reference number: GR/M82110.

References

1. Bounsaythip C., Maouche S. and Neus M. (1995), Evolutionary search techniques: Application to automated lay-planning optimisation problem, IEEE international conference on Systems, Manufacturing and Cybernetics (SMC), Vancouver, Canada, Vol. 5, pp. 4497-4503.
2. Dowsland, K. A. Dowsland, W. B. (1995), Solution approaches to irregular nesting problems, European Journal of Operations Research, 84, pp. 506-521. Elsevier.
3. Anand, S. McCord, C. Sharma, R. Balachander, T. (1999), An integrated machine vision based system for solving the non-convex cutting stock problem using genetic algorithms, Journal of Manufacturing Systems, Vol. 18, No. 6, pp. 396-415.
4. Kahng, A. (2000), Classical Floorplanning Harmful, International Symposium on Physical Design, Proceeding ACM/IEEE, pp. 207-213.
5. Adamowicz M. and Albano A. (1976), Nesting two-dimensional shapes in rectangular modules, Computer-Aided Design, Vol. 8, No. 1, pp. 27-33.
6. Burke and Kendall G. (1999), Evaluation of two-dimensional bin packing using the no fit polygon, Computers and Industrial Engineering Conference, Melbourne, Australia.
7. Cunninghame-Green, R. (1989), Geometry, shoemaking and the milk tray problem, New Scientist, Vol. 12, August, 1677, pp. 50-53.
8. Ghosh, P.K. (1993), A unified computational framework for Minkowski operations, Computer and graphics, 17(4), pp. 119-144.
9. Bennell, J.A. Dowsland, K.A. Dowsland, W.B. (2000), The irregular cutting-stock problem – a new procedure for deriving the no-fit polygon, Computers and operations research, Vol. 28, pp. 271-287.
10. Clay P. and Crispin A. (2001), Automated lay-planning in leather manufacturing, 17th National Conference on Manufacturing Research, September 2001, Cardiff, pp 257-262.
11. De Berg, M. van Kreveld, M. Overmars, M. Schwarzkopf, O. (1998), Computational Geometry Algorithms and Applications 2nd Ed. Springer Verlag, Berlin, pp. 275-281.
12. Goldberg, D. E. (1989), Genetic Algorithms in Search, Optimisation and Machine Learning, Addison Wesley, Longman Inc. ISBN 0-020-15767-5
13. Mitchell, M. (1996), An introduction to Genetic Algorithms, Massachusetts Institute of Technology Press, ISBN 0-262-13316-4.

14. Clay P., Crispin A. and Crossley S. (2000), Comparative analysis of search methods as applied to shearographic fringe modelling, Proceeding of 13th International Conference on Industrial and Engineering Applications of Artificial Intelligence and Expert Systems, June 2000, New Orleans, pp. 99-108.
15. Mühlenbein, H. and Schlierkamp-Voosen, D. (1993), Predictive Models for the Breeder Genetic Algorithm, Evolutionary Computation, Vol. 1, No. 1, pp. 25-49.

A Fast Evolutionary Algorithm for Image Compression in Hardware

Mehrdad Salami and Tim Hendtlass

Centre for Intelligent Systems and Complex Processes,
School of Biophysical Sciences and Electrical Engineering,
Swinburne University of Technology
P.O. Box 218, Hawthorn, VIC 3122 Australia
{msalami,thendtlass}@swin.edu.au

Abstract: A hardware implementation of an evolutionary algorithm is capable of running much faster than a software implementation. However, the speed advantage of the hardware implementation will disappear for slow fitness evaluation systems. In this paper a Fast Evolutionary Algorithm (FEA) is implemented in hardware to examine the real time advantage of such a system. The timing specifications show that the hardware FEA is approximately 50 times faster than the software FEA. An image compression hardware subsystem is used as the fitness evaluation unit for the hardware FEA to show the benefit of the FEA for time-consuming applications in a hardware environment. The results show that the FEA is faster than the EA and generates better compression ratios.

Keywords: Evolutionary Algorithm, Evolvable Hardware, Fitness Evaluation, Image Compression

1 Introduction

Evolutionary algorithms (EAs) were inspired from the evolutionary processes observed in the natural world 1. By following the processes of natural selection, recombination and mutation, a population of individuals can evolve to produce fitter individuals. By working on a population of individuals, there is no need to concentrate on a specific area of the solution space and all possible solutions can be considered. Therefore the EAs are suitable when there is little information about the problem at hand. Although EAs can be used for machine learning or artificial life problems, they are mostly used for optimisation and search problems 2. While the EAs have implicit parallelism in searching the problem space they become very slow when applied to very complex or hard problems. Furthermore efficient EAs require the population size to be large and run over a large number of generations, making the EA very slow. This paper uses a version of an evolutionary algorithm, a fast evolutionary algorithm (FEA). The FEA is fast even for large population sizes, allowing more generations to be processed in a given time 3.

There have been various hardware implementations of conventional evolutionary algorithms 4 5 6. However, there is little benefit to having a very fast EA in hardware if the fitness evaluation is slow. For any hardware EA if the time for fitness evaluation is more than a critical time then the advantage to be gained from a fast EA is minimal. Ironically faster EA hardware systems decrease this critical time and limit the possible application for the EA hardware.

The Fast Evolutionary Algorithm (FEA) is a method that reduces the number of evaluations by assigning a fitness value to a new child rather than evaluating it. Obviously, provided that the fitness value is not the result of a stochastic process, a parent's fitness can be assigned to a child that is identical to that parent, which may save some fitness evaluations. However, if a child is different from its parents, an algorithm is needed to assign a fitness value to the child. As this assigned fitness value is not the real fitness value, another parameter is needed to show how reliable the assigned value is thought to be. Only when this reliability falls below some threshold is the child sent for a real fitness evaluation. It has also proven useful to include a number of randomly selected true evaluations even if the child being evaluated has a reliability value above the threshold. Further, since there will be occasions when both a true fitness value and an assigned fitness value are available, it is possible to apply a compensation mechanism to correct for the tendency of the assigned fitness value that diverge from the true fitness value. For further details on the FEA see 3.

The FEA is implemented in hardware to examine the real time advantages of the system. Image compression was chosen as an application for the FEA hardware,. In image compression 7, the fitness calculation is complex and the option of reducing the number of times the fitness evaluation unit is used per generation can save a significant amount of processing.

Image compression involves classification of the current data that is then compressed using a method suitable for that class. One of the standard algorithms for the compression of images is JPEGLS 8, which can provide either lossy or lossless compression depending on the NEAR parameter. The degree of information loss is described by the picture signal to noise ratio (PSNR). The lower the PSNR the larger the information loss.

The JPEGLS algorithm has four classification parameters that control which class should be used to compress the current pixel. An evolutionary algorithm can be used to determine these parameters for any image and produce the best compression ratios. Normally one image contains a huge number of pixels and for fitness evaluation all pixels must be examined. That means each fitness calculation is very time consuming and a typical evolutionary algorithm will not be efficient at finding the best compression ratio in a reasonable time. However with the FEA, it is possible to achieve similar results in a much shorter time.

This paper describes a hardware model of the FEA implemented on an FPGA (Field Programmable Gate Arrays). Section 2 shows the internal architecture of the hardware and the timing comparison of the hardware with a similar software implementation. Section 3 displays how the FEA is used for image compression application in hardware. Later the compression performances are compared with various configuration of the FEA for four images.

2 Hardware Implementation of the Fast Evolutionary Algorithms

The main advantage of the FEA is to estimate the fitness of new individuals and quickly prepare the population of the next generation. The FEA could be used in an evolutionary algorithm application to reduce the evolution time. However, such an approach requires that the time for calculating a fitness value must be relatively large compare to the time for preparing individuals. If we assume the fast evolutionary algorithm is running on a CPU with sequential instruction processing, then at any time the CPU is either running the evolutionary algorithm, a fitness evaluation or a fitness estimation. If the time for calculating the fitness is much larger than the time for the other two tasks, then the FEA can produce better individuals faster than a typical evolutionary algorithm.

The advantage of the FEA can be utilised in a real time system, where it can be implemented in hardware, leading to a faster and more efficient system. In hardware the processes can be run in parallel.

Fig. 1 shows the structure of a system that implements a control unit with three other peripheral units. In this configuration, the estimation and evolution units can operate continuously in parallel. The fitness evaluation unit can also be run in parallel with the other units but is only activated whenever a fitness evaluation is required.

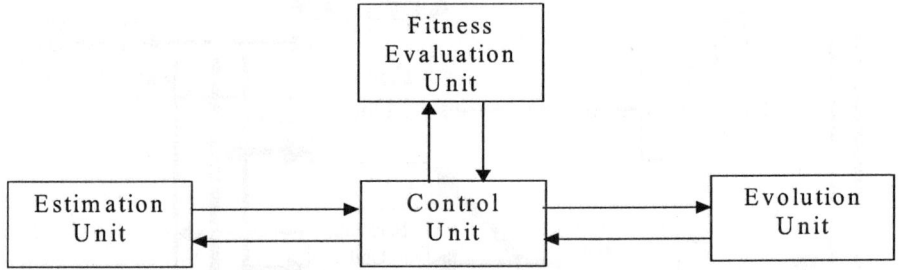

Fig. 1. The communication between three processes and the control unit

Evolution, estimation and fitness calculation can operate in parallel because they are independent of each other. While the estimation unit processes a pair of individuals, the evolution unit can be generating the next pair. As a result the time for estimation can be hidden and does not need to be compared to the time for fitness evaluation.

However these multiple processes may need access to common resources and, to avoid a conflict in the system, access to these common resources should be handled carefully. There are two main resources that can be accessed by these processes: Memory and Fitness Evaluation. To reduce the number of signals in the design, a bus line is allocated for each one of these common resources and the control unit determines the right of access.

Fig. 2 shows the block diagram of the fast evolutionary algorithm (FEA) hardware. The heart of this design is the control unit. The random unit reads the initial seed and starts generating random numbers. At each clock cycle, this unit will generate a new random number. The control unit initialises the memory using the output of the random unit to write a new random population into the memory.

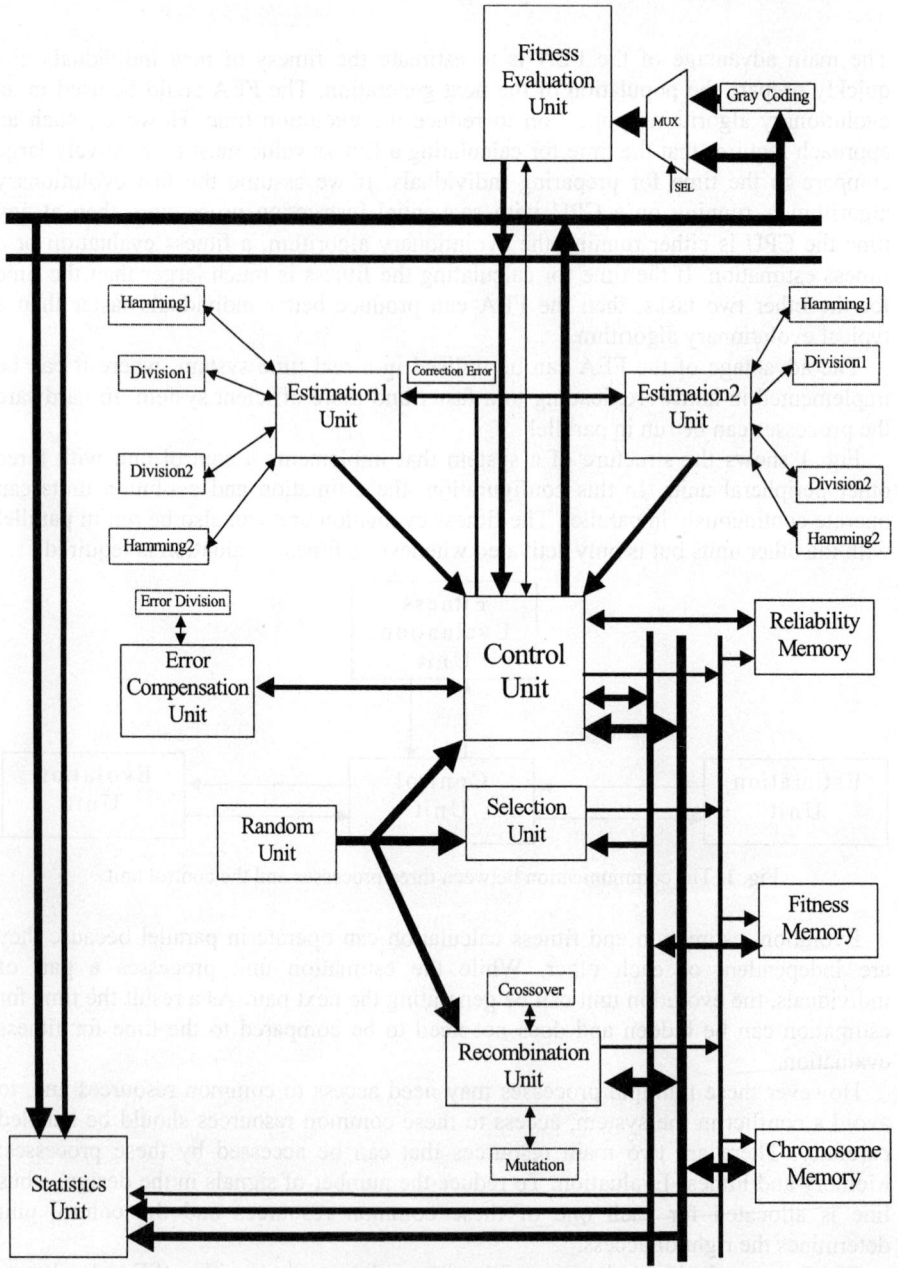

Fig. 2. The block diagram of the fast evolutionary algorithm

At any time after this initialisation there will be two separate populations in the memory. Individuals are selected from the current population using the evolutionary operators. Since the individuals are created sequentially, the new individuals must be written in a different location than the original population to allow creating the remaining individuals of the new population from the same original population. At the end of each generation, the index locations of these two populations are swapped removing the need to copy one population into the other.

The estimation units are also active during the creation of new individuals and provide the fitness and the reliability values for the newly created individuals. Since the time difference between creating two individuals in the evolutionary algorithm is very short (a few clock cycles), two estimation units are included to evaluate the fitness of the two new individuals in parallel.

Currently the FEA hardware is implemented on a Xilinx Virtex XV800 chip and is run at 50MHZ. The hardware is 50 times faster than the equivalent FEA running on a 850 MHz Pentium III processor. The next section demonstrates the performance of the FEA, compressing four typical images using a standard image compression system. The maximum fitness produced by the FEA and EA are also compared as a function of time.

3 Experiment Results for Image Compression in Hardware

The image compression algorithm is fully implemented in hardware and can coexist and operate with the fast evolutionary algorithm on the same chip. The system is applied to the four images shown in Fig. 3 in lossless (NEAR parameter=0), near lossless (PSNR=30 or NEAR parameter=20) and lossy (PSNR=20 or NEAR parameter=40) compression. The hardware unit in

Fig. 4 consists of four main blocks: context module, coding module, run length module and code write module. The coding module consists of a prediction module, a quantisation module and a Golomb coding module. The run length module consists of a run counter, a quantisation module and a Golomb coding module. The results of coding will be passed to the write module, which transfers the compressed code into memory. This system can only produce the compressed data - a similar system with a small modification is required for the decompression process. The FEA determines the best parameters for the compression process. These parameters can be added to the compressed data and at the destination the original image or data can be reproduced knowing these parameters.

The hardware performance is shown in Fig. 5 to Fig. 8 for the three FEA methods. The three methods displayed on the top of the graphs are 0, 1 and 2. Method 0 is the basic FEA without random evaluation or the error compensation mechanism. Method 1 is the basic FEA and 5% random evaluations for reducing noise in the population. Method 2 is the basic FEA with 5% random evaluations and the error compensation mechanism. The figures show that, generally, the compression ratio produced by the FEA is higher than the EA. Although both Method 1 and Method 2 are better than method 0 there is no clear general winner between these two methods.

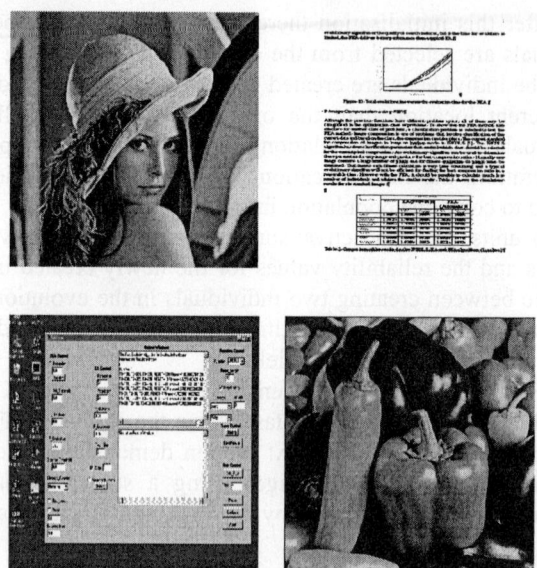

Fig. 3. The four test images (LENA, TEXT, SCREEN and PEPPER)

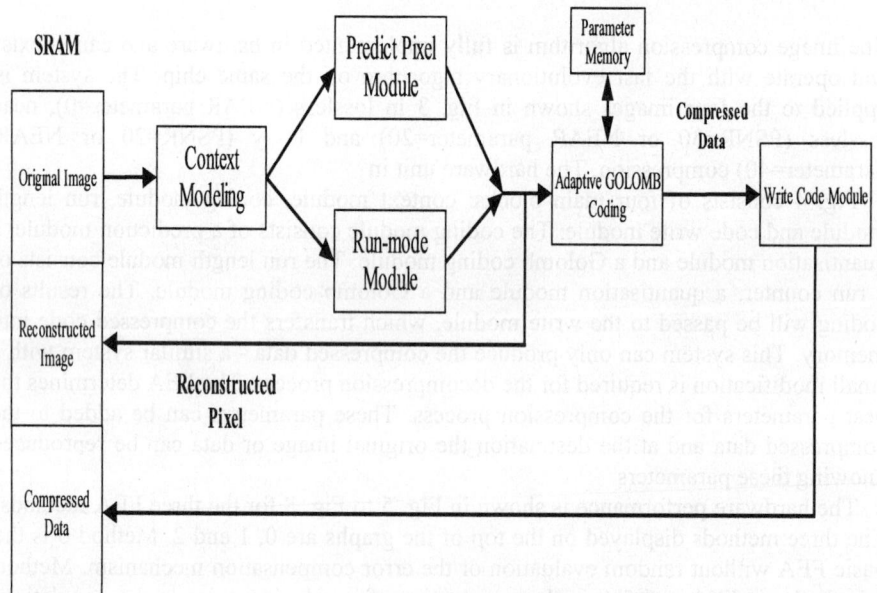

Fig. 4. The image compression hardware system

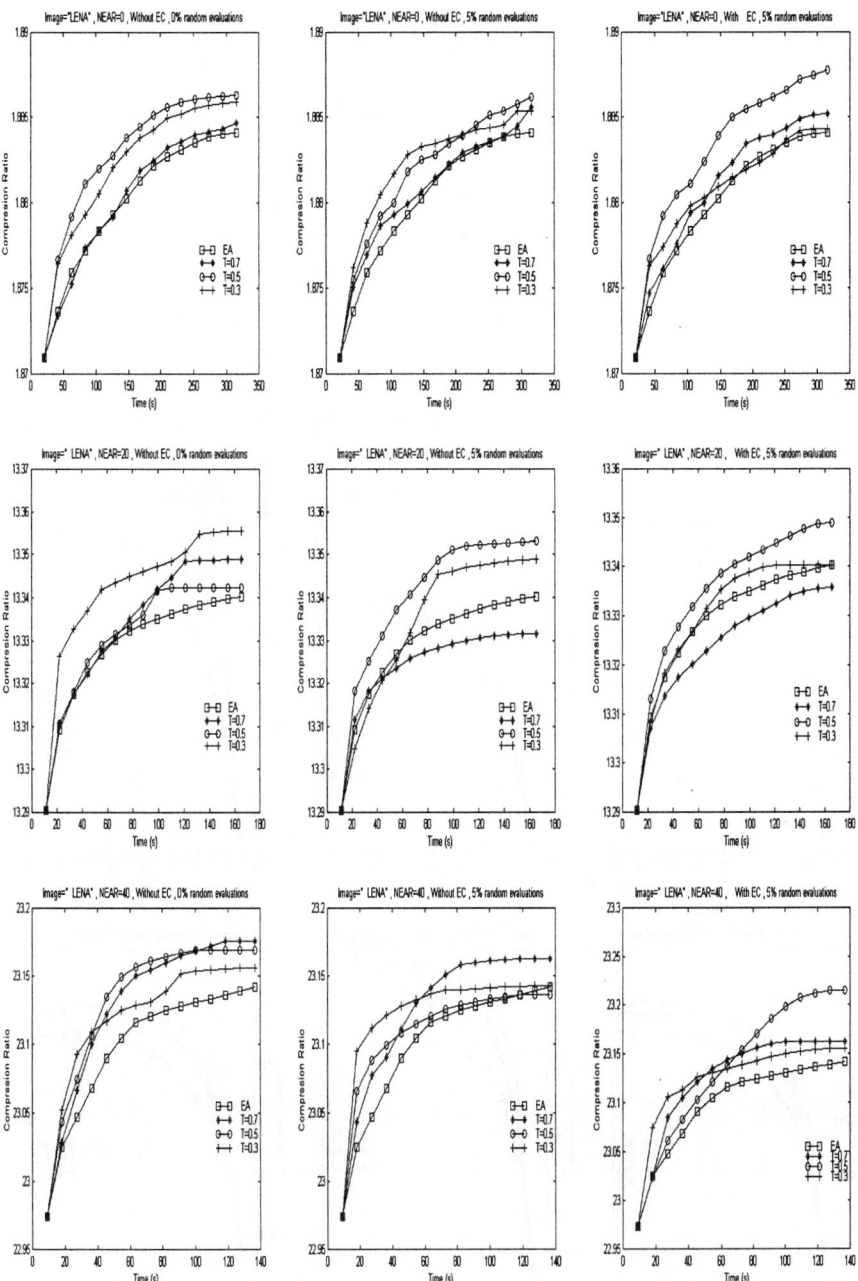

Fig. 5. The results of hardware compression of the "LENA" image by three FEA methods

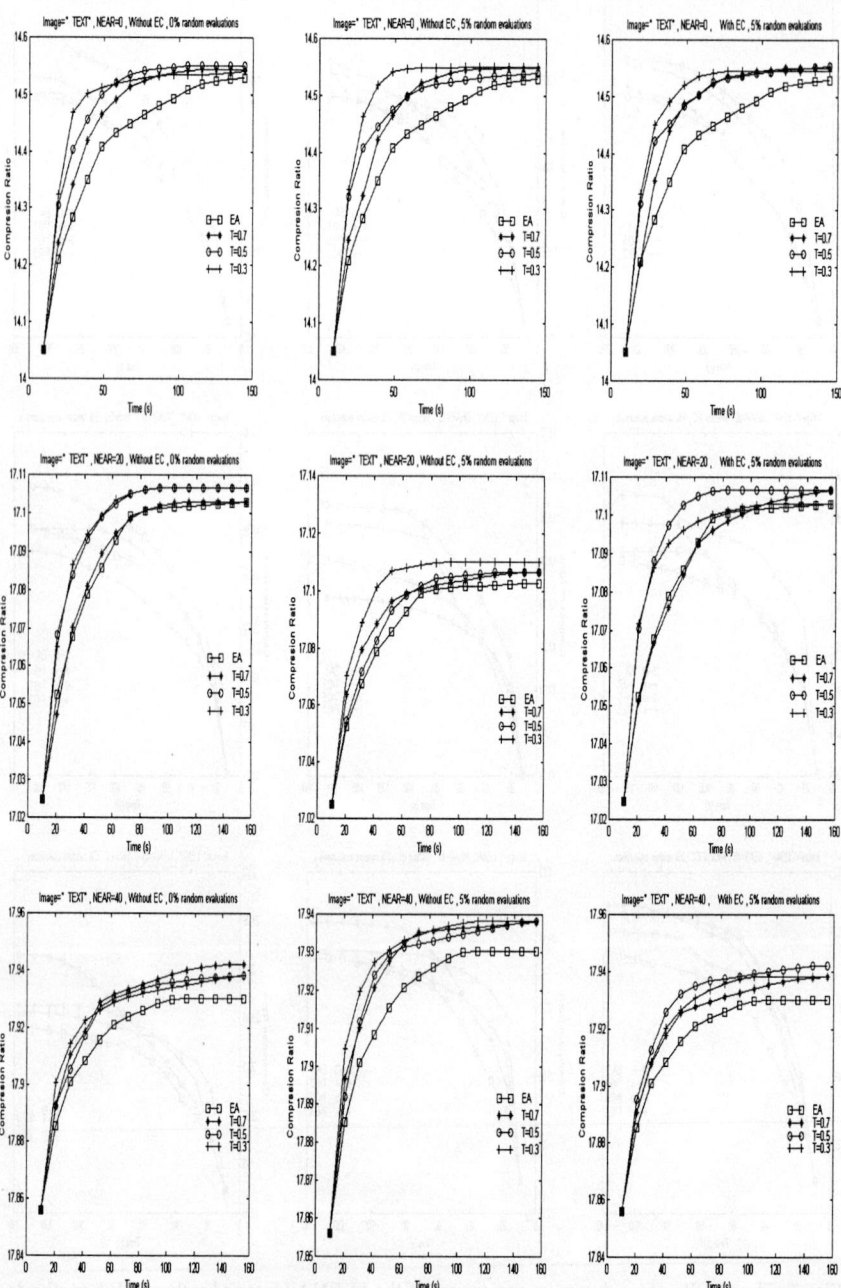

Fig. 6. The results of hardware compression of the "TEXT" image by three FEA methods

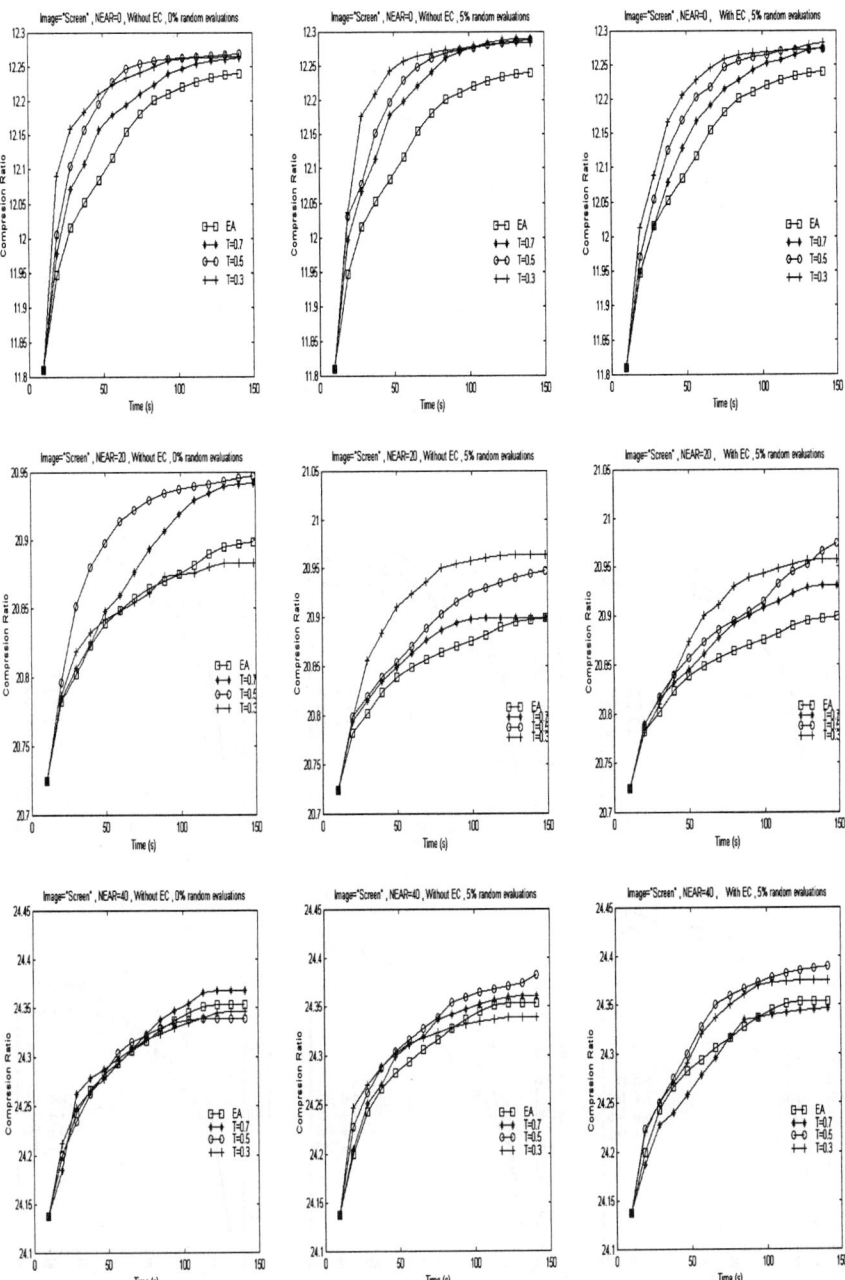

Fig. 7. The results of hardware compression of the "SCREEN" image by three FEA methods

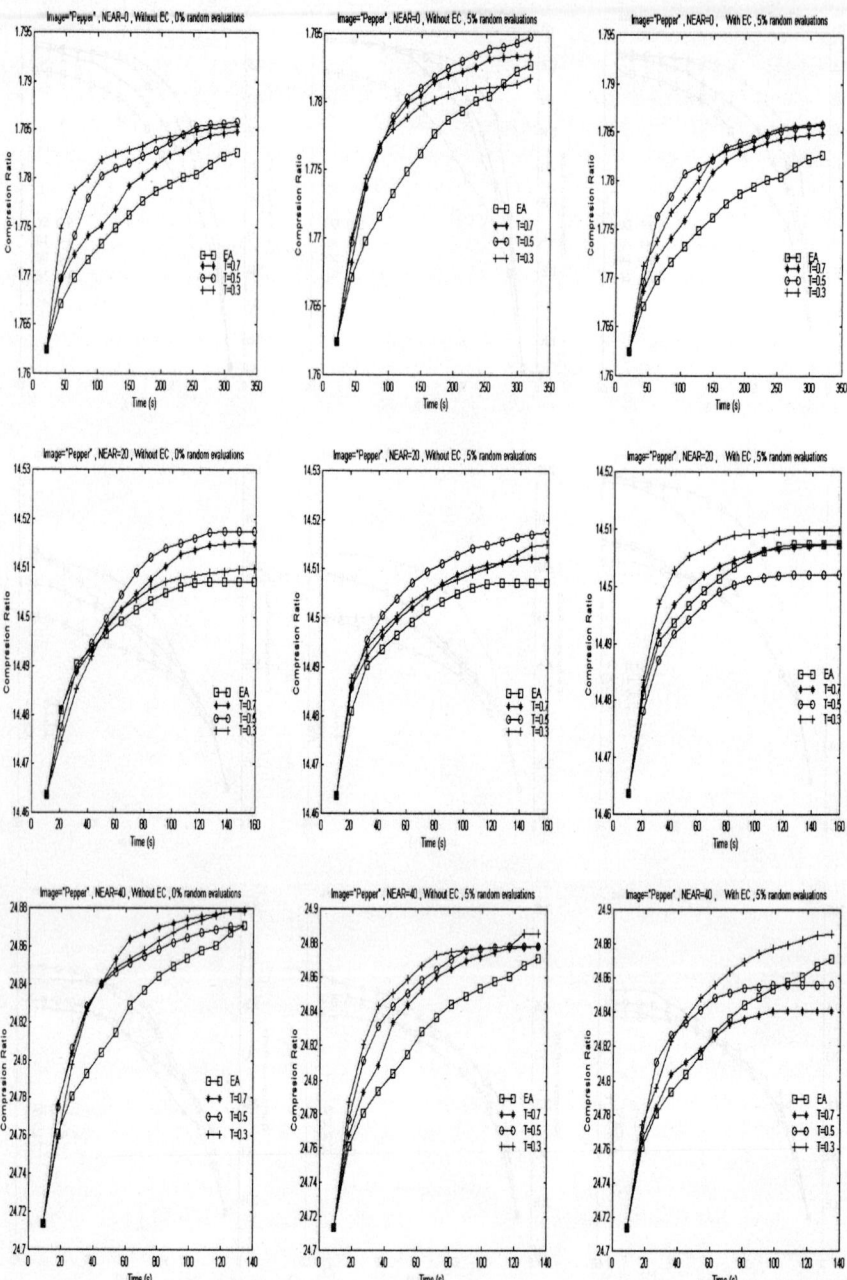

Fig. 8. The results of hardware compression of the "PEPPER" image by three FEA methods

The average time for fitness evaluation (calculating the image compression ratio) for any set of control values is between 0.15 to 0.3 seconds depending on the value of the NEAR parameter. By setting the number of evaluations to 1000, the whole evolutionary process can be completed in about 150 to 300 seconds. The time for one generation of the FEA excluding fitness evaluation time is about 100µs to 120µs depending on which method of the FEA was used. This timing situation, where the fitness evaluation is much slower than generating the next generation, is very suitable for the FEA. Of course those times, both for calculating the image compression and the FEA body itself, depend on the speed of the hardware. Each graph includes three FEA settings with different threshold values plus the EA performance. The graphs show a clear difference in real time to produce similar compression ratios between the three FEA settings and the EA. Note that the compression ratio yielded by the JPEGLS standard algorithm is generally much worse than the EA or FEA compression ratios.

4 Conclusions

Generally the most time consuming task for any evolvable hardware system is fitness evaluation. Although the hardware implementation of the evolutionary algorithm can be very fast compared to the software implementation, its speed advantage will disappear when the fitness evaluation is slow. The paper discusses a hardware implementation of a Fast Evolutionary Algorithm (FEA) that, at the cost of requiring an additional parameter, the fitness reliability, to be added to each individual, allows the fitness of any new individual to be estimated based on the fitness and reliability values of its parents. This FEA achieves a speed improvement by restricting the number of evaluations required.

To demonstrate the advantages of the FEA in real time, the system has been implemented on an FPGA chip running at 50 MHz. This hardware implementation is on average 50 times faster than the software implementation on Pentium III processors (850 MHZ).

The full advantage of the FEA is realised when the fitness evaluation is very slow, such as in image compression. In this application the time for producing a new generation in the FEA hardware is around 100µs while the fitness evaluation time for the image compression in hardware is around 0.3s. This is an ideal situation to show how the FEA obtains a faster convergence by reducing the number of fitness evaluations required.

The results also show that the FEA produces better image compression ratios than the EA at most times during evolution. The two enhancements of error compensation and random evaluation further improved the performance of the FEA. The FEA hardware also produces much better compression ratios than the standard JPEGLS algorithm.

References

1. Holland J., "Adaptation in Natural and Artificial Systems", MIT Press, Cambridge, MA, 1975.
2. Back T., "Evolutionary Algorithms in Theory and Practice", Oxford University Press, New York, 1996.
3. Salami, M, and Hendtlass T., "A Fitness Evaluation Strategy for Genetic Algorithms", The Fifteenth International Conference on Industrial and Engineering Application of Artificial Intelligent and Expert Systems (IEA/AIE2002), Cairns, Australia.
4. Spiessens P., and Manderick B., "A Massively Parallel Genetic Algorithm: Implementation and First Analysis", Proceedings of the Fourth International Conference on Genetic Algorithms, Morgan Kaufmann, San Mateo CA, pp. 279-285, 1991.
5. Salami, M., "Genetic Algorithm Processor on Reprogrammable Architectures", The Proceedings of The Fifth Annual Conference on Evolutionary Programming 1996 (EP96), MIT Press, San Diego, CA, March 1996.
6. Graham P., and Nelson B., "Genetic Algorithms in software and in Hardware – A Performance analysis of Workstation and Custom Computing Machine Implementation", Proceedings of the IEEE Symposium on FPGAs for Custom Computing Machines, pp. 341-345, 1997.
7. Salami, M., Sakanashi, H., Iwata, M., Higuchi, T., "On-Line Compression of High Precision Printer Images by Evolvable Hardware", The Proceedings of The 1998 Data Compression Conference (DCC98), IEEE Computer Society Press, Los Alamitos, CA, USA, 1998.
8. Weinberger M.J., Seroussi G., and Sapiro G., "LOCO-I: A Low Complexity, Context-Based, Lossless Image Compression Algorithm", Proceedings of Data Compression Conference (DCC96), Snowbird, Utah, pp. 140-149, April 1996.

Automatic Speech Recognition: The New Millennium

Khalid Daoudi

INRIA-LORIA, Speech Group
B.P. 101 - 54602 Villers les Nancy. France
daoudi@loria.fr

Abstract. We present a new approach to automatic speech recognition (ASR) based on the formalism of Bayesian networks. We put the foundations of new ASR systems for which the robustness relies on the fidelity in speech modeling and on the information contained in training data.

1 Introduction

Sate-of-the-art automatic speech recognition (ASR) systems are based on probabilistic modeling of the speech signal using Hidden Markov Models (HMM). These models lead to the best recognition performances in ideal "lab" conditions or for easy tasks. However, in real word conditions of speech processing (noisy environment, spontaneous speech, non-native speakers...), the performances of HMM-based ASR systems can decrease drastically and their use becomes very limited. For this reason, the conception of robust and viable ASR systems has been a tremendous scientific and technological challenge in the field of ASR for the last decades.

The speech research community has been addressing this challenge for many years. The most commonly proposed solutions to deal with real world conditions are *adaptation* techniques (in the wide sense of the term) of HMM-based systems. Namely, the speech signals or/and the acoustic models are adapted to compensate for the miss-match between training and test conditions. While the ideas behind adaptation techniques are attractive and justified, the capability of HMM-based ASR systems to seriously address the challenge seems to be out of reach (at least in our opinion).

During the last two years, we[1] have been conducting research dedicated to address the challenge from another perspective. We focus on what we believe is the core of the problem: the *robustness* in speech modeling. Precisely, our strategy is to conceive ASR systems for which robustness relies on:

- the fidelity and the flexibility in speech modeling rather than (ad-hoc) tuning of HMMs,

---

[1] The Speech Group members involved in this project are: C. Antoine, M. Deviren, D. Fohr and myself (www.loria.fr/equipes/parole).

– a better exploitation of the information contained in the available statistical data.

This is motivated by the fact that the discrepancy of HMMs in real conditions is mainly due to their weakness in capturing some acoustic and phonetic phenomena which are specific to speech, and to their "limited" processing of the training databases. In order to hope obtaining robust ASR systems, it is then crucial to develop new probabilistic models capable of capturing all the speech features and of exploiting at best the available data.

A family of models which seems to be an ideal candidate to achieve this goal is the one of *probabilistic graphical models* (PGMs). Indeed, in last decade, PGMs have emerged as a powerful formalism unifying many concepts of probabilistic modeling widely used in statistics, artificial intelligence, signal processing and other fields. For example, HMMs, mixture models, factorial analysis, Kalman filters and Ising models are all particular instances of the more general PGMs formalism. However, the use of PGMs in automatic speech recognition has gained attention only very recently [2,14,21].

In this paper, we present an overview of our recent research in the field of ASR using probabilistic graphical models. The scope of this paper is to bring this new concept (from our perspective) to the attention of researchers, engineers and industrials who are interested in the conception of robust ASR. We develop the main ideas behind our perspective and argue that PGMs are a very promising framework in ASR and could be the foundation of a new generation of ASR systems. We do not provide full algorithmic and implementation details, but we give all the necessary references to readers interested in such details[2].

2   Probabilistic Graphical Models

During the last decade, PGMs have become very popular in artificial intelligence (and other fields) due to many breakthroughs in many aspects of inference and learning. The literature is now extremely rich in papers and books dealing with PGMs in artificial intelligence among of which we refer to [4] for a very good introduction. The formalism of probabilistic graphical models (PGMs) is well summarized in the following quotation by M. Jordan [18]:

"Graphical models are a marriage between probability theory and graph theory. They provide a natural tool for dealing with two problems that occur throughout applied mathematics and engineering - uncertainty and complexity - and in particular they are playing an increasingly important role in the design of machine learning algorithms. Fundamental to the idea of a graphical model is the notion of modularity - a complex system is built by combining simpler parts. Probability theory provides the glue whereby the parts are combined, ensuring that the system as whole is consistent, and providing ways to interface models to

---

[2] All author's papers can be down-loaded from *www.loria.fr/~daoudi* or *www.loria.fr/equipes/parole*

data. The graph theoretic side of graphical models provides both an intuitively appealing interface by which humans can model highly-interacting sets of variables as well as a data structure that lends itself naturally to the design of efficient general-purpose algorithms."

More precisely, given a system of random variables (r.v.), a PGM consists in associating a graphical structure to the joint probability distribution of this system. The nodes of this graph represent the r.v., while the edges encode the (in)dependencies which exist between these variables. One distinguishes three types of graphs: directed, undirected and those for which the edges are a mixture of both. In first case, one talks about *Bayesian networks*, in the second case, one talks about *Markov random fields*, and in the third case one talks about *chain networks*. PGMs have two major advantages:

- They provide a natural and intuitive tool to illustrate the dependencies which exist between variables. In particular, the graphical structure of a PGM clarifies the conditional independencies embedded in the associated joint probability distribution.
- By exploiting these conditional independencies, they provide a powerful setting to specify efficient inference algorithms. Moreover, these algorithms may be specified automatically once the initial structure of the graph is determined.

So far, the conditional independencies semantics (or Markov properties) embedded in a PGM are well-understood for Bayesian networks and Markov random fields. For chain networks, these are still not well-understood. In our current research, given the causal and dynamic aspects of speech, Bayesian networks (BNs) are of particular interest to us. Indeed, thanks to their structure and Markov properties, BNs are well-adapted to interpret causality between variables and to model temporal data and dynamic systems. In addition, not only HMMs are a particular instance of BNs, but also the Viterbi and Forward-Backward algorithms (which made the success of HMMs in speech) are particular instances of generic inference algorithms associated to BNs [20]. This shows that BNs provide a more general and flexible framework than the HMMs paradigm which has ruled ASR for the last three decades.

Formally, a BN has two components: a directed acyclic graph S and a numerical parameterization Θ. Given a set of random variables $X = \{X_1, ..., X_N\}$ and $P(X)$ its joint probability distribution (JPD), the graph S encodes the conditional independencies (CI) which (are supposed to) exist in the JPD. The parameterization Θ is given in term of conditional probabilities of variables given their parents. Once S and Θ are specified, the JPD can be expressed in a factored way as[3][4]

$$P(x) = \prod_{i=1}^{N} P(x_i | pa(x_i)) \qquad (1)$$

---

[3] This factorization can not be obtained if the graph is cyclic.
[4] In the whole paper, upper-case (resp. lower-case) letters are used for random variables (resp. outcomes).

where $pa(x_i)$ denotes an outcome of the parents of X_i.

The Markov properties of a BN imply that, conditioned on its parents, a variable is independent of all the other variables except its descendants. Thus, it is obvious to represent a HMM as a particular BN, as shown in Figure 1. Contrarily to the usual state transition diagram, in the BN representation each node H_t (resp. O_t) is a random variable whose outcome indicates the state occupied (resp. the observation vector) at time t. Time is thus made explicit and arrows linking the H_t must be understood as "causal influences" (not as state transitions).

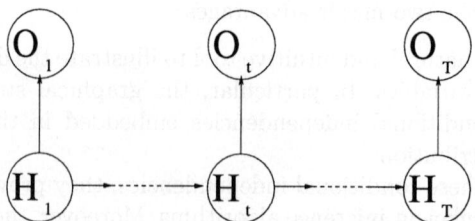

Fig. 1. a HMM represented as a Bayesian network

3 The Language of Data

When conceiving an ASR system, the only "real" material available is the training speech databases. It is then extremely important to exploit at best the information contained in these databases to build the speech acoustic models. The conception of state-of-the-art ASR systems can be decomposed in three major steps. First a front-end is decided for the parameterization of the speech signals, in general Mel Frequency Cepstral Coefficients (MFCC) are used to extract acoustic features vectors. Second, system variables are designed, in general continuous multi-Gaussian variables are used to represent the observed acoustic vectors and discrete variables are used to capture the hidden dynamics of the system. Finally, dependency relationships between system variables are *imposed* once for all in order to define the speech probabilistic model (a HMM in general), then training is done using this model. In our opinion, this last point can be a serious weakness in such systems. Indeed, these dependency relationships are motivated only by prior knowledge, no examination of data is done in order to check if they are really consistent with data. It is then more realistic to *combine* prior knowledge and data information to decide which dependency relationships are more appropriate in the modeling of each acoustic unit in the vocabulary. In the next subsections, we explain how to do so using the BNs framework.

3.1 Structural Learning of BNs

As mentioned above, the Markov properties of a BN (and a PGM in general) determine the dependency relationships between system variables. Thus, the problem of finding the appropriate dependency relationships which are consistent with data comes back to finding the graphical structure (and its numerical parameterization) which best explain data. This problem is known as *structural learning* in the BN literature.

The general principle of structural learning [15] is to introduce a scoring metric that evaluates the degree to which a structure fits the observed data, then a search procedure (over all possible structures) is performed to determine the structure which yields the best score according to this metric. If S is a given structure and D is the observed data, the scoring metric has the following form in general :

$$Score(S, D) = \log P(D|\hat{\Theta}, S) - Pen(S) \qquad (2)$$

where $\hat{\Theta}$ is the estimated numerical parameterization given the structure S. The scoring metric has two terms. The first one represents the (log)likelihood of data. The second one is a penalty term introduced to penalize overly complex structures. The consideration of this term is very important because otherwise complex structures would be always favored, resulting in untractable networks.

There are basically two approaches for penalizing complex structures. In the first one, a prior probability distribution on structures is used. This scoring metric, known as the Bayesian Dirichlet metric, gives high probabilities to more realistic structures [16]. The second approach, which is the one we use, is known as the Bayesian Information Criterion (BIC) score or the Minimum Description Length (MDL) score [17]. The BIC score, which is based on universal coding, defines a penalty term proportional to the number of parameters used to encode data:

$$Pen(S) = \frac{\log N}{2} \sum_{i=1}^{n} ||X_i, pa(X_i)|| \qquad (3)$$

where N is the number of examples (realizations) in D and $||X, Y||$ is defined as the number of parameters required to encode the conditional probability $P(X|Y)$.

The evaluation of the scoring metric for all possible structures is generally not feasible. Therefore, many algorithms have been proposed to search the structure space so as to achieve a maximum scoring structure. In [13], a structural Expectation-Maximization (SEM) algorithm is proposed to find the optimal structure and parameters for a BN, in the case of incomplete data. This algorithm starts with random structure and parameters. At each step, first a parametric Expectation-Maximization algorithm is performed to find the optimal parameters for the current structure. Second, a structural search is performed to increase the scoring. The evaluation of the scoring metric for the next possible structure is performed using the parameters of the previous structure. This algorithm guarantees an increase in the scoring metric at each iteration and converges to a local maximum.

3.2 Application to Speech Recognition

In HMM-based systems, the observed process is assumed to be governed by a hidden (unobserved) dynamic one, under some dependency assumptions. The latter state that the hidden process is first-order Markov, and that each observation depends only on the current hidden variable. There is however a fundamental question regarding these dependency assumptions: are they consistent with data? In case the answer is no: what (more plausible) dependency assumptions should we consider? We have applied the structural learning framework to learn (from training data) the appropriate dependency relationships between the observed and hidden process variables, instead of imposing them as HMM-based systems do. We have also introduced a slight but important modification in the initialization of the SEM algorithm. Namely, instead of starting from a random structure, we initialize the algorithm using the HMM structure (see Figure 1). This way we exploit prior knowledge, namely that HMMs are good "initial" models for speech recognition. More importantly, we thus guaranty that the SEM algorithm will converge to a BN which models speech data with higher (or equal) fidelity than a HMM. We refer the reader to [8] for details on the application of this strategy to an isolated speech recognition task. In the latter, the decoding algorithm is readily given by the inference algorithms associated to BNs. However, in a continuous speech recognition task, decoding requires more attention. Indeed, the SEM algorithm yields in general different structures for different words in the vocabulary. This leads in turn to an *asymmetry* in the representation of dependency in the decoding process. In [9], we have developed a decoding algorithm which deals with such asymmetry and, consequently, allows recognition in a continuous speech recognition task.

Our approach to build acoustic speech models described above has many advantages:

- It leads to speech models which are consistent with training databases.
- It guarantees improvement in modeling fidelity w.r.t. to HMMs.
- It allows capturing phonetic features, such as the *anticipation* phenomena, which can not be modeled by any Markov process (not only HMMs).
- It is implicitly discriminative because two words in the vocabulary may be modeled using completely different networks (in term of Markov properties and the number of parameters).
- It is technically attractive because all the computational effort is made in the training phase.
- It allows the user to make a control on the trade-off between modeling accuracy and model complexity.

The experiments carried out in [9,8] and the results obtained show that, indeed, this approach leads to significant improvement in recognition accuracy w.r.t. to a classical system.

4 Multi-band Speech Recognition

In the previous section we argued that data should be combined with prior knowledge in order to build speech acoustic models *a posteriori*. We applied this principle in the setting where speech is assumed to be the superposition of an observed process with a hidden one. We showed that (w.r.t. HMMs) substantial gain in recognition accuracy can be obtained using this methodology. In this section, we address the problem of modeling *robustness* from the multi-band principle perspective.

The multi-band principle was originally introduced by Harvey Fletcher [12] who conducted (during the first half of the 20th century) extensive research on the way humans process and recognize speech. He "showed" that the human auditory system recognizes speech using partial information across frequency, probably in the form of speech features that are localized in the frequency domain. However, Fletcher's work has been miss-known until 1994 when Jont B. Allen published a paper [1] in which he summarized the work of Fletcher and also proposed to adapt the multi-band paradigm to automatic speech recognition. Allen's work has then inspired researchers to develop a multi-band approach to speech recognition in order to overcome the limitations of classical HMM modeling. Indeed, in many applications (spontaneous speech, non-native speakers...) the performances of HMMs can be very low. One of the major reasons for this discrepancy (from the acoustic point of view) is the fact that the frequency dynamics are weakly modeled by classical MFCC parameterization. Another application where HMMs present a serious limitation is when the system is trained on clean speech but tested in noisy conditions (particularly additive noise). Even when the noise covers only a small frequency sub-band, HMMs yield bad performances since the MFCC coefficients are calculated on the whole spectrum and are then all corrupted.

In the classical multi-band (CMB) approach, the frequency axis is divided into several sub-bands, then each sub-band is independently modeled by a HMM. The recognition scores in the sub-bands are then fusioned with some recombination module. The introduction of multi-band speech recognition [3,11] has been essentially motivated by two desires. The first one is to mime the behavior of the human auditory system (which decomposes the speech signal into different sub-bands before recognition [1]). The second one is to improve the robustness to band-limited noise. While the ideas leading to multi-band speech recognition are attractive, the CMB approach has many drawbacks. For instance, the sub-bands are assumed mutually independent which is an unrealistic hypothesis. Moreover, the information contained in one sub-band is not discriminative in general. In addition, it is not easy to deal with asynchrony between sub-bands, particularly in continuous speech recognition. As a consequence, the recombination step can be a very difficult task. Using the BNs formalism, we present in the next subsection an alternative approach to perform multi-band speech recognition which has the advantage to overcome *all* the limitations (mentioned above) of the CMB approach.

4.1 A Multi-band Bayesian Network

The basic idea behind our approach is the following: instead of considering an independent HMM for each sub-band (as in the CMB approach), we build a more complex but uniform BN on the time-frequency domain by "coupling" all the HMMs associated with the different sub-bands. By coupling we mean adding (directed) links between the variables in order to capture the dependency between sub-bands. A natural question is: what are the "appropriate" links to add?. Following our reasoning of the previous section, the best answer is to learn the graphical structure from training data. Meanwhile, it is also logical to first impose a "reasonable" structure in order to see if this new approach is promising. If the answer is yes, then this "reasonable" structure could play the role of the initial structure (prior knowledge) in a structural learning procedure (as we did with HMMs in the classical full-band case). We build such "reasonable" structure using the following computational and physical criteria. We want a model where no continuous variable has discrete children in order to apply an exact inference algorithm (see [19]). We also want a model with a relatively small number of parameters and for which the inference algorithms are tractable. Finally, we want to have links between the hidden variables along the frequency axis in order to capture the asynchrony between sub-bands. A simple model which satisfies these criteria is the one shown in Figure 2. In this BN, the hidden variables of sub-band n are linked to those of sub-band $n+1$ in such way that the state of a hidden variable in sub-band $n+1$ at time t is conditioned by the state of two hidden variables: at time $t-1$ in the same sub-band and at time t in sub-band n. Each $H_t^{(n)}$ is a discrete variable taking its values in $\{1,...,m\}$, for

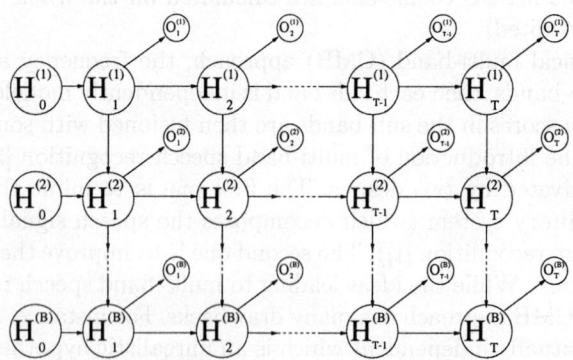

Fig. 2. B-band Bayesian network

some integer $m \geq 1$. Each $O_t^{(n)}$ is a continuous variable with a multi-Gaussian distribution representing the observation vector at time t in sub-band n ($n = 1,...,B$), B is the number of sub-bands. We impose a left-to-right topology on each sub-band and assume that model parameters are stationary. The numerical parameterization of our model is then:

$$a_{ij} \triangleq P(H_t^{(1)} = j | H_{t-1}^{(1)} = i) ;$$

$$u_{ijk}^{(n)} \triangleq P(H_t^{(n)} = k | H_t^{(n-1)} = i, H_{t-1}^{(n)} = j) ;$$

$$b_i^{(n)}(o_t^{(n)}) \triangleq P(O_t^{(n)} = o_t^{(n)} | H_t^{(n)} = i)$$

where $b_i^{(n)}$ is a multi-Gaussian density. The asynchrony between sub-bands is taken into account by allowing all the $u_{ijk}^{(n)}$ to be non-zero, except when $k < j$ or $k > j + 1$ because of the left-to-right topology.

We now stretch the advantages of such approach to multi-band ASR. Contrarily to HMMs, our multi-band BN provides "a" modeling of the frequency dynamics of speech. As opposed to the CMB approach, this BN allows interaction between sub-bands and the possible asynchrony between them is taken into account. Moreover, our model uses the information contained in all sub-bands and no recombination step is needed. A related work has been proposed in [10] where a multi-band Markov random field is analyzed by mean of Gibbs distributions. This approach (contrarily to ours) does not lead however to exact nor fast inference algorithms and assumes a linear model for asynchrony between sub-bands. In our approach, the asynchrony is learned from data. In the next subsection, we present some experiments which illustrate the power of this new approach.

4.2 Experiments

Implementation details and experiments in an isolated speech recognition task can be found in [5] and [6]. The experiments has been carried out in clean and noisy speech conditions and has led to three main results. First, in clean conditions, our approach not only outperforms the CMB one, but also HMMs. To the best of our knowledge, the only multi-band systems which out-perform HMMs in clean conditions use the full-band parameterization as an additional "sub-band". Second, through comparison to a synchronous multi-band BN, the results show the importance of asynchrony in multi-band ASR. Finally and more importantly, our approach largely outperforms HMMs and the CMB approach in noisy speech conditions.

Implementation details of our approach in a continuous speech recognition task can be found in [7]. A preliminary experiment has been carried out but only in clean speech conditions. The results obtained are analogous to those obtain in the isolated speech setting. We now present some of our latest experiments in noisy continuous speech conditions.

Our experiments are carried out on the Tidigits database. In learning we only use the isolated part of the training database where each speaker utters 11 digits twice. In test, we use the full (test) database in which 8636 sentences are uttered, each sentence contains between 1 and 7 digits. We show comparisons of the performances of a 2-band BN to HMMs (it is well-known that the performances of the CMB approach in the continuous setting are generally lower than in the isolated setting). For every digit and the silence model, the number of hidden states is seven ($m = 7$) and we have a single Gaussian per state with a diagonal covariance matrix. We use a uniform language model, i.e., $P(v|v') = \frac{1}{12}$ (eleven digits + silence). The parameterization of the classical full-band HMM is done as follows: 25ms frames with a frame shift of 10ms, each frame is passed through a set of 24 triangular filters resulting in a vector of 35 features, namely, 11 static MFCC (the energy is dropped), 12 Δ and 12 $\Delta\Delta$. The parameterization of the 2-band BN is done as follows: each frame is passed through the 16 first (resp. last 8) filters resulting in the acoustic vector of sub-band 1 (resp. sub-band 2). Each vector contains 17 features: 5 static MFCC, 6 Δ and 6 $\Delta\Delta$. The resulting bandwidths of sub-bands 1 and 2 are $[0..2152Hz]$ and $[1777Hz..10000Hz]$ respectively. The training of all models is done on clean speech only. The test however is performed on clean and noisy speech. The latter is obtained by adding, at different SNRs, a band-pass filtered white noise with a bandwidth of $[3000Hz..6000Hz]$. Table 1 shows the digit recognition rates we obtain using both models. These results illustrate the potential of our approach in exploiting the information contained in the non-corrupted sub-band.

Table 1. Digit recognition scores

Noise SNR	HMM	2-band BN
26 db	89.95%	**96.16%**
20 db	82.17%	**94.89%**
14 db	73.27%	**90.81%**
8 db	62.57%	**82.27%**
2 db	58.86%	**75.51%**

In summary, our new approach to multi-band ASR seems to be more robust than the classical approach and HMMs, both in clean and noisy conditions. The next step (which we did not carry out yet) will be to perform a multi-band structural learning using our B-band BN as an initial structure. Such procedure should increase the robustness of the resulting multi-band system.

5 Conclusion

Based on the PGMs formalism, we presented a new methodology to ASR which seems to be very promising. While the results we obtained are from the most "striking" in the literature (to the best of our knowledge), we do not claim that our perspective of applying PGMs to automatic speech recognition is the "best" one. As we have mentioned, research in this field is still new and PGMs can be applied in many different ways. Our only claim is that, at the beginning of this new millennium, PGMs seem to have a bright future in the field of ASR.

Acknowledgments

The author would like to thank C. Antoine, M. Deviren and D. Fohr for their major contributions in the implementation of the ideas presented in this paper.

References

1. J. Allen. How do humans process and recognize speech. *IEEE Trans. Speech and Audio Processing*, 2(4):567–576, 1994.
2. Jeff A. Bilmes. *Natural Statistical Models for Automatic Speech Recognition*. PhD thesis, International Compute Science Institute, Berkeley, California, 1999.
3. H. Bourlard and S. Dupont. A new ASR approach based on independent processing and recombination of partial frequency bands. ICSLP'96.
4. Robert G. Cowell, A. Philip Dawid, Steffen L. Lauritzen, and David J. Spiegelhalter. *Probabilistic Networks and Expert Systems*. Springer, 1999.
5. K. Daoudi, D. Fohr, and C. Antoine. A new approach for multi-band speech recognition based on probabilistic graphical models. In *ICSLP*, 2000.
6. K. Daoudi, D. Fohr, and C. Antoine. A Bayesian network for time-frequency speech modeling and recognition. In *International Conference on Artificial Intelligence and Soft Computing*, 2001.
7. K. Daoudi, D. Fohr, and C. Antoine. Continuous Multi-Band Speech Recognition using Bayesian Networks. In *IEEE ASRU Workshop*, 2001.
8. M. Deviren and K. Daoudi. Structural learning of dynamic bayesian networks in speech recognition. In *Eurospeech*, 2001.
9. M. Deviren and K. Daoudi. Continuous speech recognition using structural learning of dynamic bayesian networks. Technical report, 2002.
10. G. Gravier et al. A markov random field based multi-band model. ICASSP'2000.
11. H. Hermansky et al. Towards ASR on partially corrupted speech. ICSLP'96.
12. H. Fletcher. *Speech and hearing in communication*. Krieger, New-York, 1953.
13. N. Friedman. Learning belief networks in the presence of missing values and hidden variables. In *Int. Conf. Machine Learning*, 1997.
14. G. Gravier. *Analyse statistique á deux dimensions pour la modélisation segmentale du signal de parole: Application á la reconnaissance*. PhD thesis, ENST Paris, 2000.
15. D. Heckerman. A tutorial on learning with bayesian networks. Technical Report MSR-TR-95-06, Microsoft Research, Advanced Technology Division, March 1995.
16. D. Heckerman, D. Geiger, and D. M. Chickering. Learning bayesian networks: The combination of knowledge and statistical data. *Machine Learning*, 20:197–243, 1995.
17. W. Lam and F. Bacchus. Learning bayesian belief networks an approach based on the mdl principle. *Computational Intelligence*, 10(4):269–293, 1994.
18. M. Jordan, editor. Learning in graphical models. *MIT Press*, 1999.
19. S. L. Lauritzen. Propagation of probabilities, means, and variances in mixed graphical association models. *Jour. Amer. Stat. Ass.*, 87(420):1098–1108, 1992.
20. P. Smyth, D. Heckerman, and M. Jordan. Probabilistic independence networks for hidden markov probability models. *Neural Computation*, 9(2):227–269, 1997.
21. G. G. Zweig. *Speech Recognition with Dynamic Bayesian Networks*. PhD thesis, University of California, Berkeley, 1998.

Applying Machine Learning for Ensemble Branch Predictors

Gabriel H. Loh[1] and Dana S. Henry[2]

[1] Department of Computer Science; Yale University
New Haven, CT, 06520, USA
`gabriel.loh@yale.edu`
[2] Department of Electrical Engineering; Yale University
New Haven, CT, 06520, USA
`dana.henry@yale.edu`

Abstract. The problem of predicting the outcome of a conditional branch instruction is a prerequisite for high performance in modern processors. It has been shown that combining different branch predictors can yield more accurate prediction schemes, but the existing research only examines selection-based approaches where one predictor is chosen without considering the actual predictions of the available predictors. The machine learning literature contains many papers addressing the problem of predicting a binary sequence in the presence of an ensemble of predictors or experts. We show that the Weighted Majority algorithm applied to an ensemble of branch predictors yields a prediction scheme that results in a 5-11% reduction in mispredictions. We also demonstrate that a variant of the Weighted Majority algorithm that is simplified for efficient hardware implementation still achieves misprediction rates that are within 1.2% of the ideal case.

1 Introduction

High accuracy branch prediction algorithms are essential for the continued performance increases in modern superscalar processors. Many algorithms exist for using the history of branch outcomes to predict future branches. The different algorithms often make use of different types of information and therefore target different types of branches.

The problem of branch prediction fits into the framework of the machine learning problem of sequentially predicting a binary sequence. For each *trial* of the learning problem, the branch predictor must make a prediction, and then at the end of the trial, the actual outcome is presented. The prediction algorithm then updates its own state in an attempt to improve future predictions. In each round, the algorithm may be presented with additional information, such as the address of the branch instruction. All predictions and all outcomes are either 0 or 1 (for branch not-taken or branch taken, respectively). The prediction algorithm is called an *expert*.

Algorithms such as the Weighted Majority algorithm and the Winnow algorithm have been proposed for learning problems where there are several experts, or an *ensemble* of experts, and it is desired to combine their "advice" to form one final prediction. Some of these techniques may be applicable to the prediction of conditional branches since many individual algorithms (the experts) have already been proven to perform reasonably well. This study explores the application of the Weighted Majority algorithm for the dynamic prediction of conditional branches.

The Conditional Branch Prediction Problem Conditional branches in programs are a serious bottleneck to improving the performance of modern processors. Superscalar processors attempt to boost performance by exploiting *instruction level parallelism*, which allows the execution of multiple instructions during the same processor clock cycle. Before a conditional branch has been resolved in such a processor, it is unknown which instructions should follow the branch. To increase the number of instructions that execute in parallel, modern processors make a *branch prediction* and speculatively execute the instructions in the predicted path of program control flow. If later on the branch is discovered to have been mispredicted, actions are taken to recover the state of the processor to the point before the mispredicted branch, and execution is resumed along the correct path.

For each branch, the address of the branch is available to the predictor, and the predictor itself may maintain state to track the past outcomes of branches. From this information, a binary prediction is made. This is the *lookup phase* of the prediction algorithm. After the actual branch prediction has been computed, the outcome is presented to the prediction algorithm and the algorithm may choose to update its internal state to (hopefully) make better predictions in the future. This is the *update phase* of the predictor. The overall sequence of events can be viewed as alternating lookup and update phases.

The Binary Prediction Problem Consider the problem of predicting the next outcome of a binary sequence based on observations of the past. Borrowing from the terminology used in [20], the problem proceeds in a sequence of *trials*. At the start of the trial, the prediction algorithm is presented with an *instance*, and then the algorithm returns a binary prediction. Next, the algorithm receives a *label*, which is the correct prediction for the instance, and then the trial ends.

Now consider the situation where there exists multiple prediction algorithms or *experts*, and the problem is to design a *master algorithm* that may consider the predictions made by the n experts $E_1, E_2, ..., E_n$, as well as observations of the past to compute an overall prediction. Much research has gone into the design and analysis of such master algorithms. One example is the Weighted Majority algorithm which tracks the performance of the experts by assigning weights to each expert that are updated in a multiplicative fashion. The master algorithm "homes in" on the best expert(s) of the ensemble.

The binary prediction problem fits very well with the conditional branch prediction problem. The domain of dynamic branch prediction places some additional constraints on the prediction algorithms employed. Because these techniques are implemented directly in hardware, the algorithms must be amenable to efficient implementations in terms of logic gate delays and the storage necessary to maintain the state of the algorithm. In this paper, we present a hardware unintensive variant of the Weighted Majority algorithm, and experimentally demonstrate the performance of our algorithm.

Paper Outline The rest of this paper is organized as follows. Section 2 discusses some of the relevant work from machine learning and branch prediction research. Section 3 briefly explains our methodology for evaluating the prediction algorithms. Section 4 analyzes some of the existing prediction algorithms, and provides some motivation for the use of the Weighted Majority algorithm. Section 5 describes our hardware implementable approximations to the Weighted Majority algorithm, and provides experimental results. Section 6 concludes and describes directions for future research.

2 Related Work

The most relevant work to this paper falls into two categories. Machine learning algorithms for predicting binary sequences comprise the first category. The second group consists of past research in combining multiple branch predictors to build better prediction algorithms. Other areas of artificial intelligence have been applied to the problem of conditional branch prediction, such as perceptron and neural networks [5,15] and genetic algorithms [9].

Machine Learning Algorithms for Predicting Binary Sequences The problem of predicting the outcomes of a sequence of binary symbols has received much attention in the machine learning research. Littlestone and Warmuth introduced the Weighted Majority algorithm [20], which was independently proposed by Vovk [24]. The Weighted Majority algorithm works with an ensemble of experts, and is able to predict nearly as well as the best expert in the group without any *a priori* knowledge about which experts perform well. Theoretical analysis has shown that these algorithms behave very well even if presented with irrelevant attributes, noise, or a target function that changes with time [12,19,20].

The Weighted Majority algorithm and other multiplicative update master algorithms have been successfully applied to several problem domains including gene structure prediction [7], scheduling problems [3], and text processing [11] for example.

Techniques for Combining Branch Predictors Research in the computer architecture community has gone into the design of techniques that combine multiple branch predictor experts into a hybrid or ensemble predictor. The common theme among these techniques is that the master algorithm is an *expert-selection* algorithm. By expert-selection, we mean that the final prediction is formed by choosing one of the n experts based solely on the past performance of the experts. The current predictions made by the experts are not factored into the decision process.

The earliest published branch predictor that took advantage of multiple experts is the tournament predictor [21]. The tournament algorithm is only capable

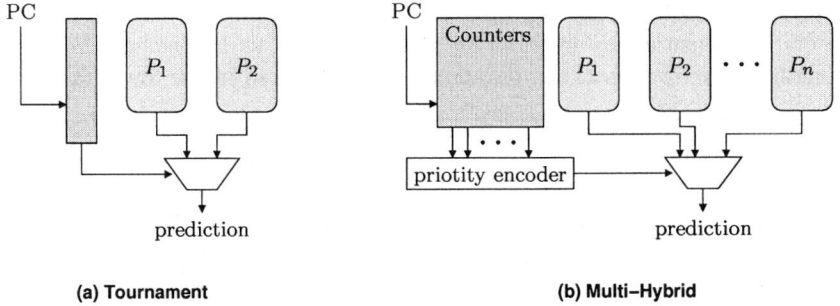

Fig. 1. The tournament predictor (a) chooses between only two experts while the multi-hybrid (b) selects from an arbitrarily sized ensemble

of choosing between two experts, E_0 and E_1. For each branch, a small counter (typically only 2 bits wide) keeps track of the past performance of the two experts. The algorithm always selects the expert that has performed the best in recent trials. The predictor is shown schematically in Figure 1a.

The multi-hybrid predictor is a generalization of the two-expert tournament scheme [10]. The multi-hybrid is not limited to two experts, but the selection process still ignores the predictions of the experts. For each branch, a small (2-bit) counter is associated with each expert. The experts that are correct more often have higher counter values. The multi-hybrid simply selects the expert with the highest counter value. The predictor is shown schematically in Figure 1b.

3 Methodology

We chose several branch predictor configurations using the best algorithms from the state of the art. From these predictors, we formed different sized ensembles based on the total required storage. Analyzing branch predictors at different hardware budgets is a common approach since the storage needed is roughly proportional to the processor chip area consumed. The branch prediction algorithms used for our experts are the global branch history gskewed [22], Bi-Mode [17], YAGS [8] and perceptron [14], and per-branch history versions of the gskewed (called pskewed) and the loop predictor [10]. Out of all of the branch prediction algorithms currently published in the literature, the perceptron is the best performing algorithm that we examined. For the master algorithms studied in this paper, the sets of predictors used are listed in Table 1.

We simulated each master algorithm with the ensembles of experts listed in Table 1. The notation X_y denotes the composite algorithm of master algorithm X applied to ensemble y. For example, multi-hybrid$_\gamma$ denotes the multi-hybrid algorithm applied to ensemble γ.

The simulation is based on the SimpleScalar 3.0 processor simulation toolset [1,4] for the Alpha AXP instruction set architecture. The benchmarks used are

Table 1. The sets of experts used in our experiments for varying hardware budgets as measured by bytes of storage necessary to implement the branch predictors. Descriptions of the parameters for the listed in the lower table

Ensemble Name	Hardware Budget	Number of Experts	Actual Size (KB)	Experts in the Ensemble
α	8KB	3	6.4	PC(7,30) YG(11,10) LP(8)
β	16KB	5	14.4	BM(13,13) PC(7,30) YG(11,10) PS(10,11,4) LP(8)
γ	32KB	6	29.9	BM(14,14) GS(12,12) PC(7,30) YG(11,10) PS(11,13,8) LP(9)
δ	64KB	6	62.4	BM(13,13) GS(16,16) PC(7,30) YG(11,10) PS(10,11,4) LP(8)
η	128KB	6	118.9	BM(13,13) GS(16,16) PC(7,30) YG(11,10) PS(13,16,10) LP(9)

Expert Name	Abbreviation	Parameters
Bi-Mode	BM	$\log_2$(counter table entries), history length
gskewed	GS	$\log_2$(counter table entries), history length
YAGS	YG	$\log_2$(counter table entries), history length
perceptron	PC	$\log_2$(number of perceptrons), history length
pskewed	PS	$\log_2$(pattern table entries), $\log_2$(counter table entries), history length
loop	LP	counter width

the integer applications from the SPEC2000 benchmark suite [23], using primarily the training input sets. The binaries were compiled on an Alpha 21264 system, using the cc compiler. The first one billion instructions of each benchmark were simulated. The conditional branch misprediction rate averaged over all of the benchmarks is the metric used to compare the performance of the different master algorithms.

4 Motivation

In this section, we analyze the performance of the multi-hybrid master algorithm and motivate the use of a "smarter" approach using results from machine learning research. Selection based master algorithms compute an index $idx \in [1, n]$, and then the master algorithm's final prediction equals the prediction of E_{idx}. Note that the computation of idx ignores the current predictions of the experts.

We collected statistics about the branches predicted by the multi-hybrid master algorithm. We classified each predicted branch based on the number of experts that were correct, and the correctness of the overall prediction of the multi-hybrid. The percentage of all branches predicted for the gcc benchmark that fall into each class is tabulated in Table 2. The trends are similar for the other benchmarks.

Table 2. Statistics collected for the multi-hybrid algorithm on gcc. Branches are classified depending on the correctness of the selected expert, and the total number of correct experts

Ensemble	α	β	γ	δ	η
Experts	3	5	6	6	6
% All Wrong	1.49	0.78	0.60	0.62	0.53
% Correct (no choice)	45.14	41.64	40.98	41.35	41.83
% Correct (w/ choice)	49.16	53.47	54.58	54.80	54.50
% Overall Wrong When:					
1 Expert Correct	2.78	1.32	1.01	0.94	0.86
2 Experts Correct	1.44	1.26	0.94	0.80	0.77
3 Experts Correct		1.05	0.83	0.68	0.66
4 Experts Correct		0.47	0.72	0.56	0.58
5 Experts Correct			0.34	0.26	0.28

The first two classes represent those cases where a selection based master algorithm has no impact on the final prediction. If all of the experts agree on the same prediction, it does not matter which expert the master algorithm chooses. The next class consists of the branches that were predicted correctly by the multi-hybrid selector when some decision had to be made. The remaining classes are situations where one or more experts disagreed with the others, and the master algorithm failed to choose an expert that made a correct prediction. These are the most interesting cases because they represent opportunities where we can apply better algorithms to improve the overall prediction rate. A key observation is that if a master algorithm was to look at the predictions made by the experts, and simply sided with the majority, then an additional 0.82%-1.52% of all branches could be predicted correctly, which translates into a 20%-30% reduction in branch mispredictions. The primary conclusion that we draw from these statistics is that there is indeed useful information in the actual predictions of the experts that can and should be used by the master algorithm.

5 Weighted Majority Branch Predictors

The statistics presented in Section 4 motivate the exploration of better master algorithms for branch prediction, particularly algorithms that take into account the current predictions of the ensemble of experts. The algorithm that we choose to use in this study is the Weighted Majority algorithm, and in particular the variant that Littlestone and Warmuth call WML in [20].

The WML algorithm assigns each of the n experts $E_1, E_2, ..., E_n$ positive real-valued weights $w_1, w_2, ..., w_n$, where all weights are initially the same. During the prediction phase of each trial, the algorithm computes q_0, which is the sum of the weights that correspond to experts that predict 0, and q_1, which is the sum of the weights for experts predicting 1. If $q_0 < q_1$, then the Weighted Majority algorithm returns a final prediction of 1. That is, the algorithm predicts with the group that has the larger total weight. During the update phase of the trial, the weights corresponding to experts that mispredicted are multiplicatively reduced by a factor of $\beta \in (0,1)$ unless the weight is already less than γ/n times the sum of all of the weights at the start of the trial, where $\gamma \in [0, \frac{1}{2})$. The multiplication by β reduces the relative influence of poorly performing experts for future predictions. The parameter γ acts as an "update threshold" which prevents an expert's influence from being reduced too much, so that the expert may more rapidly regain influence when it starts performing well. This property is useful in situations where there is a "shifting target" where some subset of experts perform well for some sequence of predictions, and then another subset of experts become the best performing of the group. This is particularly applicable in the branch prediction domain where branch behavior may vary as the program enters different phases of execution.

To apply the Weighted Majority algorithm to branch prediction, we maintain a table of weights indexed by the branch address (the *instance*). The weights corresponding to the branch in question are used to compute the weighted sums q_0 and q_1 which determine the final prediction. This allows different instances to use different sets of weights, since the best subset of experts for one branch may be different than the best subset for another. Throughout the rest of this section, the tables used in the simulations have 2048 sets of weights.

We simulated the Weighted Majority algorithm using the same ensembles used in Section 4 for the multi-hybrid algorithm. Out of the combinations of β and γ that we simulated, the best values are $\beta = 0.85$ and $\gamma = 0.499$. Since $\gamma = 0.499$ is very close to the maximum allowable value of $\frac{1}{2}$, we conclude that for the branch prediction domain the best set of experts shifts relatively frequently.

As presented, the WML algorithm can not be easily and efficiently implemented in hardware. The primary reason is that the weights would require large floating point formats to represent, and the operations of adding and comparing the weights can not be performed within a single processor clock cycle. As described, the WML algorithm decreases the weights monotonically, but the performance is the same if the weights are all renormalized after each trial. Renormalization prevents underflow in a floating point representation. Furthermore, the storage needed to keep track of all of the weights is prohibitive. Nevertheless, the performance of the WML algorithm is interesting as it provides performance numbers for an ideal case. Figure 2 shows the branch misprediction rates of the WML algorithm (for $\beta = 0.85, \gamma = 0.499$) compared against the multi-hybrid. The best branch prediction scheme consisting of a single expert (the perceptron predictor) is also included for comparison. Compared to the single predictor, WML_η makes 10.9% fewer mispredictions, and 5.4% fewer than multi-hybrid$_\eta$. A 5-11% improvement in the misprediction rate is non-trivial in the branch prediction field where the average misprediction rates are already less than 5%. These results are encouraging and motivate the exploration of hardware unintensive implementations.

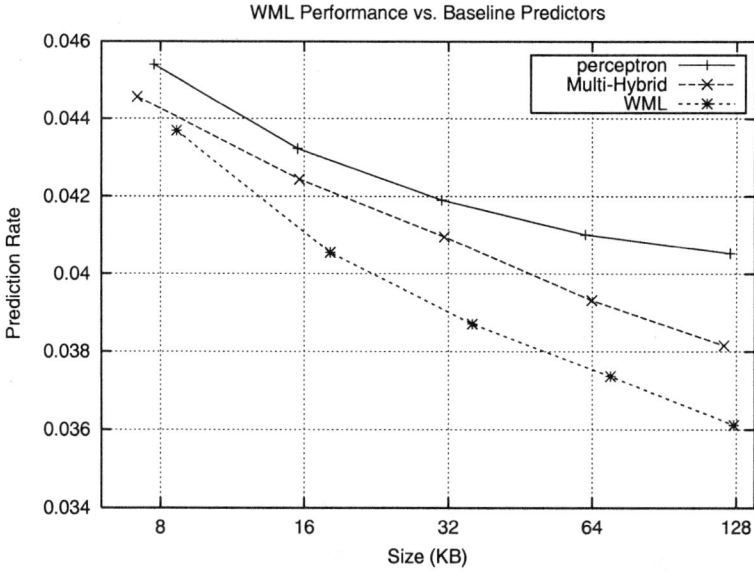

Fig. 2. The WML master algorithm yields lower misprediction rates than the selection based multi-hybrid master algorithm and the best singleton algorithm

There are several aspects of the WML algorithm that need to be addressed: the representation of weights, updating the weights, the γ update threshold, and normalization. In place of floating point weights, k-bit integers can be used instead. Multiplication is a costly operation to perform in hardware. We therefore replace the multiplication by β with additive updates. This changes the theoretical mistake bounds of the algorithm. We justify this modification because in our application domain, the size of the ensemble of experts is relatively small and so this change in the asymptotic mistake bounds should not greatly affect prediction accuracy. The limited range of values that can be represented by a k-bit integer plays a role similar to the update threshold imposed by γ because a weight can never be decreased beyond the range of allowable values. Instead of normalizing weights at the end of each trial, we use update rules that increment the weights of correct experts when the overall prediction was wrong, and decrement the weights of mispredicting experts when the overall prediction was correct. We call this modified version of the Weighted Majority algorithm aWM (for approximated Weighted Majority).

Figure 3 shows the performance of the aWM algorithm compared to the ideal case of WML. Because the smaller ensembles have smaller hardware budgets, the amount of additional resources that may be dedicated to the weights is limited, and therefore the size of the weights used are smaller. For ensembles α, β and γ, the sizes of the weights are 2, 3 and 4 bits, respectively. Ensembles

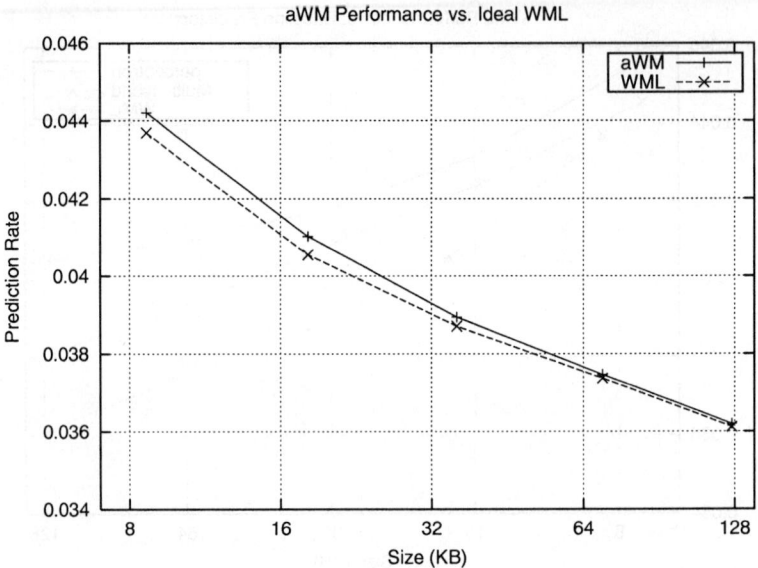

Fig. 3. Despite several simplifications made to adapt the WML algorithm for efficient hardware implementation, the modified version aWM still performs very well when compared to the ideal case of WML

δ and η each use 5-bit weights. The most interesting observation is how well the aWM algorithm performs when compared to the ideal case of WML despite the numerous simplifications of the algorithm that were necessary to allow an efficient hardware implementation.

The Weighted Majority approach to combining the advice from multiple experts has an advantage over selection based approaches. Consider a situation where the expert that has been performing the best over the last several trials suddenly makes an incorrect prediction. In a selection based approach, the master algorithm simply chooses based on the past performance of the experts and will make a misprediction in this situation. On the other hand, it is possible that when the best expert mispredicts, there may be enough experts that predict in the opposite direction such that their collective weight is greater than the mispredicting experts. For the aWM$_\eta$ predictor executing the gcc benchmark, situations where the weighted majority is correct and the expert with the greatest weight is incorrect occur over 80% more frequently than cases where the best expert is correct and the weighted majority is wrong. This confirms of our earlier hypothesis that the information conveyed by the collection of all of the experts' predictions is indeed valuable for the accurate prediction of branches.

The Weighted Majority algorithm can also be useful for the problem of determining *branch confidence*, which is a measure of how likely a prediction will be correct [13]. For aWM$_\eta$ on gcc, when q_0 and q_1 are within 10% of being equal, the overall accuracy of the predictor is only 54%. This means that when we do not have a "strong" majority, the confidence in the prediction should not be very great.

6 Conclusions and Future Work

The problem of correctly predicting the outcome of conditional branches is a critical issue for the continued improvement of microprocessor performance. Algorithms from the machine learning community applied to the branch prediction problem allow computer architects to design and implement more accurate predictors. Despite the simplification of the Weighted Majority algorithm into a form that amenable to implementation in a processor, the hardware unintensive variant achieves misprediction rates within 0.2-1.2% of an idealized version. The results show the importance of combining the information provided by *all* of the experts.

The Weighted Majority algorithm is but one possible approach to combining the predictions from an ensemble of experts. Future work includes researching more accurate algorithms and techniques which yield even simpler implementations. The research described in this paper may be applied to similar areas in the design of microprocessors. Prediction and speculation are pervasive in the computer architecture literature, for example memory dependence prediction [6,16], value prediction [18], and cache miss prediction [2]. Any of these areas could potentially benefit from applications of more theoretically grounded machine learning algorithms.

References

1. Todd M. Austin. SimpleScalar Hacker's Guide (for toolset release 2.0). Technical report, SimpleScalar LLC. http://www.simplescalar.com/docs/hack_guide_v2.pdf.
2. James E. Bennett and Michael J. Flynn. Prediction Caches for Superscalar Processors. In *Proceedings of the 30th International Symposium on Microarchitecture*, 1997.
3. Avrim Blum. Empirical Support for Winnow and Weight-Majority Based Algorithms: Results on a Calendar Scheduling Domain. In *Proceedings of the 12th International Conference on Machine Learning*, pages 64–72, 1995.
4. Doug Burger and Todd M. Austin. The SimpleScalar Tool Set, Version 2.0. Technical Report 1342, University of Wisconsin, June 1997.
5. Brad Calder, Dirk Grunwalk, Michael Jones, Donald Lindsay, James Martin, Michael Mozer, and Benjamin Zorn. Evidence-Based Static Branch Prediction Using Machine Learning. *ACM Transactions on Programming Languages and Systems*, 19(1):188–222, 1997.
6. George Z. Chrysos and Joel S. Emer. Memory Dependence Prediction Using Store Sets. In *Proceedings of the 25th International Symposium on Computer Architecture*, pages 142–153, 1998.

7. Adam A. Deaton and Rocco A. Servedio. Gene Structure Prediction From Many Attributes. *Journal of Computational Biology*, 2001.
8. Avinoam N. Eden and Trevor N. Mudge. The YAGS Branch Prediction Scheme. In *Proceedings of the 31st International Symposium on Microarchitecture*, pages 69–77, December 1998.
9. Joel Emer and Nikolas Gloy. A Language for Describing Predictors and it Application to Automatic Synthesis. In *Proceedings of the 24th International Symposium on Computer Architecture*, June 1997.
10. Marius Evers, Po-Yung Chang, and Yale N. Patt. Using Hybrid Branch Predictors to Improve Branch Prediction Accuracy in the Presence of Context Switches. In *Proceedings of the 23rd International Symposium on Computer Architecture*, May 1996.
11. Andrew R. Golding and Dan Roth. Applying Winnow to Context-Sensitive Spelling Correction. In *Proceedings of the 13th International Conference on Machine Learning*, pages 182–190, 1996.
12. Mark Herbster and Manfred Warmuth. Track the Best Expert. In *Proceedings of the 34th International Conference on Machine Learning*, pages 286–294, 1995.
13. Erik Jacobson, Eric Rotenberg, and James E. Smith. Assigning Confidence to Conditional Branch Predictions. In *Proceedings of the 29th International Symposium on Microarchitecture*, pages 142–152, December 1996.
14. Daniel A. Jiménez and Calvin Lin. Dynamic Branch Prediction with Perceptrons. In *Proceedings of the 7th International Symposium on High Performance Computer Architecture*, January 2001.
15. Daniel A. Jiménez and Calvin Lin. Perceptron Learning for Predicting the Behavior of Conditional Branches. In *Proceedings of the*, 2001.
16. R. E. Kessler. The Alpha 21264 Microprocessor. *IEEE Micro Magazine*, March–April 1999.
17. Chih-Chieh Lee, I-Cheng K. Chen, and Trevor N. Mudge. The Bi-Mode Branch Predictor. In *Proceedings of the 30th International Symposium on Microarchitecture*, pages 4–13, December 1997.
18. Mikko H. Lipasti, Christopher B. Wilkerson, and John Paul Shen. Value Locality and Load Value Prediction. In *Proceedings of the Symposium on Architectural Support for Programming Languages and Operating Systems*, October 1996.
19. Nick Littlestone. Learning Quickly When Irrelevant Attributes Abound: A New Linear-threshold Algorithm. *Machine Learning*, 2:285–318, 1988.
20. Nick Littlestone and Manfred K. Warmuth. The Weighted Majority Algorithm. *Information and Computation*, 108:212–261, 1994.
21. Scott McFarling. Combining Branch Predictors. TN 36, Compaq Computer Corporation Western Research Laboratory, June 1993.
22. Pierre Michaud, Andre Seznec, and Richard Uhlig. Trading Conflict and Capacity Aliasing in Conditional Branch Predictors. In *Proceedings of the 24th International Symposium on Computer Architecture*, 1997.
23. The Standard Performance Evaluation Corporation. WWW Site. http://www.spec.org.
24. Volodya Vovk. Universal Forecasting Strategies. *Information and Computation*, 96:245–277, 1992.

A Customizable Configuration Tool for Design of Multi-part Products

Niall Murtagh

Industrial Electronics and Systems Laboratory, Mitsubishi Electric Corporation
Tsukaguchi-honmachi 8-1-1, Amagasaki-shi, Hyogo 661-8661, Japan
niall@fas.sdl.melco.co.jp

Abstract. The configuration of complex multi-part products often requires that a human expert be available to determine a compatible set of parts satisfying the specification. With the availability of on-line web catalogs, such experts can now be supported or even substituted by intelligent web tools. This paper describes the architecture and problem representation for such a tool, built upon a constraint-based reasoning engine. The flexible strategy we employ enables a high degree of user customization of the web tool, allowing users to personalize not only the tool interface but also to edit or add to the catalogs as required, and even to change the product representation details. We illustrate the ideas by referring to the configuration of industrial products such as programmable logic controllers (PLC).

1 Introduction

Continuous improvements in the functionality of products have resulted in greater possibilities for satisfying particular customer requirements but have also meant that more expertise is required to configure the products to fulfill both user specifications and other design constraints. Where human experts are needed to carry out the configuration details, costs may increase and delivery time may be slowed. The increased sophistication of web systems provides a possible solution to this problem – intelligent interfaces and programs can enable customers to perform configuration tasks that otherwise would require a human expert. However, web-based and other computer systems often have the drawback that they are less flexible than human experts and tend to be restricted in the type of problem they can solve.

The web is already moving to overcome some of this inflexibility by allowing users to customize interfaces so that the systems show only the information the user is interested in. Web systems are also capable of giving users their own personalized databases, by either automatically storing personal data or by allowing the user to edit existing databases. But beyond this, user customization is normally not possible and, in some cases, may not be desirable. However, we propose that in the case of web-based product configuration tools, it is desirable to allow the user to adjust more than databases and interfaces: by availing of the strategies we outline in this paper, it is feasible to allow the user to freely adjust various details of the problem description itself, such as the product physical and functional makeup. This would enable customers to combine their own products and parts with the pre-arranged system catalogs. It would also allow a system user to tackle a completely new problem type, once it can be represented using the available description language. The only part of the system not open to user customization is the problem solving program, the kernel of the system.

2 Development Strategy for Customizable Web System

The initial goal for the current project was to provide an interface to facilitate access to part catalogs, and to provide assistance in selecting parts to fulfill compatibility and other requirements in order to compose a multi-part product. This was realized in the first version of our system, ACAT, in which the user interface listed specification parameters which most users would wish to constrain (cost, delivery time, functional capabilities, etc.). In response to user input for these parameters, other requirements written into the underlying program (e.g., physical and functional constraints) were processed and an adequate set of parts was selected from the catalogs. Fig. 1 illustrates the conventional data flow between client and web-server, as used in the initial version of ACAT.

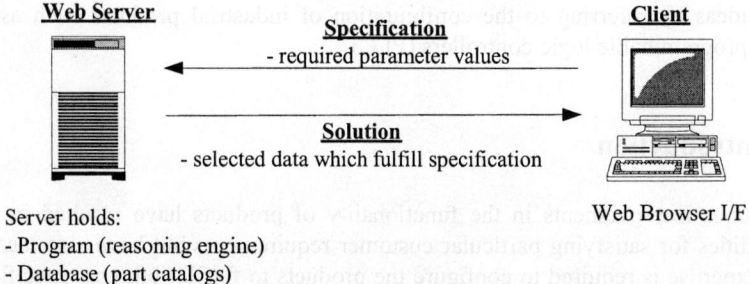

Fig. 1. Conventional Model for Data Flow between Client and Web-Server System

2.1 Steps in Adding Flexibility to the System

While the initial system proved useful where the product description and input/output details did not change, whenever a different product had to be configured or when the interface for a particular product had to be altered, program editing and re-compiling were necessary. To minimize or avoid such program alterations and recompilations, we created a more flexible system by separating the various data and reasoning layers. This was achieved in the following manner:

- Firstly, the interface was separated from the program or reasoning engine and represented in an editable data file containing lists of parts and attributes to be displayed for user input and output.
- While databases/catalogs are conventionally separate from processing programs, the catalog topologies, such as the names and number of data fields, are usually

included in the programs. We separated this topology information from the reasoning engine program in order to facilitate changing of catalogs without any program editing.

- With alterations to the interface and/or catalogs, it is quite likely that further aspects of the problem description itself may change, e.g., a newly added part may have a required relation with an existing part or parts, which must be checked. While conventional systems would have such checks written into the processing program, we separated the problem description information from the program and represented it in a data file, using a constraint language. Within the limits of the language terminology, there is no restriction as to the type of changes that are permitted or to the type of problem that can be represented and tackled.

Thus an existing problem description may be edited or completely replaced with a new problem description, while leaving the configuration program or reasoning engine untouched. Figure 2 summarizes the data flow where partial or complete customization of the system takes place.

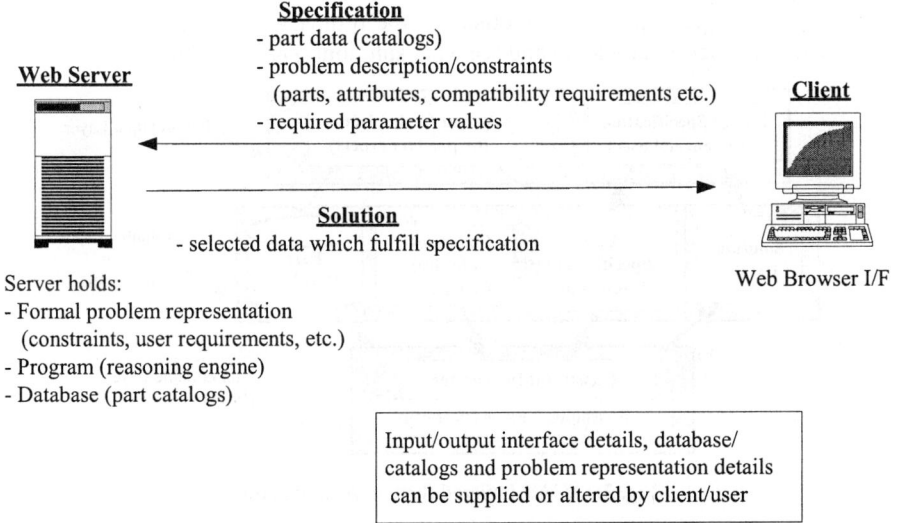

Fig. 2. ACAT Client-Server Data Flow Pattern

2.2 System Application Scenarios

The ACAT system can be used in various scenarios depending on the level of customization carried out. The user may search for a solution using the current problem description and catalogs without adjustment; the user may edit some or all of the interface, part catalogs, problem description details; or, the user may supply alternative interface, part catalogs and problem details.

Where partial or complete user customization is allowed, security and system integrity are ensured by preventing client access to the original database/catalogs and program, and by checking user supplied data before processing.

3 Architecture and Implementation

Figure 3 shows the system architecture. The bottom layer containing the reasoning program and constraint interpreter is not editable by customers, whether normal users or product experts. The layer above it, containing the problem template, is open either for editing or complete replacement by a product expert, before being used by a normal user. The non-expert user can provide the specification constraints through the user input interface, run the program, and see the proposed solution details.

The more complex functional and physical constraints are not visible to the non-expert user. Instead, this portion of the problem description is contained in a data file editable by a product expert, who can also adjust databases or part catalogs and can edit the items listed in the user input and output interface.

The top layer shows the two usage patterns: one for the product expert who can edit the problem description and another for the ordinary user who only wishes to indicate the product specification and see what solution(s) are possible.

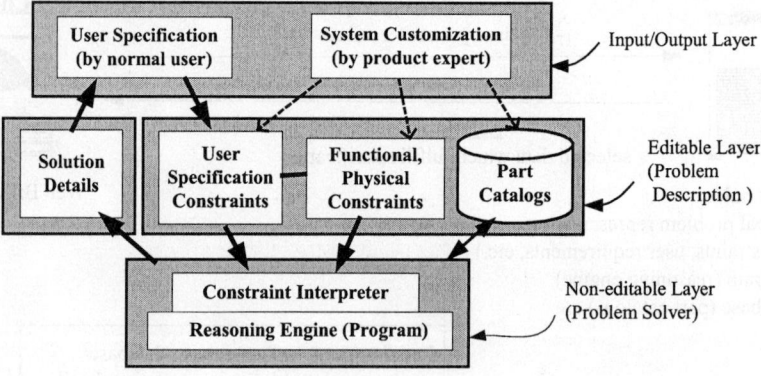

Fig. 3. ACAT Layered System Architecture

3.1 Reasoning Engine Program

The bottom layer of the system which is only accessible by the system manager contains a constraint-based problem solver which processes all the constraints to find a set of compatible parts satisfying user specification details and various other constraints. Where a large database or catalog is involved, the search space may be large and various strategies are employed to minimize the time required to find a solution, or to conclude that no solution is possible, if this is the case. To improve the efficiently of the search algorithm, the solver employs heuristics from constraint-based reasoning research, such as initial removal of infeasible values to minimize the search space, and the ordering of constraints to tackle the most highly constrained variables and therefore hardest to satisfy, first [1]. Where a solution cannot be found, the system points to constraints which must be relaxed but allows the user to make the decisions.

3.2 Detailed Problem Representation

We use a constraint language to represent the details of the problem description. ACAT contains an interpreter with constraint definitions for various types of relationships and restrictions on attributes. The format for each constraint listed in the problem description file is:

constraint-type : constraint-symbol : constrained part-attribute list

The *constraint-type* provides information for processing the constraint, e.g., whether the constraint is unary (acting on a single attribute), binary (acting on two attributes), or *n-ary* (acting on more than two attributes)

The *constraint-symbol* is defined in the constraint interpreter, where it is translated into computational form. Some examples of constraint symbols used in ACAT and their meaning are:

- *GEQ*: the first of the following part-attributes must be *greater than or equal to* the second, e.g., *GEQ Connector-Points InputUnit-Points*
- *SUMLEQX*: the sum of the following attributes, with the exception of the final attribute, must be less than or equal to the final attribute.
- *META-1*: a constraint which can alter another constraint, as defined in a procedure, e.g., if a certain constraint cannot be satisfied (such as minimum memory requirement), then add another part (such as extra memory).

Where a problem cannot be adequately represented using currently defined constraint types, the representation language can be expanded by adding new definitions to the reasoning engine and interpreter, which are written in Java/JSP. The modular structure of the program facilitates the adding of additional constraint definitions as required, so that arbitrarily complex problem representations can be tackled. However, these can only be edited or expanded by changing the reasoning engine program and interpreter and hence must be carried out by the system manager.

4 Example

While the ideas proposed here may be relevant to various types of information system, ACAT was built with a particular industrial problem in mind: the configuration of complex multi-part products such as programmable controllers, servo motors, sequencers, etc. We take as an example the configuration of a PLC (programmable logic controller). PLCs are robust and versatile control units used for various types of automated factory systems and processes. They are composed of CPUs, displays, and I/O relays and switches to control equipment. Fig. 4 shows the scenario with sample part catalogs and input and output data. The ACAT system provides the link between input, catalogs or databases and output.

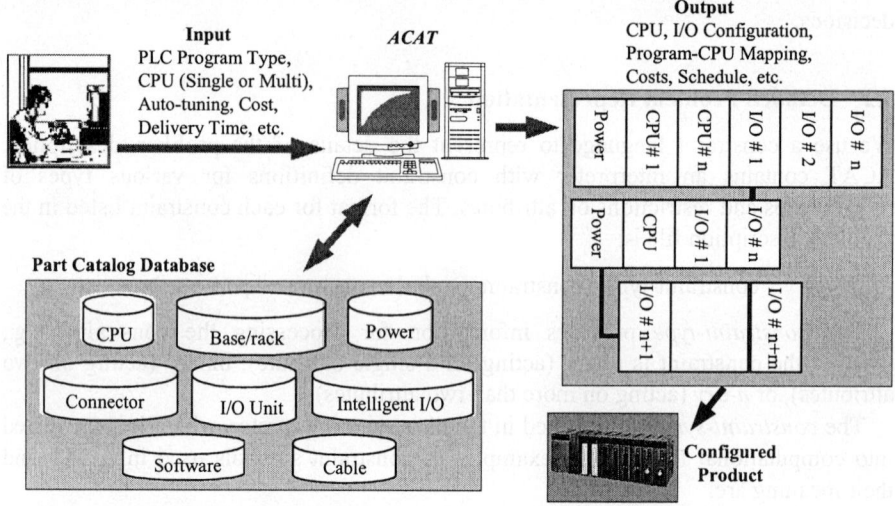

Fig. 4. Configuration of Programmable Logic Controller

A typical interface for the current version of ACAT in shown in Fig. 5, with the input on the left hand side and the output on the right. The items in this interface can be easily changed by editing the data file which contains the interface description. When a configuration is carried out, ACAT reads the input data (required attribute parameter values) and outputs a compatible set of parts and attribute values, if a solution can be found.

The input parameters (attributes) constitute some of the problem description but do not include the more complicated requirements such as constraints involving two or more parameters. Such constraints are contained in the problem representation file as described in section 3.2.

When configuring a product, the user typically provides specification details by editing the input parameter default values (left hand frame of Fig. 5) and then executes the reasoning engine. ACAT reads the catalog file(s) to determine basic information about the product to be configured – constituent part names and lists of attributes for each part. It then reads the detailed problem description file before searching the catalog(s) for candidate solutions.

Currently input parameters are, by default, *fixed* constraints which must be fulfilled. However, each input requirement can be relaxed by re-designating it as a *preference*, to represent constraints that can be automatically relaxed if no solution is found. ACAT displays suggested input parameters to be relaxed where a solution cannot be found. Optimization is possible for specified attributes with continuous numerical ranges.

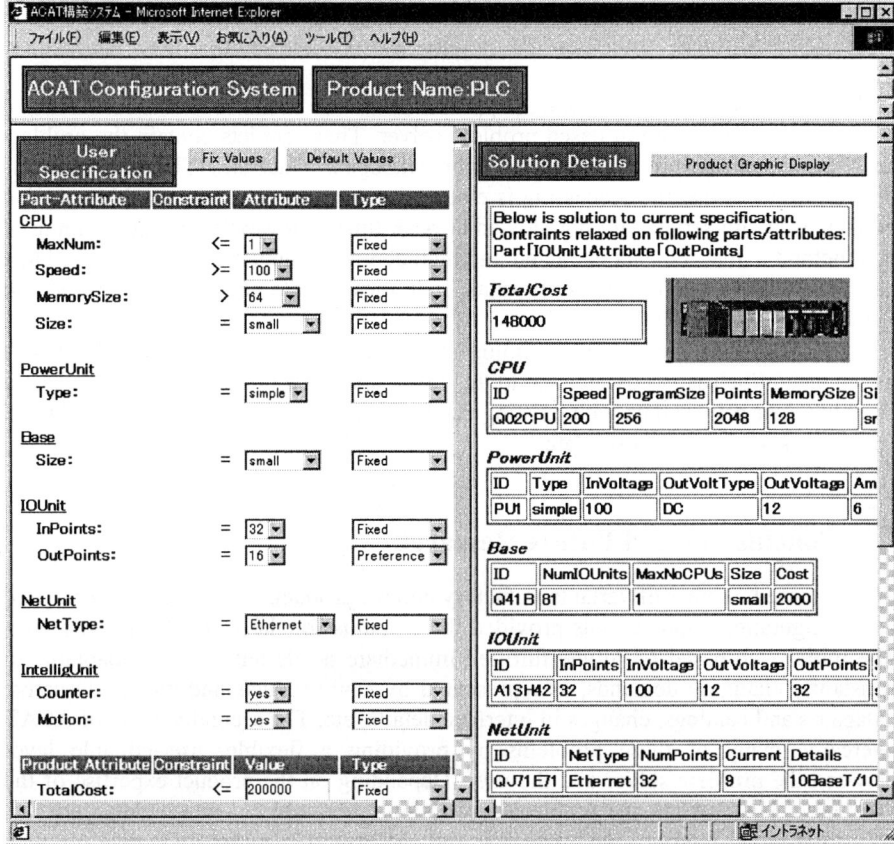

Fig. 5. Input-Output Interface of ACAT for Normal Usage

5 Related Research

Earlier work on configuration systems is reported in [2], focusing on the constraint-types and the stage in the problem solving process at which they are processed. The constraint-based representation ideas are developed in the present research to facilitate the flexible re-arrangement of problem entities.

The automatic design of interfaces, based on user profiling, data types and hardware is proposed by Penner et al. [3], and the personalization of catalog content is outlined in Ardissono et al. [4]. While it would be desirable to ultimately automate or partly automate the construction of interfaces in ACAT, we have focused in this research on enabling the personalization of the application at the more fundamental problem description level.

Commercial configuration systems include the ILOG configurator [5] consisting of a software components library which facilitates the building of large systems in the conventional model, i.e., as shown in Fig. 1. The ILOG Configurator requires the product structure to be expressed using inter-part *ports* and part hierarchies, in addition to attributes. In contrast, ACAT uses a purely attribute-based approach to express the problem and does not require any further inter-part or hierarchical descriptions.

The work done by Torrens et al. [6] proposes the use of smart clients consisting of java-applets in a constraint-based problem solver. These applets contain the problem description and relevant data extracted from catalogs or databases. The advantage this method offers is that the processing is essentially done on the client machine, so that popular web site servers are not slowed at peak times. However, the shortcoming of this method is that the data extracted and included in the applet cannot cater for large databases, i.e., it assumes that only small quantities of data need be downloaded and searched for a final solution.

ACAT, on the other hand, is aiming at processing large databases and catalogs which cannot be easily downloaded to the client, and at facilitating users who wish to combine their private data with existing catalogs on the server, or to apply the ACAT constraint solver to their own data.

6 Conclusions and Future Research

The increased number of variants in factory control products and the consequent need for configuration support tools provided the stimulus for the work described above. The initial web-based prototype fulfilled immediate needs but was not adaptable to constantly changing demands, as represented by new products and parts, alternative databases and catalogs, changes in interface details, etc. The second version of ACAT overcame many of these problems by providing a flexible, user-editable layer between the problem solver and the user. Depending on the product expertise of the user, he or she can edit the problem-description layer, add or remove data catalogs, adjust the initial interface parameter list and can provide new or alternative problem constraints.

At present, adjustment of the editable layer must be done by changing or replacing the text files which contain the problem data. While this is not overly complicated, some editor-level interface is necessary to provide error checking and to prevent non-acceptable data, e.g., checks are necessary to ensure that only defined constraint symbols be used. Another area requiring further work is that of catalog preparation: although data for most products are available in electronic form and are accessible through the web, standard format data catalogs for all parts have not yet been completed. Finally, feedback from factory managers and company sales departments giving actual user needs and desirable system functions is being collected in order to guide further improvements.

References

1. Murtagh, N.: A Product Configuration Tool using Web Catalogs, Japanese Society of Artificial Intelligence, Knowledge-Based Systems Research Meeting, SIG-KBS, Nagoya, (2001) (in Japanese)

2. Murtagh, N.: Artifact Configuration across Different Design Levels. In Mantyla, M., Finger, S., Tomiyama, T. (eds): Knowledge Intensive CAD I, Chapman & Hall, (1996)
3. Penner, R., Steinmetz, E., Johnson, C.: Adaptive User Interfaces through Dynamic Design Automation, American Association of Artificial Intelligence, Proceedings of the National Conference, (2000) 1127-1128
4. Ardissono, L., Goy, A.: Dynamic Generation of Adaptive Web Catalogs, Lecture Notes in Computer Science LNCS 1892, Springer-Verlag, Berlin Heidelberg, (2000) 5-16
5. ILOG Corporation: White Paper, http://www.ilog.com/products/configurator (2001)
6. Torrens, M., Faltings, B.: Constraint Satisfaction for Modelling Scalable Electronic Catalogs, Lecture Notes in Artificial Intelligence LNAI 1991, Springer-Verlag, Berlin Heidelberg, (2001) 214-228

Phase-to-Phase Wave Parameters Measurement of Distribution Lines Based on BP Networks

Fengling Han[1], Xinghuo Yu[1], Yong Feng[2], and Huifeng Dong[3]

[1] Faulty of Informatics and Communication, Central Queensland University
Rockhampton, QLD 4702, Australia
{f.han,x.yu}@cqu.edu.au
[2] Department of Electrical Engineering, Harbin Institute of Technology
Harbin, 150001, P.R. China
yfeng@hope.hit.edu.cn
[3] Dongfang Electrical Corporation, Yantai, P.R. China

Abstract. In order to overcome the influence of measurement errors in phase-to-phase wave parameters, this paper presents a neural network model for radial distribution lines. By providing a sample that is satisfactory and can resist measurement-error, and choosing a proper training method, the network can converge quickly. Simulation results show that this NN model can resist both amplitude-error and phase-error successfully, and can enhance the measurement precision of phase-to-phase wave parameters under working conditions.

Keywords: radial distribution lines, phase-to-phase wave parameters, neural network

1 Introduction

In recent years fault location for distribution lines has attracted much attention. Because of its complex structure and multi-branches, artificial intelligence method and steady state analysis method are generally used to locate a fault on distribution lines. When locating a fault, the artificial intelligence method generally using multi-terminal measurement or additional sensors. While one-terminal measurement steady state method is welcomed to the field worker because of its convenience. Srinivasan and St-Jacques[1] presented a fault location method by using the symmetry component method and one-terminal information. Zhu, Lubkeman and Girgis[2] presented another fault location method based on the lumped parameter model of radial transmission lines. The wave parameters of lines are calculated theoretically. This method takes the loads influence into consideration. Saha and Provoost[3] proposed a solution for the solidly ground network of radial type. Su and Xu[4] used the open-circuit and short-circuit methods to measure the wave parameters of no-branch lines. For locating a fault, the field measuring data are used. When in practical used, environment noise affects the wave parameters measurement and fault location. Sometimes this noise would cause error or even failure when locating a fault[5,6]. Neural networks are performing successfully where recognizing and matching complicated, vague, or incomplete patterns [7].

Based on the [5] and [6], this paper sets up a BP network model taking into consideration noises in the field. A perfect sample that can overcome measurement noise is provided to the network. By choosing a proper training method, the network can converge quickly. The trained network can overcome the errors in an actual measurement. This paper is organized as follows: section 2 briefly introduces the principle of fault location for distribution lines. In section 3, the wave parameters measuring errors are analysed. In section 4, the NN model is proposed to overcome the wave parameters measuring errors. Simulations are given in section 5. The main findings are summarized in the last section.

2   Principle of Fault Location for Distribution Lines

The fault location method is based on sinusoidal steady states analysis of the circuit. When a fault occurs in a distribution line, power is cut off. Assuming that only a terminal current and voltage are available, we can send a sinusoidal excitation signal through the distribution lines.

2.1   Fault Location for No-branch Distribution Line

Phase-to-phase short-circuit for no-branch line is shown in Fig.1. Fig.1(a) is a sinusoidal excitation signal generator and a sample resistor. Fig.1(b) is a no-branch transmission line.

The distributed-parameter equations for Fig.1 are :

$$\begin{cases} V_x = V_1 \operatorname{ch} \gamma x - Z_c I_1 \operatorname{sh} \gamma x \\ I_x = I_1 \operatorname{ch} \gamma x - \dfrac{V_1}{Z_c} \operatorname{sh} \gamma x \\ V_2 = V_x \operatorname{ch} \gamma(l-x) - Z_c I'_x \operatorname{sh} \gamma(l-x) \\ I_2 = I'_x \operatorname{ch} \gamma(l-x) - \dfrac{V_x}{Z_c} \operatorname{sh} \gamma(l-x) \\ V_x = R(I_x - I'_x) \\ V_2 = Z_2 I_2 \end{cases} \qquad (1)$$

where V_1 and I_1 are the beginning terminal voltage and current phasor respectively. V_2 and I_2 are the end terminal voltage and current phasor respectively. l is the line length and x is distance of the fault from the beginning point; Z_c is wave impedance and γ is propagation constant; R is fault contact resistance.

From Equ.(1), we can obtain x and R by solving the nonlinear equations. Therefore, we can locate the location of the fault.

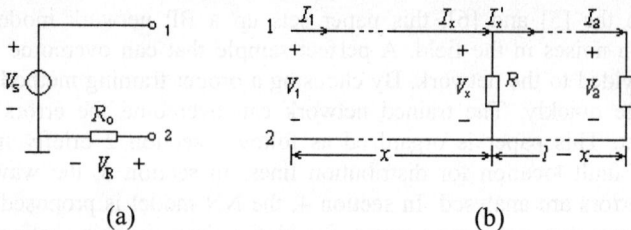

Fig. 1. Short-circuit happened at distance x away from the beginning terminal (a) Sinusoidal signal generator. (b) No-branch transmission line

2.2 Fault Location for Distribution Lines

A distribution line is shown in Fig.2.

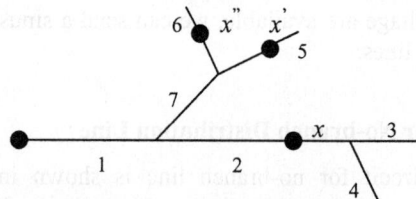

Fig. 2. Radial distribution line

Assume that only one fault happens in section k (in Fig.2, $k=1,2,\ldots,7$), and only V_1 and I_1 are measurable. We can calculate the voltage and current of beginning point of any section from the measurements of V_1 and I_1. Then x is obtained by solving the nonlinear Equ.(1). There is one answer or no answer for each section. The real fault happens at x, but by resolving the circuit equations (1), we obtain x, x' and x'' (where the electrical distance from the beginning terminal are equal). Changing the frequency of the excitation signal, we can get another set of data. Since the real location of the fault is unrelated to the frequency of the exciting signal, for two different measuring frequencies, the location of the fault will be the same under ideal circumstances. In practice, there are errors in the wave parameters of the line and the current and voltage used in location the fault, so the answer differ and we can compare the results obtained using two different frequencies and calculate the difference of x between two results. The spot where the absolute difference value of x is the smallest is the real location of fault[4].

The above method requires the model and the wave parameters of the lines to be accurate.

3 Analysis of Wave Parameters Measuring Errors

Propagation constant γ and wave impedance Z_c are wave parameters of distribution lines. They represent characteristics of the lines under working condition. When using the open-circuit and the short-circuit method to measure the wave parameters, there are errors in both the amplitude and the phase of the voltage and current respectively, so errors exist in the calculated wave parameters. In this paper simulation was done on triangle LJ120mm² lines. The distance between two lines is 60cm. The phase-to-phase wave parameters γ = 0.0050 − j0.2147 km⁻¹ and Z_c=287.2656 − j0.6661Ω are calculated theoretically. The effect of amplitude deviation of voltage and current on phase-to-phase wave parameters is shown in Table 1. The effect of phase deviation is shown in Table 2.

Table 1. Influences of voltage and current amplitude errors on phase-to-phase wave parameters

Add −5%→+5%errors	$\Delta\text{Re}(Z_c)/(\Omega)$	$\Delta\text{Im}(\gamma)/(\text{km}^{-1})$
Open-circuit voltage	Linear, +7→ −7	Linear, −0.8×10⁻⁴→+0.8×10⁻⁴
Short-circuit voltage	Linear, −7→ +7	Linear, +0.8×10⁻⁴→−0.8×10⁻⁴
Open-circuit current	Linear, +7→ −7	Linear, +0.8×10⁻⁴→−0.8×10⁻⁴
Short-circuit current	Linear, −7→ +7	Linear, −0.8×10⁻⁴→+0.8×10⁻⁴

Table 2. Influences of voltage and current phase-error on phase-to-phase wave parameters

Add −5%→+5%errors	$\Delta\text{Re}(Z_c)/(\Omega)$	$\Delta\text{Im}(\gamma)/(\text{km}^{-1})$
Open-circuit voltage	Linear, −0.004→+0.004	Linear, +0.05×10⁻²→ −0.05×10⁻²
Short-circuit voltage	Linear, −0.004→+0.004	Linear, +0.05×10⁻²→ −0.05×10⁻²
Open-circuit current	Square, max 0.077, min 0	Linear, −0.05×10⁻²→ +0.05×10⁻²
Short-circuit current	Square, max 0.077, min 0	Linear, −0.05×10⁻²→ +0.05×10⁻²

From the Table.1 and Table.2 we know that amplitude measurement errors affect the real part of the wave impedance, while phase measurement errors affect the imaginary part of the propagation constant.

4 Model, Structure and Samples of the Network

When there are no measuring errors in the voltage and current in both the end terminal open-circuit and short-circuit, the wave parameters of the lines can be determined exactly according to the theoretical calculations. It is the measuring errors that make the wave parameters inexact. So, we want to improve the original Equ. (1).

4.1 Model and Structure

When using the BP network, the input of the network is the amplitude and phase of voltage and current when the end terminal both open-circuit and short-circuit. We use the open-circuit voltage as a reference. There are a total of seven inputs. In order to simplify the network and enhance the convergence rates, we use four sub-networks to measuring the real and imaginary parts of wave impedances Z_c and propagation constant γ. The network has only one output. Simulation results show that when there are 30 units in the hidden layer, both the convergence rates and the size of the errors are acceptable.

4.2 Calculation of Wave Parameters within a Neighborhood

Based on the voltage and the current when the end terminal both open-circuit and short-circuit, a series of the original samples of wave parameters can be given by theoretical calculation. The four numbers, the real parts and the imaginary parts of the wave impedance Z_c and the propagation constant γ, have relationships. When the type of distribution lines and the erect method are chosen, the four numbers can be determined.

This paper investigates the measurement of phase-to-phase wave parameters when the lines are erected as a triangle structure. Because of the geometry symmetry, the couple are in opposition to each other. The model can be replaced by a two-lines. Let R_0, L_0, C_0 be the resistor, inductor and capacitor of unit length of the wire respectively. We can calculate the one-order parameters by using the following equations:

$$\begin{cases} R_0 = \rho/s \\ L_0 = [4.6 \lg(D_m/r) + 0.5\mu_r] * 10^{-4} \\ C_0 = 0.0241 * 10^{-6} * [\lg(D_m/r)]^{-1} \end{cases} \quad (2)$$

where ρ is the resistor rate;
  μ_r is the relative magnetic conductor rate;
  D_m is geometry average distance of three phases wire;
  r is the radii of the wire;
  s is the cross area of the wire.

Assume that the angular frequency of the sinusoidal input is ω. We can calculate the second-order parameters (propagation constant γ and wave impedance Z_c) by using the following equations:

$$\begin{cases} \gamma = \sqrt{(R_0 + j\omega L_0)(G_0 + j\omega C_0)} \\ Z_c = \sqrt{\dfrac{R_0 + j\omega L_0}{G_0 + j\omega C_0}} \end{cases} \quad (3)$$

After the structure and the type of the wire are chosen, we assume that γ and D_m varies among a neighborhood, we can calculate a series of Z_c and γ. Corresponding to these Z_c and γ of the lines, we conclude that as a set close to the actual lines, and these values are exact corresponding to a section of the lines. We define this set as **B**.

4.3 Determination of the Basic Set of Sample Set

After obtaining neighborhood **B** of phase-to-phase wave parameters, let **B** be the output of the whole neural network, then a serious of sample can be obtained according to the sinusoidal steady state equations of this section as follows:

$$\begin{cases} V_2 = V_1 \operatorname{ch} \gamma l - I_1 Z_c \operatorname{sh} \gamma l. \\ I_2 = I_1 \operatorname{ch} \gamma l - \dfrac{V_1}{Z_c} \operatorname{sh} \gamma l. \end{cases} \quad (4)$$

For the given Z_c and γ, assume V_{io} and V_{is} to be the measuring voltage when the end terminal is both open circuit and short circuit respectively. I_{io} and I_{is} can be calculated by using equation (4) and the initial conditions (the measuring value at the beginning terminal). Now Z_c, γ, V_{io}, V_{is}, I_{io}, I_{is} are the samples. After that, we change the values of V_{io}, V_{is}, l.

Repeating the above process, we can obtain all the samples. These are called basic set of the sample set **A1**. These samples are theoretically exact. There are no errors because the measured data have not been used yet.

4.4 Improvements to the Sample Set

The network trained by sample set **A1** can access to network equations (3). Both the measured amplitude and phase have errors. We can modify sample **A1** according to the errors. Extent sample **A1** as follows: Let V_{io} be the reference vector, and add 0, ±3%, ±5% relative deviation to the amplitude and phase of V_{is}, I_{io} and I_{is} while Z_c and γ are not changed. The extend sample is called sample **A2**.

Among sample **A2**, there is an exact unit without errors, it also include some samples with errors. So **A2** has the ability to resist errors. In this way the problem of parameter measurement errors can be overcome.

The extent sample set **A2** doesn't satisfied Equ.(4), we should say that the function relationship of the network is the father-set of Equ.(4).

5 Simulation

We use the Levenberg-Marquardt method, a training-function provided for BP network in the Neural Network Toolbox of MATLIB5.3. The Levenberg-Marquardt method is one that converges quickly, and it does not need much computation. The basic idea is to calculate Hessian Matrix by using Jacobian Matrix

Take a triangle erect LJ120mm^2 wires as an example, the distance between the two lines is 60cm. The connection relationship of the network is shown in Fig.2.

The wave parameter are calculated theoretically: $\gamma = 0.0050 - j0.214 \text{ km}^{-1}$, $Z_c = 287.2656 - j0.6661 \Omega$.

After the network has been trained, its ability to resist disturbance is tested. When measuring, first we add 5% random number to the amplitude and the phase of the voltage and the current of the original network, the output is obtained by theoretical calculation.

Two typical networks are tested. The imaginary part of γ is the output of network 1. The training result is shown in Fig.3. The training sample of this network is obtained by exert deviation on the phase of the voltage and the current while the end terminal is both open-circuit and short-circuit.

The real part of Z_c is the output of network 2, the training result is shown in Fig.4. The training sample of this network is obtained by exert deviation on the amplitude of the voltage and the current while the end terminal is both open-circuit and short-circuit.

Then, we do the simulation on the BP network presented in this paper.

For 1500 samples, when there is 5% deviation in the phase of voltage and current while the end terminal is both open-circuit and short-circuit, the absolute deviation of the imaginary parts γ is shown in Fig.5.

When there is 5% deviation in the amplitude of the voltage and current while the end terminal is both open-circuit and short-circuit, the real parts absolute deviation of Z_c is shown in Fig.6.

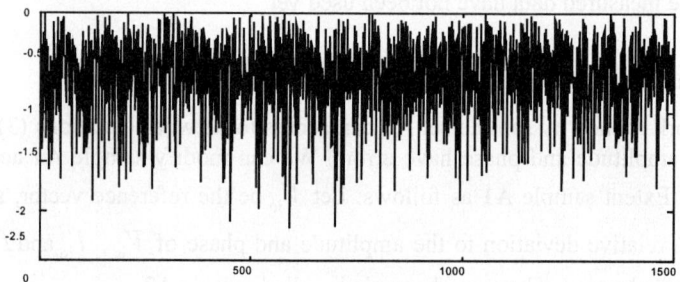

Fig. 3. Theoretical computation of γ imaginary parts when 5% errors exist in the phase. Horizontal axis: 1500 samples, Vertical axis: deviation in the imaginary part of γ

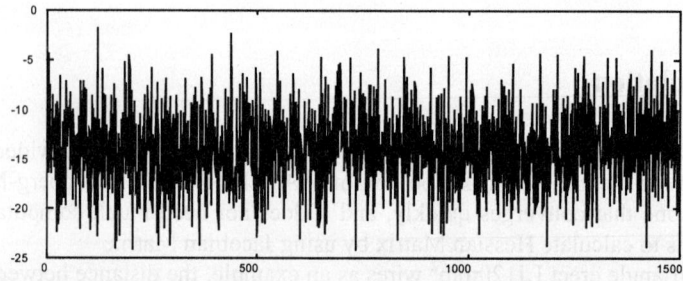

Fig. 4. Theoretical computation of Zc real parts when 5% errors exist in the amplitude. Horizontal axis: 1500 samples, Vertical axis: deviation in the real part of Z_c

6 Conclusions

The paper investigated the effect of the amplitude and the phase deviation on wave parameters measurement by using simulation. The neural networks are trained based on the amplitude and phase deviation. The networks are tested by a random deviation. The conclusion is:

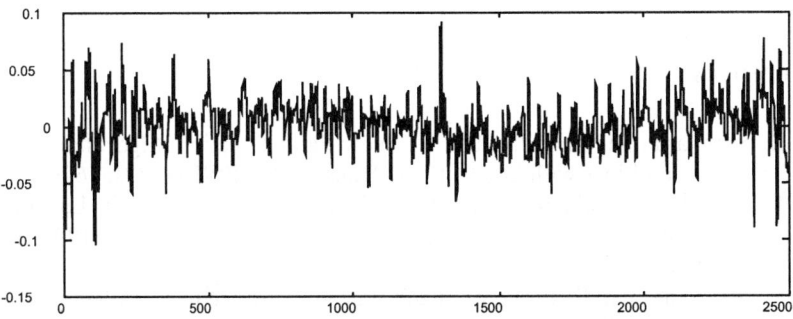

Fig. 5. Network output of imaginary part of γ which sample is obtained by 5%phase-error. Horizontal axis: 1500 samples, Vertical axis: deviation in the imaginary part of γ

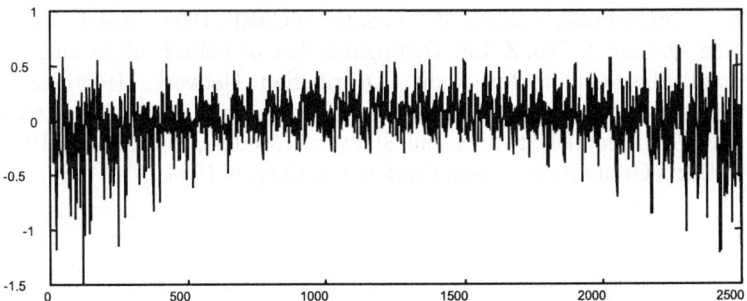

Fig. 6. Network output of real part of Z_c which sample is obtained by 5%amplitude-error. Horizontal axis: 1500 samples, Vertical axis: deviation in the real part of Z_c

(1) For the influence of the measuring deviation of the voltage and the current on wave parameters, the amplitude deviation influences wave impedance greatly while the phase deviation influences propagation constant obviously.
(2) When the amplitude of the voltage and the current exist 5% deviation, the output of the imaginary part of γ trained by this network is 2 digits more exact than the theoretical calculation.
(3) When the phase of the voltage and the current exist 5% deviation, the output of the real part of Z_c trained by this network is 1 digit more exact than the theoretical calculation.

This network is based on the classical calculation method. If there are no errors, the network can produce the phase-to-phase wave parameters exactly as that in theoretical calculation. When a certain amount of errors caused by measurement exist, the neural network model of the lines can resist amplitude and phase deviation effectively. It can enhance the measurement precision of the phase-to-phase parameters. Only with a relatively accurate wave parameters can the one-terminal measurement fault location method be put into practical use.

References

1. K. Srinivasan and A.St-Jacques. A New Fault Location Algorithm for Radial Transmission Lines with Loads. IEEE Trans., 1989, PWRD-4(3): 1676-1682
2. J. Zhu, D. L.Lubkeman and A. A.Girgis. Automated Fault Location and Diagnosis on Electric Power Distribution Feeders. IEEE Trans., 1997, PWRD-12(2):801-809
3. M. M. Saha and F. Provoost. Fault Location in Medium Voltage Networks. Proceedings of CIRED' Conference, 1999.
4. H. Su and C. Xu. A New Algorithm for single-line-to-ground fault location on 10kV or 6kV direct distribution line. Proceedings of the CSEE. 1995,15(5):423-428
5. F. Han, C. Xu, Z. Liu. Radial distribution network short-circuit point feature recognition and fault location. Proceedings of CSEE. 1997,17(6):412-415
6. F. Han, Q. Lian, C. Xu, Z. Liu. Distinguish Real or False Fault of Single Line-to-Ground Short-Circuit for Radial Distribution Network. 1998 International Conference on Power System Technology Proceedings. Aug. 1998, 260-264
7. Sarle, W.S. Neural Networks and Statistical Models. Proceedings 19th Annual SAS Users Group International Conference, Cary, N 1994,1538-1550.

Learning Capability: Classical RBF Network vs. SVM with Gaussian Kernel

Rameswar Debnath and Haruhisa Takahashi

Department of Information and Communication Engineering
The University of Electro-Communications
1-5-1 Chofugaoka, Chofu-shi, Tokyo, 182-8585, Japan
{rdebnath,takahasi}@ice.uec.ac.jp

Abstract. The Support Vector Machine (SVM) has recently been introduced as a new learning technique for solving variety of real-world applications based on statistical learning theory. The classical Radial Basis Function (RBF) network has similar structure as SVM with Gaussian kernel. In this paper we have compared the generalization performance of RBF network and SVM in classification problems. We applied Lagrangian differential gradient method for training and pruning RBF network. RBF network shows better generalization performance and computationally faster than SVM with Gaussian kernel, specially for large training data sets.

Keywords: Radial Basis Function network, Lagrangian differential gradient, Support Vector Machine.

1 Introduction

In classification problem, the task is to learn a rule from a training data set that contains attributes or covariates with class levels. Let $X \in R^d$ be covariates used for binary classification, and $Y \in \{-1, 1\}$ be class level that are related by a probabilistic relationship. Usually it is assumed that (X, Y) is an independently and identically distributed sample from an unknown probability distribution $P(X, Y) = P(X)P(X|Y)$ where $P(X|Y)$ is the conditional probability of Y given X, and $P(X)$ is the marginal probability of X.

Consider the training data set is separated by a hyperplane that is constructed by a sparse linear combination of basis functions depending on the training data points. Let $p(x) = P(Y = 1|X = x)$ is the conditional probability of a random sample being in class 1 given $X = x$. The weakness of SVM is that it only estimates $sign[p(x) - \frac{1}{2}]$, while the probability of $p(x)$ is often of interest in most practical learning tasks. Feedforward networks such as Radial Basis Function network, multilayer Perceptron provide an estimate of the probability $p(x)$. Applying Lagrangian differential gradient method as a learning rule in RBF network, the hyperplane is constructed relatively in a small hypothesis space that follows VC theory. The basic idea of Vapnik's (VC) theory is: for a finite set of training examples, search the best model or approximating function that has to be constructed to an approximately small hypothesis space. If the space is too large, model can be found that will fit exactly the data but will have a poor generalization performance that is poor predictive capability on new data. Applying Lagrangian differential gradient method in RBF network we have got better generalization performance and it is computationally faster than SVM, especially for large training data sets.

In section 2 and 3, we briefly discuss about SVM and RBF network. In section 4, the background of Lagrangian differential gradient method is presented. Then we formulate the problem of training and pruning neural networks in terms of Lagrangian differential gradient method in the binary classification problem. Experimental results are reported in section 5. Comparison of RBF network to SVM is presented in section 6. We end with conclusion in section 7.

2 Support Vector Machine (SVM)

Support Vector Machine produces a non-linear classification boundary in the original input space by constructing a linear hyperplane in a transformed version of the original input space. The input space is transformed in a high dimensional feature space whose dimension is very large, even infinite in some case by using a non-linear function which is an element of a Hilbert space while the capacity of the system is controlled by a parameter that does not depend on the dimensionality of the feature space. Given a set of points which belongs to either of two classes, SVM finds the hyperplane leaving the largest possible fraction of points of the same class on the same side, while maximizing the distance of either class from the hyperplane. The separating hyperplane can be written as:

$$f(\mathbf{x}) = \sum_{i=1}^{N} \alpha_i y_i K(\mathbf{x}_i, \mathbf{x}) + b$$

where $\alpha_i \geq 0$, $K(\mathbf{x}_i, \mathbf{x})$ is called inner-product kernel, $\mathbf{x}_i \in R^d$ is the input vector for the i-th example, and, y_i, the target value for the i-th example takes values $\{1, -1\}$, b is a bias and N is training sample size.

The parameter α_i can be found by solving the following convex quadratic programming problem:

$$\text{maximize} \quad \sum_{i=1}^{N} \alpha_i - \frac{1}{2} \sum_{i,j=1}^{N} \alpha_i \alpha_j y_i y_j K(\mathbf{x}_i, \mathbf{x}_j)$$

$$\text{subject to} \quad \sum_{i=1}^{N} \alpha_i y_i = 0,$$

$$0 \leq \alpha_i \leq C, \quad i = 1, \ldots, N$$

where C controls the trade-off between training error and generalization ability and its value is chosen by means of a validation set.

It often happens that a sizeable fraction of the N values of α_i can be zero. Only the points lie closest to the classification hyperplane including those on the wrong side of the hyperplane are the corresponding α_i non-zero. These training points $\mathbf{x}_i$'s are called support vectors. Therefore, the support vectors condense all the information contained in the training data set which is needed to classify new data points.

The inner-product kernel $K(\mathbf{x}_i, \mathbf{x}_j)$ is used to construct the optimal hyperplane in the feature space without considering the feature space itself in explicit form. The requirement of the inner-product kernel $K(\mathbf{x}_i, \mathbf{x}_j)$ is to satisfy Mercer's theorem. Common types of inner-product kernels are Gaussian kernel, polynomial kernel, and sigmoid kernel.

The bias b can be computed by taking any support vector $\mathbf{x}_j$ with $\alpha_j < C$ (training point $\mathbf{x}_j$ on the wrong side of hyperplane is the corresponding $\alpha_j = C$), and hence

$$b = y_j - \sum_{i=1}^{N} y_i \alpha_i K(\mathbf{x}_i, \mathbf{x}_j)$$

3 Radial Basis Function (RBF) Network

The construction of a Radial Basis Function (RBF) network, in its most basic form, involves three layers with entirely different roles. The input layer is made up of source nodes that connects the network to its environment. The second layer, the only hidden layer in the network, applies a nonlinear transformation from the input space to the hidden space; in most applications the hidden space is of high dimensionality. The output layer is linear, supplying the response of the network to the activation pattern applied to the input layer. When a Radial Basis Function (RBF) network is used to perform a complex pattern classification task, the problem is basically solved by transforming it into a high dimensional space in a nonlinear manner. A mathematical justification is found in Cover's theorem on the separability of patterns may be stated as:

> A complex pattern-classification problem cast in a high-dimensional space nonlinearly is more likely to be linearly separable than in a low-dimensional space.

The most common RBF network is:

$$F(\mathbf{x}) = \sum_{i=1}^{M} w_i \phi \left(\|\mathbf{x} - \mathbf{c}_i\|^2 \right) + b$$

where $\mathbf{x} \in R^d$ is the input vector, $w_i \in R$, for $i = 1, \ldots, M$ are the output weights, b is a bias, $\mathbf{c}_i \in R^d$, for $i = 1, \ldots, M$ are centers; the known data points $\mathbf{x}_i \in R^d$, $i = 1, \ldots, N$ are taken to be the centers, $\mathbf{c}_i$, and it guarantees that $M = N$, and $\phi \left(\|\mathbf{x} - \mathbf{c}_i\|^2 \right)$ is the multivariate Gaussian function defined by

$$\phi \left(\|\mathbf{x} - \mathbf{c}_i\|^2 \right) = exp \left(-\frac{1}{2\sigma_i^2} \|\mathbf{x} - \mathbf{c}_i\|^2 \right)$$

The condition $\sigma_i = \sigma$ for all i is often imposed on $F(\mathbf{x})$.

4 Our Approach in Learning Neural Networks

We applied Lagrangian differential gradient method for learning RBF network. Lagrangian differential gradient method finds the solution of a constrained optimization problem. Consider the problem of the form:

$$\text{Minimize} \quad f(x)$$
$$\text{Subject to} \quad \phi_i(x) \leq 0 \quad \text{for all } i = 1, \ldots, m$$

Then, x is modified according to the following rule:

$$\frac{dx}{dt} = -\nabla f(x) - \sum_{i=1}^{m} \delta_i(x) \nabla \phi_i(x)$$

where

$$\delta_i(x) = \begin{cases} 0, & \text{when } \phi_i(x) \leq 0 \\ k > 0, & \text{when } \phi_i(x) > 0 \end{cases}$$

is a kick-back function. When the trajectory goes to infeasible region, the constraint condition does not satisfy, it is kicked back by the value of k.

4.1 Lagrangian Differential Gradient Based Training and Pruning

In this section, we show how to apply Lagrangian differential gradient method in learning network. For simplicity of presentation we assume that network is developed for binary classification problem.

The binary classification problem is as follows. Consider two finite subsets of vectors $\mathbf{x}$ from the training set:

$$(\mathbf{x}_1, y_1), (\mathbf{x}_2, y_2), \ldots, (\mathbf{x}_m, y_m), \quad \text{where } \mathbf{x} \in R^d, y \in \{-1, 1\}$$

one subset I for which $y = 1$ and the other subset II for which $y = -1$ are separable by the hyper-surface $g(\mathbf{x}, \mathbf{w}) = 0$, where $\mathbf{x}$ is an input vector and $\mathbf{w}$ is an adjustable weight vector. The following inequalities

$$g(\mathbf{x}_i, \mathbf{w}) \geq 0 \quad \text{for } y_i = 1$$
$$g(\mathbf{x}_i, \mathbf{w}) < 0 \quad \text{for } y_i = -1$$

hold. We may rewrite the above inequalities in the equivalent form

$$y_i g(\mathbf{x}_i, \mathbf{w}) \geq 0 \quad \text{for all } i = 1, 2, \ldots, m.$$

According to VC theory, best network model or approximating function should be constructed to a relatively small hypothesis space in order to obtain good generalization. We wish to search a network that is constructed in small hypothesis space (i.e., a network with small number of weight parameters) and perfectly classifies all training examples by the following optimization problem:

$$\text{Minimize} \quad |\mathbf{w}|$$
$$\text{Subject to} \quad -y_i g(\mathbf{x}_i, \mathbf{w}) \leq 0 \quad \text{for all } i = 1, 2, \ldots, m$$

Then, **w** is modified according to the following rule:

$$\frac{d\mathbf{w}}{dt} = -\nabla(|\mathbf{w}|) + \sum_{i=1}^{m} \delta_i(\mathbf{w})\nabla\big(y_i g(\mathbf{x}_i, \mathbf{w})\big)$$

where

$$\delta_i(x) = \begin{cases} 0, & \text{when } y_i g(\mathbf{x}_i, \mathbf{w}) \geq 0 \\ k > 0, & \text{when } y_i g(\mathbf{x}_i, \mathbf{w}) < 0 \end{cases}$$

The correction $\Delta\mathbf{w}(n)$ applied to $\mathbf{w}(n)$ is defined by:

$$\Delta\mathbf{w}(n) = \eta \frac{d\mathbf{w}(n)}{dt}$$

where η is a positive constant called the learning-rate parameter and has to be chosen by the user, and n is time step (iteration). This adjustment incrementally decreases the cost function.

This optimization procedure includes pruning process as well as training. Training of network is run until all training examples are classified correctly. The value of $|\mathbf{w}|$ becomes minimum when a small number of weights have non-zero value and all other weights have zero value. The pruning process is done as in the following way:

When the training examples are classified correctly with $\mathbf{w}(n)$, the second term of Lagrangian differential becomes zero. The weights are then modified with the first term. Modification of weights only with the first term is run until there is any error found of the training examples with the current weights. During this weight modification process some weight values w_i become zero or are near to zero. Remove those weights and pruning of the network is done at this stage. When any error is found in the training examples, network is then re-trained with both term of the differential starting with the current weight vector.

The whole process of training—pruning will be continued until no further improvement in the network model is noticed, i.e., training errors remain within specific limit.

5 Experimental Results

Three experiments were used to compare the results of RBF network to SVM. For each model, the kernels were Gaussian functions with $\sigma = 1$. Sequential Minimal Optimization (SMO) algorithm was used for training SVM where the value of $C = 100$. In the first and second experiments, we tested two classes of 4 and 5-dimensional threshold-XOR function and in the third experiment, we tested two classes of 5-dimensional majority-XOR function. A N-dimensional threshold-XOR function is natural extension of the real-valued threshold function which maps $[0,1]^{N-1} \times \{0,1\} \to \{-1,1\}$. The real-valued threshold function maps $[0,1]^N \to \{0,1\}$: the output is 1 if and only if the sum of N real-valued inputs is greater than $N/2$. In N-dimensional threshold-XOR function, output is 1 if and only if the N-th input, which is binary, disagrees with the threshold function computed by the first $N-1$ real-valued inputs. A N-dimensional majority-XOR function is a Boolean function where the output is 1 if and only if the N-th input disagrees with the majority computed by the first $N-1$ inputs; majority is a boolean predicate in which the output is 1 if and only if more than half of the inputs are 1. Training and test examples were independently drawn from same distribution. A simulator run considered of training a randomly initialized network on a training set of m examples of the target function. The fraction of test examples classified incorrectly was used as an estimate of the generalization error of the network. The average generalization error for a given value of m training examples was typically computed by averaging the results of 45 simulator run, each with different random initial weights chosen uniformly from $[-1, 1]$. We restricted our attention to values m ranging from 20 to 100 training examples because we have got our expected results within this range. The generalization error of the resulting network was estimated by testing on a set of 5000 examples.

5.1 Choosing the Learning Rate Parameters of Our Proposed Method

Each weight is adjusted by an amount which is proportional to the Lagrangian differential, where the proportionality constant, η, called learning rate. Training may take a long time and convergence behavior may depend on learning rate. From the experiment, we found that different learning rate of different terms of the Lagrangian differential speed up the training process and we achieved the best results applying the following values. In the experiment correction $\Delta \mathbf{w}(n)$ applied to $\mathbf{w}(n)$ as:

$$\Delta \mathbf{w}(n) = -\eta_1 \nabla \left(|\mathbf{w}(n)|\right) + \eta_2 \sum_{i=1}^{m} \delta_i \left(\mathbf{w}(n)\right) \nabla \left(y_i g\left(\mathbf{x}_i, \mathbf{w}(n)\right)\right)$$

where η_1 is dynamic learning rate. Different η_1 values are used in the training process and in the pruning process. Network training is done using $\eta_1 = 0.00001$ and $\eta_2 = 0.009 \sim 0.09$. In the pruning process the value of η_1 is started at $0.0001 \sim 0.0006$. Increasing the number of step size (iteration), the number of weights reduces and the η_1 value increases with the reduced number of weights. All cases either training or pruning we set the sum of absolute value of weights to a fixed value (e.g., 50 in this experiment). As the number of weights reduces and each weight value increases, the update η_1 value speed up the pruning process. We set the value of $\delta_i(\mathbf{w}(n))$ as the inverse of total number of misclassification errors in the training data set obtained at iteration n.

Figures 1-3 show a comparison of generalization performance of RBF network learned by our proposed method and SVM. Table 1 shows the number of remaining weights of RBF network and support vectors of SVM.

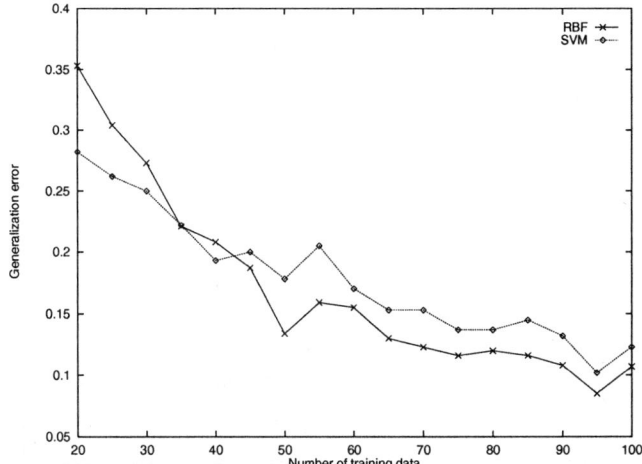

Fig. 1. Comparison of generalization error of RBF network learned by our proposed method and SVM in two classes of 4-dimensional threshold-XOR function. Training data sizes range from 20 to 100, in increment of 5. RBF network shows better generalization performance than SVM when the training size is large

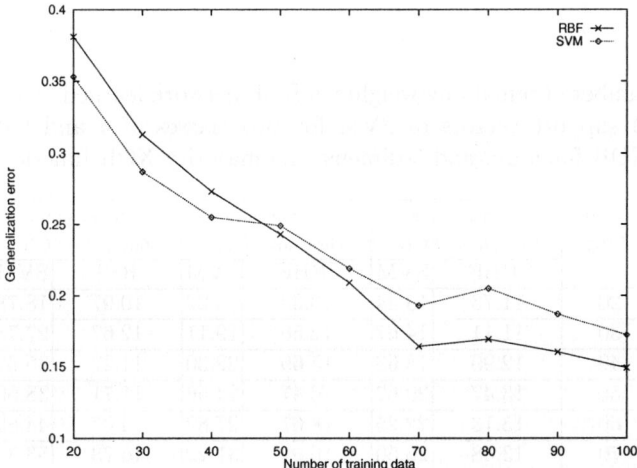

Fig. 2. Comparison of generalization error of RBF network learned by our proposed method and SVM in two classes of 5-dimensional threshold-XOR function. Training data sizes range from 20 to 100, in increment of 10. RBF network shows better generalization performance than SVM when the training size is large

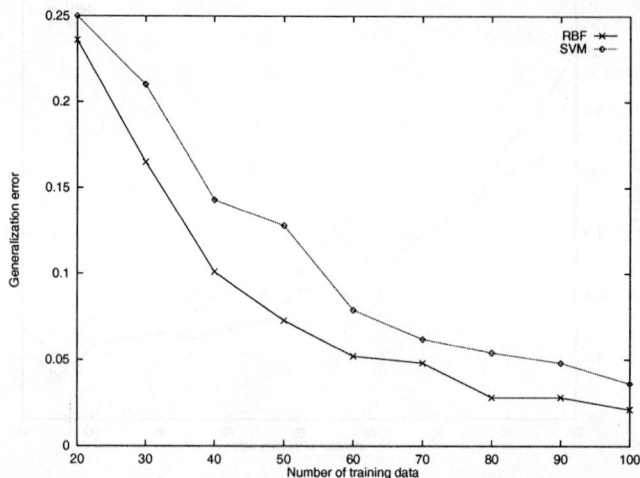

Fig. 3. Comparison of generalization error of RBF network learned by our proposed method and SVM in two classes of 5-dimensional majority-XOR function. Training data sizes range from 20 to 100, in increment of 10. RBF network shows better generalization performance than SVM in all sizes

Table 1. Number of remaining weights of RBF network learned by our proposed method and support vectors of SVM for two classes of 4 and 5-dimensional Threshold-XOR function, and 5-dimensional majority-XOR function

Training data	4-dimensional threshold-XOR		5-dimensional threshold-XOR		5-dimensional majority-XOR	
	RBF	SVM	RBF	SVM	RBF	SVM
20	11.73	13.44	10.38	15.22	10.07	18.78
30	11.11	14.67	12.56	19.11	12.67	27.78
40	12.90	18.63	15.69	22.00	11.22	35.67
50	13.47	20.67	16.47	24.56	11.71	38.00
60	13.13	22.25	18.67	27.89	14.07	44.89
70	13.48	26.50	16.09	31.22	16.73	53.33
80	14.16	27.22	16.89	32.33	14.98	57.22
90	17.16	31.00	22.49	35.56	16.64	63.00
100	21.91	32.22	25.82	35.67	21.53	68.89

6 Discussion and Comparison

The support vectors of the SVM are those which are close to the classification boundary or misclassified and usually have large values $[p(x)(1 - p(x))]$. Since training data are independently and identically distributed and may contain an amount of noise, the number of support vectors increases as the training data size increases. By fitting more and more noisy data, the machine may implement a rapidly oscillating function rather than the smooth mapping which characterizes most practical learning tasks. The input points of RBF network corresponding to remaining weights can be either close to or far from the classification boundary. Increasing training data size, number of remaining weights will not increase rapidly in RBF network using our proposed learning method like SVM.

The dividing hyperplane of SVM has smaller variation comparing to RBF network since it passes through the middle of two classes. If the number of support vectors is similar to the number of remaining weights of RBF network, SVM shows better generalization performance than RBF network. For this why, we have got better generalization performance in SVM than RBF network's when the training data size is relatively small for some data sets. When the training data size is large, variation of surface becomes small. Increasing the training data size RBF network using our proposed learning method implement much smoother function comparing to SVM. This cause better generalization performance obtained in RBF network than SVM's when training data size is large.

7 Conclusion

In this paper, we applied Lagrangian differential gradient method for training and pruning RBF network. Empirical results show that RBF network needs much smaller fraction of training data points than SVM to index kernel basis function. This gives a computational advantage over SVM when the size of the training data sets is large. RBF network with our proposed learning method shows better generalization performance than SVM, specially for large training data sets.

References

1. Leone, R. D., Capparuccia, R., and Merelli, E.: A Successive Overrelaxation Backpropagation Algorithm for Neural-Network Traning. IEEE Trans. Neural Networks, Vol. 9, No. 3, pp. 381-388 (1998).
2. Hsin, H.-C., Li, C.-C., Sun, M., and Sclabassi, R. J.: An Adaptive Training Algorithm for Back-Propagation Neural Networks. IEEE Trans on systems, man and cybernetics, Vol. 25, No. 3, pp. 512-514 (1995).
3. Castellano, G., Fanelli, A. M., and Pelillo, M.: An Iterative Pruning Algorithm for Feedforward Neural Networks. IEEE Trans Neural Networks, Vol. 8, No. 3, pp. 519-531 (1997).

4. Heskes, T.: On 'Natural' Learning and Pruning in Multilayered Perceptrons. Neural Computation, Vol. 12, No. 4, pp. 881-901 (2000).
5. Ponnapalli, P. V. S., Ho, K. C., and Thomson, M.: A Formal Selection and Pruning Algorithm for Feedforward Artificial Neural Network Optimization. IEEE Trans Neural Networks, Vol. 10, No. 4, pp. 964-968 (1999).
6. Abdel-Wahhab, O., and Sid-Ahmed, M. A.: A new scheme for training feed-forward neural networks. Pattern Recognition, Vol. 30, No. 3, pp. 519-524 (1997).
7. Karras, D. A., and Perantonis, S. J.: An Efficient Constrained Training Algorithm for Feedforward Networks. IEEE Trans Neural Networks, Vol. 6, No. 6, pp. 1420-1434 (1995).
8. McLoone, S., Brown, M. D., Irwin, G., and Lightbody, G.: A Hybrid Linear/Nonlinear Training Algorithm for Feedforward Neural Networks. IEEE Trans Neural Networks, Vol. 9, No. 4, pp. 669-684 (1998).
9. Tsukamoto, N., and Takahashi, H.: Generalization Performance: SVM vs. Complexity-Regularization. Technical Report of IEICE, Vol. 101, No. 154, pp. 49-55 (2001).
10. Fine, T. L. and Mukherjee, S.: Parameter Convergence and Learning Curves for Neural Networks. Neural Computation, Vol. 11, No. 3, pp. 747-769 (1999).
11. Zhu, J., and Hastie, T.: Kernel Logistic Regression and the Import Vector Machine (2001). For online version: http://www.stanford.edu/ jzhu/research/nips01.ps.
12. Lee, Y., Lin, Y., and Wahba, G.: Multicategory Support Vector Machines, Proceedings of the 33rd Symposium on the Interface, Costa Mesa, CA (2001).
13. Campbell, C., and Cristianini, N.: Simple Learning Algorithms for Training Support Vector Machine, Technical report, University of Bristol (1998).
14. Haykin, S.: Neural Networks-A comprehensive foundation, 2nd ed., Prentice Hall International, Inc (1999).
15. Cristanni, N, and Shawe-Taylor, J.: An Introduction to Support Vector Machines and other kernel-based learning methods (pp. 93-121), Cambridge University Press (2000).
16. Buzzi, C., and Grippo, L.: Convergent Decomposition Techniques for Training RBF Neural Networks, Neural Computation, Vol. 13, No. 8, pp. 1891-1920 (2001).
17. Cohn, D.: Separating Formal Bounds from Practical Performance in Learning Systems (1992). For online version: http://citeseer.nj.nec.com/cohn92separating.html.

Trading off between Misclassification, Recognition and Generalization in Data Mining with Continuous Features

Dianhui Wang[1], Tharam Dillon[1], and Elizabeth Chang[2]

[1] Department of Computer Science and Computer Engineering,
La Trobe University, Melbourne, VIC 3083, Australia
dhwang@cs.latrobe.edu.au
[2] Department of Computer Science and Software Engineering,
Newcastle University, Newcastle, Australia

Abstract. This paper aims at developing a data mining approach for classification rule representation and automated acquisition from numerical data with continuous attributes. The classification rules are crisp and described by ellipsoidal regions with different attributes for each individual rule. A regularization model trading off misclassification rate, recognition rate and generalization ability is presented and applied to rule refinement. A regularizing data mining algorithm is given, which includes self-organizing map network based clustering techniques, feature selection using breakpoint technique, rule initialization and optimization, classifier structure and usage. An Illustrative example demonstrates the applicability and potential of the proposed techniques for domains with continuous attributes.

1 Introduction

Given a set of pre-classified examples described in terms of some attributes, the goal of data mining for pattern classification is to derive a set of IF-THEN rules that can be used to assign new events to the appropriate classes [3]. The classification rules can be roughly categorized into crisp rules and fuzzy rules [1],[10]. The crisp rules may be expressed by hyper-box regions whose surfaces are parallel to the inputs axis [4],[8],[11],[13], polyhedron regions and ellipsoidal regions [2],[14],[16]. Existing methods for dealing with continuous features usually discretize continuous data into either two groups using a threshold value or into several groups creating bins throughout the possible range of values [7],[9]. Neural networks are able to use the raw continuous values as inputs, and as such, the method permits direct exploitation of this class of problem [6]. In this manner we are using the actual data itself directly, rather than arbitrarily specified divisions. Using supervised BRAINNE (Building Representations for AI using Neural NEtworks) we are able to extract the important attributes, and from this we are able to determine the appropriate range, without the need for arbitrary thresholds [4]. A version of supervised BRAINNE was developed to solve this continuous features problem. But it required multiple re-solutions of the supervised neural net [4]. However, difficulties occur if one wishes to extract disjunctive rules directly using this technique although it works well for direct extraction of conjunctive rules, and to overcome these hybrid BRAINN has been developed. Currently hybrid BRAINNE is only applicable to discrete variables [15]. Therefore, preprocessing for data discretization is required when the domain data is continuous or mixed-mode. To handle domains characterized by continuous features and disjunctive rules, a hybrid method utilizing an unsupervised net for clustering followed by optimization of those rules using a GA technique has been recently developed by the current authors [16], where the ellipsoidal regions in expressing rules were adopted. A drawback existing in the technique proposed is the use of all attributes in representing the classification rules. Note that the goal of data mining for classification is to construct a classifier to distinguish examples from different classes but not to build up a rule descriptor. Features selection is pretty important and necessary in the study of the pattern recognition domain [14]. This is because the issue is closely relevant to the generalization power and computational complexity of the classification system.

Unlike fuzzy classifiers, the issue of completeness arises from crisp classifiers and it can be measured by using a coverage rate [8],[11]. Theoretical study showed that a single Kohonen net may not capture all patterns concerned through unsupervised learning due to the training data distributions [17]. It is essential to extract rules with a high coverage and especially to obtain some rules for patterns with lower probability of occurrence but considerable conceptual importance. For example, mining some rules to classify heavy rain instances from a rainfall database. The event, heavy rain only occurs infrequently, but is a significant event for weather forecasting. To overcome this inherent shortcoming caused by the use of a single Kohonen net for clustering, a further discovery for meaningful and significant patterns is possible, and this can be dealt with re-training the network with a set of new data.

This paper presents a data mining approach for extracting accurate classification rules for domains characterized by continuous features and disjunctive rules. The proposed data mining algorithm here can handle classification problems with continuous data directly, and can assign a rule set covering more patterns especially for those with lower probability density and high conceptual importance. Experimental results for iris data are given.

2 Rules Representation and Feature Selection

This section defines a class of crisp rules characterized by the generic ellipsoidal regions. Each individual rule may have its own attributes, which corresponds to a lower dimensional space. The basic idea for feature selection is to directly use the information of the cluster centers without considering the shape effect as done in [14]. The discarded features from a rule are those having relative small distances from other classes in terms of the cluster centers. We use the following notation in this paper. Let $R^{m \times m}$ and R^m denote the set of $m \times m$ real matrices and the m-dimensional real vector space, respectively. If $M \in R^{m \times m} (v \in R^m)$, then $M^T (v^T)$ denotes the transpose of $M(v)$, M^{-1} the inverse of M as it is nonsingular and $Tr(M)$ the trace of M, respectively. Let S be a set with finite elements, we use $|S|$ to represent the cardinality of S, i.e., the number of elements contained in the set S. For a given positive finite matrix $M \in R^{m \times m}$, $X, Y \in R^m$, the well-known Mahalanobis distance measure is defined as $d_M(X,Y) = (X-Y)^T M^{-1}(X-Y)$, and the Euclidean distance measure, denoted by $d(X,Y)$, is given by setting M to be the identity matrix in Mahalanobis distance measure. The 2-norm of a vector $X \in R^m$, denoted by $\|X\|$, is calculated by $\|X\| = \sqrt{d(X,0)}$.

Suppose that the classification problem contains n-classes data sets, denoted by $\Theta_1, \Theta_2, \ldots, \Theta_n$, characterized by continuous attributes in an m-dimensional subspace $[0,1]^m \subset R^m$. Let

$$S_{all} = \{X : X \in R^m \cap (\Theta_1 \cup \Theta_2 \cup \ldots \Theta_n)\} \quad (1)$$

This data set above is firstly divided into a training set denoted by S_{tra} and a test set denoted by S_{tes}, respectively. The training set will be used to initialize and optimize the rules, and the test set, unseen by the learning methods, will be used to evaluate the performance of the classifier.

We use notation O_{kl} to represent the l-th cluster of the class Θ_k and $C_{kl} = [c_{kl,1}, c_{kl,2}, \cdots, c_{kl,m}]^T$ the center of the cluster O_{kl}, $A_{kl} \in R^{m \times m}$ a positive definite matrix related to this cluster. Then, a crisp type classification rule covering patterns around the cluster O_{kl} can be expressed as follows:

$$R_{kl} : IF\ d_{A_{kl}}(X, C_{kl}) \leq 1,\ THEN\ X \in \Theta_k \quad (2)$$

The classification rules expressed as above give ellipsoidal regions, which are exactly determined by the cluster center C_{kl} and the matrix A_{kl}. In the space for X, this ellipsoid represents the region that it is covered by the Rule R_{kl} for classification. The rule representation in (2) utilizes all the attributes in feature space. Therefore, it will increase the rule complexity and at the same time decrease the generalization power of the rule. To overcome these difficulties and improve the classifier performance, removal of some unnecessary attributes from feature space can be done.

Suppose that the input selected for R_{kl} is $\tilde{X} = [x_{kl_1}, x_{kl_2}, \cdots, x_{kl_H}]^T \in R^H$, where H stands for a smaller integer than m. A condensed matrix, denoted by $\tilde{A}_{kl} \in R^{H \times H}$, can be obtained by discarding some rows and columns corresponding to the removed attributes. Similarly, we denote the derived cluster center by $\tilde{C}_{kl} = [c_{kl_1}, c_{kl_2}, \cdots, c_{kl_H}]^T \in R^H$. Then, the rule R_{kl} can be rewritten as

$$R_{kl}: \text{IF } d_{\psi \tilde{A}_{kl}}(\tilde{X}, \tilde{C}_{kl}) \leq 1, \text{THEN } X \in \Theta_k \tag{3}$$

where $\psi > 0$ is an adjustable real number. Note that the set of selected attributes may be different for each rule, therefore the rules representation given in (3) is quite general. There are three steps to induce the rule in (3), they are:

- Generating a rule in (2) by the techniques proposed in [16].
- Selecting features and forming initial rules in (3).
- Finalizing the rules by adjusting ψ in (3).

2.1 Simplified Feature Selection Algorithm (SFSA)

Suppose that a set of rules with full attributes is available. The following simplified algorithm is for features selection:

- Calculate $\delta_{k_j l_j}(q) = |c_{kl,p} - c_{k_j l_j, p}|, k_j \neq k, l_j = 1,2,\cdots,q = 1,2\cdots,m$.
- Calculate $\sigma_{kl}(q) = \min\{\sigma_{k_j l_j}(q), l_j = 1,2,\cdots,\kappa_j \neq \kappa\}, q = 1,2,\cdots,m$.
- Sort $\sigma_{kl}(q), q = 1,2,\cdots,m$ such that $\sigma_{kl}(q_i) \leq \sigma_{kl}(q_{i+1})$ to form an index vector, namely $IV_{kl} = [\sigma_{kl}(q_1), \sigma_{kl}(q_2), \cdots, \sigma_{kl}(q_m)]$.
- Determine a breakpoint q_z such that $\sum_{j \geq z} \sigma_{kl}(q_j) / \sum_j \sigma_{kl}(q_j) = \beta$, where β is a threshold given by designers, usually chosen as 0.85~0.90.
- Select the contributory attributes in the antecedent of the rule as $x_{q_z}, x_{q_{z+1}}, \cdots, x_{q_m}$, and the dimensional size H in (3) becomes $m - z + 1$.

2.2 Scale ψ Determination

Theoretically, the scale ψ in (3) can be calculated by

$$\psi = \max\{d_{\tilde{A}_{kl}}(\tilde{X}, \tilde{C}_{kl})\}, \text{ s.t. } d_{A_{kl}}(X, C_{kl}) = 1 \tag{4}$$

This value gives a projection of the region in the original space to a lower dimensional space. There are two aspects to be considered when we try to get an exact solution in (4). First, it is difficult to obtain an analytical solution, indeed it is even not so easy to get a numerical approximation. Next, the obtained value of ψ needs further adjusting to ensure an acceptable performance of the rule. Based on these considerations, we here suggest a search algorithm to determine the scale ψ in (3) as follows:

- Set an initial value of the ψ to be 1, i.e., $\psi_0 = 1$.
- Let $\psi = \psi_0$, $\Delta > 0$ is a small positive real number.
- Check out the misclassification rate (MR) of the rule.
- Take $\psi = (1 + \Delta)\psi$ if $MR = 0$; otherwise take $\psi = (1 - \Delta)\psi$.
- Repeat the last step for determining the final value of ψ, which satisfies $\psi^* = \sup_\psi \{\arg MR = 0\}$.

3 Regularization Approach for Rule Refinement and Tradeoffs

The backbone of our rule induction approach is a two-phase process: a rule initialization stage followed by a rule optimization stage. In the optimization phase, it contains three parts, i.e., initial rules refinement, attributes elimination and rules finalizing. In this paper, we present a regularization theory to tradeoff misclassification rate, recognition rate and generalization ability for refining the initial rules.

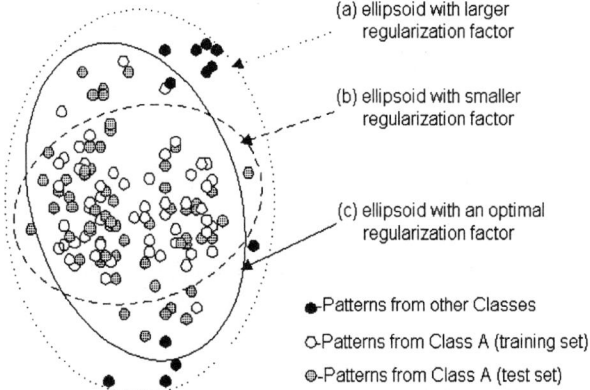

Fig.1 A graphical illustration of our regularization theory. If the regularizing factor (RF) takes a larger value, the misclassification rate will be possibly higher, see (a); On another hand, if the RF takes a smaller value, the coverage rate will be possibly lower, see (b), which also implies a possibly weaker in generalization; Case (c) corresponds to the optimal RF

The basic idea behind the regularization theory is to balance two or more possibly conflicting criteria for performance in the design of a system. In data mining for pattern classification, we mainly consider three important criteria that the system is trying to meet, i.e., misclassification rate (MR), an increased recognition rate (RR) and an improved generalization power (GP). A good classifier will have lower MR and higher RR for both training set and test set and a higher GP simultaneously. As the ellipsoidal region tends to be large enough to enclose many training patterns for a certain category of data, the misclassification rate could be increased for this class. It could lead to lower recognition for other classes. On the other hand, if the size of region is reduced and only a small number of training data for a class is enclosed, a good generalization ability of the rule cannot be expected because of over-fitting, also the recognition rate of this class could be reduced due to poor coverage. Moreover, the degree of complexity will increase with the number of rules. Therefore, a tradeoff between the size of the ellipsoidal region to achieve good generalization ability, a high coverage rate (CR) and a low misclassification rate should be addressed. Note that there is no direct way to exactly evaluate the GP during the design phase. However, this measurement can be indirectly expressed by using the size of the rule region. Recall that the trace is a reasonable measure to characterize the size of the ellipsoidal region. Thus, GP and CR for Rule R_{kl} in (2) can be approximately evaluated by

$$E_{GP+CR} = |G_k(A_{kl}, C_{kl})| + Tr(A_{kl}) \qquad (5)$$

The MR can be calculated by

$$E_{MR} = \frac{\sum_{p \neq k} |G_p(A_{kl}, C_{kl})|}{|G_k(A_{kl}, C_{kl})|} \qquad (6)$$

where the set $G_p(A_{kl}, C_{kl})$ is defined as

$$G_p(A_{kl}, C_{kl}) = \{X : d_{A_{kl}}(X, C_{kl}) \leq 1, X \in \Theta_p \cap S_{tra}\} \qquad (7)$$

Using this notation, we define a regularization function (RF) as follows:

$$F(A_{kl}, C_{kl}) = (1 - \lambda_{kl}) E_{MR} + \lambda_{kl} E_{GR+CR}^{-1} \qquad (8)$$

where $0 < \lambda_{kl} < 1$ is a regularizing factor.

The regularization function (8) above allows us to balance the importance of misclassification against the generalization ability of the rules.

4 Regularizing Data Mining Algorithm: RDMA

This section provides an outline of our rule induction techniques. Some details on the RDMA can be referred to [16]. The Kohonen net is a good tool for clustering [12], so we employ it in this work.

The Kohonen approach works best with a normalized set of input features [5, 18]. Let the original features in the real world space be $X = [x_1, x_2, \ldots, x_m]^T$. Then, we define a normalized set of features Z as follows:

$$X \to Z : Z = \frac{1}{\sqrt{1 + \|X\|^2}} [\frac{X}{\|X\|}, \|X\|]^T = [z_1, \cdots, z_m, z_{m+1}]^T \qquad (9)$$

Using transformation (9), the training set S_{tra} in the real world space can be mapped to a corresponding data set in Z-space, denoted by $\widetilde{S}_{tra}$, which will be used to initialize rules. To recover the value of a vector X in the real world from a value of a vector Z in the normalized space of feature vectors, we use the following inverse transformation:

$$Z \to X : X = \frac{z_{m+1}}{1 - z_{m+1}^2} [z_1, z_2, \cdots, z_m]^T \qquad (10)$$

Let $Net = \{(i, j)\}$ denote the set of nodes in the net, $W_{set} = \{W_{ij} : (i, j) \in Net\}$ the set of weights where W_{ij} is the weight vector at each node $(i, j) \in Net$. Using the data from training set to fire the weight set to obtain some cluster centers in Z-space. Then, we use transformation (10) to obtain the cluster centers in X-space.

4.1 Setting Initial Rules

To associate specific examples in the training data set with given clusters, it is necessary to not only have the cluster center but also to define a region around this center in which examples belong to that cluster. We do this by defining a radius that is associated with the cluster center. This radius can be determined iteratively. Once we obtain the radius associated with the cluster centers we are in a position to carry out rule initialization. This essentially consists of determining the matrix A_{kl}. The steps in carrying out this rule initialization are as follows:

- Determine the points in the Z-space associated with cluster l from class k.
- Determine the centers of clusters and the patterns associated with cluster l from class k in X-space.
- Determine the matrix A in the expression (2) using the patterns associated with cluster l.

Note that the initial rule with the ellipsoidal region is equivalent to a Gaussian classifier, where the probability distribution function of the training data within the cluster is assumed to obey the Gaussian distribution. However, the data belonging to the cluster do not necessarily obey this distribution, so we need to tune the rules by the proposed regularization function to maximize the performance.

4.2 Rule Set Optimization and Finalization

There are two aspects to rule set optimization. The first one is to optimize the individual rules for each class at the same level. The second one is to discard the redundancy so that the rule set keeps the number of rules as small as possible. Let $E_{kl}^r(p)$ stand for the set of examples (from both training set and test set) covered by rules R_{kl} for class p at level r. If there exists a relationship $E_{kl}^r(p) \subseteq E_{k_1 l_1}^r(p) \cup \cdots \cup E_{k_m l_m}^r(p)$ for some $E_{k_j l_j}^r(p), j = 1, 2, \cdots, m$, then the R_{kl} will be removed as a redundant rule. To optimize the individual rules, a simple and practical way for saving the computational burden for refining rules is to set 3~5 regularizing factors properly, and then to choose the one with the minimal criterion function value as the final regularizing factor. The finally sub-optimal A_{kl}^* and C_{kl}^* corresponding to the final regularizing factor can be obtained directly after completing rule optimization. Using the proposed techniques above, we can finalize rules.

5 Completeness, Structure and Usage

Because the rule used here is a kind of crisp rule, the performance of the classification system depends on the completeness of the rule set. Thus, the assignment of the cluster centers in the initial rules is important. To some classes, if no cluster centers can be identified by the procedures given in the last section, there will not be any rules associated with these classes. Theoretical studies show that Kohonen nets tend to learn the input distribution if the input vector is a random variable with a stationary probability density function. This result implies that it is very possible to miss necessary or even important cluster centers for some classes. In some situations one could not be expected to build a complete rule set using a single Kohonen net only. To extract further rules for these hidden patterns that may not be covered by the obtained rule set, we reorganize the training data set and then go through the whole operation described in the last section. The procedure is as follows:

- Obtain a rule set for the l-th level by using the training set S_{tra}^l with the techniques described above.
- Check out the patterns covered by the classification rules in $RS(l)$, and denote the set of these patterns by T_{tra}^l. Set a new training set and test set for the next level as $S_{tra}^{l+1} = S_{tra}^l - T_{tra}^l$.

Denote the full rule set by $CRS = \{RS(0), RS(1), \ldots, RS(N_L - 1)\}$, where N_L represents the number of the level of the classifier. As can be seen, the proposed classifier has a layered structure and it works successively in practice, i.e., we judge an unseen pattern starting from $RS(0)$ until it is classified by certain rules of a rule set $RS(\omega)$, $0 \leq \omega \leq N_L - 1$. At the same level, the rule set can be further categorized by using the number of attributes used in rules. Then, an unseen pattern will be checked out starting from the group of rules with more attributes until it is classified by certain rules. This usage of the rule set can be geometrically explained and we omit the details here. Note that the hierarchical use of the rule set is driven by our rule generation approach.

6 An Illustrative Example

With the space limitation, we here give a simple example to demonstrate our proposed techniques. The dataset consists of 150 examples of three different types of Iris plant [4]. The training set comprised of 120 examples the test set 30. The input attributes were first mapped into the range 0.0 to 1.0 using the transformation given in [4]. A 6×6 dimension Kohonen net was employed to locate the cluster centers. In Table 1, the first two rows showed that there are 148 original examples covered by the 11 rules with only 1 example from the test set classified incorrectly, and 2 examples uncovered by the rule set generated using the RDMA before attributes elimination RDMA(BAE). The second two rows indicated that all of the examples covered by the 11 rules with 1 example from the training set and 1 example from the test set classified incorrectly by the rule set generated using the RDMA after attributes elimination RDMA(AAE). This implied that the generalization capability of the rule set with attributes elimination has been improved. The corresponding results for supervised BRAINNE are 144 original examples covered, 4 examples classified incorrectly and 6 examples uncovered by the rules.

A comparison of the multi-layered neural-nets classifier with supervised BRANNE is given in Table 2, where the figures in brackets represent the results for the test set. The figures for the neural net classifier were the average values taken from the 10 runs. Results show that the proposed method gave better performance with respect to the corresponding results obtained by using both the neural-nets classifier and supervised BRAINNE for this domain.

Table 1. Statistics for the Iris Plant Data where NR: Numbers of Rules; NCE: Numbers of Covered Examples by the Rules; Lk: the k-th Level; NME: Numbers of Misclassified Examples by the Rules; DCk: Detailed NCE by the Rules for Class k

	\multicolumn{6}{c}{Training set}	\multicolumn{6}{c}{Test set}
	NR	NCE	NME	DC1	DC2	DC3	NCE	NME	DC1	DC2	DC3
L1	8	105	0	40	34	31	23	0	10	7	6
L2	3	14	0	0	5	9	6	1	0	3	3
L1	8	110	0	40	35	35	28	0	10	10	8
L2	3	10	0	1	4	6	2	1	0	1	1

Table 2. Results Comparison for the Iris Plant Data

	Misclassification Rate	Coverage Rate	Rules #
NN Classifier	0.83 (6.67)	100.00 (100.00)	3 (nodes)
BRAINNE	2.50 (3.33)	95.83 (96.67)	11
RDMA (BAE)	0.00 (3.33)	99.17 (96.67)	11
RDMA (AAE)	0.83 (3.33)	100.00 (100.00)	11

7 Conclusions

This paper considers the problems of multidimensional pattern classification characterized by continuous features and disjunctive rules. The classification system is built using crisp rules with generic ellipsoidal regions. Our main contributions in this work are: (1) develop a regularization approach for data mining which trades off misclassification rate, recognition rate and generalization ability performance; (2) present simplified algorithms for features selection and rules finalization; and (3) resolve the problem of completeness of rule set by a multiple level structure, which makes it possible to obtain classification rules for patterns with low probability of occurrence but high conceptual importance.

Representation, generation and adaptation are three essential and important aspects in building up rules based classification systems. The regularization approach proposed here aims at building a model based method to tradeoff some requested performance in the course of rule generation. We trust the framework presented in this paper is significant and valuable for classifier design. Lastly, it should be pointed out that the use of Kohonen net for clustering purposes and the form of regularization function for refining the rules are not unique.

References

1. R. L. Kennedy, Y. Lee, B. V. Roy, C. D. Reed and R. P. Lippmann, Solvong Data Mining Problems Through Pattern Recognition. Prentice Hall, PTR, Unica Technologies, Inc., (1998)
2. S. Theodoridis and K. Koutroumbas: Pattern Recognition, Academic Press, (1999)
3. A.K. Jain, P. W. Duin, and J. Mao, Statistical pattern recognition: a review, IEEE Trans.On Pattern Analysis and Machine Intelligence, 5(2000) 4-37
4. S. Sestito and T. S. Dillon, Automated Knowledge Acquisition. Australia: Prentice Hall, (1994)
5. T. Kohonen, Self-Organization and Associative Memory. Berlin: Springer-Verlag (1989)
6. R. P. Lippmann, Pattern classification using neural networks, IEEE Communications Magazine, (1989) 47-64
7. J. Y. Ching, A. K. C. Wong and K. C. C. Chan, Class-dependent discretization for inductive learning from continuous and mixed-mode data, IEEE Trans.On Pattern Analysis and Machine Intelligence, 7(1995) 641-651
8. P. K. Simpson, Fuzzy min-max neural networks-Part I: Classification, IEEE Trans. On Neural Networks, 5(1992) 776-786
9. X. Wu, Fuzzy interpretation of discretized intervals, IEEE Trans. On Fuzzy Systems, 6(1999) 753-759
10. S. Mitra, R. K. De and S. K. Pal, Knowledge-based fuzzy MLP for classification and rule generation, IEEE Trans. On Neural Networks, 6(1997) 1338-1350
11. L. M. Fu, A neural-network model for learning domain rules based on its activation function characteristics, IEEE Trans. On Neural Networks, 5(1998) 787-795
12. J. Vesanto, and E. Alhoniemi, Clustering of the self-organizing map, IEEE Trans. On Neural Networks, Special Issue On Neural Networks for Data Mining and Knowledge Discovery, 3(2000) 586-600
13. H. Lu, R. Setion and H. Liu, Effective data mining using neural networks, IEEE Trans. On Knowledge and Data Engineering, 6(1996) 957-961
14. S. Abe, R. Thawonmas and Y. Kobayashi, Feature selection by analyzing class regions approximated by ellipsoids, IEEE Trans. On SMC-Part C: Applications and Reviews, 2(1998) 282-287
15. W. F. Bloomer, T. S. Dillon and M. Witten, Hybrid BRAINNE: Further developments in extracting symbolic disjunctive rules, Expert Systems with Applications, 3(1997) 163-168

16. D. H. Wang and T. S. Dillon, Extraction and optimization of classification rules for continuous or mixed mode data using neural nets, Proceedings of SPIE International Conference on Data Mining and Knowledge Discovery: Theory, Tools and Technology III, pp.38-45, April 16-20, 2001, Orlando, Florida, USA.
17. H. Yang and T. S. Dillon, Convergence of self-organizing neural algorithm, Neural Networks, 5(1992) 485-493
18. G. G. Sutton and J. A. Reggia, Effects of normalization constraints on competitive learning, IEEE Trans. On Neural Networks, 3(1994) 502-504

Interacting Neural Modules

Garry Briscoe

Computer Science Department, University of Wisconsin Oshkosh
800 Algoma Bvld., Oshkosh Wisconsin, 54901, USA
briscoe@uwosh.edu

Abstract. Neural computation offers many potential advantages, yet little research has attempted to explore how solutions to complex problems might be achieved using neural networks. This paper explores the use of linked and interacting neural modules, each capable of learning simple finite state machines, yet capable of being combined to perform complex actions. The modules are suitable for a hardware-only implementation.

1 Introduction

Most current neural network applications regard them simply as algorithms capable of providing a learned mapping of inputs to outputs. Whether the algorithm is for a static mapping or a dynamic mapping using recurrency, these applications usually consider only a single neural module. Little work has been done on the application of neural networks to more complex tasks.

As well, current neural net applications are generally simulated on existing computational devices. This solution, of necessity, reduces the neural computation to one of serial processing. While parallel computers go some way to alleviating this problem, the degree of parallelism is limited. This results in a common criticism of neural networks—they do not scale up well to large problems because of the combinatorial problems associated with serial digital computers.

In order for neural networks to become more useful tools, these two problems will need to be addressed. Mechanisms will need to be found whereby complex application tasks can be split off into separate, simple, yet interacting neural modules in order to reduce the overall problem complexity, and hardware neural network devices, capable of massive parallelism, will need to be developed either as stand-alone systems or as extensions to existing digital computers. This paper looks at the first of these issues.

The use of neural networks for finite state machine learning has a long history as an active research area [2,4]. Most approaches use the simple recurrent network (SRN) model initially described by Elman [3], which is based on the back-propagation method of learning. While SRNs are a very powerful computational model, the task of devising effective learning procedures and designing proper architectures for a given problem is a very challenging task.

Most experiments to date using SRNs involve only small automata with a limited number of states. Obtaining solutions for complex problems that involve large numbers of states is quite a difficult problem. In general, only regular and context free languages have been explored [2,4,7], and little work has been done on higher-level languages such as context sensitive tasks.

This paper outlines a number of experiments that were conducted to investigate a different type of neural structure, one based on the recurrent linkage of self-organizing maps (SOM). The overall structure is similar to that of the SRN, but with a number of significant differences. Firstly, while both are three-layered devices, the central layer of the model proposed here is a self-organizing map. In the simulations described in the following sections, a modified Kohonen algorithm was used [5], but in a hardware-only device some equivalent neighborhood-inhibition capability would be provided [6]. Secondly, the back-propagation neural weight update algorithm used within the SRN method is replaced with simple Hebbian learning as well as the normal update rules for SOM surfaces. Thirdly, the model utilizes supervised training of the output layer in order to learn the various states of each finite state machine. An alternate model in which the states are learned in an unsupervised, implicit manner is found in [1].

2 Temporal Learning—Finite State Automata

The initial task is to build a network that is able to learn a Finite-State Automata (FSA) language. FSAs are very simple computing machines, capable of parsing finite-state grammars—the lowest on Chomsky's hierarchy of grammars. A Finite-State Transition Network (FSTN) can be regarded as a description of a language. It can also be interpreted as a specification of an FSA to recognize (or in some cases, generate) elements of the language. A transition network consists of a set of *states* connected by *arcs*. The transitions between states are directional, and occur when a new input is to be parsed (or generated). Usually there is an *initial* state and one or more *final* states.

2.1 Stopwatch Example

As a simple example, consider the case of a stopwatch. The stopwatch transition network is shown in Fig. 1 (a). The training data is a set of state transitions. For example, if the stopwatch is in the *rest* state and the **sw1** button is pressed, the watch starts the clock mechanism and moves to state *count*. A subsequent press of the **sw1** button will stop the mechanism and return the watch to the *rest* state. Pressing the **sw2** button in the *rest* state has no effect, returning to the same state, whereas pressing this button while in the *count* state will cause the hand to stop while counting continues—the *hold* state. The full set of state transitions and associated inputs is shown in the table of Fig. 1 (b).

Consider the network arrangement as shown in Fig. 2 (a). The initial layer of the network is formed by the concatenation of two vectors; an input vector representing a button press, and a vector which is a direct copy of the output layer on the previous iteration. This recurrent input represents the current state of the FSA. The second layer is a two-dimensional self-organizing map, while

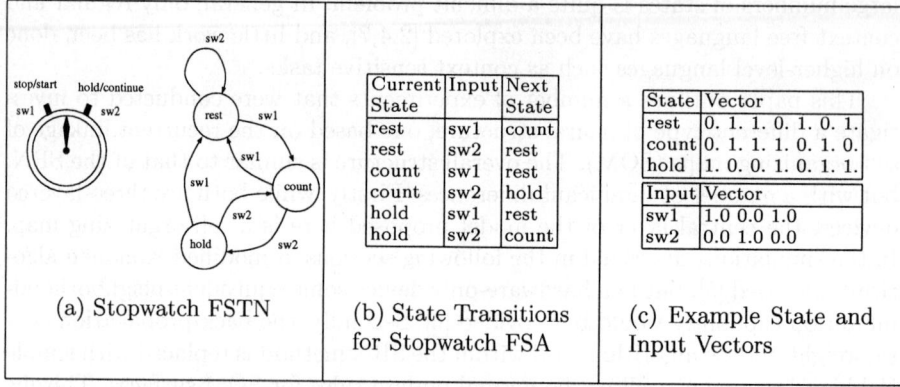

Fig. 1. Stopwatch Example

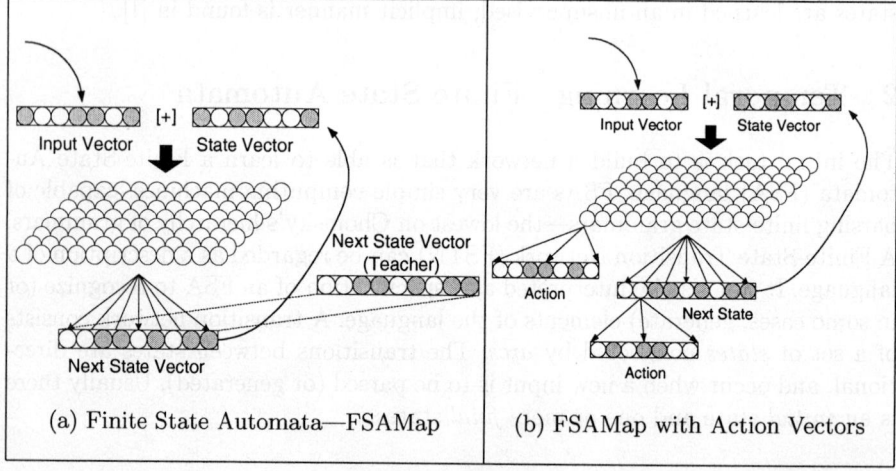

Fig. 2. FSAMap

the third layer is another vector used to represent the state that the system is to move to, given a particular current state and input button press. Note that neural weights connect every SOM neuron to every neuron in the output layer, although only a connection to a single SOM neuron is shown in the figure. The final component is a training vector that is used to update the weight connections between the SOM surface and the output layer. This teacher signal is used for supervised Hebbian learning of the output weights.

Let us refer to this overall structure as an FSAMap. To train the system, pairs of input states and button press vectors (which have been assigned random vectors containing 'bits' of either 1.0 or 0.0) are presented as inputs, and the system trained to generate the vector appropriate to the next state as its output. Possible input vector values are shown in the table of Fig. 1 (c).

The internal process followed by the FSAMap is simple: the vector corresponding to the initial state is concatenated with the vector of the first button-press input. The combined vector is mapped to the SOM and winning nodes determined. In training mode, the SOM weights are adjusted using a modified Kohonen algorithm. The outputs of the SOM surface nodes are calculated and combined with the next-state weights to calculate the next state vector. In training mode, these weights are updated via Hebbian learning to better represent the required output state vector on the next iteration. [1]

Training the FSAMap involves applying the sets of the three input vectors (*state*, *input*, and *next-state*) in order to form a set of SOM and output weights. Each input vector is combined with the current state vector, and the combined vector passed through the system to calculate a vector for the next state. This state vector is then (recurrently) used as the state vector for the next input.

Once the system is trained, it can be tested to see if it has learned the FSTN. An initial state is provided as input, and then a series of button-press inputs. The system is tracked to check that it follows the required state transitions. For example, an input sequence of: rest sw2 sw1 sw2 sw2 sw1 sw2 should follow the state transition sequence: rest rest count hold count rest rest.

Once trained, the FSAMap was able to perform the task of emulating the stop-watch FSTN to 100% accuracy; given an initial state and an input, the system was able to correctly determine the next state. This state vector, when recurrently returned and combined with another input, correctly determines the next state, and so on indefinitely.

It is important to recognize that no real distinction needs to be made between the training and recognition stages of the method. The process is continuous, with recognition and learning occurring simultaneously, albeit at a lower learning rate for later epochs. The software simulation however, was done in these two stages for convenience. An initial program performed both recognition and learning for a certain number of epochs until the system had learned. For the FSAMap, this was usually set to 200 epochs, a number that is much less than the usual number of training epochs required for SRNs. A separate program was then used to read in the learned weights for both the SOM and the output weights, and to test the learned system against a series of test inputs.

2.2 Number Example

A second FSA problem presents a more difficult example—learning to determine if a series of digits is a valid number sequence—in integer, real or scientific notation. The FSTN is shown in Fig. 3 (a). This example is both more complicated, and exhibits the effective use of an alternate (error) path to return to a particular state.

---

[1] Due to space limitations, full details of the simulations such as the algorithm and parameter settings are not given here, but may be obtained from the author. Brief details of some of the parameter settings are given in the Notes section.

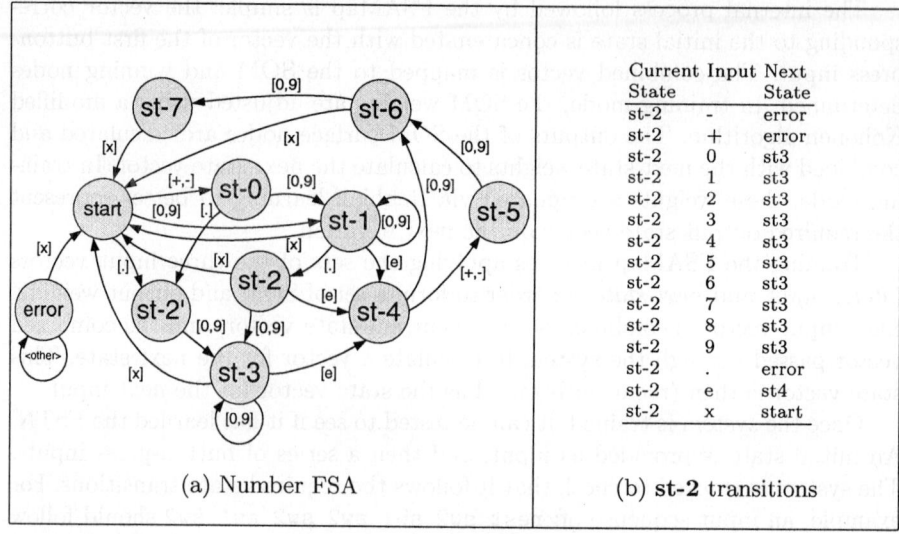

Fig. 3. Number Example

Input characters are ['-','+','0'-'9','.','e','x'] where 'x' is used to indicate the end of a sequence. The training triplets are similar to the stopwatch example; for example, an input of start 6 st-1 indicates that if the FSAMap is in state **start** and receives an input of **6**, then it should move to state **st-1**. Once trained, and given an initial state (**start**), subsequent input sequences such as:

```
6 . 0 x
- 3 5 0 e 1 x
+ . 2 x
+ 2 1 7 0 8 3 . 5 2 1 6 e - 2 1 x
```

are to be parsed correctly, and the state returned to the **start** state, while sequences such as:

```
1 . 2 e 1 0 2 x
9 . . 3 x
. x
```

should go to the **error** state upon detecting an error, before returning to the **start** state (following the input of an **x**) in order to accept another number. The Number FSA worked exactly as required. There were 165 state transitions (15 inputs for 11 states), and so represents a reasonably large FSA problem. The *error exit* state transitions provide a means of correcting for invalid inputs. If the FSAMap is in a certain state, and an input is received for which there is no valid state transition in the FSTN, then a transition will be made to an **error** state. Once in this state, all inputs except for an **x** are absorbed and the **error** state maintained. An entry of **x** from the **error** state will result in a transition to the **start** state for the entry of the inputs for the next number. The error handling is built by explicit state transitions. For example, the inputs of **-**, **+**

and . are not allowed from state **st-2**. The explicit training state transitions for state **st-2** are shown in Figure 3 (b).

2.3 Output Actions

The FSAMap architecture as described may be extended to initiate some action or a set of actions whenever a state transition occurs. The action could be to turn some device on or off, or perhaps to output some form of information.

A set of "action vectors" can be trained directly from the SOM surface (equivalent to the next-state vector arrangement), or even as another layer of nodes following the next-state vector, as shown in Fig. 2 (b). Hebbian learning is used to train the output action weights to produce appropriate action vectors.

3 Hierarchical FSAMaps

The single-level FSAMap arrangement presented so far is somewhat inflexible. If the system is in *state1* say, and it receives an input *input1*, then it will always go to the same state, *state2*. This is not always what is required—sometimes it may be necessary that the system proceed to alternate states depending on some other contextual information. This alternate requirement cannot be met with a single layered FSAMap. The next state chosen may need to be determined not only by the input, but also by some other criteria.

Context moderated behavior can be achieved with the introduction of a second FSAMap state machine as indicated in Fig. 4. The behavior of FSAMap 1 will depend on the different states of FSAMap 2, as well as the inputs and states of FSAMap 1. The network shown in Fig. 4 shows an optional external input vector for FSAMap 2. If this is used, then the two FSAMaps are seen to be symmetrical. By redrawing the FSAMaps as black boxes, hiding the internal details and just indicating that each FSAMap can have two inputs and one output (that is, given 2 inputs, the FSAMap will move to a new state and output a vector), then this symmetry can be seen more clearly in Fig. 5 (a). The multiple input mechanism allows the connection of a number of FSAMaps together, and the number of inputs per FSAMap may be extended to further increase the possibilities.

3.1 Multiple FSAMaps—Context-Sensitive Grammar Example

The Chomsky hierarchy of languages includes the *context-sensitive* grammar. An example of a context-sensitive grammar that is often cited in the computer science literature is $a^n b^n c^n$. Obviously any parser of this grammar must be able to examine the context—the number of a characters must be counted in some way in order to be able to parse the b and c components.

To implement the grammar $a^n b^n c^n$ within a hierarchical FSAMap structure, two counters and a controlling state machine are required, as is shown in Fig. 5 (b). Each counter is a simple device that learns to count up to say 5, then back

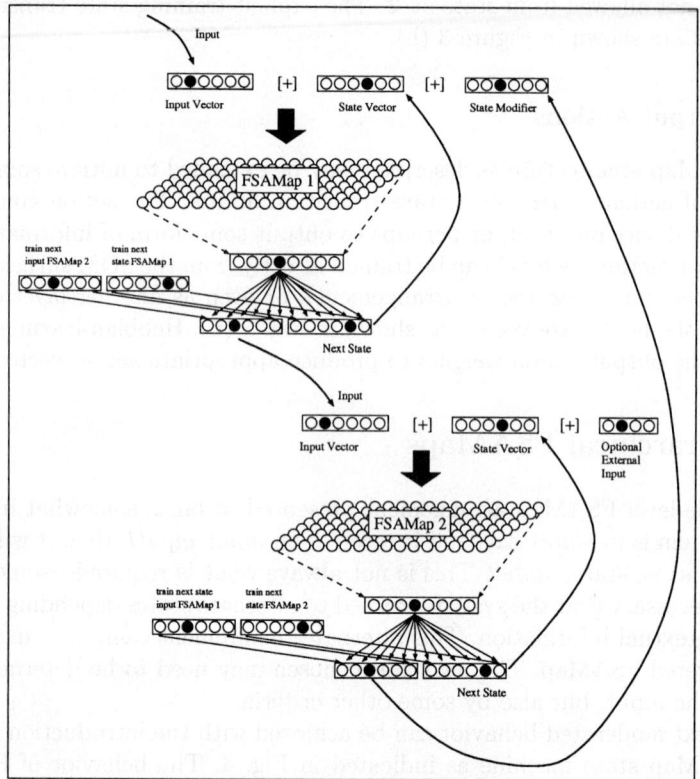

Fig. 4. Hierarchical FSAMaps

down to zero. To achieve this, 6 number states are required (**state0**, ..., **state5**) as well as an error state to indicate underflow, overflow or mismatched indices (**error**). The system could be designed to differentiate these errors and take separate actions depending upon the error type, but this was not done.

As well as the inputs *0*, *1* and *-1*, the system also required inputs *reset* (which resets the current state back to **state0**), *init* (which initializes the current state to **state1**), and *check* (which checks that the count is zero; if yes, it performs the same action as *init*; if no, it starts the error routine). The outputs for each counter (**ok** and **err**) are copied back to the main Control FSAMap.

Each FSAMap was trained individually on the state transitions and output vectors appropriate to it. Obviously, as the Counter FSAMap is used twice in the multi-level structure, the same learned weights can be used for both. Once both the Control and Counter FSAMaps were trained, their weights were used in another program which controlled the co-ordination of the three FSAMaps. Input to this program was of the form

```
start st0 st0 a a a b b b c c c a b c a a b b c c a a a b b c c c a a b b c c
```

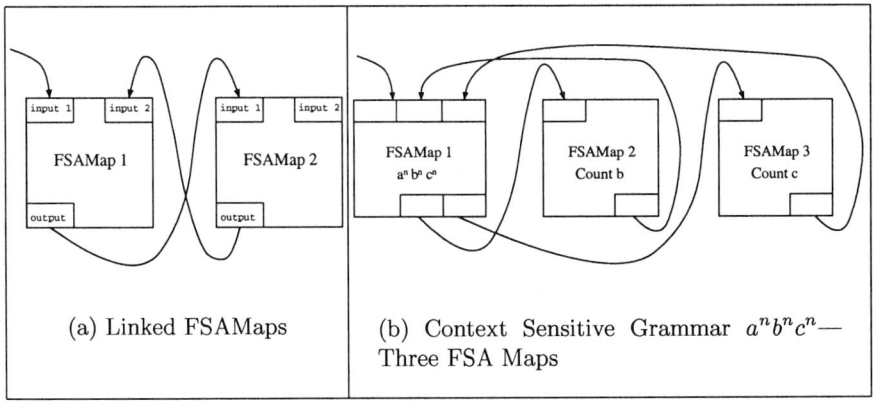

Fig. 5. Connected FSAMap Modules

The first three inputs are the initial states of each of the three components of the combined multi-level structure. Subsequent entries are input characters which are passed to the Control FSAMap at each time step. These inputs, along with the current outputs from the two Counter FSAMaps, are processed at each time step to evaluate the next state, and the two output values appropriate to the Control FSAMap. At the *same time step*, each of the outputs of the Control FSAMap are used as inputs to one of the counters, and along with the current state of the particular Counter FSAMap, is used to determine the next state and the appropriate output of these FSAMaps.

The multi-level FSAMap structure so described was able to correctly parse the input data. The above input data includes strings of $a^n b^n c^n$ for n equal to 3, 1 and 2 respectively. These were parsed correctly and the appropriate states of all of the FSAMaps were encountered. The next input of the above test is an invalid sequence (a a a b b c c c) and the combined system correctly determined that there was a missing **b** following the input of the first **c** of the sequence. The **check** input passed to the **b** Counter FSAMap causes an error exit to occur, and the subsequent **c** characters of the invalid string were absorbed via the error processing. The system was able to recover following the input of the first **a** of the next sequence, and the final string sequence was then parsed correctly.

3.2 Multiple FSAMaps—Computer Language Example

A more difficult problem is to correctly parse a C-like (programming) language. This example is used simply as a difficult context-sensitive problem, and is not intended to suggest that this technique should be used to parse computer languages. Here the input component for each state transition comprises two parts; the actual input, and the state from a second FSAMap. The contextual input from the second FSAMap provides a way of making alternate state transitions.

```
statements  :=   statement statements
              |  statement
statement   :=   read list ;;
              |  print list ;;
              |  for ( expression ; expression ; expression ) ;;
                     statement ;;
              |  expression ;;
              |  if ( expression ) ;; statement ;;
              |  while ( expression ) ;; statement ;;
              |  { statements }
expression  :=   variable logicalop expreval
expreval    :=   term
              |  term arithop expreval
              |  ( expreval )
list        :=   variable
              |  variable , list
term        :=   variable
              |  number
variable    :=   a | b | c
number      :=   0 | 1 | 2 | 3 | 4 | 5 | 6 | 7 | 8 | 9
logicalop   :=   = | == | < | <= | > | >= | ! = | ++
arithop     :=   + | - | * | /
```

```
read a , b ;;
for ( a = 1 ; a < b ; a ++ ) ;;
{
    c = a + b ;;
}
if ( c >= 3 ) ;;
    print a , c ;;
while ( c < a + b ) ;;
{
    c = 1 + a ;;
    if ( a == c * 2 - b / 3 ) ;;
    {
        c ++ ;;
    }
}
```

(a) C-like Language Description   (b) Example Code

Fig. 6. C-language Example

The C-like language is defined by the description shown in Figure 6 (a). This language is very simple, and probably not very useful, but it does illustrate a number of points concerning the use of multiple FSAMaps. Some example statements in the language (that are able to be parsed by the FSAMaps) are shown in Figure 6 (b). Note the extra semi-colons in the statements that are not found in the standard C-language. These were included to enable a better synchronization between the two FSAMaps, and to reduce the number of the state transitions in each. While this is somewhat arbitrary, the point of the exercise is not to define a suitable language for parsing by FSAMaps, but rather to illustrate the possibility of hierarchical, recurrent maps cooperating to learn difficult and context-dependent tasks.

The parsing of the statements is only considered in a very simple manner. A number of extensions could be made; for example, a count of the level of nesting (resulting from the { and } statement block brackets) could be maintained with another FSAMap counter. In this FSAMap, an input of '{' would increment a count (state $0 \Rightarrow 1$, $1 \Rightarrow 2$, etc) whereas an input of '}' would decrement the count (state $1 \Rightarrow 0$, $2 \Rightarrow 1$, etc). This extension was not included here in order to make the problem relatively simple.

Note also that the parsing has been simplified by treating a number of two character inputs as a single symbol (such as == and ! =). As well, the keywords (such as **while**) are treated as single symbols and not as individual characters that go to make up each keyword.

4 Discussion—FSAMap

This paper examined the case of a recurrent neural network being trained to learn the 'rules' of an FSA. Although the rules (the state transitions) were learned, they were supplied as data in an *explicit* form.

Within the FSAMap, the SOM layer learns to order the input vectors (composed of the input and state vectors concatenated together). So long as the input vectors remain stationary (that is, do not change over time), then the distribution of the input vectors over the SOM surface is stable, resulting in a unique representation of each input being found on the output side of the SOM whenever an input/current state vector is mapped to the surface. Weighted connections from this unique SOM output representation are then able to be trained to reproduce the vector elements of the next state in the sequence.

The FSAMap mechanism performs extremely well, and is robust yet flexible, and with the use of appropriate hardware implementation could scale up to much larger problems. It indicates that recurrent neural structures are able to handle complicated state transition procedures, and are able to take some form of response depending upon the state change. As well, using hierarchically connected FSAMaps allows for a rich form of mapping between inputs and outputs. The hierarchical FSAMap structures described in the previous sections show that complicated and involved behaviors are able to be produced by linked neural networks.

4.1 Notes

Selected System Details	stopwatch example	number example	$a^n b^n c^n$ Control	$a^n b^n c^n$ Counter	c-language Parser	c-language Context
Dimension of state vector	7	10	12	6	8	8
Number of input vectors	2	15	3	1	2	1
Number of output vectors	-	-	2	1	1	1
Number nodes on map	5 x 5	30 x 30	30 x 30	25 x 25	30 x 30	20 x 20
Epochs for map	100	100	200	200	100	100
Epochs for output weights	200	200	300	300	500	500

References

1. G. J. Briscoe. *Adaptive Behavioural Cognition*. PhD thesis, School of Computing, Curtin University of Technology, Perth, Western Australia, 1997.
2. Axel Cleeremans, David Servan-Schreiber, and James L. McClelland. Finite state automata and simple recurrent networks. *Neural Computation*, 1:372–381, 1989.
3. Jeffrey L. Elman. Finding structure in time. *Cognitive Science*, 14:179–211, 1990.
4. C. L. Giles, C. B. Miller, D. Chen, H. H. Chen, G. Z. Sun, and Y. C. Lee. Learning and extracting finite state automata with second-order recurrent neural networks. *Neural Computation*, 4:393–405, 1992.

5. T. Kohonen. Self-organized formation of topologically correct feature maps. *Biological Cybernetics*, 43:59–69, 1982.
6. H. Onodera, K. Takeshita, and K. Tamaru. Hardware architecture for Kohonen network. *IEICE Transactions On Electronics*, E76C(7):1159–1167, 1993.
7. Janet Wiles and Jeff Elman. Learning to count without a counter: A case study of dynamics and activation landscapes in recurrent networks. In *Proceedings of the Seventeenth Annual Conference of the Cognitive Science Society, Pittsburg, PA*. Cambridge, MA: The MIT Press, 1995.

The Application of Visualization and Neural Network Techniques in a Power Transformer Condition Monitoring System

Zhi-Hua Zhou, Yuan Jiang, Xu-Ri Yin, and Shi-Fu Chen

National Laboratory for Novel Software Technology
Nanjing University, Nanjing 210093, P.R.China
zhouzh@nju.edu.cn
http://cs.nju.edu.cn/people/zhouzh

Abstract. In this paper, visualization and neural network techniques are applied together to a power transformer condition monitoring system. Through visualizing the data from the chromatogram of oil-dissolved gases by 2-D and/or 3-D graphs, the potential failures of the power transformers become easy to be identified. Through employing some specific neural network techniques, the data from the chromatogram of oil-dissolved gases as well as those from the electrical inspections can be effectively analyzed. Experiments show that the described system works quite well in condition monitoring of power transformers.

1 Introduction

Effective and efficient condition monitoring is very important in guaranteeing the safe running of power transformers [7, 8]. With good condition monitoring, potential failures of the power transformers can be identified in their incipient phases of development so that the maintenance of the power transformers can be condition based in addition to periodically scheduled. Since the physical essence of the failures of the power transformers have not been clearly recognized at present, the monitors usually set up mappings between the failures and their appearances and then analyzes or predicts the potential failures with pattern recognition techniques. As an important pattern recognition technique, neural networks have been extensively applied in this area in past decades and have exhibited strong ability in modelling complex mappings between the failures and their appearances [1, 2]. However, in some cases the information on the condition of the power transformers can be visualized so that it is easy and intuitive for a human monitor to identify the potential failures, which can not only improve the accuracy of the analysis but also reduce the computational cost required by the analysis [14]. So, utilizing neural networks and visualization techniques together in a power transformer condition monitoring system seems an attracting alternative.

In this paper, visualization and neural network techniques are applied together to a power transformer condition monitoring system. In the system, the data from the chromatogram of oil-dissolved gases are visualized by 2-D or 3-D graphs. If a human monitor cannot identify potential failures from those graphs, the information on the condition of the power transformers is passed to a neural network module. The neural network module utilizes a specific paired neural network architecture, which enables it to deal with the data from the chromatogram of oil-dissolved gases as well as those from the electrical inspections. It employs some redundant input attributes to accelerate the training speed and reduce the number of hidden units in the networks. Moreover, the neural network module exploits fuzzy techniques to preprocess the primitive input data so that important features with relatively small values will not be blocked off by those with relatively big values. Experiments show that the described system attains good monitoring effect.

The rest of this paper is organized as follows. In Section 2, we briefly introduce the whole condition monitoring system. In Section 3, we describe the visualization techniques used in the system. In Section 4, we present the neural network techniques employed in the system. In Section 5, we report on some experimental results. Finally in Section 6, we conclude.

2 Sketch of the System

The data fed to the described system can be categorized into two types, i.e. the data from the electrical inspections and the data from the chromatogram of oil-dissolved gases. The former type of data include *ultrasonic measure* (abbr. UL), *abnormal sound* (abbr. AS), *direct current resistance* (abbr. DR), etc., some of which can be obtained when the power transformers are in running while some can be obtained only when the power supply are terminated. The latter type of data include the volume of H_2 in the oil-dissolved gas, the volume of CH_4 in the oil-dissolved gas, the volume of C_2H_6 in the oil-dissolved gas, etc., all of which can be obtained when the power transformers are in running.

When the system is working, the data from the chromatogram of oil-dissolved gases are visualized in 2-D or 3-D graphs at first, and a human monitor is asked to check the graphs. If he or she can identify some potential failures, the analysis process terminates. Otherwise the data from the chromatogram of oil-dissolved gases as well as those from the electrical inspections are fed to the neural network module which will return an analysis result. In summary, the analysis process of the system is depicted in Fig. 1.

3 Visualization

3.1 Visual Modelling

Visualization techniques enable the deep insight of the information by transforming the digital symbols to vivid images and/or graphs [9]. It could be used not only to enrich the interface between human and machine but also to help human recognize useful information more quickly and intuitively.

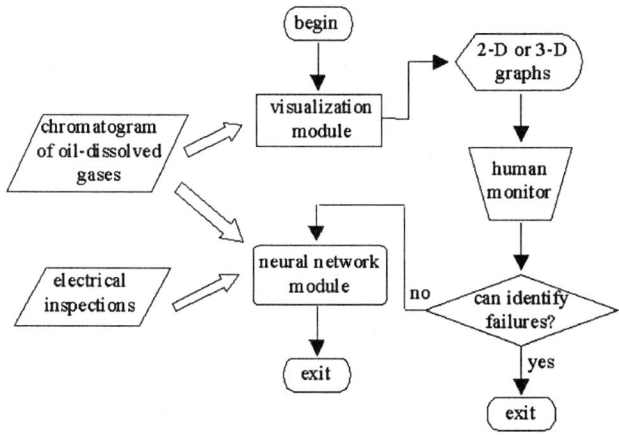

Fig. 1. The analysis process of the described system

At present treble-ratio-rule [10] is the most authoritative criterion in analyzing the data from the chromatogram of oil-dissolved gases in China. Treble-ratio-rule computes the ratios, including CH_4/H_2, C_2H_2/C_2H_4, and C_2H_4/C_2H_6, of the volumes of H_2, CH_4, C_2H_6, C_2H_4, and C_2H_2 in the oil-dissolved gases, and then performs condition monitoring according to some empirical rules. In the described system, visualization techniques are introduced to depict those ratios, which endows data with spatial properties from multi-dimensional perspectives.

Before visualization, the problem to be solved must be modelled. Nowadays, pipeline model [4] is one of the most prevailing visualization models. In the described system, the pipeline model is adapted according to the requirements of the condition monitoring of power transformers, which is shown in Fig. 2.

The functions of the components in Fig. 2 are as follows.

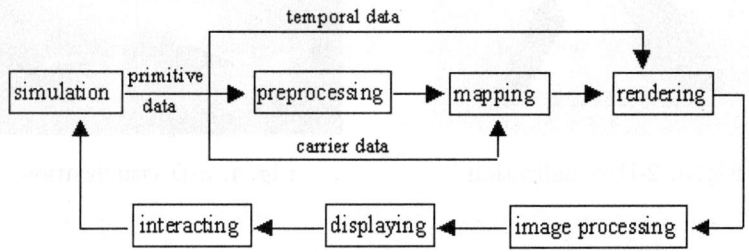

Fig. 2. Visualization model used in the described system

1) Simulation: to generate primitive data, i.e. the data from the chromatogram of oil-dissolved gases, for visualization.
2) Preprocessing: to transform and filter the primitive data, which filters out noise, extracts interested data, and derives new data such as gradients and interpolations.
3) Mapping: to provide functions of modelling and classifying. Modelling means to extract geometrical elements such as points, lines, etc. from the filtered data. Classifying is to classify voxels with different values.
4) Rendering: to derive basic geometrical elemental information.
5) Image processing: to assemble basic functions of graphics transformation, e.g. scaling, rotating, etc.
6) Displaying: to display the images in a way that diverse perspectives and queries are available.
7) Interacting: to help the human monitor track the analysis process so that the simulation can be controlled and guided.

3.2 Implementation

In the described system, the data from treble-ratio-rule of the chromatogram of oil-dissolved gases are visualized in 2-D or 3-D graphs. In 2-D visualization, the fault space is depicted as an equilateral triangle where the axes are respective the percentage of the volumes of CH_4, C_2H_4, and C_2H_2 in the oil-dissolved gases. Thus, there are six states in the fault space as shown in Fig. 3. In 3-D visualization, the fault space is depicted as an open-cube where the axes are the ratios of the five characteristic gases, i.e. H_2, CH_4, C_2H_6, C_2H_4, and C_2H_2. Thus, there are also six states in the fault space as shown in Fig. 4.

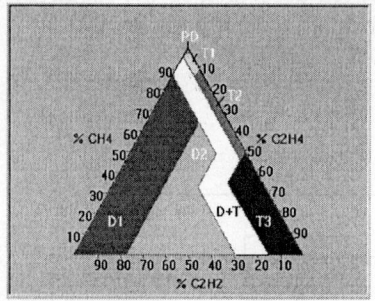
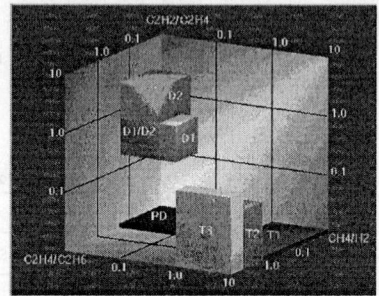

Fig. 3. 2-D visualization   **Fig. 4.** 3-D visualization

In Fig. 3 and Fig. 4, PD denotes *partial discharge*, D1 denotes *low-energy discharge*, D2 denotes *high-temperature discharge*, T1 denotes *high-temperature overheat* with $t < 300°C$, T2 denotes *high-temperature overheat* with $300°C \leq t < 700°C$, T3 denotes *high-temperature overheat* with $700°C \leq t$.

From the 2-D and 3-D graphs, some potential failures are easy to be identified. Since different measurements of the data are used in those two graphs, the analysis results could verify each other so that the accuracy of analysis is improved.

4 Neural Networks

4.1 Paired Neural Networks

In most condition monitoring systems that utilize neural network techniques, only the data from the chromatogram of oil-dissolved gases are processed while those from the electrical inspections are seldom used [5]. The reason lies in the diversity, the high cost, and the temporal relationships of the electrical inspections. On one hand, there is accurate symbolic criterion [10] in analyzing the data from the chromatogram of oil-dissolved gases so that dealing with only gas data does not well exert the power of neural techniques. On the other hand, the data from the electrical inspections is very useful in locating some potential failures that can not be identified by the analysis of gas data. Therefore building a neural network module that has the ability of dealing with the data from the chromatogram of oil-dissolved gases as well as those from the electrical inspections is of great value to condition monitoring of power transformers.

The temporal relationships of the electrical inspections lie in the orders of inspections, that is, whether an inspection B is required is determined by the result of an inspection A. For example, the *insulation resistance* need not be measured if other inspections show that the power transformer is well running. If it is measured preceding some other electrical inspections, unnecessary cost is paid because measuring the *insulation resistance* requires terminating the power supply. Such kind of orders of electrical inspections lead to the result that it is almost impossible to attain all the helpful data from the electrical inspections at the same time without unnecessary cost. However, neural networks claim that all the input data should be provided before the training begins. Thus the obstacle appears.

In order to get across the obstacle, the described system adopts a specific paired neural network architecture based on the fault set and the fault attribute set [6]. At first, features from some important online electrical inspections and those from the analysis of the chromatogram of oil-dissolved gases are selected to constitute the online fault attribute set. The failure types that could be identified from the online fault attribute set constitute the online fault set. An online monitoring neural network is built by regarding the online fault attribute set and the online fault set as the input and output attribute sets respectively. Then, features from some important offline electrical inspections, along with features from the selected online electrical inspections and those from the analysis of the chromatogram of oil-dissolved gases, are selected to constitute another fault attribute set, i.e. the offline fault attribute set. The failure types that could be identified from the offline fault attribute set constitute the offline fault set. An offline monitoring neural network is built by regarding the offline fault attribute set and the offline fault set as the input and output attribute sets respectively. Both the online and the offline neural networks are trained so that the former is applicable when the power transformers are in running while the latter is applicable when the power supply is terminated.

As for the online monitoring neural network, since obtaining the inputs does not require terminating the power supply, its inputs could be easily collected at the same time. But as for the offline monitoring neural network, obtaining the inputs may require terminating the power supply. However, the inputs of the offline monitoring neural network could be obtained at the same time when the power transformers are in periodic termination and examine.

Moreover, since the online fault set and the online fault attribute set are proper subsets of the offline fault set and the offline fault attribute set respectively, the online monitoring neural network is also applicable when the offline monitoring neural network is applicable. Therefore the accuracy of the analysis could be improved in simulating the consultation of multiple experts where the analysis result of one network is verified with that of the other network.

4.2 Redundant Input Attributes

In the described system, the percentage of the volumes of H_2, CH_4, C_2H_6, C_2H_4, and C_2H_2 in the oil-dissolved gases are included as inputs to the neural networks. In addition, since the volumes of *total hydrocarbon* ($CH_4+C_2H_6+C_2H_4+C_2H_2$, abbr. C1+C2), CO, and CO_2 are helpful in locating some potential failures, they are also included as inputs to the neural networks.

Among the features from the electrical inspections, *ultrasonic measure* (abbr. UL), *abnormal sound* (abbr. AS), and *current from iron core to earth* (abbr. CE) are included as inputs to the online monitoring neural network; *direct current resistance* (abbr. DR) and *iron core insulation resistance* (abbr. IR), along with those used for the online monitoring neural network, are included as inputs to the offline monitoring neural network.

Note that *total hydrocarbon* is a redundant attribute because it does not provide any information that cannot be derived from the attributes CH_4, C_2H_6, C_2H_4, and C_2H_2. Therefore *total hydrocarbon* can be automatically learned by the neural networks through their implicitly constructive learning ability [3]. However, explicitly providing such a helpful redundant attribute as input could not only increase the learning ability of the neural networks but also reduce the number of hidden units required in achieving good generalization ability [3]. So, total hydrocarbon is included as mentioned above. Similarly, the ratios of CH_4/H_2, C_2H_2/C_2H_4, and C_2H_4/C_2H_6 are also included as inputs to the neural networks.

4.3 Fuzzy Preprocessing

There are great differences in the value ranges of the input attributes. For example, the value of the attribute CO_2 is usually bigger than 1,000 while the value of the attribute CH_4/H_2 is usually between 0 and 1. If the primitive attribute values are directly input to the neural networks, the features with relatively small values may be blocked off by those with relatively big values. The solution is to map the value ranges of different attributes to a same interval, e.g. [0, 1].

It is easy to transform all the primitive attribute values to the interval [0, 1] through dividing them by the length of their corresponding value ranges. However, such a linear mapping may drop some important characteristics of the original probabilistic distribution of the attributes. Instead, the described system employs fuzzy techniques. Some *attention values* are obtained from senior human monitors, which is shown in Table 1.

Table 1. Some *attention values* (abbr. *a.v.*) provided by senior human monitors. The metric of gases, CE, and IR are respectively ppm, mA, and $m\Omega$

attr.	H_2	CH_4	C_2H_6	C_2H_4	C_2H_2	CO	CO_2	C1+C2	CE	IR
a.v.	100	50	100	100	3	300	5,000	150	20	1,500

Based on those *attention values*, the membership grades of the primitive attribute values against their corresponding *attention values* are computed according to a membership function, and then the membership grades are regarded as the inputs to the neural networks instead of those primitive values. The membership function used here is *Sigmoid* function shown in Eq.(1), where x denotes the primitive value while x_a denotes its corresponding *attention value*.

$$x' = \left(1 + e^{-x/x_a}\right)^{-1} \tag{1}$$

Moreover, the values of binary attributes are mapped to 0.1 and 0.9, and the values of ternary attributes are mapped to 0.1, 0.5, and 0.9. Those mappings could speed up the converging of the neural networks [13].

5 Experiments

The described system is tested on a data set comprising 704 cases, among which 528 cases are used as the training set while the rest 176 cases are used as the test set. The data set is provided by the Institute of Electric Science of Shandong Province, P.R.China. All the cases are collected from real-world power transformers used in Shandong Province.

The neural networks are trained with SuperSAB algorithm [12], which is one of the fastest variations of Backpropagation. Tollenaere [12] reported that it is 10-100 times faster than standard BP [11]. The parameters of SuperSAB are set to the values recommended by Wasserman [13], i.e. the weight-increasing factor

η_{up} is set to 1.05, the weight-reducing factor η_{down} is set to 0.2, and the upper ground of the maximum step of the k-th weight η_{ij}^k is set to 10.

In the experiments, the visualization results are checked by a junior human monitor. If he cannot identify any potential failures, the case is passed to the neural network module for further analysis. Experimental results show that the test set accuracy of the described system is 91.5%, which is better than the level of junior human monitors (usually 80.1%) and is very close to the level of senior human monitors (usually 93.2%).

For comparison, the treble-ratio-rule is also tested with the same test set. The results show that the monitoring effect of the described system is far better than that of the treble-ratio-rule. Some test cases and their corresponding results are shown in Table 2, where "*" denotes that treble-ratio-rule has not identified any potential failures. The abbreviations used in Table 2 are shown in Table 3.

Table 2. Some test cases and their corresponding results

H_2	CH_4	C_2H_6	C_2H_4	C_2H_2	CO	CO_2	UL	AS	CE	TRR result	our result	real result
29	23	40	12	4.7	140	790	1	0	1.3	*	OSP	OSP
40	15.7	1.8	35.5	6.59	446	809	1	0	1.0	HTD	OSP	OSP
69	12.9	7.2	5.6	12.4	377	554	2	1	0.9	*	OSE	OSE
44	7.3	1.6	2.2	3.1	558	1134	2	1	1.2	*	OSE	OSE
140	8.1	8.3	15	23	680	2020	0	0	0.6	*	OSE	ID
54	7	7.4	8.6	5.4	88	297	0	0	1.3	LED	ID	ID
350	1001	298	1001	7.9	131	1401	0	0	100.3	HTO	ICO	ICO
90	149	32.4	486	19.2	315	10305	0	0	501.2	HTO	ICO	ICO
747	2065	1029	4589	6.4	664	1430	0	0	0.1	HTO	TWO	TWO
428	1660	533	4094	11.4	637	4759	0	0	0.3	HTO	TWO	TWO

6 Conclusion

In this paper, visualization and neural network techniques are applied together to a power transformer condition monitoring system where a visualization module is used to identify relatively simple potential failures while a neural network module is used to identify relatively complex potential failures. Differing from some previous condition monitoring systems, the described system has the ability of dealing with the data from the chromatogram of oil-dissolved gases along with those from the electrical inspections. Experiments show that the monitoring effect of the described system is close to that of senior human monitors.

The described system has been used by the Institute of Electric Science of Shandong Province, P.R.China, for over a year, and has received much appreciation for its good monitoring effect. We believe that the success of the described

Table 3. Abbreviations used in Table 2

full description	abbr.
treble-ration-rule	TRR
high-temperature discharge	HTD
low-energy discharge	LED
internal discharge	ID
high-temperature overheat	HTO
iron core overheat	ICO
tap switch overheat	TWO
oil static electricity	OSE
oil submerged pump fault	OSP

system shows that the techniques from diverse domains of computer science, such as neural networks from computational intelligence and visualization from graphics, could greatly profit practical engineering tasks when they are adequately combined.

Acknowledgements

The comments and suggestions from the anonymous reviewers greatly improve this paper. The authors wish to thank their collaborators at the Institute of Electric Science of Shandong Province, P.R.China. This work was partially supported by the National Natural Science Foundation of P.R.China, and the Natural Science Foundation of Jiangsu Province, P.R.China.

References

1. Booth C., McDonald J. R.: The use of artificial neural networks for condition monitoring of electrical power transformers. Neurocomputing **23** (1998) 97–109
2. Cannas B., Celli G., Marchesi M., Pilo F.: Neural networks for power system condition monitoring and protection. Neurocomputing **23** (1998) 111–123
3. Cherkauer K. J., Shavlik J. W.: Rapid quality estimation of neural network input representations. In: Touretzky D., Mozer M., Hasselmo M. (eds.): Advances in Neural Information Processing Systems, Vol. **8** MIT Press, Denver, CO (1996) 45–51
4. Dyer D. S.: A data flow toolkit for visualization. IEEE Transactions on Computer Graphics and Application **10** (1990) 60–69
5. He D., Tang G., Chen H.: Neural network methods for diagnosing faults of power transformers. Automation of Electric Power Systems **17** (1993) 33–38 (in Chinese)
6. He J.-Z., Zhou Z.-H., Zhao Z.-H., Chen S.-F.: A general design technique for fault diagnostic systems. In: Proceedings of the INNS-IEEE International Joint Conference on Neural Networks (2001) **II**–1307–1311

7. Kang P., Birtwhistle D.: Condition monitoring of power transformer on-load tap-changers. Part 1: Automatic condition diagnostics. IEE Proceedings of Generation Transmission and Distribution **148** (2001) 301–306
8. Kang P., Birtwhistle D.: Condition monitoring of power transformer on-load tap-changers. Part 2: Detection of ageing from vibration signatures. IEE Proceedings of Generation Transmission and Distribution **148** (2001) 307–311
9. McCormick B. H., DeFanti T. A., Brown M. D.: Visualization in scientific computing. Computer Graphics **21** (1987) 1–14
10. Ministry of Water Resources and Electric Power of P. R.China: Guideline of Analysis and Judgement of Dissolved Gases in the Oil of Power Transformers. Standards SD187-86
11. Rumelhart D., Hinton G., Williams R.: Learning representation by backpropagating errors. Nature **323** (1986) 533–536
12. Tollenaere T.: SuperSAB: Fast adaptive backpropagation with good scaling properties. Neural Networks **3** (1990) 561–573
13. Wasserman P. D.: Advanced Methods in Neural Computing. Van Nostrand Reinhold, New York (1993)
14. Yamashita H., Cingoski V., Nakamae E., Namera A., Kitamura H.: Design improvements on graded insulation of power transformers using transient electric field analysis and visualization technique. IEEE Transactions on Energy Conversion **14** (1999) 1379–1384

Entrepreneurial Intervention in an Electronic Market Place

John Debenham

Faculty of Information Technology, University of Technology, Sydney
debenham@it.uts.edu.au

Abstract. An electronic market has been constructed in an on-going collaborative research project between a university and a software house. The way in which actors (buyers, sellers and others) use the market will be influenced by the information available to them, including information drawn from outside the immediate market environment. In this experiment, data mining and filtering techniques are used to distil both individual signals drawn from the markets and signals from the Internet into meaningful advice for the actors. The goal of this experiment is first to learn how actors will use the advice available to them, and second how the market will evolve through entrepreneurial intervention. In this electronic market a multiagent process management system is used to manage all market transactions including those that drive the market evolutionary process.

1 Introduction

The overall goal of this project is to derive fundamental insight into how e-markets evolve. The perturbation of market equilibrium through entrepreneurial action is the essence of market evolution. Entrepreneurship relies both on intuition and information discovery. The term 'entrepreneur' is used here in its technical sense [1]. This overall goal includes the provision of timely information to support the market evolutionary process.

A subset of this project is substantially complete; it is called the *basic system*. The goal of this subset is to identify timely information for traders in an e-market. The traders are the buyers and sellers. This basic system does not address the question of market evolution. The basic system is constructed in two parts:

- the e-market
- the actors' assistant

The e-market has been constructed by Bullant Australasia Pty Ltd—an Australian software house with a strong interest in business-to-business (B2B) e-business [www.bullant.com]. The e-market is part of their on-going research effort in this area. It has been constructed using Bullant's proprietary software development tools. The e-market was designed by the author. The actors' assistant is being constructed in the Faculty of Information Technology at the University of Technology, Sydney. It is funded by two Australian Research Council Grants; one awarded to the author, and one awarded to Dr Simeon Simoff.

One feature of the whole project is that every transaction is treated as a business process and is managed by a process management system. In other words, the process management system makes the whole thing work. In the basic system these transactions include simple market transactions such as „buy" and „sell" as well as transactions that assist the actors in buying and selling. For example, „get me information on the financial state of the Sydney Steel Co. by 4.00pm". The process management system is based on a robust multiagent architecture. The use of multiagent systems is justified first by the distributed nature of e-business, and second by the critical nature of the transactions involved. This second reason means that the system should be able to manage a transaction (eg: buy ten tons of steel by next Tuesday at the best available price) reliably in an essentially unreliable environment. The environment may be unreliable due to the unreliability of the network and components in it, or due to the unreliability of players—for example, a seller may simply renege on a deal.

The overall goal of this project is to investigate the evolution of e-markets. A market is said to be in *equilibrium* if there is no opportunity for arbitrage—ie. no opportunity for no-risk or low-risk profit. Real markets are seldom in equilibrium. Market *evolution* occurs when a player identifies an opportunity for arbitrage and takes advantage of it possibly by introducing a novel form of transaction.. The investigation of e-market evolution will commence at the end of 2001. The overall design framework is already complete and a brief description of it is given in Sec. 4. Section 2 describes the e-market and the actors' assistant in the basic system. Section 3 describes the multiagent process management system.

2 The Basic System

The basic system consists of the e-market and the actors' assistant. The goal of the basic system is to identify timely information for traders in the e-market. Trading in the e-market either takes place through the 'e-exchange' or through a 'solution provider'. The *e-exchange* is an open exchange where goods and services are offered on a fixed price basis or by some competitive mechanism. The *solution providers* assist in the negotiation of contracts and facilitate the development of business relationships between buyers and sellers. So the actor classes in the basic system are buyer, seller, e-exchange and solution provider. The four actor classes are shown in Fig. 1. Before describing the system itself a justification is given of this choice of four actor classes.

2.1 Actor Classes in the Basic System

For some while there has been optimism in the role of agents in electronic commerce. „During this next-generation of agent-mediated electronic commerce,.... Agents will strategically form and reform coalitions to bid on contracts and leverage economies of scale...... It is in this third-generation of agent-mediated electronic commerce where companies will be at their most agile and markets will approach perfect efficiency." [2]. There is a wealth of material, developed principally by micro-economists, on the behaviour of rational economic agents. The value of that work in describing the behaviour of human agents is limited in part by the inability of humans to necessarily behave in an (economically) rational way, particularly when their (computational) resources are limited. That work provides a firm foundation for describing the behaviour of rational, intelligent software agents whose resource bounds are known, but more work has to be done [3]. Further, new market mechanisms that may be particularly well-suited to markets populated by software agents is now an established area of research [4] [5]. Most electronic business to date has centred on on-line exchanges in which a single issue, usually price, is negotiated through the application of traditional auction-based market mechanisms. Systems for multi-issue negotiation are also being developed [6], also IBM's Silkroad project [7]. The efficient management of multi-issue negotiation towards a possible solution when new issues may be introduced as the negotiation progresses remains a complex problem [8].

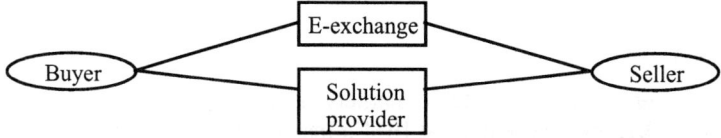

Fig 1. The four actor classes in the basic e-market system

Given the optimism in the future of agents in electronic commerce and the body of theoretical work describing the behaviour of rational agents, it is perhaps surprising that the basic structure of the emerging e-business world is far from clear. The majority of Internet e-exchanges are floundering, and it appears that few will survive [9]. There are indications that exchanges may even charge a negative commission to gain business and so too market intelligence [op. cit.]. For example, the Knight Trading Group currently pays on-line brokers for their orders. The rationale for negative commissions is discussed in Sec. 4. One reason for the recent failure of e-exchanges is that the process of competitive bidding to obtain the lowest possible price is not compatible with the development of buyer-seller relations. The preoccupation with a single issue, namely price, can overshadow other attributes such as quality, reliability, availability and customisation. A second reason for the failure Internet e-exchanges is that they deliver little benefit to the seller—few suppliers want to engage in a ruthless bidding war [op. cit.]. The future of electronic commerce must include the negotiation of complex transactions and the development of long-term relationships between buyer and seller as well as the e-exchanges. Support for these complex transactions and relationships is provided here by *solution providers*.

A considerable amount of work has been published on the comparative virtues of open market e-exchanges and solution providers that facilitate direct negotiation. For example, [10] argues that for privately informed traders the 'weak' trader types will systematically migrate from direct negotiations to competitive open markets. Also, for example, see [11] who compare the virtues of auctions and negotiation. Those results are derived in a supply/demand-bounded world into which signals may flow. These signals may be received by one or more of the agents in that world, and so may cause those agents to revise their valuation of the matter at hand. No attempt is made to accommodate measures of the intrinsic validity of those signals, measures of the significance of those signals to the matter at hand, or measures of interdependencies between those signals. That is, those models ignore the general knowledge that is required to make good business decisions. That general knowledge may be drawn from outside the context of the market place. So those results do not necessarily mean that e-exchanges will be preferred over negotiated contracts in the real e-business world. For example, the negotiation of a long term contract for some commodity may be based on an individual's 'hunch' that future political instability in a foreign country will (partly) cause the price of that commodity to rise.

The issue addressed here is limited to single issue negotiation either in an e-exchange or through a solution provider. This problem is not trivial. For example, if a company has a regular requirement for so-many tons of steel a month then will it be better off for the next 24 months *either* making offers in an e-exchange on a monthly basis *or* negotiating a 24-month contract?

2.2 The E-market

The construction of experimental e-markets is an active area of research. For example, [12] describes work done at IBM's Institute for Advanced Commerce. The basic e-market described here has been designed in the context of the complete system described briefly in Sec. 4. There are two functional components in the *basic e-market*: the e-exchange and a solution provider. The *solution provider* is 'minimal' and simply provides a conduit between buyer and seller through which long term contracts are negotiated. The *solution provider* in its present form does not give third-party support to the negotiation process.

An e-exchange is created for a fixed duration. An *e-exchange* is a virtual space in which a variety of market-type *activities* can take place at specified times. The time is determined by the e-exchange *clock*. Each activity is advertised on a notice *board* which shows the start and stop time for that activity as well as what the activity is and the *regulations* that apply to players who wish to participate in it. A human player works though a PC (or similar) by interacting with a *user agent* which communicates with a *proxy agent* or a solution provider situated in the e-market. The inter-agent communication is discussed in Sec 3. Each activity has an *activity manager* that ensures that the regulations of that activity are complied with.

When an e-exchange is created, a specification is made of the e-exchange *rules*. These rules will state who is permitted to enter the e-exchange and the roles that they are permitted to play in the e-exchange. These rules are enforced by an *e-exchange manager*. For example, can any player create a sale activity (which could be some sort of auction), or, can any player enter the e-exchange by offering some service, such as advice on what to buy, or by offering 'package deals' of goods derived from different suppliers? A high-level view of the e-market is shown in Fig. 2.

The activities in the basic e-market are limited to opportunities to buy and sell goods. The regulations for this limited class of activities are called *market mechanisms* [5]. The subject of a negotiation is a *good*, buyers make *bids*, sellers make *asks*. Designing market mechanisms is an active area of research. For example, see optimal auctions [4]. One important feature of a mechanism is the 'optimal' strategy that a player should use, and whether that strategy is „truth revealing" [11]. Mechanisms can be for single-unit (ie a single good) or multi-unit (ie a number of identical goods). Any single-unit mechanism may be extended to a multi-unit mechanism by using it to establish a price for a good, permitting the 'winner' to take as many goods as required at that price (and maybe then permitting other buyers to take further goods at that price), then putting the remaining goods up for sale to determine another price, and so on, until the process terminates.

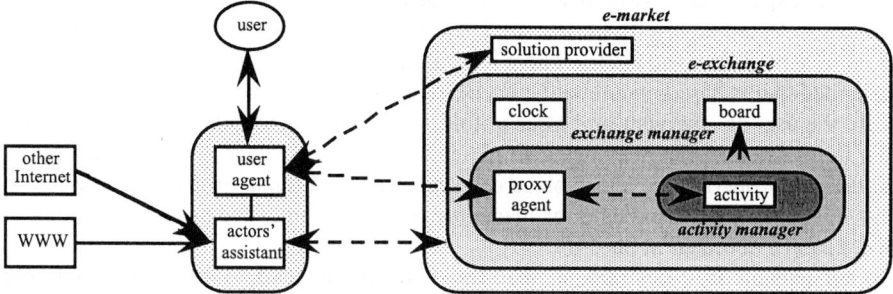

Fig. 2. High-level model of the e-market and user

To enable the activity manager to do its job, each proxy agent reveals in confidence the assets that the proxy agent possesses to the activity manager. The activity manager may 'ear mark' some amount of a buying proxy agent's money if the market mechanism requires that that amount must be set aside to support a bid, or if a buyer's deposit is required. Each player may withdraw assets from the proxy agent and so from the e-market.

3 Process Management

Fig 2. may give the false impression that all the process management system does is to support communication between the user agents and their corresponding proxy agents. All transactions are managed as business processes, including a simple 'buy order', and a complex request for information placed with an actor's assistant. For example, a simple 'buy order' placed by a user agent may cause its proxy agent to place a bid in an activity which may then provoke interaction with a number of user agents who are trying to sell the appropriate goods, and so on. Alternatively, a user agent may seek out other agents through a solution provider. In either case, any buy or sell order will be partly responsible for triggering a reaction along the entire supply (or value) chain. Building e-business process management systems is business process reengineering on a massive scale, it often named *industry process reengineering* [17]. This can lead to considerable problems unless there is an agreed basis for transacting business. The majority of market transactions are constrained by time („I need it before Tuesday"), or more complex constraints („I only need the engine if I also have a chassis and as long as the total cost is less than..). The majority of transactions are *critical* in that they must be dealt with and can't be forgotten or mislaid. Or at least it is an awful nuisance if they are. So this means that a system for managing them is required that can handle complex constraints and that attempts to prevent process failure.

E-market processes will typically be *goal-directed* in the sense that it may be known *what* goals have to be achieved, but not necessarily *how* to achieve those goals today. A goal-directed process may be modelled as a (possibly conditional) sequence of goals. Alternatively a process may be *emergent* in the sense that the person who triggers the process may not have any particular goal in mind and may be on a 'fishing expedition' [18]—this is particularly common in the entrepreneurial processes described briefly in Sec. 4. There has been little work on the management of emergent processes [19]. There a multiagent process management system is described that is based on a three-layer, BDI, hybrid architecture. That system 'works with' the user as emergent processes unfold. It also manages goal-directed processes in a fairly conventional way using single-entry quadruple-exit plans that give almost-failure-proof operation. Those plans can represent constraints of the type referred to above, and so it is a candidate for managing the operation of the system described in Sec. 2, and elaborated in Sec. 4.

One tension in e-business is that the transactions tend to be critical and the environment in which they are being processed is inherently unreliable. On one level sellers and buyers may renege on a deal, or may take an unpredictable amount of time to come to a conclusion. On another level the whole e-business environment, including the computer systems on it and the Internet itself, can not be relied on to perform on time, and, on occasions, may not operate at all. So this means that the system which manages the transactions must be able to adapt intelligently as the performance characteristics of the various parts of its environment are more or less satisfactory and more or less predictable. A more subtle complicating factor is that whereas much valuable information may be available free of charge on the World Wide Web it is not necessarily current. For example, the Sydney Stock Exchange publishes its quoted prices free of charge but with a time delay of several minutes—current information costs money. So, while working in an unreliable environment, the system for managing processes should balance time and cost tradeoffs in sub-processes whilst attempting to keep the whole process on time and within budget, and whilst delivering results of an acceptable standard.

Multiagent technology is an attractive basis for industry process re-engineering [20] [21]. A multiagent system consists of autonomous components that negotiate with one another. The scalability issue of industry process reengineering is „solved"—in theory—by establishing a common understanding for inter-agent communication and interaction. Standard XML-based ontologies will enable data to be communicated freely [22] but much work has yet to be done on standards for communicating expertise [23]. Results in ontological analysis and engineering [24] [23] is a potential source for formal communication languages which supports information exchange between the actors in an e-market place. Systems such as CommerceNet's Eco [www.commerce.net] and Rosettanet [www.rosettanet.org] are attempting to establish common languages and frameworks for business transactions and negotiations. Specifying an agent interaction protocol is complex as it in effect specifies the common understanding of the basis on which the whole system will operate.

A variety of architectures have been described for autonomous agents. A fundamental distinction is the extent to which an architecture exhibits deliberative (feed forward, planning) reasoning and reactive (feed back) reasoning. If an agent architecture combines these two forms of reasoning it is a *hybrid architecture*. One well reported class of hybrid architectures is the three-layer, BDI agent architectures. One member of this class is the INTERRAP architecture [25], which has its origins in the work of [26]. A multiagent system to manage „goal-driven" processes is described in [19]. In that system each human user is assisted by an agent which is based on a generic three-layer, BDI hybrid agent architecture similar to the INTERRAP architecture. That system has been extended to support emergent processes and so to support and the full range of industry processes. That conceptual architecture is adapted slightly for use here; see Fig 3(a). Each agent receives messages from other agents (and, if it is a personal agent, from its user) in its message area. The world beliefs are derived from reading messages, observing the e-market and from the World Wide Web (as accessed by an actor's assistant).

Deliberative reasoning is effected by the non-deterministic procedure: „on the basis of current *beliefs*—identify the current *options*, on the basis of current options and existing commitments—select the current commitments (called the agent's *goals* or *desires*), for each newly-committed goal choose a *plan* for that goal, from the selected plans choose a consistent set of things to do next (called the agent's *intentions*)". A *plan* for a goal is a conditional sequence of sub-goals that may include iterative or recursive structures. If the current options do not include a current commitment then that commitment is dropped. In outline, the reactive reasoning mechanism employs triggers that observe the agent's beliefs and are 'hot wired' back to the procedural intentions. If those triggers fire then they take precedence over the agent's deliberative reasoning. The environment is intrinsically unreliable. In particular plans can not necessarily be relied upon to achieve their goal. So at the end of every plan there is a *success condition* which tests whether that plan's goal has been achieved; see Fig 3(b). That success condition is itself a procedure which can succeed (3), fail (7) or be aborted (**A**). So this leads to each plan having four possible exits: success (3), failure (7), aborted (**A**) and unknown (**?**). In practice these four exists do not necessarily have to lead to different sub-goals, and so the growth in the size of plan with depth is not quite as bad as could be expected. KQML (<u>K</u>nowledge <u>Q</u>uery and <u>M</u>anipulation <u>L</u>anguage) is used for inter-agent communication [27].

The system operates in an environment whose performance and reliability will be unreliable and unpredictable. Further, choices may have to be made that balance reliability with cost. To apply the deliberative reasoning procedure requires a mechanism for *identifying* options, for *selecting* goals, for *choosing* plans and for *scheduling* intentions. A plan may perform well or badly. The process management system takes account of the „process knowledge" and the „performance knowledge". *Process knowledge* is the wisdom that has been accumulated, particularly that which is relevant to the process instance at hand. *Performance knowledge* is knowledge of how effective agents, people, systems, methods and plans are at achieving various things. A plan's *performance* is defined in terms of: the likelihood that the plan will succeed, the expected cost and time to execute the plan, the expected value added to the process by the plan, or some combination of these measures. If each agent knows how well the choices that it has made have performed in the past then it can be expected to make decisions reasonably well as long as plan performance remains reasonably stable. One mechanism for achieving this form of adaptivity is reinforcement learning. An alternative approach based on probability is described in [19]. In addition, an agent may know things about the system environment, and may have some idea of the *reason why* one choice lead to failure. An agent's belief in these reasons may result from communication with other agents. Such beliefs may be used to revise the „historical" estimates to give an *informed* estimate of plan performance that takes into account the *reasons why* a plan behaved the way that it did [op. cit.].

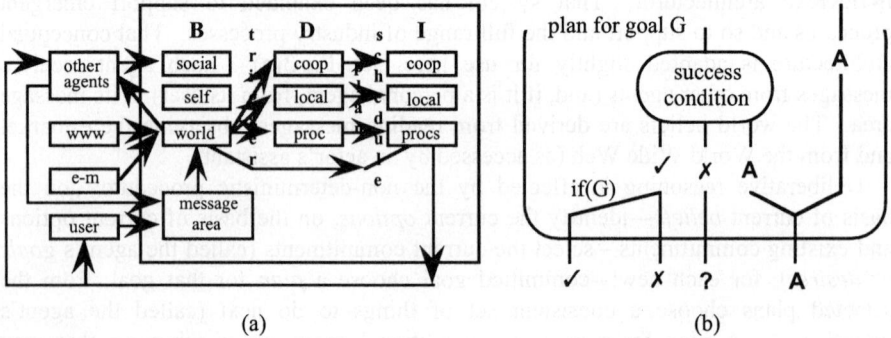

Fig 3. (a) conceptual architecture, (b) the four plan exits

4   E-market Evolution

E-markets, with their rapid, extensive, pervasive and visible flows of information provide an ideal context in which (a) to study market evolution in a context that is undergoing rapid evolution and innovation, and (b) to build systems that assist entrepreneurial intervention in actual market situations. The term 'entrepreneur' is used here in its technical sense [1]. Entrepreneurial transactions are typically more complex than the basic e-market transactions described in Sec. 2. Work on this part of the project will commence in the latter half of 2001, and the outline design is complete.

One entrepreneurial transaction is a request for information. For example, „look for Sydney-based manufacturing companies with a high asset backing that have been subject of at least one failed takeover bid in the past two years". Such a request triggers an instance of an *information process* in the user agent that will invoke information discovery and data mining methods, and perhaps hand-crafted analytical tools, in that agent's actors' assistant. These methods and tools are applied to either the e-market data or to Internet data. So there are four basic problems here. First, determining where to access the individual signals. Second, assessing some measure of confidence in the validity of those signals. Third combining those signals—which may be invalid and which may have been observed at different times and different places—into reliable advice. Fourth to do all this within tight time and cost constraints. Another entrepreneurial transaction is *entrepreneurial intervention* itself. For example, „I am offering the following unique combination of products and services with a novel payment structure.....". Such a transaction also triggers a business process that sets up the required structures within the e-market.

4.1 Actor Classes in the Full System

In the full e-market system, additional actor classes are introduced to support market evolution. *E-speculators* take short term positions in the e-exchange. This introduces the possibility of negative commission charges for others using an e-exchange. Sell-side *asset exchanges* exchange or share assets between sellers. *Content aggregators*, who act as forward aggregators, coordinate and package goods and services from various sellers. *Specialist originators*, who act as reverse aggregators, coordinate and package orders for goods and services from various buyers. The specialist originators, content aggregators and e-speculators represent their presence in the e-exchange by creating activities. This finally gives justification to the use of the term 'activity' to describe what happens in the e-exchange rather than 'sale' which might be more appropriate to the limited range of transactions in the basic system described in Sec. 2. The eight actor classes are illustrated in Fig. 4. They are a super-set of the actor classes in the basic system.

The full system will be used to study market evolution using intelligent software and the vast resources of the Internet. First, e-markets are in themselves a fruitful arena for hunting entrepreneurial opportunities—witness the proliferation of new market forms and players [28], [9] and the characterisation of the new network economy as an „opportunity factory". Second, the opportunity arises to leverage the rich flows of information in and around e-markets using smart computer systems to facilitate the discovery of potentially profitable opportunities and thereby stimulate the innovation and evolution process.

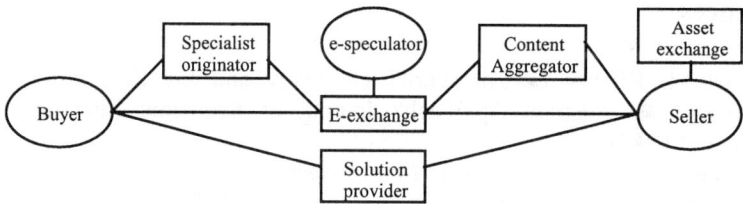

Fig 4. The eight actor classes in the full e-market system

There are valuable studies on the cognitive processes and factors driving alertness to opportunities [29]. There is also a wealth of established work in economic theory, that provides a basis for work in e-markets, including: the theory of auctions, theory of bargaining, and theory of contracts. [8] presents mechanisms for negotiation between computational agents. That work describes tactics for the manipulation of the utility of deals, trade-off mechanisms that manipulate the value, rather than the overall utility, of an offer, and manipulation mechanisms that add and remove issues from the negotiation set. Much has to be done before this established body of work may form a practical basis for an investigation into electronic market evolution [3]. Little is known on how entrepreneurs will operate in electronic market places, although the capacity of the vast amount of information that will reside in those market places, and on the Internet generally, to assist the market evolution process is self-evident. The overall aim of this project is to understand the e-market evolutionary process.

5   Conclusion

One of the innovations in this project is the development of a coherent environment for e-market places, a comprehensive set of actor classes and the use of a powerful multiagent process management system to make the whole thing work. The use of a powerful business process management system to drive all the electronic market transactions unifies the whole market operation. The development of computational models of the basic market transactions, deploying those models in the e-market place, and including them as part of the building blocks for creating a complete e-market place provides a practical instrument for continued research and development in electronic markets.

Acknowledgment

The work described herein was completed whilst the author was a visitor at the CSIRO Joint Research Centre for Advanced Systems Engineering, Macquarie University, Sydney. The contribution to the work by the members of that Centre is gratefully acknowledged.

References

[1] Israel M. Kirzner Entrepreneurial Discovery and the Competitive Market Process: An Austrian Approach" Journal of Economic Literature XXXV (March) 1997 60-85.
[2] R. Guttman, A. Moukas, and P. Maes. Agent-mediated Electronic Commerce: A Survey. Knowledge Engineering Review, June 1998.
[3] Moshe Tennenholtz. Electronic Commerce: From Economic and Game-Theoretic Models to Working Protocols. Invited paper. Proceedings Sixteenth International Joint Conference on Artificial Intelligence, IJCAI'99, Stockholm, Sweden.
[4] Milgrom, P. Auction Theory for Privatization. Cambridge Univ Press (2001).
[5] Bichler, M. The Future of E-Commerce: Multi-Dimensional Market Mechanisms. Cambridge University Press (2001).
[6] Sandholm, T. Agents in Electronic Commerce: Component Technologies for Automated Negotiation and Coalition Formation. Autonomous Agents and Multi-Agent Systems, 3(1), 73-96.
[7] Stršbel, M. Design of Roles and Protocols for Electronic Negotiations. Electronic Commerce Research Journal, Special Issue on Market Design 2001.
[8] Peyman Faratin. Automated Service Negotiation Between Autonomous Computational Agents. PhD dissertation, University of London (Dec 2000).

[9] R. Wise & D. Morrison. Beyond the Exchange; The Future of B2B. Harvard Business review Nov-Dec 2000, pp86-96.
[10] Neeman, Z. & Vulkan, N. Markets Versus Negotiations. The Hebrew University of Jerusalem Discussion Paper 239. (February 2001).
[11] Bulow, J. & Klemperer, P. Auctions Versus Negotiations. American Economic Review, 1996.
[12] Kumar, M. & Feldman, S.I. Business Negotiations on the Internet. Proceedings INET'98 Internet Summit, Geneva, July 21-24, 1998.
[13] B. Kovalerchuk & E. Vityaev. Data Mining in Finance: Advances in Relational and Hybrid Methods. Kluwer, 2000.
[14] J. Han, L.V.S. Lakshmanan & R.T. Ng. Constraint-based multidimensional data mining. IEEE Computer, 8, 46-50, 1999.
[15] Han, J. & Kamber, M. Data Mining: Concepts and Techniques. Morgan Kaufmann (2000).
[16] Chen, Z. Computational Intelligence for Decision Support. CRC Press, Boca Raton, 2000.
[17] Feldman, S. Technology Trends and Drivers and a Vision of the Future of e-business. Proceedings 4th International Enterprise Distributed Object Computing Conference, September 25-28, 2000, Makuhari, Japan.
[18] Fischer, L. (Ed). Workflow Handbook 2001. Future Strategies, 2000.
[19] Debenham, J.K.. Supporting knowledge-driven processes in a multiagent process management system. Proceedings Twentieth International Conference on Knowledge Based Systems and Applied Artificial Intelligence, ES'2000: Research and Development in Intelligent Systems XVII, Cambridge UK, December 2000, pp273-286.
[20] Jain, A.K., Aparicio, M. and Singh, M.P. „Agents for Process Coherence in Virtual Enterprises" in Communications of the ACM, Volume 42, No 3, March 1999, pp62—69.
[21] Jennings, N.R., Faratin, P., Norman, T.J., O'Brien, P. & Odgers, B. Autonomous Agents for Business Process Management. Int. Journal of Applied Artificial Intelligence 14 (2) 145—189, 2000.
[22] Robert Skinstad, R. „Business process integration through XML". In proceedings XML Europe 2000, Paris, 12-16 June 2000.
[23] Guarino N., Masolo C., and Vetere G., OntoSeek: Content-Based Access to the Web, IEEE Intelligent Systems 14(3), May/June 1999, pp. 70-80
[24] Uschold, M. and Gruninger, M.: 1996, Ontologies: principles, methods and applications. Knowledge Engineering Review, 11(2), 1996.
[25] MŸller, J.P. „The Design of Intelligent Agents" Springer-Verlag, 1996.
[26] Rao, A.S. and Georgeff, M.P. „BDI Agents: From Theory to Practice", in proceedings First International Conference on Multi-Agent Systems (ICMAS-95), San Francisco, USA, pp 312—319.
[27] Finin, F. Labrou, Y., and Mayfield, J. „KQML as an agent communication language." In Jeff Bradshaw (Ed.) Software Agents. MIT Press (1997).
[28] Kaplan, Steven and Sawhney, Mohanbir. E-Hubs: The New B2B Marketplace. Harvard Business Review 78 May-June 2000 97-103.
[29] Shane, Scott. Prior knowledge and the discovery of entrepreneurial opportunities. Organization Science 11 (July-August), 2000, 448-469.

Intelligent Auto-downloading of Images

Vikram Natarajan and Angela Goh

School of Computer Engineering, Nanyang Technological University
Singapore 639798
Tel: (065)7904929; Fax: (065)7926559
asesgoh@ntu.edu.sg

Abstract. There is a proliferation of news sites on the World Wide Web and hence, it is becoming increasingly harder for journalists to monitor the news on such sites. In response to this need, a software tool JWeb-Watch was developed to provide journalists with a suite of tools for checking the latest news on the web. One particular feature of this tool is the ability to download relevant images from these sites and discard extraneous figures. The paper reviews related software and discusses the unique features of this approach adopted.

Keywords: Internet News Sites, Images downloading

1 Introduction

The impact of the Information Revolution, known by various names such as Cyberspace, Global Information Infrastructure, Internet and the like, is being felt all across the world. The ability to communicate through the Internet has been supported by browsers and other tools that assist users to locate sources of information. In April 2000, IDC reported that since 1995, 2 billion Web pages have been created, with more than 200 million new ones added every month and another 100 million becoming obsolete in the same period [2].

One sector that the Internet has had a significant impact on is the newspaper industry. There has been a rapid transition to digital newspapers. Though traditional printed news is still very much in demand, the co-existence of its digital counterpart provides readers with many more features and services [5]. These include the ability to search, retrieve and send specific news articles from current or archived news. Many news services provide multimedia information including audio and video clips. Personalization or customization is another service unique to e-newspapers. The idea of an agent that scours the Internet for news and packaging its results has been discussed at length in [4]. The motivation for such News on Demand services arise out of the proliferation of sites offering on-line and real-time news.

This paper focuses on the needs of a specialized group of people in the newspaper industry, namely, the journalists. A tedious but essential part of a journalist's work is to monitor news as it occurs. News monitoring is a term used to depict the tracking of news. This involves the ability to gather information from news sites, manage it and determine the changes that have occurred over time. Presently, the process of news monitoring requires reading through an enormous number of news articles and downloading the relevant data that the user wishes to monitor. Once this is done, the user has to keep referring back to the web sites to check if the article has changed and what these changes are. The problems associated with such a process include:

- To monitor even a single article is time consuming and painstaking.
- As the number of articles to be analyzed increases, the effort becomes even greater.
- There might be many subtle changes (such as a change in count of electoral votes) that might not be noticed
- Saving the results for further analysis has to be done manually and needs to be indexed to ease future reference.

In order to alleviate these problems, an investigation and development of a tool named JWebWatch was undertaken. JWebWatch conceptualizes and implements an architecture that would support a journalist's work. In this paper, one of the features will be elaborated on. While other features adopted existing techniques, such as text extraction, the intelligent image downloading of images from the web site involves interesting and novel approaches. The approach can be easily applied to other applications such as financial systems that require web monitoring.

Following this introduction, the objectives of JWebWatch and its system architecture is given. This is followed by a brief review of related software. Details of the image downloading feature are then provided. This is followed by examples of results.

2 Overview of JWebwatch

The tool comprises six major modules as shown in Figure 1. These are summarised as follows:

1. Text Extraction from News Websites
   This module enables the downloading of articles from websites by extracting only the relevant text from the page. Care has to be taken not to extract unwanted text, such as quotes, flash news, latest stock prices, text logos, unwanted headings and so on. There is a tremendous amount of such unwanted and redundant text on sites such as CNN (http://www.cnn.com).
2. Summarization of Articles
   The user is able to summarize any article he/she desires. The algorithm used ensures that the summarization is robust and return results as close to a natural summary as possible. A list of keywords of the article can also be produced.
3. Images and Advertisements on websites
   Along with the text of the article, the images associated with the text can

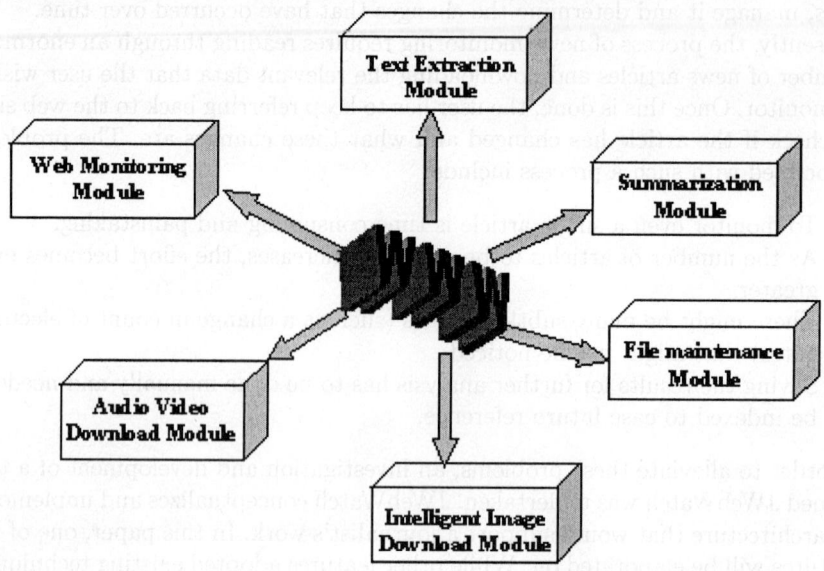

Fig. 1. An overview of JWebWatch

also be downloaded. Many websites contain a huge number of advertisements, bullets and images. For example, CNN has an average of 80 to 100 images on every web page. These advertisements and unwanted images are filtered out.

4. Audio and Video Files

   Audio and video files related to an article are also available for downloading. This is to enable the user to listen/view the file from the hard disk instead of going to the site. JWebWatch allows users to play the audio and video without having to invoke another application program.

5. Monitoring Updates on Websites

   Dynamic changes to web sites are tracked so that the changes can be highlighted automatically. This will save tedious manual checks to determine which portion of the website has been altered. The monitoring is sophisticated and error free, providing a user-friendly means of viewing the changes.

6. File Maintenance

   Articles can be stored and retrieved through a simple-to-use file system.

In order to facilitate the use of the above functions, there is a need for a simple Graphical User Interface. As most users are familiar with typical Windows applications, the Interface was designed to have a similar look-and-feel with such applications.

The relationship between the modules described above are shown in Figure 2.

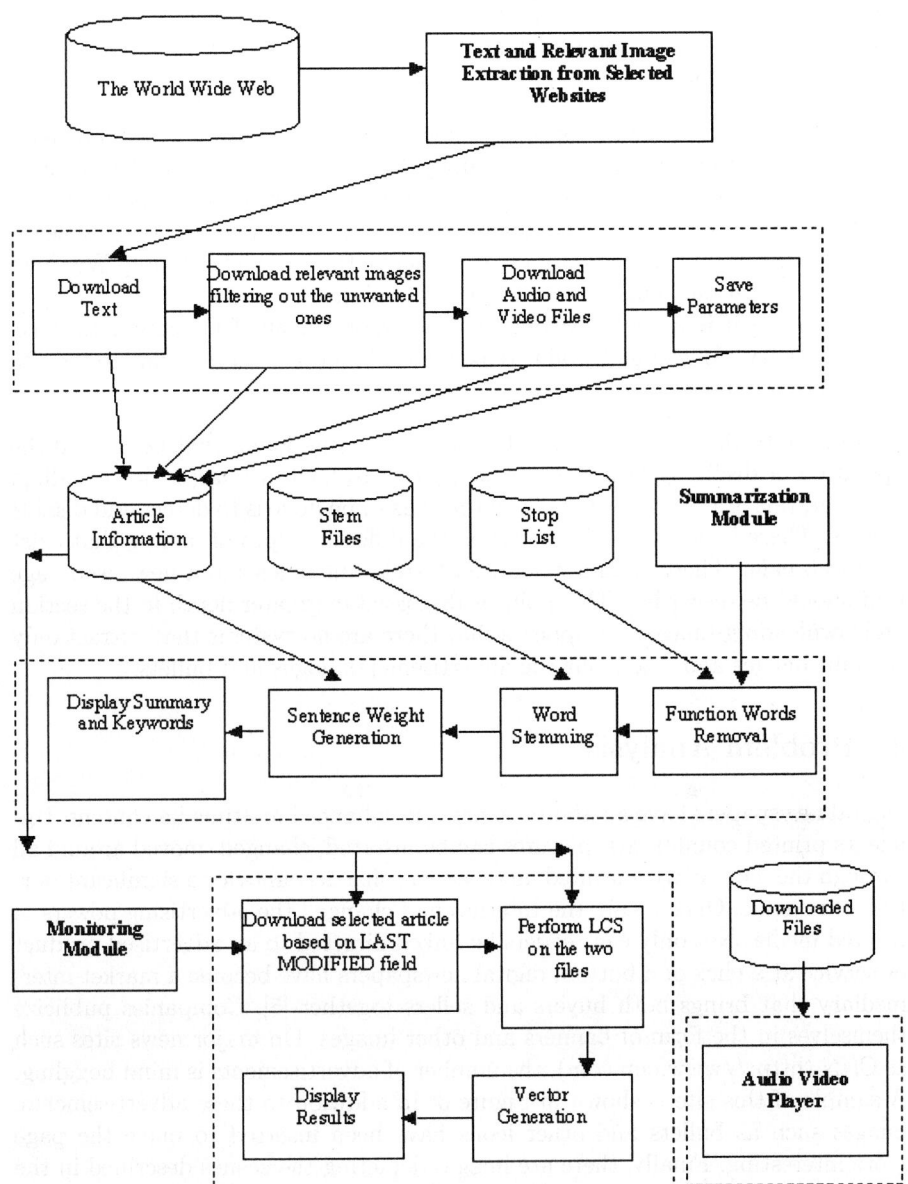

Fig. 2. Relationship between the modules comprising JWebWatch

3 Related Work

In this section, a brief survey of the image detection on web sites will be given. There is existing software relating to downloading of pertinent images and the discarding of irrelevant graphics. There are a few commercial tools that disable advertisements when a user is browsing the web.

- AdsOff is a tool that works as a background process to accelerate page loading and make browsing more enjoyable while connected to the Internet. It achieves this by filtering irrelevant content such as advertisements [1]. Features include a universal filter that works with all web browsers, a means to render web pages without advertisements, the removal of Popup Windows and Animations, and the disabling of Cookies.
- AdEater [3] is a tool developed at the Department of Computer Science, University College at Dublin. It possesses functions very similar to AdsOff described above.

In addition to the above-mentioned tools, others were found that performed the operation of disabling of images on web pages to facilitate faster downloading. However, this is a trivial task since all that has to be done is to detect and disable images. There is no distinction between the different types of images that exist on a web page. There is an abundance of irrelevant images in a news web page that should be discarded. This point is discussed in greater depth in the section on Downloading Images . It appears that there are no systems that extract only relevant images and filters out the advertisements, logos and bullets.

4 Problem Analysis

Digital newspapers have an ability to provide enhanced multimedia content. Unlike its printed counterpart, pictures can be inserted, changed, moved around all through the day. As with printed news, advertising accounts for a significant portion of revenue. Once again, the Internet had changed the advertising powers of printed media. Not only can readers be linked directed to an advertised product or service at a click of a button, digital newspapers have become a market intermediary that brings both buyers and sellers together [5]. Companies publicize themselves in the form of banners and other images. On major news sites such as CNN (http://www.cnn.com), the number of advertisements is mind boggling. A sample of this site is shown in Figure 3. In addition to these advertisements, images such as bullets and other icons have been inserted to make the page more interesting. Finally, there are images depicting the scenes described in the article. These are the pictures of interest to the journalists. There is therefore a need to be able to detect the relevant images and to only download these.

A study of various news sites was carried out. The following factors were considered important in determining the relevancy of an image:

Intelligent Auto-downloading of Images 351

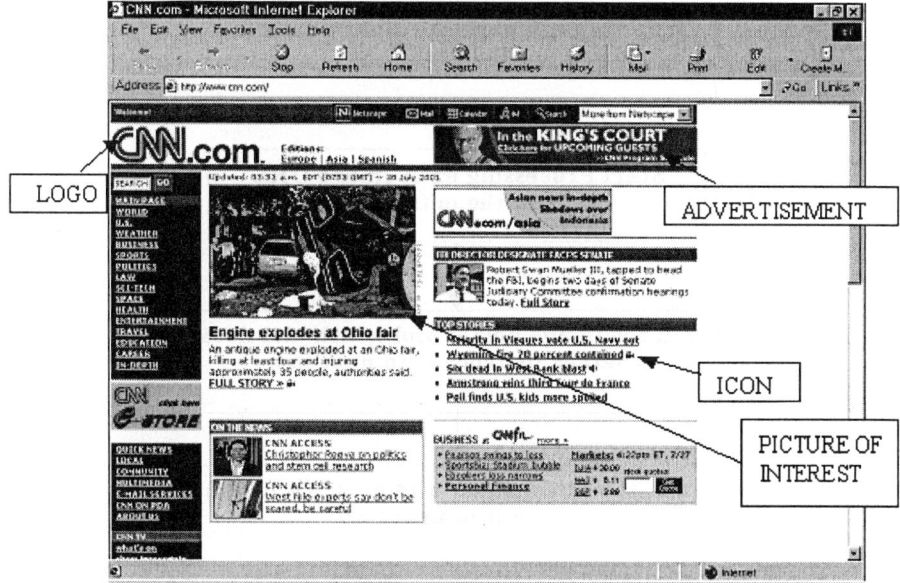

Fig. 3. Sample Web Site: cnn.com

- The domain site on which the image resides in
- Height of the image (small images are of height less than 100 pixels)
- Width of the image (small images are of width less than 100 pixels)
- The file extension of the image file
- If the height and width are not available, the file size of the image is considered
- The page the image points to (if there is any link)

5 The Intelligent Downloading Algorithm

Arising from the above analysis, an algorithm that intelligently downloads relevant images and discards other unwanted images was developed. Figure 4 contains the steps involved. The steps depict in detail the module shown as Download relevant images filtering out the unwanted ones' in Figure 2. The first step of the algorithm is to create a vector comprising image URLs. The URL vector is created by examining the img tags of the HTML page. The img element embeds an image in the current document at the location of the element's definition. The img element also has no content and is usually replaced inline by the image designated with the src attribute. Attributes such as height and width can also be specified along with the tag. However, due to the unstructured nature of HTML, these attributes are not mandatory.

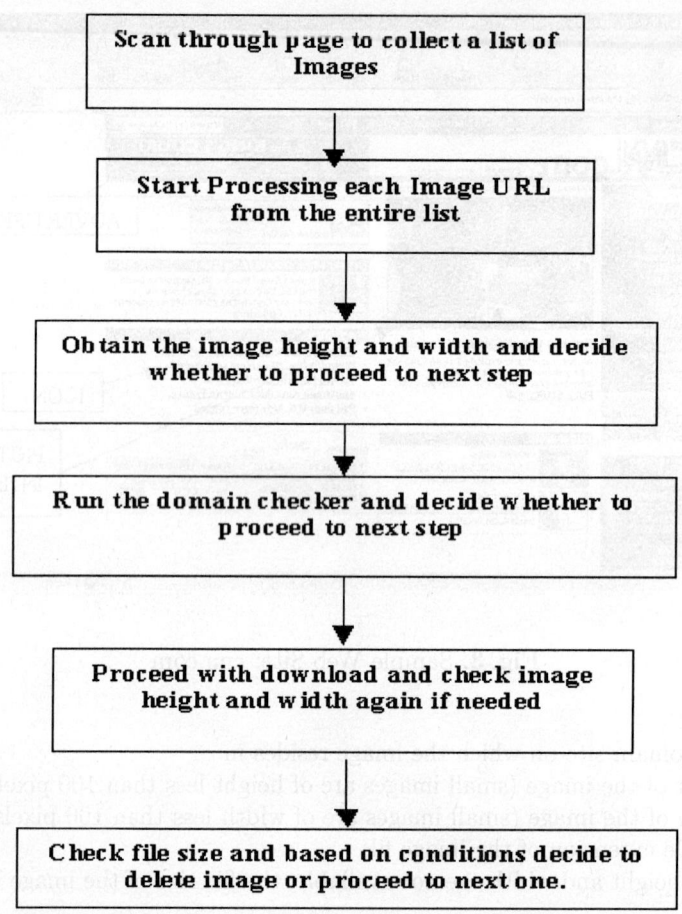

Fig. 4. Summary of the Image Downloading Algorithm

Next, the algorithm examines the height and width attributes (if present). The minimum height and width was specified to be 99 pixels based on the studies carried out on the various sites. Small images are unable to capture a reasonable amount of content and hence images of such sizes would normally be some icon or bullet. The image in Figure 5 is 120 pixels in width and 100 pixels in height.

Images that are smaller than this are virtually guaranteed to contain unimportant information. Thus, at this step, the algorithm would discard these small images that typically represent bullets and extremely small images. It should be noted that the value of 99 pixels, though stringent, was not increased as it might result in the filtering of some relevant images.

Figure 6 shows the images that are filtered out and those that remain in the list. Images (a to d) are dropped from the list. It is noted that advertisements and

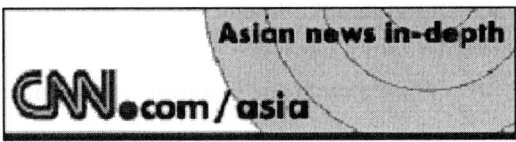

Fig. 5. Image Height and Width

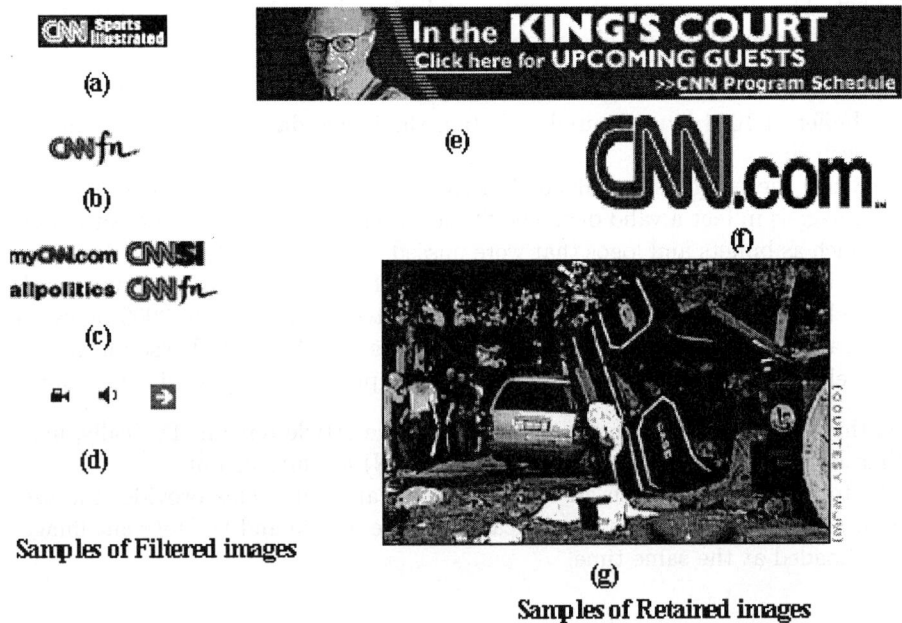

Fig. 6. Image Filtering

logos may still be retained as they can be of fairly large sizes. Hence, (e) and (f) appear in the list. The only valid image (g) in the article is also retained.

The next step consists of running the domain checker on the remaining Image URLs. The rationale behind the domain checker is as follows: Advertisements always point to external sites (or sites not in the same domain). This was an interesting fact observed by monitoring the web pages in order to detect advertisements. Image (f) in Figure 5 is a company logo. The image is not kept in an external site and hence would be retained. To solve this problem, the relative domain was also checked. A valid image (g) is stored in the same directory as the article. In such situations, the src attribute in the img tag normally points to the relative path by simply giving the name of the image of the entire URL. However images such as (f) are needed throughout every page in CNN and so are not stored in the sub-directory of each article. An example of the path in such a case is:

$$\text{http://www.cnn.com/images/...... xxx.gif}$$

Thus, by obtaining the URL of the image and checking the domain of the image with respect to the original URL of the article, further image filtering is accomplished. At this stage, more than 85 percent of the images have been filtered out. However to attain an even higher level of accuracy, an additional stage is added. This is to ensure that irrelevant images have not slipped through the checks conducted so far. The steps of this stage are as follows:

- Process the remaining image URLs by downloading them one by one. A buffer of 1024 bytes is used to obtain the image data as a stream from the web site.
- Once the image is downloaded, a check function is invoked to verify if the image is in fact a valid one. The file size is checked in order to detect images such as bullets and logos that were missed in the earlier stages due to absent height and width attributes. Image sizes of less than 2000 bytes are simply deleted. Images that are less than 5000 bytes but more than 2000 bytes are loaded and the height and width attributes are obtained. These images are deleted if the attributes do not satisfy the previously specified constraints.

At this stage, only the images that relate to the article remain. Typically, more than a hundred images (for a site such as CNN) are filtered out.
The tool is enhanced with an image-viewing capability. This provides the user with great flexibility since he/she can view the article and the relevant images downloaded at the same time.

6 Tests Carried Out

The testing methodology for this module involves running the algorithm on a number of web pages containing images. Thus a large number of images are processed in order to verify the accuracy and effectiveness of the algorithm.
In a typical CNN web page, there are about 80 images on the main page. The links occurring in every selected URL are also downloaded along with the page. Approximately 40 links are present on each page and each of these contains another 80 or so images. Thus, by simply selecting one page, approximately 3300 images are processed. Tests were carried out on 30 web pages, resulting in the algorithm being executed on 100,000 images. These extensive tests for helped to evaluate the accuracy of the module in filtering out unwanted images. Table I contains the results obtained for a few web pages. The six sites tabulated are chosen randomly from the 30 pages in order to depict the nature of the results. These results are representative of the entire set of tests conducted. In all the cases tested, the images that were relevant to the article were always downloaded. It was noticed that in some rare cases, such as pages showing stock market ratings, there are some extremely large images of charts, which might not

be useful to all users. These pages consist mainly of such images and have very little text. In this case, the images were downloaded. In general, the downloading of images was performed with high accuracy and provided extremely encouraging results.

Table 1. Results of the downloading module

Site Index No.	Total Number of Images	Relevant Images Downloaded	Unwanted Images (not downloaded)	Comments
1	72	6	66	Results were accurate
2	56	2	54	Results were accurate
3	89	1	88	Results were accurate
4	61	0	61	No relevant images to download
5	85	2	83	Results were accurate
6	94	1	93	A stock market page. One stock logo downloaded

7 Conclusion

The paper presents a tool for journalists to download and monitor web sites. The specific focus of the article has been on the design and implementation of downloading of images from the World Wide Web. Experiments have shown that the algorithm performs well in its ability to discriminate between irrelevant images and those directly related to the article. The modular design of this function enables it to be adapted to other applications requiring similar features. Further work is being carried out to extract captions from downloaded pictures. This related work is a challenging one but is seen as a good enhancement of the current intelligent image downloading feature.

References

1. AdsOff, http://www.intercantech.com
2. IDC (2000) http://www.idc.com/itforecaster/itf20010403Home.stm
3. Kushmerick, N. (1999) Learning to remove Internet advertisements, Third International Conference on AUTONOMOUS AGENTS (Agents '99) Seattle, Washington (postscript version found in http://www.cs.ucd.ie/staff/nick /home/research/ae/)
4. Maybury, M. (2000) Special Issue on News on Demand, Communications of the ACM, Vol. 43, No.2, pp.32-74
5. Palmer, J. W. and Eriksen, L. B. (1999) Digital Newspapers explore marketing on the Internet Communications of the ACM, Vol. 42, No.9, pp.33-40.

Intelligent Facilitation Agent for Online Web-Based Group Discussion System

Junalux Chalidabhongse, Wirat Chinnan,
Pichet Wechasaethnon, and Arpakorn Tantisirithanakorn

Program of Information Technology
Sirindhorn International Institute of Technology, Thammasat University
P.O. Box 22 Thammasat Rangsit Post Office, Pathumthani 12121, Thailand
junalux@siit.tu.ac.th

Abstract. Facilitation is considered one of the most important factors in the effective use of group decision support system (GDSS). However, high quality human facilitator may not be easily available. Furthermore, increase in globalization and telecommuting requires GDSS that supports dispersed group meeting, which is not conveniently implemented with human facilitators. This paper proposes an intelligent facilitation agent (IFA) that can be applied to facilitate group meeting through an online web-based discussion system. The IFA employs the power of asking questions, based on the supporting decision model associated with the problem being discussed, to structure group conversation. The experiments were set to study the effects of our IFA on group meeting through an online discussion system. The experimental results illustrate that group meetings with our IFA have higher number of generated ideas, higher discussion participation, higher amount of supporting information for decision making, and lower group distraction.

1 Introduction

Group decision support system (GDSS) has been introduced in past few years to enhance group meeting efficiency. The typical GDSS for face-to-face meeting is a decision room where the discussants use computer system to interact at the same time and the same place [1]. Facilitation has been included as one of the most critical factors for successful use of GDSS. High quality facilitators can result in high quality group outcomes and enhance GDSS effectiveness [2]. However, high quality human facilitators may not be widely available and human facilitators may produce some human bias [2]-[5]. Moreover, globalization, telecommuting, and the rapid growth of the Internet are likely to accelerate the trend of people working in dispersed groups [6]. This phenomenon causes the problem of having human facilitators to support dispersed group meeting where participants are located in different places. Thus, to avoid the problems that may occur with human facilitators, this study proposes the intelligent facilitation agent (IFA) to support group meeting via an online web-based discussion system. The task of the IFA is to facilitate group meeting through the application of discourse analysis and the power of asking questions to structure group conversation in order to encourage group participation and discourage group distraction. The IFA is one of the main components of our ongoing work on intelligent group discussion support agent (IGDSA) [7]. The next section of this paper presents the general framework of the IGDSA and its components. The third section discusses more detail of the IFA, which is the main focus of this paper. The fourth section reports the experiments and their results on the study of the effects of our system on group meeting through an online discussion system. Finally, we describe the conclusions of this study and possible future research.

2   General Framework of the System

The IFA is one of the main components of our ongoing work on the development of the IGDSA for electronic meeting. The IGDSA is a multi-agent system, which is composed of four agents: the information extraction agent, the facilitation agent, the information support agent, and the decision support agent. The framework of the IGDSA is presented in Figure 1.

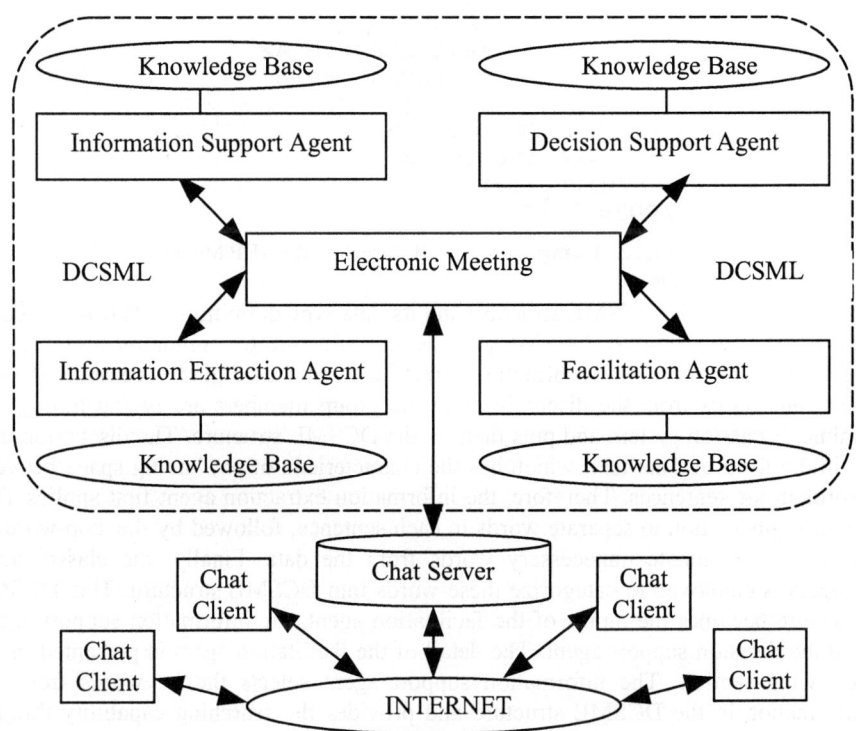

Fig. 1. Framework of Intelligent Group Discussion Support Agent

Communications among these four agents are represented in decision markup language (DCSML). DCSML is a markup language that classifies the data from the discussion in the XML format. DCSML provides the structure for extracting key decision factors from group conversation. DCSML uses start-tag and end-tag to form the tag-elements that parenthesize the key decision factors of the problem in the discussion. DCSML can support many types of problems based on the appropriate decision models associated with those types of problems. For the current version of our ongoing work on IGDSA, we focus on the problem of selection among many alternatives based on various criteria by applying the Analytic Hierarchy Process (AHP) model [8]. Some examples of the key decision factors for DCSML that is based on AHP include alternative, decision criterion, weight of relative importance among criteria, and priority of preferences of available alternatives on each criterion. An example of DCSML structure for AHP model is illustrated in Figure 2.

```
<objective>
    <discussant: Name>
        <criterion X>
            <weight>
                <criterion Y>
                <criterion Z>
                    :
            </weight>
            <priority>
                <alternative A>
                <alternative B>
                    :
            </priority>
        </criterion X>
    </discussant>
</objective>
```

Fig. 2. Example of DCSML Structure for AHP Model

The detail of the DCSML structure and its data type definition (DTD) is presented in our previous study [9]. DCSML provides the structure for communications among agents in our system. The information extraction agent (see Figure 3) retrieves the key decision factors from the discussion that the group members are typing through an online discussion system and puts them in the DCSML structure. The discussion data is in the form of Thai text, which has the characteristic of not having space between words in the sentences. Therefore, the information extraction agent first applies Thai word segmentation to separate words in each sentence, followed by the stop-wording process to eliminate unnecessary words from the data. Finally, the classification process is employed to categorize these words into DCSML structure. This DCSML structure becomes the inputs of the facilitation agent, the information support agent, and the decision support agent. The detail of the facilitation agent is presented in the following section. The information support agent selects the keywords from the information in the DCSML structure and provides the searching capability through the Internet and the knowledge base of the system to provide more supporting information for group discussion. The decision support agent keeps checking the completeness of the information in the DCSML structure and for any level of completeness of the information that is sufficient to run the decision model associated with the problem being discussed, the decision support agent will perform analysis and provide the suggested alternative to the group members based on the available information at that moment.

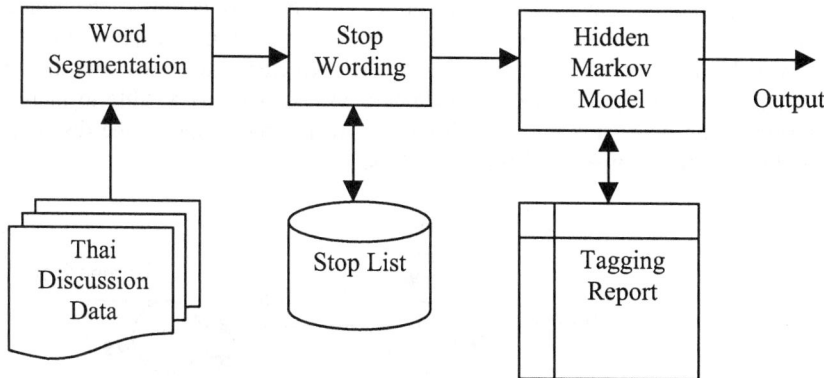

Fig. 3. Information Extraction Agent

3 Intelligent Facilitation Agent (IFA)

The IFA employs the power of asking questions to structure group conversation in order to encourage group participation within the scope of the problem being discussed to avoid group distraction. Many researches on meeting facilitation are based on adding structure to the interactions (see the review in [10]). However, those meeting facilitation systems are typically driven by human facilitators. The IFA supports group discussion by combining the capability of the information extraction agent with the power of asking questions by the facilitation agent. The facilitation agent analyzes the information in the DCSML structure, generated by the information extraction agent, to check any available information that is important to support group decision making. The facilitation agent is composed of sets of questions associated with the matching pattern of the available information in the DCSML structure. The status of the information in the DCSML structure for the AHP model can be categorized by the state diagram illustrated in Figure 4.

The IFA comprises the rule-based subsystem to choose appropriate set of questions according to the state diagram. The rules are triggered according to the matching pattern of the available information in DCSML structure at each moment. To create a variety of questions, we prepared many forms of questions for each state. Thus, for each matching, the IFA will randomly select a question from the set of questions matched with the current state of the available information in the DCSML structure and send the question to the computer screen of each group member. For example, the whole process starts from State 1 with no alternative and no decision criterion. The set of questions in this state is composed of the questions that will stimulate the group members to generate alternatives and decision criteria. At State 4 where we have only alternatives available without decision criterion, the set of questions consists of the questions that will encourage the participants to think about the criteria that can be used to evaluate the available alternatives. At State 6 where we have one criterion and more than one alternative, the set of questions comprises the questions that will encourage the discussants to compare the available alternatives based on this criterion. At State 7 and State 8 where we have more than one criterion, the set of questions include the questions that will ask the group members to compare the relative importance of the available criteria. Therefore, by keeping asking the questions based on the available key decision factors in the DCSML structure associated with the type of problem being discussed, the group members are expected to be able to increase their participation by continuing their discussion within the scope of the discussion topic and reduce the number of distracting issues.

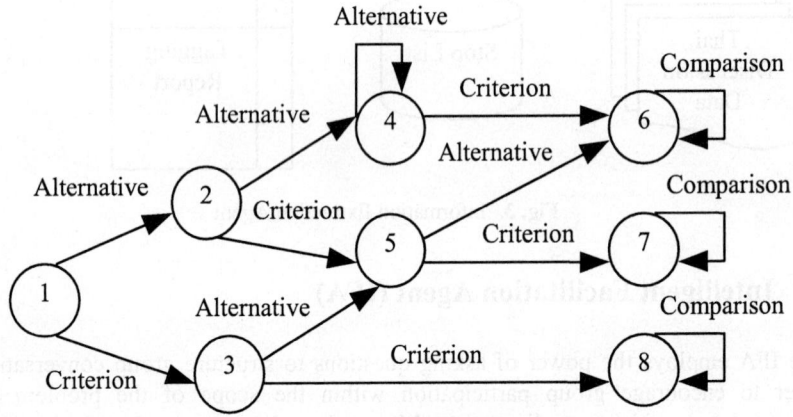

Fig. 4. State Diagram of Facilitation Agent for AHP Model

4 Experiments and Results

This paper postulates that the IFA can improve the quality of group discussion by encouraging group participation and discouraging group distraction. For our preliminary study on the development of this system, four hypotheses are formulated to investigate the possible effects of the proposed system.

H1: Group meeting with the proposed IFA results in less group distraction than group meeting without the proposed system.

H2: Group meeting with the proposed IFA results in more group participation than group meeting without the proposed system.

H3: Group meeting with the proposed IFA results in more available useful information than group meeting without the proposed system.

H4: Group meeting with the proposed IFA results in higher number of generated ideas than group meeting without the proposed system.

In order to analyze the efficiency of the system and examine the above hypotheses, we considered two types of group meeting: group meeting through an online discussion system with the proposed IFA and group meeting through an online discussion system without the proposed system. Thirty students were recruited to participate in the experiments. The participants were assigned to groups of three members. Each group participated in both the meetings with and without the proposed IFA, but on different topics. Each meeting lasted 20 minutes. The meeting conversations were saved in the files for further analysis. After each meeting, the participants received the questionnaires to assess their satisfaction levels on their group discussion. The questionnaires were designed using a five-point Likert-type scale (from the highest level of satisfaction = 5 to the lowest level of satisfaction = 1). The t-test hypothesis testing has been conducted to the questionnaire data to examine the hypotheses at the significance level (or α) of 0.05. The results of the experiments are illustrated in Table 1.

Table 1. Experimental Results of the Three Hypotheses

	Mean Value with Proposed System	Mean Value without Proposed System	T-Score	P-Value	Hypothesis Result
H1	2.74	2.10	2.91	0.003	Accept
H2	2.93	2.50	2.09	0.02	Accept
H3	3.43	2.90	2.89	0.003	Accept

From the experimental results, the first three hypotheses are accepted at the significance level of 0.05. Thus, the results have illustrated that the participants in the experiments statistically satisfy with the group discussion with the proposed support system more than the group meeting without such system on all these three criteria. Furthermore, the saved data files of meeting conversation were analyzed on the number of generated ideas and the number of distracting issues during the meeting. In our experiments, the distracting issues are defined as the issues that are not considered as the components of the DCSML structure for the discussed topic. The analysis of the meeting conversation from the saved data files shows that the meeting with the proposed IFA has 23.45% higher percentage of the number of ideas generated during the meeting and 18.56% lower percentage of the number of distracting issues than the meeting without the proposed system. Thus, the H4 hypothesis is accepted and the H1 hypothesis is again accepted with these additional results. The reasons to support these results can be noticed from the conversational data files, which illustrate that the questions from the IFA could persuade group members to come back to the discussed problem when they were about to be distracted from the issues and the questions could also help the discussants to continue their conversation when they ran out of the ideas to be discussed.

5 Conclusions

This paper presents the application of the IFA to support group meeting via an online discussion system. The proposed IFA is one of the four main agents in our ongoing work on intelligent group discussion support agent. The system applies the knowledge of natural language processing, data mining, intelligent agent, and management science to develop a group support system that can enhance group discussion efficiency. An empirical study was conducted to test the effectiveness of the proposed system. Two types of meeting were carried out: group meetings with and without our proposed support system. After each meeting, the participants filled out the questionnaires to assess their satisfaction levels on group meeting. The results of the empirical study indicate that group meeting with the proposed support system resulted in less group distraction, more group participation, more available useful information, and more generated ideas. Our future research will focus on the improvement of the IFA and examine its other effects on group interactions such as the effects on group conflict based on various discussion topics.

References

1. Turban, E.: Decision Support Systems and Expert Systems. 4th edn. Prentice-Hall International, USA (1995)
2. Anson, R., Bostrom, R., Wynne, V.: An Experiment Assessing Group Support System and Facilitation Effects on Meeting Outcomes. Management Science, Vol. 41 (1995) 189-208
3. Broome, B. J., Keever, D. B.: Next Generation Group Facilitation. Management Communication Quarterly, Vol. 3 (1989) 107-127
4. Cooks, L. M., Hale, C. L.: The Construction of Ethics in Mediation. Mediation Quarterly, Vol. 12 (1994) 55-76
5. Griffith, T. L., Fuller, M. A., Northcraft, G. B.: Facilitator Influence in Group Support Systems: Intended and Unintended Effects. Information Systems Research, Vol. 9, No. 1 (1998) 20-36
6. Tung, L., Turban, E.: A Proposed Research Framework for Distributed Group Support Systems. Decision Support Systems, Vol. 23 (1998) 175-188
7. Chalidabhongse, J., Chinnan, W.: Intelligent Group Discussion Support Agent for Electronic Meeting. In Proceedings of the International Symposium on Communications and Information Technology (ISCIT), Thailand. IEEE (2001)
8. Saaty, T. L.: The Analytic Hierarchy Process. McGraw-Hill, New York (1980)
9. Chinnan, W., Chalidabhongse, J., Tanhermhong, T.: Decision Markup Language for Discussion System. In Proceedings of the 16th International Conference on Production Research (ICPR-16), Czech Republic (2001)
10. Farnham, S., Chesley, H. R., McGhee, D. E., Kawal, R., Landau, J.: Structured Online Interactions: Improving the Decision-Making of Small Discussion Groups. In Proceedings of the ACM 2000 Conference on Computer Supported Cooperative Work, USA (2000) 299-308

TWIMC: An Anonymous Recipient E-mail System

Sebon Ku[1], Bogju Lee[2], and Dongman Lee[1]

[1]Engineering Dept., Information and Communications University (ICU), Korea
{prodo,dlee}@icu.ac.kr
[2] Computer Engineering Dept., Dankook University, Korea
blee@dankook.ac.kr

Abstract. More and more people rely on e-mails rather than postal letters to communicate each other. Although e-mails are more convenient, letters still have many nice features. The ability to handle "anonymous recipient" is one of them. This research aims to develop a software agent that performs the routing task as human beings for the anonymous recipient e-mails. The software agent named "TWIMC (To Whom It May Concern)" receives anonymous recipient e-mails, analyze it, and then routes the e-mail to the mostly qualified person (i.e., e-mail account) inside the organization. The machine learning and automatic text categorization (ATC) techniques are applied for the task. We view each e-mail account as a category (or class) of ATC. Everyday e-mail collections for each e-mail account provide an excellent source of training data. The experiment shows the high possibility that TWIMC could be deployed in the real world.

1 Introduction

Recently more and more people use e-mails rather than postal letters to communicate each other. E-mails are faster, easier to use, and more reliable than postal letters. E-mails are really substituting postal letters rapidly. Compared with e-mails, however, letters still have many nice features. Letters can be hand-written, nicely decorated, attached to other postal packages, and so on. The best feature of letters over e-mails, however, is the ability to handle "anonymous recipient". People used to send anonymous recipient letters by designating "To Whom It May Concern" inside the letters. The one who receives the letter in an organization reads the letter and routes to a qualified person. Of course, the "routing expert" knows well each of organization member's personal interest, job function, and business role. If organizations provide a special and universally accepted e-mail account such as TWIMC (To Whom It May Concern) (e.g., twimc@ancompany.com), people will be able to use anonymous recipient e-mails as in postal letters.

Another well-known task in this type happens in the call center of companies. Usually businesses run a representative e-mail account (e.g., info@acompany.com) to support customers' queries and orders. Companies spend many efforts to handle these anonymous recipient e-mails. Most of the efforts are to find the proper persons to route them. We call this problem the automatic e-mail routing (AER) problem. The AER problem can be defined as follows.

X: a set of e-mail instances
x: an e-mail instance, $x \in X$
e: an instance of e-mail account (including the anonymous account)
Ce: the target concept for e, that is, $Ce: X \rightarrow \{0, 1\}$
D: the positive and negative training data for the target concept
$D = \{<x, C(x)>\}$
H: the hypothesis of an approximated target concept by using D
The goal of the AER problem: Find $He(X) = Ce(X)$ for all e

There are companies that handle the AER problem such as Calypso Message Center (http://www.calypsomessagecenter.com) and White Pajama (http://www.whitepajama.com) in which the routing solution is integrated in the call center system. The systems use a simple rule based filtering technique where rules consist of keywords. When the predefined words are found in the incoming e-mail, then the e-mail is routed to the pre-classified agent or department. Managers can create/edit these rules so that changes in the business role can be incorporated. The problem of this approach, however, is that obtaining and maintaining the rules is not an easy task so most people do not like it. Moreover no one knows the exact business role of every organization member. Another approach is to use the existing machine learning technique to this domain. The main barrier of employing the machine learning algorithms is the high dimensional nature of e-mails, and the lack of ability handling large amount of data efficiently. Most of the applications depend on feature selection to reduce input space dimension. This step also involves complicated parameter setting. Most of all the difficulty to maintain high precision makes AER problem hard to be solved automatically.

This research aims to develop an autonomous software agent that performs the routing task that achieves the goal of the AER problem. The TWIMC system receives anonymous recipient e-mails, analyze it, and then routes the e-mail to the mostly qualified person (i.e., e-mail account) inside the organization. Existing machine learning and the automatic text categorization (ATC) techniques are applied for the task. Each e-mail account is viewed as a category (or class) of the ATC. Everyday incoming e-mail collections for each e-mail account provide an excellent source of training data. TWIMC is also the name of the anonymous recipient e-mail address in this work.

The automatic text categorization (ATC) is the problem assigning category labels to new documents based on the likelihood suggested by a training set of labeled documents [8]. This is very similar to AER problem in that both use text data and the training data are well provided. Many algorithms have been proposed or adopted to solve ATC problem. Examples include Naïve Bayes, Decision Tree, Neural-Network, RIPPER, and k-Nearest Neighbor algorithms [3], [5], [4], [6], [7]. The Support Vector Machine (SVM) is one of these algorithms that showed high performance [9], [10]. Especially the fact that SVM works well under large amount of features is very promising to the ATC and AER

Some researches also applied text categorization techniques to e-mail message categorization problem [11], [12]. However, most of them used machine-learning algorithms for personal information agent, which categorizes incoming e-mails by predefined set of "subjects" for a person. Each subject becomes a class in this problem, which is different from the AER problem. There are also many attempts to use text categorization to eliminate junk e-mails [13], [14]. In these researches text categorization techniques are used to filtering out unwanted e-mails from user's point of view. Also a new application domain of ATC with e-mails was proposed in [15] that learn authorship identification and categorization. SVM was used in this research and showed promising results. Another problem domain, customer e-mail routing, can be found in [16], that used decision tree to classify customer e-mails by predefined subjects such as address change, check order. This is similar with AER but domain area is more specific, bank system.

Joachims analyzed why SVM is suitable for ATC problem [9]. He concluded that problems with high dimensional input space, few irrelevant features, sparse vectors, and linearly separable spaces like ATC are proper domain for SVM and can be solved efficiently. The AER also has the similar characteristics with ATC problem except large amount of noisy data. In this paper SVM is mainly examined for the AER problem.

We evaluated the performance of various machine-learning algorithms such as SVM, Naïve Bayes, and K-NN. As a result SVM appeared to be a proper algorithm in the AER problem. The experimental result shows the high possibility that TWIMC could be deployed in the real world. Section 2 explains the design and implementation issues of TWIMC. In Section 3 the result of algorithm comparison is presented. Section 4 is donated to analyze the SVM result.

2 Design Considerations

Our approach that treats the e-mail account as a class and uses the incoming e-mail collections as training data has many inherent challenges. First, given an e-mail account, the diversity of e-mail data is very high, that is, they reflect many different subjects and user's interests. Second, there can be identical training data (i.e., e-mails) over different classes (i.e., e-mail accounts). Despite the high diversity, we hypothesized that the incoming e-mails for a person reflects his/her business roles that distinguishes him/her from others.

Two things were considered when designing TWIMC – efficiency and scalability. The system should be efficient in that learning the business role should be automatic and done within a reasonable time. E-mail routing process also should be done almost in real-time. The scalability means that as the number of classes in the system increases, the accuracy of the system should not be degraded much.

2.1 Preprocessing

Preprocessing is very important phase in the ATC problem and also in the TWIMC. Through this phase the learning data are prepared more adequately for the consequent processes. TWIMC performs additional preprocessing as follows. As mentioned the e-mail data in an e-mail account are very diverse. Lots of data do not reflect the user's business roles and job functions. Examples include news mails, commercial advertisement mails, and junk mails. One observation we found was that they were usually in HTML format. On the contrary most e-mails that reflects the business role were plain text. So our preprocessing phase drops all of the HTML e-mails. Another preprocessing we made is regarding the e-mails in reply form. People usually include the original message when reply. The included original message is eliminated since it is not considered as part of the incoming mail.

2.2 Indexing

In order for the learning algorithm to process e-mail data easier, e-mails are converted to other representation. The vector model with binary weighting was used [1]. In this scheme e-mail is changed to binary vector with each weight either 1 or 0. The 1 represents the word to appear in the e-mail while 0 is not. Other representation schemes such as TF and TFIDF were considered also but not used. The frequency of word is less significant. If one word appears in an e-mail, it is enough to differentiate the user's business role from others.

Normally ATC system performs the feature reduction techniques like stemming and stop-word elimination to reduce the number of features. TWIMC also adopted these techniques during indexing phase.

2.3 Feature Selection

The feature selection mechanism in ATC was also considered in TWIMC. When system has run quite a long time, the number of word that appeared at least once during parsing phase becomes up to from tens of thousands to hundreds of thousands. Most algorithms are not able to perform well in such huge feature spaces. So less significant features should be eliminated. Yiming Yang and Jan O. Pedersen compared various feature selection schemes in ATC [20]. They used Document Frequency Thresholding (DF), Information Gain (IG), Mutual Information (MI), X^2 statistic (CHI), and Term Strength (TS). The paper concluded that DF, IG and CHI showed excellent performance. Also David D. Lewis researched on this topic [19]. In ATC problem less frequently appeared words are less significant for differentiation. But care should be taken to apply feature selection scheme to AER problem directly. There are few irrelevant features in e-mails [9]. If we choose too many features, algorithms could not work efficiently. If we choose too few features, important information could be missed. So we used very simple feature selection scheme, which implementation cost is very low. The words that appeared at least two times in a parsing phase were selected as features, that is, the words appearing only once are not regarded as good features. Applying this method with DF feature selection decreased the dictionary size dramatically from 20,000 to 5,000.

2.4 Data Sampling

When building the model by using the incoming e-mail collections, TWIMC needs to select positive and negative data. The system use all the incoming e-mail collections that passed the preprocessing as positive data, but the system can control the size of positive data by date. This simple window-based method can handle the "Concept Drifting" problem [18]. The method to select negative data is as follows. Ideally all the e-mails in the collection except the positive data should be used as negative data. But this has a problem since there could be tens of thousand e-mails in the collection. It could decrease the performance of system seriously. So we used proportion-enforced sampling [17].

Pn: Expected number of negative training data
Tn: Total number of candidate negative data except designated class
Tc: Number of data in a class

Number of sampled negative data in a class: $\dfrac{(Tc)}{(Tn)} * Pn$

$$\sum_{c=0}^{c=n} \frac{Tc}{Tn} \times Pn = Pn \tag{1}$$

The number of negative data Pn is controlled by applications. We used Pn as 500 in the experiment.

2.5 The System

The system diagram of TWIMC is shown in the figure 1. The TWIMC is basically resides in the mail server. E-mails to be used as training data are preprocessed and indexed at first. During this phase e-mails are converted to binary vector representation. At the same time dictionary is built and stored after applying feature selection scheme. From time to time, the model building procedure is aroused by system-timer that was set by the user previously. This procedure converts each e-mail vector that is stored in database to a file format that is suitable to be used by the machine learning algorithms. Then the models reflecting each class are built. They are stored also in the database.

If a new TWIMC e-mail (i.e., anonymous recipient e-mail) arrives, the system performs the same processing - indexing and file converting. Then vector representation of the e-mail goes to the classifier. The classifier figures out a qualified receiver of this e-mail by using the previously built models.

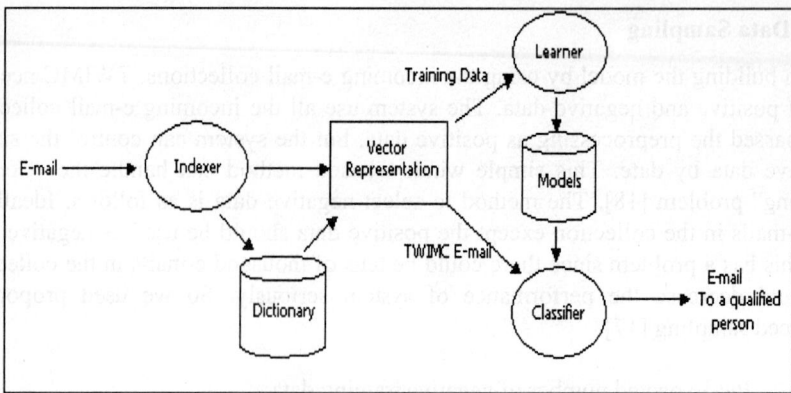

Fig. 1. System Architecture of the TWIMC

3 Algorithm Comparisons

Three machine learning algorithms are chosen to compare the performance of the system - SVM, Naïve Bayes, and k-NN. Ten-fold cross validation [2] is used for the training and classifying. Then five cross validation results are averaged for each algorithm. Existing machine learning libraries are used for the experiment, which include SVMlight and Weka [22], [23].

3.1 Performance Measures

There have been proposed many evaluation methods for the ATC problem. They were mainly from the information retrieval community such as recall, precision, F1 measure, R-precision [1]. There is a research that proposed the standard procedures for ATC [21]. Recall is regarded as the most significant single value. David proposed also macro- and micro-averaging. The macro-averaging is the value that first computes individual recall of classes and averages theses results. The micro-averaging, on the other hand, considers all decisions as a single group and computes recall. He concluded 'if categorization were used to route documents to divisions of a company, with each division viewed as being equally important, then macro-averaging would be more informative'. In TWIMC every e-mail account is not equally important. Some users have clear business roles and the others do not. We selected micro-averaging recall as the primary criterion for performance measure.

3.2 Result

The evaluation was done under the condition of nine classes and 326 e-mails (after preprocessing). The e-mail corpus that was used in the experiment is shown in Table 1.

Table 1. E-mail corpus

E-mail account	Number of e-mails
Address 1	62
Address 2	27
Address 3	36
Address 4	45
Address 5	30
Address 6	47
Address 7	41
Address 8	16
Address 9	22
Total	326

Table 2 shows the result of the experiment. This result shows that SVM outperforms the other two algorithms.

Table 2. Performance of the algorithms

Algorithms	Macro-recall	Micro-recall
SVM	96.39	96.31
Naïve Bayes	48.73	58.58
k-NN	83.07	81.28

4 Further Experiments

Since SVM showed best performance, we design further experiment to investigate how scalable SVM is as the number of classes increases. We divided the e-mail account set and corresponding data into the two. The first set belongs to the ones who have clear business roles. E-mail data in the set are relatively consistent and are not diverse compared with the other set. This set (i.e., both classes and data) is called 'strong data set'. The remaining classes and corresponding data belong to the second set that is named 'weak data set'. The e-mail data in the second set is very diverse, even by manual way the classification is difficult. Note that in real situation the anonymous recipient e-mails will be concentrated on the e-mail accounts in the first set. Hence the goal of the system like TWIMC is to categorize the strong set accurately. The difficulty is that there is large weak data set, which can be regarded as noise to TWIMC, than the strong set in real world. Table 3 is the classes and the number of data in each set.

Table 3. Strong and weak data set

Data Set	Classes	E-mails	Avg # of e-mails per class
Strong Set (Training Data)	10	1806	189.6
Strong Set (Testing Data)		194	19.4
Weak Set (Training Only)	35	1543	44.08
Total	45	3528	78.4

Table 4 was obtained by applying SVM with five ten-fold cross validation to the strong and weak data sets.

Table 4. SVM results with the strong and weak data set

Learning/Cross-validation	Calculation	Macro Recall	Micro Recall
Strong + Weak Set	Strong + Weak	41.45	59.89
	Strong	72.39	77.98
	Weak	31.61	32.61
Strong Set	Strong	86.30	92.14
Weak Set	Weak	42.64	45.36

'Learning/Cross-validation' column represents how learning and cross-validation were performed. 'Strong + Weak Set' means that strong and weak data set were used together to build models and cross-validation was performed within these models. But in 'Strong' or 'Weak' respectively, learning and cross-validation are performed on each data set alone. 'Calculation' column represents how recall is calculated. In 'Strong + Weak', cross-validation results of 'Strong' and 'Weak' data set are all included in recall calculation. But in 'Strong' or 'Weak', each 'Strong' or 'Weak' data set is included in the recall calculation exclusively. The most plausible scenario is in real situation is 'Strong + Weak Set' in learning (since we do not know which ones are strong or weak) and 'Strong' in calculation (since the anonymous recipient e-mails will be usually go to the strong set).

Note that the results of the (Strong + Weak Set, Strong) combination are lower than those of (Strong Set, Strong). It can be interpreted as the accuracy of strong data set is lowered when evaluation is done with weak data set. This leads to conclusion that with larger amount of weak data the accuracy of strong data could be degraded – the scalability problem. Note also that the SVM result in the table 2 (about 96%) is better than the one in the table 4. The reason is due the size of the data. In small data space, the algorithms tend to perform well.

5 Conclusions

We proposed a new method that automatically routes e-mails to the best-qualified recipient by using the machine learning. This has potential application areas as the anonymous recipient e-mails and the e-mail call center. In this method everyday (incoming) e-mail collections are used as training data to learn the business role of each e-mail account. In fact the e-mails are very valuable resource in which the business roles and job functionality can be extracted. From the machine learning's point of view each e-mail address is one class. This differs from previous works regarding e-mail mining that applied text categorization techniques to filter out spam mails or automatically organize personal e-mail directories. The new method also automatically handles the concept drifting problem by using the recent training data periodically. Several machine learning algorithms were examined to discriminate adequate algorithms in this new domain named AER problem. By comparing some algorithms we concluded that SVM is proper one, and we analyzed the performance of the algorithm in virtually real situation. Although there found some scalability problem when we directly applied SVM, the prediction result showed that the TWIMC could be used in the real situation. In the ATC, the scalability problem can be avoided by using conceptual hierarchy [24]. Although building the conceptual hierarchy is not an easy task in AER, we need similar approach to tackle this problem. It can be an issue of extended research.

References

1. Yates, B., Neto, R.: Modern Information Retrieval. Addison-Wesley (1999)
2. Han, J., Kamber, M.: Data mining: Concepts and techniques. Morgan Kaufmann Publishers (2001)
3. William, W. C.: Learning Rules that Classify E-Mail. Advances in Inductive Logic Programming (1996)
4. Moulinier, I., Jean-Gabriel, G.: Applying an Existing Machine Learning Algorithm to Text Categorization. Connectionist, Statistical, and Symbolic Approaches to Learning for Natural Language Processing, Springer-Verlag (1996)
5. Yang, Y.: An Evaluation of Statistical Approaches to Text Categorization. Journal of Information Retrieval, Vol. 1 (1999) 67-88
6. Han, E., Karypis, G., Vipin, K.: Text Categorization Using Weight Adjusted k-Nearest Neighbor Classification. Computer Science Technical Report TR99-019, University of Minnesota (1999)
7. Schutze, H., Hull, D.A., Pedersen, J.O.: A Comparison of Classifiers and Document Representations for the Routing Problem. Proc. of the 18th annual international ACM SIGIR conference on Research and Development in information retrieval (1995)
8. Yang, Y., Liu, X.: A Re-examination of Text Categorization Methods. Proc. of ACM SIGIR Conference on Research and Development in Information Retrieval (1999)

9. Joachims, T.: Text Categorization with Support Vector Machines - Learning with Many Relevant Features. Proc. of the European Conference on Machine Learning, Springer (1998)
10. Cristianini, N., Shawe-Taylor, J.: An introduction to Support Vector Machines and Other Kernel-Based Learning Methods, Cambridge University Press (2000)
11. Kim, S., Hall, W.: A Hybrid User Model in Text Categorization
12. Rennie, J.D.M.: ifile: An Application of Machine Learning to E-mail Filtering. Proc. of KDD-2000 Workshop on Text Mining (2000)
13. Sahami, M., Dumais, S., Heckerman. D., Horvitz, E.: A Bayesian Approach to Filtering Junk E-Mail. Proc. of AAAI-98 Workshop on Learning for Text Categorization (1998)
14. Katirai, H.: Filtering Junk E-Mail: A Performance Comparison between Genetic Programming and Naïve Bayes, Carnegie Mellon University (1999)
15. de Vel, O.: Mining E-mail Authorship. Proc. of KDD-2000 Workshop on Text Mining (2000)
16. Weiss, S.M., Apte, C., Damerau, F.J., Johnson, D.E., Oles, F.J., Goetz, T., Hampp, T.: Maximizing Text-Mining Performance, IEEE Intelligent Systems (1999)
17. Yang, Y.: Sampling Strategies and Learning Efficiency in Text Categorization. Proc. of the AAAI Spring Symposium on Machine Learning in Information Access (1996)
18. Klinkerberg, R., Joachims, T.: Detecting Concept Drift with Support Vector Machines. Proc. of 17th International Conf. on Machine Learning (2000)
19. Lewis, D. D.: Feature Selection and Feature Extraction for Text Categorization. Proc. of the Speech and Natural Langauge Workshop (1992)
20. Yang, Y., Pedersen, J.O.: A Comparative Study on Feature Selection in Text Categorization. Proc. 14th International Conference on Machine Learning (1997)
21. Lewis, D.D.: Evaluating Text Categorization. Proc. of the Speech and Natural Language Workshop (1991)
22. Joachims, T.: Making Large-Scale SVM Learning Practical, LS VIII-Report (1998)
23. Garner, S. R.: WEKA - The Waikato Environment for Knowledge Analysis. Proc. of the New Zealand Computer Science Research Students Conference (1995) 57-64
24. Grobelnik, M., Mladenic, D.: Efficient Text Categorization. Proc. of the European Conference on Machine Learning Workshop on Text Mining (1998)

Mental States of Autonomous Agents in Competitive and Cooperative Settings

Walid S. Saba

Knowledge & Language Engineering Group, School of Computer Science,
University of Windsor, Windsor, ON N9B-3P4 Canada
saba@cs.uwindsor.ca
http://santana.cs.uwindsor.ca/

Abstract. A mental state model for autonomous agent negotiation in is described. In this model, agent negotiation is assumed to be a function of the agents' mental state (attitude) and their prior experiences. The mental state model we describe here subsumes both competitive and cooperative agent negotiations. The model is first instantiated by buying and selling agents (competitively) negotiating in a virtual marketplace. Subsequently, it is shown that agent negotiations tend to be more cooperative than competitive as agents tend to agree (more so than disagree) on attributes of their mental state.

1 Introduction

Autonomous agents can potentially be engaged in a number of E-Commerce activities ranging from competitive negotiations, such as buying and selling agents negotiating the purchase/sale of a certain product, to cooperative negotiations, as in the coordination of supply-chain activities. In this paper we describe a mental state model which we argue subsumes both competitive and cooperative environments. In particular, assuming that a mental state (reflecting an agent's attitude) is given as an n-tuple $\langle a_1, ..., a_n \rangle$ where a_i are attributes from a partially ordered domain, a cooperative model is one where all the participants agree on at least one of these attributes. In the competitive setting no such constraint can be assumed. That is, we argue that the competitive and cooperative models are two extreme points in a continuum defined by the degree to which negotiating agents share their attitude (or mental state).

In the next section an abstract mental state model for agent negotiation is introduced. The model is then instantiated by buying and selling agents that competitively negotiate in a virtual market place. It is then shown that as agents agree more so than disagree on one or more attributes of their mental states, the negotiation tends to be more cooperative than competitive. While the focus of this paper is on the mental state model, our long-term goal is to build an environment for distributed, artificial and linguistically competent intelligent agents, termed DALIA (Saba and Sathi, 2001). Notwithstanding the lack of consensus on what an 'intelligent agent' is, there are clearly several challenges in building an environment where autonomous agents are situated in a highly dynamic and uncertain environment.

In addition to a number of characteristics that intelligent agents are expected to posses (see Bradshaw (1997) and Jennings and Wooldridge (1998)), in our view intelligent agents must also have a certain level of linguistic competency and must be able to perform commonsense reasoning in a highly dynamic and uncertain environment. There are clearly several challenges in formalizing this type of commonsense reasoning which will not be discussed in this paper. Instead, our focus here is on developing a mental state model, which models the attitude of agents that learn from experience in competitive and cooperative agent negotiations[1].

2 A Mental State Model for Agent Negotiation

A mental state model for agent negotiation is taken to be a 5-tuple $M = \langle K, A, I, F, G \rangle$ where:

- $K = \{k_1, ..., k_m\}$ is a set of agents
- $A = \{a_1, ..., a_n\}$ is a set of relevant attributes
- $I = \{I_1, ..., I_m\}$ is a set of agent attitudes
- $F = \{f_1, ..., f_m\}$ is a set of attribute evaluation functions
- $G = \{g_1, ..., g_m\}$ is a set of proposal evaluation functions

These sets are explained as follows. We assume that in any agent negotiation there is a set of attributes $A = \{a_1, ..., a_n\}$ that are considered 'relevant'. For example, price might be a relevant attribute for agents competitively negotiating the purchase/sale of a certain product; while publicSafety might be a relevant attribute for agents cooperatively deliberating on the issue of banning smoking in restaurants. It is also assumed that attributes take on values from a partially ordered domain - that is, the values $V(a_i)$ and $V(a_j)$ of any attributes a_i and a_j must be comparable:

$$(\forall a_i, a_j \in A)((V(a_i) \leq V(a_j)) \vee (V(a_i) \geq V(a_j))) \qquad (1)$$

Associated with each agent $k_i \in K$ is a mental state (representing an agent's attitude), $I_i = \langle imp_i(a_1), ..., imp_i(a_n) \rangle$, representing the importance that agent i assigns to each attribute; where $0 \leq imp_i(a_j) \leq 1$. For example, for some agent u, $imp_u(price) = 0.2$ and $imp_u(time) = 0.9$ might represent the fact that for u the price is not too important, but time is very crucial. We also assume, somewhat similar to (Chun and Wong 2001), that the success of a negotiation depends on agreement between agents on a subset of the set of relevant attributes, which we term the set of distinguished attributes, $DA \subseteq A$. For example, agreement on price might be what determines the success or failure of a (competitive) negotiation between buying and selling agents in a virtual

---

[1] See (Saba and Corriveau, 2001) and (Saba, 2001) for an overview of our concurrent work on natural language understanding and commonsense reasoning.

marketplace. On the other hand, while there might be several attributes that are 'relevant' in a (cooperative) deliberation on banning smoking in public restaurants, agreement on publicSafety might be the deciding factor.

For each distinguished attribute $a_i \in DA$ we assume that agents have access to a publicly available attribute value, given as a range $[pmin(a_i), pmax(a_i)]$. Using these 'publicly' available attribute values and their attitudes, agents employ a set of functions $F = \{f_1,...,f_m\}$ to compute their 'private' values, where $f_i : I \times V(A) \to V(A)$. For example, while the range $[1000, 5000]$ might be the publicly available value for the price of a PC, depending on their mental state (attitude), specific agents might assume a different range (which must be included in the publicly available range). A private value $V(a_i) = [pmin_x(a_i), pmax_x(a_i)]$ that an agent x computes is used to make proposals in the course of a negotiation. For example, an agent x might send a message $propose_x^n(y, V(a))$ to agent y, corresponding to nth proposal from x to y regarding an attribute $a \in DA$. The nth response that y sends to x concerning attribute a is a pair $r_y^{n,a} = \langle v, V_y(a) \rangle$ where $V_y(a)$ is the nth counter proposal sent from y to x, and $v \in \{accept, high, low\}$ corresponds to whether the proposal was accepted, was high or low. More formally,

$$(r_y^{n,a} = \langle v, V_y(a) \rangle) \leftarrow propose_x^n(y, V_x(a))$$

where

$$v = \begin{cases} accept & V_x(a) = V_y(a) \\ low & V_x(a) < V_y(a) \\ high & V_x(a) > V_y(a) \end{cases} \quad (2)$$

If a proposal $propose_x^n(y, V(a))$ is not accepted, an agent makes another proposal, $propose_x^{n+1}(y, g_x(r_y^{n,a}, V(a)))$, where a new attribute value that is within the range of values acceptable to x is computed as a function $g_x \in G$ of the agent's acceptable values and the previous counter proposal received from y. In the course of a negotiation a number of proposals and counter proposals are made:

$$(r_y^{1,a} = \langle v_1, V_y(a) \rangle) \leftarrow propose_x^1(y, V_x(a))$$
$$(r_y^{2,a} = \langle v_1, V_y(a) \rangle) \leftarrow propose_x^2(y, g_x(r_y^{1,a}))$$
$$...$$
$$(r_y^{n,a} = \langle v_1, V_y(a) \rangle) \leftarrow propose_x^n(y, g_x(r_y^{n-1,a}))$$

In the sequence of proposals and counter proposals shown above, x first makes a proposal to y, regarding the (distinguished) attribute a. If y does not accept the offer, x then makes a second proposal (if possible, see blow), taking y's initial response (and its mental state) into consideration, and so on. The negotiation can be in any state $s \in \{done^+, done^-, done^0\}$ depending on whether the negotiation ended with success, failure or is still in progress, respectively. The possible states a negotiation between agents x and y can be in are defined as follows, where $[\text{pmin}_x(a), \text{pmax}_x(a)]$ is the range agent x assumes for attribute a:

$$\text{done}^+ \equiv_{df} (\forall a \in DA)(\langle \text{accept}, V_y(a) \rangle = \text{propose}_x(y, V(a)))$$

$$\text{done}^- \equiv_{df} (\exists a \in DA)$$
$$([[\langle \text{low}, V_y(a) \rangle = \text{propose}_x(y, V_x(a)) \wedge (V_y(a) > \text{pmax}_x(a))] \quad (3)$$
$$\vee [\langle \text{high}, V_y(a) \rangle = \text{propose}_x(y, V_x(a)) \wedge (V_y(a) < \text{pmin}_x(a))]])$$

$$\text{done}^0 \equiv_{df} \neg \text{done}^+ \wedge \neg \text{done}^-$$

Informally, a negotiation between x and y is successful if, *every* proposal from x on the set of distinguished attributes is accepted by y. The negotiation fails if, for *some* attribute in the set of distinguished attributes, y requires a value higher than the maximum x can propose, or if y requires a value lower than the minimum x can propose. A negotiation is still in progress if neither of these states is reached. A high-level description of this process is shown in figure 1 below.

3 Negotiating with an Attitude

Our negotiation model has several common features with a number of existing models (e.g., Chavez and Maes 1996; Esmahi and Dini 1999; Kumar and Feldman 1998; Faratin et al. 1999). However, in these models the notion of a mental state of an agent, which as will be argued below plays an important role in the negotiation process, has received little attention. Specifically, we argue that an agent's mental state defines an overall context of negotiation, which consequently defines an agent's strategy and ultimately determines how a negotiation proceeds.

3.1 Basic Components of the Model

The virtual marketplace we have developed is an environment that at any point in time has a list of buyers and sellers. Buyers and sellers in the environment are also assumed to have access to two knowledge sources: (i) an ontology for domain-specific product information; and (ii) a set of general commonsense rules.

Agents are also assumed to learn from experience using case-based reasoning (CBR) techniques: negotiations along with product information, market conditions and the outcome of the negotiation are stored as experiences for use in future negotiations. A high-level description of the virtual marketplace is shown in figure 2 below[2].

The negotiation model itself is a 5-tuple $\langle K, A, I, F, G \rangle$ where

- $K = \{b, s\}$ is a set of buying and selling agents
- $A = \{\text{time}, \text{price}, \text{commitment}\}$ is a set of 'relevant' attributes
- $I = \{\langle t_b, p_b, c_b \rangle, \langle t_s, p_s, c_s \rangle\}$ are the attitudes of b and s
- $F = \{f_b(\langle t_b, p_b, c_b \rangle, \text{price}), f_s(\langle t_s, p_s, c_s \rangle, \text{price})\}$ attribute evaluations
- $G = \{g_b(I_b, \text{price}), f_s(g_s, \text{price})\}$ proposal evaluation functions

---

[2] The ontology and the learning component are not discussed here, but see (Saba, 2002).

1. **forEach** ($a_i \in DA$) { // for each distinguished attribute
2. $[pmin(a_i), pmax(a_i)] \leftarrow KB$ // retrieve public values from a product KB
3. **forEach** ($k_j \in K$) { // for each agent
4. // using one's attitude and public values, compute 'private' attribute values
5. $([pmin_j(a_i), pmax_j(a_i)] = V_j(a_i)) \leftarrow f_j(I, A)$
6. }
7. }
8. **while** (true) { // start the negotiation
9. **forEach** ($k_x \in K \wedge k_y \in K$){// for each two agents x and y ($x \neq y$)
10. **forEach** ($a_i \in DA$) { // for each distinguished attribute
11. // x makes proposal number s to y re distinguished attributes a_i.
12. $(r_s^{y,a_i} = \langle v, V_y(a_i) \rangle) \leftarrow propose_x^s(y, g_x(r_{s-1}, V_x(a_i)))$
13. // evaluate current state using (3)
14. $done^+ \equiv_{df} (\forall a \in DA)(\langle accept, V_y(a) \rangle = propose_x(y, V(a)))$
15. $done^- \equiv_{df} (\exists a \in DA)(((\langle low, V_y(a) \rangle = propose_x(y, V_x(a))$
16. $\wedge (V_y(a) > pmax_x(a))) \vee ((\langle high, V_y(a) \rangle$
17. $= propose_x(y, V_x(a)) \wedge (V_y(a) < pmin_x(a))))$
18. }
19. **if** ($done^-$) exit(failure)
20. **if** ($done^+$) exit(success)
21. }
22. }

Fig. 1. A high-level description of the negotiation process

That is, in this simple model we are considering two types of agents, a buyer b and a seller s. We further assume that the price (of a certain product), the time a negotiation takes, and the commitment level (to the purchase/sale of a certain product) are attributes that are relevant to both buyers and sellers (of course, one can easily imagine many other relevant attributes, such as the importance an agent might place on quality, brandName, etc.) Furthermore, it is assumed that price is the only 'distinguished attribute', i.e., agreement (or disagreement) on price is what determines the success (or failure) of a negotiation.

In this model an attitude $I_x \in I$ (representing a 'hidden' mental state) of an agent x is a triple $\langle t_x, p_x, c_x \rangle$ representing the importance of time, the importance of price and the commitment level of x, respectively, and where t, p and c take on values from the open interval [0,1]. For example, for an agent with attitude $\langle 1.0, 0.3, 0.8 \rangle$ time is a priority, and the commitment level is rather high although it is somewhat indifferent to the price (an ideal buyer attitude from the perspective of a seller).

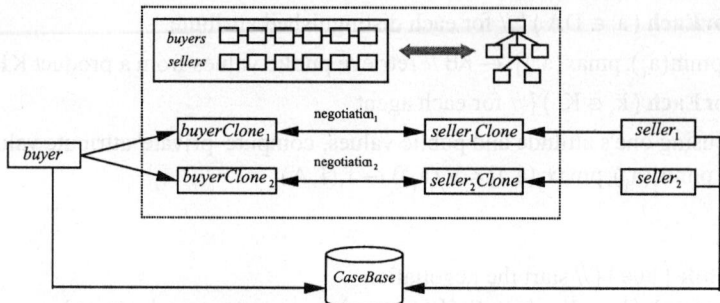

Fig. 2. Basic components of the virtual marketplace

3.2 Using the Attitude to Compute a Price Range

Using a publicly available price range, obtained form a product-specific ontology, and their own attitude, buyers and sellers evaluate their own 'private' price range. There are several functions that are plausible here. However note that the lower the minimum price range of a buyer the more time a negotiation will take. Thus, the buyer's minimum price range must be higher than the public's minimum if the importance of time is high for that buyer. Moreover, when the commitment of a buyer is high the agent should be as close to the public's maximum as possible, since a high commitment indicates an urgent need for the product. However, the importance of the price should balance the degree of commitment (in a sense, a high commitment is a form of desperation, and the importance of price provides a sobering effect). A seller reasons in exactly the opposite direction. The following two functions are one suggestion on how such reasoning might plausibly be captured:

$$f_b(\langle t_b, p_b, c_b \rangle, [mp, xp]) = [mp + (t_b)(xp - mp)/10, xp - (xp)(p_b)(1 - c_b)]$$
$$f_s(\langle t_s, p_s, c_s \rangle, [mp, xp]) = [mp + (mp)(p_b)(1 - c_b), xp - (t_b)(xp - mp)/10] \quad (4)$$

where $[mp, xp] = [p\min(price), p\max(price)]$. Based on their attitude, buyers and sellers use these functions to determine the price range they are willing to negotiate within. Note, for example, that the higher the commitment (i.e., the more disparate a buyer is), the higher the maximum they are willing to pay, and the inverse is true for sellers. Other components of the mental state have a similar effect on the strategy of buyers and sellers. Looking at the algorithm described in figure 1 and the above functions, it is clear that the attitude of agents is a crucial part of the overall negotiation process. In particular, since the attitude determines an agent's price range, it is subsequently a key in determining the result of a negotiation, as well as the time a negotiation might take.

3.3 Using the Attitude to Make (Counter) Proposals

While the attribute evaluation functions described above determine the initial price range that agents are committed to, the proposal evaluation functions that agents use to adjust their proposals based on their attitude and previous counter-offers are also crucial. The following function seem to be a plausible candidate for buyers:

$$g_b(\langle t_b, p_b, c_b \rangle, \langle v, V_s(a) \rangle)$$

$$= \begin{cases} \text{sucess}(V_s(a)) & v = \text{accept} \\ f_b(\langle t_b, p_b, c_b \rangle, [mp+\varepsilon, xp]) & v = \text{low} \wedge V_s(a) \leq xp \\ f_b(\langle t_b, p_b, c_b \rangle, [mp, xp-\varepsilon]) & v = \text{high} \wedge V_s(a) \geq mp \\ \text{failure}() & \text{otherwise} \end{cases} \quad (5)$$

Intuitively, if the response to a previous offer was not accepted, using its attitude a buyer re-computes a new price range by adjusting its minimum and maximum, if at all possible, by some increment/decrement ε. Otherwise, either an offer was accepted or the negotiation should terminate with failure. Note that the higher the increment/decrement value ε, the faster the negotiation (regardless of the result). Thus, for example, one would expect agents with an attitude $\langle t, p, c \rangle$ where $t \approx 1.0$ (i.e., where time is a high priority), to have a high value for ε. This, however, is too simplistic an analysis since a buyer might place a high priority on time as well as price, in which case the agent should not increase its minimum hastily. Moreover, since the attitude of the seller is hidden, the buyer might increase its minimum considerably, well beyond what the seller was willing to accept. Clearly, there is potentially an infinite number of possibilities that would have to be considered if an optimal price range and optimal ε are to be attained. If agents use prior (positive and negative) experiences, however, they could overtime optimize their attitude as well as the increment/decrement value ε, both of which are key in determining the success or failure of a negotiation. In our model agents learn by 'learning how to improve their attitude'. Using a case-based reasoning strategy, agents store prior experiences (cases), which include product information, their attitude at the time, as well as market conditions and the outcome of the negotiation. Agents subsequently use prior experiences (cases) to either sharpen or loosen their attitudes, based on the outcome of previous experiences. We cannot discuss this component in further detail here for lack of space. Instead we refer the reader to Saba (2002).

4 Attitudes in a Cooperative Setting

Unlike the competitive negotiation model described above, there are various scenarios where agents that have a slightly different set of goals might nevertheless negotiate in a cooperative manner. Unlike the competitive situation, agents in a cooperative setting (such as those engaged in a deliberation) must share at least one attribute value from the attributes of their mental state.

The (competitive) negotiation between buying and selling agents that was considered above could itself be more cooperative than competitive if the buyers and sellers agreed on some essential attribute of their mental state. Suppose that for some two agents b and s we have the following:

$$\begin{aligned} I_b &= \langle t_b, p_b, c \rangle \\ I_s &= \langle t_b, p_b, c \rangle \end{aligned} \quad (6)$$

That is, b and s agree on the commitment level. For example, both are highly committed (rather desperate) to the purchase/sale of a certain product, or both are rather indifferent. We claim that agreement on the commitment level attribute of the buyer's and seller's attitudes makes the negotiation process more cooperative than competitive. Recall that the buyer and seller compute their private ranges as a function of the public price range and their attitude as shown in (6), resulting in the following:

$$f_b(\langle t_b, p_b, c \rangle, [mp, xp]) = [mp + (t_b)(xp - mp)/10, xp - (xp)(p_b)(1-c)]$$
$$f_s(\langle t_s, p_s, c \rangle, [mp, xp]) = [mp + (mp)(p_s)(1-c), xp - (t_s)(xp - mp)/10] \quad (7)$$

A buyer's initial bid is its minimum while the seller's initial bid is its maximum. Thus, b's first proposal is $propose_b^1(s, mp + (t_b)(xp - mp)/10)$.

Note that in one step, what the buyer b proposed, while is the minimum possible, is almost equal to the maximum the seller s would like to get. What follows next, according to the process discussed above, is an evaluation of b's first proposal by s, where it can be assumed that the buyer always seeks a lower price and the seller always seeks a higher price. This is done as follows:

$$g_s(\langle t_s, p_s, c \rangle, \langle high, V \rangle)$$
$$= \begin{cases} sucess(V) & v = accept \\ f_s(\langle t_s, p_s, c \rangle, [mp + \varepsilon, xp]) & v = low \wedge V \leq xp \\ f_s(\langle t_s, p_s, c \rangle, [mp, xp - \varepsilon]) & v = high \wedge V \geq mp \\ failure() & otherwise \end{cases} \quad (8)$$

where

$$V = mp + (t_b)(xp - mp)/10$$

Note that since $(mp + (t_b)(xp - mp)/10) \geq mp$ the result of this step would be for the seller to decrease its maximum (by some ε) - i.e, the new maximum of the seller is $xp - \varepsilon - (t_b)(xp - mp)/10$, which would be the first counter proposal made to the buyer b. In return, b evaluates this proposal as follows:

$$g_b(\langle t_b, p_b, c \rangle, \langle low, V \rangle)$$
$$= \begin{cases} sucess(V) & v = accept \\ f_b(\langle t_b, p_b, c \rangle, [mp + \varepsilon, xp]) & v = low \wedge V \leq xp \\ f_b(\langle t_b, p_b, c \rangle, [mp, xp - \varepsilon]) & v = high \wedge V \geq mp \\ failure() & otherwise \end{cases} \quad (7)$$

where

$$V = xp - \varepsilon - (t_b)(xp - mp)/10$$

Since $(xp - \varepsilon - (t_b)(xp - mp)/10) \leq xp$, this will result in the buyer increasing its minimum yet again to a value very near the seller's maximum. This process will converge quite quickly (depending on ε), resulting in effect in a process that resembles cooperation rather than competition. The above argument could be made even more acute if the two agents are made to agree on both the commitment level and the importance of time - especially when some extreme values for t and c are chosen. For example, assume the following attitudes for agents b and s:

$$I_b = \langle 0, p_b, 1.0 \rangle$$
$$I_s = \langle 0, p_b, 1.0 \rangle \qquad (8)$$

That is, suppose both b and s agree on the commitment level (they are both rather desperate), and they both agree on the importance of time (for both time is not important at all). Now applying the functions given in (6) would result in the computation shown in (9). Note that in a single step both b and s compute the same price range, and therefore, by (7), they will reach an agreement in one step. Clearly this sounds a lot more like cooperation than competition.

$$\begin{aligned}
&f_b(\langle 0, p_b, 1 \rangle, [mp, xp]) \\
&= [mp + (0)(xp - mp)/10, xp - (xp)(p_b)(1-1)] \\
&= [mp, xp] \\
&f_s(\langle 0, p_s, 1 \rangle, [mp, xp]) \\
&= [mp + (mp)(p_b)(1-1), xp - (0)(xp - mp)/10] \\
&= [mp, xp]
\end{aligned} \qquad (9)$$

The point here is that the mental state model described above defines the boundaries of such negotiation tasks and it subsumes both competitive as well as cooperative dialogue types. That is, it seems that the competitive and cooperative models are two extreme points in a continuum defined by the degree to which negotiating agents share their attitude (mental state).

5 Concluding Remarks

A mental state model that subsumes both competitive as well as cooperative agent negotiations was presented. We first described a prototype of a virtual marketplace environment where buying and selling agents that learn from experience autonomously (and competitively) negotiate on behalf of their clients. It was consequently argued that a cooperative model is simply a model where agents agree more so than disagree on the relevant attributes of the mental state.

Acknowledgements

This research was partially supported by the National Science and Engineering Research Council of Canada (NSERC). Thanks are also due to Pratap Sathi who conducted some experiments related to this work.

References

1. Bradshaw, J. (Ed.). 1997. **Software Agents**, AAAI Press/MIT Press.
2. Chavez, A. and Maes, P. (1996), Kasbah: An Agent Marketplace for Buying and Selling Goods, In Proceedings of the 1st Int. Conf. on the Practical Application of Intelligent Agents and Multi-Agent Technology.
3. Chun, H. W., Wong, R. Y. M. (2001), Introducing User Preference Modeling for Meeting Scheduling, In N. Zhong, J. Liu, S. Ohsuga, J. Bradshaw. (Eds.), **Intelligent Agent Technology - Research and Development**, World Scientific Publishers, pp. 474-478.
4. Esmahi, L. and Dini, P. (1999), Toward an Open Virtual Market Place for Mobile Agents, In Proceedings of IEEE 8th Int. Workshops on Enabling Technologies: Infrastructure for Collaborative Enterprises.
5. Faratin, P., Jennings, N. R., and Sierra, C. (1999), Automated Negotiation for Provisioning Virtual Private Networks using FIPA-Compliant Agents, In Proceedings of AAAI Workshop on Negotiation.
6. Jennings, N. R. and Wooldridge, M. (Eds.) 1998. **Agent Technology: Foundations, Applications, and Markets**, Springer-Verlag.
7. Kumar, M. and Feldman S. I. (1998), Business Negotiation on the Internet, IBM Institute for Advanced Commerce, IBM T. J. Watson Research Center Report, Yorktown Heights, NY.
8. Saba, W. S. (2002), DALIA: Distributed and Linguistically Competent Intelligent Agents, In C. Faucher, L. Jain, and N. Ichalkaranje (Eds.), **Innovations in Knowledge Engineering**, Physica-Verlag, pp. ???-???, Springer-Verlag. (to appear summer 2002)
9. Saba, W. S. (2001), Language and Commonsense Knowledge, In Brooks, M., Corbett, D., Stumptner, M., (Eds.), AI 2001: Advances in Artificial Intelligence, Lecture Notes in Artificial Intelligence, Vol. 2256, pp. 426-437, Springer.
10. Saba, W. S. and Corriveau, J.-P. (2001), Plausible Reasoning and the Resolution of Quantifier Scope Ambiguities, Studia Logica - International Journal of Symbolic Logic, 67(2):271-289, Kluwer.
11. Saba, W. S. And Sathi, P. (2001), Agent Negotiation in a Virtual Marketplace, In N. Zhong, J. Liu, S. Ohsuga, J. Bradshaw. (Eds.), **Intelligent Agent Technology - Research and Development**, World Scientific Publishers, pp. 444-455.

An Expert System Application for Improving Results in a Handwritten Form Recognition System

Silvana Rossetto, Flávio M. Varejão, and Thomas W. Rauber

Departamento de Informática, Centro Tecnológico,
Universidade Federal do Espírito Santo,
Vitória, Espírito Santo, Brazil
{silvanar,fvarejao,thomas}@inf.ufes.br

Abstract. This paper presents a categorization of the knowledge used during the stage of post-processing of the recognized results in a system that automatically reads handwritten information from forms. The objective is to handle the uncertainty present in the semantics of each field of the form. We use grammatical rules particular of the Portuguese language and specialized information about the characteristics of the recognition system. A knowledge-based system uses the different information classes to collect evidence in order to correct the misclassified characters.

Keywords: Expert System, Handwriting Recognition, Knowledge Processing

1 Introduction

Handwritten document recognition has seen significant progress during the last few years [4]. The main goal is the correct transcription of the handwritten information to machine readable format.

Common difficulties are the segmentation and classification errors, many times due to poor input quality, incomplete training of the character classifiers and, mostly, the enormous style variety of handwritten characters. One way of handling recognition errors consists of correcting them using contextual knowledge about the domain.

This task usually includes three steps: error detection, generation of corrective hypotheses and their selection. In the case of form processing, which is one of the main applications of handwritten recognition, the contextual knowledge is provided by the meaning of the information inserted into the fields.

Benevides et al. [1] have adapted the form reading system of Garris et al. [2] to recognize a particular type of form, namely the student registration system of our university. Fig. 1 shows a sample form. The system processing is composed of two main stages: image processing and character recognition.

However, recognition errors frequently arise and the system is not able to repair them.

Our approach includes an additional post-processing stage in order to eliminate as much errors as possible. Contextual information about the form fields and commonly encountered recognition errors are used for identifying and repairing the recognition system mistakes. However, exclusive use of a knowledge type may not be enough for allowing the system to take the right decisions in most cases. It is necessary to accumulate evidences originating from different knowledge sources for dealing with uncertainty.

Since the particular form we use has typical fields for personal registration forms, we believe that our approach may also be applied to similar forms.

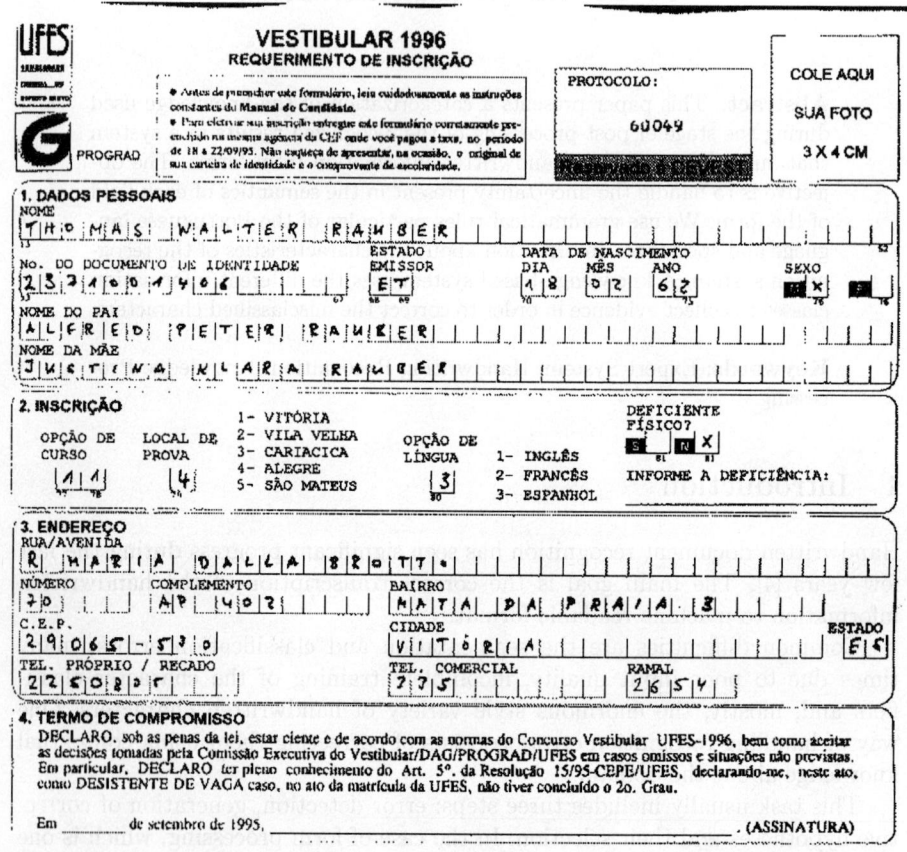

Fig. 1. Filled out form for student registration

This paper is organized in the following way: in the following section we present a short review of character recognition and post-processing, in section 3 we describe the problem of handwritten recognition and some difficulties when using knowledge for the correction of errors. In section 4 we present a knowledge categorization that can be used to improve the result of the recognition of forms. In section 5 an experiment of post-processing is presented, and finally, in section 6 we draw some conclusions of our work and outline future research.

2 Related Work

Character recognition can be considered as a classification task which categorizes visually digitized characters into their respective symbolic classes. Pattern recognition algorithms are used to extract form features of the character and classify it. Artificial neural networks have been used successfully for the implementation of character classifiers [4].

Meanwhile none of the decisions of the classifier can be considered definitely true since the majority of hypotheses are generated from local properties of the pattern. Contextual post-processing envisions to improve the locally taken decisions, exploring certain constraints and inter-object relationships. The final result using contextual information can be considered as an error prone hypothesis as well, but is usually more optimized than the exclusively local recognition [5].

Contextual post-processing techniques may use various sources of knowledge, not only about the specific domain of the document, but also about the constraints of the specific idiom, see e.g. [3].

3 Handwritten Recognition and the Use of Knowledge

One of the largest difficulties in handwritten recognition is the correct classification of the characters. Consider the recognition results obtained from the system described in Benevides et al. [1]. Figure 1 shows a filled in form and table 1 presents the recognition results of some selected fields. The table is composed of the recognized characters together with their respective degrees of confidence which are defined by the maximum activation value of the output layer of a perceptron with one hidden layer. The underlined characters indicate erroneous recognition. One way of detecting and correcting the resulting mistakes of the recognition is using the techniques of contextual post-processing. For instance, in the presented example of recognition, the semantic content of one field allow for detecting some errors. For instance, the birth date field (*"dia do nascimento"*) cannot have values higher than 31. For correcting those mistakes we need to restrict the correction possibilities and, further, select the right option from the possibilities set.

The knowledge that guides the detection and correction of the recognition mistakes is usually associated to uncertainty sources. The main uncertainty sources that may occur during the post-processing stage of the form handwritten recognition are:

Untrustable Information: Some recognized characters were not necessarily written. In some situations, the lines that delimit the fields and the internal lines that separates the characters may not be eliminated by the recognition system.

Table 1. Recognition results from the form in fig. 1

Field	Recognition results	Character confidence degree classification
"Nome"	THOMAS WALTER RAUBER	.95 .89 .63 .89 .77 .96 .97 1.00 .99 .98 .87 .76 .75 1.00 .69 .99
"Dia do nascimento"	48	.65 1.00
"Mês do nascimento"	02	.99 .78
"Nome do pai"	ALFRED PETE<u>G</u> RAU<u>Z</u>ER	.50 .99 .99 .99 .98 .91 .82 .98 .99 .93 <u>.28</u> .95 .91 .98 <u>.24</u> .99 .74
"Curso"	40	.48 .33
"Local da prova"	4	.96
"Cidade"	VIT<u>G</u>RIA	1.00 .82 1.00 <u>.61</u> .83 .99 .91
"Estado"	E<u>I</u>	.89 .26

In these situations they will be considered characters. Another untrustable information source is the inappropriate segmentation of isolated characters. They will be classified as two independent characters.

Incomplete Information: It is not possible to know in advance neither the complete set of values of some fields, nor a general formation rule for them. Even if we can find formation rules for the field values, there is often many exceptions to these rules. Therefore, if a word hypothesis was not found in the set, or does not obey the formation rule, it does not imply that the hypothesis is wrong. Moreover, the reverse is also true. If a word hypothesis is in the set, or follows the basic formation rule, we cannot guarantee that the word has been correctly recognized [5]. Actually, the word may be mistaken by another word in the value set or that also follows the formation rule.

Imprecise Information: The confidence degree of the individual character recognition (see [1]), does not assure the character correct identification. There are situations where the confidence is high and the recognition is wrong and, vice versa, where the confidence is low and the recognition is right.

Thus, the conclusions and decisions taken to detect and correct the mistakes based on the above mentioned information may also be untrustable, incomplete or imprecise.

4 Solution for Form Post-processing

A solution to reduce the uncertainty in the post-processing consists of using the different information sources altogether and all additional knowledge available to accumulate evidences for reducing or even eliminating the uncertainty of the decisions taken during the post-processing stage.

In order to make it easy the construction of a knowledge-based system capable of achieving this purpose, it is essential to classify and describe the different knowledge sources that may be used. It is important to emphasize that our classification is by no means complete or the only one possible. It was defined specifically for the development of our knowledge-based system. Nevertheless, this does not mean that it may not be useful in other domains or applications. Following, we show our classification:

1. *Domain knowledge*
   (a) Semantic knowledge
      − Lexicon of fields knowledge
      − Inter-field relationships knowledge
      − Form application context knowledge
   (b) Grammatical knowledge
      − Phonological knowledge
      − Morphological knowledge
      − Syntactical knowledge
   (c) Recognition System knowledge
      − Confidence degree knowledge
      − Delimiting line errors knowledge
      − Classification errors knowledge
      − Segmentation errors knowledge
2. *Control knowledge*
   (a) Errors detection knowledge
   (b) Generation of corrective options knowledge
   (c) Options selection knowledge

4.1 Domain Knowledge

One domain knowledge source is the semantic knowledge related to the form fields. It includes the lexicon of each field, the semantic relationships between the fields and the information regarding the form application context. Another information source are the language grammatical rules used for filling out the form. Finally, there is also knowledge about the recognition system characteristics such as the most common segmentation and classification flaws and the context where they usually happen.

Semantic Knowledge Semantic knowledge is related to the meaning of the information that should be filled out in each field.

Lexicon of fields knowledge: Each form field has associated constraints restricting the way it can be filled in. Some of these constraints are independent of the content of other fields. These constraints are lexical knowledge since they are based on the vocabulary of each field. For instance, intervals for the allowed numeric values may be defined for the fields "birth date", "place of the exam", "code of the grade" and "language".

Inter-field relationships knowledge: There are relationships among the fields which allow that the recognized information of one field supports the correction and validation of the information in another field. As an example we can consider the fields which contain the surname of the candidate and his or her parent's surnames. The inheritance habits of surnames provide a valuable tool to improve the results.

Form application context knowledge: The form application context may provide preferences among the values of each field. For instance, in the case of the student registration form, there is a larger possibility that the candidates live at neighboring states of the state where the University is located.

Grammatical Knowledge Phonological, morphological and syntactical knowledge may be associated to the fields with non numerical values.

Phonological knowledge: The phonological knowledge allows for inspecting the correct formation and representation of the written words. For instance, in Portuguese, there is an orthographic rule specifying that we should necessarily use the "m" consonant before the "p" and "b" consonants, such as, in the "campestre" word. The same rule specifies that we should use the "n" consonant before the other consonants, such as in the "cansado" word. Among other uses, this type of knowledge may be used for identifying that the character before a "p" or "b" is a "m" instead of a "n".

Morphological knowledge: The morphological knowledge allows for inspecting the word structure, formation and flexioning mechanisms and their division in grammatical classes. In case of substantives, for example, the morphological knowledge defines the rules of gender and number formation. One application of morphological knowledge is the gender inspection of words.

Syntactical knowledge: Syntactic information is associated to the word concordance. Examples (for Portuguese) are relationships of the gender (male or female) and singular and plural. In Portuguese the articles that precede the nouns vary conforming the gender and if the words are in singular or plural. For instance the surname "das Santos" may not exist, being the correct name "dos Santos".

Recognition System Knowledge Knowledge about the recognition system behavior may be a valuable source of information for identifying and correcting errors. It allows for error discovery even in situations where a supposed valid information was recognized, i.e. the information satisfied the field constraints but was not really correct.

Confidence degree knowledge: The confidence degree of each character classification is a knowledge source provided by the recognition system. Besides using this information for identifying possible errors, the recognition system may suggest characters to be used for correction.

Delimiting line errors knowledge: Some character recognition errors may have a significant rate of incidence due to the features of the recognition system. For instance, instead of recognizing the letter "O", the recognition system often recognizes the letter "Q" when the handwritten character crosses the limits of the underlying line that delimits the field.

Classification errors knowledge: Some classification errors may be related to common improper character writing. One example is the recognition of the letter "H" instead of "A" because of some discontinuity of the character strokes.

Segmentation errors knowledge: There may also be problems with the system segmentation process. For instance, the letters "I" e "J" are usually recognized instead of the letter "U" due to the improper segmentation of the character. The knowledge about classification and segmentation errors may be represented by a *confusion matrix*, i.e. a data table indicating the most common mutual errors of the recognition process [6].

4.2 Control Knowledge

The control knowledge should make use of the domain knowledge for accomplishing the basic subtasks of the post-processing stage: error detection, generation of the corrective options and selection of those options.

Errors detection knowledge: The error detection task consists of verifying if the form field fulfilled information is valid. The error detection knowledge is field-specific and involves all domain knowledge sources related to the specific field. The lexicon of the fields, their relationships and the grammatical rules enable the detection of invalid field values. The form application context and the recognition system knowledge may be used for identifying possible errors on the values assumed to be valid.

Generation of corrective options knowledge: Initially, the set of corrective options of a field value is composed of all other possible field values. However, according with the kind of error detected, it is possible to restrict this set using knowledge about the corrective procedures applicable on the specific context where the error occurred.

Options selection knowledge: Correcting errors are more difficult than detecting them. Whenever there is more than one corrective option, additional knowledge sources should be used for accumulating evidences about each corrective option and ordering them in a proper way. For instance, if there is a grammatical error in the occurrence of the consonant "m" in a recognized word, we may consider two corrective options: changing the consonant "m" or the subsequent letter in the word. For choosing a corrective option we may use other sources of information, such as: the confidence degree in the recognition of each letter and the similarity of the corrected words with the words in the lexicon.

5 Experimentation and Analysis

In order to verify the viability of our proposal we realized a constrained experiment which considers a subset of the fields of the form and the rules for detecting errors, generate and select the correction options. The fields under consideration are the name of the candidate, the name of his or her parents and the address information (street, quarter, city and state).

5.1 Rule Description

In our experiment, we have only used a subset of the error detection, correction and selection rules. We have applied rules for identifying grammatical mistakes in the use of the consonants "m", "n", "q" and "h", vocalic and consonantal encounter mistakes, nominal agreement between prepositions and name words, character segmentation and field separation lines mistakes. In addition, we have also verified if the word confidence (average of its character confidences) is below an established threshold (0.5) and if the corrected word belongs to the field lexicon.

Corrective options are associated to the different kinds of mistakes. Thus, each type of mistake has an associated rule that generates specific corrective options. For instance, a rule for generating corrective options for the "m" consonant mistake change the "m" letter to the most common letter confused with "m" by the recognition system, change the "m" subsequent letter for vowels; and change the "m" subsequent letter for the consonants "p" and "b". Since we may have many corrective options for the same word, each option has an associated evidence degree representing its possibility of being correct. The evidence degree is a numeric value estimated experimentally or subjectively proposed by the knowledge engineer. In our experiment, the evidence degrees were mostly set subjectively.

The rules for choosing the corrective option consider the whole subset of detected errors and corrective options. Some of these rules are: (a) If the field has an associated lexicon, select the corrective option that makes the word more similar to a word in the lexicon; (b) Select the corrective option suggested by the larger number of correction rules; (c) Select the corrective option with larger evidence whenever there are distinct corrective options suggested by the same number of correction rules.

In our experiment, we have used lexicons for some fields. We have associated a set of common names and surnames to the name fields. We have also associated common words, such as "president", "saint" and "coronel", to the address field lexicons. The only field with a complete lexicon is the state field (the lexicon is compound by the brazilian states abbreviation).

5.2 Description of the Experiment

The CLIPS[1] programming environment was used for the implementation of the experiment. It allows the association of object orientation and production rules in problem solving, besides being a modifiable public domain system. Each form field was implemented as a class with a set of attributes and methods capable of representing a subset of the domain knowledge. The rules were used to represent the control knowledge. Figure 2 highlights the considered fields of a form and tables 2 and 3 show the recognition and post-processing results, respectively. In the tables only the fields that were processed are considered and the words that

---

[1] http://www.jsc.nasa.gov/~clips/CLIPS.html

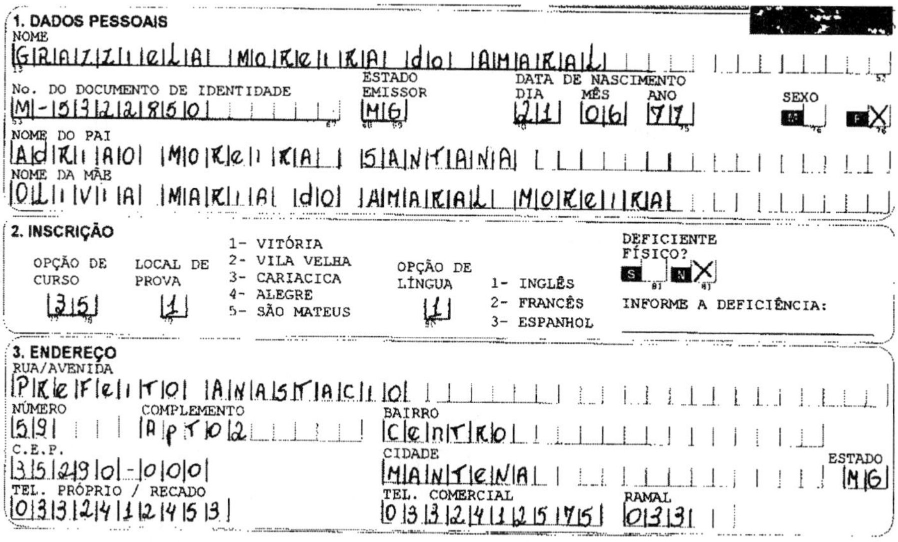

Fig. 2. Example of a recognized form

were corrected or considered correct without post-processing are emphasized in boldface.

5.3 Discussion

In the example above 17 words were processed and only one was recognized correctly. After the post-processing step the number of correct word incremented to ten. In order to outline how the post-processing was done in detail we consider the second word of the father's name. In this word grammatical errors were detected related to the use of the letter 'M'. It was considered similar to the fifth word of the mother's name and the second word of the candidate's name. The hypotheses that were inferred include the words MOXCLXA and MEXCLXA (possibilities for the correction of letter 'M') and the word MOREIRA (selected correction option for the fifth word of the mother's name). Since MOREIRA is the only option found in the dictionary of the last names it was selected.

6 Conclusion

Handwriting recognition suggests the use of the available contextual knowledge to deal with hard cases of character classification. After the character recognition stage of the recognition process, our approach applies a form specific contextual knowledge for performing the necessary corrections. The form specific knowledge is based on intra-field and inter-field lexical, morphological, syntactical and semantic constraints. Our approach consists of accumulating evidences from different knowledge sources for dealing with exceptions to the general rules and

Table 2. Recognition results of the form in fig. 2

Field	Result of the recognition process
Name	GRAIJMCLL MQKEIKA JQ RMBLAL
Name of father	AHXPAO MCXCLXA SANYANA
Name of mother	OJMVKA RMIARKA JO AMAKAL MORCIRA
Street	PRRFRITO AHASTATIO
Quarter	CRRTKD
City	MANYRNA
State	MG

Table 3. Results of the post-processing of fig. 2

Field	Result of Post-Prcessing
Name	GRAIJMCLL **MOREIRA DO AMARAL**
Name of father	AHXPAO **MOREIRA SANTANA**
Name of mother	OJMVKA RMIARKA **DO AMARAL MOREIRA**
Street	**PREFEITO** AHASTATIO
Quarter	CRRTKD
City	MANYRNA
State	**MG**

uncertainty. The principal drawback of the proposed methodology is the dimension of the rule set that have to be defined to cover reasonably well the possible correction hypotheses.

References

1. R. M. Benevides, S. Rossetto, and T. W. Rauber. Adaptação de um sistema de reconhecimento de formulário. I. International Seminar on Document Management, Curitiba, Brasil, Nov. 10-11, 1998.
2. M. D. Garris, J. L. Blue, G. T. Candela, P. J. Grother S. A. Janet, and C. L. Wilson. NIST form-based handprint recognition system. Technical Report NISTTR 5959, U. S. Dept. of Commerce, NIST, Gaitherburg, MD 20899-001, 1997.
3. Michael Malburg. Comparative evaluation of techniques for word recognition improvement by incorporation of syntactic information. Proceedings of the Fourth International Conference on Document Analysis and Recognition (ICDAR'97), 1997.
4. R. Plamondon and Sargur N. Srihari. On-line and off-line handwriting recognition: A comprehensive survey. *IEEE Transactions on Pattern Analysis and Machine Intelligence*, 22(1):63–84, January 2000.
5. J. Schürmann, N. Bartneck, T. Bayer, J. Franke E. Mandler, and M. Oberländer. Document analysis — from pixels to contents. *Proceedings of the IEEE*, 80(7):1101–1119, July 1992.
6. Kazem Taghva, Julie Borsack, Bryan Bullard, and Allen Condit. Post-editing through approximation and global correction. UNLV Information Science Research Institute, Annual Report, University of Nevada, Las Vegas, 1993.

A Knowledge-Based System for Construction Site Level Facilities Layout

K.W. Chau and M. Anson

Department of Civil and Structural Engineering,
Hong Kong Polytechnic University, Hunghom, Kowloon, Hong Kong
cekwchau@polyu.edu.hk

Abstract. The choice of good construction site layout is inherently difficult, yet has a significant impact on both monetary and time saving. It is desirable to encapsulate systematically the heuristic expertise and empirical knowledge into the decision making process by applying the latest artificial intelligence technology. This paper describes a prototype knowledge-based system for the construction site layout, SITELAYOUT. It has been developed using an expert system shell VISUAL RULE STUDIO, which acts as an ActiveX Designer under the Microsoft Visual Basic programming environment, with hybrid knowledge representation approach under object-oriented design environment. By using custom-built interactive graphical user interfaces, it is able to assist designers by furnishing with much needed expertise in this planning activity. Increase in efficiency, improvement, consistency of results and automated record keeping are among the advantages of such expert system. Solution strategies and development techniques of the system are addressed and discussed.

1 Introduction

Construction site layout, involving the assignment of locations for the temporary facilities, is an important planning activity. It is recognized that a good layout has a significant impact on cost, timeliness, operational efficiency and quality of construction, which manifest on the larger and more remote projects. Improper layout can result in loss of productivity due to excessive travel time for laborers and equipment, or inefficiencies due to safety concerns. In spite of its potential consequences, construction site layout generally receives little advanced planning, and hardly any planning during construction. It is often determined in an ad hoc manner at the time the siting requirement arises. In the evaluation process, it involves many decisions to be made by the designer based on empirical rules of thumb, heuristics, expertise, judgment, code of practice and previous experience. It is desirable to encapsulate systematically this knowledge into the decision making process by applying the latest artificial intelligence technology.

With the recent advent in artificial intelligence (AI), a knowledge-based system (KBS) is able to furnish a solution to this decision making process through incorporating the symbolic knowledge processing. During the past decade, the potential of AI techniques for providing assistance in the solution of engineering problems has been recognized. KBS are considered suitable for solving problems that demand considerable expertise, judgment or rules of thumb, which can be broadly classified into the following categories: interpretation; design; diagnosis; education; and planning. Areas of early applications of KBS technology include medical diagnosis, mineral exploration and chemical spectroscopy. KBS have made widespread applications in different fields and are able to accomplish a level of performance comparable to that of a human expert [1-5].

This paper describes a prototype KBS for the construction site level facility layout, SITELAYOUT, which has been developed using a commercially available microcomputer-based expert system shell VISUAL RULE STUDIO [6]. It is intended not only to emulate the reasoning process followed by site practitioners, but also to act as an intelligent checklist that comprises site objects and activities for expert advice to its users. The blackboard architecture with hybrid knowledge representation techniques including production rule system and object-oriented approach is adopted. Its knowledge base comprises representations of the site and the temporary facilities to be located, as well as the design knowledge of human experts. It opportunistically applies varying problem solving strategies to construct the layout incrementally. Through custom-built interactive graphical user interfaces, the user is directed throughout the planning process. Solution strategies and development techniques of the system are also addressed and discussed.

2 The Construction Site Layout Problem

Site layout design are engineering problems that are solved daily on each construction site. Temporary facilities are essential to support construction activities, but do not constitute the completed project. These facilities are diverse in nature depending on the size and location of the project. They range from simple laydown areas, material handling equipment, access roads, warehouses, job offices, fabrication shops, maintenance shops, batch plants to lodging facilities, etc. These may be on site during part or all of project construction. Very often, after the bidding stage and the startup of the project, continuous advance planning is seldom carried on. The detailed site layout is left to the day-to-day scheduling of foremen and superintendents. Inappropriate location then requires material handling that could have been avoided.

The choice of good layout is inherently difficult since it is constrained by both the site topography and the location of the permanent facilities. The nature of the problem is such that no well-defined method can guarantee a solution. In the literature, some guidelines have been proposed for laying out construction sites [7-9]. Construction site layout can be delimited as the design problem of arranging a set of predetermined facilities on the site, while satisfying a set of constraints and optimizing layout objectives. Good layout demands fulfilling a number of competing and yet often conflicting design objectives. Some objectives are to maximize operation efficiency by promoting worker productivity, shortening project time, reducing costs and to maintain good employee morale by providing for employee safety and job satisfaction, to minimize travel distance and time for movement of resources and to decrease material handling time. Besides, legal obligations may impose safety and permit constraints, and technical and physical limitations cannot be neglected. Managers who set out to meet several objectives must prioritize them, which is a nontrivial and highly subjective task.

The site layout problem involves multiple sources of expertise with often conflicting goals, requiring scheduling of activities with leveling of multiple resources and space consideration. Project managers, superintendents and subcontractors may jointly agree upon the location of major pieces of material and equipment, by using their past experience, trial and error, insight, preference, common sense and intuition. Mathematical optimization techniques such as operation research have been employed to aid layout design [10]. However, they did not gain general acceptance in industry due to the difficulty to learn and use, complexity and unavailability of the computer implementations, and the requirement of a quantifiable objective.

The designer solves the problem by employing two techniques, namely, iconic representation and decomposition. The iconic representation of the problem's physical components, such as engineering drawings, templates or three-dimensional models, is useful in allowing the designer's perceptual vision to operate on it. Human can only manipulate a smaller number of symbols in the mind at any one time. This limited symbolic processing power forces a decomposition of the problem into sub-problems, which are then dealt with relatively independently. The search for a solution is then the search for a succession of facility selection and location that achieves a satisfying solution to the layout problem. The layout problem can then be viewed as a search through a tree whose nodes are partial layouts. The search proceeds by stepping from state to state through the problem space, starting from an initial state comprising an empty job site and a list of facilities to be located, and ending at a state comprising a layout of all facilities satisfying the set of constraints.

3 Features of SITELAYOUT

A prototype KBS, SITELAYOUT, has been developed to assist construction practitioners in their complex task of designing spatial layout planning. It allows site management to continuously monitor and update the plan as site activities proceed. It integrates domain heuristics and layout strategies, reasons about spatial and temporal designs, and dynamically changes its behavior based on design strategy. SITELAYOUT uses a plan-generate-test strategy. The solution will depend entirely on the strategies the user wants to implement, the objects and constraints involved in the layout. SITELAYOUT manipulates the facilities, extracts information from the site layout, generates alternative locations for the facilities, tests constraints, selects a location and updates the layout. Site objects are first assembled into partial layouts and constraints are successively applied to identify feasible positions for them. Several partial layouts are combined to form a complete solution. The system is compiled and encrypted to create a run-only system. The user can always overrule any design options and recommendations provided by the system. It thus plays the role of a knowledgeable assistant only. Its strength lies in its flexibility for deciding what actions to take. It allows for users to fine-tune the system to their preference style, yet to let them benefit from its thoroughness, speed, and overall effectiveness for designing a site layout.

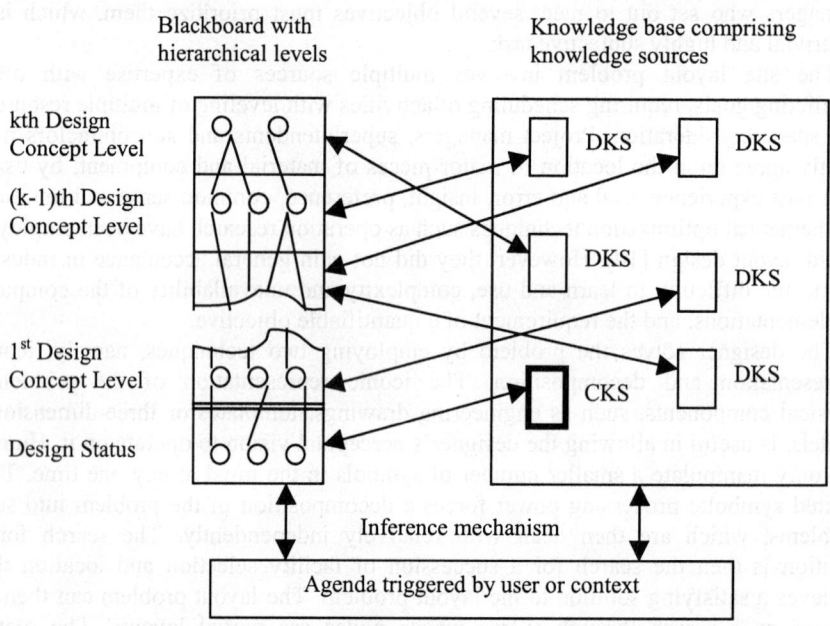

Fig. 1. Blackboard architecture of the prototype expert system (DKS and CKS denote Domain Knowledge Source and Control Knowledge Source respectively)

3.1 Development Tool

In order to facilitate development of the KBS on site layout planning, expert system shell containing specific representation methods and inference mechanisms is employed. This system has been implemented with the aid of a microcomputer shell VISUAL RULE STUDIO, which is a hybrid application development tool under object-oriented design environment. VISUAL RULE STUDIO acts as an ActiveX Designer under the Microsoft Visual Basic 6.0 programming environment. Production rules as well as procedural methods are used to represent heuristic and standard engineering design knowledge. By isolating rules as component objects, separate from objects and application logic, it produces objects that can interact with virtually any modern development product and thus rule development becomes a natural part of the component architecture development process.

3.2 System Architecture

The blackboard architecture has been successfully applied in solving a wide range of domain problems, such as speech recognition, signal processing, and planning [11]. A blackboard system consists of diverse knowledge sources that communicate through a blackboard and are controlled by an inference mechanism. Figure 1 shows the blackboard architecture of the prototype system. The blackboard serves as a global data structure, which facilitates this interaction. This architecture is adopted since the reasoning with multiple knowledge sources is essential to solve the site layout problem. Besides, the actual layout design follows from quite opportunistic decisions, which are often made incrementally. Many factors contributing to the decisions can be unpredictable based on personal preference depending on the condition. Since not all information is initially available or relevant, the initial layout plans are often modified as construction proceeds.

Apart from the usual components in a typical KBS, namely, knowledge base, inference mechanism, session context, user interface, knowledge acquisition and explanation modules, it also incorporates database. The database chosen is Microsoft Access due to its popularity as a user-friendly relational database within industry, reasonable cost, and Visual Basic support by means of Visual Basic for Applications. The system can retrieve information from a library of partial arrangements, which were gleaned from actual site conditions.

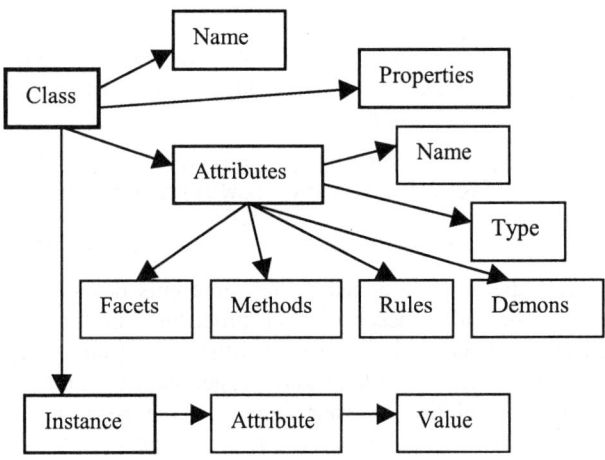

Fig. 2. Structure of VISUAL RULE STUDIO components

3.3 Object-Oriented Programming

Under the declarative knowledge representation environment, objects are used to encapsulate knowledge structure, procedures, and values. An object's structure is defined by its class and attribute declarations within a RuleSet. Object behavior is tightly bound to attributes in the form of facets, methods, rules, and demons. Figure 2 shows the structure of VISUAL RULE STUDIO components. Each attribute of a class has a specific attribute type, which may be compound, multi-compound, instance reference, numeric, simple, string, interval, and time. Facets provide control over how the inference engines process and use attributes. Methods establish developer-defined procedures associated with each attribute. The set of backward-chaining rules that conclude the same attribute is called the attribute's rule group. The set of forward-chaining demons that reference the same attribute in their antecedents is called the attribute's demon group. SITELAYOUT combines expert systems technologies, object-oriented programming, relational database models and graphics in Microsoft Windows environment. By defining various types of windows as different classes, such as Check Box, Option Button, List Box, Command Button, Text Box, etc., they can inherit common characteristics and possess their own special properties.

3.4 Knowledge Base

In this prototype system, the knowledge is represented in object-oriented programming and rules. Its knowledge base contains representations of the permanent facilities, the temporary facilities and the design knowledge of the expert.

Reasoning knowledge, both heuristic and judgmental, including the constraints between the objects, is represented as rules. Knowledge represented in the IF/THEN production rules with confidence factors can be assigned either automatically, or in response to the user's request. Site planning strategies expressed in rules form a natural representation and are easily understood. These rules are a formal way of specifying how an expert reviews a condition, considers various possibilities, recommends an action and ranks the potential layout solutions. The following is a typical example of the production rules.

```
Rule to find ScoreRebarSubcontractor: 3 of 10
IF RebarSubcontractor.DistanceToWorksArea >= 200
AND RebarSubcontractor.Area >= 100
THEN RatingScore.ScoreRebarSubcontractor:= Low CF 85
```

The objects define the static knowledge that represents design entities and their attributes, which can be either descriptive or procedural in form. This allows a description of physical objects including the facilities and of other requisite abstract objects such as the locations. Procedural knowledge, such as numerical processing and orientation of the facilities, is represented in the form of object-oriented programming. Generic construction objects are structured in a hierarchical knowledge base with inheritance properties. Besides, it comprises a blackboard together with two sets of knowledge sources, namely, Domain Knowledge Sources and Control Knowledge Sources.

Blackboard. Objects inherit properties from the class to which they belong. Each one of the objects of the representation such as points, sides, polygons, facilities and, at the highest level of the hierarchy, the site itself are represented as objects. It describes facilities by their type, possible zoning requirement, geometry, dimensions, duration on site, and mobility. In order to keep track of the design status and related information, SITELAYOUT uses a hierarchical level named Design Status that has attributes whose values change in time to represent the different states of the layout process. These states are partial layouts obtained from the previous state by the selection of a location for a certain facility. This unit keeps track of the facility being located, the alternative locations generated and the alternative selected at the previous level of the design process.

Domain Knowledge Sources. Diverse Domain Knowledge Sources, functioning independently and cooperatively through the blackboard, encode the actions to take for constructive assembly of the layout. They add particular domain specific objects to the arrangement, modify their attributes and roles, call the constraint satisfaction system to generate the appropriate possible locations for that object, and display objects on the computer screen. They encompass all construction management domain knowledge necessary to design site layouts. The expert's design knowledge consists of heuristic and rules of thumb acquired through years of experience, which is represented as a set of rules. Constraints in SITELAYOUT are desired qualities of the layout due to relationships amongst the facilities, the work area and the outside world emanating from the functionality of the facilities. They include zoning constraint, adjacency constraint, distance constraint, non-overlap constraint, access constraint, spatial constraint, position constraint, view constraint and preference constraint. When several constraints relate to an object, it satisfies them according to the heuristic ordering that reflects the order in which a human expert would satisfy them. It first picks the largest object and positions the other objects in order of decreasing of size by meeting their constraints relative to the former. The problem-solver in SITELAYOUT selects through the rules the constraints that need to test the instantiated locations for the facility at hand, instantiate them and test them.

Control Knowledge Sources. The Control Knowledge Sources involve meta-level knowledge, which establishes the problem solving strategy and controls the execution of the Domain Knowledge Sources. They delineate the strategy in selecting which action to take next and in what order to take for different site planning problems. Besides, they modify dynamically the strategy the problem-solving system uses to choose a knowledge source for execution on each problem-solving cycle.

3.5 Inference Mechanism

The inference engines control the strategies that determine how, from where, and in what order a knowledge base draws its conclusions. These inference strategies model the reasoning processes an expert uses when solving a problem. The Control Knowledge Sources evaluate the Design Status and decide what action should be performed mainly in data-driven forward chaining mechanism. The knowledge representations of the Domain Knowledge Sources, however, need both forward and backward chaining inference mechanism to arrive at the solution. At each cycle the knowledge sources recommend separate actions for which the preconditions are satisfied. Based on the rated heuristic scores, the scheduler then selects the best action and proposes it for execution. The system asks the user to accept or override the recommended action. The scheduler then executes the action, which may be to perform detailed design or change strategy. This cycle is repeated until a finite set of feasible solutions satisfying all constraints is found. In this way, it mimics human opportunistic and incremental reasoning about problem-solving actions. The solution tree is pruned intelligently beforehand, by selecting, at each step, the best possible action, and to limit an object's possible locations on site to a small finite set. The evaluation of a particular design can then be rated.

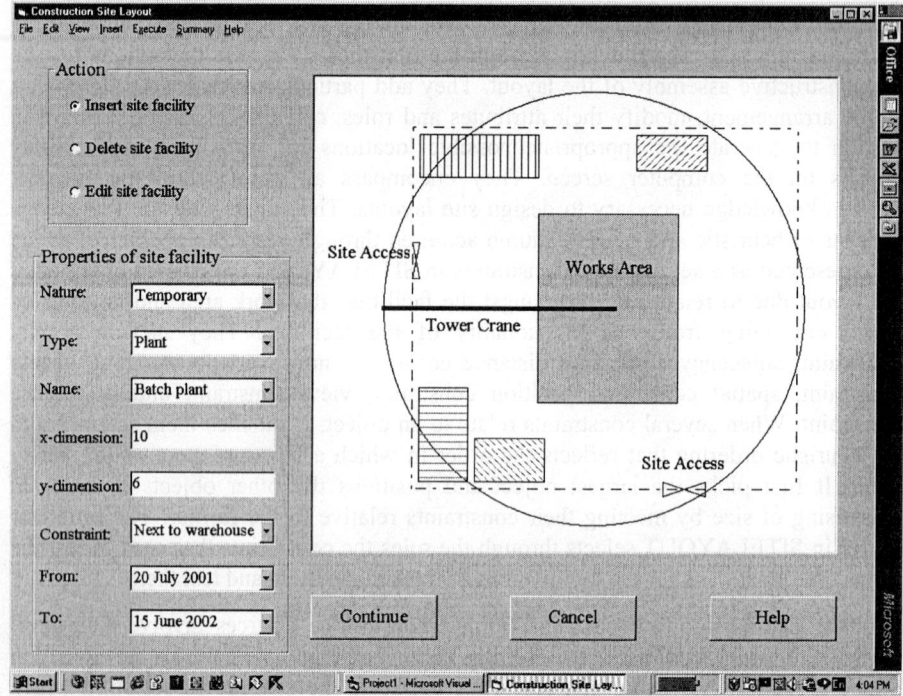

Fig. 3. Screen displaying interactive user interface of SITELAYOUT

3.6 User Interfaces

The system offers a friendly user interface such that a combination of mouse and keyboard can be used to navigate the application. Whilst input data entries are kept at minimum, they are provided by the user mostly through selection of appropriate values of parameters from the menus and answers to the queries made by the system. The input data provided by the user will be rejected if it is not within the specified range. The system provides multi-window graphic images combined with valuable textual information, which is extremely valuable to novice designers.

A set of temporary facilities has been represented into the system. The user needs to indicate which one is needed for the problem at hand and specify the size of each. In order to pick one feasible alternatives of the layout to be the solution, the choice is based on some evaluation functions taking into account user preferences and other criteria. Users can observe SITELAYOUT perform its preferred actions or they can make their own changes on the position of an object to the layout on the interactive display. That information is then sent back to the Domain Knowledge Sources which can monitor incoming information and update the blackboard for further reasoning. A computer graphics screen with multiple colors is a natural representation of possible object positions in layout design and a powerful medium for human-machine communication. The output is displayed graphically on the screen, indicating the location of every facility on the site, which facilitates expert user interface, visualization, critique and feedback. Figure 3 shows a typical screen displaying interactive user interface of SITELAYOUT.

3.7 Knowledge Acquisition Facilities

Knowledge plays an important role in a KBS, yet the major difficulty in designing a realistic site layout is the acquisition of necessary data. In order to acquire knowledge, it is better to work with the expert in the context of solving particular actual problems, instead of directly posing questions about rules. The knowledge used has been acquired mostly from written documents such as code of practice, textbooks and design manuals and conversations with several site practitioners.

3.8 Explanation Facilities

HELP command buttons provide definitions of a variety of parameters involved in order that the user can select the appropriate options. Their primary functions are to aid the user to comprehend the expert's approach to the problem of construction site layout, and to gain synthetic experience. In fact, the explanation facility is one of the distinct differences between the conventional computer programs and KBS, which is designed to explain its line of reasoning for acquiring an answer.

4 Conclusions

A prototype microcomputer KBS, SITELAYOUT, which assists in making decision on the construction site layout problem, was developed and implemented. It is shown that the hybrid knowledge representation approach combining production rule system and object-oriented programming technique is viable with the implementation of blackboard system architecture under a Windows platform for this domain problem. It comprises a detailed checklist of all activities and objects on site. The knowledge base is transparent and can easily be updated, which renders the KBS an ideal tool for incremental programming. Besides, its explanation facilities are capable of offering valuable information to the user, which can lead to a more efficient planning procedure. By using custom-built interactive graphical user interfaces, it is able to assist designers by furnishing with much needed expertise and cognitive support in a planning activity that when overlooked, results in cost overrun and schedule delays. Increase in efficiency, improvement, consistency of results and automated record keeping are among the advantages of such expert system.

Acknowledgement

The work described in this paper was substantially supported by a grant from the Research Grants Council of the Hong Kong Special Administrative Region (Project No. *PolyU 5060/99E*).

References

1. Chau, K.W.: An Expert System for the Design of Gravity-type Vertical Seawalls. Engineering Applications of Artificial Intelligence **5(4)** (1992) 363-367
2. Chau, K.W., Albermani, F.: Expert System Application on Preliminary Design of Liquid Retaining Structures. Expert Systems with Applications **22(2)** (2002) 169-178
3. Chau, K.W., Chen, W.: An Example of Expert System on Numerical Modelling System in Coastal Processes. Advances in Engineering Software **32(9)** (2001) 695-703
4. Chau, K.W., Ng, V.: A Knowledge-Based Expert System for Design of Thrust Blocks for Water Pipelines in Hong Kong. Journal of Water Supply Research and Technology - Aqua **45(2)** (1996) 96-99
5. Chau, K.W., Yang, W.W.: Development of An Integrated Expert System for Fluvial Hydrodynamics. Advances in Engineering Software **17(3)** (1993) 165-172
6. Rule Machines Corporation: Developer's Guide for Visual Rule Studio. Rule Machines Corporation, Indialantic (1998)
7. Handa, V., Lang, B.: Construction Site Planning. Construction Canada **30(3)** (1988) 43-49
8. Handa, V., Lang, B.: Construction Site Efficiency. Construction Canada **31(1)** (1989) 40-48
9. Rad, P.F., James, B.M.: The Layout of Temporary Construction Facilities. Cost Engineering **25(2)** (1983) 19-27
10. Moore, J.: Computer Methods in Facility Layout. Industrial Engineering **19(9)** (1980) 82-93
11. Engelmore, R., Morgan, T.: Blackboard Systems. Addison_Wesley, Wokingham (1988)

A Decision-Support System
to Improve Damage Survivability of Submarine

Dongkon Lee[1], Jaeyong Lee[2], and K. H. Lee[1]

[1] Korea Research Institute of Ships and Ocean Engineering/Kordi, Korea
{dklee, khlee}@kriso.re.kr
[2] Korea Naval Sea Systems Command, Korea
james-y-lee@hanmail.net

Abstract. Any small leakage in the submarines can lead to serious consecutive damages since it operates under high water pressure. Such leakage including damages on pipe and hull eventually incur human casualties and loss of expensive equipments as well as the loss of combat capabilities. In such cases, a decision-making system is necessary to respond immediately to the damages in order to maintain the safety or the survival of the submarine. So far, human decision has been the most important one based on personal experience, existing data, and any electronic information available. However, it is well recognized that such decisions may not be enough in certain emergency situations. The system that depends on only human experience may cause serious mistakes in devastating and scared situations. So it is necessary to have an automatic system that can generate responses and give advice the operator how to make decisions to maintain the survivability of the damaged vessel. In this paper, a knowledge-based decision support system for submarine safety is developed. The domain knowledge is acquired from the submarine design documents, design expertise, and interviews with operator. The knowledge consists of the responses regarding damage on pressure hull and piping system. Expert Elements are deduced to obtain the decision from the knowledge base, and for instance, the system makes recommendations on how the damages on hull and pipes decision and whether to stay in the sea or to blow. It is confirmed that developed system is well simulated to the real situation throughout sample applications.

1 Introduction

The submarines are very important in modern warfare. The number of submarines in the world's navies continues to increase and the number of major surface combatants decreases in recognition of their cost and vulnerability. Submarines are the one element of naval forces that, due to their ability to remain undetected for long periods of time, can operate in waters dominated by enemy air and surface forces. It provides us forces with various options in strategy and tactics but provide a serious menace to the enemy because of its stealth ability.

On Saturday, August 12, the giant Russian nuclear submarine Kursk - carrying a crew of 118 - sank in the waters of the Barents Sea after the incident that Russian officials described as a "catastrophe that developed at lightning speed." More than a week later divers opened the rear hatch of the sub but found no survivors[1]. It means any small leakage in the submarines can lead to serious consecutive damages since it operates under high water pressure. Such leakage including damages on pipe and hull itself eventually can incur human casualties and loss of expensive equipments as well as the loss of combat capabilities. Therefore, rapid and appropriate response to hull damage and local leakage under combat situation as well as under normal operation is very important. However, decision-making capability in the emergency situation is very limited compared to normal condition in general. The final decision-making in emergency case is made by heuristic knowledge and training based on the given situation. It always bases its inferences upon heuristic factor rather than numerical analysis results.

In this paper knowledge based system that can be support decision making process for damage and leakage of the submarine is developed. Domain knowledge for emergency response of the submarine is obtained from operators and designers based on design manuals and documents. The obtained knowledge is divided into two categories, which are pressure hull and pipe damage, through refining process. It is stored in knowledge base as rule and is adapted to backward chaining for inference. The developed system simulates given situations well and will be a useful tool for submarine operator. To install system at submarine, interface with sensors in the submarine is required further research is needed in order to make this interface.

2 Applications of Knowledge Based System in Military Vessels

Nowadays, expert and knowledge-based decision making system is available in many engineering fields such as shipbuilding, ship navigation, armor, and battle control system.

The CLEER (Configuration assessment Logics for Electromagnetic Effects Reduction) system is developed to prevent electro-magnetic interference(EMI) and to improve electro-magnetic compatibility between electronic equipments such as antenna and radar of the surface naval ship[2]. If electro-magnetic interference is occurred between electronic equipments, it can lead to serious hindrance in operation. The CLEER is to find proper installation location for the equipment when designing a ship. The system that can be used for reassigning the mission is developed when one more subsystems of the autonomous underwater vehicle(AUV) fails[3]. The rational behavior model for AUV control consists of 3 level. The knowledge-based system is adapted at strategic level which is upper most stage.

The application research for design itself and design components in system configuration is relatively active compare to other fields. A systematic procedure in the form of an interactive decision support system(DDS) incorporating artificial intelligence(AI) and other modeling tools such as a spreadsheet and a database management system to provide effective decision support at early stage of naval ship design has been developed[4]. The system contains ship design heuristics and related models that can aid a ship designer in rapidly evaluating design modifications. The system for design of antenna control system[5] and Phalanx test bed[6] are also developed.

The monitoring system has been developed to reduce operating and maintenance cost by increasing operation ability and reducing unnecessary maintenance[7]. The knowledge based system that can be used to establish maintenance plan and part exchange by monitoring and diagnosis of gas turbine takes signals from several sensors as input, and results were derived by performance curves of gas turbine and expertise based on sensor data.

A Knowledge base approach to the threat evaluation and weapon assignment process for a single warship attacked by anti-ship missile has been explored using SmallTalk 80/Humble[8]. The research described the design of the knowledge base system, the design of the testing environment for it, the evaluation of the performance of the knowledge base system in the environment.

The expert system for improvement of damage survivability, Hull Structural Survival System(HSSS), was developed and installed at surface war ship[9]. The HSSS has been designed to ensure maximum portability and to minimize life cycle costs. To these ends, it has been split into three computer software configuration items (CSCIs). The Analysis CSCI contains all the necessary algorithms and calculation routines to determine sea state, stability status and hull girder strength, and any other necessary data. The Intelligent Decision Aid (IDA) CSCI contains twinned rules-based expert systems and a conflict resolution function that work together to determine a course of action in the post-damage environment.

The knowledge-based system for military vessels is studied and applied to design, equipment monitoring, risk reduction and damage control. Most of relate research is for surface naval ships. Yet, there has been no literature of knowledge based system for submarine's leakage.

3 System Implementation

3.1 Knowledge Acquisition

The damage survivability for war ship is requires not only requires design knowledge and procedure of commercial vessel but also characteristics of war ships. For example, it must consider a state of hostilities, a plan of operations and patrol. Rapid and appropriate response is the most important for leakage in case of submarines especially since it operates under high water pressure.

This paper considers conventional submarine with diesel generator. The propulsive power is supplied from battery and it does not have an oxygen generator. To stay in the water under damaged condition, therefore, propulsive power for moving capability and oxygen for crew life is a limitation factor.

The possibility of damage increase rapidly by accumulated structural fatigue because changes of water pressure by repeated diving and surfacing cause deformation in many parts of pipe line and welding line of the submarine. And there still exists the possibility leakage due to the hull crack and pipe line damage by the explosive impact of a depth charge.

In this paper, the domain knowledge is acquired from document of submarine design, design expertise, and interview with operator[10]. The knowledge for leakage response consists of damage of pressure hull and piping system. Heuristic knowledge extracted from operators may contain large amounts of experience and technical know-how, but it is not so easy to arrange these pieces in any orderly fashion as in mathematical formulas. Also, this knowledge has to be tested and redundant knowledge eliminated before it can be added to the knowledge base. Fig. 1 and 2 are showing the decision tree of the extracted knowledge for hull and piping system respectively. There are three kinds of decision-making criteria and data in each node of the decision tree as follows;

- User(operator) input
- Signal from sensors
- External computer program

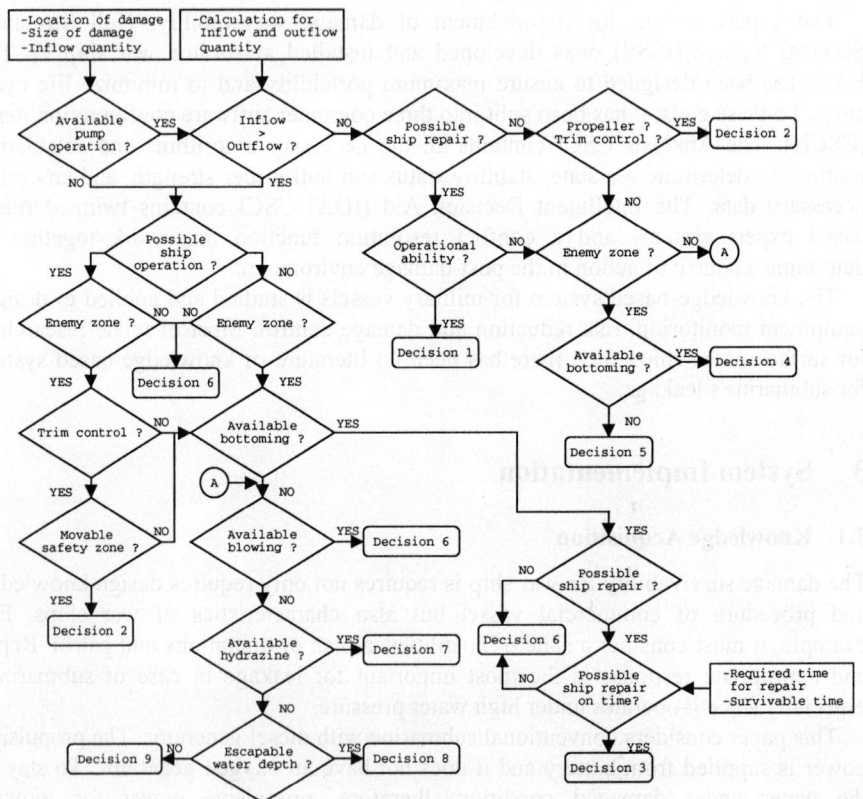

Fig. 1. The decision tree for hull damage

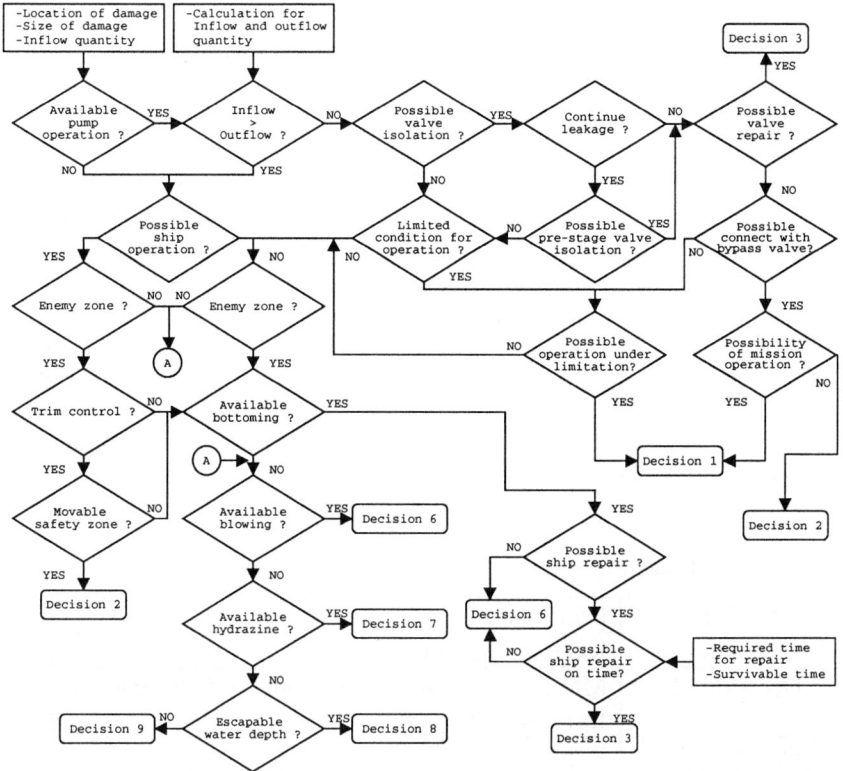

Fig. 2. The decision tree for piping system damage

The details of the above items are summarized as follows:

Input Items by Sensor Signal
1. Operational ability : Measured electrical signal of the propulsive motor
2. Enemy line : Electrical signal from ESM and noise library
3. Pump condition : Electrical signal such as READY or POWER AVAILABLE sign of pump
4. Depth(from surface to submarine) : Signal from depth sounder and depth gauge
5. Escapable depth : Digital signal from depth sounder

Input Items by user(Operator)
1. Leakage type : There are two types of leakage; pressure hull and piping line system. Great number of sensors is required for detection of leakage. Therefore, it is detected and confirmed by operator's eye
2. Connecting possibility of bypass piping system : The piping system in the submarine is constructed with network structure to prevent damage. The damaged part will be by-passed by valve control.

3. Status of leakage : The operators confirm whether or not it keeps on leaking after repair and closing intermediate valve.
4. Required repair time and possibility : The operators decide whether or not the repair is possible and its time.
5. Possibility of valve closing : The operators identify location and damage condition of valves that they want to isolate.
6. Move to safety water depth : It is decided in due consideration of distance, reaching time, available battery capacity and environmental condition by the operators.
7. Dive down to seabed : It is decided in due consideration of water depth to seabed, its shape and condition, and posture of vessel by the operators.
8. Possibility of a given mission continuation : The operators decide whether or not the submarine to sustain for her mission under leakage or malfunction of some parts.
9. Operation under limited condition : It is decided in due consideration of a given mission and current status of vessel by the operators.
10. Restriction of operation : Decided by the operators

Table 1. Type of conclusions from knowledge based system

No. of end leaf	Conclusions
Decision -1	Keep up mission
Decision -2	Make a move for safety zone and call backup forces
Decision -3	Repair
Decision -4	Bottoming
Decision -5	A stay in the water
Decision -6	Blowing
Decision -7	Hydrazine
Decision -8	Escape with suit
Decision -9	A distress buoy

Input Items by External Computer Program
1. Quantity of inflow and outflow of water : The inflow is calculated based on water depth and sectional area of damage. The outflow is determines based on the pump capacity and water depth.
2. Remained survival time : The remained survival time is obtained from calculating program that consider relationship between increase of internal air pressure and CO_2, decrease of O_2, decrease of body temperature in flooded zone, and a physiological limitation of the crew.
3. Blowing condition : The compressed air is important means to rise at flooded condition. Te decide whether to use it or not, the compressed air amount, number of usage times, rising speed to the surface and rising locus are calculated, and results are provided to operator for decision-making. The similar data in blowing condition is calculated and provided.

4. Use of hydrazine : The hydrazine is for emergency situation or when the compressed air has been exhausted.
5. Trim control : Trim moment by weight shift and ballast control is calculated. There are two ballast tanks at stern and bow.

Table 1 show the end leafs of the decision tree. The end leafs mean the final conclusion by inference in given situation by knowledge based system.

3.2 System Implementation

Previous attempts to develop expert systems to help decision processes can be grouped into two types. One attempt was to develop an expert system specialized for the specific domain. This attempt was to develop a small but efficient system which utilizes the characteristics of the application domain and selectively including the functions that are needed in that application domain. The second attempt was to use the general purpose expert system development tool. This allows flexible integration of the tool with external application programs such as calculation programs, database and so on. With general purpose expert system development tool, development time can be reduced because it provides a rich set of functions. In this paper, general-purpose expert system development tool, Expert Element, is used. It is programmed in C and uses object-oriented concepts. It has the following characteristics.

- Easy to interface with external programs because of the open architecture concept. Application programs developed in conventional languages such as C or Fortran can control the Expert Element, or the Expert Element can control the external application programs.
- To represent knowledge of the real world into the knowledge base, the object-oriented paradigm is adapted. Objects and rules comprise a hybrid knowledge representation.
- Knowledge base developed on different hardware can be shared without any modification.
- The system environment is as follow:
- OS : Windows NT
- Development tool for knowledge based system : Expert Element
- API : C

The obtained knowledge is expressed as rules in Expert Element. The rules consist of pressure hull and piping line damage.

The knowledge can be divided into static knowledge and dynamic knowledge. The static knowledge express objects and the dynamic knowledge express expertise to solve problem. To implement knowledge base, therefore, the objects of knowledge must be defined. The various knowledge representations such as class, object, property, method and rule are supported in Expert Element. Generally, objects represent real domains, and these objects may possess common feature. Class is a grouping or generalization of a set of objects. The objects are the smallest chunk of information and instance of class.

The rule is produced to implement decision support system based on defined the objects. The implemented rule in this paper is divided into two types of categories.

One is rules to solve problem and the other is meta-rules to control rules. The meta-rule is rule of rules and used to control whole decision making process. Fig. 3 shows a part of implemented knowledge base. It has two tree structures for pressure hull and piping line damage. Decision tree is constructed by cross-reference concept.

The external calculation program is developed for inflow and outflow quantity through damaged part with depth. The inflow quantity of seawater can be obtained by following expression;

$$Q = Ca \cdot A \cdot \sqrt{2gH} \qquad (1)$$

Where, Q = inflow quantity of seawater(m^3/sec)
Ca = frictional coefficient
A = sectional area of damaged part(m^2)
g = acceleration of gravity(m/sec^2)
H = water depth(m)

It is impossible for the sensors to be attached on the whole part of the hull to detect leakage. And it is not easy to find damaged parts by crews. Therefore the inflow quantity is obtained from above expression. To do this, installed sensors on bottom such as level float contact switch and pressure sensor measure the inflow water.

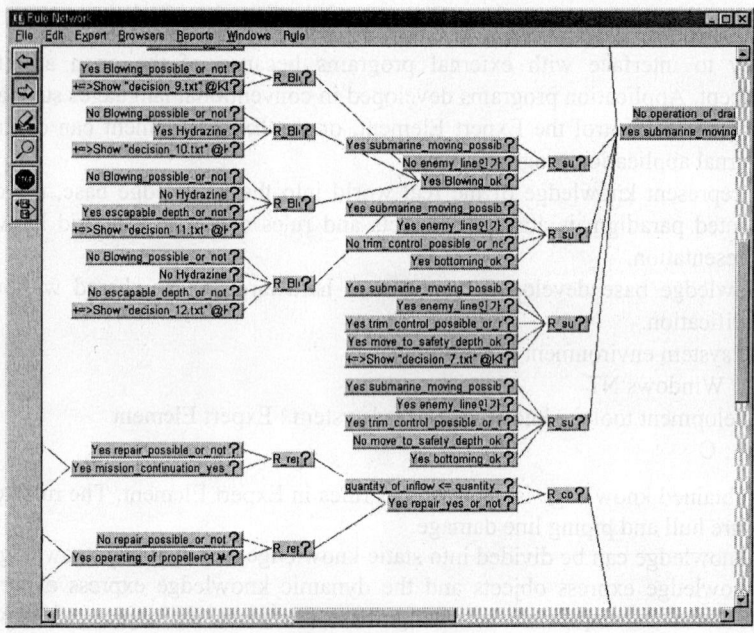

Fig.3. Knowledge base

4 Application Examples

To verify the system, the results are analyzed by test run using the potential input data. There are tested for some sample cases. In conclusion, the results well simulate for given input data and have good accuracy in comparison to decision criteria of the operator.

4.1 Piping Line Damage

Fig. 4 shows the result of application example for piping line damage. It contains inference procedure, fired rules and conclusion. The given situation is as follow:

- Piping line damage
- Pump is in malfunction
- The inflow quantity is less than the outflow.
- A cutoff of valve and bypass system is available.
- No more leakage
- Operation for mission is limited.

The conclusion for given situation is „make a move for safety zone and call backup forces".'

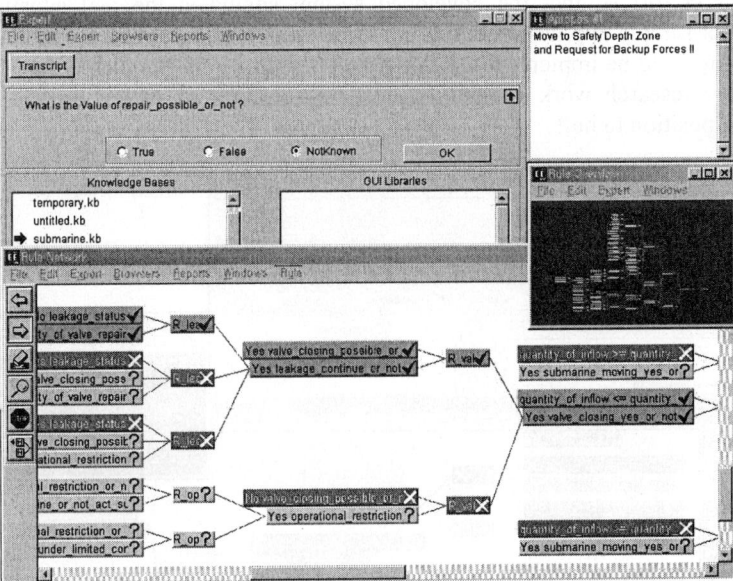

Fig. 4. The result of application example for piping line damage

4.2 Pressure Hull Damage

Fig. 5 shows the result of application example for pressure hull damage. It also contains inference procedure, fired rules and conclusion. The given situation is as follow:

- Pressure hull damage
- Pump is in malfunction
- Ship operation is not available.
- It is in enemy zone
- It is not available for bottoming and blowing.

The conclusion for given situation is „move to surface by hydrazine".

5 Conclusions

Knowledge-based response system was developed for submarine safety under damaged conditions of pressure hull and piping line system. The domain knowledge was acquired from document of submarine design, design expertise and interview with operator. The obtained knowledge was implemented to knowledge base according rule type after evaluation and validation process. The backward reasoning method for inference was adapted and graphic user interface was also developed.

It was confirmed that the developed system simulated the real situation well throughout sample application and it could be used for education of operators. And the system could be implement to submarine after functional supplementation. To do that, more research work is required such as selection of proper sensors and its attaching position to hull.

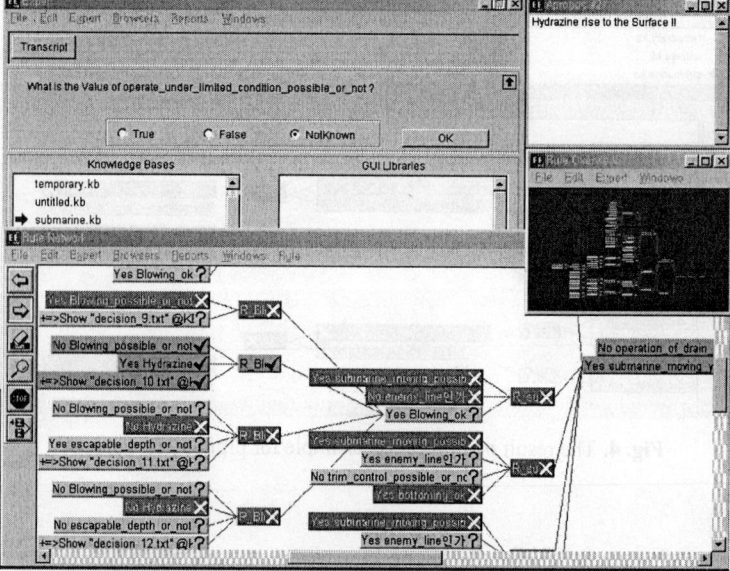

Fig. 5. The result of application example for pressure hull damage

References

1. http://www.cnn.com/SPECIALS/2000/submarine/
2. H. H Zhou, B. G Silverman, J. Simkol, CLEER : An AI System Developed to Assist Equipment Arrangements on Warship" May 1989, Naval Engineers Journal
3. R. B. Byrnes, et al "The Rational Behavior Software Architecture for Intelligent Ships" Mar. 1996, Naval Engineer Journal
4. Chou Y.C., Benjamin C.O., "An AI-Based Decision Support System for Naval Ship Design", Naval Engineers Journal, May 1992
5. J. H. Graham, R. K. Ragade, J. Shea "Design Decision Support System for the Engineering/Re-engineering of Complex Naval System, July 1997, Naval Engineer Journal
6. J. G. Shea "Virtual Prototyping using Knowledge-Based Modeling and Simulation Techniques" May 1993, Naval Engineer Journal
7. J. R. Hardin, et al "A Gas Turbine Condition Monitoring System" Nov. 1995, Naval Engineer Journal
8. R L. Carling "A Knowledge-Base System for the Threat Evaluation and Weapon Assignment Process", Jan. 1993, Naval Engineer Journal
9. J. Garber, J. Bourne, J. Snyder "Hull Structural Survival System" 1997, Advanced Marine Enterprise Ltd.
10. Scott, A.C, Clayton, J.E, Gibson, E.L "Practical Guide to Knowledge Acquisition" 1991, Addison-Wesley Publishing Company

On the Verification of an Expert System: Practical Issues

Jorge Santos[1], Zita Vale[2], and Carlos Ramos[1]

[1] Departamento de Engenharia Informática, Instituto Superior de Engenharia do Porto,
Rua Dr. António Bernardino de Almeida, 431, 4200-072 Porto – Portugal
{jorge.santos|csr}@dei.isep.ipp.pt

[2] Departamento de Engenharia Electrotécnica, Instituto Superior de Engenharia do Porto
Rua Dr. António Bernardino de Almeida, 431, 4200-072 Porto – Portugal
zav@dee.isep.ipp.pt

Abstract. The Verification and Validation (V&V) process states whether the software requirements specifications have been correctly and completely fulfilled. The methodologies proposed in software engineering showed to be inadequate for Knowledge Based Systems (KBS) validation, since KBS present some particular characteristics[1]. Designing KBS for dynamic environments requires the consideration of Temporal knowledge Reasoning and Representation (TRR) issues. Albeit, the last significant developments in TRR area, there is still a considerable gap for its successful use in practical applications.
VERITAS is an automatic tool developed for KBS verification, it is currently in development and is being tested with SPARSE, a KBS used in the Portuguese Transmission Network (REN) for incident analysis and power restoration. In this paper some solutions are proposed for still open issues on Verification of KBS applied in critical domains.

1 Introduction

The Verification and Validation is part of a wider process denominated Knowledge Maintenance [2]. The correct and efficient performance of any piece of software must be guaranteed through the Verification and Validation (V&V) process. It seems obvious that Knowledge Based Systems should undergo the same evaluation process.

It is known that knowledge maintenance is an essential issue for the success of the KBS since it must guarantees the consistency of the knowledge base (KB) after each modification in order to avoid KBS incorrect or inefficient performance.

This paper addresses the Verification of Knowledge-Based Systems in general, particularly focussing on VERITAS an automatic verification tool used in the verification of SPARSE. Section 2 describes SPARSE, a KBS used to assist the Portuguese Transmission Control Centres operators in incident analysis and power restoration, describing its characteristics and pointing the implications for Verification work. Section 3 presents VERITAS, an automatic verification tool. Some emphasis is given to the tool architecture and to the method used in anomaly detection. Finally, some conclusions and future work are presented.

Albeit there is no general agreement on the V&V terminology [3], the following definitions will be used for the rest of this paper.

- **Validation means building the right system** [4]. The purpose of validation is to assure that a KBS will provide solutions with same (or higher) confidence level as the ones provided by domain experts. Validation is then based on tests, desirably in the real environment and under real circumstances. During these tests, the KBS is considered as a "black box", meaning that, only the input and the output are really considered important.

- **Verification means building the system right** [4]. The purpose of verification is to assure that a KBS has been correctly designed and implemented and does not contain technical errors. During the verification execution the interior of KBS is examined in order to find any possible errors, this approach is also called "crystal box".

- **Anomaly** is a symptom of one (or multiple) possible error(s). Notice that an anomaly does not necessarily denotes an error [5]. Rule bases are drawn as a result of a knowledge analysis/elicitation process, including, for example, interviews with experts or the study of documents such as codes of practice and legal texts, or analysis of typical sample cases. The rule base should reflect the nature of this process, meaning that if documentary sources are used the rule base should reflect knowledge sources. Regarding that some anomalies are desirable an intentionally inserted in KB.

2 SPARSE

Control Centres (CC) are very important in the operation of electrical networks receiving real-time information about network status. CC operators should take (in a short time) the most appropriate actions in order to reach the maximum network performance.

In case of incident conditions, a huge volume of information may arrive to these Centres. The correct and efficient interpretation by a human operator becomes almost impossible. In order to solve this problem, some years ago, electrical utilities began to install intelligent applications in their Control Centres. These applications are usually KBS and are mainly intended to provide operators with assistance, especially in critical situations.

In the beginning, SPARSE [6] started to be an expert system (ES). It was developed for the Control Centres of the Portuguese Transmission Network (REN). The main goals of this ES were to assist Control Centre operators in incident analysis allowing a faster power restoration. Later the system evolved to a more complex architecture, which is normally referred as a Knowledge Based System (see **Fig. 1**).

One of the most important components of SPARSE is the knowledge base (KB).

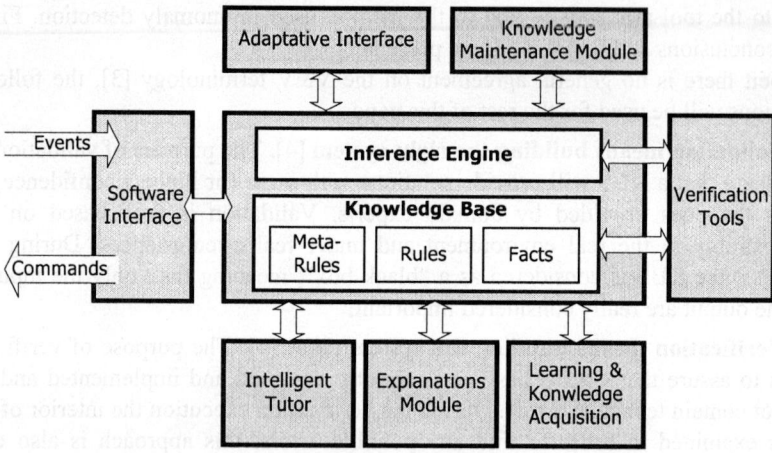

Fig. 1. SPARSE Architecture

The SPARSE KB uses rules with the following appearance:

```
'RULE' ID : 'DESCRIPTION' :
[
  [C1 'AND' C2 'AND' C3] 'OR'
  [C4 'AND' C5]
]
⇒
[A1,A2,A3].
```

Conditions considered in the LHS[1] (C1 to C5 in this example) may be of one of the following types:

- A fact, typically these facts are time-tagged;
- A temporal condition.

The actions/conclusions to be taken in RHS[2] (A1 to A3 in this example) may be of one of the following types:

- Assertion of facts (conclusions to be inserted in the knowledge base);
- Retraction of facts (conclusions to be deleted from the knowledge base);
- Interaction with the user interface.

The rule selection mechanism uses triggers (also called meta-rules) with the following structure:

trigger(Fact,[(Rule$_1$, Tb$_1$, Te$_1$),...,(Rule$_n$, Tb$_n$, Te$_n$)])
standing for:

- Fact – the arriving fact (external alarm or a previously inferred conclusion);
- Rule1,...,n – the rule that should be triggered in first place when fact arrives;

---

[1] Left Hand Side
[2] Right Hand Side

- `Tb1,...,n` – the delay time before rule triggering, used to wait for remaining facts needed to define an event;
- `Te1,...,n` – the maximum time for trying to trigger the each rule.

The inference process, (roughly speaking) relies on the following cycle (see **Fig. 2**). SPARSE collects one "message" from SCADA[3], then the respective "trigger" is selected and some rules are scheduled. The temporal window were the rule x could by triggered is defined in the interval [Tbx,Tex]. The "scheduler" selects the next rule to be tested, (the inference engines tries to prove its veracity). Notice that, when a rule succeeds, the conclusions (on the RHS) will be asserted and later processed in the same way as the SCADA messages.

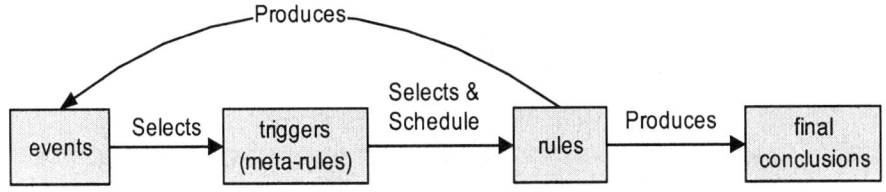

Fig. 2. SPARSE main loop

3 VERITAS

There were clearly two main reasons to start the verification of SPARSE. First, the SPARSE development team carried out a set of tests based on (previously collected) real cases and some simulated ones. Although these tests were very important for the product acceptance, the major criticism that could be pointed to this technique is that they only assure the correct performance of SPARSE under the tested scenarios.

Second, the tests applied in the Validation phase, namely the field tests, are very expensive because they required the assignment of substantial technical personnel and physical resources for their execution (e.g. transmission lines). Obviously, it is impossible to run those tests for each KB update. Under these circumstances, an automatic verification tool could offer an easy and inexpensive way to assure knowledge quality maintenance, assuring the consistency and completeness of represented knowledge.

A specific tool, named VERITAS [7] (see **Fig. 3**) has been developed to be used in the SPARSE verification. It performs structural analysis in order to detect knowledge anomalies. Originally, VERITAS used a non-temporal knowledge base verification approach. Although it proved to be very efficient in other KBS verification, some important limitations concerning SPARSE were detected.

The main goal of VERITAS is the anomaly detection and reporting, allowing the users decide whether reported symptoms reflect knowledge problems or not.

---

[3] Supervisory Control And Data Acquisition, this systems collects messages from the mechanical/electrical devices installed in the network

Basically, anomaly detection consists in the computation of all possible inference chain that could be produced during KBS performance. Later, the inference chains (expansions) are tracked in order to find out if some constraint is violated. Some well-known V&V tools, as KB-REDUCER[8], COVADIS[10] and COVER [9] used similar approaches. Notice that an expansion is an inference chain that could be generated during system functioning. For verification purposes, sometimes, an "expansion" could be explicitly computed and stored.

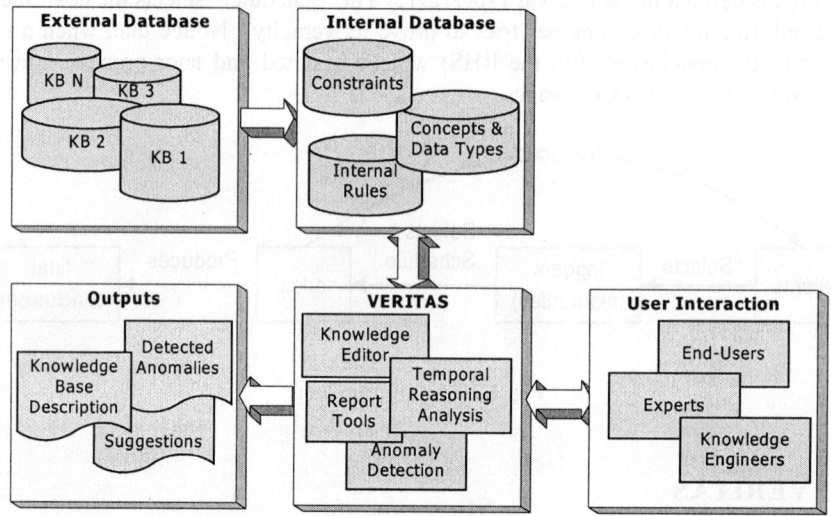

Fig. 3. VERITAS Architecture

After a filtering process, which includes "temporal reasoning analysis" and "variable evaluation" the system decides when to report an anomaly and some repair suggestion.

VERITAS is knowledge-domain and rule-grammar independent, due to these properties (at least theoretically), VERITAS could be used to verify every rule-based systems.

3.1 Some Verification Problems

As it was stated before, the anomaly detection relies in the computation of all possible inference chains (expansions) that could be entailed by the reasoning process. Later, some logical tests are performed in order to detect any constraints violation.

SPARSE presents some features (see **Fig. 4**) that made the verification work harder.

These features demand the use of more complex techniques during anomaly detection and introduce significant changes in the number and type of anomalies to detect. The most important are:

Variables evaluation: In order to obtain comprehensive and correct results during the verification process, the variables evaluation is crucial, especially for the temporal variables that represent temporal concepts. Notice that during anomaly detection (this type of verification is also called static verification) it is not possible to predict the values that a variable will have.

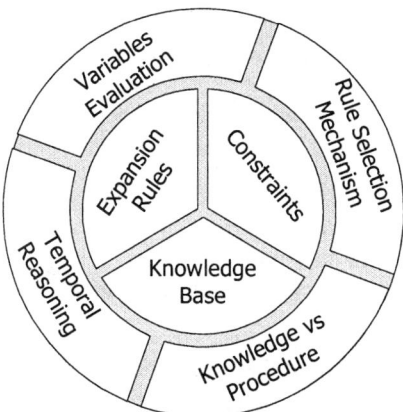

Fig. 4. The verification problem

Rule triggering selection mechanism: This mechanism was implemented using both meta-rules and the inference engine. When a "message" arrives, some rules are selected and scheduled in order to be later triggered and tested. In what concerns verification, this mechanism not only avoids some run-time errors (for instance circular chains) but also introduces another complexity axis to the verification. Thus, this mechanism constrains the existence of inference chains. For instance, during system execution, the inference engine could be able to assure that shortcuts (specialists rules) would be preferred over generic rules.

Temporal reasoning: This issue received large attention from scientific community in last two decades (surveys covering this issue could be found in [11] [12]). Despite the facts that "time" is ubiquitous in the society and the natural ability that human beings show dealing with it, a widespread representation and usage in the Artificial Intelligence domain remains scarce due to many philosophical and technical obstacles. SPARSE is an "alarm processing application" and its most difficult task is reasoning about events. Thus, it is necessary to deal with time intervals (e.g. temporal windows of validity), points (e.g. events occurrence), alarms order, duration and the presence or/and absence of data (e.g. messages lost in the collection/transmission system).

Knowledge versus procedure: When using a "programming" language to specify knowledge, sometimes knowledge engineers could be tempted to use "native" functions. For instances, the following sentence in PROLOG: X is Y + 10 is not an (pure) item, in fact is a sentence that should be evaluated in order to get the X value. It means the verification method needs to know not only the programming language syntax but also the meaning (semantic) in order to evaluate the functions. This step is particularly important when some variables values are updated.

3.2 Variables Evaluation

In order to obtain acceptable results during anomaly detection the variables evaluation is a crucial aspect during expansion computation. The used technique relies on variable name replacing in the original rule by a string that uniquely represents the argument (variable). Later, during rule expansion generation this information will be used to check possible matches.

Different match types could be used during evaluation (see **Fig. 5**), notice that "**Optimal**" match corresponds to the one that is obtained during system functioning:

Open: In this type of match, free variables could be instantiated with terminals,, consider the following example: disconnector(X), disconnector(open), where X could take "open" value. In this type of match, a large number of expansions is considered and consequently a considerable number of "false" anomalies could be detected;

Close: In this case it is not allowed that free variables would match with terminals (in opposition to the "open" previous example). The reduced number of rule expansions generated will make the detection process simpler but the trade-off is that some anomalies could remain undetected;

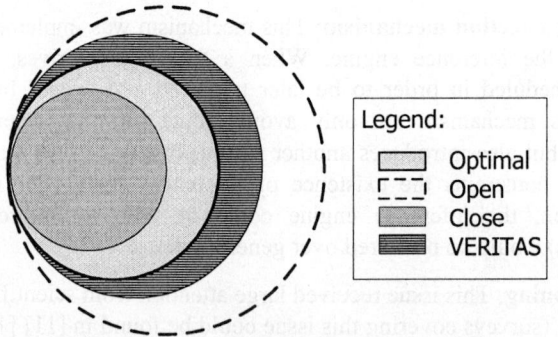

Fig. 5. Relation between match type and expansions number

VERITAS: The used approach intends to obtain the best of the two previous kinds of matches, being "**Open**" in order to allow detection of all possible anomalies but at same time generate the minimum number of expansions. How is that achieved? Firstly, a free variable just could be replaced by valid values in the knowledge domain, previously defined in the database (e.g. breaker(X), in the SPARSE domain could have the following values: open, close, moving). Secondly, the expansion rules are computed as late as possible and kept in a graph structure, meaning that, in the graph (**Fig. 6**) the following pairs would be kept (where <--> stands for matches):

breaker(X) <--> breaker(open)   (1)

breaker(close) <--> breaker(X)   (2)

But later, during expansion rules computation, a similar process to PROLOG instantiation mechanism is used, so that each variable instantiation is reflected in the following transitions. This means that an expansion that uses both relations would not be generated, meaning that, after x becomes instantiated with open it could not be later instantiated with close.

3.3 Anomaly Detection

As it has been described in section 2, SPARSE has some specific features. Regarding these features the used technique is a variation of common Assumption-based Truth Maintenance System (ATMS) [14]. Namely, the knowledge represented in the meta-rules had to be considered in rule expansions generation.

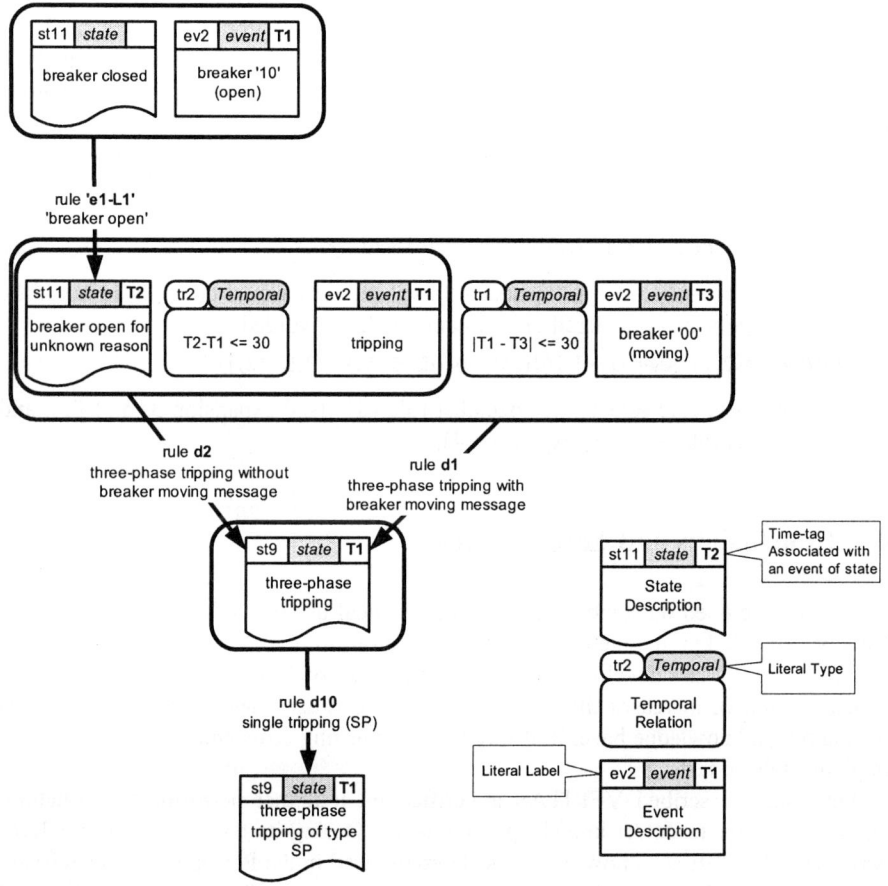

Fig. 6. Hypergraph representation

In the previous work of the authors [7], during rule expansions calculation, one data structure was created for each possible inference chain. Currently, each "literal" keeps information about the list of it predecessors, more precisely, the list of all rules that could infer this literal. It means that it is possible with a graph-visiting algorithm to compute the desired expansions. This approach showed to be more time-efficient (during anomaly detection, since some tests do not need to know all rule expansions) and data flexible (e.g. circular chains became possible to exist in the verification database).

VERITAS shows the existing relations between the rules in the KB using a directed hypergraph type representation (see **Fig. 6**). This technique allows rule representation in a manner that clearly identifies complex dependencies across compound clauses in the rule base and there is a unique directed hypergraph representation for each set of rules [13]. In the following example, a set of rules ('e1-L1',d1,d2,d10) is considered.

The VERITAS verification tool is able to detect anomalies existing in KB in a first step using logic tests. The detected "logic" anomalies could be grouped in three groups: redundancy, circularity and ambivalence (a complete classification could be found in [7]). Later, the verification system performs temporal reasoning analysis which allows filtering previously detected anomalies, providing really accurate information, namely by preventing false anomaly reporting.

For instances, in the previous example, two expansions would be generated:

$$(st11+ev2) \rightarrow \text{'e1-L1'}:(st11+tr2+ev2+tr1+ev2)\rightarrow d1:(st9)\rightarrow d10:(st9) \quad (3)$$

$$(st11+ev2) \rightarrow \text{'e1-L1'}:(st11+tr2+ev2)\rightarrow d2:(st9)\rightarrow d10:(st9) \quad (4)$$

If only logical analysis is performed a "false" anomaly would be reported but if the following triggers are considered, this situation will be avoided.

'TRIGGER'(ev2,[breaker open],[(d1,30,50),(d2,31,51),(d3,52,52)]).

Thus, the meta-rule selects the specialist rule (d1) (see expansion 3) and first and later the general rule (d2) (see expansion 4).

4 Conclusions and Future Work

This paper focussed on some aspects of the practical use of KBS in Control Centres, namely, knowledge maintenance and its relation to the Verification process. It becomes clear to the development team that the use of Verification tools (based on formal methods), increases the confidence of the end-users and eases the process of maintaining a knowledge base. It also reduces the testing costs and the time needed to implement those tests.

This paper described VERITAS a verification tool that performs the structural analysis in order to detect knowledge anomalies. During its tests the SPARSE KBS was used. SPARSE is a knowledge based system used in the Portuguese Transmission Network (REN) for incident analysis and power restoration. Some characteristics that mostly constrained the development and use of verification tool were pointed, namely, temporal reasoning and variables evaluation.

The problem of verifying temporal properties on intelligent systems take its roots in the method used to codify temporal reasoning. The verification task will be harder if is it difficult to distinguish between the domain knowledge and temporal knowledge. The solution appears to be some method that allows to integrate temporal ontologies (domain independent) with domain ontologies in a easy mode. Moreover, the verification tools could focus just the integration aspects since that temporal ontology could be previously verified.

Acknowledgements

The authors of this paper would like to acknowledge PRAXIS XXI Research Program, FEDER, the Portuguese Science and Technology Foundation and Innovation Agency for their support of SATOREN Project and Project PRAXIS/3/3.1/CEG/2586/95.

References

[1] O.Garcia and Y.Chien - "Knowledge-Based Systems: Fundamentals and Tools", IEEE Computer Society Press, 1991
[2] T.Menzies - "Knowledge Maintenance: The State of the Art, The Knowledge", Engineering Review, 1998
[3] T.Hoppe and P.Meseguer – "On the terminology of VVT", Proc.European Workshop on the Verification and Validation of Knowledge-Based Systems, Logica Cambridge, 3-13, Cambridge, 1991
[4] B.Boehm, "Verifying and validating software requirements and design specifications", IEEE Software, 1(1), 1984
[5] A.Preece and R.Shinghal "Foundation and Application of Knowledge Base Verification", Intelligence Systems, 9:683-701, 1994
[6] Z.Vale, A.Moura, M.Fernandes and A.Marques, "SPARSE - An Expert System for Alarm Processing and Operator Assistance in Substations Control Centers", ACM (Association for Computing Machinery) Press, Applied Computing Review, 2(2):18-26, December 1994
[7] J.Santos, L.Faria, C.Ramos, Z.Vale and A.Marques - "VERITAS – A Verification Tool for Real-time Applications in Power System Control Centers". Proc.12th International Florida AI Research Society (FLAIRS'99). Eds. AAAI Press, 511-515. Orlando, 1999
[8] A.Ginsberg – "A new approach to checking knowledge bases for inconsistency and redundancy", Proc.3rd Annual Expert Systems in Government Conference, IEEE Computer Society, Washington D.C., pp.102-111, October, 1987
[9] A.Preece – "Towards a methodology for evaluating expert systems", Expert Systems (UK), 7(4), 215-223, November 1990
[10] M.Rousset – "On the consistency of knowledge bases: the COVADIS system", Proceedings of European Conference on Artificial Intelligence, pp.79-84, Munchen, 1988
[11] L.Vila – "A Survey on Temporal Reasoning in Artificial Intelligence", AI Communications, 7:4-28, March 1994

[12] "Spatial and Temporal Reasoning", Eds Oliviero Stock, Kluwer Academic Publishers, 1997
[13] M.Ramaswary and S.Sarkar – "Global Verification of Knowledge Based Systems via Local Verification of Partitions", Proc. 4th European Symposium on the Validation and Verification of Knowledge Based Systems (Eurovav'97), 145-154. Leuven, Belgium, 1997
[14] J.Kleer – "An assumption-based TMS", Artificial Intelligence (Holland), 28(2), 127-162, 1986
[15] A.Preece, R.Shinghal, and A.Batarekh "Verifying expert systems: a logical framework and a practical tool" Expert Systems with Applications, 5(2):421-436, 1992

DOWNSIZINGX : A Rule-Based System for Downsizing the Corporation's Computer Systems

J. L. Mitrpanont and T. Plengpung

Department of Computer Science, Faculty of Science, and
Mahidol University Computing Center,
Mahidol University, Rama VI Rd., Bangkok 10400, THAILAND
ccjlm@mahidol.ac.th
thammskp@scb.co.th

Abstract. Downsizing the corporation's computer systems is still an important practice of many organizations. The feasibility of downsizing and breakthroughs in computer technology reinforce this concept, and cause corporations to downsize for cost reduction and to increase the efficiency and effectiveness of business operations. The problems of many processes and related complex factors need to be considered. The product of this research, „*A Rule-Based System for Downsizing the Corporation's Computer Systems (DOWNSIZINGX),*" is a tool that uses a model of a knowledge base that provides a visual programming or window interface with an expert system shell. The system provides recommendations to support the decision-making of downsizing the computer systems. It uses CLIPS as an expert system shell and makes inferences about knowledge by using the forward chaining and the depth first strategy as a conflict resolution technique. A user interacts with the system through Visual Basic Interface connected to CLIPS via CLIPS OCX. The explanation facility allows the user to trace back the recommendations to find what factors were involved and to assist effective decision-making with reduced risks. The prototype was developed corresponding to the model and tested on some business decision scenarios and case studies. The results show that the system performs all functions correctly and also recommends situations effectively. Moreover, the system demonstrates the flexibility of the implementation of the user interface through Visual Basic programming tool which will enable and enhance more friendly interface for the users.

1 Introduction

Cost reduction and „well operate" are the significant issues of many corporations, especially mainframe-based corporations [1]. With the rapid change of information technology, new hardware and software, including new architecture such as client-server, network computing and others, have convinced the corporation's leaders and boost the necessity of downsizing approach [2] [3] and [4]. However, implementing the downsizing have both benefits and pitfalls [5] due to its complexity contributed from many factors and processes that managers must be concerned. Therefore, managers need a specific tool to support them in the decision-making processes to reduce risk and impact that may be resulted from implementation. However, most original expert system shells are not flexible enough when they are used to develop the applications. They provide command lines syntax and not very user friendly. Additionally, at present most people have more skills of visual programming so it will be beneficial to apply it in designing and implementing many applications of knowledge base and expert system that use a production rule.

In this paper, we present a knowledge base architecture that provides visual programming interface with expert system shell and develop „A Rule-Based System for Downsizing the Corporation's Computer Systems" as the prototype. DOWNSIZINGX comprises many factors that are related to area in which the management must make a decision. *Such factors as corporation's objective, policy, value, customer, competitive situation, economy, risk, government's policy, technology, impact of the project, computer system, corporation's structure, finance, and human resource are provided.* The knowledge component of DOWNSIZINGX is also presented. The techniques of *the Structured Situation Analysis, the Decision Table Analysis and Dependency Diagram are used to demonstrate the overall knowledge design* before they are developed with the production rules system. See details in [6]. The system provides an interface for the expert to define rules together with questions, answers and recommendations through CLIPS Window Interface. A set of questions and valid answers specific to each module are defined. As a result, we developed approximately 2000 if-then rules. The DOWNSIZINGX Inference Mechanism is implemented by CLIPS and invoked through Visual Basic programming via CLIPS OCX. The inference engine of CLIPS uses forward chaining process to derive the recommendation results, and uses the depth first strategy as the conflict resolution strategy in selecting the rules to process.

Additionally, DOWNSIZINGX also provides the explanation facility. A user can trace the output results back to find the related information. The „Why and How questions" that the user wants to know are resolved. Furthermore, the system provides the print function to show the intermediate results that present the internal mechanism the system works. So, the user can understand more about the functionality of the knowledge-base expert system as well.

2 The Knowledge Base Model with CLIPS Engine

In order to develop the DOWNSIZINGX system, the architecture of the knowledge base model interfacing with CLIPS engine [7] has been designed. Figure 1 demonstrates this architecture.

Knowledge Base. The knowledge base contains knowledge necessary for understanding, formulating, and solving problems. It includes two basic elements:

- Facts, such as the problem situations and theories of the problem.
  Rules that direct the use of knowledge to solve specific problems.

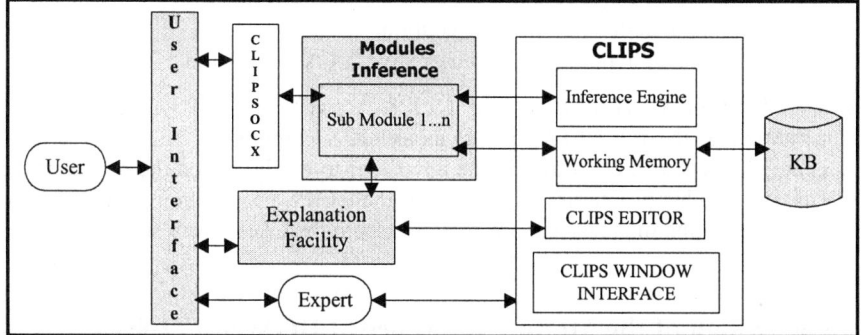

Fig. 1. The Architecture of Knowledge Base Model with CLIPS Engine

An Explanation Facility. The Explanation Facility acts as the explanation of the system by using Visual Basic programming. It has an ability to trace and explain the conclusion of some questions, such as the following:

- How was a certain conclusion reached?
- What rules made that conclusion?
- What facts are the input facts or what facts are the derived facts?
- What questions made the attribute of resource?
- Why was a certain alternative rejected?

Module Inference. The Module Inference comprises of several basic inference modules that are developed on CLIPS programs to manage the input data from the user, and control mechanism in rule interpreter of all rules in each module. It interacts with inference mechanism of CLIPS to execute the modules, and then compares the known fact to match the recommendation before it issues to the user. This module must be designed specific to the problem domain, see details in section 3.

CLIPS Environment. The CLIPS environment comprises four major components [7], i.e., CLIPS Windows Interface, CLIPS editor, CLIPS Inference engine, and Working Memory.

- *CLIPS Windows Interface.* The CLIPS Window Interface performs as acquisition subsystem. It accumulates and transforms the problem-solving's expertise from knowledge sources to a computer program.
- *CLIPS Editor.* The CLIPS editor is used to edit rules, facts, and write programs.
- *CLIPS Inference Engine.* The 'brain' of the KBS model is the inference engine, also known as the control structure or the rule interpreter. This component is essentially a computer program that provides a methodology for reasoning about information in the knowledge base and in the working memory, and for formulating conclusions.
- *Working Memory.* The area of working memory sets aside for the description of a current problem, as specified by the input data; it is also used for recording

intermediate results. The working memory records intermediate hypotheses and decisions. Three types of decisions can be recorded on the working memory:
Plan: how to attack the problem.
Agenda: potential actions awaiting execution.
Solution: candidate hypotheses and alternative courses of action.

CLIPS ActiveX or CLIPS OCX. The CLIPS ActiveX or CLIPS OCX [8] is an implementation of the full CLIPS engine as an OCX component. All of the CLIPS features are enabled. The OCX allows the developer to easily embed the power of the CLIPS engine into his programs. The details of the CLIPS programming environment are documented in the manuals and are available from the CLIPS web site [7].

3   The Design of the Module Inference of DOWNSIZINGX

In this section, the module inference of DOWNSIZINGX is presented. Based on the knowledge acquisition of downsizing, the system is divides into two major modules:

Initial Screening. Initial screening is broken down into *readiness, value and technology*. When we downsize the corporation's computer systems, the first aspect to be considered is what benefits or values we will get if we do it. Ultimately, we have to consider whether the feasibility of the current technology can reinforce this process successfully. Readiness is a factor that divided into internal and external factors.

- *Internal factors*. We consider the structure of the corporation, objective and available resources. These factors must be analyzed and recommendation should be given. The resource factors of a corporation comprise of human resource, finance and computer systems. Human resource is important because a corporation will succeed only when its personnel are ready. Finance is very important too because a corporation cannot do anything if it has no funding. The last section of resources is the computer system. We determine how well it can support business operation, capacity, and state. These factors are used as criteria when downsizing is considered.
- *External factors*. We analyze four factors that affect a corporation. They are economy, competitive situations, customers and government's policy.

Final Screening. Final screening is broken down into *risk, organization's policy and impact*. The detail of each part can be found in [6].

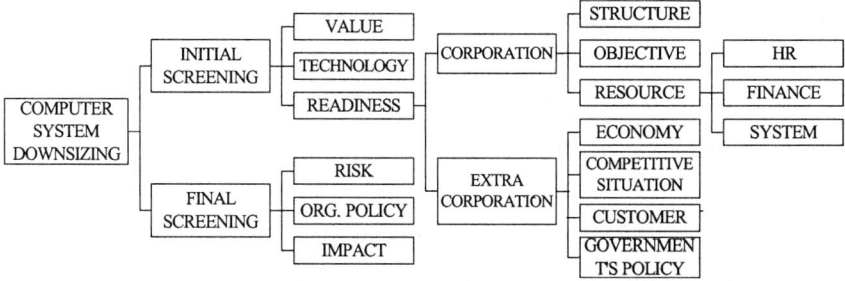

Fig. 2. Structured Situation Diagram of Main Modules of DOWNSIZINGX

After the knowledge acquisition of DOWNSIZINGX system has been done, the components of DOWNSIZINGX module inferences are developed and comprised of 14 modules as depicted in Figure 3.

DOWNSIZINGX Module Inference			
Human Resource	Government's Policy	Impact	Risk
Finance	Technology	Competitive Situation	Customer
Economy	Computer System	Value	
Corporation's Objective	Corporation's Policy	Corporation's Structure	

Fig. 3. The components of DOWNSIZINGX

Below is an example of Technology module inference.

```
;;The HANDLE-RULES module;;
(defmodule HANDLE-RULES(import MAIN ?ALL)(export ?ALL))
(deftemplate HANDLE-RULES::rule
   (multislot name)multislot if)multislot then))
(defrule HANDLE-RULES::throw-away-ands-in-antecedent
   ?f <- (rule (if and $?rest))
   => (modify ?f (if ?rest)))
(defrule HANDLE-RULES::throw-away-ands-in-consequent
   ?f <- (rule (then and $?rest))
   => (modify ?f (then ?rest)))
(defrule HANDLE-RULES::remove-is-condition-when-satisfied
   ?f <- (rule (if ?attribute is ?value $?rest))
   (attribute (name ?attribute)(value ?value))
   => (modify ?f (if ?rest)))
(defrule HANDLE-RULES::remove-is-not-condition-when-satisfied
   ?f <- (rule (if ?attribute is-not ?value $?rest))
   (attribute (name ?attribute) (value ~?value))
(modify ?f (if ?rest)))
(defrule HANDLE-RULES::perform-rule-consequent
   ?f <- (rule (if)(then ?attribute is ?value $?rest))
   => (modify ?f (then ?rest))
```

```
(assert (attribute (name ?attribute)(value ?value))))

(defmodule RULES (import HANDLE-RULES ?ALL)
          (import HANDLE-QUESTIONS ?ALL)
          (import MAIN ?ALL))
(defrule RULES::startit => (focus HANDLE-RULES))
(deffacts the-ds-rules

(rule (name rule-enable-downsizing-1)
(if high_capacity is yes and
 new_architecture is yes and compatibility is yes)
 (then enable_downsizing is yes))
(rule (name rule-enable-downsizing-2)
(if high_capacity is yes and cheaper is no and
 new_architecture is no)
 (then nable_downsizing is no))
(rule (name rule-enable-downsizing-3)
(if high_capacity is yes and cheaper is yes and
 new_architecture is yes)
 (then enable_downsizing is yes))
(rule (name rule-enable-downsizing-4)
(if high_capacity is no and cheaper is no)
 (then enable_downsizing is yes))
(rule (name rule-enable-downsizing-5)
(if high_capacity is yes and cheaper is yes and
 new_architecture is no and compatibility is no)
 (then enable_downsizing is no))
(rule (name rule-enable-downsizing-6)
(if high_capacity is yes and cheaper is no and
 new_architecture is yes and compatibility is yes)
 (then enable_downsizing is yes))
(rule (name rule-enable-downsizing-7)
(if high_capacity is no and
 new_architecture is no and compatibility is no)
 (then enable_downsizing is no))
(rule (name rule-enable-downsizing-8)
(if high_capacity is no and
 cheaper is yes and new_architecture is yes and
 compatibility is yes)
 (then enable_downsizing is yes))
(rule (name rule-enable-downsizing-9)
(if high_capacity is no and cheaper is yes and
 new_architecture is yes
 and compatibility is no)
 (then enable_downsizing is no))
```

4 DOWNSIZINGX System Design and Implementation

The system of DOWNSIZINGX consists of four major processes as shown in Figure 4. The details of system design and user interfaces are describes in [9].

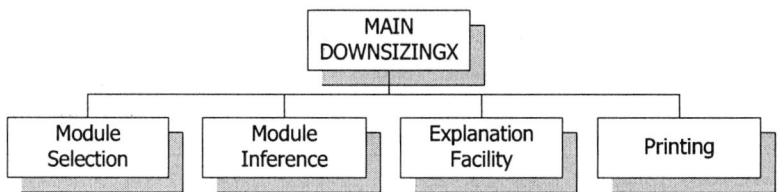

Fig. 4. Four Main Processes of DOWNSIZINGX

Examples of input and output design of DOWNSIZINGX are shown in Figure 5 through Figure 8 as follows.

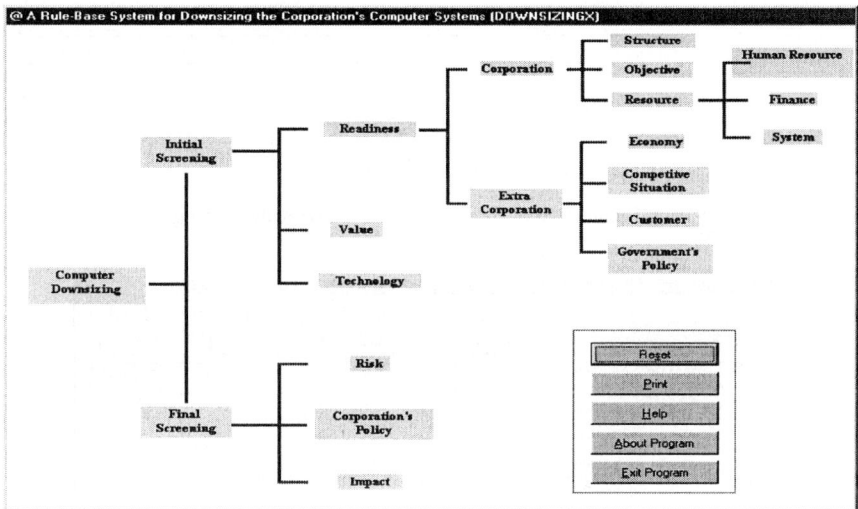

Fig. 5. User Interface of DOWNSIZINGX

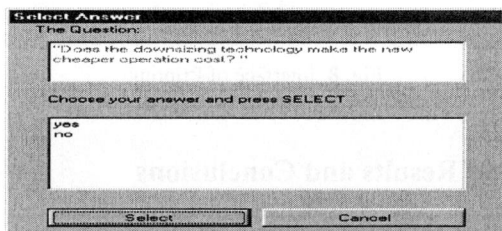

Fig. 6. Interface of Question and Answer

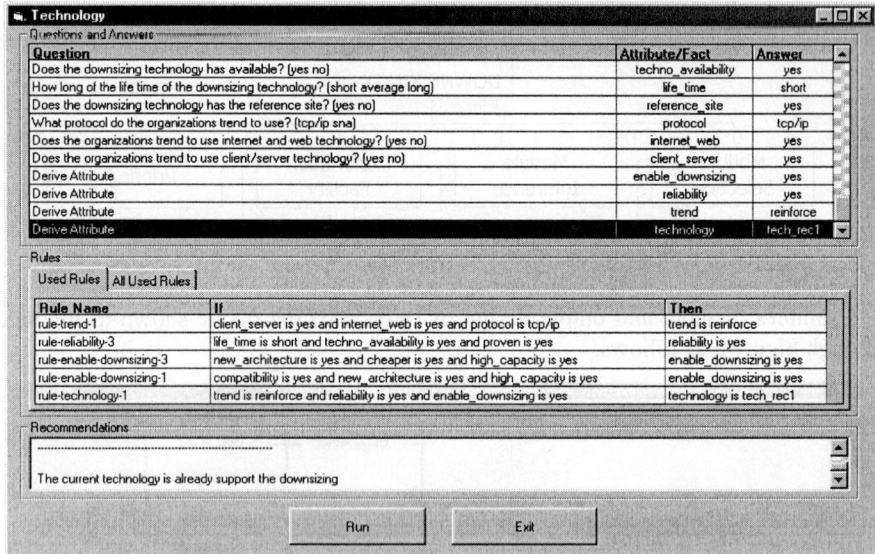

Fig. 7. Interface of Recommendation and Rule Tracing

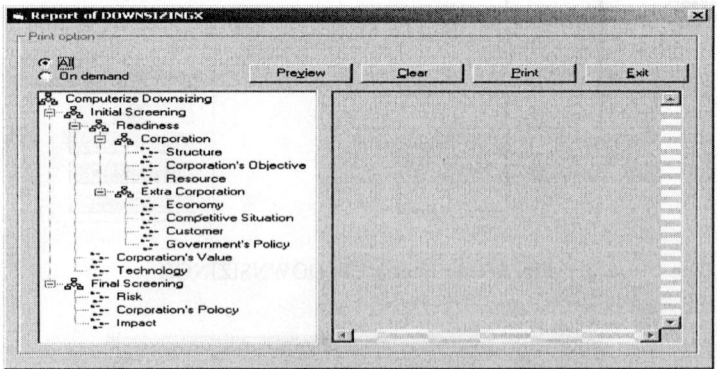

Fig. 8. Interface of Printing

5 Experimental Results and Conclusions

The prototype was tested with some business scenarios. The inference mechanisms are invoked through the Visual Basic programming interface and the recommendation results are verified. In addition, the correctness was measured by testing all functions of DOWNSIZINGX, and compared the results between running program via CLIPS Windows Interface versus running them through Visual Basic by using CLIPS OCX.

Below is a partial result of an example of business scenario which emphasizes on the recommendation obtained from the „System" module of the DOWNSIZINGX prototype, see Figure 2. However, more details can be found in [9].

Scenario 1. Use DOWNSIZINGX prototype to recommend the Computer System Situation

In this scenario, the manager of company A wants to know the capability and performance of the current computer system to consider whether downsizing the computer system would be necessary. He receives the data from the computer department as summarized in Table 1, and then he uses the DOWNSIZINGX to help him analyze the data.

Table 1. Information about the Computer System of Company A

Items	Characteristic or Feature
1. CPU IBM Mainframe 9672 R36	4 processors, 160 MIPS. Y2K passed. Fiber channel
- Avg. CPU Utilization	36 %
- Peak CPU Utilization	100 %
2. Applications	20 Applications running per day.
3. No. of downtime in this month	1
Interval time to up	20 Min.
4. Cost	
Investment Cost	High
Operation Cost	High
Maintenance Cost	High
6. Performance of Applications	
Deposit	3 sec/transaction
Loan	10 sec/transaction
Back Office	10 sec/transaction
6. Transactions per day	700,000
7. Time of use	2 years
8. Vendor support	24 Hr per day

The prototype system recommended that the current computer system of company A should be upgraded because of cost overruns the budget and high cost of operation occurs. However, more considerations on the other factors must be evaluated. Figure 9 demonstrates the recommendation interface of the „Computer System" module from the DOWNSIZINGX prototype. More experiments can be found in [6].

In summary, we have presented the development of the DOWNSIZINGX prototype system. The knowledge base of the system is acquired mainly from experts, online documents and textbooks [6]. The design of the system was based on an alternative knowledge model of interfacing the expert system shell with Visual Basic programming tools to provide the flexibility in implementation of knowledge base and expert system applications that use the production rules. The system demonstrates the flexibility of the implementation of the user interface as well. However, to provide more support on the decision-making and improve the operation of the organization, the characteristic of object oriented programming such as reusability should be considered in order to provide a better methodology for developer to implement other knowledge base and expert system.

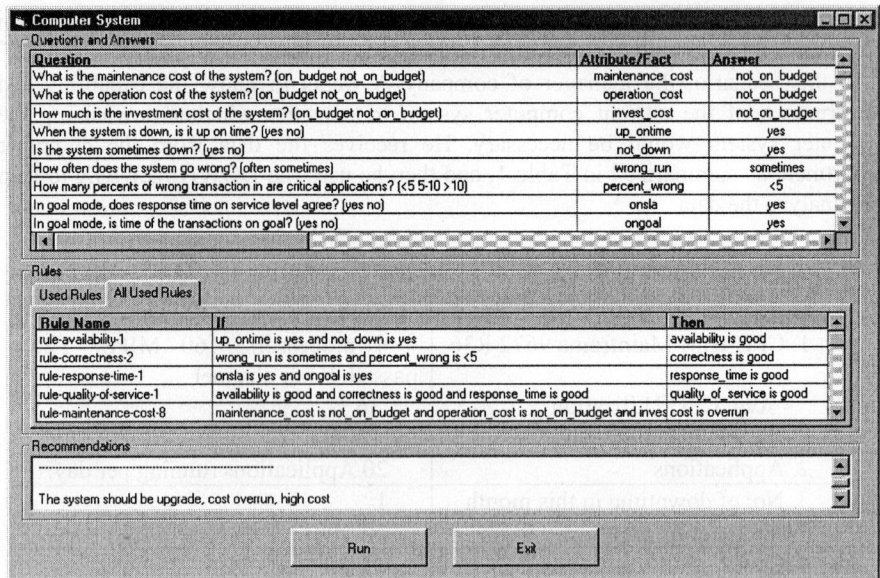

Fig. 9. Interface of recommendation of the „Computer System" inference module of the DOWNSIZINGX prototype

References

1. M. H. Azadmanesh, Daniel A. Peak: The Metamorphosis of the Mainframe Computer: The Super Server. [Online] available from:
   http://hsb.baylor.edu/ramsower/acis/papers/peak.htm (1995)
2. S. Guengerich: Downsizing Information Systems. Camel IN : Sams (1992)
3. D. Trimmer: Downsizing Strategies for Success in the Modern Computer World. Cambridge : Addison-Wesley (1993)
4. J. Hansen: Rightsizing a Mainframe Administrative System Using Client/Server. [Online] available from: http://www.cause.io/orado.edu (1993)
5. H. Fosdick: How to avoid the pitfalls of Downsizing. Datamation Magazine. May 1 (1992) 77-80
6. T. Plengpung: DOWNSIZINGX: A Rule-Based System for Downsizing the Corporation's Computer System. Master Thesis, Department of Computer Science, Faculty of Science, Mahidol University, Bangkok, Thailand (2001)
7. Gary Riley: CLIPS. [Online] available from:
   http//www.ghg.net:clips/CLIPS.html (2001)
8. M. Tomlinson: CLIPS OCX. [Online] Available from:
   http://ourworld.compuserve.com/homepages/marktoml/clipstuf.htm (1999)
9. J.L. Mitrpanont and T. Plengpung: DOWNSIZINGX: A Rule-Based System for Downsizing the Corporation's Computer System. MU-TR-44-004, Technical Report, Computer Science, Mahidol University, Bangkok, Thailand (2001)

Credit Apportionment Scheme for Rule-Based Systems: Implementation and Comparative Study

Nabil M. Hewahi and H. Ahmad

Computer Science Department
Islamic University of Gaza, Palestine
nhewahi@mail.iugaza.edu

Abstract. Credit Apportionment scheme is the backbone of the performance of adaptive rule based system. The more cases the credit apportionment scheme can consider, the better is the overall systems performance. Currently rule based systems are used in various areas such as expert systems and machine learning which means that new rules to be generated and others to be eliminated. Several credit apportionment schemes have been proposed and some of them are even used but still most of these schemes suffer from disability of distinguishing between good rules and bad rules. Correct rules might be weakened because they are involved in an incorrect inference path (produces incorrect conclusion) and incorrect rules might be strengthen because they are involved in an inference path which produces correct conclusion. In this area a lot of research has been done, we consider three algorithms, Bucket Brigade algorithm (BB), Modified Bucket Algorithm (MBB) and General Credit Apportionment (GCA). The algorithms BB and MBB are from the same family in which they use the same credit allocation techniques where GCA uses different approach.

In this research, we make a comparison study by implementing the three algorithms and apply them on a simulated "Soccer" expert rule-based system. To evaluate the algorithms, two experiments have been conducted.

Keywords: Rule-Based Systems, Credit Apportionment Scheme, Adaptive Systems

1 Introduction

Rule-based systems are very common systems used in expert systems. The standard rule structure is (IF <condition> THEN <action>) which we shall consider in our research. The problem of assigning responsibility for system behavior was recognized and explored in pioneering work in machine learning conducted by Samuel [14] and termed the credit assignment problem in [12].

The credit apportionment (credit assignment) problem has been one of the major areas of research in machine learning. Several attempts have been made to develop appropriate schemes for rule-based systems [2][3][5][7][9][11]. A very well known credit apportionment scheme is the Bucket Brigade algorithm (BB) proposed by Holland [9] for the classifier system of genetic algorithms. It may most easily envisioned as information economy where the right to trade information manufactures (the environment) to information consumer (the effectors). Two main functions are performed by BB: 1. Assign credit to each individual rule   2. Resolves the conflict resolution problem [9][10].

As noted by Westerdale [16], the BB algorithm is a sub-goal reward scheme, that is, a production p is rewarded for placing the system into a state, which satisfied the condition part of another production q, provided that q leads to payoff. However, if the subgoals are not well formulated, it may be that q obtains payoff in spite of the help of p rather than because of it (i.e., other states satisfying q's condition may be more describe). Even under these circumstances, p gets rewarded. Westerdale [16] suggested an alternate scheme, called genetic reward scheme in which the payoff, which enters the system, is divided in some other way. Antonisse [2] explored the adaptation of BB to production system where he addresses the problem of unsupervised credit assignment. He concluded that the BB is an inexpensive algorithm and stable method that appears to be convergent. However according to Antonisse the main limitations of BB is that it does not give absolute measure of performance for rules in the system and the propagation of credit along paths of inference is proportional to the lengths of these paths, rather than constant. Huang [11] realized that the usefulness of the action conducted by a rule is a function of the context in which the rule activates and scalar-valued strengths in BB algorithms are not appropriate approximator for the usefulness of rule action. He therefore proposed an algorithm called context-array BB algorithm in which more sophisticated rule strength such as array-value strengths are used. In order to handle the default hierarchy formation problem in the standard bucket brigade algorithm, Riolo [13] suggested simple modification to the standard bucket brigade algorithm. The modification involves biasing the calculation of the effective bid so that general classifiers (those with low bid ratios) have much lower effective bids than do more specific classifiers (those with high bid ratios). Dorigo[4] address the same problem and proposed an apportionment of credit algorithm, called message-based bucket brigade , in which message instead of rule are evaluated and a rule quality is considered a function of the message matching the rule conditions, of the rule specificity and of the value of the message the rule tries to post.

Spiessens [15] has shown that the BB algorithm is not very efficient in providing accurate predictions of the expected payoff from the environment and presents an alternative algorithm for a classifier system called PCS. The strength of a classifier in PCS is reliability measure of the production of this classifier.

The problem of not adequately reinforced long action sequences by BB has been discussed by Wilson [17] and an alternative form of the algorithm was suggested. The algorithm was designed to induce behavior hierarchies in which modularity of the hierarchy would keep all the BB chains sort, thus more reinforce able and more rapidly learned, but overall action sequences could be long. The problem of long action chains was mentioned for the first time by Holland [10] and he suggested a way to implement "bridging classifier" that speeds up the flow of strength down a long sequence of classifiers.

Classifier systems has difficulty forming long chains of rules and can not adequately distinguish overly general from accurate rules. As a result, the classifier system does not learn to solve sequential decision problems very well. Derrig and Johannes [3] proposed a credit assignment scheme called Hierarchical Exempler Based Credit Allocation (HEBCA) to replace the BB algorithm. HEBCA reduces the delay in reinforcing early sequence classifiers. It allocates a classifier strength as the average of its matching exemplers.

To modify the BB algorithm to be suitable for pattern matched systems, it has to be noticed that there are three main differences between the classifier system and production system namely, knowledge representation, probabilistic control, and parallel versus sequential orientation [2].

Hewahi and Bharadwaj [5] modified the BB algorithm to be suitable for production systems (rule-based system). They developed a credit assignment algorithm, which is an extension of the BB algorithm and called Modified Bucket Brigade algorithm (MBB). Hewahi and Bharadwaj [5] showed that the BB algorithm is suffering from weakening the strength of correct rules which are initiated by incorrect rules. The modified version of BB algorithm solved this problem by appending rule status to each rule in addition to its strength. The rule status helps in deciding the strategy for bid payment. Although the MBB algorithm solved the above mentioned problem but still the algorithm does not consider the case in which incorrect rules are strengthen because they are involved in an inference of a path which produces correct conclusion.

Hewahi [8] has developed a new algorithm for credit assignment applied to production systems called General Credit Apportionment (GCA). GCA solves the problems that might be created by BB and MBB algorithms by using a different approach and considering the system overall performance.

2 Credit Apportionment Scheme

In this paper, we shall consider and study three algorithms, to wit, BB, MBB and GCA. BB has been chosen because it is a well reputed algorithm. MBB has been also chosen to notice its effect and its ability to overcome the problems created by BB. GCA is using another approach that is expected to recover most of the problems posed by BB and MBB.

To have more information about BB, you can refer to [9][10]. One of the main problems of BB is the long action chains. Actually this is not our problem in this research. To understand the problem posed by Hewahi and Bharadwaj [5] created by BB and therefore they proposed MBB to solve it, let us consider the following set of rules:

1. IF A THEN B
2. IF B THEN C
3. IF C & D THEN E
4. IF E & R THEN G
5. IF G THEN M
6. IF G & F THEN N
7. IF N & C THEN K
8. IF N THEN Y
9. IF C & F THEN H

We further assume that the rules 1,2,3,5,7 and 8 are correct rules, the remaining rules are incorrect. We consider the input B,D,R,F. The inference of the rules would be as shown in Figure 2.1.

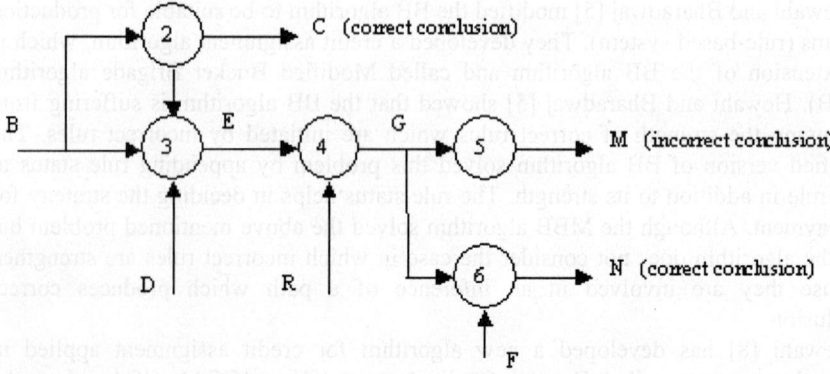

Figure 2.1 : Rules inference when the input is B,D,R,F

We assume that C and N are correct conclusions and M is incorrect conclusion. The incorrect conclusion of M is produced because rule 4 is incorrect. The correct conclusion of N might be produced because the rules 4 and 6 are incorrect and the correct conclusion comes accidentally. The problem in these two previous cases is how to know the incorrect rule in the solution path so the credit of the correct rule does not decrease.

2.1 Bucket Brigade (BB)

To implement the algorithm, each rule is assigned a quantity called its strength (credit). The BB adjusts the strength to reflect the rule's overall usefulness to the system. The strength is then used as the basis of a computation. Each rule in the matching list makes a bid, the one with highest bid is fired and its strength is decreases by the amount of the bid which is added to the strength of the rule that initiated the currently fired rule. The bid formula is $B_i = S_i * ACC_i * b$ where S_i is the strength of rule, ACC_i is the rule accuracy which should be 1 in our case and **b** is a small factor less than 1 (in our case is 0.5).

2.2 Modified Bucket Brigade (MBB)

The main problem addressed before designing the MBB algorithm is that how to modify the standard BB algorithm so that it is able to weaken wrong rules which affect good ones. This would eventually disqualify wrong rules from entering the conflict set. Thereby prohibiting them from initiating any other rule. To tackle this problem, a rule status is associated with every rule in addition to its strength. The rule status shows whether the rule is correct or incorrect. The following five cases for rule status are considered:

: unknown status (initial status of every rule).
C : Correct.
NC : Not correct.
D : Doubtful (its credit would be changed to C or NC)
X : Confirmed correct rule.

The MBB algorithm uses a similar technique for bid distribution as that employed for the standard BB algorithm except that the rules with NC status would pay their bid to the rule responsible for their activation, but would not receive any payoff from the rules they activate. However, if the rule with NC status is a final node, it may get payoff from the environment, provided it produces correct conclusion. Also under the proposed algorithm, all intermediate actions are stored because some rules may be activate by more than one condition (needs more than one intermediate rule to get fired) which may not be satisfied at the same time. For more clarification and details refer to [5] [6].

2.3 General Credit Apportionment (GCA)

The intuition behind GCA is based on the overall system performance and each rule's performance. The idea is that whenever the rule participates in a correct path (leads to correct conclusion), it means that this rule might be correct and if the same rule participates in incorrect conclusion, it means that this rule might be incorrect. To know the status of the rule (correct or incorrect) we consider the ratio of the number of the times the rule participates in correct paths to the total number of the times the rule participates in solution paths, and the ratio of the number of the times the rule participates in incorrect paths to the total number of the times the rule participates in solution paths [8]. For any rule to be evaluated if it is a good rule (correct rule) or bad rule (incorrect rule) must be involved in various solution paths.

The GCA algorithm is giving more opportunity to the rules which participate in correct paths by always adding larger value of ratio to their weights. Similarly, it deducts the smaller value of the ratio from the rules weights if they participate in incorrect conclusion. The rule which always participate in incorrect paths, are considered incorrect rule and their weights will be continuously decrease. The idea of the algorithm is to give the rules more time to prove if they are correct or not. This means that incorrect rule participates in one correct path is assumed to be correct until the system proves the reverse [8]. One problem could be raised, if a correct rule participate in an incorrect path most of the times, then the weight of the rule will be degraded very quickly. To solve this problem we use the system performance, the formula of the system performance is

SP = Ctr_i / Ntr_i, where
SP : system performance .
Ctr_i : Number of the times the ith rule participate in a correct path.
Ntr_i : Number of the times the ith rule fired.
For more clarification and details, refer to [8].

3 Implementation

To have real results, an expert system called "soccer" has been built consisting of 90 rules. "Soccer" is an expert system concerned with the rules and regulation of the soccer game. The Soccer expert system is useful for helping the soccer game referees and interested people in knowing the rules and regulations of the game and to take the correct judgments about all the cases of the game specially those cases that might cause confusion. For more details about soccer game, refer [19][20][21]. The rules structure is of the form IF <condition> THEN <action>.

A C++ simulation program for the "soccer" expert system was developed. The object-oriented approach is used to simplify the representation of the rules and paths. A path is a sequence of fired rules producing appropriate results for the given facts. The implemented program deals with four main components, Rules, Paths, Credit Computation and print reports. For more details refer to [1][18].

3.1 Rules
The rule structure used is of the form IF <condition> THEN <action>.
An example of the rules used is

IF the PLINTH is broken
THEN stop playing until it is repaired.
For more details refer to [1][18].

3.2 Paths
A solution path is represented in the form $r_1, r_2,...,r_n, r_m,...,r_k$, where r_m is the rule fired by initialization of r_n or any other given fact. To represent the output produced by a given path, we use the form O_i: Solution path, where "O_i" is the ith output given a set of facts.

We might have more than one output given the same set of facts. To check the system we have formed 59 problems to be solved by the system that produces 171 solution paths. This illustrates that with the same set of given facts, many solution paths might be followed. An example of the problems given to the system is as shown below:

Facts : Football damaged
O_1: 3, 79, 83, 88, 87
O_2: 3, 8

Figure 3.2 shows the relation between some chosen rules, each circle refers to a rule, and the number in a circle refers to the number of the rule. The rules within the square fire at the same time. In Figure 3.2, rule 13,14,15,16,17,18,19 or 20 initializes rule 10, rule 10 initializes rule 83, rule 83 initializes rule 88, rule 88 initializes rule 87 which is a final rule. Similarly, other rules in the Figure follow the same method. It is to be noted that rules 5 and 6 do not initialize any other rules; hence, they are starting and final rules.

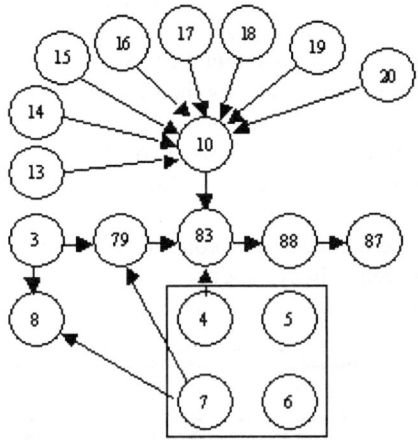

Figure 3.2 show the relation between rules 3,4,5,6,7,8,10,13,14,15,16,17,18,19, 20,79,83,87 and 88

3.3 Computing Credit of Rules

There is one master function to compute the credit of the rules. It calls three sub functions. These three sub functions are related to the three algorithms BB, MBB and GCA. Each function stores the result of the computation in array of objects with size 90 with a specific class form. The array names are BBN, MBBN and GCAN for BB, MBB and GCA respectively.

4   Experimental Results

Two main experiments are conducted to show the performance of the three algorithms, and to compare their results. The starting credit for the rules is 10. We fabricated rules 3 and 85 to become incorrect within the related paths. This makes path 1, 2,58,75,77,80,83,86,134 and 171 produce incorrect results. Since rules 3 and 85 are incorrect, they will affect the credit of other rules that have relation with them such as rules 8,31,32,42,43,44,45,54,70,79 ,83,87 and 88. To know the identity of those rules and others, refer [1][18].

Other paths could be affected by the incorrect rules but still produce correct conclusion such as path6 and path 7 which contain rule 7 as shown in Figure 3.2 [1][18].

4.1 Experiment I

This experiment is based on choosing the solution paths randomly. We take the credit of the rules after 100,300,500,700,1000,1300,1500,1700 and 2000 iterations.

Figure 4.1 shows the experiment on rule 3. Figure 4.1(b) shows the performance of the three studied algorithms on rule 3. In BB algorithm, the credit of rule 3 is fluctuating because the rules initialized by (following) rule 3 in the solution paths are correct (rules 8 and 79). Those rules sometimes participate in correct paths and other times participate in incorrect paths, thereby their credit is fluctuating (Up-Down), rule 3 always loses half of its credit and gains half of the credit of the followed rule.

In MBB algorithm, its credit decreases up to 300 iterations, then remains on 0 credit. Because the status of rule 3 becomes Not_Correct, it will not gain any credit from the followed rules.

In GCA algorithm, its credit decreases continuously, because it always participates in incorrect paths, so its credit always goes down.

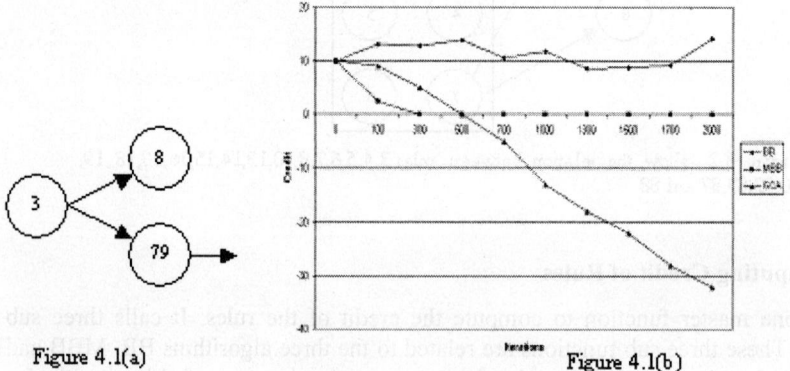

Figure 4.1(a)   Figure 4.1(b)

Figure 4.1(a) shows the relation between rule 3 and other rules, (———▶....) in the Figure means there are followed rules. Figure 4.1(b) shows the performance of rule 3 using the three credit assignment algorithms.

Figure 4.2 shows the experiment on rule 8. In BB as in MBB (typical performance), the credit of rule 8 is fluctuating because rule 8 always is a final rule. Rule 8 sometimes participates in correct paths and other times in incorrect paths, its credit is fluctuating (Up- Down).When rule 8 participates in correct path it always loses half of its credit and gains 10.

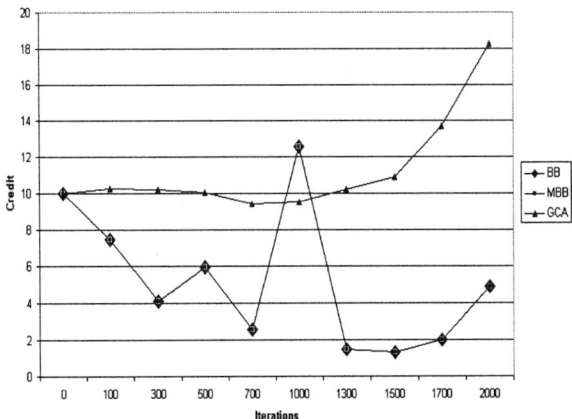

Figure 4.2 shows the performance of rule 8 using the three credit assignment algorithms

In GCA algorithm, its credit increases then decreases and then increases continuously because rule 8 sometimes participates in correct path and other times in incorrect path. In the iterations between 300 and 1000 choosing of incorrect paths is greater than correct paths so the credit of the rule decreases.

Figure 4.3 shows the experiment on rule 83. Figure 4.3(b) shows the performance of the three studied algorithms on rule 83. Figure 4.3(b) is considered to be a portion of a larger figure and it prepared in this way to cover the confusion that might occur. The confusion could happen because the values produced by GCA algorithm are much higher than the values produced by the tow other algorithms. Through BB or MBB, we get similar results.

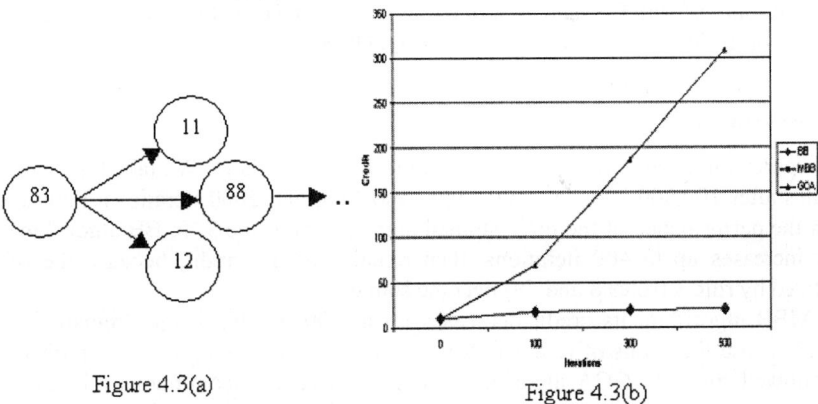

Figure 4.3(a)　　　　　　　　　Figure 4.3(b)

Figure 4.3(a) shows the relation between rule 83 and other rules, (⸺▶⸺…) in the Figure means there are followed rules. Figure 4.3(b) shows the performance of rule 83 using the three credit assignment algorithms

The credit of rule 83 is fluctuating because the rules initialized by (following) rule 83 in the solution paths are correct (11,12,88). Rules 11 and 12 are final rules and participate in correct paths while rule 88 sometimes participates in correct paths and other times in incorrect paths, thereby its credit is fluctuating (Up-Down), rule 83 always loses half of its credit and gains half of the credit of the rules 11,12 and 88. The credit of the rule 83 by algorithms BB and MBB is considered to be quite good.

In GCA algorithm, its credit increases continuously, because rule 83 sometimes participates in correct paths, and other times in incorrect paths, the rule credit increases by large value and decreases by small value. In this Figure the SP effect appears where the SP is firstly applied after 1000 iterations.

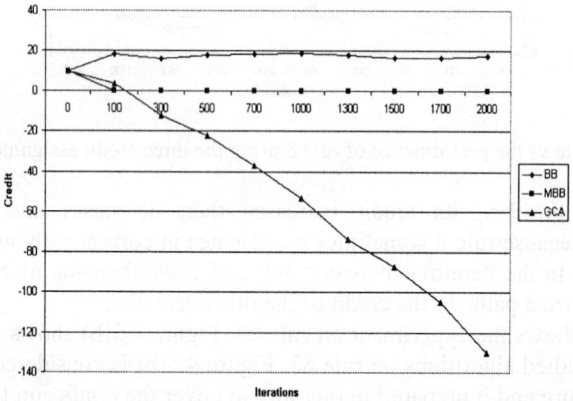

Figure 4.4 shows the performance of rule 85 using the three credit assignment algorithms

Figure 4.4 shows the experiment on rule 85. It is obvious from the figure that MBB and GCA outperform BB and could recognize the incorrectness of rule 85 very quickly. For more analysis and results, refer to [1][18].

4.2 Experiment II

This experiment is based on choosing the solution paths serially. We take the credit of the rules after 100,300,500,700,1000,1300,1500,1700 and 2000 iterations. Figure 4.5 shows the performance of the three studied algorithms on rule 3. In BB algorithm, its credit increases up to 400 iterations, then remains on 15 credit, because the rules initialized by rule 3 (rules 8 and 79) become stable.

In MBB algorithm, its credit decreases up to 300 iterations, then remains on 0 credit, because the status of rule 3 is Not_Correct so it will not gain any credit from the followed rules. In GCA algorithm, its credit decreases continuously because it always participates in incorrect path, so its credit goes down. Despite that the problems (solution paths) are chosen serially. MBB still shows better results than BB. GCA also shows good performance.

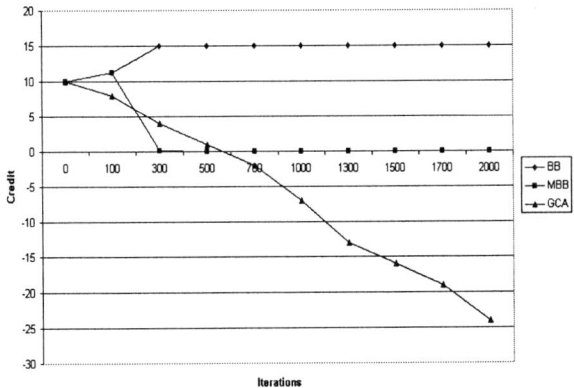

Figure 4.5 shows the performance of rule 3 using the three credit assignment algorithms

Figure 4.6 shows the performance of the three studied algorithms on rule 8. In BB as in MBB (typical performance), the credit of rule 8 increases up to 300 iterations then becomes stable on 13.33. Because rule 8 is a final rule and participates in correct paths as well as in incorrect paths, its credit is fluctuating (Up-Down), When rule 8 participates in correct path, It always loses half of its credit and gains 10.

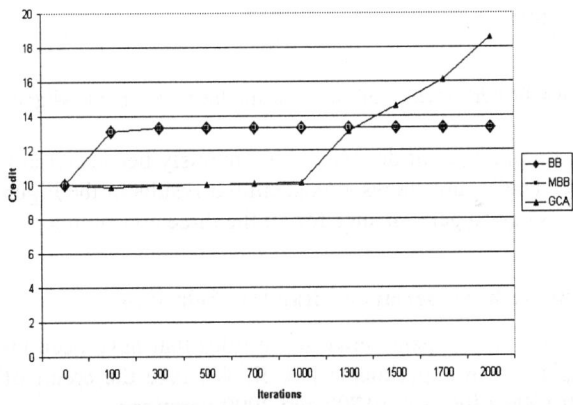

Figure 4.6 shows the performance of rule 8 using the three credit assignment algorithms

In GCA algorithm, its credit increases then decreases and then increases continuously. Because the rule some times participates in correct paths, and other times in incorrect paths, in the iterations between 1 and 300 the choosing for incorrect paths is greater than the choosing of correct paths so the credit of the rule decreases. SP effect appears here, where after 1000 iterations the increasing in credit is faster than before 1000 iterations. In case of serial testing for BB and MBB shows better performance than random testing. For GCA, in both the experiments have shown good results.

Figure 4.7 shows the performance of the three studied algorithms on rule 85. In BB algorithm, the credit of rule 85 is fluctuating because the rules initialized by (following) rule 85 in the solution paths are correct (83). Rule 83 sometimes participates in correct paths and other times participates in incorrect paths, thereby their credit is fluctuating (Up-Down), rule 85 always loses half of its credit and gains half of the credit of rule 83.

In MBB algorithm, its credit decreases up to 100 iterations, then remains on 0 credit. Because the status of rule 85 is Not_Correct, it will not gain any credit from the followed rules.

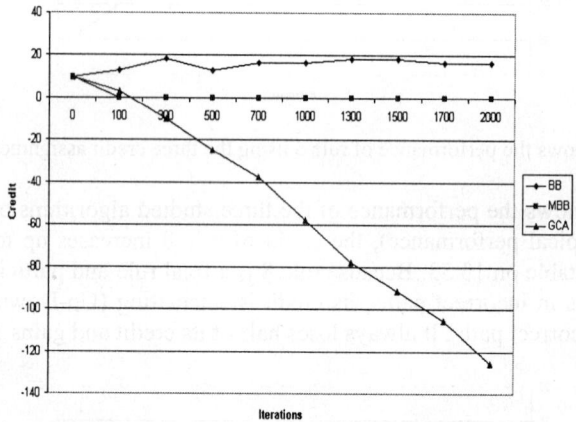

Figure 4.7 shows the performance of rule 85 using the three credit assignment algorithms

In GCA algorithm, its credit decreases continuously because it always participates in incorrect path . The Figure shows almost similar results of the Figure 4.4 of random choosing and shows good performance for all the three algorithms.

4.3 Comparison between Serial and Random Selection

This experiment is a comparison between the rules that have been taken serially and randomly as explained in experiments I & II. We take the credit of the rules after 100,300,500,700,1000,1300, 1500,1700 and 2000 iterations.

The comparison between the two experiments shows that more or less the results obtained by experiment I & II are the same for each type of credit apportionment scheme. This says that GCA outperforms the two algorithms in both the experiments. MBB still performs better than BB in both the experiments. For more details and results, refer to [1][18].

5   Conclusion

An overview of credit apportionment schemes for rule-based systems is presented. We have studied the algorithms BB, MBB and GCA for credit apportionment scheme to make a comparison among their performance. A simulation program has been implemented to simulate the three chosen algorithms on a „soccer" expert rule-based system. The results which have been obtained from the simulation program were presented in figures. Two experiments have been conducted, random choice of the problems (solution paths) and serial choice of the problems, also a study between the two experiment is presents.

To summarize our conclusions, we could say that we have reached to the following points

1. GCA gives excellent result on all the cases.
2. The algorithm BB weakens wrong rules, which affect good ones.
3. The algorithms BB and MBB have a problem, which is long action sequences.
4. All the three algorithms perform well if we have well structured rules.
5. MBB outperforms BB in discovering incorrect rules, which will stop the degradation of the credit of correct rules.
6. GCA in general outperforms the other two algorithms.
7. In random choice, if a certain correct rule participates in incorrect path more than in a correct paths (but not none), the GCA would still perform better than the other two algorithms despite that the rule credit might be decreased (this is because of SP factor).
8. BB algorithm using random choice of problems when presenting the credit of incorrect rule shows fake better performance than in serial choice because at some moment the credit of the rule using random choice might reach or even exceed the credit of the rule using serial choice. Moreover, BB in both the cases shows bad performance.
9. In GCA, correct rules might get weakened if it always participates in incorrect paths. This does not mean that the algorithm is not so good because the assumption of this statement is that most of the used rules are incorrect.
10. MBB and GCA using random or serial selection of the system problems show similar and good performance.
11. In general and despite of the type of credit apportionment scheme, it is recommended to use serial selection of the system problems so that all get the same opportunity

The future work would be

1. Applying GCA to a real adaptive system
2. One of the problems that still can not be solved totally by any of the three algorithms is, if more than one incorrect rule participates in a solution path and still the produced result is correct.
3. Since the speed of computer machines nowadays is very fast, a study can be done on the effect of that on BB and MBB to neglect the problem of long action chains.
4. Since the GCA algorithm is more complex than MBB in terms of time, a study could be done to explore whether the time lost by GCA is still less than the time needed for rule to be discovered correct or incorrect using MBB.

References

1. B.H.Ahmad, „Credit Apportionment Scheme for Rule-Based Systems: Implementation and Comparative Study ‚‚, B.Sc dissertation, Islamic University of Gaza, (2001).
2. H.J.Antonisse, "Unsupervised Credit Assignment in Knowledge-Based Sensor Fusion Systems" IEEE Trans .on Sys, Man and Cyber., sep./Oct, vol.20, no.5. (1990), 1153-1171.
3. D.Derrig, and D.J. Johannes, "Hierarchical Exemplar Based Credit Allocation for Genetic Classifier Systems", Proceedings of the 3rd annual Genetic Programming Conference, Univ. of Wisconsin, Madison, Wisconsin, Morgan Kaufman Publishers San Francisco, California ,USA,July 22-25,(1998),622-628.
4. M.Dorigo,"Message-Based Bucket Brigade: An Algorithm for the Apportionment of Credit Problem "in Lecture Notes in Artificial Intelligent, Machine Learning, Springer Verlag, New York, (1991), 235-637.
5. N.M.Hewahi and K.K.Bharadwaj, "Bucket brigade algorithm for Hierarchical Censored Production Rule-Based systems ", International Journal of Intelligent Systems, USA, Vol.11, John (1996) ,197-225.
6. N.M.Hewahi, "Genetic Algorithms Approach for Adaptive Hierarchical Censored Production Rule-Based System", School of Computer and Systems Sciences Jawaharlal Nehru University New Delhi, India, Jan. (1994).
7. N.M.Hewahi," Fuzzy system as a credit assignment scheme for rule based systems", Proceedings of 3rd Scientific Conference on Computers and Their Applications, Amman, Jordan, Feb. 11-13, (2001).
8. N.M.Hewahi, "Credit Apportionment Scheme for Rule Based Systems", Journal of Islamic University of Gaza,(2001) , vol.9, No.1, part2, 25-41.
9. J.H.Holland, "Properties of the bucket brigade algorithm", in J.Grefenstette (Ed.), Proceedings of Inter.Conf. on Genetic Algorithms and Their Applications, Carnegie-Mellon University, Pittsburg, PA, July (1985).
10. J.H.Holland, "Escaping Brittleness: The Possibility of General-Purpose Learning Algorithm Applied to Parallel Rule-Based Systems ", in R.S.Michalski, J.G.Carbonell, and T.M.Michalski, J.G.Carbonell, and T.M.Mitchell (Eds.), Machine Learning: Artificial Intelligence Approach, vol.2,(1986).
11. D.Huang, "A Framework for the Credit-Apportionment Process in Rule-Based System". IEEE Trans. on Sys.Man and Cyber, May/June (1989) , Vol.19, no.3, 489-498.
12. M.Minsky, "Step Toward Artificial Intelligence", in A.E.Feigenbaum and J.Feldman, Eds., Computers and Thought, McGraw-Hill, New York, (1961).
13. R.Riolo," Bucket Brigade Performance: I. Long Sequences of Classifier ", in J.Grefenstette (ED.), Proceeding of 2nd Inter.Conf. on Genetic Algorithms, July 1987.
14. A.L.Samuel, "Some Studies in Machine Learning Using The Game of Checkers", IBM Journal of Res. and Devel. , No.3, (1959).
15. P.Spiessens, "PCS: A Classifier System that builds a predictive internal world model", in 9th European Conf. in AI, Stockholm, Sweden, ECA190, August 6-10, (1990) , 622-627.

16. T.H.Westardale, "A Reward Scheme for Production Systems with Overlapping Conflict Sets", IEEE Trans. Sys. Man and Cyber., vol. SMC-16, no.3, May/June(1986), 369-383.
17. S.W.Wilson, "Hierarchical Credit Allocation in Classifier System ", in M.S.Elzas, T.I. Oren, and B.P. Zeiglar (Eds.), Model. and Simul. Methodol. Knowledge Systems' Paradigms, Elsevier Science Publishers B.V. (North Holland), (1989), 352-357.
18. www.basemhaa.8m.net
19. www.drblank.com/slaws.html
20. www.fifa2.com/scripts/runisa.dll?s7:gp::67173+refs/laws
21. www.ucs.mun.ca/~dgraham/10tg/

An Adaptive Web Cache Access Predictor Using Neural Network

Wen Tian, Ben Choi, and Vir V. Phoha

Computer Science, College of Engineering and Science
Louisiana Tech University, Ruston, LA71272, USA
Wen_tian2@yahoo.com
pro@BenChoi.org
phoha@coes.latech.edu

Abstract. This paper presents a novel approach to successfully predict Web pages that are most likely to be re-accessed in a given period of time. We present the design of an intelligent predictor that can be implemented on a Web server to guide caching strategies. Our approach is adaptive and learns the changing access patterns of pages in a Web site. The core of our predictor is a neural network that uses a back-propagation learning rule. We present results of the application of this predictor on static data using log files; it can be extended to learn the distribution of live Web page access patterns. Our simulations show fast learning, uniformly good prediction, and up to 82% correct prediction for the following six months based on a one-day training data. This long-range prediction accuracy is attributed to the static structure of the test Web site.

1 Introduction

Use of the Internet is rapidly increasing, and the increase in use motivates the increase in services, and vice versa. All indications are that the use of Internet, services offered, and new applications on the Internet will grow exponentially in the future, resulting in an explosive increase in traffic on the Internet. However, the Internet infrastructure has not kept pace with the increase in traffic, resulting in greater latency for retrieving and loading data on the browsers. Thus, Internet users experience slow response times, especially for popular Web sites. Recently, Web caching has emerged as a promising area to reduce latency in retrieval of Web documents.

In this paper, we present methods and the design of an intelligent predictor for effective Web caching to reduce access latency. Our predictor uses back-propagation neural network to improve the performance of Web caching by predicting the most likely re-accessed objects and then keep these objects in the cache. Our simulation results show promise of successfully capturing Web access patterns, which can be used to reduce latency for Internet users.

Our motivation to use neural networks to predict web accesses follows. The distribution of Web page requests is highly nonlinear and show self-similarity in Web page requests. Since, Neural nets are capable of examining many competing hypotheses at the same time and are more robust than statistical techniques when underlying distributions are generated by non-linear process, it is then natural to use neural nets to predict Web page accesses. Another motivation to use Back-propagation weight update rule in our study is that the Web access logs provide the history of accesses, which can be divided into training examples (test data) to train, and test the predicting power of a supervised learning neural net (see Section 3.2 for our strategy to segment the Web log into training, validation, and test data and our design of NN predictor). Thus, we have all the basic ingredients to build a successful neural network predictor.

The salient features of our approach are: (1) Our predictor learns the patterns inherent in the past Web requests to predict the future Web requests; (2) It adapts to the changing nature of Web requests; (3) It is not dependent on the underlying statistical distribution of Web requests; (4) It is robust to noise and isolated requests; and (5) It can be implemented in hardware or in software.

This paper is organized as follows. In the next subsection, we provide the definition of the symbols used in this paper. In Section 2, we briefly review the related work in caching replacement algorithms. In Section 3, we propose an adaptive Web access predictor using back-propagation neural network to predict the most likely re-accessed objects. In Section 4, we present our simulation results. Finally, in Section 5, we provide conclusion and suggest future work.

1.1 Terminology

This section provides definitions of the symbols used in this paper.

- α A threshold value to determine whether the desired value should be 1 or 0.
- β A threshold value to control learning and prediction granularity.
- W_T The training window
- N_T The size of the training window
- W_P The prediction window
- N_P The size of the prediction window
- W_B The backward-looking window in learning phase
- W_F The forward-looking window in learning phase
- W_{F2} The forward-looking window in predicting phase
- N_{F2} The size of the window W_{F2}
- $W_{i,j}$ The weight value from node i to node j in the neural model
- η The learning rate for back-propagation weight update rule.
- d   The desired output
- t   Time step in the neural net training phase

2 Related Work

Numerous cache replacement algorithms have been proposed by Web caching researchers. The most popular algorithms are: LRU (Least Recently Used) [3], LFU (Least Frequently Used) [3], GDS (Greedy Dual-Size) [14], and LFUDA (LFU with Dynamic Aging) [2], and others reported in [1][2][3] [4] [7] [11] [13].

LRU deletes the objects that have not been requested for the longest time. LFU replaces objects with the lowest access counts; however, this algorithm tends to fill the cache up with frequently accessed old objects. LFUDA is a variant of LFU that uses dynamic aging to accommodate shifts in the set of popular objects. The new feature prevents previously popular objects from polluting the cache by adding an age factor to the reference counts when a new object is added to the cache. GDS takes the size and the cost for retrieving objects into account, this algorithm assigns a value, V, to each object in the cache, the V value is set to the cost of retrieving the object from the origin server, divided by its size. When the cache is full, the object with the smallest V value is replaced.

Few of these algorithms use artificial intelligence techniques to predict Web accesses. Our work builds on the work of Foong, Hu and Heisey [5] who use logistic regression to build an adaptive Web cache. The drawbacks of their approach are: extensive computation required to fit logistic curve for small updates; difficulty of learning access patterns, and absence of a scheduler. Instead, we use a neural network [6] [9] [15], which can be implemented in hardware for real time response, and has fast learning ability. We also present a separate scheduler for training and prediction. We introduce our approach in the following section.

3 Our Design of a Web Access Predictor

In this section, we present our design of a Web access predictor. A schematic of our intelligent predictor is given in Fig. 1. The predictor has two modules: preprocessing module and processing module. Both of these modules can run as background processes in a Web server.

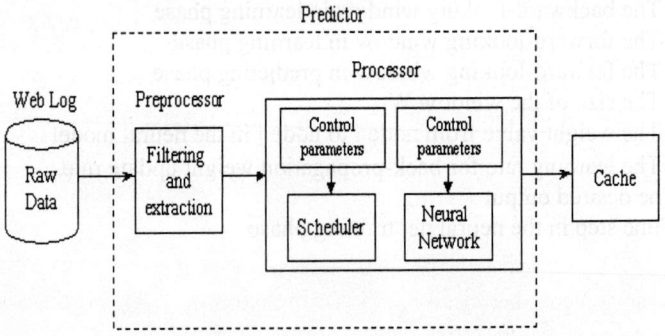

Fig. 1. Schematic of an intelligent predictor

The preprocessor in Fig. 1 handles the filtering and preparation of data in a format to be fed to the processor. The processor consists of a core and a moving window protocol. The design of these components is provided in the follow subsections.

3.1 Preprocessor

For testing our design, we use Web log files from Boston University, spanning the timeframe from November 21, 1994 through May 8, 1995 [8]. The raw data from the log file is in the following form:

<machine name, timestamp, user id, URL, size of the document, object retrieval time in second>

We remove any extraneous data, and transform the useful data into a vector of the form $\{URL, <x_1, x_2, x_3, x_4, x_5>\}$.

X_1 : Type of document
X_2 : Number of previous hits
X_3 : Relative access frequency
X_4 : Size of document (in bytes)
X_5 : Retrieval time(in seconds)

The heuristic to choose value for X_1 is as follows. The values assigned are based on the relative frequency of the type of the files, since there are more image files than HTML files, we assign a higher value to image files and a lower value to HTML files. Since there are few other file types, to penalize learning of other file types, we assign a negative value for all other file types. The following chosen values are somewhat arbitrarily but these chosen values gave us good results in our simulations:

X_1 = 10: HTML files
   15: image files
   -5: all other files

We also consider the file entries having the same URL but having different file size and different retrieval time as the updated version of previous occurrence of such file. Moreover, if a file entry has size of zero and the retrieval delay is also zero, which means the request was satisfied by the internal cache, then we obtain the file size and the retrieval time from the previous occurrence of the file having the same URL. Next, we can get the document size (X_4) and retrieval time (X_5) directly from the raw data. The process for extracting the value of X_2, and X_3 are provided in Section 3.2.1. These values of X_1, X_2, ... X_5 are provided as training vectors for the next stage.

3.2 Processor

The second stage consists of training our neural network using the data obtained from preprocessing. After the network has been trained, we use the network to predict if an object will be re-accessed or not. Fig. 2 shows the design of our predictor using a scheduler and a neural segment.

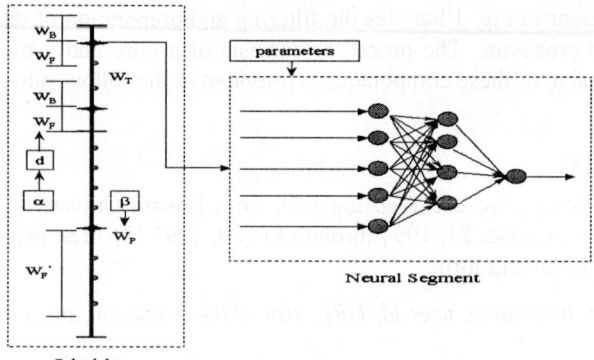

Fig. 2. Design of a neural network architecture to predict cache re-access probability

3.2.1 Training the Predictor Using Back-Propagation Algorithm

As indicated in Fig. 2, the scheduler selects the training data and the prediction windows. The training data was selected using training window W_T, which has size of N_T. We also specify a backward-looking window W_B and a forward-looking window W_F (as was done in [5]). The window W_B and W_F are sliding windows related to current entry. The W_B is the previous log accesses from the current entry while W_F is the following log accesses from the current entry.

Getting Input Vectors

We use the following method to obtain the input vectors for each access log entry.
For retrieving X_1, X_4, X_5, see Section 3.1.

X_2 (previous hit), is equal to how many times an entry has been accessed in window W_B.

X_3 (relative access frequency) is equal to the total number of accesses of a particular URL in window W_T divided by N_T.

After getting the inputs X_1, X_2, X_3, X_4, X_5, we normalize their values to make them in the similar range. Since these variables have different range, such as the size may be 10000 bytes, where as retrieval time may be 2 seconds; so, we scale these entries by dividing each item by a respective constant to scale them to the same range.

Getting Desired Output

The desired output is a binary variable. If the object in W_T has been re-accessed at least α times in W_F, then we assign value one as the desired output, otherwise zero. Higher value of α results in fewer pages to be assigned value one but more accurate prediction, while smaller α value results in more pages to be assigned value one but less accurate prediction.

We modify the standard back-propagation training algorithm [12] [10] [15] to adopt to Web caching. Figure 2 contains our architecture of a feed-forward multi-layer neural network, with five input nodes (corresponding to X_1, X_2, X_3, X_4, X_5), four hidden nodes, and one output node to predict. An outline of back-propagation algorithm follows:

```
function BP-Net(network, training data, α)
returns a network with modified weights
inputs: network, a multilayer network with initialized weight
values between -1 to 1, training data, a set of input/output,
desired output pairs, η is the learning rate
 repeat
  for each e in training data do
   normalize the input
   /* compute the output for this example */
   calculate the input values for the hidden layer
       using sigmoid function
   calculate the output for the output layer
       using sigmoid function
   /* update the weights leading to the
       output layer */
   /* $x_i$ are input values and $y_j$ s are output values,
       $d_j$ is the desired output */
   $W_{i,j}(t+1) \leftarrow W_{i,j}(t) + \eta \times \delta_j \times x_i$
   where $\delta_j = y_j(1-y_j)(d_j-y_j)$,
       $x_i$ is the input in hidden layer
   /* update the weights in the subsequent layers */
   $W_{i,j}(t+1) \leftarrow W_{i,j}(t) + \eta \times \delta_j \times x_i$
   where $\delta_j = x_j(1-x_j)\sum_k \delta_k W_{jk}$,
       k is over all nodes in the layers above node j
   /* calculate the mean square err between
       new weights and old weights */
   $Err \leftarrow (W_{i,j}(t+1) - W_{i,j}(t))^2$
  end
 until network has converged
 return network
```

3.2.2 Prediction Stage

After the network has been trained, it was used to predict future access patterns. We selected W_p as the prediction window that consists of N_p entries. The prediction results were compared with the available test log entries. In this stage, we specify a sliding forward-looking window W_{F2} and a threshold β. W_{F2} contains the next N_{F2} access entries related to the current entry. The following pseudo-code shows the method for testing the prediction.

```
boolean correct-prediction(input vector)
    calculate the actual output V with input vector;
    calculate the number of hits H the object has been
       re-accessed in window $W_{F2}$;
    /* the predictor predicts the object will
       be re-accessed */
    if ((V >= 0.6) && (H >= •))
       return true;    // correctly predicted
    /* the predictor predicts the object will not be
       re-accessed */
    if ((V < 0.6) && (H < •))
       return true;    // correctly predicted
    else
       return false;
```

4 Simulation Results

In our simulations, we choose $N_T = 500$, $N_P = 5000$, $\alpha = 3$, $N_B = N_F = 50$ and $N_{F2} = 200$. We use log file of one day as the training data. The network converged after 8000 iterations, the mean square error (MSE) value is less than 0.005. Then we use the network to predict access patterns for the next six months. Tables 1 and 2 show the results having β value ranging from 1 to 5. In the tables, learning performance represents the percentage of correct prediction on the data that were part of the training data while prediction performance represents the percentage of correct prediction on data that were not part of the training data.

Table 1. Sample of Simulation Result 1

Learning data: December 1994, prediction data: January 1995

MSE	LEARNING PERFORMANCE (testing)	β	PREDICTION PERFORMANCE
0.00188	80%	1	61%
0.00444	83%	2	67%
0.00417	86%	3	79%
0.00270	85%	4	82%
0.00326	80%	5	79%

Table 2. Sample of Simulation Result 2

Learning data: December 1994, prediction data: May 1995

MSE	LEARNING PERFORMANCE (testing)	β	PREDICTION PERFORMANCE
0.00401	80%	1	61%
0.00331	85%	2	72%
0.00336	82%	3	71%
0.00358	83%	4	75%
0.00042	82%	5	76%

The simulations were done for more than twenty times [16], Tables 1 and 2 are a representative sample of our simulation results. From the tables, we can see that as the β value increases the prediction accuracy increases and it stabilizes for $\beta > 3$. So we choose $\beta = 4$ as the threshold value for this Web site, and we achieve up to 82% percentage of correct prediction (even though the prediction data are six months from the training data). Figure 3 shows the prediction percentage for the next six months for different values of β (we use the training data from one day, then predict the next six months).

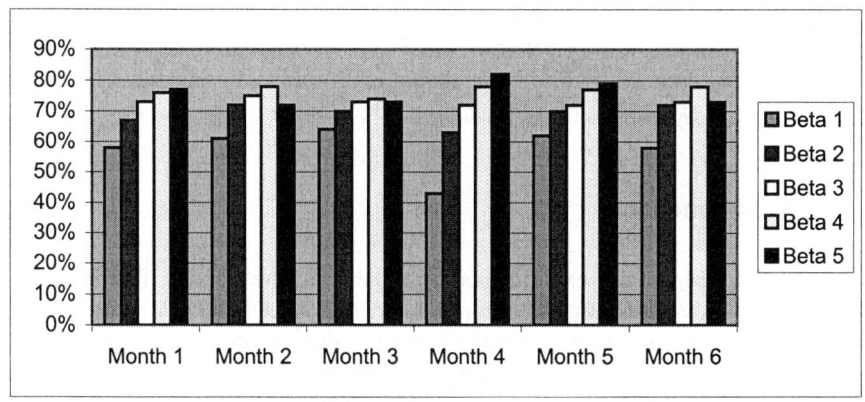

Fig. 3. . Simulation results with different β value

From Fig. 3, we can see that the prediction appears to be uniform (approximately ranging from 60% to 82%, except for month 4 for β = 1) over a period of six months, we believe this is because the structure of Web site has not changed much over the period of data collected from the Web logs [8].

5 Conclusion and Future Work

In this work, we have successfully presented a novel technique to predict future Web accesses based on history of Web accesses in a Web site. Based on this idea, we have built an intelligent predictor that learns the patterns inherent in the access history of Web pages in a Web site and successfully predicts the future accesses. Our simulations show that the predictor learns the access patterns and we have been able to predict up to 86% accuracy for data on which the network has already been trained and up to 82% accuracy on new and never seen before data. We have also presented heuristics to control the granularity of training data and prediction of the accesses. The success of this technique has opened up many new areas of research. Some of the future areas of further exploration are listed below.

- Develop heuristics to predict the short and long-range period for which current training gives good future prediction. Successful estimate of this time period will help develop strategies to retrain the network.
- Explore other architectures for building the predictor, such as, Self Organizing Feature maps, Adaptive Resonance Theory, Recurrent neural networks, and Radial Basis Function neural network.
- Implement our technique as part of server software, so that the system can be tested in live environment. At present, we have used data collected from log files, which is a static data. We would like to test this technique in a real time and live environment.
- At present, for a significant number of new Web accesses, the method requires retraining. We would like to explore building a dynamic model of a perceptron,

so that only a small part of weights may be changed by adding or deleting nodes. Thus, the training for new patterns would require weight updates for only part of the network. This type of network should learn on the fly without updating the complete set of weights, resulting in no need to completely retrain the network.

Acknowledgements

The access log used in our experiments is available at The Internet Traffic Archive (http://ita.ee.lbl.gov/index.html) sponsored by ACM SIGCOMM. It was collected by the Oceans Research Group (http://cs-www.bu.edu/groups/oceans/Home.html) at Boston University for their work, "Characteristics of WWW Client Traces", which was authored by Carlos A. Cunha, Azer Bestavros, and Mark E. Crovella.

6 References

[1] Ramon Caceres, Fred Douglis, Anja Feldman, Gideon Glass, and Michael Rabinovich, *Web proxy caching: the devil is in the details*, 1997~2001 NEC Research Institute.
[2] John Dilley and Martin Arlitt, *Improving Proxy Cache Performance: Analysis of Three Replacement Policies*, IEEE Internet Computing, November-December 1999, pp. 44- 50.
[3] Duane Wessels, *Web Caching*. O'Reilly, 2001.
[4] Li Fan, Pei Cao, and Quinn Jacobson, *Web prefetching between low-bandwidth clients and proxies: potential and performance,* Available at http://www.cs.wisc.edu/~cao/papers/prepush.html (last accessed September 18, 2001.)
[5] Annie P. Foong, Yu-Hen Hu, and Dennis M. Heisey, *Logistic Regression in an Adaptive Web Cache*, IEEE Internet Computing, September-October 1999, pp. 27-36.
[6] Freeman J. and Sakura D. Neural Networks: Algorithms, Applications, and Programming Techniques. Addison-Wesley, 1991.
[7] Dan Foygel and Dennis Strelow, *Reducing Web latency with hierarchical cache-based prefetching,* proceeding of the 2000 international workshop on parallel processing in IEEE 2000.
[8] Internet Traffic Archive, available at http://ita.ee.lbl.gov/html/traces.html (last accessed November 4, 2001.)
[9] J. Hertz, A. Krogh, and R. G. Palmer, *Introduction to theory of Neural Computation*, Addison-Wesley, Reading, Mass., 1991.
[10] Jacobs R. Increased rates of convergence through learning rate adaptation. Neural Networks 1, 295-307, 1988.
[11] Ludmila Cherkasova, Improving WWW Proxies Performance with Greedy-Dual-Size-Frequency Caching Policy, HP laboratories report No. HPL-98-69R1, April, 1998.

[12] Ronald W. Lodewyck and Pi-Sheng Deng, *Experimentation with a back-propagation neural network*, Information and Management 24 (1993) Pp. 1-8.
[13] Evangelos P. Markatos and Catherine E. Chronaki, *A Top-10 approach to prefetching on the Web*. Technical Report 173, ICS-FORTH. Available from http://www.ics.forth.gr/proj/arch-vlsi/www.html (last accessed June 27, 2001.)
[14] N. Young, *On-line caching as cache size varies*, available in the 2nd Annual ACM-SIAM Symposium on Discrete Algorithms, pp. 241-250, 1991.
[15] R. P. Lippmann, *An Introduction to Computing with Neural Nets*, IEEE ASSP Magazine, Vol.4, No.2, Apr.1987, pp. 4-22.
[16] Wen Tian, *Design Of An Adaptive Web Access Predictor Using Neural Network*, MS Report, Computer Science Department, Louisiana Tech University, 2001.

A Designated Bid Reverse Auction for Agent-Based Electronic Commerce

Tokuro Matsuo and Takayuki Ito

Center of Knowledge Science,
Japan Advanced Institute of Science and Technology
1-1 Asahidai, Tatsunokuchi-machi, Nomi-gun, Ishikawa 923-1292, Japan
{t-matsuo,itota}@jaist.ac.jp,
http://www.jaist.ac.jp/~{t-matsuo,itota}

Abstract. Internet auctions are seen as an effective form of electronic commerce. An internet auction consists of multiple buyers and a single seller. We propose an alternative, the REV auction, in which a buyer can select sellers before conducting the auction. There are several advantages to our mechanism. First, the seller selection mechanism enabled us to reflect the buyers' preference. Second, the seller's evaluation mechanism effectively maintains seller quality. Third, our mechanism can avoid consulting before bidding. We implemented an experimental e-commerce support system based on the REV auction. Experiments demonstrated that the REV auction increased the number of successful trades.

1 Introduction

As the Internet develops it has become an increasingly prosperous network for many types of commerce. Internet auctions have been a particularly effective form of electronic commerce[eBa][yah]. Internet auctions have made rapid progress in recent years, and we can find many investigations of internet auctions [Turban 00][Yokoo 00]. Internet auctions have become a promising field to apply agent technologies [Ito 01]. On many auction sites, sellers take the lead in trades, while buyers have no authority except for estimating value. An alternative is the reverse auction. Here, a buyer displays their desired goods and along with a desired price, and sellers competitively decrease their price until a sale is made. For instance, "priceline.com" conducts reverse auctions [pri].

There are currently two types of reverse auctions, defined in terms of the way a buyer purchases goods. One involves applying to an auction site for goods that a buyer hopes to purchase. First, the buyer answers questions from the auction site on the goods desired, expected price, quantity, etc. Next, the auction site searches among sellers for the buyer's objective. For example, on priceline.com, buyers must provide details on the characteristics or content of goods or services. Therefore, there is a limit on the range of goods shown to the buyer, and there is the potential that a buyer's preferences may not be reflected completely. Further, the mechanism is not robust against collusion. The other type of reverse auction involves showing goods that a buyer hopes to purchase and identifying potential sellers [eas]. But there is no guarantee that multiple sellers arrive soon after a buyer shows the goods and desired price.

To solve some of the problems with existing reverse auctions we propose the REV (Reverse Extended Vickrey) auction protocol. The REV auction is a novel protocol in which the buyer nominates sellers and the nominated sellers compete on price. We conducted experiments to verify that the REV auction can consummate more deals than the Vickrey auction protocol and the open bidding protocol. The experimental results confirmed that the REV auction can facilitate deals more effectively that the Vickrey auction protocols.

The rest of the paper is organized as follows. Section 2 describes of existing auction types. In section 3 we define the REV auction. Section 4 presents the experimental results. In section 5 we show examples of user interfaces, and discuss the advantages of the REV auction. And in section 6 we provide some final remarks and describe future work.

2 Auction Primer

In this paper, it is assumed that an agent's utility function is quasi-linear. By the quasi-linear utility, if a buyer with the evaluation value b^* buys one unit of a good at the paying price, p, we assume his or her utility is defined as b^* - p. Similarly, if a seller with the evaluation value s^* sells one unit of the good at a price, p, we assume his or her utility is defined as p - s^*. If a buyer cannot obtain a unit, or a seller cannot sell a unit, we assume a utility of 0.

If the mechanism is individually rational, each participant never suffers any loss by participating in the exchange. In a private value auction, individual rationality is indispensable [Yokoo 01]. We assume that the evaluation value for each buyer or seller is independent of the other participants' evaluation value. Such a good is called a private value good. Particular examples of private value goods are taste and hobby goods aimed at resale. Curios, such as pictures, are usually classified as private value.

A typical auction protocol is classified as shown in Table 1 [Endo 00]. In an auction called the English protocol, a bid value is exhibited and a bidder can increase a bid value freely. When bidders no longer desire a change of value the auction is awarded to the highest bidder and the bid value is paid. Another auction, called the Dutch protocol, uses an auctioneer to decrease the price until stopped by a buyer who declares that his price has been reached. The auction is awarded to the buyer at the price in effect at the time the buyer called the auctioneer to stop.

At first price sealed bid auctions, each bidder offers a bid, without telling the other bidders the bid value. The auction is awarded at the highest value by the bidder who gave the highest bid. At auctions called second price sealed bid, each bidder offers a bid, without revealing the bid value. The auction is awarded at the second value to the bidder who gave the highest bid. The second price sealed bid is also called a Vickrey auction, and was first proposed by W. Vickrey. The following are considered the key features of the auction. A successful bidder pays the evaluation value from the second-place bidder. Therefore, the auction is awarded to the successful bidder independent of the price that the successful bidder pays. In general, it is known in game theory that bidding based on one's true evaluation value is a dominant strategy.

Table 1. Auction protocols

	Open-bid	Sealed-bid
	English	Vickrey(second price)
	Dutch	first price

3 REV Auction

3.1 Reverse Auctions and Designated Auctions

In the reverse auction multiple sellers compete on goods and the evaluation value shown by the buyer. A designated bidding system is required by the public works office of Japan, and is one form of reverse auction. In designated bidding, since the auctioneer nominates a bidder based on the quality of their work, it can prevent poor companies from making a successful bid prior to bidding. The result of an auctioneer's examination of the standard of technical requirements, the right to bid is granted only to bidders accepted by aptitude.

Other features of designated auctions differ from typical bidding mechanisms. By nominating a superior contractor, it becomes possible to assure the quality of work, to some extent. The disadvantage is the increased probability of collusion, because the number of bid participants is restricted [McMillan 91][Saijo 94]. Since the bidders are identified in advance, bidders are specifically nominated based on the quality of past performance. Therefore, it is hard for companies from other areas and new companies to join the auctions. Collusion prevents any fear of competition, and can lead to kick-backs and side payments (in economics). Therefore, in theory the open bidding auction mechanism is better than the designated auction mechanism [Baba 00]. On the Internet, however, since buyers can solicit nominations via World Wide Web, sellers can maintain a collusion. Therefore, we applied the designated auction mechanism to the Internet auction.

3.2 REV Auction Mechanism

In our REV auction a buyer selects sellers who have substitute goods and the buyer's preferred goods. The nominated sellers submit sealed bids. The designated aspect of the auction can eliminate sellers who have inferior goods by pre-selecting a seller. First, sellers register their goods in our auction server, and then input the evaluation value of the goods to the agent. Second, the buyer selects the goods according to his preference and purpose from the groups of sellers registered previously. The highest successful bid price (buyer's evaluation value) is also shown at the server. Simultaneously, 1 point is assigned to the nominated sellers. This point is added to a seller at the server whenever it is nominated. The number of points indicates how well the seller is trusted. Third, when nomination and presentation are completed, the buyer and sellers begin to bid. When the bidding is completed, if the second price is lower than the buyer's evaluation value, then the buyer pays the second price. If the buyer's evaluation value is between first and second prices, the buyer pays the successful bidder his evaluation value.

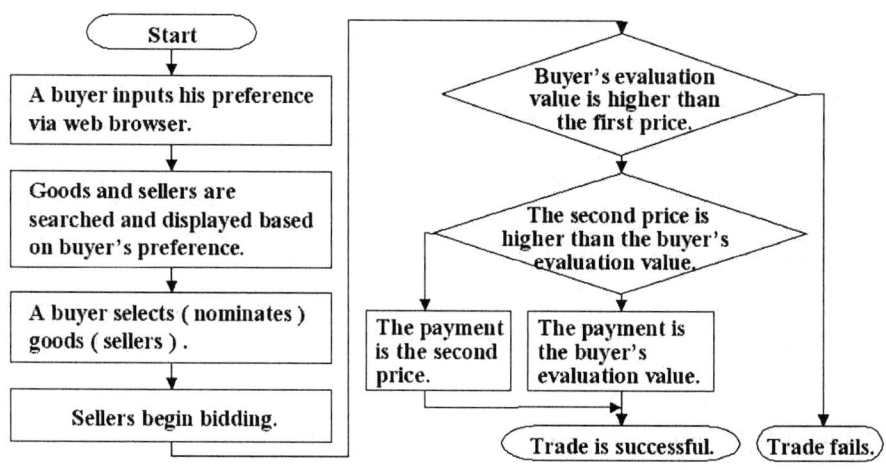

Fig. 1. REV auction flow chart

Suppose that they are n($n \gg 0$) sellers who are already registered on our server. Each seller has a single item. Suppose also that there exists one buyer. The buyer nominates m($m \leq n$) sellers with goods that suit the buyer's preferences. The buyer shows an evaluation value, b. The set of sellers whom the buyer nominates is as follows.

$$\{S_1, S_2, \cdots, S_i, \cdots, S_m\}$$

Let us represent the sellers' declared evaluation values as $s_i (i = 1, 2, \cdots, m)$. Seller agents bid for the evaluation value in the sale. The bids are sorted in a sequence from small/cheap bid-value order to simplify their description as follows.

$$s_1 \leq s_2 \leq \cdots \leq s_m$$

In the winner determination algorithm S_1 represents the smallest bid value. When a seller's agent, S_1, wins, the buyer's payment and the utility of sellers and the buyer are determined as follows.

1. If $s_2 \leq b$, the buyer pays the price, s_2, to the agent S_1.

   We assume a quasi-linear utility. Therefore, a buyer, S_1's, utility is $s_2 - s_1$ and the winner's utility is $b - s_2$.

2. If $s_1 \leq b \leq s_2$, in other words a buyer's valuation is between the first and second prices, the buyer pays price b to the seller.

   Obviously, the buyer's utility is 0, and the seller S_1's utility is $b - s_2$.

3. If $b<s_1$, then this trade fails.

In the above, since the second price is used to determine a successful bidder's payment in case 1 this auction is a second-price, sealed-bid double auction. If the buyer's evaluation value is between the first and second prices, the buyer pays the successful bidder the buyer's evaluation value.

3.3 An Example of REV Auction

In this section, we show concrete examples of the REV auction mechanism. Suppose there are three sellers, $i = 1, 2, 3$, and a buyer. A seller, i, shows his utility as $s_i (i = 1, 2, 3)$.

Case 1. A buyer pays the second price.

Buyer's evaluation value: $6
Seller's evaluation value: $\{s_1, s_2, s_3\} = \{\$3, \$5, \$7\}$

If the successful bidder is seller 1, the buyer pays $5 to seller 1. $5 was the second price. The surplus is $6 - $3 = $3. Seller 1 gains utility, $5 - $3 = $2. The buyer's utility is $6 - $5 = $1.

Case 2. A buyer pays his evaluation value.

Buyer's evaluation value: $ 4
Seller's evaluation value: $\{s_1, s_2, s_3\} = \{\$3, \$5, \$7\}$

The successful bidder is seller 1. The buyer pays $4 to seller 1. And $4 is the buyer's evaluation value. The surplus is $4-$3 = $1. Seller 1 gains utility of $1, and the buyer's utility is zero.

Case 3. No trade.

Buyer's evaluation value: $ 2
Seller's evaluation value: $\{s_1, s_2, s_3\} = \{\$3, \$5, \$7\}$

The buyer's evaluation value is lower than sellers'. If the buyer trades, his utility is below zero, $2-$3 = -$1. This trade does not satisfy individual rationality. Therefore, in this case, no trade is consummated.

4 Experiments

As one of the features of a REV auction, when the second price matches a buyer's evaluation value, the buyer's evaluation value is used as the payment. It is expected that the REV auction has a higher rate of realizing deals than the Vickrey auction and the open bidding.

We conducted a deal-making simulation using the REV auction protocol. A buyer's evaluation value and the sellers' bid values are assumed to be goods of private value. Their values are given based on a uniform distribution.

We conducted an experiment to present the REV auction's effectiveness. First, Fig. 2. shows the result of the experiment and is a graph showing successful trades in the REV and Vickrey auctions. The horizontal axis shows the number of sellers who are nominated by a buyer. The vertical axis shows the rate of successful trade. We conducted 10000 dealings for each setting and calculated its average. In the experiment, we assumed there were 2 to 20 sellers. For each case, we compared the REV auction with the Vickrey auction.

When the REV auction protocol is used, the rate of deal closings is always higher than in the equivalent Vickrey auction. In the REV auction, when a buyer nominates two sellers, the probability that a deal is consummated is 66%. In the Vickrey auction, when a buyer nominates two sellers, the probability of a deal is 33%. Therefore, the success of the REV auction is 33% higher than the Vickrey auction in terms of the rate of deals. When a buyer nominates 10 sellers, the REV auction closure is 9% higher than the Vickrey auction. For 20 nominated sellers the REV auction is 5% higher. When the REV auction is compared with the Vickrey auction for a small number of sellers nominated, an obvious difference appears in the rate of successful trades. We insist that the number of deals that the REV auction realizes is always larger than the number of deals that the Vickrey auction realizes. Furthermore, when a buyer nominates between 50 and 100 sellers, we also confirmed that the REV auction had a higher rate of deal formation than the Vickrey auction.

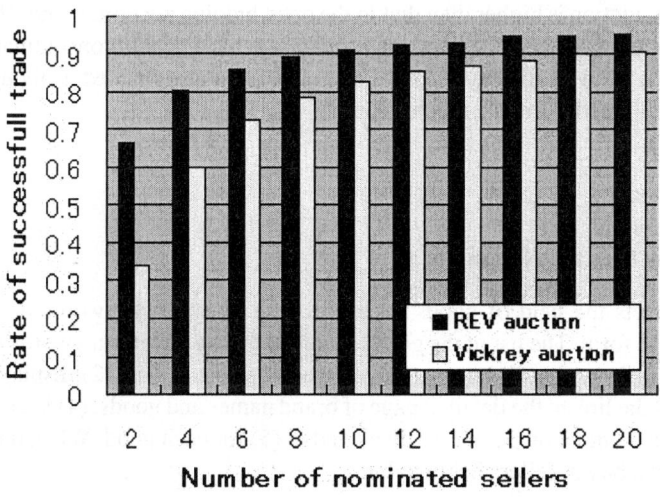

Fig. 2. Rate of successful trades in the REV and Vickrey auctions

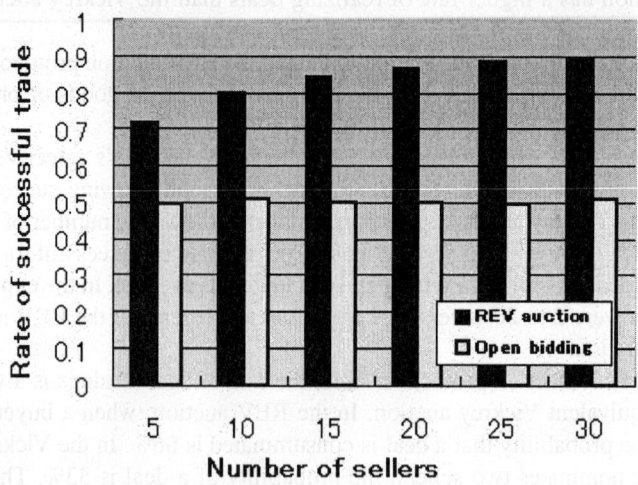

Fig. 3. Rate of successful trades in the REV auction and the open bidding

Second, Fig. 3. shows the result of the experiment and is a graph showing successful trades in the REV and the open bidding. The open bidding is an auction without nomination. The horizontal axis shows the number of sellers who bids in the REV auction and the open bidding. The vertical axis shows the rate of successful trade. We conducted 10,000 dealings for each setting and calculated its average. In the experiment, we assumed there were 5 to 30 sellers. For each case, we compared the REV auction with the open bidding. When the REV auction protocol is used, the rate of successful trades of our auction is higher than that in the open bidding auction. In the REV auction, when a browser displayed 5 sellers, the probability that a deal is consummated is about 70%. When a browser displayed 30 sellers, a deal is consummated is about 90%. Fig. 3. shows REV auction enable much trade.

5 Discussion

5.1 A User Interface Example

A buyer selects the field of goods for a purchase or inputs keywords describing the goods into the form. The list of the goods searched from a buyer's keywords is displayed in Fig. 4: (1) shows check boxes a buyer uses to select goods, (2) displays names of goods, (3) is the link to the detailed page of brand names and goods, (4) is the exhibitor's name. The accumulation points are displayed in (5) for each good. When the button (6) is pushed by a buyer, sellers begin to bid.

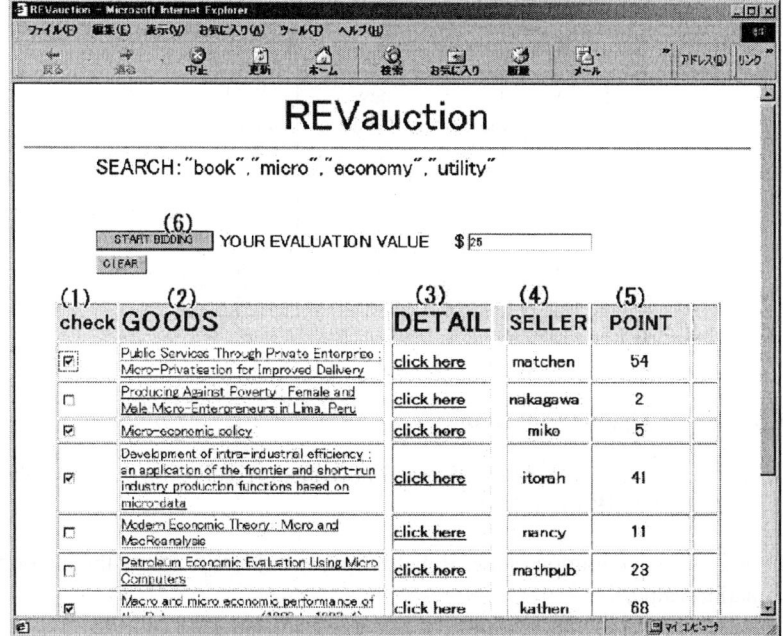

Fig. 4. An example of a user interface

5.2 A Buyer's Preference

Among the significant advantages of the REV auction are its abilities to reflect a buyer's preferences and to guarantee the quality of sellers. This advantage is achieved by having the buyer nominate sellers and by competition among the sellers. In many auctions, competition occurs only on price. In the designated auction, a buyer will not nominate sellers who have inferior quality goods. Sellers who are not nominated can not bid. Therefore, sellers must sell goods of high quality. The REV auction enables selection of goods according to the preference of the buyer. Furthermore, in the REV auction, a buyer can nominate sellers who sell goods in different categories.

Table 2 shows an example of the nomination mechanism in the REV auction. Suppose a buyer nominates sellers who have goods that can be substituted. In this case, the goods have the same value for him. Consider a buyer wants to buy a certain book in the REV auction. The CD-ROM, and online data have the same content as the book. The buyer takes the following into consideration. Is the edition new or not, is the author famous, is the price low, the contents, the size and weight. A buyer's evaluation value shows $8. If there are goods priced less than $8, the deal is a success.

In this Example, we can observe that a buyer can nominate sellers who sell the book, a CD-ROM or online data. The REV auction enables us to have multiple sellers compete simultaneously. In the REV auction, a buyer can include their own initiative in the auction.

Table 2. An example of the nomination

Item	Book1	Data CD-ROM	Online data
edition	new	new	old
author	famous	not famous	famous
price	high	low	high
contents	good	good	bad
size	large	small	-
weight	heavy	light	-
point	12	15	9
value	8	8	8

5.3 Robustness Against Collusion

In general, the designated auction is not robust against collusion for the following reasons[McMillan 92][Kajii 00].

- The goods have common value. Rebates enable bidders to know the auctioneer's evaluation value.
- If auction is repeated, there will be an incentive for sellers to collude.
- Retaliation from other colluding sellers prevents a seller from avoiding collusion.

In general, consultation is difficult without all the bidders consenting in an auction. In the REV auction, collusion is difficult for the following reasons.

- The sellers nominated do not necessarily belong to the same category.
- The sellers nominated are not known beforehand.
- Another sellers newly appear.

5.4 Rewards for Nomination

The REV auction has a positive evaluation system based on rewards for nomination. In The REV auction, when sellers are nominated by a buyer, 1 point is assigned to the nominated sellers as a reward. Based on these points, a buyer can nominate new sellers for themselves. Currently, on some auction sites, there are systems where buyers can evaluate sellers [McMillan 91]. Whenever a buyer evaluates a certain seller negatively, the seller's utility is decreased. On the other hand, in the REV auction, when sellers are nominated, the sellers' points can only increase. A buyer can not evaluate sellers negatively. Therefore, there is individual rationality for sellers in the REV auction. The point system is superior to the evaluation system on traditional auction sites.

6 Conclusions and Future Work

In this paper, we proposed the REV auction to solve problems in typical reverse auctions. The advantages of the REV auction are described as follows. First, since a buyer selects sellers, the quality of goods and the buyer's preferences can be reflected. Second, collusion is difficult. Third, the REV auction has a positive evaluation system. The point system helps the buyer to select sellers who perform well. In our experiments, we showed that the REV auction can realize more trades than the Vickrey auction.

Our future work includes extending the REV auction protocol to cases where group buying exists and to cases where the double-auction mechanism is used. In the REV auction, we will investigate a buyer's preference which is represented by MAUT (Multi Attribute Utility Theory).

References

[eas]	http://www.easyseek.net/
[eBa]	http://pages.ebay.com/
[pri]	http://tickets.priceline.com/
[yah]	http://auctions.yahoo.co.jp/
[Baba 00]	Baba Y.: Mineika no ookushon(in Japanese), "Financial Review", Policy Research Institute of the Ministry of Finance Japan, No.53 (2000)
[Endo 00]	Endo T.: Theory of Auctions, p.8, The Mitsubishi Economic Research Institute (2001)
[Ito 01]	Ito T., Shintani T.: Implementation Technologies for Multiagent Systems and Their Applications, Journal of the Japanese Society for Artificial Intelligence, Vol.16, No.4, pp. 469-475 (2001)
[Kajii 00]	Kajii A., Matsui A.: Micro-Economics; Strategic Approaches, Nippon-Hyoron-Sha, pp.113-115(2000)
[McMillan 91]	McMillan, J.: Dango: Japan's Price-Fixing Conspiracy, Economics and Politics, Vol. 3, pp. 201-218 (1991)
[McMillan 92]	McMillan J.: Games, Strategies, and Managers, Oxford University Press (1992)
[Saijo 94]	Saijo T., Une M., and Yamaguchi T.: Dango experiments, forth coming in Journal of the Japanese and International Economics (1994)
[Turban 00]	Turban E. , Lee J. , King D. , Chung H. M.: Electronic Commerce: A Managerial Perspective, Pearson Education(2000)
[Yokoo 00]	Makoto Yokoo: Internet Auctions: Theory and Application, Journal of the Japanese Society for Artificial Intelligence, Vol.15, No.3, pp.404-415 (2000)
[Yokoo 01]	Yokoo M., Sakurai Y., and Matsubara S.: Robust Double Auction Protocol against False-name bids, The 21st IEEE International Conference on Distributed Computing Systems(ICDCS-2001) (2001)

Design of a Fuzzy Usage Parameter Controller for Diffserv and MPLS

K. K. Phang, S. H. Lim, Mashkuri Hj. Yaacob, and T. C. Ling

Faculty of Computer Science and Information Technology
University of Malaya, 50603 Kuala Lumpur, Malaysia
{phang,mashkuri,tchaw}@fsktm.um.edu.my
shong@siswazah.fsktm.um.edu.my

Abstract. This paper describes a fuzzy usage parameter control (UPC) mechanism in DiffServ (DS) and Multiprotocol Label Switching (MPLS) networks based on fuzzy logics. Current research treats MPLS as the Internet's solution to high performance network. In DS network, UPC is an important factor in ensuring the sources conforms to the negotiated service level agreement (SLA). Most of the UPC techniques proposed are based on conventional crisp set which are inefficient when dealing with the conflicting requirements of UPC. Simulation results show that the proposed fuzzy scheme outperforms conventional techniques in terms of packet loss ratio, higher selectivity and lower false alarm probability.

1 Introduction

One of the most demanded features of the Internet is to provide a high level of Quality of Service (QoS) and performance assurance to the users. Unfortunately the current Internet only supports best effort IP forwarding, where all packets are treated uniformly regardless of the relative importance and timeliness requirement of any of the packets. MPLS [1,2] and DS [3,4] are generally regarded as versatile solutions to address this problem. MPLS provides a solution for high-speed switching and packet forwarding. DS, on the other hand, is intended to enable scalable service discrimination in the Internet without the need for per-flow state and signaling at every hop and hence highly scalable. DS uses the concept of per hop behavior (PHB) in the scheduling and dropping of packets. PHB is applied by a router to all the packets, which are to experience the same DS service. Three PHB groups are defined: 1) The default best-effort PHB 2) Expedited Forwarding PHB (EF) and 3) Assured Forwarding PHB Group (AF). The EF PHB is intended for low loss, low latency, low jitter, assured bandwidth, end-to-end service through DS domains; whereas, the AF PHB Group is intended for low loss (high assurance) traffic without strict delay or jitter requirements.

The integration of DS and MPLS addresses issues related to scalability and QoS provided by the traditional best effort IP routing and the overlay model [1]. It offers better quality of service, scalability and traffic management. The complex, dynamic and unpredictable nature of DS and MPLS traffic presents a challenge to conventional, statistical approaches in various traffic management functions.

The term UPC is adopted from [5] to designate a mechanism that performs the metering and policing of the traffic, which is part of the traffic conditioning functions in DS [4]. Various approaches have been proposed for the UPC. The most prevalent candidate among the various policing mechanism is the Leaky Bucket (LB) mechanism. Other approaches are window-based mechanisms, including the Moving Window (MW), the Jumping Window (JW), the Triggered Jumping Window (TJW), and the Exponential Weighted Moving Average (EWMA) mechanisms [6,7].

Fuzzy logic is based on the idea of fuzzy sets and linguistic variables with the use of expert knowledge [8]. It has been shown that fuzzy logic is efficient in handling complex, nonlinear, and dynamic systems such as a computer network. Various fuzzy logic based policers have been proposed in the literature for ATM network. In [9], a fuzzy policer which uses a window-based control mechanism is proposed. In [10], a fuzzy UPC based on dual leaky bucket is proposed. In [11], a fuzzy policer based on the concepts of modified leaky bucket and moving window is proposed. Simulation results in all the mentioned systems show that fuzzy policers are much better than the conventional approaches.

This paper proposes a fuzzy traffic policer that is different from the discussed previous works [9,10,11,12,13] in three ways. First, it applies the policer in a DS and MPLS network. Next, it polices the AF PHB traffic. Finally, it polices the mean transmission rate and burst size separately by using two rule bases. The paper is organized as follows. Section 2 describes the difficulties of using non-fuzzy model in UPC. Section 3 provides details of the proposed scheme; Section 4 and 5 discuss the relevant parameters of the proposed fuzzy controller; Section 6 and 7 present the simulation results and conclusions.

2 Difficulties with Current Models Using Crisp Model

The problems with all these approaches come from several conflicting requirements of the policing function. These basic requirements include high selectivity, low false alarm probability, high responsiveness and simplicity of implementation [9]. In addition, the long-term congestion (violation) within each AF PHB should be minimized, while allowing short-term violations resulting from burst [4,9]. It is hard to fulfill all these requirements at the same time. For example, a low false alarm probability will result in lower responsiveness (to allow slight violation margin). Furthermore, a high selectivity will involve high complexity in implementation, thus, conflicting with the simplicity requirement. Due to the dynamic behavior of various types of traffic source, even high selectivity itself is difficult to achieve.

In Leaky Bucket, two parameters are involved (the bucket size and the leak rate) [4,5]. In policing an average rate where certain burst is allowed, the leak rate is set to the negotiated rate and the bucket size has to be large enough to lower the false alarm probability. But with a large bucket size, the responsiveness to real violations will be slow. In order to increase the responsiveness by reducing the bucket size, the leak rate has to be increased. But this will reduce the capability to detect long-term violation to the average rate.

The limitations of conventional policing mechanisms lead to alternate solutions. Fuzzy logic offers intuitive approach based on the mathematical foundation of Fuzzy Set Theory. A UPC mechanism, which is more "natural," is preferable.

3 The Proposed Scheme

In our proposed scheme, the metering and policing of the AF PHB [4] traffic are implemented using the following three parameters: the Peak Transmission Rate (PTR), the Sustained Transmission Rate (STR) and the Maximum Burst Size (MBS). PTR is the maximum rate at which packets will be sent. STR is the long-term average transmission rate. MBS, on the other hand, is the maximum number of packet that can be sent continuously at peak rate. The drop precedence within each AF class is implementation using different values for the leaky bucket size. For AF PHB with low lost precedence, a larger bucket size is used; whereas for traffic with high drop precedence, a smaller bucket size. Both recent and long-term behaviors of traffic sources are taken into consideration. The source is free to transmit at a rate below or equal to the STR at any time. Occasionally, it can transmit beyond the STR but not exceeding the PTR. This means that the AF PHB service allows bursty traffic. The size of burst is determined by the negotiated MBS.

To implement a UPC function to monitor the AF traffic, the UPC mechanism must be able to monitor the compliance of both the STR and the MBS. For any incoming packet, we can test whether the packet conforms to the traffic profile by using a continuous leaky bucket algorithm. It is a fact that even a fully compliant connection may occasionally contain non-conforming packets, due to many factors [4,5].

The main concern with the UPC function is the ability to avoid false alarm without sacrificing the responsiveness. This conflict, together with the weaknesses of the conventional methods in traffic policing, leads to the use of fuzzy logic in UPC. The proposed Fuzzy AF Policer sets out to minimize this conflict by considering both recent and long-term traffic patterns before making any decision.

To monitor the STR, the incoming transmission rate has to be measured. On the other hand, to monitor the MBS, the burst length of incoming packets has to be measured. Due to the bursty nature of a source, it is almost certain that from time to time, there will be burst length that exceeds the MBS unless the negotiated value is much larger than needed. But the latter case will cause inefficient resource utilization. In this situation, the policer should not simply drop the packet, but will take into consideration the long-term burst length and the long-term transmission rate. If one or both of these two long-term parameters are well below the negotiated value, there is no reason to penalize the short-term violation, unless the violation continues.

Another issue arises from the fact that the negotiated STR is an upper-bound value of the mean transmission rate [14]. This means that for any AF connections, the long-term mean rate must not exceed the STR. Consider a traffic source that violates the negotiated rate for a considerable amount of time, so that the long-term mean rate is well above the negotiated value, but then stops the violation by transmitting at a low rate. During the violation period, it will be penalized by dropping packets. But immediately after the violation period (the long-term average is still well above the negotiated STR), should it still be penalized (with a smaller magnitude)? During this short after-violation period, the connection is basically still non-compliant, and therefore the QoS needs not be met.

At this point, it is clear that the above situations require a more flexible policing mechanism than the conventional leaky bucket. Therefore, a fuzzy controller would be a good candidate. With proper input parameters and rule base, all the above problems could be minimized.

4 Input and Output Parameter of the Fuzzy Controller

Four input parameters are used for the proposed Fuzzy AF Policer.

1. **Recent Burst Length (RBL)** This is the most currently measured burst length of an incoming traffic.
2. **Long-term Burst Length (LBL)** This is a running average of all the measured burst lengths since the beginning of the connection. If at any moment, there is a total of N bursts, then

$$LBL = \frac{\sum_{i=1}^{N} Burst\ length\ of\ i^{th}\ burst}{N} \quad (1)$$

3. **Recent Transmission Rate (RTR)** To measure the current transmission rate, a window is needed. The window size is determined by both the MBS and the STR, and can be adjusted by an external factor, determined by the policing policy.

$$RTR = \frac{Amount\ Transmitted\ in\ last\ window}{Window\ length} \quad (2)$$

4. **Long-term Transmission Rate (LTR)** The long-term transmission rate is calculated as the average transmission rate from the beginning of the connection.

$$LTR = \frac{Amount\ Received}{Total\ Connection\ Time} \quad (3)$$

With the above parameters defined, the input parameters for the Fuzzy Traffic Policer are as follows.

$$Ratio\ of\ RBL\ to\ MBS\ (rRBL) = \frac{RBL}{MBS} \quad (4)$$

$$Ratio\ of\ LBL\ to\ MBS\ (rLBL) = \frac{LBL}{MBS} \quad (5)$$

$$\text{Ratio of } RTR \text{ to } STR \ (rRTR) = \frac{RTR}{STR} \tag{6}$$

$$\text{Ratio of } LTR \text{ to } STR \ (rLTR) = \frac{LTR}{STR} \tag{7}$$

There is one output parameter, the Drop Rate (DR). It is a value from [0,1] where 0 means total pass, and 1 means total drop. Each of the input and output parameters is a fuzzy variable, which is in turn, is described by a set of membership functions. They are defined below (where $T(x)$ gives the set of membership functions for fuzzy variable x):

$$T(rRBL) = \{Low(L_{RBL}), Medium(M_{RBL}), High(L_{RBL})\}$$

$$T(rLBL) = \{Low(L_{LBL}), Medium(M_{LBL}), High(L_{LBL})\}$$

$$T(rRTR) = \{Low(L_{RTR}), Medium(M_{RTR}), High(L_{RTR})\}$$

$$T(rLTR) = \{Low(L_{LTR}), Medium(M_{LTR}), High(L_{LTR})\}$$

$$T(DR) = \{Drop \ (D), Between \ Pass \ \& \ Drop \ (B), Pass \ (P)\}$$

The membership functions of all of the input parameters are similar; they take the form of the graph in Fig. 1. The membership functions for the output parameters are depicted in Fig. 2.

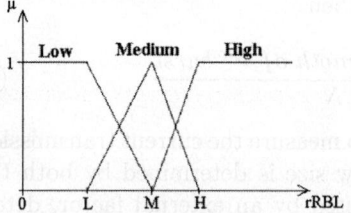

Fig. 1. Membership functions for fuzzy variable rRBL

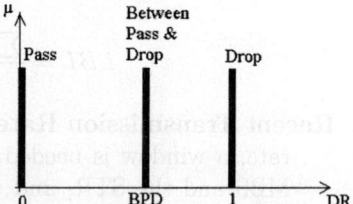

Fig. 2. Membership functions for Drop Rate

5 The Fuzzy Rule Base

Two fuzzy rule bases are used to generate the final output. The first rule base takes 3 inputs and produces 1 output, while the second rule base takes 2 inputs and produces 1 output. The first rule base is for policing of the MBS. Its inputs are the rRBL, rLBL and rLTR. The output is DR. With these three inputs, the rule base has 27 rules. All the rules are in the following canonical form:

$$IF \ x_1 \ is \ A_1 \ AND \ x_2 \ is \ A_2 \ AND \ x_3 \ is \ A_3 \ THEN \ y \ is \ B$$

Table 1. Rule base for policing the MBS

Rule	rLBL	rRBL	rLTR	DR	Rule	rLBL	rRBL	rLTR	DR
1	L	L	L	P	15	M	M	H	P
2	L	L	M	P	16	M	H	L	P
3	L	L	H	P	17	M	H	M	B
4	L	M	L	P	18	M	H	H	D
5	L	M	M	P	19	H	L	L	D
6	L	M	H	P	20	H	L	M	D
7	L	H	L	P	21	H	L	H	D
8	L	H	M	B	22	H	M	L	D
9	L	H	H	D	23	H	M	M	D
10	M	L	L	P	24	H	M	H	D
11	M	L	M	P	25	H	H	L	D
12	M	L	H	P	26	H	H	M	D
13	M	M	L	P	27	H	H	H	D
14	M	M	M	P					

Where A1, A2 and A3 are members of rRBL, rLBL and rLTR respectively and B is DR. Table 1 summarizes the rule base.

It has to be noted here that this rule base is for the policing of the MBS, which is the maximum allowed burst length (at peak transmission rate). Therefore, the long-term transmission rate is only used to assist decision-making. So even when it is high, a penalty is not necessarily imposed. But when the long-term burst length is high, the action is almost certainly a Drop.

The second rule base (Table 2) is for policing of the STR. The inputs are rRTR and rLTR. The output is again DR.

Table 2. Rule base for policing the STR

Rule	rLTR	rRTR	DR	Rule	rLTR	rRTR	DR
1	L	L	P	6	M	H	B
2	L	M	P	7	H	L	D
3	L	H	B	8	H	H	D
4	M	L	P	9	H	H	D
5	M	M	P				

With two rule bases, there will always be two output values, both are Drop Rate (DR). Because each rule base monitors different parameters (one for STR and another for MBS), their output is not related. In the implementation, both values are added arithmetically to obtain the final Drop Rate. This means that the violation of one of them is enough to get some penalty. If both are violated, then a heavier penalty will be imposed.

6 Simulation Results

On-off source is used to model bursty traffic. Each traffic source has two states: on-state and off-state. During the on-state, packets are transmitted at the peak transmission rate (PTR). During the off-state, no packets are transmitted. The mean duration for the on-state and the off-state are carefully selected to achieve a particular average transmission rate (STR). Also, the duration of the on-state determines the burst length (amount transmitted during a burst). The on-off model with geometrically distributed on-time and exponentially distributed off-time is used.

The steady state response of the policer can be interpreted as the "seriousness of penalty" corresponding to different amount or level of violation. The evaluation of the steady state response is based on the percentage of packets dropped by the policer.

Packetized voice source are used to evaluate the steady state response. Packetized voice represents sources with low burstiness characterized by frequent short bursts. Their parameters are listed in table 3.

Table 3. Source parameters for packetized voice

	Packetized Voice
Peak bit rate:	32 Kbps
Mean bit rate:	11.2 Kbps
Mean burst length:	2650 bytes

The UPC typically sits at the edge of the network, that is, at the edge routers. To evaluate the performance of the proposed fuzzy policer, simulations are carried out by using the NIST ATM/HFC Network Simulator [15]. Several new components are developed and incorporated into the simulator. The values defining the membership functions for all the fuzzy sets used are given in Tables 4 and 5. These values are used throughout the simulations for both the steady-state and the transient response. They can be fine-tuned to suit particular policies.

Table 4. Membership function boundary values for the input parameters

Parameters	L	M	H
rRBL	0.5	1	1.5
rLBL	0.5	1	1.5
rRTR	0.8	1	3
rLTR	0.8	1	1.2

The steady state response is evaluated using sources with constant on-off state duration. This is to ensure that the results are not affected by unintended

Table 5. Member function boundary values for the output parameter

Parameter	Pass	BPD	Drop
DR	0	0.5	1

violations of the negotiated parameters caused by the randomness in the distribution. The results for violation of both STR and MBS for each type of traffic source are shown in Figures 3 and 4.

From the steady state response simulation result graphs, it is obvious that the fuzzy policer outperforms the leaky bucket policer. When the level of violation increases, the penalty imposed by the leaky bucket increases very slowly while the penalty imposed by the fuzzy policer increases very quickly.

The significance of this response is that once a connection has been identified as strictly non-compliant, no QoS has to be committed to it, including the Packet Loss Ratio. As specified in [5], when a non-compliant connection has been identified and the violation level is high, the switch may initiate a mechanism to disconnect the source totally. With the fuzzy policer, the packet drop percentage approaches 100% when the violation level is constantly high (above 20% for STR (i.e. at $x = 1.2$) and above 50% for MBS (i.e. at $x = 1.5$)), this has the same effect of 'shutting down' the connection without actually disconnecting it. Of course, in actual environment, explicitly disconnecting a non-compliant connection is still desirable to free some resource.

The transient response of the policer is its behavior during a sudden change of an incoming source traffic, especially when the change is a violation to the negotiated traffic parameters. The behavior of the policer is sometimes in an uncertain state of whether to penalize or not to penalize the source. This is especially true when the source behavior is unintentional or just a false alarm.

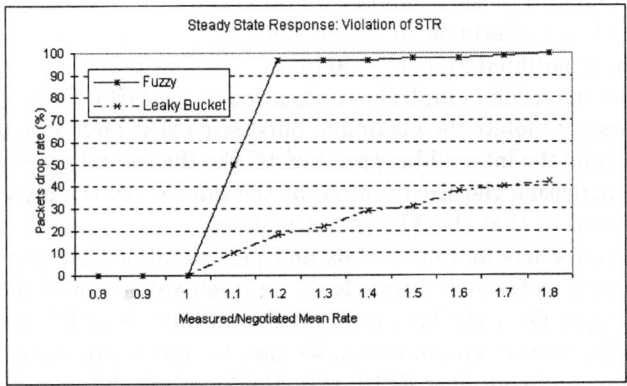

Fig. 3. Steady state simulation results: Cell drop rate for various measures/negotiated mean rate

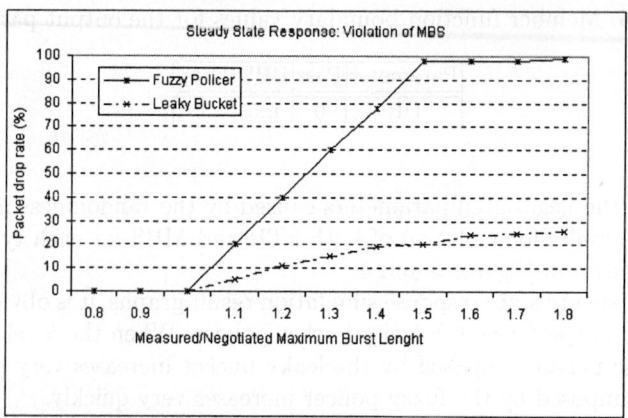

Fig. 4. Steady state simulation results: Cell drop rate for various measured/negotiated maximum burst length

A serious penalty in this situation will greatly reduce the quality of the network service.

Using a source that is compliant at the beginning of the connection, and suddenly starts to violate one of the negotiated parameters after a certain point tests the transient response of the proposed fuzzy policer. The packetized voice source is used, with roughly same parameters as shown in table 3, but the magnitude is increased to the megabit range for ease of evaluation. Also, the on-off period is set to constant to eliminate any unwanted disruption when evaluating the transient response due to the randomness.

The number of packets dropped by the policer is used as a measurement of the amount of penalty imposed by the policer. Again, both the behaviors of the leaky bucket and the fuzzy policer are evaluated and compared.

Figures 5 and 6 show the result of the transient-response test. The negotiated STR (that is, the long-term mean transmission rate) is equivalent to a bit rate of 3.5 Mbps. The negotiated MBS is 2650 bytes. The first graph shows violation of the STR, while the second graph shows violation of the MBS. For the first graph, the source does not violate the maximum burst size (MBS) but violates the mean rate by shortening the interval between bursts. For the second graph, the MBS is violated but increasing the size of burst, at the same time, the interval between bursts is adjusted so that the STR is not violated.

The violation starts at time $t = 6s$ and persists after that point. Both the long-term bit rate and recent bit rate is shown (their values are on the left y-axis) for the first graph. Only the long-term maximum burst length is showed in the second graph. In the first graph, we can see that the recent bit rate suddenly rises above 3.5 Mbps and continues. With this, the long-term bit rate also increases slowly toward the recent bit rate. The situation is also the same for the second graph. The following discussion will refer to the first graph. The behaviors for both violations are similar.

Notice that there is a period (roughly between $t = 6s$ to $t = 10s$) where the long-term bit rate is still below 3.5 Mbps. Strictly speaking, by definition of the STR, the source is still compliant. But if the situation persists, then sooner or later, the STR will be violated.

From Figs. 5 and 6, we see that the leaky bucket displays constant behavior throughout the entire transient period, which causes a constant packet drop rate (hence the straight line). But the fuzzy policer starts slowly, especially during that uncertain period. But once the violation persists, and the long-term bit rate is violated, quick action is taken so that more packets are dropped.

The slow-start behavior in this situation is desirable because if the sudden violation is only a false alarm (that is, the source quickly return to the previous state), the number of dropped-packets will be small. On the other hand, when the STR is definitely being violated, fast increase of packet drop rate is necessary as a warning sign to the source. These behaviors show that the fuzzy policer is much more flexible than the leaky bucket in handling dynamic situations.

To evaluate the behavior of the policers in an 'actual' false alarm, the simulation is run again the source returns to the previous state after only a very short time (500ms). With this, the STR is never violated throughout the whole period. The result is shown in Fig. 7. This time, we see that the fuzzy policer managed to drop only a very small number of packets, and it stops dropping packet just after that violation period. Because the STR is never actually violated, the fuzzy policer will not continue the penalty like what happened in the previous test. Please note that the short delay for the fuzzy policer to stop is due to the use of measurement windows, which is typical in any windowing mechanism. This period is random in nature but is upper-bounded by the window length. By

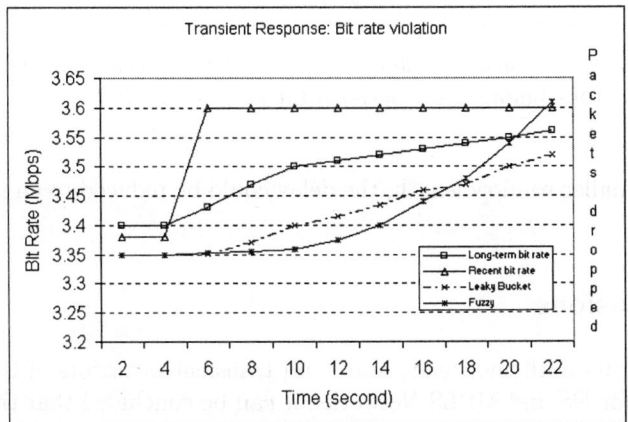

Fig. 5. Transient response: Sudden violation of the negotiated rate (3.5 Mbps) and violation persists

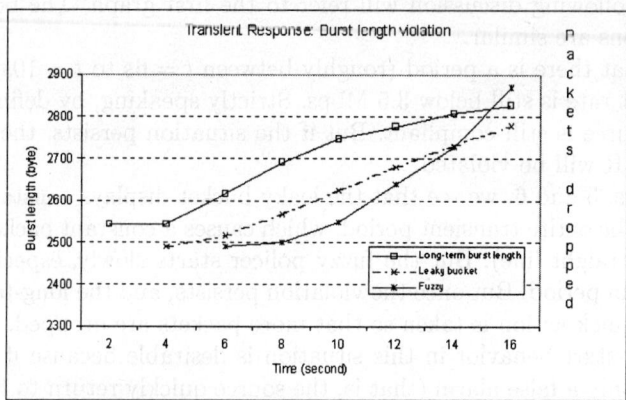

Fig. 6. Transient response: Sudden violation of the negotiated maximum burst size (2650 bytes) and violation persists

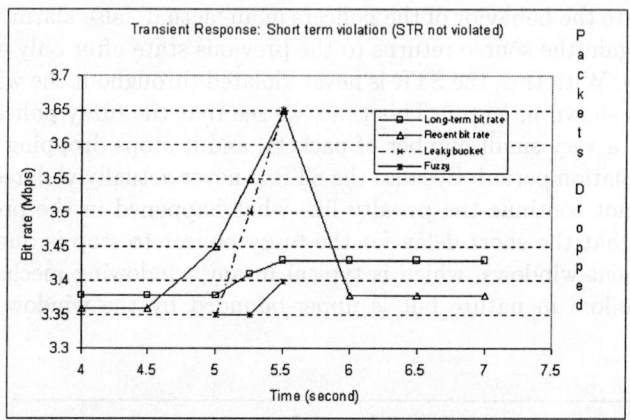

Fig. 7. Transient response: Sudden violation of the negotiated bit rate (3.5Mbps) but violation stops before the long-term bit rate is violated

choosing a smaller window length, the delay could be reduced to an insignificant value.

7 Conclusions

After evaluating both the steady state and transient behaviors of the proposed Fuzzy UPC for DS and MPLS Networks, it can be concluded that the proposed scheme achieves higher effectiveness and selectivity than the leaky bucket algorithm, with low probability of false alarm. Possible future work includes UPC for other PHB groups in DS, including the EF and even the best-effort group.

References

1. Lawrence, J.: Designing multiprotocol label switching networks. IEEE Communications Magazine, Vol. 39 Issue. 7, July (2001)
2. Girish, M. K., Zhou, B., Hu, J. Q.: Formulation of the Traffic Engineering Problems in MPLS Based IP Networks. Proc. Fifth IEEE Symposium on Computers and Communications (2000) 214–219
3. Zhang, G., Mouftah, H. T.: End-to-end QoS guarantees over diffserv networks. Proc. Sixth IEEE Symposium on Computers and Communications 2001 (2001) 302–310
4. Heinanen, J., Baker, F., Weiss, W., Wroclawski, J.: Assured Forwarding PHB Group, RFC 2597, June (1999)
5. ATM Forum: Traffic Management Specification, Version 4.1, af-tm-0121.000, March (1999)
6. Butto, M., Cavallero, E., Tonietti, A.: Effectiveness of the Leaky Bucket Policing Mechanism in ATM Networks. IEEE Journal on Selected Areas in Communication, Vol. 9, No. 3, April (1991) 335–342
7. Rathgeb, E. P.: Modeling and Performance Comparison of Policing Mechanisms for ATM Networks. IEEE Journal on Selected Areas in Communication, Vol. 9, No. 3, April (1991) 325–334
8. Asai, K.(Ed): Fuzzy Systems for Information Processing. IOS Press (1995)
9. Catania, V., Ficili, G., Palazzo, S., Panno, D.: A Comparative Analysis of Fuzzy Versus Conventional Policing Mechanisms for ATM Networks. IEEE/ACM Transactions on Networking, Vol. 4, No. 3, Jun (1996) 449–459
10. Cheng, R. G., Chang, C. J.: Design of a Fuzzy Traffic Controller for ATM Networks. IEEE/ACM Transactions on Networking, Vol. 4, No. 3, Jun (1996) 460–469
11. Wu, X., Ge, L.: The application of fuzzy logic in real-time traffic control in ATM networks. ICCT '98, Vol. 1 (1998) 467–471
12. Chang, C. J., Chang, C., Eul, Z., Lin, L.: Intelligent Leaky Bucket Algorithms for Sustainable-Cell-Rate Usage Parameter Control in ATM Networks. IEEE, Proceedings 15th International Conference on Information Networking, Jun (2001)
13. Pitsillides, A., Ahmet Sekercioglu, Y., Ramamurthy, G.: Effective Control of Traffic Flow in ATM Networks Using Fuzzy Explicit Rate Marking (FERM). IEEE Journal on Selected Areas in Communications, Vol. 15, No. 2, Feb (1997) 209–225
14. Jain, R.: Congestion Control and Traffic Management in ATM Networks: Recent Advances and A Survey. Dept. of Computer and Information Science, The Ohio State U., August (1996)
15. Golmie, N., Mouveaux, F., Hester, L., Saintillan, Y.,Koenig, A., Su, D.: The NIST ATM/HFC Network Simulator: Operation and Programming Guide. High-Speed Networks Technologies Group, NIST, US Dept. of Commerce, December (1998)

A Tool for Extension and Restructuring Natural Language Question Answering Domains

Boris Galitsky

Knowledge-trail, Inc. 9 Charles Str Natick MA 01760
bgalitsky@knowledge-trail.com
http://www.knowledge-trail.com
http://dimacs.rutgers.edu/~galitsky/NL

Abstract. In this report, we present the system that allows various forms of knowledge exchange for users of the natural language question answering system. The tool is also capable of performing the domain restructuring by domain experts to adjust it to a particular audience of customers. The tool is implemented for financial and legal advisors, where the information is extremely dynamic by nature and requires fast correction and update in natural language. Knowledge management takes advantage of the technique of semantic headers, which is applied to represent the poorly structured and logically complex data in the form of textual answers. Question answering is performed by matching the semantic representation of a query with the ones of the answers. The issues of domain extension via understanding of natural language definitions of new entities and new objects are addressed.

1 Introduction

Due to the tremendous amount of information available online on one hand and to the constantly increasing demand for the customer support in a variety of domains on the other hand, the role of natural language (NL) question answering (Q/A) systems has dramatically increased in the last few years [1,7]. In financial business and legal domains, Q/A system becomes valuable when it can be promptly updated by emerging information and quickly adjusted to a particular customer environment [5] . The users must be able to use NL to share the impression of the received advice with the other users, as well as to supply additional up-to-date information. At the same time, in parallel, the experts releasing the Q/A system need to restructure the content to better fit a particular customer audience.

In this paper, we present the methodology of dynamic modification of the Q/A domain, built either manually or automatically. We have developed the suite of financial domains, including tax, insurance, real estate, investments [7], as well as legal, psychological and medical advisors. Details on the syntactic and semantic analysis in our system, as well as the capabilities of semantic header technique for representation of the poorly-structured knowledge can be found elsewhere [5]. Representation of mentioned above set of domains is much less structured and requires much richer set of entities than that of domains with database knowledge representation on one hand and with the statistical question-answer matching machinery on the other hand [2,10]. Therefore, the complexity of domain creation and support tools grows correspondingly. Naturally, the higher the domain complexity, the higher flexibility and automation degree of domain adjustment tools is required. We believe that the capability of domain extension in natural language allows approaching the practical limit of flexibility for an expert Q/A system.

In this paper, we present the following three components of the knowledge management system (Fig.1), oriented to the semantic headers type of domains [5]:

(1) *Domain Restructuring Component* (DRC), capturing structural information from the experts and acquiring marketing information from the users, asking questions;

(2) *Advice Acquisition Component* (AAC), capturing additional domain knowledge from the experts and users, willing to share their experience.

(3) *Definition Understanding Component* (DUC) that allows the users and domain development personnel to introduce new objects and new entities to the encoded Q/A domain.

In spite of the thorough commercial evaluation of semantic header Q/A technique [4], our interactive domain development tools have been used by rather limited number of customers. In the future, we expect much higher contribution of these tools in the creation and management of Q/A domains.

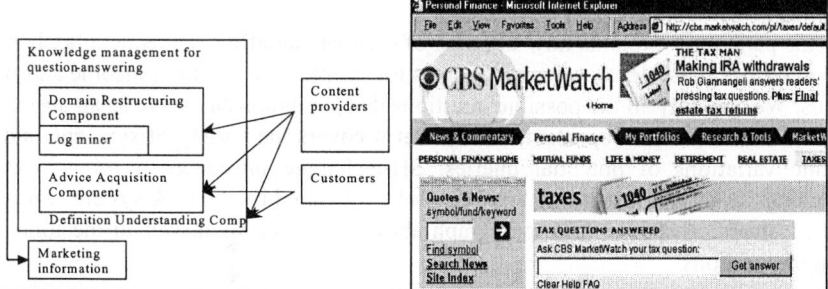

Fig.1. *(On the left)* Knowledge management architecture. Domain Restructuring Component is in use by the content provider and domain expert personnel, adjusting the question answering for a particular audience. Customers share their experience via Advice Acquisition Component, new information automatically becomes available for the other customers. *(On the right)* TAX advisor, deployed by CBS Market Watch (tax season of 2000). Customers are encouraged to ask questions about individual tax and provide an answer quality feedback. The log of question and answers is then processed by domain expert and is incorporated in a new version of Q/A, together with the question-answer pairs, obtained by the AAC

The process of domain extension can be naturally considered from the viewpoints of knowledge acquisition as the transfer and transformation of potential problem solving expertise from some knowledge source to a program. In particular, the system of ontological analysis [8] use three main categories for structuring domain knowledge: the static ontology, which consists from domain entities with objects and attributes, dynamic ontology, which defines the states that occur in problem solving (here, Q/A) and epistemic ontology, which describes knowledge that guides and constrains state transformation in the domain. It is worth mentioning such rule-based knowledge acquisition systems as COMPASS [9], which examines error messages derived from the switch's self-test routines, and EMYCIN [11], domain-independent framework for constructing and running consultation programs. The essential feature of our approach is that a rule-based system merges the semantic and domain knowledge to produce a tool for developing, modification and extension of NL Q/A domain by an expert, content provider and end user in addition to a knowledge engineer.

2   Technique of Semantic Headers for Knowledge Representation

The technique of semantic headers (SH) is intended to resolve the problem of converting an abstract textual document into a form, appropriate for answering a question and easy annotation of a new answer. SH technique is based on logic programming, taking advantage of its convenient handling of semantic rules on one hand, and explicit implementation of the domain common-sense reasoning on the other hand. Only the data, which can be explicitly mentioned in a potential query, occurs in semantic headers. Under SH technique, the domain coding starts with the set of answers (the content). As an example, we consider an answer from the tax domain: „*The timing of your divorce could have a significant effect on the amount of federal income tax you will pay this year...* „

This paragraph explains how the time of divorce can affect someone's tax liability and describes possible tax saving strategies for people with different income and filing status. We consider all the possible questions this paragraph can serve as an answer to, and build their formal representation so that it covers the variety of syntactic and semantic variations of potential queries. SH-technique introduces a robust way of matching the formal representation of answers, obtained by the Q/A system, with precoded semantic headers. The paragraph above serves as an answer to the following kind of questions

> *What are the tax issues of divorce? How can timing your divorce save a lot of federal income tax? I am recently divorced; how should I file so I do not have a net tax liability? Can I avoid a marriage penalty? How to take advantage of the married filing jointly status when I am getting divorced?*

Below is the list of semantic headers for the answer above (we use the logic programming presentation style).

```
divorce(tax(_,_,_),_):-divorceTax.
divorce(tax(_,_,_), time):-divorceTax.
divorce(tax(file(_),liability,_), _):-divorceTax.
divorce(file(joint),_):-divorceTax.
```

Then the call to divorceTax will add the paragraph above to the current answer, which may consist from the multiple pre-prepared ones. Under the acquisition of new knowledge, semantic headers are automatically added to the knowledge base. Their compatibility with the existing semantic headers needs to be verified; if necessary, the metapredicates of the backtracking control (*var/nonvar* below) are redistributed between the updated semantic headers [5,7]. A generic set of semantic headers for an entity *e*, its attributes $a_1, a_2, ...$ and other entities $a_1, a_2, ...$ looks like the following:

---

e(A):-var(A), answer(#). This is a very general answer, introducing (defining) the entity *e*. It is not always appropriate to provide a general answer (e.g. to answer *What is tax*), so the system may ask a user to be more specific:

e(A):-var(A), clarify([$a_1, a_2, ...$]). If the attribute of *e* is unknown, clarification procedure is initiated, suggesting to choose an attribute from the list $a_1, a_2, ...$ to have a specific answer about *e(a_1)* instead of just for *e*.

e(A):-nonvar(A), A = a_1 , answer(#). The attribute is determined and the system outputs the answer, associated with the entity and its attribute (# is the answer id).

e(e_1(A),e_2):-nonvar(A), A≠a_1 , e_2(_). Depending on the existence and values of attributes, an embedded expression is reduced to its innermost entity that calls another SH.

e(A,#). This semantic headers serves as a constraints for the representation of a complex query *e_1(A,#), e_2(B,#)* to deliver just an *answer(#)* instead of all pairs for e_1 and e_2. It works in the situation where e_1 and e_2 cannot be mutually substituted into each other.

---

3 Restructuring Domain Knowledge

There is a variety of quite similar question answering domains, associated with a particular topic of world knowledge. These domains are common in presenting the essential issues, explaining the basic concepts and links between them. Marketing targets of the content providing companies, employing the question answering engines, motivate the difference between these domains. Therefore, each of these domain contains individual topics and a peculiar, biased prospective of presentation. DRC is oriented to the knowledge management for such domain, to efficiently build the specialized domains given a generic one.

Natural language understanding problem is posed as recognition of the most relevant answer, given a query [5]. Therefore, all answers (more precise, associated sets of semantic headers) are subject to hierarchical classification. Proper categorization of answers and establishing their taxonomy is the essential step in knowledge representation for Q/A. DRC is designed to visualize a Q/A domain in a form, convenient to modify the meaning of participating entities meanings modification, as well as to distribute overall knowledge through the answers.

Q/A domain restructuring toolkit (Fig.2) presents the domain classification graph with the nodes, which can be moved to provide a clearer classification schema. The nodes are assigned with entities and the edges represent the facts of their mutual substitution to form the semantic headers. Answers are associated with the sets of paths on this graph. Modifying the graph, a domain expert changes the current way of answer classification. One can think of this procedure as an analogous to shifting the boundaries between the areas of a feature space to improve the pattern recognition accuracy. In case of Q/A, such modification performs the redistribution of accents (links to the questions) between the answers. DRC is capable of reading/writing the knowledge representation code (the list of semantic headers) to match current domain graph. The visual representation of this graph can be edited manually by the content manager to modify the domain taxonomy. As a result, modified semantic headers are created to represent the updated links between the entities and their attributes.

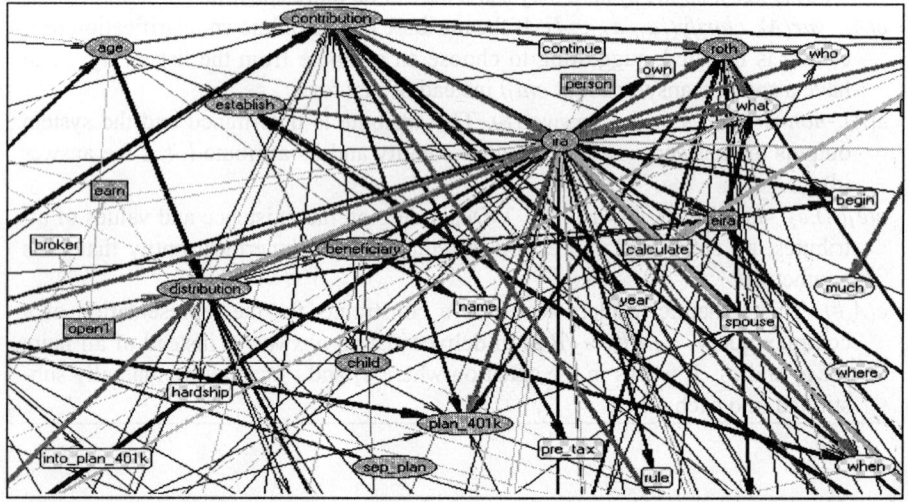

Fig.2. The fragment of the screenshot of the Domain Restructuring Component

Each customer question and its formal representation are accumulated (logged) by the Log Miner unit of DRC (Fig.1). The totality of such formal representations can be visualized as a graph with the same nodes (or a subset of nodes) as the domain graph (shown as ovals). There are edges that are present in the domain graph and are not present in the logging graph (with rectangle shape nodes), and vice versa. It means that there are natural language expressions, encoded by the corresponding links between entities, which are not in use while querying. At the same time, there are links between entities, used by customers and not prepared in advance by the knowledge engineers or content personnel. The latter is essential information, concerning the possibly missing marketing features of a product or a service that is the subject of Q/A. Also, comparing the edge thickness (the number of links) for the domain and log graphs helps to estimate the domain issues, most important from the customers' prospective.

4 Acquiring New Domain Knowledge from the Users

In this section, we present the Advice Acquisition Component (AAC), which is a means to extend the Q/A domain by obtaining information „from the field", based on the personal experience of the members of users' community. AAC allows knowledge capturing in NL, not requiring the efforts of knowledge engineers to encode this information manually using the low-level code.

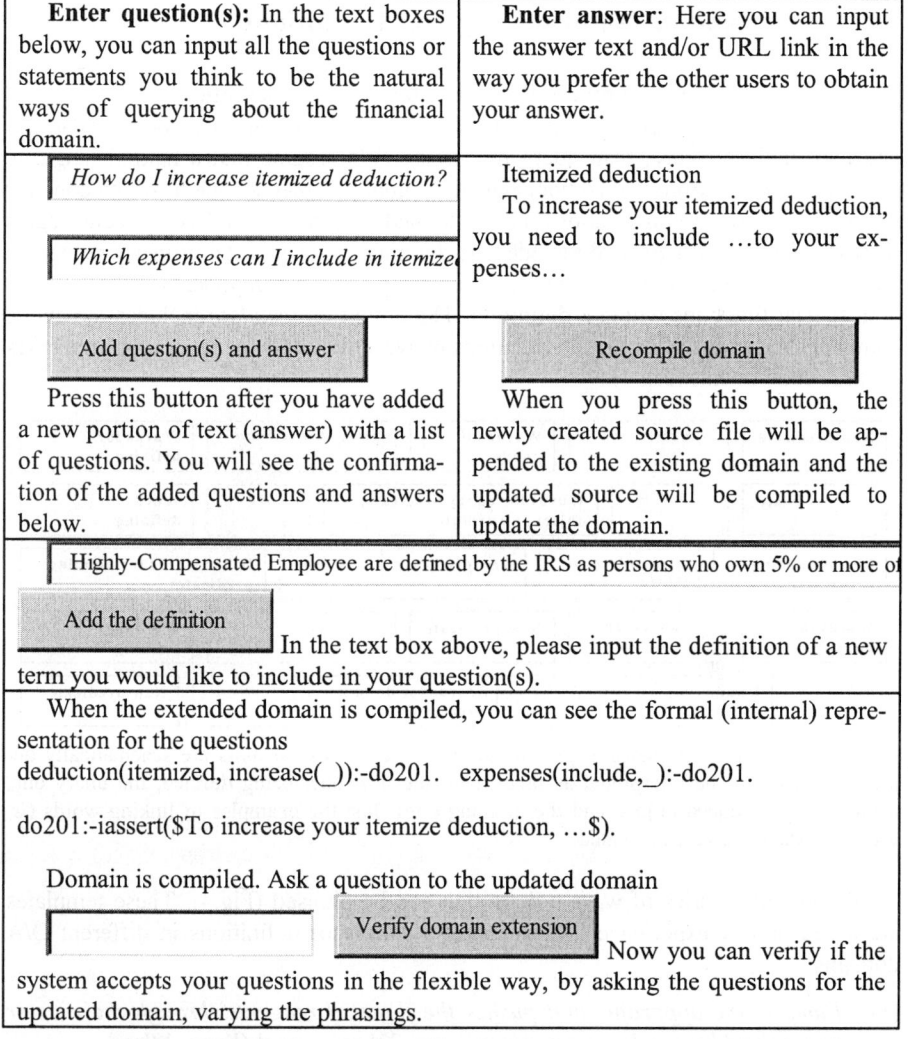

Fig.3. Screenshot of the Advice Acquisition Component for the tax domain

AAC processes the *combination* of the answer (the textual representation of advice with possible URL links to more detailed references) and a set of questions or statements. These questions or statements are supposed to be natural from the user's prospective to generate the assigned document or URL links as a reasonable answer (association). Processing of canonical questions or the automatic annotation of answers occurs similar to that of the queries under Q/A. In addition, the current set of SHs for an entity is analyzed in respect to completeness and consistency. Machinery of default logic for pragmatic processing that follows the semantic analysis may be required to eliminate unnecessary entities or to add implicitly assumed ones [6].

5 Understanding Natural Language Definitions

DUC (Fig.1) performs the processing of natural language definitions of new entities (predicates) that is connected with significantly more difficulties of both syntactical and pure logical nature, than query processing. We assume that a natural language definition consists from two components of a complex sentence, one of which contains the predicate (word combination), being defined, and the second component begins from *if, when, as only as,* and includes the defining predicates. To properly recognize the definitions, we use the templates, including the above components and the words with special functions within a definition. The definition *template* includes the above components and specifies the interrelations between the sentence words for predicates of these components (Fig.4).

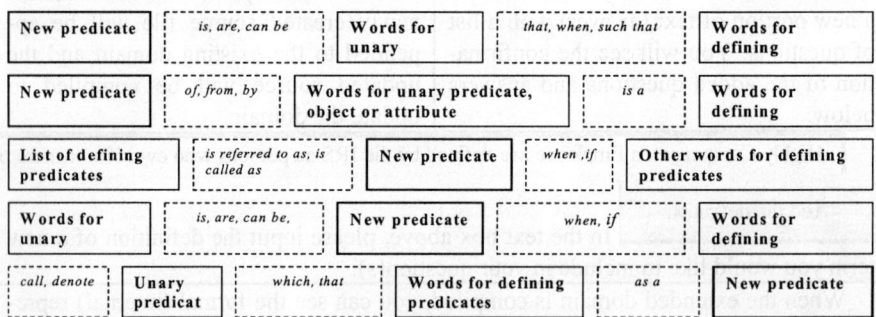

Fig.4. The syntactic templates to process the definitions. Lexical units are schematically depicted, which will be interpreted as predicated: the new one, being defined, the unary one, naming the introduced object, and the defining ones. Just the examples of linking words (*is, are, that, when,* etc.) are presented

We present a series of ways a definition can be phrased (Fig.4). These templates are based on our experience of processing a variety of definitions in different Q/A domains.

(1) *Pump is the apparatus that pushes the flows of reagents through the reactor vessels and heatexchangers: pump(Xapp, Xflow) :- push(Xapp, Xflow), (reactor(Xapp, Xflow) ; heatexchanger(Xapp, Xflow)).*

(2) *Maximum pressure of a pump is such pressure value that a flow stops flowing: maximum_pressure(pump, Xvalue):- pressure(pump, Xvalue), flow(stop). In other words, a pressure „becomes maximum" if the flow stops.*

(3) *Stopping the engine under excessive temperature is referred to as its emergency shut-off when the valve is being closed:*
emergency_shut_off(Xapp):- engine(Xapp, stop), temperature(Xapp, Xflow, Xtemp), valve(Xapp, close).

(4) *We call an apparatus that exchanges energy between the flows as a heatexchanger.* We take into account that *energy* and *temperature* are synonyms here and ignore plural for *flows*, approximating the reality in our definition:
heatexchanger(Xapp, xflow):- apparatus(Xapp), temperature(Xapp, xflow, xtemp), flow(Xflow).

(5) *A person informs another person about a subject if he wants the other person to know that subject, believes that she does not know the subject yet and also believes that that other person wants to know the subject.* This definition of a mental entity has neither unary nor main defining predicates; also, it consists from the metapredicates (the last argument ranges over the arbitrary formula).
inform(Who, Whom, What) :-want(Who, know(Whom, What)), believe(Who, not know(Whom, What)), believe(Who, want(Whom, know(Whom, What))).

(6) *SEP IRA is a retirement plan for self-employed individuals without becoming involved in more complex retirement plans.*
ira(sep):- retirement(plan, self_employed), not retirement(plan, complex).

The last item is the definition for semi-structured domains, where knowledge is not fully formalized. This definition introduces a new clause, linking existing entities or introducing a new one to grow the formalized component of domain representation.

To form a new definition, NL expression undergoes the steps, depicted at Fig.5.

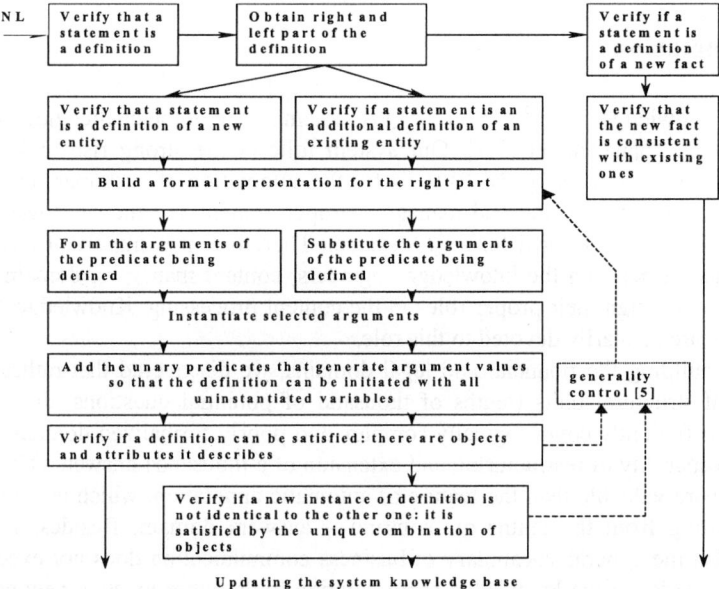

Fig. 5. Sketch of the algorithm for understanding a new definition

There are the groups of semantic rules with the following functions, in addition to the ones deriving formal representation of a query:

- Obtaining the variables that link the predicates in the defining part;
- Obtaining the variables that constitute the arguments of the predicate being defined (choosing the *main* predicate among the defining ones). This main predicate is chosen such that it verifies the value, generated by the unary predicate, syntactically linked to one being defined;
- Assign the *neutral* roles to the rest of variables;
- Conversion of the translation formula of the defining part into the defining part of a clause.

Main semantic patterns (clause templates) of definitions are depicted at Fig. 6. The rectangles denote the predicate names and the squares above – their arguments. An arrow from a square (variable) of one predicate to a square (variable) of another predicate depicts the equality of corresponding value such that the first variable generates its value and the second verifies it. Bi-directional dotted arrow path means that the backtracking is expected when the definition clause is initiated. The most typical definition templates for definitions of the fully formalized domains are depicted at (1-3). In the mental states domain, multiple mental metapredicates simultaneously inherit the arguments to the new predicate (4). In the partly formalized domains, where the arguments serve the purpose of separating answers, the arguments of a new predicate are instantiated by the objects or attributes from the definition, and semantics of the predicate being defined is not inherited by the defining predicates (5).

6 Conclusions

Handling knowledge representation for Q/A system is one of the challenging problems of knowledge management [3,4]. Our system follows the strong rise of interest in natural language processing (NLP) tools for the automation of customer support systems. The role of choosing and managing proper content and the right way of providing access to it is critical for a Q/A system. Therefore, it is important to organize the interaction between the knowledge engineers, content managers, domain experts and testers to assign their proper roles in the content processing. Knowledge management tools are primarily devoted to this role.

While building the financial and legal domains, we discovered that rather limited number of structural units (tenths of thousand of potential questions, thousands of answers) sufficiently cover logically complex and poorly formalized domains. Therefore, the capability of restructuring and extension of a thousands-answers Q/A domain is much more valuable than the feature of automatic annotation, which becomes profitable starting from the tenths of thousand – answers domain. Besides, it is well known, that the general vocabulary of business communication does not exceed four thousand words, so the knowledge representation tools rather meets a new combination of known words than a new word itself.

Analysis of customer satisfaction showed that even the rate of 70% of correct answers satisfies more than 95% of customers, who prefer to use Q/A instead of the keyword search, menu-based representations, browsing the FAQ lists, etc. More than 50% percent of customers expressed the interest to contribute to the domain content as soon as the Advice Acquisition toolkit becomes commercially available, possibly defining new terms via Definition Understanding Component.

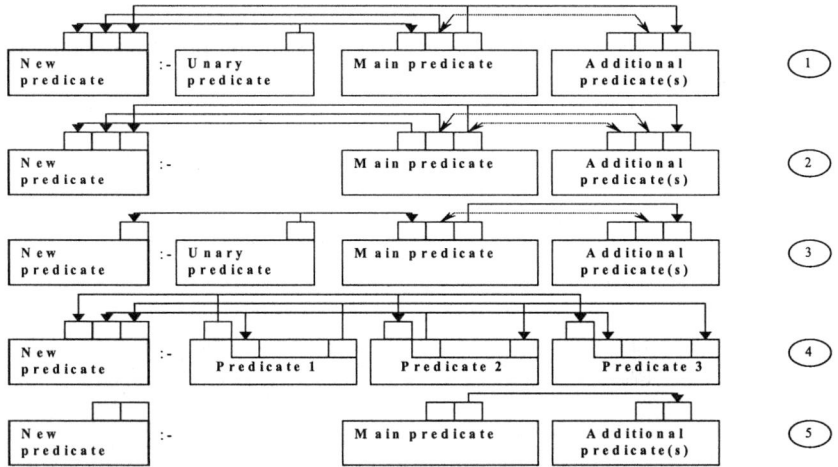

Fig. 6. The semantic (clause) templates for definitions of new entities

References

1. Answer Logic, Inc. White Paper www.answer-logic.com (2000).
2. Allen, J.F. Natural language, knowledge representation, and logical form. In Bates, M. & Weischedel, R.M., eds. Challenges in NLP pp146-175. Cambridge Univ. Press. (1993).
3. Tiwana, A. 1999 Knowledge management toolkit: practical techniques for building a knowledge management system. Prentice Hall (1999).
4. Liebowitz, J. Knowledge management handbook. CRC Press (1999).
5. Galitsky, B. 2000 Technique of semantic headers: a manual for knowledge engineers DIMACS Tech. Report #2000-29, Rutgers University (2000).
6. Ourioupina, O., Galitsky, B. Application of default reasoning to semantic processing under question-answering. DIMACS Tech. Report #2001-15, Rutgers University (2001).
7. Galitsky, B. Semi-structured knowledge representation for the automated financial advisor. In Monostori, L., Vancza, J., Ali, M., eds. Engineering of Intelligent Systems LNAI 2070: 14th IEA/AIE Conference, p.874-879 (2001).

8. Alexander J.H., Freiling M.J., Shulman S.J., Staley J.L., Rehfuss S. Messick S.L. Knowledge level engineering: ontological analysis in AAAI Natl Conf pp. 963-68 (1986).
9. Prerau D.S. Developing and managing expert systems. Reading, MA Addison-Wesley (1990).
10. Turmo, J. and Rodrigues, H. Selecting a relevant set of examples to learn IE-rules. In Monostori, L., Vancza, J., Ali, M., eds. Engineering of Intelligent Systems LNAI 2070: 14th IEA/AIE Conference (2001).
11. van Melle W.J. System aids in constructing consultation programs. Ann Arbor MI : UMI Research Press (1981).

Effective Retrieval of Information in Tables on the Internet*

Sung-Won Jung, Kyung-Hee Sung, Tae-Won Park, and Hyuk-chul Kwon

AI Lab. Dept. of Computer Science, Pusan National University,
San 30, Jang-geon Dong, 609-735, Busan, Korea
{swjung,skhastro,twnpark,hckwon}@pusan.ac.kr

Abstract. Information retrieval services on the Internet have tried to provide Internet users with information processed as the way users want. These information retrieval systems merely show html pages that include the index word abstracted from the user's query. There are a number of difficulties to be overcome in the present technology which could be solved by considering semantics in html documents. However it can heighten the precision of retrieval results by using the structural information of the document as an alternative source of information. The tabular form, which appears on ordinary documents, usually has the most relevant information. Based on the similarity to a Web's html document, we try to improve the precision of results by analyzing the table on html documents. Our main purpose here is to do table parsing and construct a dictionary of table indexes for applying to our information retrieval system and thus enhance the accuracy.

1 Introduction

It is the ultimate goal of the information retrieval systems to offer suitable information to the user, to grasp the user's exact intention. To achieve this goal, we need to analyze the semantics of html documents. However, it is quite difficult with present technology to apply semantics to the retrieval system. Another method to grasp the user's intention is to analyze the structural information of documents. The structural information of documents is predictable from the writing pattern of the people. For example, in the majority of documents the authors represent their subject as a title and they indent each paragraph to distinguish hierarchy among paragraphs. This structural information depends on the category that the document belongs to. There can be various kinds of structural information. In this paper, we make use of the tabular form that appears in documents. We use a structured table in documents to convey our subject clearly. Therefore, the table form is more obvious than the plain text form. As the

---

* This work has been supported by Korea Science and Engineering Foundation. (Contract Number : R01-2000-00275)

table form is structured by the writer's intention, it may contain more important information.

It is easier to use a table than to use the other structural information in documents for the following reasons. First, a table in documents can be distinguished without great effort. Using the fact that a table in html documents begins with the <TABLE> tag and ends with the </TABLE> tag, we can discover the boundary of the table contents to some degree. Second, it is relatively easy to abstract the meaning from a table. Unlike other structural information in documents, the table has index words, something that serves to guide or facilitate reference. We can generally get the table index from the uppermost row and the leftmost column.

We wish to improve the accuracy of information retrieval by using the characters of a table described above. If the table analysis is added to current ranking models such as Vector Space Model or P-norm Model[1,2,3,4,5], the accuracy of retrieval result would be better about specific queries.

2 What We Have to Overcome

Current information retrieval systems rank related documents based on similarities between documents and user's query[1,2]. This usually goes through the following process. Internally inverted files are created from the indexes, which result from parsing the documents. An inverted file is a sorted index of keywords, with each keyword having links to documents containing that keyword. When an Internet user inputs a certain query, the current systems select the documents which have or relate to the query words. And the related documents are ranked by Vector Space model[1,2,5] or P-norm Model[3,4], etc. Additionally, they use Hyperlink information of html documents[7] and Relevance Feedback method[8, 9]. The current retrieval systems that place great importance on index words have the following limitations.

First, current systems do not draw the meaning from html documents. To retrieve information more accurately, semantics of html documents should be considered. Current information retrieval systems measure the similarity between html documents and user's query based on the term frequency and document frequency. Term frequency (TF) is a frequency of a term or an index word in each document. Document frequency (DF) is the frequency of documents which contain the term. Current systems accept the assumption that documents are related to the index word in proportion to the term frequency of the documents. However, this assumption is only of limited validity.

Second, current systems do not distinguish relatively weighted indexes or keywords from general indexes in html documents. Because there can be more important keywords even in one document, people tend to remember the important part when they read a certain document. If all keywords in a document are given the same weight, it is difficult to retrieve what the users want exactly. Current systems consider term frequency and document frequency of the indexes by an alternative method, but it does not increase the matching accuracy enough to satisfy all users.

Third, current systems do not reflect structural information in html documents. We can grasp the writer's intention to some degree if we consider structural information of documents. As the writer uses titles of each paragraph, indentations and tabular forms to convey his intention clearly, they have significance in a document. But current information retrieval systems ignore the structural information when they extract index words from documents.

To overcome the above limits, the system presented in this paper focuses on analyzing and processing the tabular form. When an author writes a document, it may be more effective to represent the document with a tabular form rather than to describe it in the usual manner. In some html pages, the whole document is consisted of one table for a special purpose. When a table in a document makes the main component, it is more effective to extract indexes from the table than the plain text. The system improves accuracy of information retrieval by assigning weights on the indexes, which are extracted from tables. For example, if a user wants to know about 'the yearly GNP of Korea', showing the pages that contain a table having the words, 'GNP' and 'Korea' as indexes is more adequate than showing the pages that contain the words in text.

However, when a table in a document is the minor component, it is more effective to assign weights on the keywords, which are extracted from the <CAPTION> tag or from the sentences before and after the table, instead of using the table indexes, because a 'minor' table in itself has little significance and tends to be related to the other part of the document. And most of the minor tables have titles or captions, which summarize the table contents.

3 Implementation

To implement the Table Parser, the system presented here assumes three aspects in common sense.

1. If a table occupies over 90 percents of an html page, the table has significance.
2. A meaningful table has table indexes, which represents relevant rows and columns.
3. When a table that appears in an html page is a 'minor' table, then the title of the table has more importance than the table indexes.

With these assumptions, the system focuses on the following points.

1. Distinction of a meaningful table from a 'decorative' one.
2. Extraction of table indexes, which are worth representing contents of the table, in variant shapes of tables.
3. Method of applying the information of table indexes to information retrieval.

There are two types of tables in html documents, decorative and non-decorative (i.e. meaningful) tables. To distinguish the latter from former, we classify a meaningful table as one that contains indexes. Therefore, to extract indexes from a table is the most important issue in the table parsing. If we draw exact table indexes, we can easily distinguish a meaningful table from a decorative table and reprocess the table to obtain table information based on these indexes.

3.1 Selection of a Meaningful Table

To distinguish a meaningful table from a decorative table, the system uses a person's way of writing html tags related to the table in html documents. To begin with, the system regards the horizontal and vertical iteration of the table contents. If a Web designer uses a table merely to decorate the page, the shape of the table would be complicated and the contents would not repeat. The system notices the repetition of tags such as <TR>, or table row element, <TH> and <TD>, or data cell elements. The <TH> tag is located in the table header and <TD> tag in the table body. The system analyzes tables which repeat over the whole page except for partially repeated tables. During the analysis, the width and the height of the table are counted.

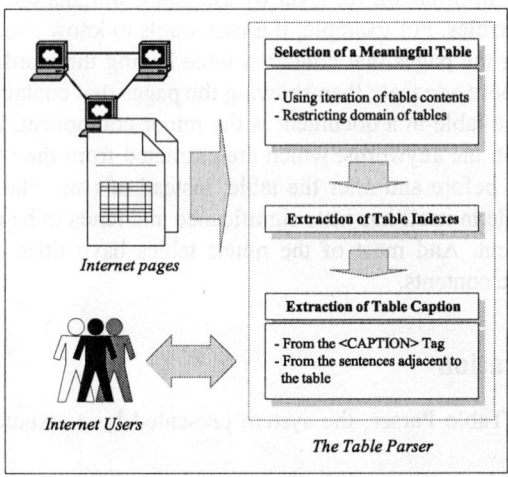

Fig. 1. The architecture of the table parser

The system excludes tables that have extremely complex shapes, tables that are nested over twice, and pages in which there are more than four tables with different shapes which have no explanatory sentences.

The system excludes simple tables for similar reasons. When an html page is divided into frames, through which more than one html file can be displayed at a time, we often use a table to arrange sentences. The system excludes one-dimensional tables, which have only one row or one column, and two dimensional tables with less than two rows and three columns. In summary, the system processes tables of relatively regular shapes only.

3.2 Extraction of Table Indexes

After filtering tables on the html pages, the system extracts the table indexes, which represent the table contents, from the meaningful tables. It is not a simple problem to pull out indexes from a table because tables on the Internet appear in a variety of forms. As Web designers use various tabular forms according to their purposes, the number of rows and columns, and the position of indexes are also various in each table.

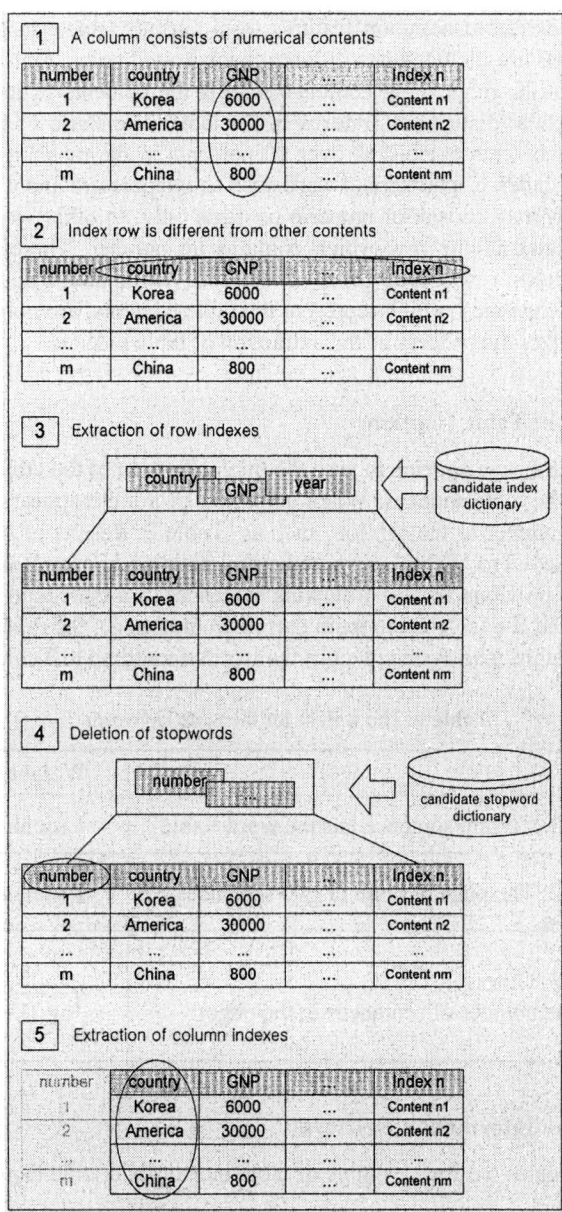

Fig. 2. Extraction of table Indexes

Because it is impossible to consider all kinds of tables on the html pages, the research extracted common features among relatively regular tables. The system supposes that table indexes are located in the upper rows and left columns. Generally row indexes are more important than column indexes, therefore the system first scans or parses a table from the uppermost row downward for some rows.

The research extracted common features from various tables. In general, the cell where index words are included has different background color, font-family and font-weight style from the other table contents. And if the contents of a table consist of numbers, it is easy to distinguish indexes in the table. The Ratio of the tables which contain numeric contents extends to over 80 percents in the meaningful table on the Internet. If a row index is a numerical word such as temperature and price, the column related to the row may consist of numeric contents only. In other words, the indexes are probably located in the row which contains no number. Therefore, the system regards the uppermost row among the rows that does not contain a number as indexes. And most of table indexes, which represent the table contents, are nouns of one word. Figure 2 summarizes the process of the extraction of table indexes.

3.3 Extraction of Table Caption

If a table in an html page appears in text, the title or caption of the table becomes more significant than the table indexes. When a caption of a table appears in text, people usually attach a number to the caption, such as 'Table 2: Results of Implementation'. In this case, the keyword 'Table' can be a helpful notation to extract the table caption. Additionally, the previous and the following sentences of a table tend to represent the table contents. And if a table is the main part in an html page, the remaining sentences may explain the table. The system weighs the words according to Table 1.

Table 1. The weight for the table keywords

Cases	Weight (W)
Case 1: Certain sentence has the word 'table'	1 (default)
Case 2: The page has one or two sentences	1 (default)
Case 3: Others (n : the number of sentences in the page)	$\dfrac{1}{\log_2(1+n)}$

3.4 Applying on Information Retrieval

The system calculates the final weights of table indexes and table caption using equation (1).

$$\text{Final-Weight} : W_f = c \log_2 (N_{index}) + W \qquad (1)$$

where

N_{index} = the number of indexes in a document
$c = 0.1$ (a correction constant)
W = the value from Table 1

This is similar to using Hyperlink information for information retrieval. Based on equation (1) and Vector Space Model, the system applies final similarity as equation (2).

Final- Similarity : Sim(Query, Doc) = W_f + Sim (by Vector Space Model) (2)

4 Experimental Results

We implemented the system with a Sun Enterprise 3500 connected to the Internet by a T1-level LAN. We used previous works on technology which was related to Information Retrieval. It was studied in the Artificial Intelligence Laboratory at Pusan National University.

4.1 Statistics for Tables on the Internet

First, we analyzed the property of html documents that contained a table. As a result, it was found that about 10% of the whole html documents had a 'major' table, which the system can process. We selected one thousand documents among them and analyzed these documents on two categories. Table 2 shows the results.

Table 2. Statistics for Tables in one thousand sample documents

Table Category	Number of html documents	Percentage
The table contains numerical contents	726	72.6 %
Index row and column have decorations	823	82.3 %
Each Index is one word noun	856	85.6 %

4.2 Accuracy Improvement

The system implemented to weight the table form information with the established Vector Space Model. Because the documents with valid table were only about 10% of all documents, the accuracy of retrieval seldom improved.

Table 3. Accuracy improvement by the table parsing

Criteria	Current system	Method 1	Method 2	Method 3
Extraction accuracy	-	62.3%	74.3%	86.7%
Retrieval accuracy	81.5 %	82.4%	83.1%	83.4%

Method 1 – Current information retrieval system with processing the table which contains numerical contents

Method 2 – Method 1 with processing the table whose index row and column have decorations

Method 3 – Method 2 with processing the table whose indexes consist of one word noun

Table 3 shows the result of accuracy experiment. The extraction accuracy means the precision when the system extracts meaningful tables from a thousand html documents, which contain a meaningful table, by using the method that this paper suggests. The retrieval accuracy means the precision of retrieval with the extracted meaningful tables. The retrieval accuracy of current system is about 81.5% and it slightly increases up to 83.4% when the table parsing technology is applied. However, if a query is retrieved only in the meaningful tables, the results are much improved.

5 Conclusions

Information retrieval systems have had tried to improve performance. Their effort was the start of the traditional Boolean Model and now is expressed variously such as Vector space Model, P-norm Model, and Hyperlink Information Model spotlighted recently. The analysis method for the table we described above is one of those methods. This research is intended to improve the accuracy of information retrieval system by giving a meaning in the html table since current information retrieval systems ignore the fact that a table in an html page contains significant information.

We proposed various ways to distinguish between a relevantly meaningful table and just a decorative one. Having done this, a method to effectively abstract information from the table indexes was presented, and this was applied to the systems. This research can be applied to the previous works. When an Internet user tries to get a specific table, this research can provide better results.

6 Future Works

Giving meaning to a table and applying this to information retrieval systems are difficult processes to achieve. As characteristic of html, its structure and presentation are not detached, there are lots of cases where that people tend to use a decorate table due to preference. (Also, there is a semantic representation form that people basically prefer.) Html document source can be different even if the semantic representation form is similar in appearance. Therefore the analysis result can be completely mistaken. For the exceptional matters, since the html tables have various shapes, it needs to be handled case-by-case. However, we have not drawn many cases because the research period was short. In the near future, this system here will cover more cases to improve the quality of information retrieval. The system will also investigate the relationship between table caption and table indexes, and other structural information on the Internet pages.

References

1. Kobayashi, M., Takeda, K. : Information Retrieval on the Web. ACM Computing Surveys (2000) 144-173
2. Salton, G., McGill, M. J. : Introduction to Modern Information Retrieval. McGraw-Hill, New York (1983)
3. Fox, E. A. : Extending the Boolean and Vector Space Models of Information Retrieval with P-norm Queries and Multiple Concept Types. Dissertation Cornell University (1983)
4. Smith, M. E : Aspects of the P-norm Model of Information Retrieval : Syntactic Query Generation, Efficiency, and Theoretical Properties. Dissertation Cornell University (1990)
5. Salton, G., Fox, E. A., Wu, H. : Extended Boolean Information Retrieval. ncstrl.cornell (1982) 82-511
6. Frakes, W. B., Baeza-Yates, R. : Information Retrieval : Data Structures & Algorithms. Prentice-Hall, New Jersey (1992)
7. Golovchinsky, G. : What the Query Told the Link : The Integration of Hypertext and Information Retrieval. Hypertext97 Proceeding, Southampton UK (1997) 67-74
8. Salton, G., Buckely C. : Improving Retrieval Performance bye Relevance Feedback. JASIS (1990) 288-297
9. Buckely, C., Salton, G. : The Effect of Adding Relevance Information in a Relevance Feedback Environment. SIGIR (1994) 292-301

A Fitness Estimation Strategy for Genetic Algorithms

Mehrdad Salami and Tim Hendtlass

Centre for Intelligent Systems and Complex Processes,
School of Biophysical Sciences and Electrical Engineering,
Swinburne University of Technology
P.O. Box 218, Hawthorn, VIC 3122 Australia
{msalami,thendtlass}@swin.edu.au

Abstract: Genetic Algorithms (GAs) are a popular and robust strategy for optimisation problems. However, these algorithms often require huge computation power for solving real problems and are often criticized for their slow operation. For most applications, the bottleneck of the GAs is the fitness evaluation task. This paper introduces a fitness estimation strategy (FES) for genetic algorithms that does not evaluate all new individuals, thus operating faster. A fitness and associated reliability value are assigned to each new individual that is only evaluated using the true fitness function if the reliability value is below some threshold. Moreover, applying some random evaluation and error compensation strategies to the FES further enhances the performance of the algorithm. Simulation results show that for six optimization functions, the GA with FES requires fewer evaluations while obtaining similar solutions to those found using a traditional genetic algorithm. For these same functions the algorithm generally also finds a better fitness value on average for the same number of evaluations. Additionally the GA with FES does not have the side effect of premature convergence of the population. It climbs faster in the initial stages of the evolution process without becoming trapped in the local minima.

Keywords: Genetic Algorithms, Evolutionary Algorithms, Fitness Evaluation

1 Introduction

Genetic Algorithms (GAs) are very effective at finding global solutions to a variety of problems 1,2. This optimisation technique performs especially well when solving complex problems because they don't impose many of the limitations of traditional optimisation techniques. Due to its evolutionary nature, a GA will search for solutions without regard to the specific inner working of the problem. This ability lets the same general-purpose GA routine perform well on large, complex problems. The GA searches the problem space in different directions simultaneously instead of one direction as in the gradient-based methods 3. However the time for performing the

GA processes is generally longer than times for gradient-based methods. This paper considers a version of a genetic algorithm that uses a fitness estimation strategy (FES) allowing it to run fast even for large population sizes thus allowing more evolution to be performed in a given time.

A genetic algorithm involves several operators: crossover, mutation and selection. The time to perform the crossover or mutation processes is proportional to the chromosome length of the individuals in the population. Generally even for large chromosome lengths the time taken by the crossover and mutation operators is very short. This is also true for most selection operators that only require reading from or writing into a population. In any genetic algorithm, the fitness of new individuals must be determined before storing them in the next population. The fitness evaluation is problem specific and a black box to the genetic algorithm that has, and needs, no information about the fitness evaluation process, including how long it will take.

The goal of any GA is to find the individual with the best fitness value. Normally there is no time limit in achieving this goal and the GA is allowed to run for any number of generations to achieve this. However, if there is a time limit in which to find the best fitness value, the fitness evaluation time will be very critical.

If the fitness evaluation process is slow, the GA has to reduce the number of evaluations made while still achieving the best possible fitness value. Making an approximation function of the fitness function that is faster than the actual fitness function to process, and using this faster approximation function can achieve this. There has been work on estimating the fitness value of the children by building a model of the objective function 4. However this model has to be updated regularly to reflect the current estimated fitness of the various parts of the objective function. Although this approach might be useful, the computation time to build the model and update it regularly can easily become enormous if an accurate model is required.

Ratle 5 used an approximation function built using the Kriging method to reduce the number of fitness evaluations. The approximation function was built using the results of a number of true fitness evaluations and then updated after certain number of generations. Although the calculation of the approximation function may take less time than an actual fitness evaluation, the time for building and maintaining the Kriging method approximation function should also be considered as it may defeat the purpose of the approach. The approximation time depends on the complexity of the approximation function and the number of samples used for approximation. Accurate approximation would be time consuming because it requires complex functions and large number of samples. While low order Kriging functions and fewer samples can be used for less complex fitness functions, this might affect the evolution operators and cause premature convergence.

The solution proposed here to achieve the reduction in the number of evaluations while avoiding premature convergence is to assign a fitness value to a new child rather than evaluate it. Obviously, provided that the fitness value is not the result of a stochastic process, a parent's fitness can be assigned to a child that is identical to that parent, which may save some fitness evaluations. However, if a child is different from its parents, an algorithm is needed to assign a fitness value for the child. As this assigned fitness value is not the real fitness value, another parameter is needed to show how reliable the assigned value is thought to be. Only when this reliability falls below some threshold is the child sent for a real fitness evaluation.

This paper assumes that the time to perform an actual fitness evaluation has already been reduced as far as practicable by some combination of avoiding redundant evaluations **Fehler! Verweisquelle konnte nicht gefunden werden.**, approximating true evaluation by using sampling data 5, 8 and exploiting parallel processing 9. It describes a Fitness Estimation Strategy (FES) for GAs that achieves a speed improvement by restricting the number of evaluations required. While a traditional GA selects parents for the next generation on the basis of the true fitness, the GAs with FES selects based on the true fitness for some individuals and the assigned fitness for the rest. As long as the difference between the true fitness and the assigned fitness is not too large the parental population, selected by the two methods, will be very similar and so will the final results obtained. The advantage of the FES is the reduced number of true evaluations required.

The next section describes the FES for the GAs in details. Section 3 shows two enhancements to the basic FES. In Section 4, simulation results for six optimisation functions will be presented. Section 5 shows the results of the GAs with FES when applied to the same function as used in 5.

2 The Fast Estimation Strategy

It should be possible to estimate the fitness of the children produced using the known fitness values of the parents. The accuracy of the child's fitness is likely to be less than the accuracy of the parents' fitness. However, it may well be possible to assign a fitness value close enough to the real fitness value that the breeding process bias towards improved performance is not compromised. In this case using fitness values generated from the parental fitness values rather than true evaluation should not be unrealistic, at least for a few generations. However, in time, the difference between the assigned and true fitness will become large enough to compromise the breeding process. Before this happens individuals must be truly evaluated again to realign their fitness values with reality. In order to know when to do this, another parameter is associated with each fitness value to show how reliable or accurate the assigned value is believed to be. The reliability of the assigned fitness of a child is less than the reliability of the best of its parents. If the child's fitness reliability drops below a threshold, the assigned fitness is discarded and the fitness of the child is actually evaluated, updated and recorded as again being fully reliable.

The closer the assigned fitness values are to the real fitness values the better the genetic algorithms with fitness estimation strategy might be expected to work. However, since the actual fitness value can be replaced in the breeding process by the fitness rank of individuals in the population 10, precise fitness evaluation is not essential - any algorithm that produces the correct rank order list of individual fitness would be adequate. As a consequence, the genetic algorithm with FES will still perform reasonably even with imprecise fitness values.

The reliability value, R, varies between 0 and 1 and can also be used to distinguish between evaluated and assigned fitness values. A value of 1.0 for R means the fitness value was actually evaluated, a value less than this means that the fitness value was estimated. The value of R for assigned fitness values depends on two factors: the reliability of parents and how close parents and children are.

The process for generating consecutive populations with the Fast Estimation Algorithm is as follows:

- An initial random population is generated. Each individual consists of three elements: a genetic string (the chromosome), a fitness value (*f*) and a reliability value (*R*). For each random individual in the initial population, the fitness will be calculated using the true fitness function and the *R* value will be set to 1.0.
- Two individuals will be selected using a selection procedure. These two individuals are the parents for two new individuals. The two parents can be identified as P1=($p1, f_1, R_1$) and P2=($p2, f_2, R_2$), where f_1 is the fitness and R_1 the reliability for the chromosome *p1* and f_2 is the fitness and R_2 the reliability for the chromosome *p2*. The parents will undergo crossover and mutation processes to generate two new children (*c1* and *c2*). For each child, a fitness value *f* and a reliability value *R* must be determined.
- First the similarity *S* between the children and their parents must be calculated. The difference *D* between a child *c1*, with the genetic string consisting of the binary values c_0 to c_{l-1}, and a parent, with the genetic string p_0 to p_{l-1} is defined as a number $D(c1,p1)$ between 0 and 1 given by

$$\left(\sum_{i=0}^{l-1} \text{abs}(c_i - p_i)\right)/l \quad (1a)$$

and the similarity S between them, also a number between 0 and 1, is given by

$$1 - D(c1,p1) \quad (1b)$$

- Let S_1 be the similarity between *c1* and *p1* and S_2 the similarity between *c1* and *p2*. If *p1* and *c1* are identical (S_1 is equal to 1) then *c1* will take $f=f_1$ and $R=R_1$. Conversely, if *p2* and *c1* are identical (S_2 is equal to 1) then *c1* will take $f=f_2$ and $R=R_2$. If neither of the above conditions applies, then the *f* and *R* values for *c1* will be calculated from:

$$f = \frac{S_1 R_1 f_1 + S_2 R_2 f_2}{S_1 R_1 + S_2 R_2} \quad \text{and} \quad R = \frac{(S_1 R_1)^2 + (S_2 R_2)^2}{S_1 R_1 + S_2 R_2} \quad (2)$$

The fitness and reliability values for *c2* are assigned similarly.

- If the *R* value for a child is less than a threshold value (*T*) the assigned fitness for that child will be replaced by a fitness calculated using the true fitness function and the *R* value will be set to 1.0. After this process is completed, *c1* and *c2* together with their fitness and reliability values are transferred into the next generation.

Note that in this algorithm if $S_1=S_2$, the contribution from each parent to the child's fitness value will be proportional to that parent's own *R* value. For threshold values close to 1.0, the algorithm involves little or no fitness assignment and effectively results in a conventional genetic algorithm. Conversely, a threshold value close to zero means running the evolutionary process almost exclusively by fitness assignment.

If T is very small the fitness values in the population are not generally correct and the selection process may not work properly. The estimation strategy will fail if the individuals selected differ significantly from those that would have been chosen if the real fitness of each individual were used.

3 Two Enhancements to the Basic Fast Estimation Strategy

The algorithm described in the previous section is the core method for the FES. This basic FES uses a fixed threshold value and uses the unchanged assigned fitness value for the selection process. The two enhancements to the basic FES discussed below add some random fitness evaluation and an error compensation strategy.

3.1 Random Fitness Evaluation

The GA with FES reduces the number of evaluations and for small threshold values the fitness values of most individuals in the population will be estimated. Such an approach, similar to a greedy optimisation algorithm, might affect the selection process of the GA and cause the population to converge to a sub-optimum point. No evaluation will happen with a threshold value of zero, and the best fitness in population will never change. However for threshold values closer to 1.0, the algorithm would have enough true evaluations to ensure that the selection operator will work properly.

To improve the performance of the algorithm when the threshold is small, a number of randomly chosen individuals in the population are evaluated regardless of their reliability values. A Probability of Evaluation (P_E) between 0 and 1 is chosen. When the reliability value of the child is more than the threshold value a random number is obtained. If this is less than P_E the fitness of the child is truly evaluated. As a result, a minimum number of true evaluations will happen to every generation. The value of P_E must be small enough to make sure it does not dominate the threshold value in the FES. For the simulations in this paper P_E is set to 5%.

3.2 Error Compensation

In the above algorithm, the fitness value assigned to a child is always between the fitness values of the parents. In a traditional genetic algorithm the children may have a better fitness than their parents. Provided the threshold is greater than zero, some true evaluations still happen during the evolution. Thus there will be occasions when both a true and an estimated fitness value are available. Using these, a compensation system can be introduced to minimize the difference between the true and estimated values. **Fig. 1** shows a block diagram of a compensation system that converts a raw assigned value into a compensated assigned value. Whenever a true fitness value is calculated an error value (the difference between the compensated assigned value and the true fitness value) is used to update the compensation system. The true fitness is used for the fitness of individuals when available; otherwise the compensated assigned value is used.

The compensating system used is very simple as it is only intended to apply an approximate correction. The running average of the previous error values, adjusted to emphasize recent errors, is used to compensate the assigned value. The CAV (Compensated Assigned Value) is derived from the AV (Assigned Value) and the compensation error (CE) by

$$CAV = AV + CE/M \tag{3}$$

where M is the number of true evaluations performed. CE is initialized to zero and M to one. After the real evaluation *i*, the value of M will be incremented and the CE will be updated by

$$CE = CE + (FV_i - AV_i) \qquad (4)$$

where FV_i is the true fitness value. After performing N real evaluations, the values of CE and M are both divided by two. This has the effect of emphasizing the contributions from the next errors to be received. Since the assigned values are on average higher than the true fitness values, the CE and hence the correction applied are normally negative.

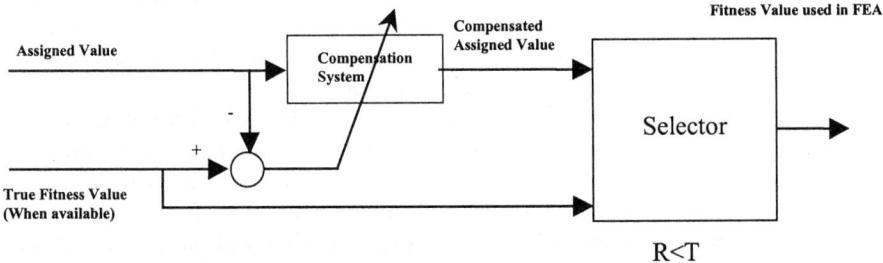

Fig. 1. The error compensation system for the FES

4 Simulation Results

The GA with FES has been applied to the following six functions taken from 11 12. In each case 16 bit precision was used.

The first one is Shekel's Foxholes function:

$$F_1(x) = \sum_{j=1}^{25} \frac{1}{j + \sum_{i=1}^{2}(x_i - a_{ij})^6}, \text{ where } a_{ij} \text{ is a constant}, -65 < x_i < 65 \qquad (5)$$

The second one is the Six-hump Camel-Back function:

$$F_2 = 10*(4x_1^2 - 2.1x_1^4 + \frac{1}{3}x_1^6 + x_1 x_2 - 4x_2^2 + 4x_2^4 + 2)^{-1}, \quad -65 < x_i < 65 \qquad (6)$$

The third one is the Michalewicz's function:

$$F_3 = \sum_{i=1}^{10} \sin(x_i) \sin^{20}\left(\frac{ix_i^2}{\pi}\right) \quad -10*\pi < x_i < 10*\pi \qquad (7)$$

The fourth one is the Griewank's function:

$$F_4 = 1 - \sum_{i=1}^{10} \frac{(x_i - 100)^2}{4000} + \prod_{i=1}^{10} \cos\left(\frac{x_i - 100}{\sqrt{i}}\right), \quad -600 < x_i < 600 \qquad (8)$$

The fifth one is the Sphere Model function:

$$F_5(x) = -\sum_{i=1}^{30}(x_i - 1)^2 \qquad -5.12 < x_i < 5.12 \qquad (9)$$

The sixth one is Rastrigin's function:

$$F_6(x) = (-3000) - \sum_{i=1}^{30} x_i^2 - 10\cos(2\pi x_i) \qquad -5.12 < x_i < 5.12 \qquad (10)$$

The first two functions are two-dimensional. The third and fourth functions are 10-dimensional. The last two functions are 30-dimensional parameters and test the FES for functions with large bit strings. As 16 bits are used for each parameter, the chromosome used for the last two functions consists of a total of 480 bits. The GA is set to find the maximum fitness value for all functions.

All parameters common to the conventional GAs and the GA with FES were set to the same value for both algorithms, although the actual values were a function of the objective function being used as shown in Table 1. These values were selected to give the best conventional GA performance. Uniform crossover (rate = 0.95) and tournament selection (window size=2) were used, together with an elitism strategy. Both algorithms used a generational recombination strategy. All simulation results presented in this section are averaged over 100 runs with different random starting points.

Table 1. The values for three parameters used in the GA and the GA with FES

Function	F_1	F_2	F_3	F_4	F_5	F_6
Bit mutation rate	0.03	0.03	0.001	0.001	0.001	0.001
Population Size (P)	64	64	128	128	256	256
Terminal Evaluation (E_t)	2500	2000	10000	3500	20000	20000

Fig. 2 shows graphs of fitness versus the number of true evaluations for the six functions defined in Equations 5 to 10. The figure shows the results for the FES with error compensation strategy and 5% random evaluations. The results for other FES configurations such as 'without error compensation' or 'without random evaluation' show that the FES with random evaluation and error compensation produces the best performance.

Although the above figure shows the difference in the fitness values for any number of evaluations, it does not show how many times the GA with FES was better than the GA among those 100 runs. To report the difference between the GA and the GA with FES, the number of times that the GA with FES was better than the GA was recorded for three points. The first point is at one third of the total evaluations, the second point is at two thirds of the total evaluations and the last point is at the end of all evaluations.

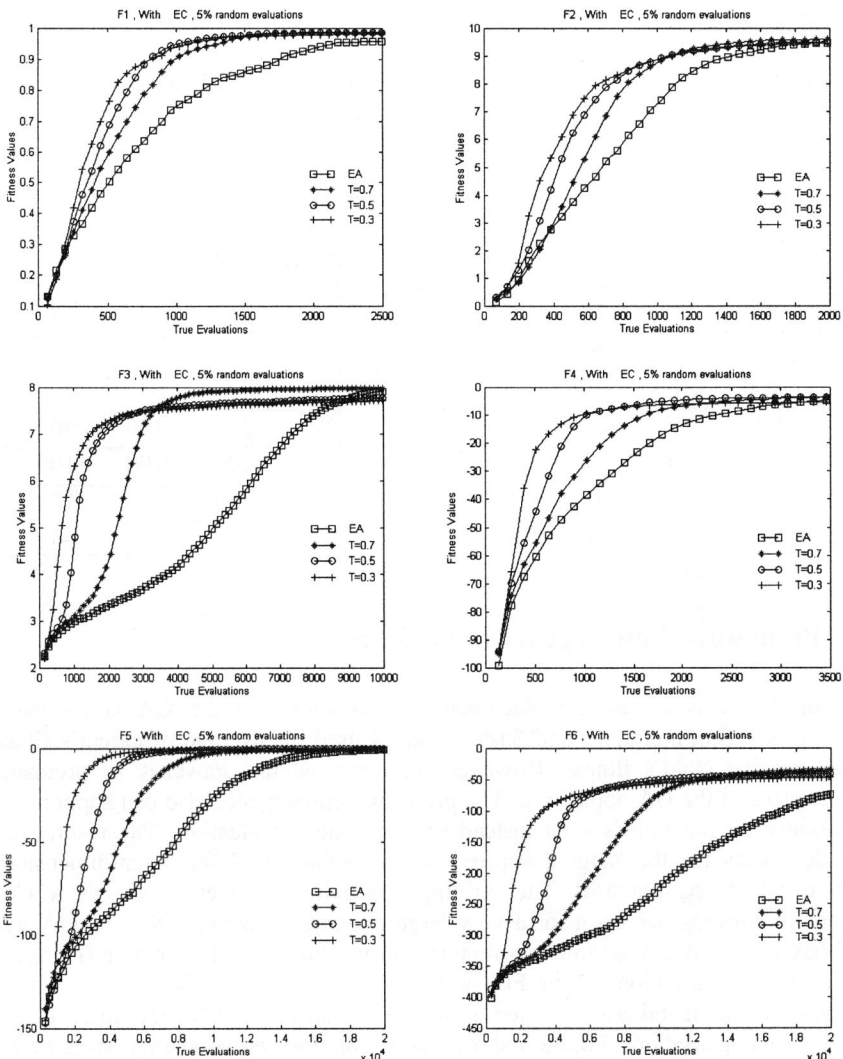

Fig. 2. The comparison of the average of the maximum fitness values in population between GA and GA with FES with three threshold values for six optimisation functions. The GA with FEA generally creates better fitness values for the same number of evaluations

Table 2 shows the three measured values for six functions for three threshold values. For the first measurement point, the best threshold is 0.3 for all functions and by increasing the threshold value the performance of the GA with FES will reduce. Therefore, for the fastest convergence of the GA with FES, the threshold must be small. The measurement at the second point shows similar performance for all threshold values. On the other hand for the final test point the higher threshold values produce slightly better performance. For that point, the best threshold is 0.7 and in some cases the performance of the smaller threshold values is very poor. It means that for the larger number of evaluations, the threshold value of the FES should not be small.

The table also shows that the performance enhancements of the FES are very good for the more complex functions F_5 and F_6. For those functions larger population sizes and more evaluations were used because of the complexity.

Table 2. Proportion of runs for which GA with FES outperform GA for the three testing points

Functio	1/3 of evaluations			2/3 of evaluations			3/3 of evaluations		
	0.3	0.5	0.7	0.3	0.5	0.7	0.3	0.5	0.7
F_1	87	84	74	79	78	74	74	53	77
F_2	83	77	68	68	65	76	37	28	95
F_3	100	100	100	95	98	100	17	8	89
F_4	99	97	85	96	100	97	25	100	98
F_5	100	100	100	100	100	100	100	100	100
F_6	100	100	100	100	100	100	100	100	100
Overall	97	93	88	89	90	91	59	65	93

5 Premature Convergence in the FES

One of problems of using replacement fitness values in the GA is premature convergence. Smith et al 13 and Sastry et al 14 used the average of parent's fitness values for the child's fitness. However they reported the drawback of premature convergence of the GA population. The premature convergence also can happen if an approximation function is used instead of true fitness evaluation. The results in 5 include results for the Kriging approximation method used for a ten-dimensional version of the F_6 function. The Kriging approximation method shows a clear premature convergence even for a very large population size of 200. The GA with FES has also been applied to a ten-dimensional version of the F_6 function (named F_7 here). Two configurations of the FES were tested. The first configuration is without random evaluation and error compensation, the second is with 5% random evaluation and error compensation. Figure 3 shows the results for two population sizes of 200 and 64.

Using the first configuration of the FES and a population size of 64 gives results that are similar to the results in 5 which shows strong premature convergence in the population. With the population increased to 200, the results still show premature convergence for the threshold values of 0.3 and 0.5, but better convergence for T=0.7. Using the second configuration, the results are much better and there is no sign of premature convergence. As might be expected, the population of 200 has slower convergence rate per evaluation compared to the population of 64 and requires more than 3000 evaluations to convergence to the optimum point of zero.

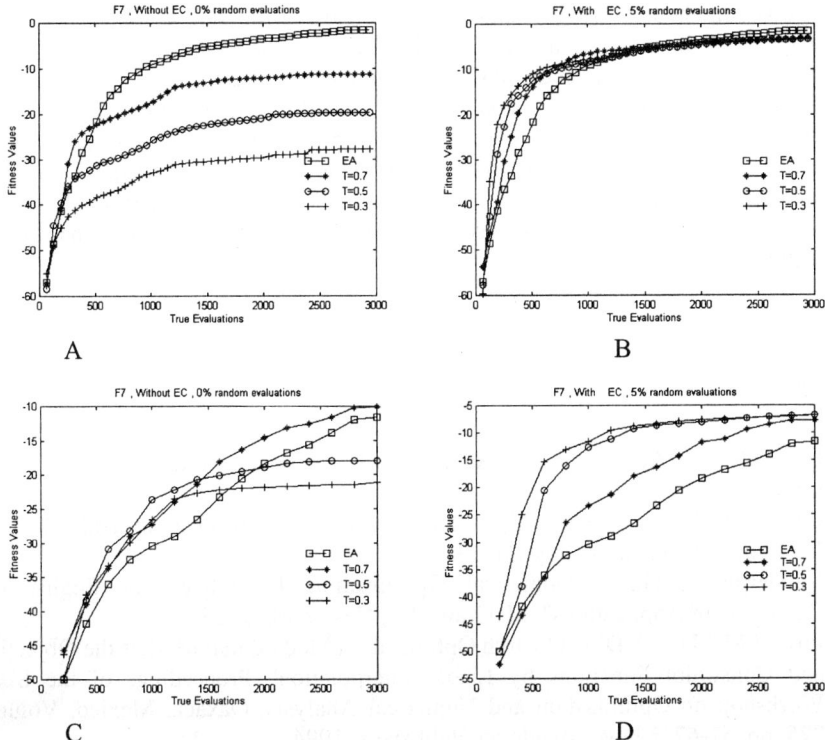

Fig. 3. The performance of the GA with FES for the F_7 function. Premature convergence can be avoided by increasing the population size threshold for the basic FES method or using two enhancement methods for the FES. (A= Without error compensation and no random evaluation for population=64, B= With error compensation and 5% random evaluation for population=64, C= Without error compensation and no random evaluation for population=200, D= With error compensation and 5% random evaluation for population=200)

6 Conclusions

This paper introduces a fitness estimation strategy (FES) whose use in a genetic algorithm results in faster operation of the algorithm. This requires an additional parameter, the fitness reliability to be added to each individual. Two enhancements are added to the FES to make the estimation more accurate. Firstly a small percentage of random evaluation is added to the FES. Secondly, an error compensation system is applied to the FES. The results presented cover six optimisation functions and show that the FES generally takes less time to reach the same solution as the GA as a result of the reduced number of evaluations.

The number of evaluations required by the GA with FES was reduced compared to a traditional genetic algorithm to reach the same fitness value even though in general better fitness values were achieved. For the earlier stages of evolution the performance advantage of the FES was more recognisable than at the final point.

While other methods of fitness approximation showed the side effect of premature convergence of the population, the FES with the two enhancements displayed a very similar behavior to the conventional GA without the side effect.

The results also show that a small threshold of 0.3 is more effective than other threshold values in achieving a faster climb to good fitness values, which may be important if only limited time is available to achieve as good a result as possible. However, if the obtaining the best result is most important, a larger threshold value of 0.7 is more appropriate. This result will still be obtained faster than by a conventional GA.

References

1. Holland J., "Adaptation in Natural and Artificial Systems", MIT Press, Cambridge, MA, 1975.
2. Goldberg D.E., "Genetic Algorithms in Search, Optimization and Machine Learning", Addison-Wesley Co, 1989.
3. Vanderplaats, G.N., "Numerical Optimization Techniques for Engineering Design: with Applications", McGraw-Hill, New York, 1984.
4. Powell M.J.D., "A Direct Search Optimization Method that Models the Objective and Constraint Functions by Linear Interpolation" Proceedings of the Sixth Workshop on Optimisation and Numerical Analysis, Oaxaca, Mexico, Volume 275, pp. 51-67, Kluwer Academic Publishers, 1994.
5. Ratle, A., "Optimal Sampling Strategies for Learning a Fitness Model", Proceedings of the 1999 Congress on Evolutionary Computation, Vol. 3, pp. 2078-85, Piscataway, NJ, 1999, IEEE Press.
6. Bäck T., Fogel D.B., Michalewicz Z., "Handbook of Evolutionary Computation", Oxford University Press, Oxford, 1997.
7. Fitzpatrick J.M., Grefenstette J.J., "Genetic Algorithm in Noisy Environment", Machine Learning, Vol 3, pp. 101-20, 1988.
8. Hammel U., Bäck T., "Evolution Strategies on Noisy Functions: How to Improve Convergence Properties", Parallel Problem Solving from Nature (PPSN III), Lecture notes in computer science 866, Springer-Verlag, Berlin, pp. 159-68, 1994.
9. Grefenstette J.J., "Robot Learning with Parallel Genetic Algorithms on Network Computers", The Proceedings of the 1995 Summer Computer Simulation Conference (SCSC95), Ottawa, CA, pp. 352-57, 1995.
10. Baker J.E., "Reducing Bias and Inefficiency in the Selection Algorithm", Proceedings of the Second International Conference on Genetic Algorithms, Lawrence Erlbaum Associates, Hillsdale, NJ, 1987.
11. Marin J. and Sole R.V., "Macroevolutionary Algorithms: A New Optimization Method on Fitness Landscape", IEEE Transaction on Evolutionary Computation, Vol. 3, No. 4, pp. 272-286, 1999.
12. Eiben A.E. and Bäck T., "An Empirical Investigation of Multi-Parent Recombination Operators in Evolution Strategies, Evolutionary Computation, Vol. 5, No.3, pp. 347-365, 1997.

13. Smith R., Dike B., and Stegmann S., "Fitness Inheritance in Genetic Algorithms", Proceedings of the ACM Symposium on Applied Computing, pp. 345-350, New York, NY, 1995.
14. Sastry K., Goldberg D.E., and Pelikan M., "Don't Evaluate, Inherit", IlliGAL Technical Report No. 2001013, University of Illinois at Urbana-Champaign, Urbana, Illinois, January 2001.

Derivation of L-system Models from Measurements of Biological Branching Structures Using Genetic Algorithms

Bian Runqiang[1,2,3], Phoebe Chen[1,2,4], Kevin Burrage[1,2], Jim Hanan[2], Peter Room[2], and John Belward[1,2]

[1] Advanced Computational Modelling Centre, Department of Mathematics
University of Queensland, Brisbane, Qld 4072, Australia
{bianrq,kb,jab}@maths.uq.edu.au
[2] Centre for Plant Architecture Informatics, Department of Mathematics
University of Queensland, Brisbane, Qld 4072, Australia
{jim,peter.room}@cpai.uq.edu.au
[3] Department of Automation, Tianjin University of Technology
Tianjin 300191, P. R. China
bianrq@public.tpt.tj.cn
[4] Faculty of Information Technology, Queensland University of Technology
Qld 4001, Australia
p.chen@qut.edu.au

Abstract. L-systems are widely used in the modelling of branching structures and the growth process of biological objects such as plants, nerves and airways in lungs. The derivation of such L-system models involves a lot of hard mental work and time-consuming manual procedures. A method based on genetic algorithms for automating the derivation of L-systems is presented here. The method involves representation of branching structure, translation of L-systems to axial tree architectures, comparison of branching structure and the application of genetic algorithms. Branching structures are represented as axial trees and positional information is considered as an important attribute along with length and angle in the database configuration of branches. An algorithm is proposed for automatic L-system translation that compares randomly generated branching structures with the target structure. Edit distance, which is proposed as a measure of dissimilarity between rooted trees, is extended for the comparison of structures represented in axial trees and positional information is involved in the local cost function. Conventional genetic algorithms and repair mechanics are employed in the search for L-system models having the best fit to observational data.

1 Introduction

Lindenmayer systems (or L-systems) were proposed by Aristid Lindenmayer in 1968 as a mathematical representation of biological development and they have been widely used in modelling and simulation of branching biological objects. L-systems find applications in modelling the architectural development of plants [1], [2], [3], corals [4], and patterns of cell division in embryos [5]. For example, Frijters showed that L-systems can be applied to formalize the different florescence states of *Hieracium murorum* [6] and Renshaw simulated the root structure and canopy development of sitka spruce *Picea sitchensis* with L-systems [7]. Software environments called vlab (for Unix) and L-studio (for Windows) with powerful functions have been developed to support the modeling of plants using L-systems. They allow the user to enter and edit an L-system and interpret it to generate 3D images that can be viewed from any perspective [1], [8], [9].

To derive an L-system model from data on real biological objects involves difficult mental work and time-consuming manual procedures. Work has been done on the data collection and analysis of cotton, sorghum, pinaster and other plants [10], [11]. There is a need for software to automate the derivation of L-system productions (rewriting rules) from measurement data [1] and some work has been reported on the use of genetic algorithms for this purpose [12].

In this paper, a method based on genetic algorithms is presented for automating derivation of L-systems which are used to model the branching structure of definite biological objects. Previous work in this area is reviewed in section 2 of the paper; the main concepts of L-systems are outlined in section 3; in section 4, we present the genetic algorithm strategy and the repair mechanism employed. Test results are reported in section 5 and conclusions are given in section 6.

2 Previous Work in the Field

Ochoa described the evolution of plant branching structures represented by L-systems [13]. Evolution was simulated using a genetic algorithm and fitness evaluation focused on five factors: (a) positive phototropism, (b) bilateral symmetry, (c) light gathering ability, (d) structural stability and (e) proportion of branching points. The weighting average of the five factors was regarded as the fitness

$$Fitness = \frac{aw_a + bw_b + cw_c + dw_d + ew_e}{w_a + w_b + w_c + w_d + w_e} \quad (1)$$

where w_a, w_b, w_c, w_d, w_e are weighting parameters for tuning the effect of each component on the final fitness function. Some operating strategies of genetic algorithms were modified to adapt to the problem.

In Curry [12], a user interface was described that allows the user to guide a genetic algorithm evolving plant models expressed as parametric L-systems. The user interface displays 9 typical plant structures and the user may visually evaluate and choose two of them to crossover or choose one to mutate until the system produces the desired structure.

In Jacob [14], a method is described for inference of axiom (or starting string) and L-system productions (or rewriting rules) by genetic algorithms. The characters that may appear in the L-axiom and L-productions are encoded in symbolic form and a repair mechanism is considered to adjust the randomly generated individuals. The approach is used in inference of the fractal structures of the quadratic Koch island. The reference structure, generated from a sequence of turtle commands, is represented in unit-length line segments in an equidistant grid. Individuals in the population are compared to the line structure and overlapping as well as disjunct line segments are taken into account to define the similarity, i.e., the fitness.

Jacob also studied the evolution of plantlike structures [14]. Simple inflorescent plants were considered and L-systems were used in combination with genetic algorithms. Because no target fractal could be used to evaluate the fitness of a structure, the number of blooms and a measure of the overall extension of the evolved structures was determined with a fitness function.

The present paper presents a method for derivation of L-system models of branching structures based on genetic algorithms. The method searches for an L-system that will generate a branching structure that matches observational data for a particular real plant. In contrast to this, in the methods of Ochoa [13] and Jacob [14], genetic algorithms are used to search for L-systems for generalized plant-like structures according to general attributes such as bilateral symmetry, structural stability, and abundance of flowers, not a particular target structure. The work of Jacob deriving L-system models for a quadratic Koch island is not relevant to plants because a special grid and unit length line segment on the grid were employed to evaluate fitness [14]. The interface of Curry can be used to find an L-system model for a given plant architecture, but to the operation is not automated and depends on choices made by the user [12].

3 L-systems

Branching structures are a significant characteristic of many biological objects, such as plants, corals, etc., and mathematical descriptions of tree-like structures are needed for modeling purposes [15].

A rewriting mechanism can be used to model the development of branching structures. In the mechanism, a rewriting rule replaces a predecessor edge by a successor axial tree in such a way that the starting node of the predecessor is identified with the successor's base and the ending node is identified with the successor's extremity. Such a rewriting mechanism is employed in L-systems.

An L-system can be defined as a triple $G=(V, \omega, P)$, where V is a set of symbols, ω is the axiom or starting string, and P is a set of productions or rewriting rules. A production $(a, \chi) \in P$ is written as a --> χ, the letter a and χ represent the predecessor and successor and of this production respectively. With a definite axiom, a set of productions and a given step, an L-system can form a string composed of symbols in V, which is called an L-string. Given the L-system G, an axial tree T_2 is directly derived from a tree T_1, denoted $T_1 => T_2$, if T_2 is obtained from T_1 by simultaneously replacing each symbol in T_1 by its successor according to the production set P. A tree

T is generated by G in a derivation of length n if there exists a sequence of trees T_0, $T_1, \ldots, T_n$ such that $T_0 = \omega$, $T_n = T$ and $T_0 => T_1 => \ldots => T_n$. For example, the structure shown in Fig. 1 is produced by the following L-system.

Fig. 1. An example of plant-like structure [16]

$G_1 = (V_1, \omega_1, P_1)$
$V_1 = \{X, F, +, -, [,]\}$
$\omega_1 = X$
$P_1 = \{X \text{-->} F[+X][-X]FX, F \text{-->} FF, + \text{-->} +, - \text{-->} -, [\text{-->} [,] \text{-->}]\}$

To create Fig. 1 from the above L-system, the number of steps n is 7 and the lateral branch angle δ is 25.7°. In the example, when step $n=1$, the L-string = F[+X][-X]FX; and when step $n=2$, the L-string = FF[+F[+X][-X]FX][-F[+X][-X]FX]FFF[+X][-X]FX.

4 Derivation of L-systems and Genetic Algorithm Strategy

Genetic algorithms are optimization algorithms based on Darwinian natural selection and they are an important part of computational intelligence. The essential unit in a genetic algorithm is a gene (bit or symbol). A series of genes composes a chromosome (individual) according to an encoding rule. An evolving population consists of the individuals. Three operations including crossover, mutation and selection are carried out on individuals. The evaluation of fitness is decided according to the particular problem.

Genetic algorithms were used to derive and optimize L-system axioms and productions. For the matching of L-systems with specific branching structures, an algorithm was introduced to transform L-systems into branching structures.

In the Genetic algorithm used in the derivation of L-systems in this paper, a symbolic encoding strategy is employed. Each individual is composed of an axiom and several productions that are generated with a random combination of symbols from the variable set V. In order to guarantee that every individual was not meaningless in an L-system, a repair mechanism was introduced before fitness was evaluated. Conventional genetic algorithm operators were used for crossover, selection, and modification of the mutation operator.

4.1 Encoding and Repair Mechanism

There were three alternative forms for encoding: binary bit, real number, and symbol. Though binary bit encoding can be used, symbolic encoding is simpler [13], [14]. The symbols in L-systems acted directly as genes in the genetic algorithm.

Individuals in the genetic algorithm were composed of an axiom and several production successors. Because of the possibility of different lengths of axioms and productions, a special symbol '0' was appended to the L-system symbol set V, which produced no operation in an L-system but took the character's position in an individual.

We divided the L-system symbol set V into several subsets. Symbols such as F, X, etc., constituted a variable symbol subset V_v; '+' and '-' were sorted in the direction symbol subset V_d; the branch symbol subset was composed of '[' and ']', and a '[' and a ']' were start and end of branch symbols V_b; the symbol '0' represented a space V_{null} and could be involved in any subset.

Not all possible permutations of symbol set elements are syntactic in L-systems, e.g., ++-F, F[+F][, etc., and this fact introduces the problem of constraints. Three approaches: searching domain, repair mechanism, and a penalty scheme, are used to cope with the problem of constraints.

The searching domain approach is to embed the condition of constraints in the system by confining the searching space of an individual. This approach guarantees that all individuals are valid and that the constraint will not be violated. Another approach for handling constraints is to set up a penalty scheme for invalid individuals such that they become low performers. The constrained problem is then transformed to an unconstrained condition by associating the penalty with all the constraint violations. This can be done by including a penalty to adjust the optimized objective function.

Unfortunately the above two methods are not appropriate for our situation. The first method suits only the situation when constraints have an apparent numerical boundary. When an individual does not conform to the L-system syntax, we cannot interpret it to generate an architecture, so we have to assign a lowest fitness to it. Therefore the penalty is meaningless for the evaluation of the individuals. Consequently, we used the repair mechanism here. The repair mechanism corrected individuals so that they became valid if any condition of the constraint was violated. This was achieved by modifying some genes randomly within the valid solution space, or backtracking toward the parents' genetic material.

The repair mechanism was used to make the axiom and productions comply with L-system syntax, before the evaluation of fitness. The following cases were considered to be illegal:

(1) Wrong number or wrong position of branch symbols, including

- Brackets that do not appear in pairs, e.g., F[+F]F];
- Left brackets do not appear to the left of the right branch symbols, e.g., F[+F]]-F[;

(2) More than one direction symbol appears in front of a variable symbol, e.g., ++-F;
(3) No variable symbols between a pair of branch symbols, e.g., [-].

In the repair process, the numbers of left and right brackets N_l, N_r were tested. If $e = N_l - N_r > 0$ and e was even, $1/2e$ left brackets were changed to right ones, which made them equal; if e was odd, a left bracket was added or deleted according to a definite probability (e.g., 0.5) at a random position, which made e become even, and $1/2\ e$ left brackets were changed to right ones to make them equal. If $e = N_l - N_r < 0$, a similar operation was applied to the redundant right brackets to match the left ones. When N_l was equal to N_r, but there was no right bracket to match one or more left bracket(s) to the right, and the left brackets were exchanged with the right ones.

When there was more than one direction symbol adjacent to another, as in illegal situation (2), the redundant symbols were replaced by randomly selected symbols from the set $V_v \cup V_{null}$.

To repair strings as in illegal situation (3), variable symbols were added into the pair of brackets. If the branch was long enough (length ≥ 2), only the first direction symbol remained unchanged, and the others were replaced with randomly selected characters $c \in V_v \cup V_{null}$. If the branch was not long enough, symbol(s) $c \in V_v \cup V_{null}$ were appended.

4.2 Operators

Crossover, mutation and selection are the three fundamental operators in genetic algorithms. In the derivation of L-systems, a two-point crossover operator and a proportion selection operator were used.

In mutation, a symbol selected randomly according to the probability of mutation p_m was changed to any symbol $c \in V$. When the selected symbol was a branch symbol $c \in V_b$, a special treatment was introduced: at a random position another branch symbol was placed to match it.

Another kind of operator applied in the derivation of L-systems involved symbols moving in an individual. An L-system is order sensitive and by normal crossover and mutation, it is difficult to change the symbol order. A series of randomly selected symbols was moved to an also randomly selected position in an individual as a part of mutation. The space symbols in an individual were replaced by variable or direction symbols in a mutation operator to increase a branch's length. The decreasing length mutation was used to add space symbols randomly into a branch or a sub-branch.

Because crossover and mutation may make some strings illegal, individuals were repaired after these operations.

4.3 Fitness Evaluation

To evaluate the fitness of an individual composed of an axiom and set of productions, the branching structure generated from the L-system was compared with the target structure and a distance was calculated between them. The smaller the distance, the higher the individual's fitness.

Similar to the axial trees, we defined the order of L-strings. The symbols that were not involved in any pair of brackets constituted order zero string. In an n-order string, the symbols that are embraced only by a pair of brackets within the n-order string constitute a string with order $n+1$. For example, suppose S_n=F[+F]F[-F[+F][F]] F be a n-order string, S_n can be separated into three strings: one n-order string $s_{n,1}$=FFF, and strings with order $n+1$ $s_{n+1,1}$= +F, $s_{n+1,2}$= -F[+F][F]. $s_{n,1}$ is a special string without brackets, corresponding to an axis (without sub-branches) in a branching structure, we call it a axial string. To distinguish it from other strings, we noted it as $a_{n,1}$. Comparing it with the branching structure of S_n in Fig. 2, we see that the L-string is correspondingly matched with an axial branch. $s_{n,1}$=FFF corresponds to the n-order axis; $s_{n+1,1}$=+F corresponds to axis with order $n+1$; and $s_{n+1,2}$= -F[+F] corresponds to a branch with order $n+1$. Similar to axial trees, we defined strings without higher order axis as axial strings, and strings with higher order axis as branch strings or strings when no confusion may arise.

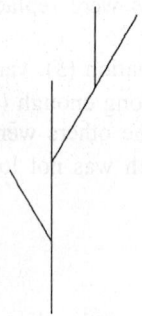

Fig. 2. Branching structure of S_n

Suppose L-string T represents a string with n-orders of strings, and $s_{i-1,j}$ is one of (i-1)th order strings, $1<i<n$, $1<j<m$, where m is the number of (i-1)th order strings. By classifying and summing classified symbols in string $s_{i-1,j}$, we can obtain one (i-1)th order axial string a_i and k strings with ith order $s_{i,1}$, $s_{i,2}$, ..., $s_{i,k}$. Suppose s_{ij} is an arbitrary ith order string, and c is an arbitrary symbol in L-system symbol set, i.e., $c \in V$ and $l(c)$ is the length of the string, then we can obtain the length of the axial string of a_{ij}, $l(a_{ij})$, and its position on the (i-1)th order axial string $a_{i-1,j}$, $p(a_{ij})$, where we use '<' to present a symbol before another one, by the following equations.

$$l(a_{i,j}) = \sum_{v \in V_v, v \in a_{i,j}} l(c) \qquad (2)$$

$$p(a_{i,j}) = \sum_{v \in V_v, v \in a_{i-1,j}, v < a_{i,j}} l(c). \qquad (3)$$

With the above equations, the lengths and positions of axes of different order in a tree were calculated and a database was built based on the calculated information. The angle was determined by each axial string's direction symbol. The fitness of an individual was evaluated as the inverse of the distance between the L-string structures and the target branching structure as

$$f = M_c - D(T_{L-system}, T_{target}) \qquad (4)$$

$$M_c = \sum_{i=1}^{m}\sum_{j=1}^{n} b_i(j), \tag{5}$$

where m is the number of target structure's branches, and n is the number of each branch's parameters. Here $D(T_{\text{L-system}}, T_{\text{target}})$ is the distance between an L-system structure with the reference structure. The distance is obtained by comparing the calculated information with the target measurement data as in Table 1 and Table 2.

5 Tests and Results

5.1 Test Conditions

For these tests, the axioms were predefined and the numbers of steps were considered as known. In a single variable L-system test 'F' was used as the variable symbol, and 'F' and 'X' were used as variable symbols in a double variable L-system test. In the single variable L-system test, the axiom was 'F', and in the double variable L-system test, the axiom was 'X'. An 'F' produces a segment of line and an 'X' produces a point in the interpretation of L-systems into branching structures. In these tests, L-productions were allowed to have at most 14 symbols.

The genetic algorithm parameters were crossover probability $p_c=0.8$, mutation probability $p_m=0.05$, population $n=20$, and the generation of each test was 100. The individual length was equal to the production length of 14.

5.2 Single Variable L-systems

A branching structure produced with an L-system, shown in Fig. 3, is used for the single variable test. The measurement data are shown in Table 1.

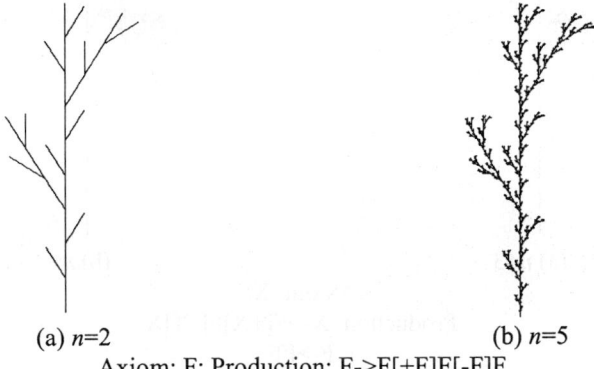

(a) $n=2$ (b) $n=5$
Axiom: F; Production: F->F[+F]F[-F]F

Fig. 3. A simple, one-variable L-system and the branching structure it generates after (a) 2 and (b) 5 iteration [16]

Table 1. Database of the L-system shown in Fig. 3(a)

axis no.	position	length in Fs	angle	sub-axes
1	-1	9	0	8
2	1	1	1	0
3	2	1	-1	0
4	3	3	1	2
5	4	1	1	0
6	5	1	-1	0
7	6	3	-1	2
8	7	1	1	0
9	8	1	-1	0
10	3,1	1	1	0
11	3,2	1	-1	0
12	6,1	1	1	0
13	6,2	1	-1	0

We executed the search program 50 times and obtained the following results: the longest search generation was 38; shortest search generation was 1; average search generation was 10.8. All the searches successfully gave the optimal result.

5.3 Double Variable L-systems

One L-system with two variables was used to test the program. The L-axiom and L-productions and the branching structure of the L-system are shown in Fig. 4. The attribute database of the branching structure is listed in Table 2.

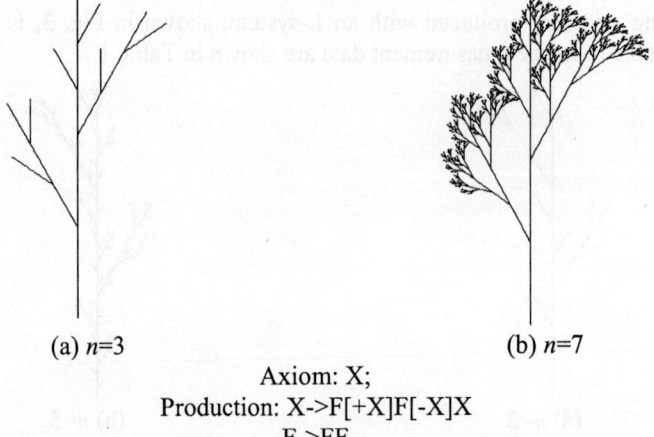

(a) $n=3$ (b) $n=7$

Axiom: X;
Production: X->F[+X]F[-X]X
F->FF

Fig. 4. L-system and its structure

Table 2. Database of L-system shown in Fig. 4 (a)

axial no.	Position	length in Fs	angle	sub-axes
1	-1	14	0	4
2	4	6	1	2
3	8	6	-1	2
4	10	2	1	0
5	12	1	-1	0
6	4,2	2	1	0
7	4,4	2	-1	0
8	8,2	2	1	0
9	8,4	2	-1	0

The results of 50 executions show that the exact fit was not found 17 times; in the found situations, the longest generation was 97; the shortest generation was 32; average generation was 53.5. The double-variable L-system search results were not as good as the single-variable. One reason is that two variables make the search space wider and another reason is that the individual length of 14 symbols for the searching of 2 symbols made the search less efficient.

To our knowledge, this is the first attempt at using genetic algorithms in this type of application. Using Matlab, each simulation on data in Table 1 took approximately 38 seconds and each simulation on data in Table 2 took approximately 738 seconds on a Pentium II.

6 Conclusion

We have demonstrated the feasibility of using a genetic algorithm method to automate the derivation of L-system specifications for simple branching structures. Realistic models of more complex biological objects, often require more complicated L-systems to be employed such as parametric, context sensitive and stochastic L-systems. Further studies on the derivation of these kinds of L-systems for modelling biological structures will be valuable. Along with search space becoming larger, the definition of individual length and the efficiency of genetic search will need more effective methods to guarantee that optimal results will be found.

References

1. Room, P., Hanan, J., Prusinkiewicz, P.: Virtual plants: new perspectives for ecologists, pathologists and agricultural scientists. Elsevier trends journals, 1 (1996) 33-38
2. ColladoVides, L., GomezAlcaraz, G., RivasLechuga. G., GomezGutierrez, V.: Simulation of the clonal growth of Bostrychia radicans (Ceramiales-Rhodophyta) using Lindenmayer systems. BIOSYSTEMS, 1 (1997) 19-27

3. Gautier, H., Mech, R., Prusinkiewicz, P., Varlet-grancher, C.: 3D architectural modeling of aerial photomorphoginisis in white clover (trifolium repens L.) using L-systems. Annals of Botany, 85 (2000) 359-370
4. Kaandorp, J. A.: Fractal modelling : growth and form in biology. Springer-Verlag, Berlin New York (1994)
5. De Boer, M. J. M.: Analysis and computer generation of division patterns in cell layers using developmental algorithms. PhD thesis, Uijks Universiteit Utrecht (1989)
6. Frijters, D.: An automata-theoretical model of the vegetative and flowering development of *Hieracium Murorum*. Biological Cybernetics, 24 (1976) 1-13
7. Renshaw, E.: Computer simulation of sitka spruce: spatial branching models for canopy growth and root structure. IMA Journal of Mathematics applied in Medicine &Biology, 2 (1985) 183-200
8. Hanan, J.: Virtual plants-integrating architectural and physiological models. Environmental Modelling & Software, 1 (1997) 35-42
9. Prusinkiewicz, P., Hanan, J. , Mech, R.: (2000) An L-system-based plant modeling language. In: Nagl, M., Schurr, A., Munch, M. (eds.): Lecture Notes in Computer Science, Vol. 1779. Springer-Verlag, Berlin (2000) 395-410
10. Kaitaniemi, P., Hanan, J., Room, P.: Virtual sorghum: visualisation of partitioning and morphogenesis. Computers and Electronics in Agriculture, 3 (2000) 195-205
11. Danjon, F., Bert, D., Godin, C., Trichet, P.: Structural root architecture of 5-year-old Pinus pinaster measured by 3D digitising and analysed with AMAPmod. Plant and Soil, 217 (1999) 49-63
12. Curry, R.: On the evolution of parametric L-systems. Research Report of Department of Computer Science, University of Calgary, Alberta, Canada (1999)
13. Ochoa, G.: On genetic algorithms and Lindenmayer systems. In Problem Solving from Nature- PPSN V. Lecture Notes in Computer Science, Vol. 1498. Springer-Verlag, Berlin Heidelberg New York (1998) 335-344
14. Jacob, C.: Illustrating evolutionary computation with mathematica. Morgan Kaufmann (2001)
15. Chen, P., Colomb, B. M.: Database Technologies for L-system Simulations in Virtual Plant Applications on Bioinformatics. to appear in the Int. Journal of Knowledge and Information Systems. Springer-Verlag, Berlin (2002)
16. Prusinkiewicz, P., Lindenmayer, A.: The algorithmic beauty of plants. Springer-Verlag, New York (1990)

Evolving a Schedule with Batching, Precedence Constraints, and Sequence-Dependent Setup Times: Crossover Needs Building Blocks

Paul J. Darwen[1]

School of Computer Science and Electrical Engineering, The University of Queensland
Brisbane 4072 Australia
darwen@ieee.org

Abstract. The Travelling Salesman Problem (TSP) has a "big valley" search space landscape: good solutions share common building blocks. In evolutionary computation, *crossover* mixes building blocks, and so crossover works well on TSP. This paper considers a more complicated and realistic single-machine problem, with batching/lotsizing, sequence-dependent setup times, and time-dependent costs. Instead of a big valley, it turns out that good solutions share few building blocks. For large enough problems, good solutions have essentially nothing in common. This suggests that crossover (which mixes building blocks) is not suited to this more complex problem.

1 Introduction: A Difficult Scheduling Problem

A common complaint in industry is that academic research on schedule optimization mostly studies idealized, simplified problems with limited relevance to the real world. The Travelling Salesman Problem (TSP) is the simplest of a large range of schedule optimization problems.

This paper studies a scheduling problem that resembles the TSP, but with the addition of realistic features that make it harder:

- Jobs can be done in batches, to reduce setup times.
- Those setup times are sequence-dependent.
- Jobs have precedence constraints, so that some jobs must be done before others (sometimes called "multi-level" constraints [5, page 230] [13]).
- Payoffs are time-dependent, i.e., the cost of a job depends not only on its order, but also when it is done.

That is, the scheduling problem studied in this paper is basically TSP but with extra features that are commonly found in real-world manufacturing. These extra features make it a harder problem.

1.1 Evolutionary Computation: So Many Flavours

Evolutionary computation comes in a bewildering assortment of different approaches: evolutionary programming [6], genetic algorithms [7], and genetic programming [14] are all similar. In brief, they maintain a population of trial solutions in a way that is loosely analogous to biological evolution:

1. Randomly initialize the population of trial solutions.
2. An *evaluation function* evaluates those trial solutions.
3. The poor solutions are erased.
4. The surviving, better solutions are copied.
5. The copies are mixed and matched using variation operators loosely based on *mutation* and *recombination*.
6. Go to step 2, repeat until a suitable solution is found.

Much of the difference between the various flavours of evolutionary computation lies in what variation operators they use [3]. In particular:

- genetic algorithms [7] emphasize recombination (or *crossover*), which mixes the many building blocks contained in a large population.
- evolutionary programming [6] emphasizes mutation, and so tends not to require as big a population.

So which approach is better? It depends on the problem. The No Free Lunch theorem [19] proves that the best optimization algorithm for a particular problem depends on the attributes of that problem — there is no single algorithm that is better for all possible problems. The algorithm must suit the problem.

For the problem studied here, it turns out that the best solutions do *not* share building blocks, so crossover may not be the best operator.

1.2 Previous Scheduling Research on Similar Problems

Perhaps because it is so difficult, the single-machine scheduling problem with all of these complexities has not been studied before. Even the single-machine problem with batching and sequence-dependent setups is rarely studied [8].

Hurink [9] presents a nice way to solve the one-machine batch scheduling problem. Unfortunately, it doesn't deal with precedence constraints, sequence-dependent setup times, or time-dependent costs.

Jordan [11] looks at scheduling with batching, including sequence-dependent setup times. Unfortunately, although he does consider setup times, he doesn't deal with precedence constraints or time-dependent costs.

A recent review of scheduling with genetic algorithms [4] mentioned only one study that considered batch sizing, that of Lee, Sikora, and Shaw [15]. Unfortunately, it too ignored precedence constraints and time-dependent costs.

The closest that previous research comes to touching the problem studied in this paper is the work of Kimms [12] [13]. This looked at single- and n-machine scheduling, with both batching and with precedence constraints. However, these studies ignored sequence-dependent setup costs [12, page 87]. With sequence-dependent setups, the problem becomes too hard for the enumerative methods that work so well on sequence-independent setup costs [8].

Many previous studies have avoided NP-completeness by using sequence independent setup times, or by having all jobs take the same amount of time. Salomon [18] shows that without setup costs, some of these problems are polynomially solveable, but setup-dependent costs or times make it NP-hard.

To summarize, this paper is about a scheduling problem with several realistic features, but previous research has only one of those features or (in the case of Kimms [13]) just two of them. This paper is the first to simultaneously consider all four of batching, precedence constraints, sequence-dependent setup times, and time-dependent costs.

1.3 Global Landscape Properties

Some combinatorial optimization problems have a "big valley" structure in the fitness landscape of the search space. These include the Travelling Salesman Problem (TSP) [2] and the flowshop scheduling problem [16]. These show a strong correlation between a solution's fitness, and the number of differences between that solution and the global optimum. Basically, a search landscape with a "big valley" means all the best solutions have similar building blocks. This makes crossover work well, because it passes building blocks around.

This paper considers a single-machine scheduling problem like TSP, with additional realistic features. Will this problem also have this helpful correlation?

Of course, there is no way of knowing what exactly are the global best solutions in the trial problems studied here, because this paper is the first to consider this kind of problem. Instead the best solution found for each problem is used instead. For the smaller problems, we can be quite confident that the best solution found is close to the global optimum. For the largest problem studied here (2013 jobs), only a few runs were done, so the global optimum is almost certainly better by a considerable margin.

2 Experimental Setup

2.1 The Test Problems

The test problems used in this paper are simplified, toy-sized versions of a scheduling problem used by a Brisbane company. This paper uses randomly-generated problems of different sizes: 116, 215, 672, and 2013 jobs.

The precedence constraints, of what jobs must be done before what, are described by what's called a *gozinto* graph (sounds like "goes into"). When each job has at most a single successor job (i.e, no job has two or more future jobs waiting for it), then the problem's gozinto graph is said to be *convergent*. A useful property is that any gozinto graph can be converted into one that is convergent [10, page 28]. So without loss of generality, the trial problems used here have convergent precedence constraints.

The time-dependence of costs is here only done by depreciation, which over the lifetime of the problem is considerable. More elaborate (and realistic) time dependencies are problem-specific. In order to keep this study as general as possible, this simple time dependence was used.

Sequence-dependent setup times were calculated in a rather arbitrary way, that simply took into account certain attributes of the various jobs in a problem. This is simple, but again not very realistic. More realistic setup times would be problem-specific, so this simple approach provides more generality.

To calculate a schedule's fitness, the time-dependent cost (or profit) of each job as it is completed is summed. Only jobs that result in a finished product being shipped give a profit, others cost money. The time-dependence means that it's better to ship finished products sooner, and avoid the carrying costs of half-finished inventory.

2.2 A Genetic Algorithm Representation

An elegant way to represent a schedule is as a *priority list*. An individual's genome contains a priority for every job in the problem. To evaluate that individual, jobs are done one at a time on the single machine. To choose the next job, of all feasible jobs (i.e., satisfying precedence constraints) the one with the highest priority is chosen.

The priority list approach is the best evolutionary method so far discovered for Muth and Thompson's job shop benchmarks [4, page 95]. It's also called *random key* representation [1].

The beauty of this approach is that a priority list is always feasible, because the priority list itself has no constraints. Constraints are embodied in the evaluation, which only chooses from feasible jobs. This allows for all kinds of messy operators, without the fragility of permutation representations.

There are problems. Random key representation is good for searching the relative ordering, but it can struggle with absolute ordering [17, page 3].

As for TSP, a simple greedy hill-climber is used. The priority of jobs are repeatedly swapped in order to find which swap gives the biggest improvement. Then that swap is implemented. This greedy hill-climber is called repeatedly until no further improvements are possible, and it's at a local optimum. This same greedy hill-climber is used to find the local optima in Section 3.1, as well as in the genetic algorithm runs in Section 3.

2.3 A Distance Metric, and Operators

Reeves [16] makes the point that the search landscape depends on what distance metric you use to decide which solutions are closer. To say that the local optima are clustered close together begs the question of what metric is used that makes them seem closer together.

Following Boese [2], the distance metric counts the number of common *links*: if one schedule has job a followed by job b, and another solution also has job a

followed by job b, then they share that link irrespective of the absolute order of job a. So for an n-job problem, there are $n - 1$ links. In the rest of this paper, the "distance" between two schedules is the number of links they have different.

The genetic algorithm (GA) here used two-point crossover and uniform crossover. As it's desirable to try and reduce the disruption to batches of jobs, a customized mutation operator simultaneously raised and lowered the priorities of all the jobs in a batch, instead of just one job.

3 Results

In the rest of this paper, the distance between two schedules is the number of links in which they differ, as outlined in Section 2.3. Also, without knowing what the global optimum of a problem is (as this is the first paper to consider problems of this type), the best solution found is used instead. For the three smaller problems of 116, 215, and 672 jobs this should be so close to the true global optimum to make little difference, but for the 2013 job problem the global optimum may be better than the best solution used here.

3.1 Local Optima: Random, and during GA Runs

In TSP, random local optima have a strong correlation between the distance from the global optimum, and the value of that local optima [2, chapter 7]. For the problem studied here, Figure 1 shows no such correlation. Where TSP has a strong linear correlation, our problem gives random scatter.

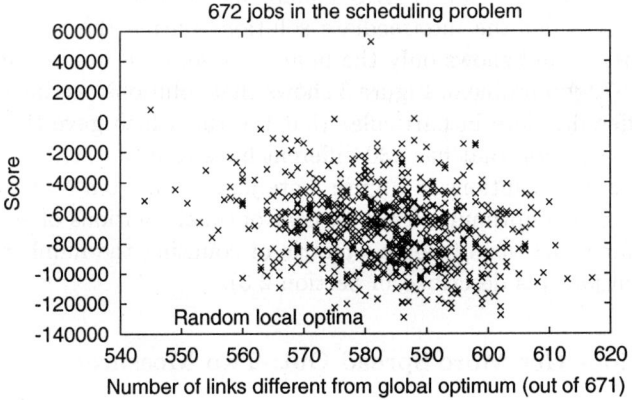

Fig. 1. Hill-climbing from random starting points gives this distribution of local optima. There appears to be no correlation between fitness and distance from the best solution, in stark contrast to TSP's "big valley"

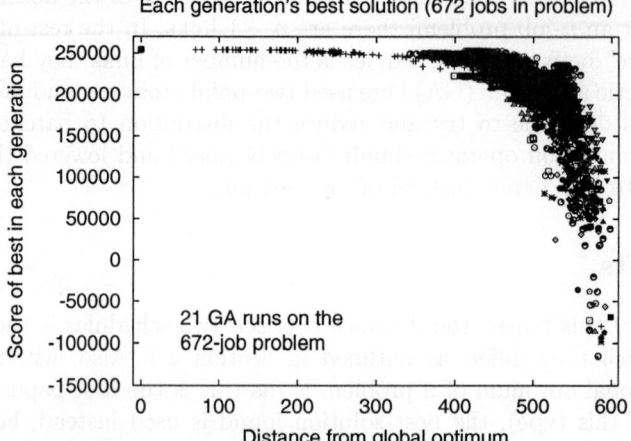

Fig. 2. Instead of random local optima, this shows a selection of the local optima found during the course of 21 different runs. Instead of a linear correlation, it gives a shape like a pistol, or a map of Florida

Instead of *random* local optima, by just walking uphill from anywhere, what about following a GA population during a run? Figure 2 shows various local optima found during 21 separate GA runs on the 672-job problem. Figure 2 shows that these 21 different GA runs share a common distribution of solutions, but no "big valley". Instead, it gives a shape like a pistol.

Figure 3 cuts out the intermediate solutions during the course of the GA runs in Figure 2, and shows only the final, best solution from each of those 21 runs on the 672-job problem. Figure 3 shows that solutions can be equally good, but very different. Note in particular that the run which gave the second-best score found a solution that has 430 different links (out of 671).

And those best solutions in Figure 3 are just as far from each other as they are from the global optimum. That is, near-optimal solutions of similar quality are spaced far apart, using Boese's metric of counting the number of different links between jobs (as described in Section 2.3).

3.2 Solutions Get More Spread Out: Two Measures

Not only are the best solutions far apart, but this spread gets wider for larger problems, i.e., the Florida shape in Figure 2 gets a narrower peninsula further to the left for larger problems. This section uses two ways to get a quantitative handle on just how widely spread is the distribution of local optima.

For the first way, consider the random local optima. Figure 4 shows the average distance from those random optima to the global optimum, as a fraction of the number of jobs in the problem.

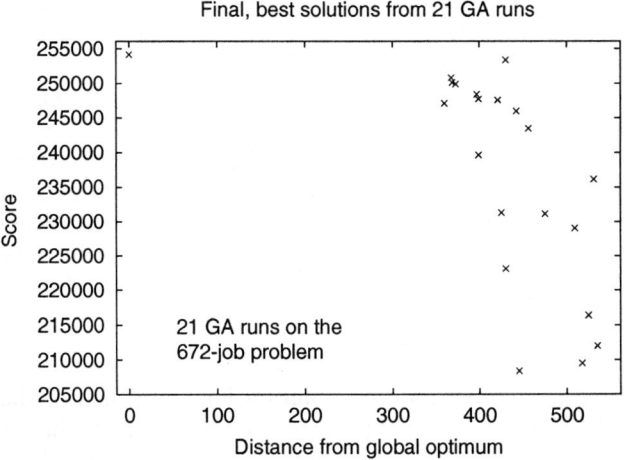

Fig. 3. For 21 GA runs of the 672-job problem, this shows the final, best solution from each run. Instead of a linear correlation like in TSP, here solutions can have many differences, but still have very similar payoffs

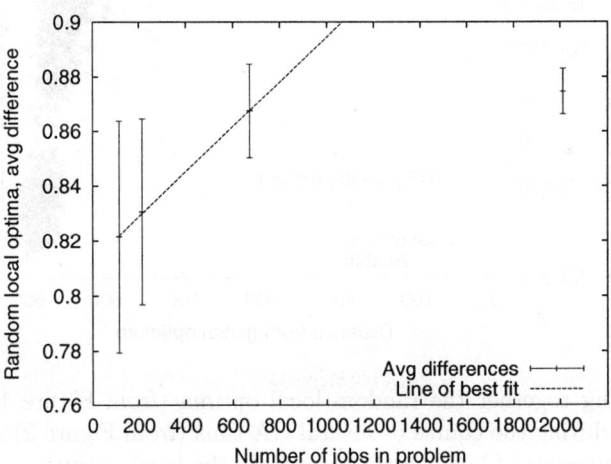

Fig. 4. As problem size increases, the average number of differences between a random local optimum and the global optimum, as a *fraction* of the problem size, appears to increase linearly for the smaller problems. Note that the error bars get smaller as the problems get bigger, in accordance with the "central limit catastrophe"

Figure 4 suggests the general trend is that random local optima are increasingly distant from the global optimum, as problem size increases. The standard deviation also gets smaller, in accordance with a sampling artifact that Boese calls the "central limit catastrophe" [2].

The second, more principled, method to measure how scattered solutions are will make use of *all* the local optima from the various GA runs, as well as the random local optima. Then, a least-squares fit passes a curve through all those points. Finding where that curve has a gentle slope (i.e., close to the global optimum) gives a measure of how widely spaced are the best local optima.

Figure 5 shows all the local optima (both 1000 random ones, plus those accumulated during GA runs) for the 672-job problem. The function chosen to fit that distribution is

$$l(x) = f_{\text{best}} - e^{cx} \qquad (1)$$

Here, f_{best} is the fitness of the best solution found (in lieu of the true global optimum). Doing a least-squares fit on the constant c yields the line of best fit, as in Figure 5. The handful of GA runs for the 2013-job problem gave such a sparse distribution that it was not used.

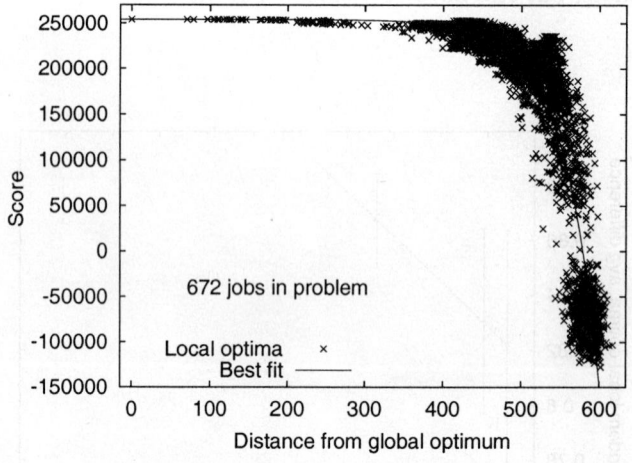

Fig. 5. Putting together the random local optima (from Figure 1) plus those accumulated during the course of several GA runs (from Figure 2) allows a best fit to give a measure of how widely spread are the local optima

The spread of the distribution is taken to be where the line of best fit has a slope of 45 degrees, i.e., a derivative of -1. This is chosen as it is among the better local optima from the GA runs, instead of down among the low-quality random local optima. Taking the derivative of Equation 1 and solving gives $x = c\ln(c)$.

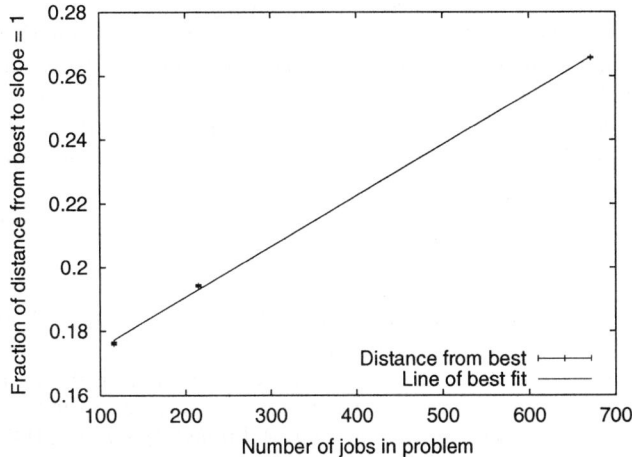

Fig. 6. Instead of measuring the number of differences from the best solution to the average of the random local optima, as in Figure 4, this takes the best-fit versions of Equation 1 and measures the distance to where that best-fit has a slope of -1 (i.e., 45 degrees downhill). The bigger the problem, the further apart are the best solutions

Figures 4 and 6 show the general trend: the larger the problem, the less that good solutions have in common. Just for interest's sake, if we extrapolate the line in Figure 4, it would reach 100% around problems of size 5269 jobs. While extrapolating like this is unreliable, the broad conclusion is that on large problems, equally good solutions can share few building blocks.

4 Discussion and Conclusions

Solving the TSP is made easier by the "big valley" structure of that problem. However, our realistic scheduling problem lacks that "big valley" structure. Instead, solutions with essentially the same fitness can nonetheless be very different, as for example the two best solutions in Figure 3.

Section 3.2 uses two different methods to show that (at least for the small-scale problems studied here) the distances between near-global optima get wider with increasing problem size. Extrapolating the second, more principled, method of Figure 6 predicts that for problems larger than about 5300 jobs, near-optimal solutions have practically nothing in common.

That claim may seem bold, but in Figure 3 the two best solutions have similar fitness but are very different — and that's on only 672 jobs. Commercial-sized problems range from 40,000 jobs to 200,000 jobs. In this search landscape whose peaks have so little in common, crossover and building blocks don't really help.

Another possibility is that some uninvented measure of commonality will show good solutions do share something. Reeves [16] points out that the search landscape depends on what distance metric you choose. TSP's metric in Section 2.3 (which follows Boese [2]) may simply be inappropriate for this more complicated scheduling problem, as it considers only the *relative* order of jobs; if job B follows job A that's a link in common. Future work will consider a distance metric using the *absolute* order of jobs.

References

1. James C. Bean. Genetic algorithms and random keys for sequencing and optimization. *ORSA Journal on Computing*, 6(2):154–160, 1994.
2. Ken D. Boese. *Models for Iterative Global Optimization*. PhD thesis, Computer Science Department, University of California at Los Angeles, 1996.
3. Ken De Jong. *Evolutionary Computation: A Unified Approach*. MIT Press, 2001.
4. Christos Dimopoulos and Ali M. S. Zalzala. Recent developments in evolutionary computation for manufacturing optimization: Problems, solutions, and comparisons. *IEEE Transactions on Evolutionary Computation*, 4(2):93–113, July 2000.
5. Andreas Drexl and Alf Kimms. Lot sizing and scheduling — survey and extensions. *European Journal of Operational Research*, 99(2):221–235, June 1997.
6. David B. Fogel. *Evolutionary Computation: Toward a New Philosophy of Machine Intelligence*. IEEE Press, New York, 1995.
7. David E. Goldberg. *Genetic Algorithms in Search, Optimization, and Machine Learning*. Addison-Wesley, Reading, Massachusetts, 1989.
8. Knut Haase and Alf Kimms. Lot sizing and scheduling with sequence-dependent setup costs and times and efficient rescheduling opportunities. *International Journal of Production Economics*, 66(2):159–169, June 2000.
9. Johann Hurink. An exponential neighbourhood for a one-machine batching problem. *OR-Spektrum*, 21(4):461–476, 1999.
10. Carsten Jordan. *Batching and Scheduling: Models and Methods for Several Problem Classes*, volume 437 of *Lecture Notes in Economics and Mathematical Systems*. Springer, 1996.
11. Carsten Jordan. A two-phase genetic algorithm to solve variants of the batch sequencing problem. *International Journal of Production Research*, 36(3):745–760, 1998.
12. Alf Kimms. Multi-level, single-machine lot sizing and scheduling (with initial inventory). *European Journal of Operational Research*, 89(1-2):86–99, 1996.
13. Alf Kimms. A genetic algorithm for multi-level, multi-machine lot sizing and scheduling. *Computers and Operations Research*, 26(8):829–848, July 1999.
14. John R. Koza. *Genetic Programming: on the programming of computers by means of natural selection*. MIT Press, Cambridge, Massachusetts, 1992.
15. In Lee, Riyaz Sikora, and Michael J. Shaw. A genetic algorithm-based approach to flexible flow-line scheduling with variable lot sizes. *IEEE Transactions on Systems, Man, and Cybernetics — Part B: Cybernetics*, 27(1):36–54, February 1997.
16. Colin R. Reeves. Landscapes, operators and heuristic search. *Annals of Operational Research*, 86:473–490, 1999.
17. Franz Rothlauf, David Goldberg, and Armin Heinzl. Network random keys: A tree network representation scheme for genetic and evolutionary algorithms. Technical Report IlliGAL 2000031, Illinois Genetic Algorithms Laboratory, University of Illinois at Urbana-Champaign, 2000.

18. Marc Salomon. *Deterministic Lotsizing Models for Production Planning*. Lecture Notes in Economics and Mathematical Systems. Springer-Verlag, 1991.
19. David H. Wolpert and William G. Macready. No Free Lunch theorems for optimization. *IEEE Transactions on Evolutionary Computation*, 1(1):67–82, 1997.

The Development of the Feature Extraction Algorithms for Thai Handwritten Character Recognition System

J. L. Mitrpanont and Surasit Kiwprasopsak

Department of Computer Science, Faculty of Science, and
Mahidol University Computing Center
Mahidol University, Rama VI Rd., Bangkok 10400, THAILAND
ccjlm@mahidol.ac.th
surasitq@hotmail.com

Abstract. This paper presents the development of feature extraction algorithms for the recognition of off-line Thai handwritten characters. These algorithms are used to exploit prominent features of Thai characters. The decision trees were used to classify Thai characters that share common features into five classes then 12 algorithms were developed. As a result, the major features of Thai characters such as an end-point (EP), a turning point (TP), a loop (LP), a zigzag (ZZ), a closed top (CT), a closed bottom (CB), and a number of legs were identified. These features were defined as standard features or the "Thai Character Feature Space." Then, we defined the 5x3 standard regions used to map these standard features, result in the "Thai Character Solution Space," which will be used as a fundamental tool for recognition. The algorithms have been tested thoroughly by using of more than 44,600 Thai characters handwritten by 22 individuals from 100 documents. The feature extraction rate is as high as 98.66% with the average of 93.08% while the recognition rate is as high as 99.19%[1] with the average of 91.42%. The results indicate that our proposed algorithms are well established and effective.

1 Introduction

Thai handwritten character recognition is complicated than some others since there are many complex characteristics such as different line levels, combination of consonants, vowels and tonal symbols. Typically, many Thai characters are similar to each other. Each can be differentiated from the others by a slightly different feature. For example, there is a little distinction among character 'ก', 'ถ' and 'ภ'. Although many approaches have been proposed [3], [4], [5], and [6], the problem of off-line handwritten character recognition of Thai characters still has very few solutions. The major cause is based on the difficulty of the recognition of handwritten characters and

---

[1] The comparison to other researches is not stated since they are studied in different environments.

the various noises emerged in the writing environment. Moreover, there are a very limited number of researches that concrete enough to extract most of the significant features of Thai handwritten characters. In specific, most of them concentrated on the extraction of only a few features mainly for Thai Optical Character Recognition (OCR) system. Recently, there are many OCR researches [7], [8] and [9] for the Thai language, but unfortunately, there is no commercial software available for both off-line and on-line system. A few products are designed for off-line Thai OCR such as the system of NECTEC's ARNThai [9].

In general, Thai characters consist of 44 consonants, 21 physical vowels, four tone marks, three special symbols and 10 numerals as described in Kiwprasopsak [2]. These characters are organized in four horizontal lines structure as shown in Figure 1.

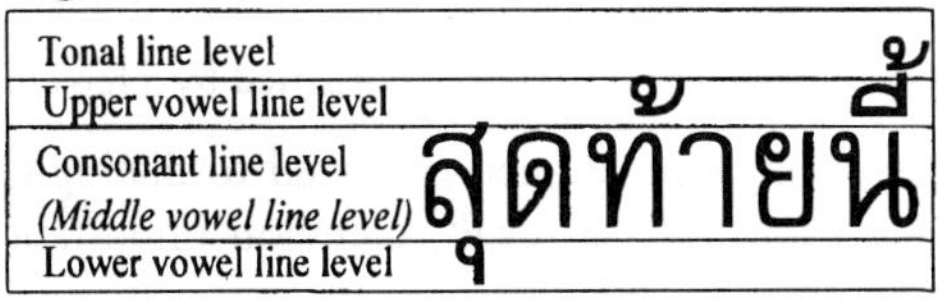

Tonal line level	
Upper vowel line level	ํ ่
Consonant line level (Middle vowel line level)	สุดท้ายน
Lower vowel line level	ุ

Fig. 1. An organization of Thai characters. A writing of Thai word can be organized into tonal line level, upper vowel line level, consonant line level or middle vowel line level, and lower vowel line level by using the terms of Airphaiboon and Kondo [7]. A consonant line level is a major part of Thai characters

Our research emphasizes on developing the algorithms for feature extractions of non-cursive or isolated Thai characters, which consist of 44 consonants, as well as 10 vowels and two special characters in the middle line of a character as shown in Figure 2. This means that upper vowels, lower vowels, other special characters and also Thai numeric characters are beyond the scope of the research. However, the extension of the algorithms could be done later.

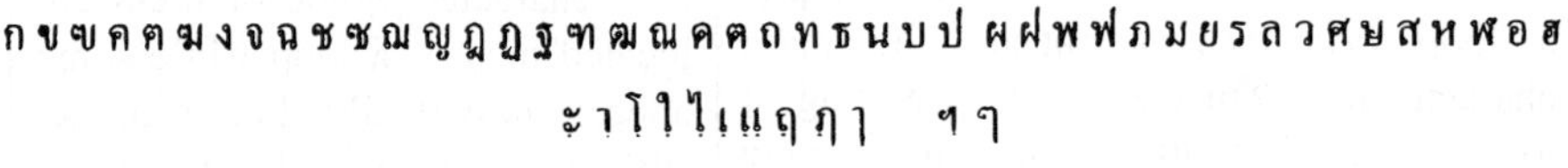

กขฃคฅฆงจฉชซฌญฎฏฐฑฒณดตถทธนบปผฝพฟภมยรลวศษสหฬอฮ
ะาโใไแเฤฦๅ ๆๆ

Fig. 2. 44 Thai consonants, 10 vowels and two special characters in the consonant line level or middle vowel line level within the scope of the research

The paper is organized into six parts. This section provides introduction to Thai character system. In section 2, basic features of Thai character are identified. In section 3, Thai character feature space and solution space are presented. The proposed algorithms are described next. Finally, experimental results and conclusions are given, respectively.

1.1 Identifying Basic Features of Thai Characters

Firstly, *"a leg"* or a number of legs is used as a group separator of Thai characters. Each Thai character consists of a certain number of leg(s) in both upper and lower parts of the character. It consists of at least one leg at the upper part and also at least one at the bottom. Consequently, we have classified Thai characters into five groups by using this leg characteristic as shown in Figure 3.

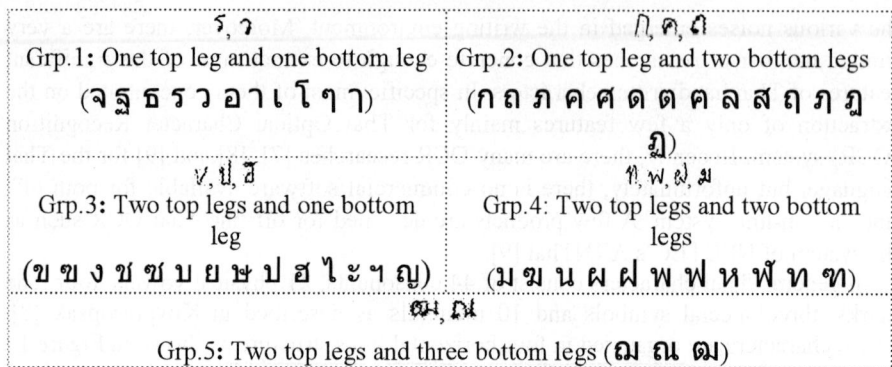

Fig. 3. Five groups of Thai characters, classified by a leg characteristic

After that, the decision trees have been designed to distinguish each character as described in Kiwprasopsak [2]. As a result, common important features are clearly distinguished as summarized in Table 1.

Table 1. Important features of Thai characters

• An End-Point (EP) An end-point is a point that is an edge or an end of a line. It is one of the most important features of Thai characters that exists in almost every character.	• A Turning Point (TP) A turning point (TP) is a turning position of a line from down to up or up to down.
• A Loop (LP) A loop can be classified into two categories: inner and outer loops. An inner loop is a loop that is located inside the main character body. In contrary, an outer loop is a loop that located outside the main character body. In case of handwriting, a loop is normally written with an incomplete loop.	• A Zigzag (ZZ) A zigzag is a curved line that generally moves down/up or up/down. It exists in different areas of a character. Some of them are located next to a loop while some appear individually. For example, 'ฆ' has a zigzag head at the upper left.
• A Closed Top (CT) A closed top indicates whether a character has a closed line at the top or not. For example, 'ค' has a closed top.	• A Closed Bottom (CB) A closed bottom indicates whether a character has a closed line at the bottom or not. For example, 'บ' has a closed bottom.

2 Defining the Thai Character Feature Space and Solution Space

This section provides the formal definition of both the proposed feature space and solution space of Thai characters. The implication of these concepts is significant, i.e., to define two standard frameworks used throughout this paper.

First, the feature of Thai characters can be defined as a set of the following properties:

- Features = {EP, TP, LP, ZZ, Height, Width, NumberOfUpperLegs, NumberOfLowerLegs, CT, CB}
- LP = {Inner, Outer}
- Height = {Normal, Long (Top or Bottom), Short}
- Width = {Normal, Thin, Wide}
- CT = {True, False}
- CB = {True, False}
- NumberOfUpperLegs = {1, 2}
- NumberOfLowerLegs = {1, 2, 3}

Second, the standard 5 x 3 matrix is defined to map each character into a standard format. Features of a character will be extracted and mapped into this matrix. It contains 15 standard regions (R1, R2, ..., R15) as shown in Figure 4.

	Left	Center	Right
Upper-Most	R_1	R_2	R_3
Upper	R_4	R_5	R_6
Middle	R_7	R_8	R_9
Lower	R_{10}	R_{11}	R_{12}
Lower-Most	R_{13}	R_{14}	R_{15}

Fig. 4. Standard regions for Thai characters

These standard regions consist of Upper-Most, Upper, Middle, Lower and Lower-Most in the horizontal direction, and consist of Left, Center and Right in the vertical direction. Based on this optimized format, the set of standard template features for Thai characters has been mapped to the standard region template generating and mapped to the standard region template generating the *"Thai Character Solution Space."* The incoming character is divided into 15 regions. Then, features are extracted and mapped before they are compared with the Solution Space for the recognition. See Mitrpanont and Kiwprasopsak [1] for more details.

3 The Development of the Feature Extraction Algorithms

Our recognition system consists of four major processes: preliminary processing, feature extraction, recognition and post-processing processes. In this paper, we emphasized on the feature extraction algorithms. However, there are a few issues about the preliminary processing worth to mention. For instance, the functions included in this process are noise reduction, reference line finding and character bitmap extraction algorithms but not the thinning process. We have found that it creates serious problems in Thai handwritten character recognition. General thinning algorithms, such as those of Zhang and Suen [10], and Arcelli [11] rely on English language in which the filled part of a character is insignificant. For example, 'A' does not differ from 'A' because the filled loop in the lower parts is not an important feature. Therefore, the skeletonized patterns of 'A' and 'A' are the same.

Unfortunately, for Thai handwritten characters, a filled loop and an incomplete loop frequently occur, but the general thinning algorithms cannot solve this problem.

From our testing, the thinning algorithms tend to generate an indistinguishable skeletonized pattern in the case of Thai handwritten characters as shown in Figure 5 (a), (b) and (c).

| (a) Character 'ก' | (b) Character 'ถ' | (c) Character 'ภ' |

Fig. 5. Examples of indistinguishable skeletonized patterns of Thai characters resulted from thinning process

In the case of a filled loop or an incomplete loop writing, the use of the well-known thinning algorithms [10] and [11] will make some important features of Thai characters disappear, such as an inner or an outer loop. The patterns of 'ถ' and 'ภ' are the same as 'ก'. Our proposed approach, therefore, does not rely on the thinning process and used original binary patterns of a character. We considered a loop as the prominent feature of Thai characters. See details in Kiwprasopsak [2].

3.1 Feature Extraction Algorithms

After the completion of the preliminary processing, the arrays of isolated characters are obtained, and then the set of feature extraction processes is performed. The logic of the feature extraction process is shown in Figure 6.

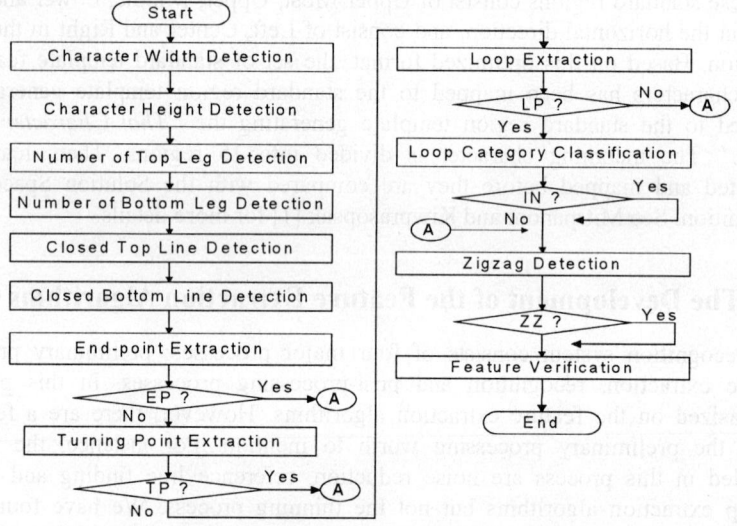

Fig. 6. The logic of the Feature Extraction processes

Our developed feature extraction processes consist of 12 algorithms. However, we will describe only three major algorithms, which are the End-point, Turning Point and Loop Extraction algorithms, in this paper. The rest are described in Mitrpanont and Kiwprasopsak [1]

3.1.1 The End-Point, Turning Point and Loop Extraction Algorithms

The End-point (EP), Turning Point (TP) and Loop (LP) extraction algorithms consists of two phases, which are Preliminary and Verification phases. The result from the Preliminary phase should be one of the following features: an EP, a LP (a complete loop), or a TP, as shown in Figure 7. However, this result is just a rough classification and is not certain enough to be able to recognize the feature because it can also represent other features. For instance, in some cases with noises, an end-point may be a filled loop while a complete loop may be a turning point, and a turning point may represent an incomplete loop. The Verification phase is later used for fine extraction of the feature.

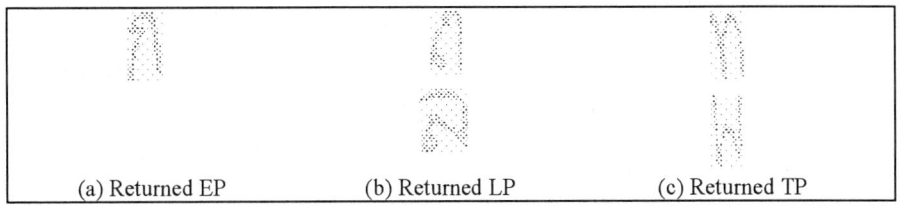

(a) Returned EP   (b) Returned LP   (c) Returned TP

Fig. 7. Examples of returned result from the Preliminary phase (at the lower left region)

The feature returned from the Preliminary phase and its candidate is summarized as follows:

- An end-point (EP) can either be the EP or the LP (filled loop).
- A loop (LP) can either be the LP or the TP.
- A turning point (TP) can either be the TP or the LP (incomplete loop) or the EP.

To extract an EP, a TP, a LP, obviously, we need to determine various patterns of each character. Therefore, we will introduce these patterns in the beginning and then these patterns will be frequently used in the rest of the algorithms.

- *General Pattern*

  Let P denotes pixel pattern: $P = \{p_{i,j} \mid i = 1 \rightarrow CharHeight,\ j = 1 \rightarrow CharWidth\}$

  Given $p_{i,j}$, the neighbors of $p_{i,j}$ are as follows:

  $X = \{x_{m,n} \mid m = i-1 \rightarrow i+1,\ n = j-1 \rightarrow j+1\}$ where: $m \neq i$ and $n \neq j$

X_3	X_2	X_1
X_4	$p_{i,j}$	X_0
X_5	X_6	X_7

  Fig. 8. The general 3 x 3 window for feature extraction

- *End-point (EP) Pattern*

  Let $p_{i,j}^{EP}$ denotes an end-point pattern, when given $p_{i,j}$:

  $p_{i,j}^{EP}$ consists of two subsets: $p_{i,j}^{TEP}$ and $p_{i,j}^{BEP}$. Where $p_{i,j}^{TEP}$ is a top end-point, and $p_{i,j}^{BEP}$ is a bottom end-point. Their patterns are:

  $p_{i,j}^{TEP} = \{p_{i,j} \mid P_{i-1,j=1,3} = 0\}\ p_{i,j}^{BEP} = \{p_{i,j} \mid P_{i+1,j=1,3} = 0\}$

- *Left-Most (LM) and Right-Most (RM) Patterns*

  Let $p_{i,j}^{LM}$ denotes left-most pixel pattern, when given $p_{i,j}$:
  $$p_{i,j}^{LM} = \{p_{i,j} \mid p_{i,j-1} = 0, p_{i,j+1} = 1\}$$
  Let $p_{i,j}^{RM}$ denotes right-most pixel pattern, when given $p_{i,j}$:
  $$p_{i,j}^{RM} = \{p_{i,j} \mid p_{i,j-1} = 1, p_{i,j+1} = 0\}$$

- *Connection Patterns (CP):*

  Let $p_{i,j}^{CP}$ denotes a connection pattern, when given $p_{i,j}$:

  $p_{i,j}^{CP}$ consists of two subsets: $p_{i,j}^{TCP}$ and $p_{i,j}^{BCP}$, where $p_{i,j}^{TCP}$ is a top-to-bottom connection pattern, and $p_{i,j}^{BCP}$ is a bottom-to-top connection pattern, respectively. Their patterns are as follows:
  $$p_{i,j}^{TCP} = \{p_{i,j} \mid p_{i+1,j-1} = 1 \text{ or } p_{i+1,j} = 1 \text{ or } p_{i+1,j+1} = 1\}$$
  $$p_{i,j}^{BCP} = \{p_{i,j} \mid p_{i-1,j-1} = 1 \text{ or } p_{i-1,j} = 1 \text{ or } p_{i-1,j+1} = 1\}$$

- *Junction Patterns (JP):*

  Please refer to Mitrpanont and Kiwprasopsak [1] for more details of Junction Patterns.

- *Joining Patterns (JoP)*

  Let $p_{i,j}^{JoP}$ denotes a joining pattern, when given $p_{i,j}$:

  $p_{i,j}^{JoP}$ consists of two subsets: $p_{i,j}^{TJoP}$ and $p_{i,j}^{BJoP}$, where $p_{i,j}^{TJoP}$ is a top-to-bottom connection pattern, and $p_{i,j}^{BJoP}$ is a bottom-to-top connection pattern, respectively. These patterns are as follows:
  $$p_{i,j}^{TJoP} = \{p_{i,j} \mid p_{i+1,j-1} = 3 \text{ or } p_{i+1,j} = 3 \text{ or } p_{i+1,j+1} = 3\}$$
  $$p_{i,j}^{BJoP} = \{p_{i,j} \mid p_{i-1,j-1} = 3 \text{ or } p_{i-1,j} = 3 \text{ or } p_{i-1,j+1} = 3\}$$

- **Algorithm 1: An End-Point Extraction Algorithm**

At the beginning, the Preliminary phase determines whether there is a Junction Pattern (JP) in the intended region or not. If there is no JP at all, the rough result will be an EP. Then, the Verification phase will determine whether it is really an EP or a filled LP by comparing line thickness with the average line thickness. The algorithm is provided below:

```
Preliminary Phase:
NextPosition ← Func. Search_FirstBlackPosition
If JunctionPositionDetermination (NextPosition)=False, then
     PreliminaryResult ← EP
End If

Verification Phase:
Remark: Determine a filled loop by checking LineThickness with
AverageLineThickness
```

```
If contiguous outstanding gDensityGraph_i > CharHeight *
LOOP_THRESHOLD, then
     VerificationResult ← LP (Filled Loop)
Else
     VerificationResult ← EP
End If

Function JunctionPositionDetermination
ReturnValue ← False
While (PN <> EP) AND (PN <> JP)
Begin
     P_{i,j} ← NextPosition
     LM_Position ← Func. Search_LM (P_{i,j})
     P_{i,j} ← LM_Position
     While (P_{i,j} <> RM)
     Begin
          P_{i,j} ← PROCESSED_PIXEL
          If PN = JP, then
               gNextLeftPosition ← FirstPosition of 01 in
                    the Next Line
               gNextRightPosition ← FirstPosition of 10
                    in the Next Line
               ReturnValue ← True
          Else If PN = EP, then
               ReturnValue ← False
          Else If PN = CP, then
               NextPosition ← CP
          End If
          j ← j+1
     End While
     RM_Position ← RM
     gDensityGraph_i ← RM_Position-LM_Position
End While
End Function
```

Remark: *Func. Search_BlackPosition and Search_LM are described in Kiwprasopsak [2].*

- **Algorithm 2: A Turning Point Extraction Algorithm**

A Turning Point Extraction algorithm is used to extract a TP in a given region. Firstly, the Preliminary phase determines a TP by searching for a Junction Pattern (JP) and non Joining Position (JoP). Then, the Verification phase checks its size to determine whether it is really a TP or a LP (incomplete).

```
Preliminary Phase:
NextPosition ← Func. Search_FirstBlackPosition
If JunctionPositionDetermination (NextPosition)=True, then
     If JoiningPositionDetermination()=False, then
               PreliminaryResult ← TP
     End If
End If

Verification Phase:
Remark: Verify TP size (is it really a TP or an incomplete LP)
Left2RightRatio=LeftSideLength / RightSideLength
Right2LeftRatio=RightSideLength / LeftSideLength
If (Left2RightRatio > 2) Or (Right2LeftRatio > 2), then
     VerificationResult ← LP (Incomplete Loop)
Else
     VerificationResult ← TP
End If
```

```
Function JoiningPositionDetermination
    Call Proc. Process_LeftSideFromJP
    If Process_RightSideFromJP (NextPosition)=True, then
            RetValue ← True
    Else
            RetValue ← False
    End If
    Return RetValue
End Function

Remark: Procedure Process_LeftSideFromJP and Process_RightSideFromJP
are described in Kiwprasopsak [2].
```

- **Algorithm 3: A Loop Extraction Algorithm**

A Loop Extraction algorithm is used to extract a LP in a given region. It also consists of two phases: Preliminary and Verification. Firstly, the Preliminary phase determines a LP by searching for a Junction Position (JP) and Joining Position (JoP). Then, the Verification phase checks its size to determine whether it is really a LP or a TP.

```
Preliminary Phase:
NextPosition ← Func. Search_FirstBlackPosition
If JunctionPositionDetermination (NextPosition)=True, then
    If JoiningPositionDetermination()=True, then
            PreliminaryResult ← LP
    End If
End If

Verification Phase:
Remark: Analyze loop size and its characteristics (it is really a
loop or a turning point)
LoopHeight ← Abs (EndLoopPosition-JunctionPosition)
LoopWidth ← RightMostOfLoop-LeftMostOfLoop
If (LoopHeight > (TopLine-BaseLine) / 2) And (LoopWidth >
(RightMostOfChar-LeftMostOfChar)) /2), then
    VerificationResult ← TP
Else
    VerificationResult ← LP (Complete Loop)
End If
```

4 Experimental Results and Conclusions

Our experiments were tested with various handwritings and with different and random Thai character set. Therefore, 50 original distinct documents have been written by 22 individuals generating 100 documents, which contain 44,616 characters. They are used to verify and validate the correctness of our proposed algorithms. The vowels and the tonal marks at the upper or lower level of characters are eliminated manually because they are beyond the scope of the research.

To evaluate the performance of the algorithms, two criteria are used, i.e., feature extraction rate and character recognition rate. *Feature Extraction Rate* is used to evaluate the correctness of the proposed feature extraction algorithms while *Character Recognition Rate* is used to evaluate the correctness of the character recognition system.

The feature extraction and recognition rates from the experiments are summarized as shown in Figure 9. The result is in between 83.47% to 98.66% with the average of

93.08%. The results of the character recognition are up to 99.19% with the average of 91.42%. For more details and performance analysis in each feature, please see Kiwprasopsak [2].

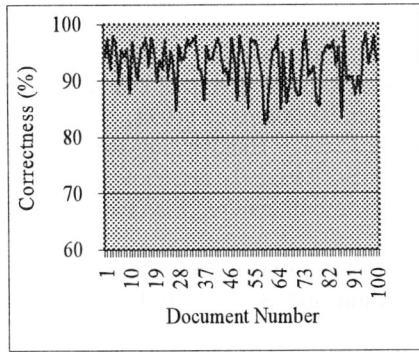
(a) Average feature extraction rate

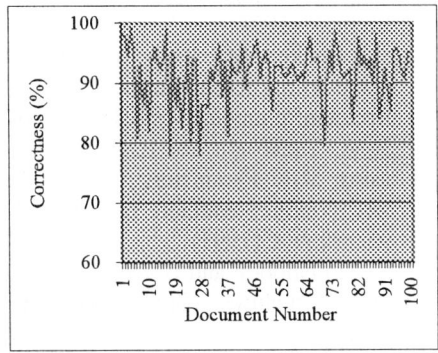
(b) Average character recognition rate

Fig. 9. Average feature extraction and character recognition rates

From the experiments, it demonstrates that our feature extraction algorithms provide a very effective recognition rate to the off-line Thai handwritten characters recognition. In addition, an incomplete feature such as an incomplete loop and a filled loop can also be extracted. This achievement came from many reasons as follows:

- Firstly, standard features of Thai handwritten characters are well defined as *"Thai Character Feature Space,"* which was effectively used for distinguishing one character from another.
- Secondly, the *"Thai Character Solution Space"* was well designed to be used as fundamental tool for recognition of all Thai characters.
- Finally, our algorithms are well established and concrete enough to tackle not only a well writing but also some incomplete writing of Thai characters.

In summary, our algorithms provide significant improvements in Thai handwritten characters and Thai OCR applications and will be used as core algorithms for extracting features for more advanced researches in the future. However, the accuracy of leg detection greatly affects the accuracy of other feature extraction, since our approach relies on number of legs detection. We found that the major error was contributed from the writing behavior with different leg styles.

References

1. J. L. Mitrpanont and S. Kiwprasopsak: The Recognition of Thai Handwritten Characters Using Feature-based Approach. MU-TR-44-005, Technical Report, Computer Science, Mahidol University, Bangkok, Thailand (2001)
2. S. Kiwprasopsak: The Recognition of Thai Handwritten Characters Using Feature-based Approach. Master Thesis, Department of Computer Science, Faculty of Science, Mahidol University, Bangkok, Thailand (2001)

3. R. M. Bozinovic and S. N. Srihari: Off-line Cursive Script Word Recognition. IEEE Trans. on Pattern Analysis and Machine Intelligence, Vol. 11. (1989)
4. A. W. Senior: Off-line Cursive Handwriting Recognition using Recurrent Neural Networks. Ph.D. Thesis, Department of Engineering, University of Cambridge, England (1994)
5. H. Bunke, M. Roth and E. G. Schukat-Talamazzini: Off-line Cursive Handwriting Recognition using Hidden Markov Models. [Online] available from ftp://iamftp.unibe.ch/pub/TechReports/1994/iam-94-008.ps.gz (1994)
6. B. A. Yanikoglu and P. A. Sandon: Off-line Cursive Handwriting Recognition using Style Parameters. Technical report PCS-TR93-192, Dartmount College, NH, [Online] available from http://www.cs.dartmouth.edu/reports /abstracts/TR-93-192.ps.Z (1993)
7. S. Airphaiboon and S. Kondo: Recognition of Handprinted Thai Character using Loop Structures. IEICE Trans. on Information and System, Vol. E79-D (9). (1996) 1296-1340
8. C. Kimpan and S. Walairacht: Thai Characters Recognition. Proceedings of the Symposium on Natural Language Processings in Thailand, (1993) 196-260
9. C. Tanprasert and T. Koanantakool: Thai OCR: A Neural Network Application. IEEE TENCON – Digital Signal Processing Applications. (1996) 90-95
10. T. Y. Zhang and C. Y. Suen: A Fast-Parallel Thinning Algorithm for Thinning Digital Patterns. Communications of the ACM, Vol. 27. No. 3 (1994) 236-239
11. C. Arcelli: A Condition for Digital Points Removal. Signal Processing, Vol. 1. (1979) 283-285

Route Planning Wizard: Basic Concept and Its Implementation

Teruaki Ito

Department of Mechanical Engineering, University of Tokushima
2-1 Minami-Josanjima, Tokushima 770-8506, Japan
ito@me.tokushima-u.ac.jp

Abstract. Route planning is one of the design problems and studies in various application areas, such as building/factory layout design, robotics, automobile navigation, VLSI design, etc. Route planning is to design an appropriate route from various candidates in terms of various perspectives, which is a time-consuming and difficult task even to a skilled designer. The author has proposed an approach of genetic algorithm (GA) to pipe route planning, and has reported the basic idea and its prototype system. Although the prototype system can generate a candidate route after the convergence of route planning process, its performance was found to heavily rely on the parameters and constraint conditions. For better performance, the previous paper proposed heuristics which was developed to narrow the search space and to improve the performance of GA engine as a preprocessor. Considering several issues we had in the past research, the paper proposes our new approach for chromosome generation, which partitions the design space, put random nodes to each partition, pick up nodes for connection, generates connection routes, set up network using these node, design routes from the network. Since we redesigned definition of chromosome from flexible length to fix length, GA operations became simpler and easier, and calculation time for design was drastically reduced. We have also modified and extended several functions in GUI modules, and implemented a prototype system called *Route Planning Wizard*. This paper describes basic ideas and implementation for our route planning method, then presents some experimental results using road roadmap data and maze problem to show the validity of our approach.

1 Introduction

The author has proposed an approach of genetic algorithm (GA) to pipe route planning [12] to be used as a part of interactive system [5]. The goal of the approach is to create a computer-based design system with a powerful problem-solving engine for interactive use [4]. The system is designed to enable a designer to make consideration of requirements in a collaborative and interactive manner, and to evaluate the impact of design alternatives for route planning in terms of various perspectives[1][11].

The author has reported the basic idea and its prototype system [6][7]. Although the prototype system can generate a candidate route after the convergence of route planning process, its performance was found to heavily rely on the parameters and constraint conditions. As for designers who use design support systems as a tool to work on design tasks, thoughts of designers are effectively activated by nice system interactions [14]. The interactions may stimulate the brain of designers to produce some innovative ideas [10][13]. Therefore, improvement of performance in the prototype system was the key-point in the next step of our research. For better performance, the previous paper proposed heuristics which was developed to narrow the search space and to improve the performance of GA engine as a preprocessor [8][9].

Considering our research results in the past, we have modified the fundamental part of our approach, which is the definition of chromosome. Internal procedures and GUI modules were also modified to meet the requirements The paper first reviews the basic idea of GA-based route planning method, then clarifies the problems we faced, describes our solution to the problems, and presents a prototype system, which is called Route Planning Wizard. The paper also shows some of the experimental results for route planning using the prototype system, and concluding remarks follow.

2 Route Planning Method Using Genetic Algorithm

As an optimal search method for multiple peak functions, GA stemming from the generation of the evolution of living things is applied to various optimization problems and its validity has been verified so far [2][3][15]. This section presents a brief summary of GA-based route planning approach.

2.1 Definition of Chromosome

In order to use GA in route planning applications as one of the optimization problems, we have represented a route from a starting point to a destination point using a character string, which is regarded as a design parameter.

In our approach, a working space for pipe route planning was represented by a cell-combination model, and the space is divided into the cells of MxN. A route is represented using a combination of cells connecting a starting cell and a destination cell. To represent direction of a route path, a set of unit vector {r, u, l, d, o} is defined, each vector represents right, up, left, down and stop, respectively, and character string of {1, 2, 3, 4, 0} corresponds to each vector. Using information on the cells which compose the route, each individual is coded. For example, the gene type for a route is expressed using symbols including {1, 2, 3, 4, 0}, where zero means the current point already reached the destination cell.

To generate a route based on this definition, we applied our original two ideas, namely eye mechanism and zone concept.

2.2 Fitness Function

In pipe route planning, as an example of route planning applications, high priority is given to the shorter route path. In addition to this, a route must go along the wall and obstacles as closely as possible, avoiding a diagonal path, the most appropriate route is designed. Using the concept of spatial potential, the degree of access to the wall or the obstacles is quantitatively calculated, and used as a part of objective function for the generation of a pipe route using GA [5].

The scheme (1) shows the basic idea for fitness function applied in our approach. Since the evaluation criteria for any designed route is closely related to its application area, the function is modified for each case in a trial-and-error manner.

2.3 Eye Mechanism

When a route is generated, obstacles and passed cell should be avoided. To do so, eye mechanism was applied. When a cell proceeds to any adjacent cell, eye mechanism checks if it is an obstacle or already passed cell, and it decides which cell should be connected. Just like eyes watch carefully before you go forward, eye mechanism works to find out an appropriate cell to go.

2.4 Zone Concept

We define the concept of "zone" to give chromosome the tendencies in the direction of a route path from a current position towards the goal point. To determine a zone, coordinates of current cell and goal cell are used, and priority vector is set to each zone. If the priority is set too high, however, all chromosomes have the same tendency in direction and a route path cannot be appropriately generated. The priority is set in a trial-and-error manner to generate chromosomes having variety of route paths. In this case, careful considerations are also taken so that variety of route path can be generated, otherwise all the route paths would only go straight in the right direction.

2.5 Dynamic Crossover Method

A route must connect the two points, namely, the starting and goal points. The length of genotype is variable and not fixed, which means that a route path is like a rubber band which is fixed with a pin in each of the terminals, and it smoothly expands.

To avoid generating genes including obstacles, potential value of each cell is checked to see if the cell in on the obstacle. If it is, a vector is repeatedly generated until it does not on the obstacle. In this way, we could obtain those genes which does not contain obstacle cells in the earlier generations. As a result, a wide range of search area is considered in GA.

Although two-points crossover generated appropriate route paths, some of them seem to be a locally optimized route paths. To avoid that, we applied dynamic selection ratio based on the minimum fitness value, average fitness value and the number of cells on obstacles. The first selection ratio of 40% is used until all of the individuals become obstacle free. Then the ratio is set down to 3% and to study all the possible routes. When the convergence status becomes a certain level, the ratio is set back up to 40%. If the difference between the average fitness value and the minimum fitness value is below 5, we assumed that the convergence is going to terminate. In this way, we excluded those individuals including obstacles in the earlier generations, we tried to take time to find the most appropriate route path without converging to a locally optimized route. When individuals are likely to converge to an appropriate route path, convergence speed is accelerated. In addition to the distinction using the fitness value, a certain number is subtracted from the fitness value to distinguish each individual more effectively.

2.6 Modification on Intermediate Points Setting

Most of the routes in the initial individuals have the tendency to go straight to the goal point. To generate initial individuals more randomly, we defined intermediate points to be passed in the route and applied in our approach.

Design area is divided into 4 sub-areas, or upper-right, upper-left, lower-left, and lower-right, each of which contains randomly distributed several intermediate points. Initial individuals are generated by way of an intermediate point. Since the arbitrary cell is randomly selected, the route paths cover the overall working space.

3 Reconsideration and Redefinition

Based on the method described in the previous section, we have implemented a prototype system, and verified our approach in pipe route planning application. As we have reported in the previous papers, feasibility of our approach has been reported but several points are still remained for future studies.

- *Fitness Function*: We have defined a fitness function considering piping route planning problems. Even if we limit the application area to pipe route planning, to define an appropriate function is not an easy task. Since we are planning to extend our idea to more general areas, further consideration is required.
- *Performance*: Although route planning is available, system performance was not good enough to use it for interactive applications. We need to improve the performance not only for interactive use but also for 3D applications.
- *Crossover problems*: Inheritance problem is observed in route generation. Inheritance problem means that child does not inherit properties of parents.
- *Representation of route*: Since the length of chromosome is not fixed, it is very flexible to represent any route or conduct any crossover. However, internal procedure is very complicated, which may be one of the reasons for performance problems.
- *Internal Procedure*: Definition of chromosome using flexible length makes it complicated to make internal procedure for GA operations.
- *User Interface*: User interface provides basic function but it was not enough to deal with more practical design. Further development is required.

3.1 Redefinition of Chromosome

The original definition of route path with several heuristics works fine as reported in the previous papers. However, several problems remained unsolved, such as low performance in GA procedures, local optimization for route design, inappropriate design for route alternatives, etc. Although several heuristics were proposed and proved to be effective in some cases, we have come to the decision that some drastic counter measures should be taken to really solve the problems. As a result, chromosome for routes was redesigned.

Figure 1 shows the revised definition of chromosome. A chromosome is composed of a fixed number of cells, which correspond to the connecting or terminal node in a graph structure. This example shows that chromosome_A is composed of red-colored 6 nodes, namely {S, 0, 1, 4, 7, G}, where S is connected to 0, then followed by 1, 4, 7, and finally G. The series of these nodes shows one route starting from S towards G.

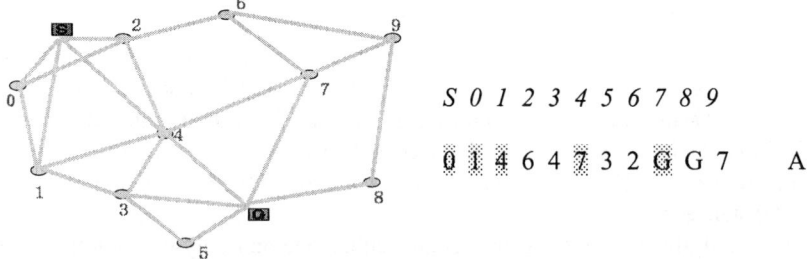

Fig. 1. Definition of chromosome

To represent a route based on this definition, both a start point and a goal point are assigned. Then the design space is partitioned using a specific number for both horizontally and vertically. A specified number of nodes are randomly generated in each partition. Connectivity of each node to adjacent nodes is examined using some parameters, such as node distance, and connection network is established. Starting from node S, the adjacent node is randomly selected from adjacent candidate nodes. Then, this adjacent node selection procedure is carried out until the connection reaches to the final node G. If any cell is open in the chromosome table, any connecting node is randomly assigned to these cells.

3.2 GA Operations

An example of crossover operation is shown in Figure 2. Two individuals A1 and A2 are the parents. A1 and A2 represent route S-0-1-4-7-G and route S-2-6-7-4-1-3-5-G, respectively. First, either of the starting node number is picked up and assigned to that of child B1. In this case, 0 is assigned from S of A1. Then its connecting node 0 is checked, and its value is picked up either from A1 or A2. In this case, both values are the same, or 1. In this way, connecting nodes are traversed until it reaches to G. Finally, B1 representing route S-0-1-3-4-7-G is generated.

GA operations for the redefined chromosome are also modified but its basic idea is the same as that of original definition. Since the length of chromosome is fixed, internal operation becomes simpler. As a result, performance was significantly improved.

S 0 1 2 3 4 5 6 7 8 9

0 1 4 6 4 7 3 2 G G 7   (A1)

2 1 3 6 5 1 G 7 4 9 8   (A2)

0 1 3 6 4 7 G 2 G G 8   (B1)

Fig. 2. Crossover operation between parents A1 and A2 to yields B1

3.3 GA-Based Approach vs. A* Algorithm

To evaluate the performance of GA-based approach, we compared the route design results with those of A* algorithm. As for the definitions of parameters in GA-based approach, Dx/Dy are vertical/horizontal partition sizes, N is the number of nodes in each partition, D is the connection distance between nodes. A* algorithm applied fitness function $f(x) = g(x)+h(x)$, where $g(x)$ is the cost to the point x and $h(x)$ is the estimated distance to the goal.

A* algorithm always finds the best route, either one-and-only-one solution problem, or more than one solution problem.

Figure 3 shows the results for a maze problem with one and only solution. Both approaches find the solution, but GA-based approach required less time.

Figure 4 shows the result for a maze problem with more than one solution. A* always finds the best solution, or the route with minimum length, whereas GA-based approach does not. However, GA-based approach finds a candidate route within less time which is required for A* algorithm.

4 Route Planning Wizard

4.1 System Overview

The idea was implemented as a prototype system called *Route Planning Wizard*. Figure 5 shows a snapshot of GUI window during design operation. With a series of simple GUI-based operations, candidate routes generation is carried out. This figure shows a pipe route planning for a factory layout.

After loading of a map for factory layout, design space is configured, such as setting up of cell size, modeling and modification of design space, etc. Upper left section window in Figure 5 shows the modeled design space. Setting up for starting/goal points is also available in this section window. The next step is route generation using GA engine. Considering the starting/goal points, the engine starts candidate routes, some of which are shown in the lower section of the window in Figure 5. The appropriate route which is selected from the candidates are displayed in the upper right section of the window. The operation is very simple and easy.

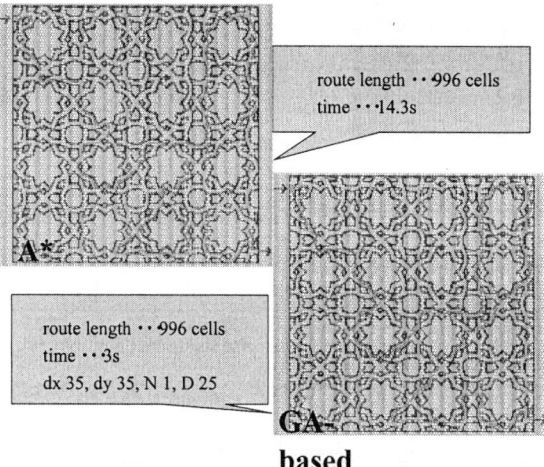

Fig. 3. Route planning for a maze problem with one and only one solution

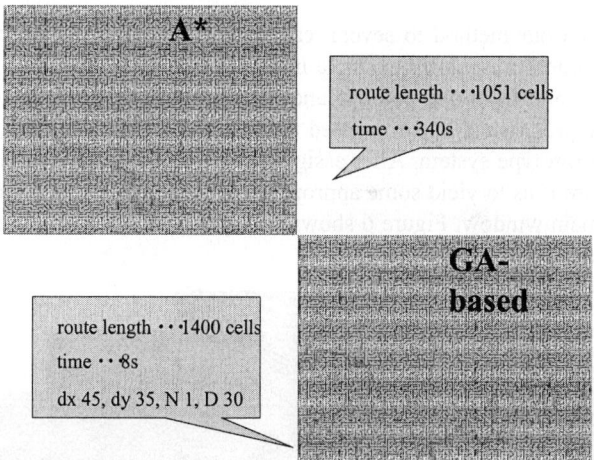

Fig. 4. Route planning for a maze problem with more than one solution

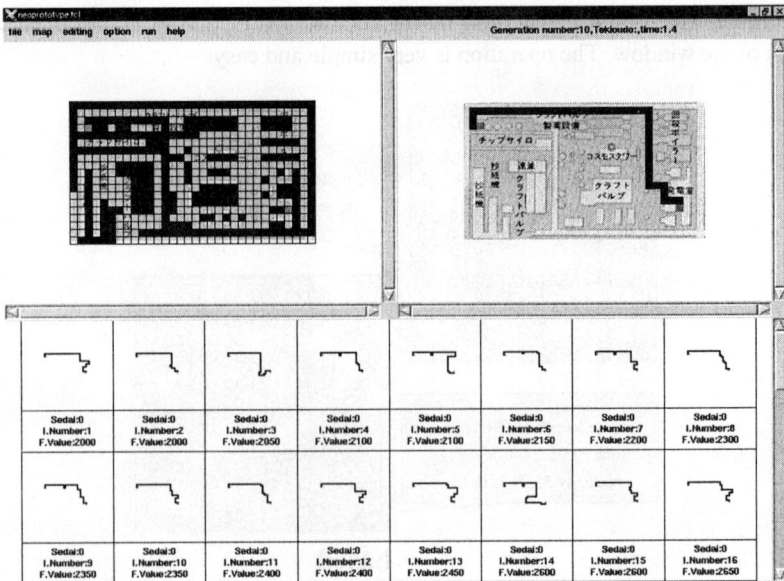

Fig. 5. A snapshot of design operation

4.2 Route Generation Example

We have applied our method to several case studies to show the validity of our approach. This section presents an example in application to a city road map to find out an appropriate route from one point to another. The original map data was modified by some image processing, and converted to our specific data structure, which can be loaded to our prototype system. After assignment of start/goal points, GA-based route design procedure runs to yield some appropriate routes, and the final routing design is shown in the main window. Figure 6 shows a snapshot of some stages in route design for the road map data.

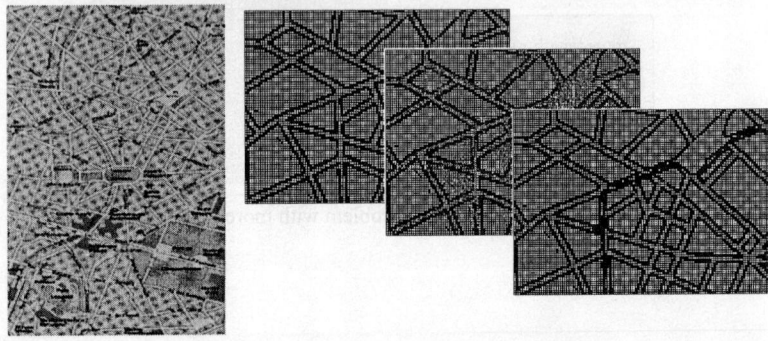

Fig. 6. Route design example to a road map data

5 Concluding Remarks

This paper presented the basic idea of GA-based route planning method, including definition of chromosome, fitness function, eye mechanism, zone concept, dynamic crossover method, and intermediate point setting. Considering several issues we had in the past research, the paper proposes our new approach for chromosome generation, which partitions the design space, put random nodes to each partition, pick up nodes for connection, generates connection routes, set up network using these node, design routes from the network. Since we redesigned definition of chromosome from flexible length to fix length, GA operations became simpler and easier, and calculation time for design was drastically reduced. To evaluate the performance of GA-based approach, we compared the route design results with those of A* algorithm, and obtained the better performance of GA-based approach.

We have also modified and extended several functions in GUI modules, and implemented a prototype system called *Route Planning Wizard*. The paper presented some examples of route design to show the validity of the approach. As some of the results are shown in this paper, appropriate routes can be designed in an interactive manner. However, results sometimes depend upon parameters, such as partition size, number of nodes in each partition, node distance to be connected, etc. Further consideration should be taken to these problems to improve the performance and quality of route design.

References

1. Finger, S., Fox, M. S., Printz, F. B. and Rinderle, J. R., „Concurrent Design", Applied Artificial Intelligence, 6, pp.257-283, 1992.
2. Goldberg, D. E., Genetic Algorithm in Search, Optimization and Machine Learning, Addison-Wesley Publishing Co., 1989.
3. Goldberg, D. E., „Genetic and Evolutionary Algorithms Come of Age", Communications of the ACM, 37(3), pp.113-119, 1994.
4. Hancock, P. A. and Chignell, M. H. (eds.), Intelligent Interfaces: Theory, Research and Design, Elsevier Science Publishers B. V., 1989.
5. Ito, T., „Towards a conceptual design support system for pipe route planning", European Simulation Symposium, Passau, Germany, pp.473-477, 1997.
6. Ito, T., „Piping Layout Wizard: Basic Concepts and its Potential for pipe route planning", IEA/AIE-1998, Castellon, Spain, Vol.1, pp.438-447, 1998.
7. Ito, T., „A genetic algorithm approach to piping route path planning," Journal of Intelligent Manufacturing, Vol.10, No.1, pp.103-114, 1999.
8. Ito, T., „Heuristics for route generation in pipe route planning", European Simulation Symposium, Hamburg, Germany, pp.178-182, 2000.
9. Ito, T., „Implementation and and evaluation for GA-based Pipe Route Planning Method", European Simulation Symposium, Marseille, France, pp.462-466, 2001.

10. Newman, W. and Lamming, M., Interactive System Design, Addison-Wesley, 1995.
11. Prasad, B., Concurrent Engineering Fundamentals, Vol.1, Prentice Hall, 1996.
12. Takakuwa, T., Analysis and Design for Drainpipe Networks, Morikita Shoten, 1978. (in Japanese)
13. van Dam, A., „Post-WINP user interfaces", Communications of the ACM, 40(2), pp.63-67, 1997.
14. Wegner, P., „Why interaction is more powerful than algorithms", Communications of the ACM, 40(5), pp.80-91, 1997.
15. Yamamura, M. and Kobayashi, S., „Toward Application Methodology of Genetic Algorithms", Journal of Japanese Society for Artificial Intelligence, 9(4), pp.506-511, 1994 (in Japanese).

The Design and Implementation of Color Matching System Based on Back Propagation

HaiYi Zhang[1], JianDong Bi[2], and Barbro Back[1]

[1]Turku Centre for Computer Science TUCS
Department of Information Systems, Abo Akademi University
Lemminkäinengatan 14 B, Fin-20520 Turku, Finland
{Zhang.haiyi,B.Barbro}@abo.fi
[2]Department of computer Science, Harbin Institute of Technology
HeilongJiang, 150001, China
bijiana@public.hr.hl.cn

Abstract. First, we implemented the iterative self-organizing data analysis techniques algorithm (ISODATA) in Color Matching Method (CMM). Then, the BP algorithm and the neural network structure in CMM are presented. We used four methods in the CMM to enhance network efficiency. Finally, we made quantitative analysis for the network learning procedure.

1 Introduction

As the computer technology and office automation is developing, monochromatic printing is not satisfying. Color printing with images and texts becomes popular. How to make the images among scanners, cameras, screens and color printers consistent is a common research task of many color technological corporations and science research agencies [1]. However, to our knowledge no company has achieved this aim yet.

The current color matching methods are based on simple chromatics theory or color space conversion relation [13,4]. More and less there exist faults, such as readjustment of parameters depending on the experience, inferior printing quality, slower printing speed, etc.

We propose a new method, the color matching method based on back propagation (BP_CMM). Our method clusters the abundant examples using the iterative self-organizing data analysis techniques algorithm (ISODATA) ahead of Artificial Neural Network (ANN) training, along with color difference theory and physical vision characteristics. Finally the BP algorithm is applied to complete conversion between two different color spaces.

We also put forward some problems that are needed to study deeply. They are the evaluation of experimental results and combination of BP algorithm and chromatics theory, which makes the initial value setting more significant and improves the precision of conversion of CMM.

The color matching method based on back propagation is a new one. It accomplishes the nonlinear conversion between two different color spaces. Therefore BP_CMM presents a new method to reach the goal of color matching and color consistency between colors.

2 Iterative Self-Organizing Data Analysis Techniques Algorithm and Its Application in CMM

2.1 Iterative Self-Organizing Data Analysis Techniques Algorithm (ISODATA)

Iterative Self-Organizing Data Analysis Techniques Algorithm [2,17] is similar to K Average Algorithm [8,6]. That is, the clustering centre is decided by an iterative operation of the sample average, but ISODATA Algorithm adds some probing steps and interactive functions between man and computer. The algorithm can assimilate the experience from the intermediate results. It can divide one class into two and merge two classes into one in an iterative procedure, called the self-organizing feature [12]. So the algorithm possesses heuristic characteristics. ISODATA Algorithm is a self-organizing complex dynamic clustering algorithm. It includes setting of the initial values, fission processing, merging and iteration.

2.2 Select Samples and Their Conversion in CMM

Selecting samples heavily influences the learning system and its results. If the selections are proper, then the samples can represent all features so that they can reflect the whole change trend. If the samples selected are not proper, the learning results only can represent partly features so that the clustering results are poor. Selecting training samples is a key point for a learning system.

2.2.1 Selecting Samples in CMM

In the color matching, the total number of sample for RGB (Red/Green/Blue) space is approximately 16700000 (256x256x256). Such huge number makes it very difficulty to learn all the samples, even impossible. So, selecting proper samples must be done for learning. Based on the knowledge of chromatics [5,16], we selected 800 group data as the training samples in our test. It consists of 400 group single colour data and 400 group mixed colour data. In the processing of printing a picture, vibrating mostly influences printing quality of the picture. There exist permeating and covering phenomenon in printing for some vibrating ways. So, the mixed color training samples are also important part. Such selecting training the samples well reflects the features of the whole samples by test proving.

2.2.2 The Sample Conversion in CMM

Using the ISODATA algorithm the sample clusters based Lab space are adopted. It is very difficult to compare the similarity between two color spaces in RGB color space [13]. So, we must convert a corresponding color space. There is a standard conversion between RGB spaces to Lab spaces, which provided by CIE [1,7]. We can calculate the similarity between two spaces in Lab colour spaces by colour difference formula we can calculate colour difference ΔE. The color difference in CIELab color spaces may be calculated in some formula. The conversion from RGB colour space to Lab colour space is obtained with experience formula, across the conversion of XYZ color space.

2.3 Application of Iterative Self-Organizing Data Analysis Techniques Algorithm in CMM

2.3.1 The Analysis for Iterative Self-Organizing Data Analysis Techniques Algorithm

ISODATA algorithm is a Self-Organizing Data complex dynamic clustering algorithm. It includes several procedures such as setting up initial values, fission processing, merging process and iterative.

It can automatically determine the reasonable clustering number, but it is a dynamic clustering algorithm, It possesses the selections of setting initial values and similar measure method and the determination of clustering rule functions, which a general dynamic clustering algorithm possess in common. Here, we depicted how to solve the three problems in CMM.

1. Setting initial values

    According to the selections of training samples and the knowledge related to the chromatics theory, we set up the pre-select initial values in ISODATA algorithm:

    (1) Expected the cluster centre number K the parameter specified according to the real printing required precision. The higher the precision is required, the larger K is. K is set as 40, related to a data file, in experimental of CMM.

    (2) The smallest sample number θ_n in each cluster θ_n is set as 1 in CMM. This can guarantee that there is only one sample in a cluster, avoiding that in this case the sample is forced to put it into the cluster, which is close to it. In this way the "color change" phenomenon is avoided in CMM printing for such case.

    (3) The smallest distance between two cluster centres, that is, merging parameter θ_c, θ_c is set as 1. The reason why it is set as 1 is that in chromatics theory, the colour difference between colours is less than 1, the eyes of human being can not distinguish it. In other word, that is, two colours in printing can be become one. The corresponding clustering is same class.

    (4) The standard deviation parameter θ_s of the sample distance distributions in a cluster, Here, θ_s is set the value of θ_c, that is, $\theta_s = 1$

    (5) Allowed merging cluster pair number L in each iteration According to the class number clustering, the value L is set as 10 in the experiment.

    (6) Allowed maximum iterative number I In the dynamic cluster of ISODATA, normally the maximum iterative number I is set, to avoid infinitely iterating. But this constrain sometime may affect to the quality of clustering. So, the value I is set as 10000 in our real application in CMM.

2. The selection of similar measure method According to the knowledge related to the chromatics theory, similar measure method of ISODATA dynamic clustering algorithm is using the color difference formula as distance similar measure dynamic clustering algorithm. In chromatic theory, the color difference ΔE is calculated with a formula [5,14]

    The color difference reflects similar level of two colors. Normally when ΔE ≈ 1 the viewer cannot distinguish the color difference. Just because of this, we can use it efficiently to print in the color printing. That is, for the two colors they can view as one color in printing, and it cannot affect the view result for viewers. It also means that the color difference of color changing around 1 is optimal. According to this character the gate limited is 1 in the clustering analysis, in this way the sample in the same cluster represents same color for the viewers, so they can be clustered as the same class. At same time, the units used in RGB, XYZ and Lab color spaces are unified for each sub vector. Two problems of dynamic clustering are solved.

3. Determining cluster criterion functions In the CMM experimental the terminal condition is that the clustering centres are no change. The cluster criterion functions may be selected as sum-squared-error criterion [15].

3 The Experimental Result Analysis for the Sample Clustering

3.1 The Algorithm Testing

For the correctness of testing cluster, here we take 9 examples, which have obviously classed features to test. The 9 examples are respectively (0, 0), (1, 1), (0, 2), (2, 0), (6, 6), (7, 5), (7, 6), (7, 7), (8, 6). We can obviously obtain the classes of the samples. That is, Class 1: (0, 0), (1, 1), (0, 2), (2, 0), Class 2: (6, 6), (7, 5), (7, 6), (7, 7), (8, 6).

In the clustering process, we take initial values respectively as: $N = 2$, $\theta_n = 2$, $\theta_s = 1$, $\theta_c = 1.5$, $L = 2$, $I = 100$. The program is running 8 times in practice, the running results are same as the real situation. The prototype differentials are respectively $\delta_1 = (0.83, 0.59)^T$, $\delta_2 = (0.63, 0.45)^T$; The clustering centres are respectively (0.75, 0.75), (7.00, 6.00).

3.2 The Experimental Results in CMM

In CMM, 800 group data were clustered, 276 group data were obtained after clustering. The number of samples is reduced quickly. The reduced ratio is almost 3:1. The paper [18] gave the final clustering results.

4 The BP Algorithm and Its Application in Colour Matching

BP algorithm is widely used in neural network models. From the structure point of view, a BP net is a typical multiplayer network. It is divided into an input layer, a hidden layer and an output layer. The full connection manner is employed normally between layers. There is no connection at same layer.

4.1 The BP Network Algorithm

The BP algorithm belongs to δ learning category. It is a supervised learning algorithm [10,9]. Its input learning samples are $x^1, x^2, ..., x^p$, related to the teachers $t^1, t^2, ..., t^p$. The learning algorithm is to modify the weights and threshold with the differences between input $y^1, y^2, ..., y^p$ and $t^1, t^2, ..., t^p$ so that y^p is as close to desired t^p as possible.

The whole network learning procedure in BP network is divided into two stages: The first stage is that the calculations are executed from the bottom of the network to up. If the structure of the net and the weights are set and input knowing learning samples, we can calculate the neuron outputs in each layer with the formula. The second stage is modifying the weights and threshold, by calculating and modifying from the place which is closest input layer to bottom, from knowing the error of the closest input layer to modify the weights of output layer, then following the formula to modify the weights of each layer, the two procedures were executed in turn repeatedly, until reach the converge.

4.2 The Application of BP Algorithm in CMM

4.2.1 BP Network Structure in CMM

According to the detailed requirements for colour matching, we used BP network structure [5,6], which has three layers for implementation, that is, an input layer, a hidden layer and an output layer. For the input layer and output layer it is easy to determine the number of neuron according to the real requirements of printing. Currently, the colour printers basically have following three printing manners: CMY (Cyan/Magenta/Yellow), CMYK (Cyan/Magenta/Yellow/Black) and CMYKcm. The corresponding neuron numbers of output layer in BP networks are respectively 3, 4 and 6. The input layer is the data of RGB space, so the neuron number of input layer is 3. It is not easy for us to determine the neuron number normally in the hidden layer. If the neuron number is more then networks converged ratio is slow, but if the neuron number is less, then the networks is limited local extreme minimum. In the some literatures, some empirical formulas are given in order to determine the neuron number in the hidden layer [11,3], for example, the hidden neuron number is half of the sum for input layer neuron number and output layer neuron number. But it is not satisfied in the practice application. After referencing the empirical formulas and testing repeat in the laboratory, we believe that it is suitable for that the hidden neuron number is a little bit greater than the empirical value in CMM. For instance, in the CMYK printing manner, it is set as 6, and we can obtain the BP network structure.

The situations of the converge is shown in the Table 1 when the hidden layer neuron number are reactively 2, 6 and 10 for 20 training samples. Here the iterative number unit is 1000. From this table we can see that the hidden neuron number is greatly affected for network converge ratio.

Table 1. The affection of the hidden node number for the network converge ratio

The hidden node number	Learning sample number	The average of iterative number
2	20	583
6	20	119
10	20	307

4.2.2 The Standard (Prototype) Processing of Training Data

In the CMM the Sigmoid function is used as an encourage function in BP network, its form is shown as $f(x) = \dfrac{1}{1+e^{-x}}$. For this function we may calculate when $x = 14$, $f(x) \approx 1.000001$, when $x = 15$, $f(x) \equiv 1$. the data is shown [6,19] in color matching. From this data we can see that when $x = 15$, $f(x)$ must be one. In such case we cannot adjust the weights. So we have to standardize this data, for implementation of CMM, the testing data are respectively divided with 100 so that the maximum value is less than 15. In this way $f(x)$ can not reach 1, avoiding the weight "stopping" appearance.

4.2.3 Four Methods to Enhance BP Network Efficiency

In order to enhance the efficiency of the BP networks, four methods were used in CMM during the realizing the BP algorithm.

1. The processing for derivative of the Sigmoid function;
2. The variant step η (n_0) ;
3. Adding a dynamic item;
4. Gradually learning.

4.2.4 The Implementation for BP Algorithm

BP network algorithm is a complex iterative algorithm, which modify the weights and the threshold by iterative repeat so that the difference error between the practice output of the networks and expecting output is smallest. It consists of the forward chaining procedure and the backward chaining procedure.

5 The Quantitative Analysis of Network Learning Process

Since the neural networks have the characteristics, which the network converge ratio is slow and it is easy fail into the local minimum, it is quite important to know the states when the neural networks are training; in additional, from user point of the view, the neural network is a black box, so it is necessary for us to check the results.

5.1 Learning Progress Situation Analysis

In the learning process in CMM, the average square root RMS error reflects the learning capability. RMS error is defined as:

$$E_{RMS} = \sqrt{\frac{\sum_{p=1}^{P}\sum_{l=1}^{m}\left(t_l^p - y_l^p\right)^2}{pm}}$$

Here, p stands for the pair number in the training set, M stands for unit number of the network output layer

In the learning process of BP networks, following the conjugate –gradient algorithm, RMS error should be gradually decreased.

When the network oscillatory the initial parameter matrix of the network is:

$$\begin{matrix}
0.21 & 0.82 & -0.37 & 0.28 \\
-0.56 & -0.50 & -0.48 & 0.32 \\
0.84 & 0.89 & -0.37 & 0.95 \\
-0.38 & 0.15 & -0.03 & -0.65 \\
0.65 & 0.72 & -0.06 & -0.69 \\
0.76 & -0.77 & -0.06 & -0.69
\end{matrix}$$

and

$$\begin{matrix}
0.03 & -0.64 & 0.46 & 0.53 & 0.41 & 0.30 & -0.81 \\
-0.10 & 0.29 & -0.71 & 0.27 & 0.21 & 0.38 & -0.48 \\
0.84 & -0.84 & -0.96 & -0.85 & -0.55 & -0.89 & -0.15 \\
0.87 & 0.94 & -0.19 & 0.83 & -0.83 & 0.25 & 0.55
\end{matrix}$$

when the network converges, the initial parameter matrix of the network is

$$\begin{matrix}
0.48 & -0.59 & -0.75 & -0.34 \\
-0.29 & 0.61 & 0.86 & 0.78 \\
-0.29 & 0.23 & -0.90 & -0.46 \\
0.52 & -0.09 & 0.97 & 0.31 \\
-0.52 & -0.76 & 0.24 & 0.81
\end{matrix}$$

and

$$\begin{matrix}
0.41 & 0.25 & -0.37 & 0.72 & 0.28 & 0.73 & -0.49 \\
0.15 & 0.15 & 0.32 & 0.41 & -0.95 & -0.13 & 0.35
\end{matrix}$$

$$\begin{matrix} -0.95 & 0.99 & 0.42 & 0.39 & 0.13 & 0.90 & 0.81 \\ 0.86 & 0.95 & 0.35 & 0.38 & -0.76 & -0.03 & -0.61 \end{matrix}$$

The parameter matrixes above, 6 x 4 matrix is the parameter matrix from the input layer to the hidden layer, 4 x 7 matrix is the parameter matrix from the hidden layer to the output layer,

5.2 The Network Capability Testing

1. The correctness testing for the network training After clustering 276 samples are used as the training samples to network training, after iterative 81635 000 times, the network converges. The parameter matrix is obtained as following:

$$\begin{matrix} -2.59 & 4.08 & 0.24 & 7.53 \\ 2.01 & -6.13 & -2.71 & -19.34 \\ 2.91 & -2.30 & -1.64 & -2.31 \\ -3.25 & -11.43 & -22.86 & -64.00 \\ 1.68 & 0.96 & 1.43 & -1.00 \\ 4.55 & 18.11 & -5.49 & 18.34 \end{matrix}$$

and

$$\begin{matrix} 2.47 & 3.49 & -8.00 & 24.82 & -1.59 & -2.13 & 0.57 \\ -1.74 & 2.00 & -1.23 & -0.62 & -2.01 & -1.95 & 2.99 \\ -1.60 & -2.00 & -1.63 & -0.71 & -3.12 & -1.62 & 2.01 \\ -1.46 & -2.00 & -1.20 & -0.46 & -2.87 & -1.98 & 2.23 \end{matrix}$$

In order to check the correctness for the network training, we take these 276 samples to test, the testing results are completely correct. RMS error is 0.1, which satisfied the requirements practically.

The capability testing of the network generalization

We take the training samples, which are not used in the clustering process, as the data of training samples to test the example, totally 524 examples. Taking these 524 examples as the inputs, we can obtain corresponding practice outputs, then we obtain the corresponding RMS error is almost 1.6. Which basically satisfied the requirements.

The data file of the testing results is given [18]. It includes the data that we used to check the correctness and wideness.

6 Summary

The paper achieved the conversion from RGB space to CMY(K) space with BP algorithm, compared to other conversion, it has following advantages:

1. Based on the color matching method of BP network we realized non-linea conversion from RGB to CMY(K). RGB space is an adding color system space but CMY(K) is subtracting color system space. So there is a non-linea relation be-

tween them. The BP algorithm can deal with such non-linea relation .In this way we overcome the shortcoming with non-linea manual adjusting.
2. The CMM based on BP network possesses generality. It can be used to convert not only from RGB to CMY(K) but also from RGB to RGB and CMY(K) to RGB. At same time, it does not rely on any equipment. It has no special processing for any particular equipment. So it can be widely used.
3. The CMM based on BP network possesses the stronger learning ability. It can learn different weights for different papers, ink and vibrating mode, it can realize the automatically adjust for different cases.
4. The CMM based on BP network is brand new method in CMM. It gives us new way to convert from a color space to other space. But, as a new method, it has the following shortcomings:
5. The network learning time is long Since the neural networks have the characterises, which the network converge ratio is slow, the disadvantage of the BP learning algorithm is that the training time normally grows exponentially as number of nodes increases. Even though in the paper the enhancing network ratio methods are used, for example, the samples clustering before learning, the network converge problem is not solved essentially.
6. The knowledge related to Chromatic theory is less applied The CMM based on BP network adopted the less knowledge of chromatic theory, when we set the initial values and train the network without many considerations about the plenty chromatic theory relations, This results in setting the network parameter without the purpose. It makes the network spend a long time to run. It affects the result accuracy in some sense.

From describe above, the CMM based on BP network, for printing quality of color pictures there is obviously enhancement. It also increases the automatic learning ability of the system. But there are some shortcomings, which we mentioned above. So much more research is needed for those areas.

Acknowledgements

The authors would like to thank Mr. XingQi Wang for his large experimental work and some useful ideas for this project.

References

1. Hong Bozhe. "The color adjustment of four-color printer." Japanese picture Committee. 56(2): 112-122, 1993.
2. B. Hayes-Roth. An architecture for adaptive intelligent systems. Artificial Intelligence, 72:329-365,1995.

3. B. Hayes-Roth, K. Pfleger, P. Lalanda, P. Morignot, and M. Balabanovic. A domain-specific software architecture for adaptive intelligent systems. IEEE Transactions on Software Engineering, 21(4):288-301,1995.
4. Mark D. Fairchild. "A Novel method for the determination of color matching functions." Color Research and Application. 14(3):122-131, 1989.
5. K. I. Funahashi. "On the approximate realization of continuous mapping by neural networks." Neural Networks. (2):183-192, 1989.
6. A. K. Jain., R. C. Duber. "Algorithms for clustering data." Englewood Cliffs. NJ: Prentice-Hall. P42-59, 1988.
7. James R. Coakley and Carol E. Brown "Artificial neural networks in Accounting and Finance: modelling issues" International Journal of Intelligent Systems in Accounting, Finance & Management 9, 119-144 (2000).
8. R. P. Lippman An introduction to computing with neural nets. IEEE ASSP Magazine April, 4-22 (1987).
9. Y. T. Mcintyre-Bhatty "Neural network analysis and the characteristics of market sentiment in the financial markets", Expert Systems, September 2000, vol. 17, no. 4,p191-198.
10. M. Mulholland., D. B. Hibbert, P. R. Haddad and P. Parslov "A comparison of classification in artificial intelligence, induction versus neural networks" http://www.emsl.pnl.gov:2080/docs/incinc/alt_nns/Mmdoc.html.
11. B. D. Ripley "Neural Networks and related methods for classification." Journal of the Royal. Stat. Society. 56(3):409-437, 1994.
12. Carlos Serrano-Cinca "Self organizing neural networks for financial diagnosis" Decision Support Systems 17, p227-238, 1996.
13. Xiao Sheng, Cun Jing, Xi Duo. "High Accyrate Color Transformation based on UCR." Japan Hardcopy 94 papers collection. 177-180, 1994.
14. G. Spencer, P. Shirley, K. Zimmerman etc. "Physically-based Glare Effects for computer Generated Images." Proceedings ACM SIGGRAOH'95. p325-334 (1995).
15. S. Sugiura and Makita T. "An improved multilevel error diffusion method." Journal of Imaging Science and Technology. 39(6):495-501, 1995.
16. Sunil Vadera and Said Nechab. The MD Shell and its use to develop Dust-Expert. Expert System. August 1995,Vol.12,No.3:231-237
17. V. Vemuri "Artificial neural networks: an introduction." Artificial neural networks: Theoretical Concepts. P1-12, 1988.
18. Xingqi Wang "research and implementation of color matching method based on back propagation" Master thesis of Harbin Institute Technology, China, p20-38, 1999.
19. Hugh J. Watson and Robert I. Mann. Expert System: Past, Present, and Future. Journal of International System management. 1998, Vol. 5, No. 4:39-46

Component-Oriented Programming as an AI-Planning Problem

Debasis Mitra and Walter P. Bond

Department of Computer Sciences, Florida Institute of Technology
Melbourne, Florida, USA
{dmitra,pbond}@cs.fit.edu

Abstract. A dream of the software-engineering discipline is to develop reusable program-components and to build programs out of them. Formalization of a type of component-oriented programming (COP) problem (that does not need any non-trivial effort for gluing components together) shows a surprising similarity to the problem of *Planning* as defined within the Artificial Intelligence (AI). This short paper explores the possibility of solving the COP by using AI-planning techniques. We have looked into some closely related AI-planning algorithms and suggested directions on how to adopt them for the purpose. Other important related issues like the target specification languages and other relevant research disciplines are also being touched upon here.

Keywords: Planning; Component-oriented programming; Intelligent software-engineering

1 Introduction

The success story of modern engineering lies with the capability of designing any target object using well-understood components. Unfortunately this is still mostly a dream within the Software Engineering discipline. As a result, programming is still considered as primarily an art. A successful reuse of the software components - as a standard practice - would transform this area from an art to an engineering discipline. A preliminary approach in this direction was to develop the standard library of routines and add it to the environment of a programming language. Common Lisp and C languages have used this technique quite extensively. In the recent years C++ has extended the technique by standardizing its template library (STL) that contains a repertoire of higher-level objects. JAVA language from its inception has incorporated a huge set of API's in a very organized way. However, the concept of reusable components demands an even higher level of architecture. A recent movement toward that direction comes from the introduction of the JAVA Beans. It provides the programmer to store and reuse components in an appropriate environment. However, the maintenance of the library of components is almost left to the user programmer. Such environments (like BDK or Bean Development Kit) provide very little help for writing a program by utilizing such components. This paper proposes a framework for intelligent usage of components for the purpose of developing a program.

There are some domains that expect the programs to be developed out of the existing components only. Numerical computation, particularly in the scientific and engineering area, is a domain of this nature. Often the user (who is typically a scientist or an engineer) writes a script code developing a model for computation that he or she wants to perform. The script code is nothing but an ordering of some of the library routines, with some parameters instantiated within the latter ones. An example where such numerical computation is done is in the area of seismic data processing, typically within the petroleum exploration-industry. In that domain a script code is being written for ordering some signal processing and data handling routines from a software library. The script code is subsequently pre-processed for developing a program (which primarily calls the library routines) for processing data.

The types of domains described above deploy an extreme situation of component-oriented programming, where the "glue" codes between components in a program are mostly trivial and the pre-processor automatically generates them. However, even in this situation the user is expected to know great details about the individual routines as well as their relationships with each other. The interface modules (graphical or otherwise) or the pre-processors do not provide much help to the user in writing the script code or in checking its consistency. In a nutshell, these domains deploy "component-oriented programming" but do so quite manually. Our proposal is to apply the AI-planning techniques for the purpose of automatically (or semi-automatically) developing programs using reusable components from a library. This proposal should be distinguished from the classical works in *automatic programming*, where the "components" were of very low level – of that of the level of individual statements in a third generation programming language.

Planning is a core area within the Artificial Intelligence [9]. The central problem there is to order (partial or linear) some operators from a given set, in order to achieve some goal. World states are represented in some language. A finite number of primitives are provided as a starting state of the world and the goal state of the world. Both are described in the same language. Each operator has a set of preconditions and a set of effects or post-conditions, again both are described using the same Planning-language. Planning algorithms search through the operator set in order to develop a sequence (or a *partial order*) of some of the operators so that the world changes from the input *start-state* to the input *goal-state*. In the AI-planning problem, if we replace the operators with some program-components, then the same problem could be viewed as a program development-problem, where the program is developed by only using components. Planning is a heavily researched area and some good progress has been made in the last few decades that could be taken advantage of towards the component-oriented programming. This concept paper explores that possibility.

Section 2 defines the problem formally. In a following section we describe the Planning problem and mention a few related AI-planning techniques. In section 4 we discuss a feasible software specification language that may be modified toward an AI-planning language for our purpose. Section 5 puts forward a proposal toward solving the component-oriented programming problem. A short section 6 mentions some other research works where the developed methodologies are strongly related to the proposed framework here and so, has a strong relevance in the research. The paper is concluded in the following section.

2 The Problem Definition

A *component-library* constitutes a set of *components* $C = \{c_1, c_2, ...\}$. Each component c_i is recursively defined as either a simple element from a set S of *software pieces*, or an n-ary directed graph $\{V, E\}$, where the nodes in V are components from C, and E is a set of n-ary edges between some nodes. The semantics of an edge could be a simple component hierarchy, or could be a temporal linkage (before/after chain) between the sub-components within the component. Each component also has a specified *required-environment* r_{ci} needed for its existence in a solution (program), and creates a *target-environment* t_{ci} after it is executed. Each of these environments could be as simple as the input/output parameters-list for a component/subroutine. Each c_i is also associated with a set of *properties* $p_i = \{p_{i1}, p_{i2}, ...\}$.

The component-oriented programming problem, in our restricted sense, is COP=(A, R, T). The problem is to elaborate (or follow) a given software architecture A by creating a graph with nodes c_i's from C, when some input-environment R and target-environment T are being provided apriori, such that the whole target-environment T is satisfied. The architecture A itself may be a graph with its nodes being a higher-level components, or may be a broad-level description, using the language that is used for specifying the sets of properties p_i's of the components. Architecture A is a list of predicates that need to be satisfied in a solution by the p_i's of the components. Nodes in the architecture A may be instantiated from C in such a way that the target environment T is satisfied. Initial environment R can be used to satisfy the components' required environments r_{ci}. Also, a component c_1's output t_{c1i} can be used to satisfy another component c_2's required-environment r_{c2i}, when c_2 is located *downstream* in the solution (a partial order) compared to c_1. A solved COP may be subsequently added to the library C as a newly added component.

Example: A data processing situation using filters. S is a simple suit of filter routines, C is a library of elements of S and some other composite filters developed by solving problems. Each c_i can be applied under some assumption (r_{ci}) about the data that needs to be processed, and will produce some quality (t_{ci}) in the filtered data. Each filter c_i also has a description $p_i = \{p_{i1}, p_{i2}, ...\}$. A problem instance COP is a broad architecture (A) about the requirement of the type of filters needed in a specific order, some information on the quality of the input data (R) and the required quality of the output data (T). A solution will be appropriately ordered (linear or partial) chain of components from the library C for achieving the required data-quality T such that the "chain" (solution) follows the input broad-level architecture A for the solution.

3   Relation to the AI-Planning

A Planning problem [9] involves a library of operators C={$c_1, c_2, \ldots$}, each c_i has a set of preconditions r_{ci} and a set of post-conditions t_{ci}. A planning problem P=(R, T) with a set of *start-state* predicates R and *goal-state* predicates T, is to create a directed graph A out of the operators from C such that all elements of T are achieved. All of R, T, r_{ci}, t_{ci} come from a set of world state predicates specified in a Planning-language. As posed here, P is a sub-problem of COP described in the previous section where the architecture/descriptions A is a *null* graph (this could be loosely interpreted as *no-suggestion* on how to solve the problem). Also, descriptions p_i for each component c_i is a *null* set in the planning problem, i.e., the operators in a typical planning problem do not have any "description" but only a symbolic name.

Actually COP problem is an easier version of the planning problem because of the extra constraints in A. Existence of a non-null initial A can be considered as a plan shell to start with. Additional descriptions p_i's in c_i's help in the problem-solving process by possibly pruning the search space. In both the problems the objective is to develop a final graph A. In this sense COP problem is a sub-problem of P, where P is enhanced with additional input A and p_i's. Hence, both the problems COP and P are equivalent to each other.

Some of the planning strategies that we can preclude for COP are dynamic planners, conditional planners or reactive planners [9]. These types of planners involve a capability to change the plan at run time, which is not warranted in the COP problem. The later problem is very much of a static nature. A first pass observation suggests that we need a partial-order planning [12] because the resulting plan need not be a linear chain of components in COP. Secondly, we need to deploy some type of hierarchical planning for utilizing some component-hierarchy that may be available in the component library C. Below, we will discuss some of the existing planning frameworks that suit the COP problem.

GraphPlan [1] generates a partial-order plan where one or more operators are allocated in each time-step, total plan being a sequence of a finite number of time-steps. The algorithm works by propagating mutual-exclusion constraints between instantiated operators (and state-predicates) from one stage to the next, which prevents operators being "chained" next to each other. The GraphPlan could be adopted toward hierarchical plan generation for solving the COP problem. Extension of the GraphPlan has been made by Do and Kambhampati [3] by using constraint satisfaction (CSP) approach. Thus, the propagation of mutual-exclusion constraints could be made more efficient by using different heuristics like dependency-directed back-jumping or forward-checking etc.

Consider a somewhat more specific data-filtering scenario of the example in the previous section. Suppose, there are two sets of filter routines within the component library, one set is onboard a satellite and another set is for using on ground (in an online satellite-generated data processing scenario). A simple architecture (A) is given in a problem instance that states that the ground filters should follow the satellite filters. This constraint could be translated as mutex constraints between the operators while solving the problem. These mutex constraints between components are not static as considered in the "GrpahPlan," but are created dynamically for the problem instance (in A/p_i's). In case in the initial architecture A is reversed, i.e., the input description requires that the satellite filters should follow the ground filters, then the mutex constraints would be reversed and a different solution will be achieved accordingly.

O-Plan [11] is the Open Plan Architecture in which hierarchical planning could be done with a mixed-initiative between the system and the human. This feature is useful for component-oriented programming with programmer interaction. Also, the framework provides opportunity for incorporating "descriptions" as in 'A' and p_i's in the COP. For these reasons O-Plan is also a closely related framework to our problem, even though it is developed primarily for a dynamic environment, e.g., emergency management. A particularly interesting model within the O-Plan is the <I-N-OVA> model, where the planning is done by using constraint manipulation and where the design rationale is being captured systematically. Formalizing such "design rationale" could constitute a step forward toward handling the "descriptions" as in 'A' and p_i's in the COP.

4 Requirement/Component Specification Language

One of the major problems in this research is to develop a software architecture-specification language. The language should be able to describe the components (p_i, r_{ci}, t_{ci}) and the COP problem (A, R, T) uniformly. It also needs to be a Planning language at the same time. Hence, it must be a hybrid between two such languages.

Currently the most popular object-oriented software architecture-specification language is the *unified modeling language* or UML [8]. It is a standardized visual language with semantics attached to the icons. While the visual interface could be very useful for specifying component-based architecture of a target program, the propositional nature of UML makes it somewhat inflexible (cannot handle first-order formulas) and thus, not so suitable as a target for adopting AI-planning languages. Most of the current-generation Planning languages are based on the first-order predicate logic that allows a better expressiveness.

A good candidate for the functional specification of component-oriented architecture of any software is the Z-notation (mainly from the Programming Research Group at the University of Oxford, see in Shaw and Garlan, [10]). Z-notation has a first-order type syntax and semantics. All the three aspects of the COP problem, A/p_i, R/r_{ci}, and T/t_{ci}, could be described in such a modified Z-language.

5 A Scheme for Doing Component-Oriented Programming

In Mitra [5] we have proposed a relational data model-based algorithm for helping the programmer to choose appropriate components (stored in a relational database) for an input architecture of the target software. The algorithm (CPRAO) does constraint propagation (very similar to the one required in the map-coloring problem) utilizing relational algebraic operators like *project* and *join*, relevant to the relational data model. At each stage of the iteration it lets user instantiate an element in the architecture (or a node in the graph) out of a set of valid components suggested by the algorithm, and then propagates the constraint (by utilizing relational operators) in order to filter the set of valid components (domains for the nodes) on the adjacent nodes. For a chosen target problem domain of component-oriented programming, we have used matching of the set of input/output parameter lists of adjacent components as the basis of constraint propagation. In the COP problem as stated in this paper, these i/o parameters could be considered as the set of required-environments r_{ci} and the set of target-environments t_{ci} for each component. The CPRAO algorithm of Mitra [5] could be adopted for the purpose of constraint propagation within a planning algorithm. A practical advantage of the CPRAO algorithm is that it allows us to use a back-end relational database for storing the component library.

The satisfaction of the initial architecture/descriptions A in the COP problem has to be achieved by the set of descriptions p_i's of the instantiated components. This can be done by deploying a technique like the unification/resolution as in the Logic Programming. Hence our scheme is to combine two different schemes of search: (1) Planning, for the purpose of chaining the required and target environments of the components by choosing appropriate components, and (2) Logic Programming, for the purpose of satisfying the original description in a first-order like requirement specification language (used for specifying A and p_i's). The two search techniques will complement each other and thus, enhance the aggregate efficiency of the algorithm.

Example of such combination already exists in the TAL-planner or the Temporal-action logic Planner from the Linkopings University-group in Sweden (Kvarnstrom and Doherty, [4]), which apparently is one of the fastest running Planner at this moment. In this planning language each of the planning operators is optionally extended with two parameters t1 and t2, indicating respectively the start and the end times of the intervals for applications of the operators. The planning algorithm uses these two special variables for unification/resolution directing the search for satisfying the temporal constraints. Our scheme is to use the same technique as that used in the TAL for the purpose of satisfying the architectural description A with the properties p_i's of the components/operators, instead of satisfying only the temporal constraints as in TAL.

The scheme is to run the GraphPlan algorithm [1] for the COP-problem, or the GraphPlan's extension GP-CSP algorithm [3]. At each stage of elaborating the Plan Graph in the algorithm, the scheme will use the component-descriptions p_i in order to satisfy the input descriptions A, which either leads to pruning of some branches (components) of the search tree or to guide the search toward the maximal satisfaction (by choosing one or more components). This step will deploy some type of resolution algorithm as in the TAL planner. The mutual-exclusion-relationships needed by the GraphPlan algorithm will be created by running the CPRAO algorithm over the environment parameters.

6 Related Works in other Areas

(1) There are attempts to develop databases for programs [7]. This type of work typically addresses whole programs and not components. However, the ontology developed for describing a program for the purpose of designing a database is relevant to our work. (2) In a workshop on the area of Intelligent Software Engineering at AAAI-

1999 conference Fischer and others from the NASA Ames Research Center have proposed a framework for retrieval and adaptation of components. Their work in this area is a close parallel to our proposed scheme. (3) The CAD (computer-aided design) area is also involved with the type of domains we are targeting. Components in CAD software simulate geometrical objects. Such software is supposed to allow a user to assemble components in a drag-and-drop oriented visual environment in order to develop a model for subsequent construction (and also possibly for simulation of its behavior). Constraint processing is a strongly relevant issue there and schemes have been proposed (Bhansali and Hoar, [2]) for the purpose. The proximity of the CAD and the COP areas is self-evident. (4) Memon et al [6] have tried to develop a mechanism for generating test-cases for checking functionalities of graphical user interfaces. They utilized an AI-planning approach that is somewhat similar to that proposed here for solving the COP-problem.

7 Conclusion

In this short paper we have proposed a framework for doing component-oriented programming automatically (or semi-automatically) using the Planning techniques from the artificial intelligence area. The work is primarily based on the observation that the two problems, COP and AI-Planning, are very similar to each other. We have formulated the two problems in a similar framework here to show this symmetry. We have identified some planning techniques and proposed how to modify them to solve the COP problem. We expect that such modifications to the AI-planning algorithms will actually enhance the efficiency of those algorithms, because the additional information (description or architecture of the required solution) will provide more guidance to the search procedure. Ostensibly, this also points to a new direction in the AI-planning research. We have also touched upon some related works in other areas.

References

1. Blum, A. V., and Furst, M. L.: Fast planning through planning graph analysis. Artificial Intelligence journal, (1997) 90, pp. 281-300.
2. Bhansali, S. and Hoar, T.H.: Automated software synthesis: An application in mechanical CAD. IEEE Transactions on Software Engineering, (1998) 24(10).
3. Do, M. B., and Kambhampati, S.: Planning as constraint satisfaction: Solving the planning by compiling it into CSP. Artificial Intelligence journal, (2001) 132, No. 2, pp. 151-182.
4. Kvarnstrom, J., and Doherty, P.: TAL planner: A temporal action logic based forward chaining planner. Annals of Mathematics and Artificial Intelligence journal (2001).
5. Mitra, D.: Interactive modeling for batch simulation of engineering systems: A constraint satisfaction problem. Lecture Notes in Artificial Intelligence (Proceedings of the IEA/AIE-2001), (2001) 2070, pp. 602-611, Springer.

6. Memon, A. M., Martha, E. P., and Soffa, M. L.: Plan generation for GUI testing. Proceedings of the Fifth International Conference on Artificial Intelligence Planning and Scheduling (AIPS-2000), pp. 226-235, Breckenridge, Colorado, USA.
7. Paul, S. and Prakash, A.: A query algebra for program databases. IEEE Transactions on Software Engineering, (1996) 22(3).
8. Rumbaugh, J., Jacobson, I., and Booch, G.: The Unified Modeling Language Reference Manual. Addison-Wesley, Reading, Massachusetts (1999).
9. Russell, S. and Norvig, P.. Artificial Intelligence: A Modern Approach. Prentice Hall, Englewood Cliffs, New Jersey (1995).
10. Shaw, M., and Garlan, D.: Software Architecture: Perspectives on an Emerging Discipline. Printice-Hall, Inc., New Jersey, USA (1996)..
11. Tate, A.: Representing plans as a set of constraints – the <I-N-OVA> model. Proceedings of the AIPS-96 conference, Edinburgh, UK (1996).
12. Weld, D.: An introduction to partial-order planning. AI Magazine, AAAI Press (1994)

Dynamic CSPs
for Interval-Based Temporal Reasoning

Malek Mouhoub and Jonathan Yip

Department of Math & Computer Science, University of Lethbridge
4401 University Drive, Lethbridge, AB, Canada, T1K 3M4
phone: (+1 403) 329 2557; fax: (+1 403) 329 2519
{mouhoub,yipj9}@cs.uleth.ca

Abstract. Many applications such as planning, scheduling, computational linguistics and computational models for molecular biology involve systems capable of managing qualitative and metric time information. An important issue in designing such systems is the efficient handling of temporal information in an evolutive environment. In a previous work, we have developed a temporal model, TemPro, based on the interval algebra, to express such information in terms of qualitative and quantitative temporal constraints. In order to find a good policy for solving time constraints in a dynamic environment, we present in this paper, a study of dynamic arc-consistency algorithms in the case of temporal constraints. We show that, an adaptation of the new AC-3 algorithm presents promising results comparing to the other dynamic arc-consistency algorithms. Indeed, while keeping an optimal worst-case time complexity, this algorithm has a better space complexity than the other methods.

Keywords: Temporal Reasoning, Dynamic Arc Consistency, Planning, Scheduling

1 Introduction

Many applications such as planning, scheduling, computational linguistics[18,9], data base design[15] and computational models for molecular biology[8] involve managing temporal constraints. In linear planning, for example, during the search process the planner must order the set of actions forming the plan by imposing a collection of appropriate ordering constraints. These constraints are essential to guarantee the consistency of the resulting plan, that is, to guarantee that if the actions are executed starting at the initial state and consistently with these constraints, then the goal will be achieved. In nonlinear planning[5,17] where the actions in a plan are partially ordered, maintaining the consistency of the ordering constraints is required, for example, when the planner attempts to establish a subgoal by reusing an action already in the plan under construction. Reasoning about constraints that prevent an action A from lying within a certain interval between two other actions A_1 and A_2 is also important in planners such as UCPOP[16]. The development of a domain-independent temporal reasoning system is then practically important. An important issue in designing such systems is the efficient handling of qualitative and metric time information. Indeed, the separation between the two aspects does not exist in the real world. In our daily life activities, for example, we combine the two type of information to describe different situations. In the case of scheduling problems, we can have qualitative information such as the ordering between tasks and quantitative information describing the temporal windows of the tasks i.e earliest start time, latest end time and the duration of each task.

In a previous work[13], we have developed a temporal model, TemPro, based on the interval algebra, to express such information in terms of qualitative and quantitative temporal constraints. TemPro translates an application involving time information into a binary Constraint Satisfaction Problem[1] where constraints are temporal relations, we call it Temporal Constraint Satisfaction Problem (TCSP)[2]. Managing temporal information consists then in solving the TCSP using local consistency algorithms and search strategies based on constraint satisfaction techniques.

The aim of our work here is to solve a TCSP in a dynamic environment. Indeed, in the real world, when solving a TCSP we may need to add new information or relax some constraints when, for example, there are no more solutions (case of over constrained problems). In this paper, we will mainly focus on maintaining the arc-consistency dynamically. In a previous work, we have used arc-consistency algorithms[13] to reduce the size of the TCSP representing the initial problem, by removing some values that do not belong to any solution. Indeed, an arc-consistency algorithm removes all inconsistencies involving all subsets of 2 variables belonging to the set of variables of the problem. In a dynamic environment we need to check if there still exist solutions to the problem every time a constraint has been added or removed. Adding temporal constraints can easily be handled by the arc-consistency algorithms we have used, we have just to put in this case the new constraint in the lists of constraints to be checked. However, constraint relaxation cannot be handled by these algorithms. Indeed, when we remove a constraint, these algorithms cannot find which value, that has been already removed, must be put back and which one must not. We must then use incremental arc-consistency algorithms instead (called also dynamic arc-consistency algorithms). Some dynamic arc-consistency algorithms have already been proposed in the literature. We also present in this paper a new dynamic arc-consistency algorithm which is a modification of a recent arc consistency algorithm[20] in order to handle dynamic constraints. Comparisons tests of the different dynamic arc consistency algorithms were performed on randomly generated dynamic temporal constraint problems. The results show that

---

[1] A binary CSP involves a list of variables defined on finite domains of values and a list of binary relations between variables.

[2] Note that this name and the corresponding acronym was used in [7]. A comparison of the approach presented in this paper and our model TemPro is described in [13].

the new algorithm we propose has better performance than the others in most cases.

The rest of the paper is organized as follows: in the next section, we will present through an example, the different components of the model TemPro and its corresponding resolution methods. Maintaining dynamic arc-consistency in the case of temporal constraints is then presented in section 3. Section 4 is dedicated to the experimental comparison of the different dynamic arc-consistency algorithms. Concluding remarks and possible perspectives of our work are then presented in section 6.

2 Knowledge Representation

Example 1 : Consider the following typical temporal reasoning problem[3] :

1. *John, Mary and Wendy* **separately** *rode to the soccer game.*
2. *It takes John* **30 minutes**, *Mary* **20 minutes** *and Wendy* **50 minutes** *to get to the soccer game.*
3. *John* **either started or arrived** *just as Mary started.*
4. *John* **either started or arrived** *just as Wendy started.*
5. *John left home* **between 7:00 and 7:10**.
6. *Mary and Wendy* **arrived together** *but* **started at different times**.
7. *Mary arrived at work* **between 7:55 and 8:00**.
8. *John's trip* **overlapped** *the soccer game.*
9. *Mary's trip took place* **during** *the game or else the game took place* **during** *her trip.*

The above story includes numeric and qualitative information (words in boldface). There are four main events: John, Mary and Wendy are going to the soccer game respectively and the soccer game itself. Some numeric constraints specify the duration of the different events, e.g. *20 minutes is the duration of Mary's event*. Other numeric constraints describe the temporal windows in which the different events occur. And finally, symbolic constraints state the relative positions between events e.g. *John's trip overlapped the soccer game*.

Given these kind of information, one important task is to represent and reason about such knowledge and answer queries such as: "is the above problem consistent?", "what are the possible times at which Wendy arrived at the soccer game?", ... etc.
To reach this goal, and using an extension of the Allen algebra[1] to handle numeric constraints, our model TemPro transforms a temporal problem involving numeric and symbolic information into a temporal constraint satisfaction problem (TCSP) including a set of events $\{EV_1, \ldots, EV_n\}$, each defined on a discrete

---
[3] This problem is basically taken from an example presented by Ligozat, Guesgen and Anger at the tutorial: Tractability in Qualitative Spatial and Temporal Reasoning, IJCAI'01. We have added numeric constraints for the purpose of our work.

domain standing for the set of possible occurrences (time intervals) in which the corresponding event can hold; and a set of binary constraints, each representing a qualitative disjunctive relation between a pair of events and thus restricting the values that the events can simultaneously take. A disjunctive relation involves one or more Allen primitives.

Table 1. Allen primitives

Relation	Symbol	Inverse	Meaning
X precedes Y	P	$P^\smile$	XXX YYY
X equals Y	E	E	XXX YYY
X meets Y	M	$M^\smile$	XXXYYY
X overlaps Y	O	$O^\smile$	XXXX YYYY
X during y	D	$D^\smile$	XXX YYYYYY
X starts Y	S	$S^\smile$	XXX YYYYY
X finishes Y	F	$F^\smile$	XXX YYYYY

The initial problem of figure 1 corresponds to the transformation of the temporal reasoning problem, we presented before, to a TCSP using the model TemPro. Information about the relative position between each pair of events is converted to a disjunction of Allen primitives. Indeed, Allen[1] has proposed 13 basic relations between time intervals: starts (S), during (D), meets (M), overlaps (O), finishes (F), precedes (P), their converses and the relation equals (E) (see table 3 for the definition of the 13 Allen primitives). For example, the information "John either started or arrived just as Wendy started" is translated as follows: $J\ (S \vee M)\ W$. In the case where there is no information between a pair of events, the corresponding relation is represented by the disjunction of the 13 Allen primitives (since this constraint is not considered during the resolution process, it does not appear on the graph of constraint as it is the case in figure 1 concerning the relation between Wendy and the Soccer game).

The domain of each event corresponding to the set of possible occurrences (we call it SOPO) that each event can take is generated given its earliest start time, latest end time and duration. In the case of Wendy's event, since we do not have any information about the earliest and latest time, these parameters are set respectively to 0 and the constant *horizon* (time before which all events should be performed). After a symbolic → numeric pre-process, these parameters are then set to 5 and 60 respectively.

Solving a TCSP consists of finding an assignment of a value from its domain to every variable, in such a way that every constraint is satisfied. Since we are

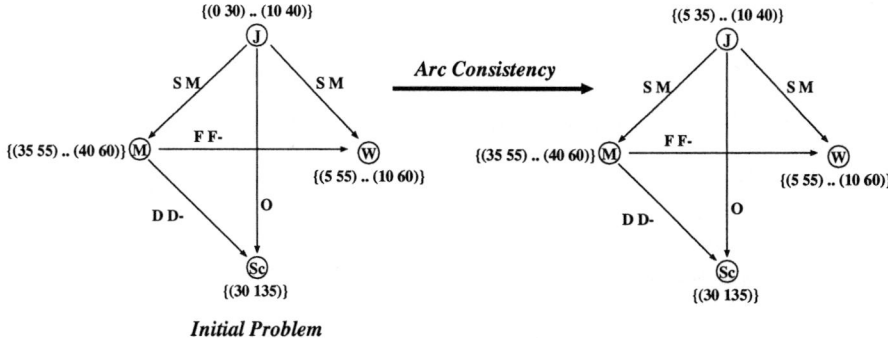

Fig. 1. Applying arc-consistency to a temporal problem

dealing with a constraint satisfaction problem, deciding consistency is in general NP-hard[4]. In order to overcome this difficulty in practice, we have developed[13] a resolution method performed in two stages. Local consistency algorithms are first used to reduce the size of the TCSP by removing some inconsistent values from the variable domains (in the case where arc-consistency is applied) and some inconsistent Allen primitives from the disjunctive qualitative relations (in the case where path consistency is performed). A backtrack search is then performed to look for a possible solution. When applying AC-3 to our temporal problem (see figure 1) the domain of event J is reduced.

3 Dynamic Maintenance of Local Consistency for Temporal Constraints

3.1 Dynamic Constraint Satisfaction Problem

A dynamic constraint satisfaction problem (DCSP) P is a sequence of static CSPs $P_0, \ldots, P_i, P_{i+1}, \ldots, P_n$ each resulting from a change in the preceding one imposed by the "outside world". This change can either be a restriction (adding a new constraint) or a relaxation (removing a constraint because it is no longer interesting or because the current CSP has no solution). More precisely, P_{i+1} is obtained by performing a restriction (addition of a constraint) or a relaxation (suppression of a constraint) on P_i. We consider that P_0 (initial CSP) has an empty set of constraints.

---

[4] Note that some CSP problems can be solved in polynomial time. For example, if the constraint graph corresponding to the CSP has no loops, then the CSP can be solved in $O(nd^2)$ where n is the number of variables of the problem and d is the domain size of the different variables

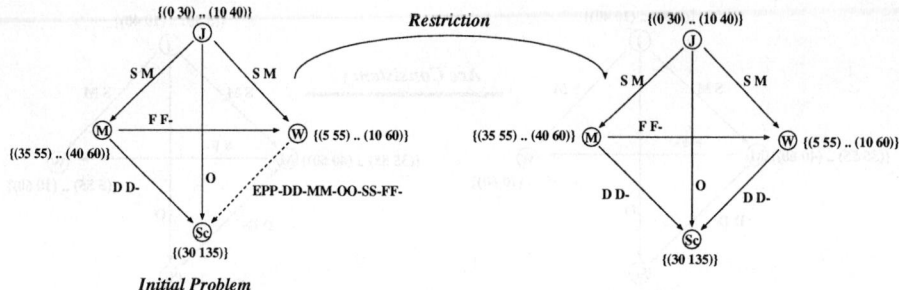

Fig. 2. A restriction in a DTCSP

3.2 Dynamic Temporal Constraint Satisfaction Problem

Since a TCSP is a CSP in which constraints are disjunctions of Allen primitives, the definition of a dynamic temporal constraint satisfaction problem (DTCSP) is slightly different from the definition of a DCSP. Indeed in the case of a DTCSP, a restriction can be obtained by removing one or more Allen primitive from a given constraint. A particular case is when the constraint is equal to the disjunction of the 13 primitives (we call it the universal relation I) which means that the constraint does not exist (there is no information about the relation between the two involved events). In this particular case, removing one or more Allen primitives from the universal relation is equivalent to adding a new constraint. Using the same way, a relaxation can be obtained by adding one or more Allen primitives to a given constraint. A particular case is when the new constraint has 13 Allen primitives which is equivalent to the suppression of the constraint.

Figure 2 shows a restriction on the problem of example 1 obtained by removing some Allen primitives from the constraint between Wendy's event and the soccer game. This restriction is equivalent to the addition of the following information to our problem: *10. Wendy's trip took place during the game or else the game took place during her trip.*

3.3 Dynamic Arc-Consistency Algorithms

The arc-consistency algorithms we have used to solve a TCSP (Mouhoub, 1998) can easily be adapted to update the variable domains incrementally when adding a new constraint. In our example, adding the new constraint (see figure 2) will lead to an arc inconsistent TCSP which leads to an inconsistent TCSP. Let us assume now that to restore the arc-consistency we decide to relax the TCSP by adding one or more Allen primitives to a chosen constraint (one of the 10 constraints of our problem). In this case, the arc-consistency algorithms are unable to update the variable domains in an incremental way because they are not able to determine the set values that must be restored to the domains. The only way, in this case, is to reset the domains, add all the constraints (including the updated one) to the "unconstrained" TCSP (TCSP with no constraints) and then perform the arc-consistency algorithm. To avoid this drawback, dynamic arc-consistency algorithms have been proposed. Bessière has proposed DnAC-4[2] which is an adaptation of AC-4[12]. This algorithm stores a justification for each deleted value. These justifications are then used to determine the set of values that have been removed because of the relaxed constraint and so can process relaxations incrementally. DnAC-4 inherits the bad time and space complexity of AC-4. Indeed, comparing to AC-3 for example, AC-4 has a bad average time complexity[19]. The worst-case space complexity of DnAC-4 is $O(ed^2 + nd)$ (e, d and n are respectively the number of constraints, the domain size of the variables and the number of variables). To work out the drawback of AC-4 while keeping an optimal worst case complexity, Bessière has proposed AC-6[3]. Debruyne has then proposed DnAC-6 adapting the idea of AC-6 for dynamic CSPs by using justifications similar to those of DnAC-4[6]. While keeping an optimal worst case time complexity ($O(ed^2)$), DnAC-6 has a lower space requirements ($O(ed + nd)$) than DnAC-4. To solve the problem of space complexity, Neveu and Berlandier proposed AC|DC[14]. AC|DC is based on AC-3 and does not require data structures for storing justifications. Thus it has a very good space complexity ($O(e + nd)$) but is less efficient in time than DnAC-4. Indeed with its $O(ed^3)$ worst case time complexity, it is not the algorithm of choice for large dynamic CSPs. More recently, Zhang and Yap proposed an new version of AC-3 (called AC-3.1) achieving the optimal worst case time complexity with $O(ed^2)$ ([20])[5]. We have modified this algorithm in order to solve dynamic CSPs as we believe the new algorithm (that we call AC-3.1|DC) may provide better performance than DnAC-4 and DnAC-6.

3.4 AC-3.1|DC

Before we present the algorithm AC3.1|DC, let us recall the algorithm AC-3 and the new view of AC-3 (called also AC-3.1) proposed by Zhang and Yap[20].

Mackworth[10] has presented the algorithm AC-3 for enforcing arc-consistency on a CSP. The following is the pseudo-code of AC-3 in the case of a TCSP. The worst case time complexity of AC-3 is bounded by $O(ed^3)$[11]. In fact this complexity depends mainly on the way line 3 of the function $REVISE$ is implemented. Indeed, if anytime the arc (i,j) is revised, b is searched from scratch then the worst case time complexity is $O(ed^3)$. Instead of a search from scratch, Zhang and Yap[20] proposed a new view that allows the search to resume from the point where it stopped in the previous revision of (i,j). By doing so the worst case time complexity of AC-3 is achieved in $O(ed^2)$.

Function $REVISE(i,j)$
1. $REVISE \leftarrow false$
2. **For** each interval $a \in SOPO_i$ **Do**
3.   **If** $\neg compatible(a,b)$ **for each interval** $b \in SOPO_j$ **Then**

---

[5] Another arc consistency algorithm (called AC-2001) based on the same idea as AC-3.1 (and having an $O(ed^2)$ worst case time complexity) was proposed by Bessière and Régin[4]. We have chosen AC-3.1 for the simplicity of its implementation

4.   remove a from $SOPO_i$
5. $REVISE \leftarrow true$
6.   End-If
7. End-For

Algorithm AC-3
1. Given a TemPro network $TN = (E, R)$
   (E: set of events,
   R: set of disjunctive relations between events)
2. $Q \leftarrow \{(i,j) \mid (i,j) \in R\}$
   (list initialized to all relations of TN)
3. **While** $Q \neq Nil$ **Do**
4. $Q \leftarrow Q - \{(i,j)\}$
5.   **If** $REVISE(i,j)$ **Then**
6. $Q \leftarrow Q \sqcup \{(k,i) \mid (k,i) \in R \wedge k \neq j\}$
7.   **End-If**
8. **End-While**

In the case of constraint restriction, AC-3.1|DC works in the same way as AC-3.1. The worst-case time complexity of a restriction is then $O(ed^2)$. The more interesting question is whether AC-3.1|DC's time complexity can remain the same during retractions. Indeed, if we use the same way as for AC|DC [14], one major concern is that during the restrictions, the AC-3.1 algorithm keeps a Resume table of the last place to start checking for consistency from. Unfortunately, during retractions, this Resume table may prove useless as values in the domain of nodes are restored. Our attempt was to follow an idea observed from the DnAC6 algorithm. Instead of replacing values in the node in the order they were deleted, the algorithm should place these values to be restored at the end of the list of values for that node, thereby keeping the Resume table intact. More precisely, the constraint relaxation, of a given relation (k, m) for example, is performed in 3 steps:

1. An estimation (over-estimation) of the set of values that have been removed because of the constraint (k, m) is first determined by looking for the values removed from the domains of k and m that have no support on (k, m),
2. the above set is then propagated to the other variables,
3. and finally a filtering procedure based on AC-3.1 is then performed to remove from the estimation set the values which are not arc-consistent with respect to the relaxed problem.

Since the time complexity of each of the above steps is $O(ed^2)$, the worst-case time complexity of a relaxation is $O(ed^2)$. Comparing to AC|DC, AC3.1|DC has a better time complexity. Indeed, the main difference between AC3.1|DC and AC|DC is the third step. This later step requires $O(ed^3)$ in the case of AC|DC (which results in a $O(ed^3)$ worst case time complexity for a restriction). In the case of AC3.1|DC, the third step can be performed in $O(ed^2)$ in the worst case because of the improvement we mentioned above. Comparing to DnAC-4 and DnAC-6, AC-3.1|DC has a better space complexity ($O(e+nd)$) while keeping an optimal worst-case time complexity ($O(ed^2)$).

4 Experimentation

In order to compare the performance of the 4 dynamic arc-consistency algorithms we have seen in the previous subsection, in the case of temporal constraints, we have performed tests on randomly generated DTCSPs. The criterion used to compare the above algorithms is the computing effort needed by an algorithm to perform the arc consistency. This criterion is measured by the running time in seconds required by each algorithm. The experiments were performed on a SUN SPARC Ultra 5 station. All the procedures are coded in C/C++. 3 classes of instances, corresponding to 3 type of tests, were generated as follows:

case 1: actions correspond to additions of constraints. $C = N(N-1)/2$ (constraints are added until a complete graph is obtained).

case 2: actions can be additions or retractions of constraints.
$C = N(N-1)/2$ additions $+ N(N-1)/4$ retractions (the final TCSP will have $N(N-1)/4$ constraints).

case 3: this case is similar to case 1 but with inconsistent DTCSPs. Indeed in the previous 2 cases the generated DTCSPs are consistent. In this last case constraints are added until an arc inconsistency and thus a global inconsistency is detected (the inconsistency is detected if one variable domain becomes empty). Retractions are then performed until the arc-consistency is restored.

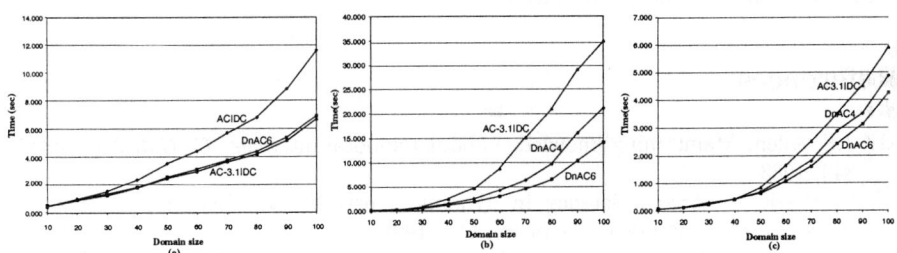

Fig. 3. Experimental Tests on random DTCSPs

4.1 Results

Figure 3a) shows the results of tests corresponding to case 1. As we can easily see, the results provided by DnAC-6 and AC-3.1|DC are better than the ones provided by AC|DC and DnAC-4 (which do not appear on the chart). Since DnAC-6 requires much more memory space than AC-3.1|DC, this latter is the algorithm of choice in the case of constraint additions. Figure 3b) and 3c) correspond to case 2 and case 3 respectively. DnAC-4 and DnAC-6 have better performance in this case than AC3.1|DC and AC|DC (the running time of AC|DC is very slow comparing to the other 3 algorithms). However, since AC3.1|DC does not require a lot of memory space, it has less limitations than DnAC-4 and DnAC-6 in terms of space requirements especially in the case of problems having large domain sizes.

5 Conclusion and Future Work

In this paper we present a comparative study of dynamic arc-consistent algorithms in the case of temporal constraint problems. The results shown demonstrate the efficiency of AC-3.1|DC, which is an adaptation of the new AC-3 algorithm for dynamic constraints, comparing to other algorithms. Indeed, while keeping an optimal worst-case time complexity, AC-3.1|DC requires less memory space than DnAC-4 and DnAC-6.

One perspective of our work is to look for a method to maintain path consistency when dealing with dynamic temporal constraints. Indeed, as we have shown in[13], path consistency is useful in the filtering phase to detect the inconsistency when solving temporal constraint problems and also in the case where the numeric information is incomplete. The other perspective is to handle the addition and relaxation of constraints during the backtrack search phase. For example, suppose that during the backtrack search a constraint is added when instantiating the current variable. In this case, the instantiation of the variables already instantiated should be reconsidered and the domains of the current and future variables should be updated.

References

1. J.F. Allen. Maintaining knowledge about temporal intervals. *CACM*, 26(11):832–843, 1983.
2. C. Bessière. Arc-consistency in dynamic constraint satisfaction problems. In *AAAI'91*, pages 221–226, Anaheim, CA, 1991.
3. C. Bessière. Arc-consistency and arc-consistency again. *Artificial Intelligence*, 65:179–190, 1994.
4. C. Bessière and J. C. Régin. Refining the basic constraint propagation algorithm. In *Seventeenth International Joint Conference on Artificial Intelligence (IJCAI'01)*, pages 309–315, Seattle, WA, 2001.
5. D. Chapman. Planning for conjunctive goals. *Artificial Intelligence.*, 32:333–377, 1987.
6. R. Debruyne. Les algorithmes d'arc-consistance dans les csp dynamiques. *Revue d'Intelligence Artificielle*, 9:239–267, 1995.
7. R. Dechter, I. Meiri, and J. Pearl. Temporal constraint networks. *Artificial Intelligence*, 49:61–95, 1991.
8. C. Golumbic and R. Shamir. Complexity and algorithms for reasoning about time: a graphic-theoretic approach. *Journal of the Association for Computing Machinery*, 40(5):1108–1133, 1993.

9. C. Hwang and L. Shubert. Interpreting tense, aspect, and time adverbials: a compositional, unified approach. In *Proceedings of the first International Conference on Temporal Logic, LNAI, vol 827*, pages 237–264, Berlin, 1994.
10. A. K. Mackworth. Consistency in networks of relations. *Artificial Intelligence*, 8:99–118, 1977.
11. A. K. Mackworth and E. Freuder. The complexity of some polynomial network-consistency algorithms for constraint satisfaction problems. *Artificial Intelligence*, 25:65–74, 1985.
12. R. Mohr and T. Henderson. Arc and path consistency revisited. *Artificial Intelligence*, 28:225–233, 1986.
13. M. Mouhoub, F. Charpillet, and J.P. Haton. Experimental analysis of numeric and symbolic constraint satisfaction techniques for temporal reasoning. *Constraints: An International Journal*, 2:151–164, Kluwer Academic Publishers, 1998.
14. B. Neuveu and P. Berlandier. Arc-consistency for dynamic constraint satisfaction problems: An rms free approach. In *ECAI-94, Workshop on Constraint Satisfaction Issues Raised by Practical Applications*, Amsterdam, 1994.
15. M. Orgun. On temporal deductive databases. *Computational Intelligence*, 12(2):235–259, 1996.
16. J. Penberthy and d. Weld. Ucpop: A sound, complete, partial order planner for adl. In B. Nebel, C Rich, and W. Swartout, editors, *Third International Conference on Principles of Knowledge Reprentation and Reasoning (KR'92)*, pages 103–114, Boston, MA, 1992.
17. J. Penberthy and d. Weld. Temporal planning with continuous change. In B. Nebel, C Rich, and W. Swartout, editors, *Twelfth National Conference of the American Association for Artificial Intelligence (AAAI-94)*, pages 1010–1015, Seattle, WA, 1994.
18. F. Song and R. Cohen. Tense interpretation in the context of narrative. In *AAAI'91*, pages 131–136, 1991.
19. R. J. Wallace. Why AC-3 is almost always better than AC-4 for establishing arc consistency in CSPs. In *IJCAI'93*, pages 239–245, Chambery, France, 1993.
20. Yuanlin Zhang and Roland H. C. Yap. Making ac-3 an optimal algorithm. In *Seventeenth International Joint Conference on Artificial Intelligence (IJCAI'01)*, pages 316–321, Seattle, WA, 2001.

Efficient Pattern Matching of Time Series Data*

Sangjun Lee, Dongseop Kwon, and Sukho Lee

School of Electrical Engineering and Computer Science,
Seoul National University, Seoul 151-742, Korea
{freude,subby}@db.snu.ac.kr

Abstract. There has been a lot of interest in matching and retrieval of similar time sequences in time series databases. Most of previous work is concentrated on similarity matching and retrieval of time sequences based on the Euclidean distance. However, the Euclidean distance is sensitive to the absolute offsets of time sequences. In addition, the Euclidean distance is not a suitable similarity measurement in terms of shape. In this paper, we propose an indexing scheme for efficient matching and retrieval of time sequences based on the minimum distance. The minimum distance can give a better estimation of similarity in shape between two time sequences. Our indexing scheme can match time sequences of similar shapes irrespective of their vertical positions and guarantees no false dismissals. We experimentally evaluated our approach on real data(stock price movement).

1 Introduction

Time series database is a set of data sequences of real numbers; each number represents a value at a time point. Examples include stock price movement, exchange rate, weather data, biomedical measurement, etc. Similarity search in time series databases is essential in many applications, such as data mining and knowledge discovery[1,2]. Although the sequential scanning can be used to perform the similarity search, it may require enormous processing time over large time series databases. To reduce the search space, an efficient indexing scheme is required. Indexing, however, on such multi-dimensional data doesn't seem to be easy because of dimensionality curse in making index structure using a spatial access method such as R-tree[3]. The most popular methods perform feature extraction as dimensionality reduction of time series data, and then use a spatial access method to index the time series data in the feature space.

Many techniques have been proposed to support the fast retrieval of similar time sequences based on the Euclidean distance[4,5,6]. However, the Euclidean distance as a similarity measurement has the following problem: it is sensitive to the absolute offsets of the time sequences, so two time sequences that have similar shapes but with different vertical positions may be classified as dissimilar. A shortcoming of the Euclidean distance is demonstrated in Figure 1. Two time sequences B and Q are the same in shape, because time sequence B can

---

* This work was supported by the Brain Korea 21 Project in 2001

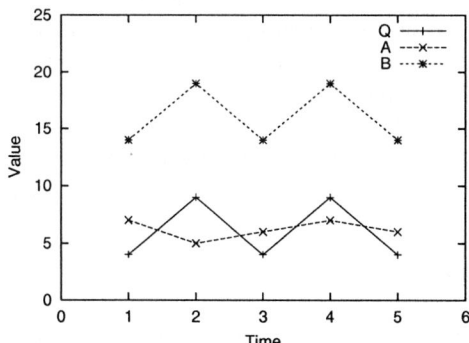

Fig. 1. Shortcoming of the Euclidean distance

be obtained by shifting time sequence Q upward 10 units. However, they can be classified as dissimilar by the Euclidean distance. From this example, the Euclidean distance is not a suitable similarity measurement when the shape of a time sequence is the major consideration.

The minimum distance[7,8,14] is often used for time sequence matching to overcome the shortcoming of the Euclidean distance above mentioned. The minimum distance is the Euclidean distance between two time sequences, neglecting their offsets. The minimum distance can be a more suitable similarity measurement when we would like to retrieve similar time sequences in shape from time series databases irrespective of their vertical positions. The minimum distance is defined as follows.

Definition 1 (Minimum Distance). *Given a threshold ϵ, two time sequences $A = \{a_1, a_2, \ldots, a_n\}$ and $B = \{b_1, b_2, \ldots, b_n\}$ of equal length n are said to be similar in shape if*

$$D_{minimum}(A, B) = (\sum_{i=1}^{n} |a_i - b_i - m|^2)^{1/2} \leq \epsilon$$

$$where \quad m = \sum_{i=1}^{n}(a_i - b_i)/n$$

To support minimum distance queries, most of previous work has the preprocessing step of vertical shifting that normalizes each time sequence by its mean before indexing[9,17,20]. The vertical shifting, however, has the additional overhead to get the mean of a sequence and to subtract the mean from each element of the sequence.

In this paper, we propose a novel and efficient indexing scheme for time series, called the segmented sum of variation indexing(SSV-indexing). Since the variation between two adjacent elements in a time sequence is invariant under vertical shifting, we can use this property to reduce dimensionality of time sequence data. Our indexing scheme can match time sequences of similar shapes without vertical shifting and guarantees no false dismissals.

The remainder of this paper is organized as follows. Section 2 provides a survey of related work. In Section 3, our proposed approach is described. Section 4 presents the experimental result. Finally, several concluding remarks are given in Section 5.

2 Related Work

Various methods have been proposed for fast matching and retrieval of similar time sequences in time series databases. The main focus is to speed up the search process. An indexing scheme called *F-index*[4] is proposed to handle sequences of the same length. The idea is to use Discrete Fourier Transform(DFT) as a dimensionality reduction technique. The results of [4] are extended in [5] and the *ST-index* is proposed for sequence matching of the different length.

In [8], authors propose to use Discrete Wavelet Transform(DWT) for dimensionality reduction and compare this method to DFT. They argue that Haar wavelet transform performs better than DFT. However, the performance of DFT can be improved using the symmetry of Fourier Transforms and DWT has the limitation that it can only be defined for time sequences with a length of the power of two.

In [9], the authors show that the similarity retrieval will be invariant to simple shifting and scaling if sequences are normalized before indexing. In [12], authors present an intuitive similarity measurement for time series data. They argue that the similarity model with scaling and shifting is better than the Euclidean distance. However, they do not present any indexing method. In [15], the authors propose a definition of similarity based on scaling and shifting transformations.

In [10], authors present a hierarchical algorithm called *HierarchyScan*. The idea of this method is to perform correlation between the stored sequences and the template in the transformed domain hierarchically. In [7,14], a definition of sequence similarity based on the slope of sequence segment is discussed.

In [11], authors propose a set of linear transformations such as moving average, time warping, and reversing. These transformations can be used as the basis of similarity queries for time series data. The results of [11] are extended in [16] and authors propose the method for processing queries that express similarity in terms of multiple transformations instead of a single one. In [13,18], authors use time warping as distance function and present algorithms for retrieving similar time sequences under this function. However, a time warping distance does not satisfy triangular inequality and can cause false dismissals.

In [17,20] the authors introduce new dimensionality reduction technique of time sequence by using segmented mean features. The same concept of independent research is proposed in [19]. They show that the segmented mean features can be used in arbitrary L_p norm. Since, however, the segmented mean features cannot support minimum distance queries, vertical shifting is required in preprocessing raw data to process minimum distance queries.

3 Proposed Approach

The problem we focus on is the design of efficient retrieval of similar time sequences in time series databases based on the minimum distance. We will now introduce the segmented sum of variation indexing(SSV-indexing) and show that it guarantees no false dismissals.

3.1 Feature Extraction as Dimensionality Reduction

Our goal is to extract features that capture the information on the shape of a time sequence, and that will lead to a feature distance definition satisfying the lower bound condition of the minimum distance. We propose to use the sum of variation as the feature of a segment. Suppose we have a set of time sequences of length n. The proposed feature extraction method consists of two steps. First, we divide each time sequence into m segments of equal length l. Then, we extract the sum of variation feature from each segment. The sum of variation is invariant under vertical shifting, so the segmented sum of variation features can be indexed to support minimum distance queries.

3.2 Lower Bounding of the Minimum Distance

In order to guarantee no false dismissals, we must construct a distance measure $D_{feature}(FA, FB)$ defined in the feature space, which has the following property.

$$D_{feature}(FA, FB) \leq D_{minimum}(A, B)$$

The distance between feature vectors $D_{feature}(FA, FB)$ is given as follows.

$$D_{feature}(FA, FB) = \frac{(\sum_{i=1}^{m} |FA_i - FB_i|^2)^{1/2}}{2 \cdot \sqrt{l-1}}$$

We must show that the distance between feature vectors is the lower bound of the minimum distance between original sequences. Before going into the main theorem, we will present some lemmas used in the main theorem.

Lemma 1. *Given two $2-point$ segments $A = (a_i, a_{i+1})$ and $B = (b_i, b_{i+1})$, variations of A and B are $FA = |a_{i+1} - a_i|$ and $FB = |b_{i+1} - b_i|$, respectively. Then, the following inequality holds.*

$$||a_{i+1} - a_i| - |b_{i+1} - b_i||^2 \leq 2 \cdot (|a_{i+1} - b_{i+1} - m|^2 + |a_i - b_i - m|^2)$$

$$\text{where} \quad m = \sum_{i=1}^{n}(a_i - b_i)/n$$

Proof. We know that

$$\||a_{i+1} - a_i| - |b_{i+1} - b_i|\|^2$$
$$= \||(a_{i+1} - m_a) - (a_i - m_a)| - |(b_{i+1} - m_b) - (b_i - m_b)|\|^2$$

where $m_a = \sum_{i=1}^{n} a_i/n$ and $m_b = \sum_{i=1}^{n} b_i/n$

Then, we have

$$\||(a_{i+1} - m_a) - (a_i - m_a)| - |(b_{i+1} - m_b) - (b_i - m_b)|\|^2$$
$$\leq |((a_{i+1} - m_a) - (a_i - m_a)) - ((b_{i+1} - m_b) - (b_i - m_b))|^2$$
$$= |((a_{i+1} - m_a) - (b_{i+1} - m_b)) - ((a_i - m_a) - (b_i - m_b))|^2$$
$$\leq (|(a_{i+1} - m_a) - (b_{i+1} - m_b)| + |(a_i - m_a) - (b_i - m_b)|)^2$$
$$\leq 2 \cdot (|(a_{i+1} - m_a) - (b_{i+1} - m_b)|^2 + |(a_i - m_a) - (b_i - m_b)|^2)$$

$$\therefore \||a_{i+1} - a_i| - |b_{i+1} - b_i|\|^2$$
$$\leq 2 \cdot (|(a_{i+1} - m_a) - (b_{i+1} - m_b)|^2 + |(a_i - m_a) - (b_i - m_b)|^2)$$
$$\leq 2 \cdot (|a_{i+1} - b_{i+1} - m|^2 + |a_i - b_i - m|^2)$$

Lemma 2. *Given two l − point segments $A = (a_1, a_2, \ldots, a_l)$ and $B = (b_1, b_2, \ldots, b_l)$, the sum of variation of A is $FA = \sum_{i=1}^{l-1} |a_{i+1} - a_i|$ and that of B is $FB = \sum_{i=1}^{l-1} |b_{i+1} - b_i|$. Then, the following inequality holds.*

$$|FA - FB|^2 \leq 2 \cdot (l-1)(\sum_{i=1}^{l-1} |a_i - b_i - m|^2 + \sum_{i=2}^{l} |a_i - b_i - m|^2)$$

where $m = \sum_{i=1}^{n}(a_i - b_i)/n$

Proof. We know that

$$|FA - FB|^2 = (\sum_{i=1}^{l-1} |a_{i+1} - a_i| - \sum_{i=1}^{l-1} |b_{i+1} - b_i|)^2$$
$$= \{\sum_{i=1}^{l-1}(|a_{i+1} - a_i| - |b_{i+1} - b_i|)\}^2$$

The following inequality holds by Cauchy-Schwartz's inequality and Lemma 1.

$$\{\sum_{i=1}^{l-1}(|a_{i+1} - a_i| - |b_{i+1} - b_i|)\}^2$$

$$\leq (l-1)\sum_{i=1}^{l-1}||a_{i+1} - a_i| - |b_{i+1} - b_i||^2$$

$$\leq 2 \cdot (l-1)(\sum_{i=1}^{l-1}|a_i - b_i - m|^2 + \sum_{i=2}^{l}|a_i - b_i - m|^2)$$

$$\therefore \quad |FA - FB|^2 \leq 2 \cdot (l-1)(\sum_{i=1}^{l-1}|a_i - b_i - m|^2 + \sum_{i=2}^{l}|a_i - b_i - m|^2)$$

Theorem 1. $D_{feature}(FA,FB)$ is the lower bound of $D_{minimum}(A, B)$, it guarantees no false dismissals.

$$D_{feature}(FA, FB) \leq D_{minimum}(A, B)$$

Proof. Based on Lemma 2, we can get the following inequality.

$$\sum_{i=1}^{m}|FA_i - FB_i|^2$$

$$\leq 2 \cdot (l-1)(\sum_{i=1}^{l-1}|a_i - b_i - m|^2 + \sum_{i=2}^{l}|a_i - b_i - m|^2)$$

$$+ 2 \cdot (l-1)(\sum_{i=l}^{2(l-1)}|a_i - b_i - m|^2 + \sum_{i=l+1}^{2(l-1)+1}|a_i - b_i - m|^2) + \ldots$$

$$+ 2 \cdot (l-1)(\sum_{i=(m-1)l+(2-m)}^{m(l-1)}|a_i - b_i - m|^2 + \sum_{i=(m-1)l+(3-m)}^{m(l-1)+1}|a_i - b_i - m|^2)$$

$$= 2 \cdot (l-1)(\sum_{i=l}^{n-1}|a_i - b_i - m|^2 + \sum_{i=2}^{n}|a_i - b_i - m|^2)$$

$$\leq 2^2 \cdot (l-1)(\sum_{i=1}^{n}|a_i - b_i - m|)^2$$

From the above result, we can get the final result.

$$D_{feature}(FA, FB) = \frac{(\sum_{i=1}^{m} |FA_i - FB_i|^2)^{1/2}}{2 \cdot \sqrt{l-1}}$$

$$\leq (\sum_{i=1}^{n} |a_i - b_i - m|^2)^{1/2} = D_{minimum}(A, B)$$

$$\text{where} \quad m = \sum_{i=1}^{n}(a_i - b_i)/n$$

3.3 Query Processing

We present the overall process of our indexing scheme in this section. Before a query is performed, we shall do some preprocessing to extract feature vectors from time sequences, and then to build an index. After the index is built, the similarity search can be performed to select candidate sequences from a database.

Preprocessing

- Step 1. Feature Extraction as Dimensionality Reduction : Each time sequence is divided into m segments. Then, feature vectors are extracted from time sequences using the method mentioned in Section 3.1 for all time sequences in a database.
- Step 2. Index Construction : We build a multidimensional index structure such as R-tree using the feature vectors extracted from time sequences.

Index Searching and Postprocessing After an index structure has been built, we can perform the similarity search against a given query sequence. The searching algorithm consists of two main parts. The first is for candidate selection and the other is for postprocessing to remove false alarms. Some non-qualifying time sequences may be included in the results of candidate selection because the $D_{feature}(FQ, FD)$ is the lower bound of $D_{minimum}(Q, D)$. The actual minimum distance between query sequence and candidate sequences are computed and only those within the error bound are reported as the query results. The implementation of the segmented sum of variation indexing is described in Algorithm 1.

4 Performance Evaluation

To measure the effectiveness of our indexing scheme, we compared the SSV-indexing with the sequential scanning in terms of the number of actual computations required. We evaluated the search space ratio to test the filtering power of removing irrelevant time sequences in the process of index searching. The search space ratio is defined as follows.

Algorithm 1: SSV-Indexing

```
begin
    Input: query time sequence Q, error bound ε
    Output: data time sequences within error bound ε

    Result ← NULL;
    Candidate ← NULL;

    // project the query sequence Q into the feature space,
    FQ ← FeatureExtraction(Q);

    // candidate selection using a Spatial Access Method,
    // if (D_feature(FQ, FD_i) ≤ ε) then D_i can be a candidate,
    Candidate ← FeatureSpaceSearching(FQ,SAM,ε);

    // postprocessing to remove false alarms,
    for all C_i ∈ Candidate do
        if ComputeMinimumDistance(Q, C_i)≤ ε then
            Result ← C_i ⋃ Result;
        else
            Reject C_i;
        end
    end
    return Result;
end
```

$$search\ space\ ratio = \frac{the\ number\ of\ candidate\ sequences}{the\ number\ of\ sequences\ in\ a\ database}$$

For the experiment, we have used the real sequence data obtained from Seoul Stock Market, Korea[1]. The stock data were based on their daily closing prices. We collected 2000 stock sequence data of average length 128. We have run 25 random queries over real dataset to find similar time sequences based on the minimum distance. The query sequences are randomly selected from the database. The sequential scanning searches an entire database. It computes the minimum distance between query sequence and all data sequences in a time series database. As the size of a time series database increases, the sequential scanning becomes deteriorated in performance. Therefore, an efficient indexing scheme is quite required to reduce the search space. Figure 2 shows the experimental result that there is a significant gain in reducing the search space by the SSV-indexing.

5 Conclusion

In this paper, we considered the problem of efficient matching and retrieval of time sequences based on the minimum distance. The minimum distance can give a better estimation of similarity in shape between two time sequences than

---

[1] This data can be retrieved from http://www.kse.or.kr/kor/stat/stat_data.htm

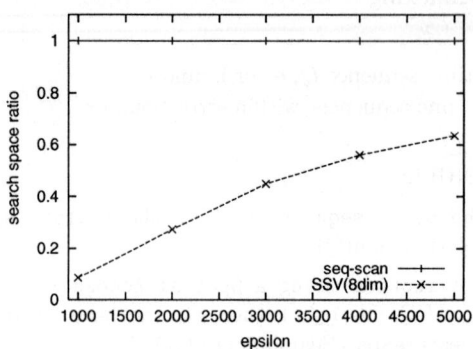

Fig. 2. Search Space Ratio: SSV-indexing vs. Sequential Scanning

the Euclidean distance. To support minimum distance queries, most of previous work has the preprocessing step of vertical shifting that normalizes each time sequence by its mean. In this paper, we have proposed a novel indexing scheme that supports minimum distance queries without vertical shifting in time series databases. We introduced the lower bound of the minimum distance to filter out dissimilar time sequences without false dismissals. We also showed via experiment that our indexing scheme could reduce the search space efficiently.

References

1. R. Agrawal, T. Imielinski and A. Swami: Database Mining: A Performance Perspective. IEEE TKDE, Special issue on Learning and Discovery in Knowledge-Based Databases 5-6(1993) 914-925
2. Usama M. Fayyad, Gregory Piatetsky-Shapiroa and Padhraic Smyth: Knowledge Discovery and Data Mining: Towards a Unifying Framework. In Proc. of the Second International Conference on Knowledge Discovery and Data Mining(1996) 82-88
3. A. Guttman : R-trees: A Dynamic Index Structure for Spatial Searching. In Proc. of SIGMOD Conference on Management of Data(1984) 47-57
4. Rakesh Agrawal, Christos Faloutsos and Arun N. Swami: Efficient Similarity Search In Sequence Databases. In Proc. of International Conference on Foundations of Data Organization and Algorithms(1993) 69-84
5. Christos Faloutsos, M. Ranganathan and Yannis. Manolopoulos: Fast Subsequence Matching in Time-Series Databases. In Proc. of SIGMOD Conference on Management of Data(1994) 419-429
6. Davood Rafiei, Alberto O. Mendelzon: Efficient Retrieval of Similar Time Sequences Using DFT. In Proc. of International Conference on Foundations of Data Organization and Algorithms(1998)
7. Kelvin Kam Wing Chu, Sze Kin Lam and Man Hon Wong: An Efficient Hash-Based Algorithm for Sequence Data Searching. The Computer Journal 41-6(1998) 402-415
8. Kin-pong Chan, Ada Wai-chee Fu: Efficient Time Series Matching by Wavelets. In Proc. of International Conference on Data Engineering(1999) 126-133

9. Dina Q. Goldin, Paris C. Kanellakis: On Similarity Queries for Time-Series Data: Constraint Specification and Implementation. In Proc. of International Conference on Principles and Practice of Constraint Programming(1995) 137-153
10. Chung-Sheng Li, Philip S. Yu and Vittorio Castelli: HierarchyScan: A Hierarchical Similarity Search Algorithm for Databases of Long Sequences. In Proc. of International Conference on Data Engineering(1996) 546-553
11. Davood Rafiei, Alberto O. Mendelzon: Similarity-Based Queries for Time Series Data. In Proc. of SIGMOD Conference on Management of Data(1997) 13-25
12. Gautam Das, Dimitrios Gunopulos and Heikki Mannila: Finding Similar Time Series. In Proc. of European Conference on Principles of Data Mining and Knowledge Discovery(1997) 88-100
13. Byoung-Kee Yi, H. V. Jagadish and Christos Faloutsos: Efficient Retrieval of Similar Time Sequences Under Time Warping. In Proc. of International Conference on Data Engineering(1998) 201-208
14. Sze Kin Lam, Man Hon Wong: A Fast Projection Algorithm for Sequence Data Searching. Data and Knowledge Engineering 28-3(1998) 321-339
15. Kelvin Kam Wing Chu, Man Hon Wong: Fast Time-Series Searching with Scaling and Shifting. In Proc. of Symposium on Principles of Database Systems(1999) 237-248
16. Davood Rafiei: On Similarity-Based Queries for Time-Series Data. In Proc. of International Conference on Data Engineering(1999) 410-417
17. Eamonn J. Keogh, Michael J. Pazzani: A Simple Dimensionality Reduction Technique for Fast Similarity Search in Large Time Series Databases. In Proc. of Pacific-Asia Conference on Knowledge Discovery and Data Mining(2000) 122-133
18. Sanghyun Park, Wesley W. Chu, Jeehee Yoon and Chihcheng Hsu: Efficient Searches for Similar Subsequences of Different Lengths in Sequence Databases. In Proc. of International Conference on Data Engineering(2000) 23-32
19. Byoung-Kee Yi, Christos Faloutsos: Fast Time Sequence Indexing for Arbitrary Lp Norms. In Proc. of International Conference on Very Large Data Bases(2000) 385-394
20. Eamonn J. Keogh, Kaushik Chakrabarti, Sharad Mehrotra and Michael J. Pazzani: Locally Adaptive Dimensionality Reduction for Indexing Large Time Series Databases. In Proc. of SIGMOD Conference on Management of Data(2001) 151-162

A Multi-attribute Decision-Making Approach toward Space System Design Automation through a Fuzzy Logic-Based Analytic Hierarchical Process

Michelle Lavagna and Amalia Ercoli Finzi

Politecnico di Milano - Dipartimento di Ingegneria, Aerospaziale
Milano-Italy

Abstract. This paper presents a method to automate the preliminary spacecraft design by applying both a Multi-Criteria Decision-Making (MCDM) methodology and the Fuzzy Logic theory. Fuzzy logic has been selected to simulate the human thinking of several experts' teams in making refined choices within a universe of on-board subsystem solutions according to a set of a given sub-criteria generated by the general "maximum product return" goal. Among MCDM approaches, the Multi-Attribute (MADM) has been chosen to implement the proposed method: starting from the Analytic Hierarchical Process (AHP), criteria relative importance is evaluated by a combination-sensitive weight vector obtained through a multi-level scheme. Uncertainty intrinsic in technical parameters is taken into account by managing all quantities with the interval algebra rules. Comparison between simulation results and existing space systems showed the validity of the proposed method. The results are really encouraging as the method detects almost identical combinations, drastically reducing the time dedicated to the preliminary spacecraft design. The suggested preliminary spacecraft configuration is also the nearest -according to an Euclidean metric in the criteria hyperspace- to the optimum detected by the MODM approach.

1 Introduction

This work faces the problem of automating the preliminary spacecraft design (the so-called phase-A)in terms of the on-board subsystem alternative selection, driven by a prefixed criteria vector. Such a tool may represent a great support for the designers' team to reduce the initial effort, and to lead further decisional processes devoted to the design refinement.

The space system design is intrinsically a complex multi-leveled process that involves several fields of scientific and engineering knowledge because of the numerous on-board devices devoted to really different tasks. As a consequence several iterative trade-offs must be done to achieve a final compromise between real device performance (e.g.propulsor specific impulse, battery capacity, solar cell efficiency, material stiffness, etc.) and desired mission objective satisfaction.

That dynamic process, at the state of the art, is managed by different teams of engineers -expert in each subsystem field - with the contribute of the scientists, expert in the on board payload instruments: by iteratively pruning existing as well as new solutions in each subsystem field to answer the mission requirements the teams converge to a first preliminary spacecraft configuration in terms of on-board subsystem set, launcher, ground network, operation requirements.

An approach - new for the space field - devoted to obtain good designs quicker, is on going at the European Space Agency, at the Jet Propulsion Laboratory, at Astrium and at Alenia Spazio[1],[9] (i.e. *Concurrent Engineering* approach).

In the field of a complete automation of the space system design process a limited number of works can be found, mostly based on the Genetic Algorithm technique [3],[4], [10].

In order to define a proper theoretic approach towards automation, two main characteristics must be highlighted in any complex system design process: a highly analytic aspect related to the lower level of designing each subsystem and simulating its performance; a heuristic aspect strictly related to the higher level of comparing intermediate solutions for each subsystem according to the criteria and constraint sets.

The first one is greatly answered by existing sophisticated software packages based on well-known mathematical models; the second one is left to the system engineers'expertise in addressing choices by pruning some solutions, by changing alternatives in one field (i.e. power supply) rather than in an other one (i.e. communications), by forcing the design direction. The pruning, changing and forcing processes can be classified, from a theoretical point of view, as a constrained Multi-Criteria Analysis driven by a deep domain knowledge .Hence,in order to automate the subsystem selection human inference method must be captured.

Multi-Criteria Analysis are faced by the Decision-Making branch that can be split into the Multi-Attribute (MADM) and the Multi-Objective (MODM) Decision - Making depending on whether the number of options is finite and predetermined or not [12].

Here a MADM-AHP (Analyic Hierarchical Process) approach is proposed as the better fitness to the actual spacecraft design process as actual solutions for each on board subsystem device are limited; moreover,the AHP - proposed by Saaty [11] - is here applied to obtain a better mapping of the complex system to be represented.

The MODM approach has also been considered to further validate the method according to optimality features [7].

The core of the proposed method is the expert system settled to simulate designers' option selection mental process: human logic is made of natural language propositions inferred each others on the basis of an acquired rule base; moreover, it does not answer the non-contradiction and the third excluded principles, that is it cannot be completely represented by a Boolean logic model: for those reasons the engineers' expertise has been here captured by models based on the Fuzzy Logic (FL) Theory – a solution to the Approximate Reasoning problems

– that not only translates natural language propositions in a mathematical formulation but also allows Multi-Valued logic([5], [15]).
In the following the proposed method is firstly presented. Attention is then focused on the FL application by reporting a module implementation, as an example. Simulations run to obtain validation are then proposed. Some final remarks are, finally, given.

2 The Proposed Method

As already explained, the problem has been approached with a MADM methodology in an Approximate Reasoning domain. Within MADM problems a main criteria vector $\underline{G}$(1xm) and the vector $\underline{Y}$(qx1) of possible alternatives (options) representing the solutive domain is either directly given by the user or computed by a pre-processing module.
In particular the $\underline{G}$ and $\underline{Y}$ elements title, respectively, the columns and the rows of the so-called decisional matrix; the fulfillment of that matrix is normally given by the domain-experts in a natural language formulation (e.g. *option y_i is quite relevant according to the criterion g_j*,etc.)[11],[2].
Within the spacecraft design, the $\underline{G}$ criteria vector collects both technical and financial requirements for the mission to be designed, that is both the engineers' and customers' point of view; the $\underline{Y}$ vector represents the feasible spacecraft configuration space generated by all possible combinations of each $\underline{Z}_j$ sub-space element ($j \in [1\ p]$ =on-board subsystems); the $\underline{Z}|_j(X,R)$ vector represents the set of actual possible devices to accomplish the j-th subsystem tasks; the $\underline{X}$ and $\underline{R}$ vectors are, respectively the technical variables to be defined by the sizing process and the current mission parameters.
Hence, the $\underline{Y}(\underline{Z}(X,R))$ is a (qx1) vector ($q = \prod_{i=1}^{p} n(j)$, n(j)=number of considered alternative devices for the j-th subsystem).
The MADM algorithms available in literature differ, basically, in the decisional matrix fulfillment, in the natural-language $\rightarrow$ numerical formulation translation and in the numerical decisional matrix management to achieve the final $\underline{Y}$ element ranking, according to the given $\underline{G}$vector [2],[12]. As the current decisional problem is quite articulated and a multi-leveled criteria net has to be created to simulated the human reasoning, the so called Analytic Hierarchical Process (AHP) has been selected to be faithful to the real process as much as possible[11],[13]. According to that theory, the main criterion (or each element of the main criteria vector) is further decomposed into several dependent inner levels made up of related sub-criteria: hence,the decisional problem to be faced is mapped with an increased granularity according to each of its nested aspects. Each decomposition level generates a local $\underline{G}_i$ criterion vector ($i \in [1\ k]$ =no. of decomposition levels) the elements of which have to be ranked according to their importance for the parent criterion: a $\underline{w}_i$ weight vector is, then, attached to the $\underline{G}_i$ criteria vector in the current i-th level. In order to better understand the current application to the space system design domain tab.1 shows the decomposed decisional matrix to be filled.

The main criterion (e.g.scientific data amount, telecommunication coverage,

Table 1. AHP: spacecraft design decomposed decisional matrix

1	Product Return			
2	Technical (w'_T)		Non-Technical($w_{NT'}$)	
3	Gross Mass(w_m)	Required Power(w_P)	Cost(w_C)	Reliability(w_R)
4	o_{1m}	o_{1P}	o_{1C}	o_{1R}
...	...	...	...	...
q	o_{qm}	o_{qP}	o_{qC}	o_{qR}

etc), is hierarchically split into the financial and technical aspects to be considered within each feasible system configuration;a further decomposition is done adding a level 3: the global mass and the required electric power are selected as the two main technical criteria; the cost and the reliability are taken into account within the financial domain.

In particular, low wet mass, low required power, low costs and high reliability are aimed.

The $\underline{w}_i$ ($i = [1\ k]$) weight vectors together with the $\underline{O}$ option matrix elements have to be given either by the user or by the domain expert; finally, by a bottom-up procedure on the decisional matrix the global ranking of the $\underline{Y}$ elements is computed. Eq.1 gives the formulation for the aforementioned iterative process.

$$L_{s-1,j} = \sum_{i=1}^{m(s)} w_i \cdot L_{s,ij}(\underline{x},\underline{r}) \forall s \in [1\ k-1];\ \forall j \in [1\ q] \quad (1)$$

$L_{s,ij}$ = Index of relevance of the visited option y_j according to the g_i criterion in the s-th level
s = current decomposition level [1 k]
$m(s)$ = No. of criteria in the current s-th level
w_i = Relevance of the criterion g_i in the current s-th level according to the parent criterion in the $s-1$ level

According to tab.1, for $s = 4 \rightarrow L_{4,ij} = o_{ij}$.
According to the AHP, (m+k-1) matrices of pairwise comparison are created: the first m(qxq) matrices, one for each option row, are filled to obtain the final absolute $\underline{O}$ (qxm(k)) option matrix [11]; the remaining k-1 (m(s)xm(s)) matrices give - within each decomposition level - each criterion absolute weight versus its parent criterion. By eventually computing the eigenvector of the maximum eigenvalue for each aforementioned matrix, the required $\underline{O}_i$ and $\underline{w}_s$ arrays to compute the final L_j index are obtained.

The aforementioned (m+k-1) matrix of pairwise comparison elements reflects the human expertise according to the current domain, and they are formally represented by qualitative propositions; according to perception analysis Saaty and

Lootsma proposed different scales to map the qualitative into quantitative[8]. In the current work, the Lootsma's exponential scale has been eventually selected as it gave simulation results with a global better mapping of actual configurations.

To gain automation in the $\underline{Q}_i$ ($\forall i \in [1\ m(k)]$) and $\underline{w}_s$ ($\forall s \in [1\ k-1]$) fulfillment, different methodologies have been defined: the proposed method to manage the $\underline{Q}$ (qx4) matrix and to consider the $\underline{X}$ vector element uncertainty together with its impact on the final cardinality computation for the $\underline{Y}$ options vector have been already exposed in [7].

In the following attention is focused on the k $\underline{w}_s(1xn(s))$ computation as it represents the core of the proposed method and it involves both the AHP and the FL theory.

3 The Criteria Weight Vector Computation

Usually, the $\underline{w}_s(1xn(s))$ vectors are set by the user as input and they remain fix throughout the process: then the ranking of the $\underline{G}_s$ elements according to their importance with respect to the parent criterion in the (s-1)th level, is stiff and highly user-dependent. However some problems, could be better modeled with variable weights according to the related $\underline{G}$ element values. In the spacecraft design domain, for example, an a priori fixed hierarchy between technical criteria cannot be set as both can make a solution unfeasible; they also cannot be considered as equal-weighted: within each visited combination both over-loaded and over-powered solutions should be highlighted: the worst criterion value should drive the combination position definition in the final global ranking of the whole feasible spacecraft configurations. On the contrary, it would not make sense to assume variable weights for the criteria related to no-strictly technical aspects: they are maintained combination-free, in the sense that their importance can correctly be considered to-be-defined by the spacecraft customer. Hence, weights related to the technical aspects are maintained variable in order to better simulate the designers' behavior in judging visited configurations from the technical point of view. To this end a second decisional matrix is created, dedicated to the second level technical weight $w_{t,j}$ computation($\forall j \in [1\ q]$ according to tab.1); tab.2 shows the decisional matrix devoted to the technical weight computation: a further level is added to the tab.1 scheme: the main criterion is decomposed till a fourth sub-criteria level for the technical branch, while decomposition stops to the third level for the non-technical branch; the fourth technical sub-criterion set represents the single on-board subsystem mass and power criteria; levels 5-q represent single possible configurations. With the former architecture, the user is asked just for the first two vectors in eq.2, while the technical weight vector $\underline{w}_T$ (see eq.2) is automatically computed by the expert module of the algorithm. To this end, a backward AHP procedure (eq.1) is applied to the weight matrix in tab.2.

$$w_{PR} = [w_{T'}\ w_{NT'}];\ w_{NT} = [w_c\ w_R];\ w_T = [w_m w_p] \qquad (2)$$

Table 2. Spacecraft design: Weight decisional matrix

1	Product Return						
2	Technical($w_{T'}$)					Non-Technical($w_{NT'}$)	
3	Gross Mass(w_m)		Required Power(w_P)			Cost(w_C)	Reliability(w_R)
4	$m_1(w_{m1})$	...	$m_p(w_{mp})$	$P_1(w_{P1})$	...	$P_p(w_{Pp})$	
5	wm_{11}	...	wm_{p1}	wP_{11}	...	wP_{p1}	
...	...	...	...	...	...	...	
q	wm_{1q}	...	wm_{pq}	wP_{1q}	...	wP_{pq}	

In particular, the fourth level weight vector is obtained by the eigenvector computation of the settled matrices of pairwise comparison .Tab.3 shows, as an example, the matrix of pairwise comparison for the mass criterion, considering four on-board subsystems; a similar matrix is created for the power technical criterion.

Actually, the matrices of pairwise comparison to be settled are more than two

Table 3. Subsystem pairwise comparison according to the mass

	Chem Propul (CP)	EPS	Thermal	TT&C
CP	1	slightly more relev	strongly more relev	more relevant
EPS	slightly less relevant	1	quite more relevant	slightly more relev
Thermal	strongly less relevant	quite less relevant	1	less relevant
TT&C	less relevant	slightly less relevant	more relevant	1

(mass and power), as there exist particular alternative device combinations that invert the nominal subsystem relative importance according to a fixed criterion. To implement the natural language → numbers translation an exponential scale has been used [8],[14] (see tab.4. By applying the classic AHP, the fourth level $\underline{w}$(1xp) criteria weight vector is obtained.

The $\underline{w}$(2xpxq) matrix(levels 5-q)is computed by several Fuzzy Logic blocks, two for each on board subsystem, devoted to model the qualitative dependencies between some selected technical parameters of the visited device (a $\underline{X}$ and $\underline{R}$ sub-vectors) and the related human score according to the current criterion (e.g. *IF the solar cell efficiency is high and the density is low THEN a photovoltaic power subsystem is light*, etc). A Mamdani approach is applied, as no quantitative relationship is available between inputs and output to be mapped by a

Table 4. Subsystem pairwise comparison according to the mass: Numerical formulation

	Chemical Propulsion	EPS	Thermal	TT&C
Chemical Propulsion	1	5	100	50
EPS	0.2	1	20	10
Thermal	0.01	0.05	1	0.5
TT&C	0.02	0.1	2	1

Sugeno scheme[15],[5]. The membership shapes and numbers, as well as the base of rules for the inference motor have been set trying to map both available data and system engineers' expertise.

By applying a Weighted Sum to each row from five to q of the matrix in tab.2 the $\underline{w}_T|_j$ vector in eq.2 is obtained; hence, by applying twice the eq.1, to the third level first and, consequently, to the second level of the matrix in tab.1 the final score is obtained.

3.1 The Fuzzy Logic Role: The Electric Power Subsystem (EPS) Example

Fuzzy logic plays a fundamental role in the $\underline{w}$ (2xqxp) element definition and it is the core of the human thinking simulation within the proposed method.

As an example the power supply and storage subsystem management is going to be analyzed. The electric power subsystem alternative scoring with respect to the fourth level sub-criteria is quite articulated as a further decomposition can be done within the framework of the electric energy suppliers: solar panels+batteries and completely autonomous electric sources. Within the current work, only the first solution is presented [6].

Their elements have to be ranked only according to the fourth level mass criterion as they, obviously, do not ask any power.

Within the (SA+BA)class, a further decomposition is done as solar panel and battery cell alternatives answer to different concurrent tasks.

A MISO architecture is implemented for the solar arrays correspondent Fuzzy Logic module with the following input technological quantities: the panel efficiency (η), the specific mass (ρ) and the power demand (P).

The power demand and the panel efficiency are modeled by similar memberships in shape and numbers, given in tab.5.

For the mass density model only two qualities have been identified in order to sharply distinguish between light and heavy power devices; analytic specifications are not given for the sake of brevity.

The $w_{sa_{mass}}$ output has a simple gaussian modelization in order to soften the input effects. Fig.3.1 shows the trend of the $w_{SA,mass}$ versus the power and the efficiency quantities.

The battery module gets- as input vector $\underline{X}$ the specific energy and the space-

A Multi-attribute Decision-Making Approach toward Space System Design

Table 5. Memberships modeling the P and the η inputs for the EPS w_{mass}

$P_{i\,range} = [0\ \max_i P_i]$	$\eta_{is\,a\,range} = [0\ 0.3]$	
low	medium	high
$\mu(w_j) = e^{(\frac{x_i}{0.2123\,\max_i x})^2}$	$e^{(\frac{x_i - 0.5\,\max_i x}{0.2123\,\max_i x})^2}$	$e^{(\frac{x_i - \max_i x}{0.2123\,\max_i x})^2}$

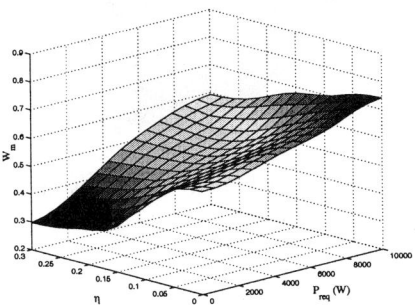

Fig. 1. Trend of the weight according to the mass criterion within the power supply-solar array solution: fuzzy model

craft power demand; the required power is mapped into its correspondent fuzzy set as explained for the power supply module (tab.5) while the energy density is represented by three triangular memberships tuned on the basis of the existing battery cells (Nichel Cadmium, Nichel-Hydrogen and Lithium-ion). Fig.3.1 shows the particular battery device $w_{iba_{mass}}$ versus its technical paramters.

To evaluate the relative importance of the current solution (SA+BA) within the whole EPS subsystem domain (level four),a simple weighted sum (as in the classic WSM-MAUT) is applied.

$$w_{EPS_{mass}} = \sum_{j}^{2} a_j b_{j,i} \quad a_2 = \neg a_1 = \neg \gamma \quad \underline{b}_i = [w_{i-ba_{mass}}\ w_{i-sa_{mass}}]^T \quad (3)$$

As the γ parameter is devoted to balance the relative importance of the solar arrays mass and the battery mass within the current configuration, the logic reasoning is simulated by considering determinant mission parameters (i.e. the sun distance D and the eclipse time/orbit T_{ecl}).

The γ output is mapped in the fuzzy domain by three gaussian memberships on $0-1$, while the inputs have dedicated modules. The obtained surface for the γ parameter is given in fig.3.1.

Tab.6 gives the selected operators within the Fuzzy Logic environment.

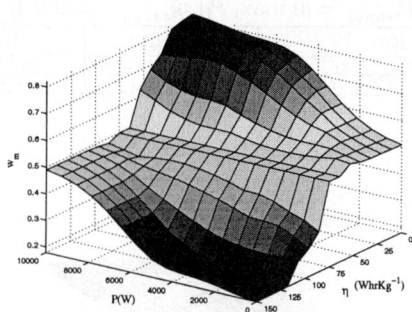

Fig. 2. Trend of the weight according to the mass criterion within the power storage-battery solution: Fuzzy model

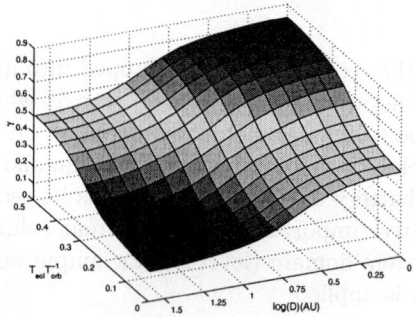

Fig. 3. Trend of the weight according to the mass criterion to balance solar arrays versus battery importance: Fuzzy model

Table 6. Selected Fuzzy Operators

Operation	t-norm	t-conorm	inference	de-fuzzification
Operator	min	max	min	centroid

4 Simulation Results

In the following results obtained by the proposed MADM approach are given for the Mars Global Surveyor NASA mission assumed as a reference to evaluate the method. Results related to the MODM approaches are discussed in [7].
Tab.7 gives a comparison among the real mission and the first suggested configuration, in the criteria domain. As it is possible to notice, the global mass is definitely better than the real one, mainly due to a fuel saving thanks to the selected propulsion module(electric ruther than chemical); for the same reason the power demand is slightly increased. For sake of brevity a complete dissertation on the whole solutive space cannot be carried on, but the obtained cardinality correctly perceives - through the technical weight trend - changes in the criteria values due to different device combination.

Table 7. MGS first case simulation results: First Suggested Configuration Criteria values

Criterion	Real Mission	LLV	MLV
Dry Mass(Kg)	670	324.628-475.778	358.713-369.445
Fuel Mass (Kg)	360.5	215.277-370.2907	245.151-288.545
Power Demand (W)	980	1007.118-1518.113	1166.987-1312.625
Global Cost (M$)	150	133.036-506.405	199.505-285.334
System Reliability	-	0.8469-0.9328	0.8755-0.9064
Mass weight	-	0.50785	
Power weight	-	0.57463	

5 Conclusions

The paper presents a method to automatically manage the MCDM of the spacecraft design obtaining a very preliminary configuration.The fuzzy logic theory has been used to implement the inference motor of linguistic dependencies, which leads the human behavior in making choices. A MADM-AHP approach has been assumed as a baseline to manage the multi-criteria decisional domain. The code drastically lowers the amount of time devoted to a preliminary design, and gives a complete cardinality of the feasible configurations risen from the input alternatives for on board-subsystem devices. Comparison with already designed probes gave satisfactory results as the suggested configurations is definitely similar, from the criteria and the on board subsystem device point of view, to the real spacecraft.

References

1. M. Bandecchi, B. Melton, F. Ongaro, *Concurrent Engineering Applied to Space Mission Assessment and Design*, ESA Bulletin, September 1999
2. A. De Montis, P. de Toro, B.Droste-Franke, I.Omann, S.Stagl, *Criteria for quality assessment of MCDA Methods*, Third Biennial Conference of the European Society for Ecological Economics, Vienna, May 3-6, 2000
3. A. S. Fukunaga, *Application of an Incremental Evolution Technique to Spacecraft Design Optimization*, 0-7803-3949-5/1997 IEEE
4. A. S. Fukunaga, S. Chien, D. Mutz, R. Sherwoood, A. Stechert, *Automating the Process of Optimization in Spacecraft Design*, Jet Propulsion Laboratory Internal Report
5. G. J. Klir, B. Yuan, *Fuzzy Sets and Fuzzy Logic-Theory and Applications*, Prentice Hall PTR-New Jersey 1995
6. W. J. Larson, J. R. Wertz, *Space Mission Analysis and Design*, Space Technology Library-Kluwer Academic Publishers,1993
7. M. Lavagna, A. E. Finzi, *Space System Design Automation through a Fuzzy Multi-Criteria Decision Making Approach*Proceedings of the International Symposium in Space Flight Dynamics, 1-6 December, 2000-Pasadena-CA
8. F. A. Lootsma, *The French and American School in Multi-Criteria Decision Analysis*, Operation Research, 24/3, pp. 193-305
9. A. Martelli, *The DSE System*, Proceedings of DASIA Conference, Nice-FR, Sept. 2001
10. T. Mosher, *Spacecraft Design Using a Genetic Algorithm Optimization Approach*, 0-7803-4311-5/98, IEEE-1998
11. T. L. Saaty, *A scaling Method for Priorities in Hierarchical Structures*, Journal of Mathematical Psychology, 15(3), pgg. 234-281,1977
12. E. Triantophyllou *Multi-Criteria Decision Making: An Operations Research Approach*Encyclopedia of Electrical and Electronics Engineering, J. G. Webster, Ed. ,John Wiley & Sons, N. Y. Vol. 15, pp. 175-186 (1998)
13. E. Triantophyllou *Using The Analytic Hierarchy Process for Decision Making in Engineering Applications: some Challenges* International Journal of Industrial Engineering: Applications and Practice, vol. 2, no. 1 pp. 35 -44, 1995
14. E. Triantaphyllou, F. A. Lootsma, P. M. Padalos, S. H. Mann, *On the Evaluation and Application of Different Scales for Quantifying Pairwise Comparisons in Fuzzy Sets*, in Journal of Multi-Criteria Decision Analysis, vol. 3 no. 3pp. 133-15, (1994)
15. R. R. Yager, D. P. Filev, *Essentials of Fuzzy Modeling and Control*, John Wiley & Sons Inc. 1994

A Case Based System for Oil and Gas Well Design

Simon Kravis[1] and Rosemary Irrgang[2]

CSIRO Mathematical and Information Sciences
[1] PO Box 664, Canberra City Act 2601
[2] Locked Bag 17 North Ryde NSW 1670

Abstract. A case base system for a complex problem like oil field design needs to be richer than the usual case based reasoning system. The system described in this paper contains large heterogeneous cases with metalevel knowledge. A multi level indexing scheme with both preallocated and dynamically computed indexing capability has been implemented. A user interface allows dynamic creation of similarity measures based on modelling of the user's intentions. Both user aiding and problem solution facilities are supported, a novel feature is that risk estimates are also provided. Performance testing indicates that the case base produces on average, better predictions for new well developments than company experts. Early versions of the system have been deployed into oil companies in 6 countries around the world and research is continuing on refining the system in response to industry feedback.

1 Introduction

Case Based Reasoning (CBR) systems store data and knowledge as a set of cases and reuse the prior knowledge to solve new problems. CBR systems can learn by storing the results of a new problem case for future use. Smythe [11] asserts that in the CBR literature it has become apparent that the "single shot" CBR in which a single case is used to solve a problem, is not adequate for complex tasks. In design and planning tasks, multiple cases are often needed to solve different parts of a complex real world problem. This paper documents a new approach to multiple case reuse involving predefined case decomposition and recomposition into a synthetic new case. Statistical methods have been incorporated to provide both mean value and risk estimates for time and cost of the operations contained in the new case.

Optimal, safe drilling of oil and gas wells is a complex operation requiring knowledge across a range of disciplines. Wells can be 10 kilometres long and may be drilled under thousands of metres of seawater. Well trajectories frequently must follow a tortuous path or track a thin horizontal reservoir column. Problems such as borehole instability can delay operations for weeks with offshore rig rates costing over $200,000 US per day. Unexpected overpressure zones can lead to a blowout and even loss of the rig. Drillers learn from each well drilled in an area but with high turnover of staff this knowledge is frequently lost to the company.

The case based software system developed, Genesis, was designed to capture expert drilling knowledge to support intelligent planning of new oil and gas wells. The core of the system is a global case base of data, information and knowledge contributed by participating companies, in the belief that capturing and sharing experience will lead to better well designs in the future with significant savings in time and cost.

Kolodner [9] believes that reasoning is more a process of "remember and modify" than "decompose and recompose". This system can be used to retrieve well instances and reuse them with modification, however a key module supports decomposition of all the wells into predefined small components. These are then recomposed into a synthetic new well case using a large number of analogous wells or sections of wells from the case base. The system functions both as an assistant and a problem solver.

Kitano [8] asserts that the vast majority of case based reasoning systems have been built as task-specific domain problem solvers and are detached from the mainstream processes of a corporation. Issues that are not addressed in such systems include:

1. Integration of the case base with existing company database systems
2. Security Control: In real applications, cases contain confidential data and secure access is required
3. Scalability: Collected cases can increase dramatically over time and in real life applications, complex multimedia data may need to be stored. Indexing will need to be constantly refined as the system grows and application needs evolve.
4. Speed: Fast case retrieval and adaptation is essential

The four points noted above have been addressed in Genesis. Firstly, the case base is stored in a relational database system. While this is not ideally suited to a case based system, it can be integrated with existing company databases, and techniques to add the additional indexing required for case based retrieval have been developed. Security has always been a key requirement as the system contains valuable confidential oil field data, this is supported by the database and use of encryption where required. A two layer index system has been adopted for retrieval of cases based on text concepts. The index is created and stored in the database. If retrieval times degrade, further layers of the index can be added to increase the speed.

The following section describes the case base design process, case contents and structure. Section 3 covers details of methods used for retrieval of cases. Section 4, on adaptation, details how cases retrieved can be used for problem solution. Section 5 reports results of performance testing, followed by discussion of research directions.

2 Case System Design

Facilitated industry workshops were held with oil companies to design a system to suit their needs. A case based system was favoured by the participants, as most appropriate for the capture and reuse of drilling knowledge and data. A common theme was that people believe that a case store must be active, it must contain problem solving processes to reuse the knowledge for a new design. Engineers wanted additional knowledge added to their normal design processes, for example displayed on plots used for planning, or used for risk assessment. This project attempts to capture and preserve company knowledge and to make it available in an active, easy to use form. Some innovative techniques designed to achieve these goals include:

- Modelling of user intent so that aspects of the information relevant to the user are retrieved and tailored to suit the users' plans.
- Algorithms for automated extraction of knowledge indexes from text reports, well summaries and from the drilling database.
- Use of Drilling Grammars to describe and recognise complex indexing concepts
- Machine learning to iteratively refine the concepts
- Metadata store of generalisations of common problems and solutions
- Techniques for information sharing while respecting company confidentiality

An important aspect of the design was that engineers want to feel in control. They are not prepared to allow a black box system to design their new wells with complete autonomy. Access to the original stored information for checking is required together with the ability to select or remove cases and add and delete operations at any stage.

2.1 Contents of Each Case

Each oil well represents one case instance. A well has attributes such as location, geology, maximum depth and trajectory type (e.g. vertical or deviated). Data and knowledge about each well is captured and stored in a database. Wells that have been already drilled are stored as historic data, new designs as planned wells. Large volumes of numeric data are stored. For example, multiple channel electronic log data from down-hole instruments may be recorded every centimetre for a 10,000 metre well. Log data is used in calculation of variables important in indexing and similarity matching, such as rock strength and rate of penetration.

Well completion reports, mud logger's reports and other text information such as lessons learned can be automatically input (in word or HTML format) and these provide valuable overview knowledge. Images may also be included in a case store. For example, images of worn drilling bits are recorded for further analysis and classification. Industry reports often include Excel spreadsheets and these can be stored as part of each case. Access to oil industry experts for knowledge acquisition is limited. This led to the development of automated tools to extract knowledge and indexes for the case base from industry reports, published papers and end of well reports. Concept definitions are stored separately from the case base indexes and are read in by the software as part of an index creation run. Companies can have concept definitions tailored to their own requirements or can use the Genesis definitions. This feature proved worthwhile when the system was deployed into a Portuguese speaking company, as concept definitions in the local language could be used to generate new Portuguese indexes without any change to the software.

When a new planned well has been designed, the new case is incorporated into the system as a planned case. After drilling, the actual data is stored as a new historic well and can be compared to the planned well for evaluation of the new design case. Post-analysis graphics features are provided to highlight strengths and weaknesses of the new case synthesis. For example, as shown in Fig. 4, the actual time depth plot of a new well can be plotted on a graph of the planned well.

2.2 Case Structure

Three levels of case structure are implemented at present. Level 1 is a metalevel covering groups of cases. Knowledge relevant to a whole basin or to a formation (geology) type is stored at this level. One index includes all wells with borehole instability problems, and another stores the main problems and solutions common to each basin, with links to the original wells from which the metadata was extracted. Level 2 indexes well attributes (e.g. location) or groups of attributes. Level 3 indexes on a decomposition of each well into a set of defined drilling phases and operations.

Fig. 1. User interface operation, retrieving wells similar in location and trajectory

3 Retrieving Similar Cases

Selection of the best cases for adaptation has been found to depend on the planned user application. For the purposes of planning a new well, the most useful cases usually come from the same geological region, with similar well path deviations and using similar hole sizes. However if a new technology such as slimhole (drilling with a very small hole) is planned it may be better to use wells with the new technology even if they are from a remote location. Users can input details of their proposed new well such as location, final hole size and approximate planned trajectory. Similarity measures are then built dynamically using a weighting derived from the user selection.

The most common retrieval method in Genesis is based on a similarity match between the user's new planned well and well cases in the database. Users can specify which features of each case are most relevant to the new well plan or system defaults can be used. An example is shown below in which the user has specified that he would like to use cases from similar locations and with similar well trajectories.

This is done by first clicking on the location image, then selecting an area defined by a latitude and longitude rectangle, location is accorded a priority of 1 in this example. Clicking the trajectory image selects trajectory as the second most relevant factor. A similarity metric using only the two user selected attributes and weighted for his priority choice is used to find the 10 wells most similar to the new planned well. Results of the search are shown in the list box: Pothos #1 is the most similar stored well in this example.

3.1 Similarity Metrics for Numeric Data

All numeric attributes are scaled to lie in the range 0 to 1, and may be generated using higher-level reasoning rather than simply scaling or application of a user-defined function. Some rules and simple drilling knowledge are applied when generating each component of the metric. For example, relevant geology requires that retrieved wells are in the same area as the planned well. Filtering on attributes is also available, for example the search can be restricted to a region such as the Cooper Basin. For n Numeric attributes, an n-dimensional weighted distance metric of the form below is used:

$$Sim_{k,j} = 1 - \sqrt{\sum_{i=1}^{n} w_i \left(\frac{f(X_{k,i} - X_{j,i})}{SF} \right)} \quad . \quad (1)$$

where

- X = selected attribute in the planned and stored wells.
- i = the attribute index
- k = new planned well index
- j = case base index
- w = weighting factor, derived from the user selection.
- SF = scaling factor.

User modelling is used for the weightings which are calculated to sum to 1. The scaling factor SF is required to compensate for the vastly differing numeric ranges of variables used in the similarity metric. Distances between wells can be many kilometres, whereas the numbers used for trajectory similarity measures are generally ratios such as TVD/MD (Total Vertical Depth/Measured Depth) and range between 0 and 1. Weighting factors are derived from the user profile selection, in the example shown we have only 2 weights, location is weighted as 2/3 and trajectory as 1/3.

3.2 Retrieval Based on Natural Language Query

Complex non numeric concepts require a higher level of abstraction and similarity matching. As an example, users may want to know what were the main problems found in the basin, and select cases solving similar problems. An example is shown in Fig 2, the user asks questions and the system searches for known indexing concepts in the word string to use for retrieval of appropriate cases. The system has an automated high level indexing system based on concepts of interest to engineers drilling oil and gas wells. Information on main problems, for example is extracted automatically from oil company reports.

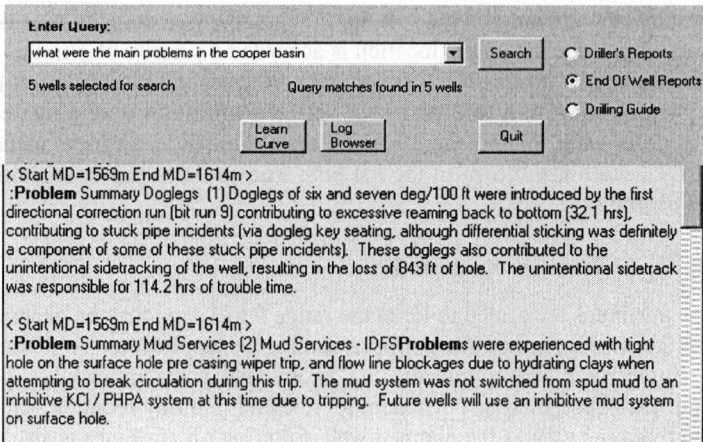

Fig. 2. Result of a natural language retrieval. Concept "main problems" used

Text pattern matching techniques are used to find relevant information in these reports. One criterion for the concept "main problems" for example, is that a relatively large amount of time is wasted because of the problem. In the example in Fig. 2, the unintentional sidetrack was responsible for 114.2 hours of trouble time. Extraction of time and depth information is also automatic based on depth and time pattern matching definitions. Higher level concepts can be combined with numeric similarity matching to retrieve similar wells. Once we have found cases with dogleg or sidetrack problems for example, then from this subset, we can search for the most similar cases based on numeric criteria. Further details of the text extraction methods can be found in [7]. A fallback index is implemented for use where no predefined concept can be found in a natural language question. This index is not based on domain knowledge but indexes on ngrams or groups of contiguous letters [12].

3.2.1 Machine Learning to Refine Retrieval Concepts

Techniques to automatically refine the drilling concepts with user feedback using CART [2] algorithms have been implemented. Users can select natural language retrieval text and rate it as relevant, irrelevant or not sure. The information is stored and run periodically to refine existing concepts. Results so far are promising but have highlighted the need to add context and more complex drilling knowledge to the existing simple concept definitions, to improve retrieval rates much above the 75% currently achieved [7].

4 Case Adaptation

Although cases can be retrieved for single case reuse with modification, the most popular feature of the system is adaptation via decomposition of multiple cases and recomposition to produce a risked complete new planned well instance. Time and costs for the new well are also estimated based on data from similar well components.

These feed automatically into an Authorization for Expenditure (AFE) spreadsheet. Feedback studies during project development and deployment into industry have consistently indicated that engineers don't have time to browse through cases and would like the system to deliver information as part of their normal work flow. Ideally the system can automate existing tasks such as well cost estimation and AFE production.

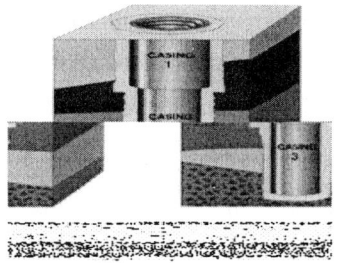

Fig. 3. Schematic [1] showing surface, intermediate and production casing phases

4.1 Classifying Drilling Activities

4.1.1 Drilling Phases

Fig. 3 illustrates three well phases. These generally correspond to hole/casing sizes. The largest hole and casing size is at the surface. Generally drilling operations for similar hole sizes at similar depths are analogous, so decomposition of well cases into components of similar size is useful for synthesis of the new well. On the drilling rig, drillers produce text reports at regular intervals each day. These reports, known as morning reports, also contain the current drilling depth and time for each event. For decomposition of each well into generic component operations, a standard set of hierarchical classifications from the text has been devised. Operations are grouped into activities such as drilling and further classified as planned operations or responses to trouble. Table 1 gives further details of the classification.

Table 1. Activity Classifications

Descriptor	No. of Codes	
Class	6	Class of activity (Drilling, Evaluation and Completion) and whether they are planned or in response to trouble.
Major Operation	16	Broad description for the activity (e.g. Drilling, Rig Move, Coring)
Operation	31	More detailed description of activity (e.g. Rotary Drilling, Drilling with Motor, Running Casing)
Trouble Type	20	Type of trouble (e.g. Wait on Weather, Stuck Pipe, Equipment Failure)
Phase	40	Phase of well construction. (e.g. Pre-Spud, Conductor, 17 inch, End-Of-Well). Each change of well diameter of more than one inch is taken to be a separate phase.

Combinations of pairs of these primitive descriptor values can be grouped together to form classification groups. Formation related problems such as swelling clays and abrasive formations are identified using pattern matching of text fields in end of well reports or driller's morning reports. Generic formation classifications such as these are stored in records abstracted from a number of wells.

A useful aspect of the decomposition of wells into small primitive operations and statistical distributions of each operation is that it is possible to use confidential well information and metadata without revealing details of individual wells.

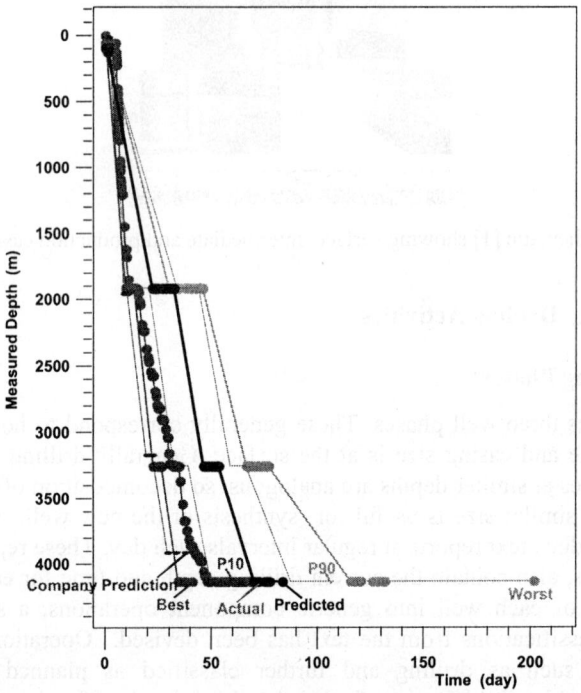

Fig. 4. Time versus depth plot showing well curve predicted by Genesis (blue) and actual well (grey dots). The red curve is the company prediction. Best (magenta) and worst (green) predictions and P10 and P90 curves are also displayed

5 Results of Performance Testing

Performance testing of a case based system requires some benchmark performance for comparison. A key output from the system is a risked prediction of time and costs for a new planned well based on relevant case base information from stored historical and planned well cases, this output was chosen for performance validation. Prior predictions by company experts of time and costs for a group of new wells was available and this data was used to compare Genesis predictions for the same wells based on cases available before commencement date of the new planned wells. Table 2 shows a summary of results from a comparison of Genesis and company estimates. Means have been calculated for the Genesis and company predicted - actual time estimates. Results show that Genesis has underestimated times by a mean of 3.6 days, the companies by 19 days. To compare the percentage error, standard deviations were calculated for both Genesis and company estimates, the ratio (Genesis SD/Company SD) is calculated. The result 0.66 indicates a 34% improvement in uncertainty could be gained using the Genesis predictions as well as removing most of the systematic underestimation bias. Further details of the performance tests can be found in [5].

Table 2. Comparison of Genesis and Company Time Predictions

Well	Actual	Plan	Genesis P10	Genesis Mean	Genesis P90	Genesis Mean - Actual	Company Plan - Actual	Genesis % Error	Company % Error
NWS-8	15.83	9.38	13.25	13.88	14.29	-1.96	-6.46	-12.37	-40.79
NWS-9	25.83	17.08	20.83	27.92	35.42	2.08	-8.75	8.06	-33.87
NWS-18	15.08	16.76	13.67	14.50	16.33	-0.58	1.68	-3.87	11.12
NWS-20	9.67	11.33	10.33	10.92	12.50	1.25	1.67	12.93	17.24
NWS-21	10.63	10.75	7.29	8.96	10.42	-1.67	0.13	-15.69	1.18
NWH-B	43.67	20.83	24.58	42.50	62.08	-1.17	-22.83	-2.67	-52.29
ME-68	150.00	62.92	87.92	98.33	119.17	-51.67	-87.08	-34.44	-58.06
ME-69	57.50	43.75	46.67	68.33	97.92	10.83	-13.75	18.84	-23.91
ME-70	72.50	40.00	62.00	83.00	107.00	10.50	-32.50	14.48	-44.83
Mean difference						-3.60	-19.00		
Mean absolute differences						9.07	19.43		
Std Dev						18.69	28.15		
Ratio of Genesis to Company Std Dev						0.66			

6   Research Direction: Creative Case Reasoning

A criticism of case based systems is that most tend to adapt and reuse old solutions in routine ways, producing robust but uninspired results. Oil engineers have also expressed concern that Genesis contains only technology from the past while rapid developments in new technology may provide better solutions to new problems. Two main initiatives are underway to address this problem. Collection of data from leading edge companies working at the technical limit of drilling is one such initiative. Companies viewing analogous wells that use advanced technology may more easily be persuaded to try the new devices or operations on their own wells. A common situation for drilling engineers is to decide whether to use a more expensive bit which may be able to drill an entire phase if it does not encounter a particular type of rock or fail prematurely. A case base of performance and problems with this type of bit will add valuable support for this decision. Cases can be synthesized to show predicted well times and costs using new technology, and also highlighting increased risks where these are significant. Another project is directed towards the automated production of summary information from recent papers and industry publications highlighting technologies or new solutions that could be substituted for old methods contained in historic cases.

7 Conclusion

A rich case based system has been produced for oil and gas well design. Cases contain heterogeneous information at varying levels of abstraction. Unlike many conventional CBR methods, problems are not solved by single case adaptation but can use multiple cases at varying levels of generalisation and decomposition. A novel feature is that the use of multiple cases also allows risk estimates to be calculated. Automated indexing of text reports and documents allows addition of high-level knowledge to the system with low maintenance. Companies using Genesis report large reductions in well design time as a new well can now be designed in hours rather than days. Empirical testing showed that the case system also outperformed company experts in predicting the time taken to drill new wells.

References

1. Baker, R.,: A primer of Oil Well Drilling, U of Texas, (1996)
2. Breiman, L., Friedman, J.H, Olshen, R..A., Stone, C . J.: Classification and Regression Trees, Wadsworth (1984)
3. Apte, C., Damerau,F. and Weiss, S. H.: Automated Learning of Decision Rules for Text Categorization, ACM Transactions on Information Systems, Vol. 12, No. 3,(July 1994.)
4. Harabagiu, Sanda M, Moldavan Dan I.: TextNet - A Text Based Intelligent System,: Natural Language Engineering 3 (2/3): 171-190 (1997)
5. Irrgang, R., Irrgang H., Kravis, S., Irrgang S., Thonhauser, G., Wrightstone A., Nakagawa, E., Agawani, M., Lollback, P., Gabler, T., Maidla, E.,: Assessment of Risk and Uncertainty for Field Developments: Integrating Reservoir and Drilling Expertise: SPE Annual Technical Conference and Exhibition, New Orleans, USA, SPE 71419, Oct, (2001)
6. Irrgang, R.; Kravis, K., Maidla, E. E., Damski, C.; Millheim, K., A Case Based System to Cut Drilling Costs: SPE Annual Technical Conference and Exhibition, October (1999)
7. Irrgang, Rosemary, Kravis, S., Agawani, Mamdouh and Maidla, Eric: Automated Storage of Drilling Experience: Capture and Re-use of Engineering Knowledge. In Petrotech New Delhi, (1999)
8. Kitano, H. and Shimazu, H.: The Experience Sharing Architecture: A Case Study in Corporate-Wide Case-Based Software Quality Control,: In Case-Based Reasoning, Experiences, Lessons and Future Directions, Leake, D. (ed) MIT Press, Massachusetts USA, (1996)

9. Kolodner, J and Leake D.: A Tutorial Introduction to Case-Based Reasoning,: In Case-Based Reasoning, Experiences, Lessons and Future Directions, Leake, D. (ed) MIT Press, Massachusetts USA, (1996)
10. Rama D. V. and Srinivasan, Padmini: An Investigation of Content Representation using Text Grammars,: ACM Transactions on Information Systems, Vol. 11 No.1, January (1993)
11. Smythe, B., Keane, M., Cunningham, P.: Hierarchical Case-Based Reasoning Integrating Case-Based and Decompositional Problem-Solving Techniques for Plant-Control Software Design, In: IEEE Transactions on Knowledge and Data Engineering, Vol. 13, No 5, September/October (2001)
12. Teufel, Bernd: Informationsspuren Zum Numerischen und Graphischen Vergleich Von Reduzierten Naturlichsprachlichen Texten, Informatik-Dissertationen ETH Zurich, NR. 13, (1989)

Ant Colony Optimisation Applied to a Dynamically Changing Problem

Daniel Angus and Tim Hendtlass

Centre for Intelligent Systems and Complex Processes
School of Biophysical Sciences and Electrical Engineering
Swinburne University of Technology
VIC 3122 Australia
{thendtlass,182921}@swin.edu.au

Abstract. Ant Colony optimisation has proved suitable to solve static optimisation problems, that is problems that do not change with time. However in the real world changing circumstances may mean that a previously optimum solution becomes suboptimial. This paper explores the ability of the ant colony optimisation algorithm to adapt from the optimum solution to one set of circumstances to the optimal solution to another set of circumstances. Results are given for a preliminary investigation based on the classical travelling salesperson problem. It is concluded that, for this problem at least, the time taken for the solution adaption process is far shorter than the time taken to find the second optimum solution if the whole process is started over from scratch.

Keywords: Meta-heuristics, Optimisation, Ant Colony Optimisation.

1 Introduction

Solving optimisation problems presents considerable challenges to researchers and practitioners alike. These often intractable problems usually arise in large industries such as telecommunications, transportation and electronics where even slight increases in solution quality can translate to increased company profit, lower consumer prices and improved services. As a result of this, numerous optimisation techniques have been studied, developed and refined.

The traditional operations research techniques of branch and bound, cutting planes and dynamic programming have been widely used, but can be computationally expensive for large and complex problems. As a result, newer meta-heuristic search algorithms such as simulated annealing [9], tabu search [6] and genetic algorithms [7] have been applied as they generally find good solutions in a moderate amount of computational time.

To further complicate the problem, external factors can, and often will, alter the resources available to perform the task, with the result that a previously optimal, or at least efficient, solution can suddenly become sub-optimal or even impossible. Most methods require that the whole optimisation process be redone with the new resource constraints, which may be a very time consuming task.

The most recent trend is to use algorithms inspired by evolutionary concepts. This paper uses an ant colony optimisation algorithm (ACO) [4] to solve a dynamic optimisation problem. ACO is a collection of meta-heuristic techniques. They are multi agent systems in which agents (ants) have very little individual intelligence but evolve a collective intelligence about a problem over time using simulated chemical markers. Unlike evolutionary algorithms, ACO does not use generational replacement, but intra-generational interaction.

Real ants do not retreat to their nest and start over if something (e.g. a foot) blocks their current efficient path, rather they adapt the path to suit the new constraint. This paper shows how an ACO algorithm can be structured so that it can adapt to a change in constraints

2 The ACO Algorithm

ACO is modeled on the foraging behaviour of Argentine ants. The seminal work by Dorigo [1] showed that this behaviour could be used to solve discrete optimisation problems. This section gives a brief overview of the ant colony mechanics using the Ant Colony System (ACS) meta-heuristic and the travelling salesman problem (TSP) together with other applications. Those readers familiar with the concepts of ACO and TSP may skip Section 2.1.

2.1 Algorithm Description

It is convenient to describe ACS with the TSP metaphor. Consider a set of cities, with known distances between each pair of cities. The aim of the TSP is to find the shortest path that traverses all cities exactly once and returns to the starting city. The ACS paradigm is applied to this problem in the following way. Consider a TSP with N cities. Cities i and j are separated by distance $d(i,j)$. Scatter m ants randomly on these cities ($m \leq N$). In discrete time steps, all ants select their next city then simultaneously move to their next city. Ants deposit a substance known as 'pheromone' to communicate with the colony about the utility (goodness) of the edges. Denote the accumulated strength of pheromone on edge (i,j) by $\tau(i,j)$.

At the commencement of each time step, Equations 1 and 2 are used to select the next city s for ant k currently at city r. Equation 1 is a greedy selection technique favouring cities which possess the best combination of short distance and large pheromone levels. Equation 2 balances this by allowing a probabilistic selection of the next city.

$$s = \begin{cases} \arg\max_{u \in J_k(r)} \left\{ \tau(r,s) \cdot [d(r,s)]^\beta \right\} & \text{if } q \leq q_0 \\ \text{Equation 2} & \text{otherwise} \end{cases} \quad (1)$$

$$p_k(r,s) = \begin{cases} \dfrac{\tau(r,s)^\alpha [d(r,s)]^\beta}{\sum_{u \in J_k(r)} \tau(r,u)[d(r,u)]^\beta} & \text{if } s \in J_k(r) \\ 0 & \text{otherwise} \end{cases} \quad (2)$$

Note that $q \in [0,1]$ is a uniform random number and q_0 is a parameter. To maintain the restriction of unique visitation, ant k is prohibited from selecting a city which it has already visited. The cities which have not yet been visited by ant k are indexed by $J_k(r)$. The parameter α controls the importance of the pheromone and is typically taken to be unity. It is typical that the parameter β is negative so that shorter edges are favoured, unless otherwise stated, in this paper a value of -4 is used. Linear dependence on $\tau(r,s)$ ensures preference is given to links that are well traversed (i.e. have a high pheromone level). The pheromone level on the selected edge is updated according to the local updating rule in Equation 3.

$$\tau(r,s) \leftarrow (1-\rho) \cdot \tau(r,s) + \rho \cdot \tau_0 \qquad (3)$$

Where:

ρ is the local pheromone decay parameter, $0 < \rho < 1$.
τ_0 is the initial amount of pheromone deposited on each of the edges.

Upon conclusion of an iteration (i.e. once all ants have constructed a tour), global updating of the pheromone takes place. Edges that compose the best solution are rewarded with an increase in their pheromone level. This is expressed in Equation 4.

$$\tau(r,s) \leftarrow (1-\gamma) \cdot \tau(r,s) + \gamma \cdot \Delta\tau(r,s) \qquad (4)$$

$$\Delta\tau(r,s) = \begin{cases} \frac{Q}{L} & \text{if } (r,s) \in \text{ globally best tour} \\ 0 & \text{otherwise.} \end{cases} \qquad (5)$$

Where:

$\Delta\tau(r,s)$ is used to enforce the pheromone on the edges of the the solution (see Equation 5).
L is the length of the best (shortest) tour to date while Q is a problem dependent constant (set to 5 for this work).
γ is the global pheromone decay parameter, $0 < \gamma < 1$.

2.2 Algorithm Modification for Dynamic Problems

It is assumed that the algorithm knows when the constraints have been changed. If the algorithm is running continuously and it has been some time since the current solution being used was found the pheromone levels on the edges associated with that solution may be high. Even so some ants will still be exploring alternate paths owing to the stochastic component of the ant decision making process. The speed of adaption may be compromised if very high levels of pheromone are allowed to remain. If all edges on the best path have their pheromone set to a small number while the pheromone level at all other edges is set to zero, all information regarding the relative merits of alternate paths would have been removed. A suitable compromise between these two extreme actions is to normalise pheromone levels on a city's edges. That is for city i the j pheromone levels P_{ij} are replaced by $\frac{P_{ij}}{P_{imax}}$ where P_{imax} is the maximum pheromone level on any of city i's edges.

3 The Test Problem Used

According to Dorigo and Di Caro [4], the TSP is a popular test problem for ACO methods because a) it is relatively easy to adapt the ACO meta-heuristics to this problem and b) it is a problem that is intuitively understood and as such it is one of the most studied problems in the combinatorial optimisation literature. In addition, this class of meta-heuristics is relatively new and research has therefore been directed at defining and exploring its features using well understood problems (such as the TSP). Many papers have solved the TSP using ACO methods including the seminal Dorigo and Gambardella [2,3], Dorigo, Maniezzo and Colorni [5] and Stützle and Dorigo [8] (this paper contains a summary of all ACO applications to the TSP to date). The Burma 14 data set is a useful small test data set for TSP algorithms. It consists of 14 pairs of data points, as shown in table 1.

Table 1. The coordinates of the Burma 14 data set

City	X	Y	City	X	Y
0	16.47	96.10	7	17.20	96.29
1	16.47	94.44	8	16.30	97.38
2	20.09	92.54	9	14.05	98.12
3	22.39	93.37	10	16.53	97.38
4	25.23	97.24	11	21.52	95.59
5	22.00	96.05	12	19.41	97.13
6	20.47	97.02	13	20.09	94.55

It has a large but manageable number of solutions, many of which have lengths that are very close to the absolute minimum path length (see figure 1). With this data set, non-exhaustive search methods have a high probability of finding a good but not optimal answer. However, if they can do this in a reasonably short time, they may be suitable methods for real life applications that require as good an answer as can be obtained in the available time.

The best path solution is shown in figure 2.

4 Results and Discussion

The results of 100 trials of an ACO solving this data set, each from a different random starting position, are shown in figure 3. For these results, as for all others unless otherwise stated, α was taken to have a value of unity and β a value of minus four. Only two times in one hundred trials was the minimum path found, with the second shortest path being found twelve times. However the speed with which the solutions (all of which were in the top 0.0084% of all results) were found is remarkable, being an average of 90 milliseconds (110 iterations on

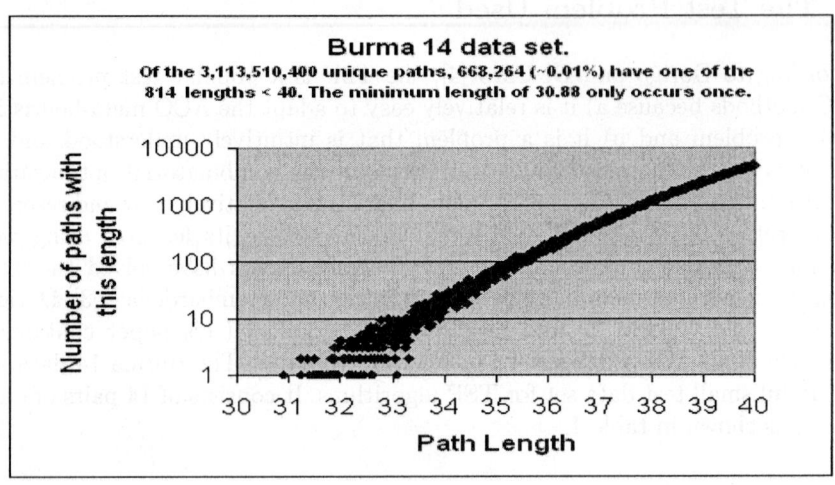

Fig. 1. The frequency of all possible paths with a length less than 40

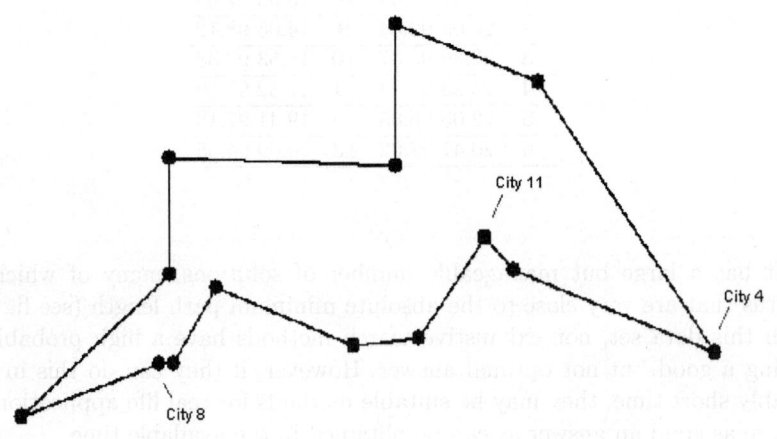

Fig. 2. The shortest path for the B14 data set

average). For comparison, using the same machine, an efficient exhaustive search took two and a half hours.

To simulate a changing environment the ants were allowed to first find a good (but not necessarily ideal) path for the full data set. When stability was achieved (arbitrarily chosen to be when 300 iterations had passed without any change in best path found) one or more cites was removed from the path and the ants were required to adapt the now nonviable path to a new path that met

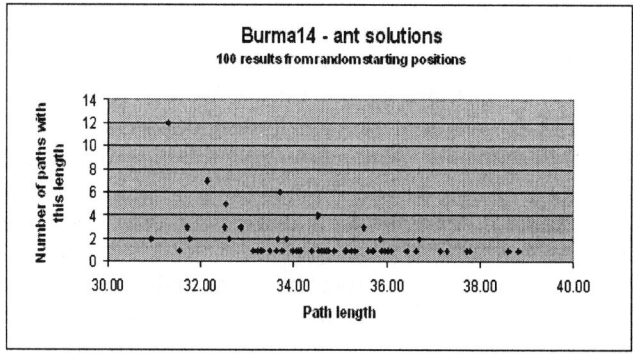

Fig. 3. The path lengths found using ant colony optimisation

the altered constraints. Cities 4, 8 and / or 11 were removed from the zero based list. The removal of city 8 requires a very small change to the best path, city 11 a modest change and city 4 a large change. In addition pairs of cities and all three cities can be removed. Table 2 gives information about the original data set and these reduced data sets.

Table 2. The data sets used for these experiments

Data set	Number of paths	Shortest length	Path lengths below threshold
B14	3,113,510,400	30.88	668,285 (0.02%) < 40
B14 - 4	239,500,800	25.06	8172 (0.0034%) < 30
B14 - 8	239,500,800	30.87	2129 (0.009%) < 35
B14 - 11	239,500,800	30.25	3475 (0.00145%) < 35
B14 - 4 & 8	19,958,400	25.05	3084 (0.015%) < 30
B14 - 4 & 11	19,958,400	24.72	4387 (0.022%) < 30
B14 - 8 & 11	19,958,400	30.24	1317 (0.0066%) < 35
B14 - 4 & 8 & 11	1,814,400	24.71	1555 (0.086%) < 30

4.1 Adapting a Very Good 14 City Solution to a Reduced Set of Cities

After finding a very good (but probably not ideal) solution to the 14-city problem one or more cities were removed from the list of cities to be visited by the ants. Before the ants were allowed to continue, the pheromone levels at each city were normalised relative to the path segment involving that city with the highest pheromone concentration. Without normalisation the very high pheromone levels on what was the best path would inhibit the ants from exploring other options, leading to path adaptation times that approached the time required to solve the reduced problem from scratch. Reducing the pheromone levels increased the relative importance of visibility while still retaining access to the collective wisdom that had been previously developed concerning the value of path segments. Path segments that used to lead to a now removed city could no longer be followed by an ant and so decisions that would have lead to travel to the removed city would now be based on visibility and the remaining (probably) low pheromone levels.

The average results obtained from 100 runs (with random initialisation) adapting to a smaller city set after first establishing a good path for the 14 city set are shown in table 3.

Table 3. Average results adapting to a shorter path

City / cities removed	Average iterations taken to adapt	Average adapted path length	Average path reduction	Optimum path reduction
8	21	34.73	1.15	0.01
11	22	34.81	1.32	0.63
8 & 11	22	34.74	1.63	0.64
4	20	29.31	5.95	5.82
4 & 8	20	29.25	5.77	5.83
4 & 11	18	29.26	6.26	6.16
4 & 8 & 11	22	29.39	5.92	6.17

As can be seen the adaptation rate was very high, significantly faster than finding the result by starting over. The average path reduction for the last three cases, in which the path reduction was significant, was comparable to the difference between the relevant ideal shortest paths. At times the average reduction exceeded this, as a result of the average result for the smaller path problem being a better approximation to the ideal than the average result from the longer path problem. This is particularly apparent for the first three results in which the expected path reduction is so small as to be lost in the noise component within the results.

4.2 Adapting a Very Good Less than 14 City Solution to a 14 City Solution

Tests were also conducted in which, after first establishing a good path for a smaller city set, one or more cities were added to the list. Again stability on the smaller path was declared after 300 iterations without improvement in the best path length and the pheromone levels were normalised before the ants were started on the revised problem. The average results obtained from 100 runs (with random initialisation) are shown in table 4 below.

Table 4. Average results adapting to a longer path, without adding pheromone to the new path segments

City / cities added	Average iterations taken to adapt	Average adapted path length	Average path increase	Optimum path increase
8	15	41.29	6.11	0.01
11	15	39.13	4.66	0.63
8 & 11	10	41.13	6.72	0.64
4	15	41.03	13.40	5.82
4 & 8	10	44.94	17.38	5.83
4 & 11	10	40.22	13.06	6.16
4 & 8 & 11	15	44.31	16.68	6.17

Although the adaptation rate is again less than starting over, the time taken to adapt to a larger city list is more than the time taken to adapt to a smaller city list. The reason for this is that the new city has no pheromone to it from any other city. Even though the pheromone levels have been normalised there is still some tendency for the ant not to include the new city until all other cities have been visited. However, once an ant has visited the city, the probability that that path segment will be included in future tours increases sharply.

Even though the pheromone levels have been normalised, the lack of any pheromone on any of the possible new path segments reduces the tendency to explore these segments. To counter this the pheromone on every new path segment was initialised to one immediately after the pheromone normalisation step. The average results obtained from 100 runs under these conditions (with random initialisation) are shown in table 5 below. Note that the time taken has actually increased, but that the extra exploration leads to a shorter path.

Table 5. Average results adapting to a longer path, with added pheromone on the new path segments

City / cities removed	Average iterations taken to adapt	Average adapted path length	Average path reduction	Optimum path reduction
8	18	37.27	2.24	0.01
11	23	38.03	4.10	0.63
8 & 11	44	36.85	2.89	0.64
4	24	37.90	10.36	5.82
4 & 8	35	37.17	9.48	5.83
4 & 11	20	36.98	9.63	6.16
4 & 8 & 11	34	37.19	10.23	6.17

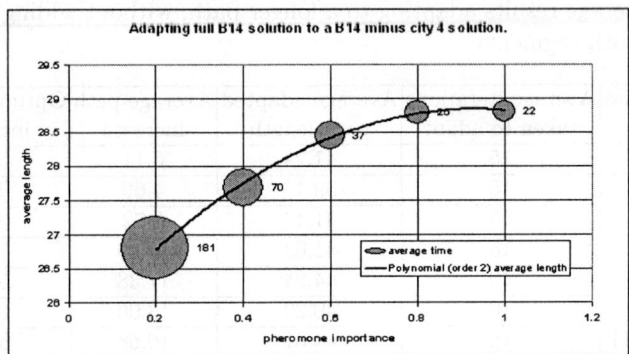

Fig. 4. The relationship between the time to find the path, the quality of the path found and the pheromone importance when adapting to a shorter path

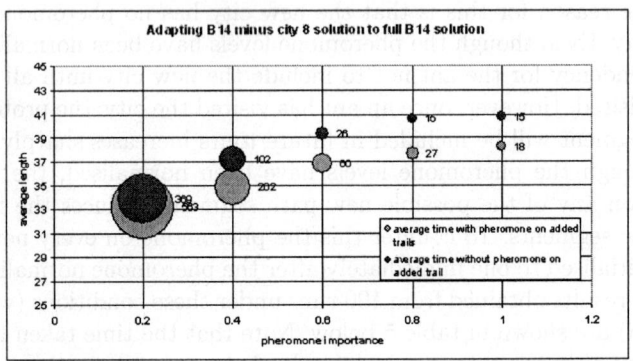

Fig. 5. The relationship between the time to find the path, the quality of the path found and the pheromone importance when adapting to a longer path

4.3 Solution Quality

Time is not the only measure of success, the quality of the solution is also important. It should be noted that the results found, while good, were not in general optimal. Varying the relative importance of visibility and pheromone shows an inverse relationship between the time taken and the quality of the solution found. This relationship appears to hold when finding both the original path and adapting to a new path using ACO. Figure 4 shows this relationship when adapting a solution to the full B14 city set to a solution that omits city 8, while figure 5 shows the relationship when adapting a solution that omits city 8 to a solution for the full B14 city set. For the results in these figures the value of β was left at minus four while the value of α was changed to the values shown on the graphs.

5 Conclusion

Although the results above are obtained using a small, although non-trivial, TSP data set, they give a clear indication that ants can adapt a path in a shorter time than would be required to find that new path starting over. As the size of the TSP city set is increased, ACO shows an improved performance compared to alternate techniques. It might be anticipated that the adaptation advantage shown by ants on the B14 city set will be at least maintained on larger data sets.

Varying the importance of the pheromone compared to the city visibility allows a trade off to be made between the time taken to reach a solution and the quality of that solution. This may prove a useful control parameter in the context of needing to find as good a solution as possible to a problem within a definite and restricted time.

No matter if starting from scratch, or if adapting a previously found solution to suit a modified set of constraints, the quality of all the solutions found for this data set are within the top 0.01% of all possible solutions. This augurs well for the use of such an approach to real life problems with dynamic constraints, for example in scheduling applications.

References

1. Dorigo, M. (1992) Optimization, Learning and Natural Algorithms, PhD Thesis, Dipartimento di Elettronica, Politechico di Milano, Italy.
2. Dorigo, M. and Gambardella, L. (1997) "Ant Colony System: A Cooperative Learning Approach to the Traveling Salesman Problem", IEEE Transactions on Evolutionary Computing, 1, pp. 53-66.
3. Dorigo, M and Gambardella, L. (1997) "Ant Colonies for the Traveling Salesman Problem", Biosystems, 43, pp. 73-81.
4. Dorigo, M. and Di Caro, G. (1999) "The Ant Colony Optimization Meta-heuristic", in New Ideas in Optimization, Corne, D., Dorigo, M. and Golver, F. (eds), McGraw-Hill, pp. 11-32.
5. Dorigo, M., Maniezzo, V. and Colorni, A. (1996) "The Ant System: Optimization by a Colony of Cooperating Agents", IEEE Transactions on Systems, Man and Cybernetics - Part B, 26, pp. 29-41.
6. Glover, F. and Laguna, M. (1997) Tabu Search, Kluwer Academic Publishers, Boston: MA, 442 pages.
7. Goldberg, D. (1989) Genetic Algorithms in Search, Optimization and Machine Learning, Addison Wesley: Reading, MA, 412 pages.
8. Stützle, T. and Dorigo, M. (1999) "ACO Algorithms for the Traveling Salesman Problem", in Evolutionary Algorithms in Engineering and Computer Science, Miettinen, K., Makela, M., Neittaanmaki, P. and Periaux, J. (eds), Wiley.
9. van Laarhoven, L. and Aarts, E. (1987) Simulated Annealing: Theory and Applications, D Reidel Publishing Company: Dordecht, 186 pages.

A GIS-Integrated Intelligent System for Optimization of Asset Management for Maintenance of Roads and Bridges

M. D. Salim[1], Tim Strauss[2], and Michael Emch[2]

[1] Construction Management, University of Northern Iowa
Cedar Falls, Iowa 50614-0178, USA
salim@uni.edu

[2] Department of Geography, University of Northern Iowa
Cedar Falls, Iowa 50614-0406, USA
{tim.strauss,mike.emch}@uni.edu

Abstract. Geographic information systems (GIS) and Artificial Intelligence (AI) techniques were used to develop an intelligent asset management system to optimize road and bridge maintenance. In a transportation context, asset management is defined as a cost-effective process to construct, operate, and maintain physical capital. This requires analytical tools to assist the allocation of resources, including personnel, equipment, materials, and supplies. One such tool, artificial intelligence (AI), is the creation of computer programs that use human-like reasoning concepts to implement and/or improve a process or task. This paper presents "heuristic" or "experience" based AI methodologies to optimize transportation asset management procedures. Specifically, we outline and illustrate a GIS-based intelligent asset management system using the case study of snow removal for winter road and bridge maintenance in Iowa, USA. The system uses ArcView GIS to access and manage road and bridge data, and ART*Enterprise, an AI shell, for the user interface.

1 Introduction

"Asset management," a concept of increasing importance to transportation engineers and planners, has several interrelated definitions. The U.S. Federal Highway Administration defines it as "a systematic process of maintaining, upgrading, and operating physical assets cost-effectively" [1]. Similarly, the Asset Management Systems Working Group of the Organization for European Cooperation and Development sees asset management as a tool through which "governments can improve program and infrastructure quality, increase information accessibility and use, enhance and sharpen decision-making, make more effective investments, and decrease overall costs" [2]. A subcommittee of the American Public Works Association has defined it as "the activities of deciding how to use society's resources to develop, operate, and maintain our infrastructure to achieve the highest possible returns" [3]. Nearly all such definitions, however, acknowledge the critical role of information systems and analytical tools that assist in the allocation of transportation-related resources, including personnel, equipment, materials, and supplies. In particular, analytical tools "facilitate the discussion underlying the decision-making process by providing the ability to articulate the impact of choosing one alternative over another through engineering and economic-based 'what-if' analyses" [2]. The development of such tools has been identified as a key area of future research in asset management [4,5].

There is much potential for the application of information science methodologies in a transportation asset management framework. For instance, asset management can be combined with artificial intelligence (AI) techniques like fuzzy sets, neural networks, genetic algorithms, and simulated annealing, and mathematical optimization methods, including linear programming or integer programming. Although powerful, these algorithms and methods can be difficult to program and slow to converge to a solution, and sometimes convergence to a solution never happens. This can hinder their application to many applied transportation management contexts, such as winter road maintenance, in which rapid feedback is required to adjust dynamic changing conditions.

An alternative to these techniques is to develop and implement a knowledge-based expert system (ES). The ES requires knowledge engineering, which is the implementation of decision-making "heuristic rules" drawn from interviews with human experts. These rules can easily be implemented in a computer environment known as expert systems shell. This paper describes a paradigm for implementing, in a transportation asset management context, knowledge engineering heuristic rules in an expert systems shell. As a case study, we describe how a geographic information system (GIS) database was integrated with an intelligent system for the maintenance of roads and bridges. The system uses the ArcView GIS software package to access and manage road and bridge data, and ART*Enterprise, an artificial intelligence shell, for the user interface.

2 Review of Literature

A considerable amount of literature on transportation asset management has emerged. Recent statements by U.S. Federal Highway Administration [1, 2], Nemmers [4], Gray-Fisher [6], and Vanier [7] have outlined the general case for applying asset management tools in the transportation construction and infrastructure maintenance industries [1,2,4,6,7]. From a more applied perspective, Ruben and Jacobs [8] and Spalding [9] have explored the use of asset management for the supply of construction materials [8,9], and Iraqi et al. [10] have described the implementation of computer resources for construction-site management and the use of real-time site-based sensors for monitoring potential construction logistics problems.

Several agencies' efforts at implementing transportation asset management principles have been documented recently [11,12,13]. In addition, there have been several recent case studies using knowledge engineering in construction related problems. For instance, Hung and Jan [14], and Jia [15] described the implementation of knowledge engineering based paradigms for construction problems [14,15], and Melhem and co-workers described the application of such systems to steel bridge construction [16]. The implementation of an expert system for asset management requires careful consideration and selection of appropriate algorithms; Tarek and Razavi compared several AI based algorithms with more traditional mathematical approaches [17].

Moreover, there are many similarities between construction logistic problems and transportation planning and maintenance problems. Strandberg described the implementation of a system-wide computer-based logistics program for the Chicago Transit Authority [18]. By integrating routing tasks, asset management functions, and maintenance scheduling, this system has shown great promise in decreasing the Chicago Transit Authority's overall costs. Hart described a similar system that includes real-time global positioning system (GPS) data [19].

3 System Architecture

The provision of high-quality road transportation services to the public on a consistent basis requires informed decision making at a variety of levels, from the planning stages, to construction of the roadway, and finally to its operation and maintenance. The allocation of resources at each level of the asset management process entails unique considerations, but common issues emerge throughout all levels. For instance, construction logistics and transportation infrastructure maintenance problems share several characteristics, and techniques developed for one can be applied to the other. Both involve complex supply chain issues, since the materials and equipment come from multiple sources and require extensive task coordination. Both involve coordination of multiple, interdependent subtasks. Both are highly time, cost, and resource sensitive. Decision making for both problems needs to be agile in response to unpredictable circumstances such as weather. Finally, and most importantly, both problems involve large numbers of variables, large numbers of constraints, and complex interactions. As such concerns make both of these problems intractable by conventional mathematical optimization and AI techniques, they require novel methods.

Figure 1 shows an overview of the expert system developed for the case study outlined below. The creation of an expert system begins with a series of detailed interviews of experts in the application area, in this case transportation managers, supervisors, planners, and engineers. The questioning takes both the form of qualitative discussions and quantitative surveys. Knowledge engineering is the process of analyzing this information and developing a knowledge base. The knowledge base consists of a set, often very large, of predicate-calculus format rules:

FORMAT:
IF (Input Variables → Input Tests) THEN (Predicates → Output Variables)

EXAMPLE:
IF (Climate = Torrential Rain) THEN (Work Location = Inside)

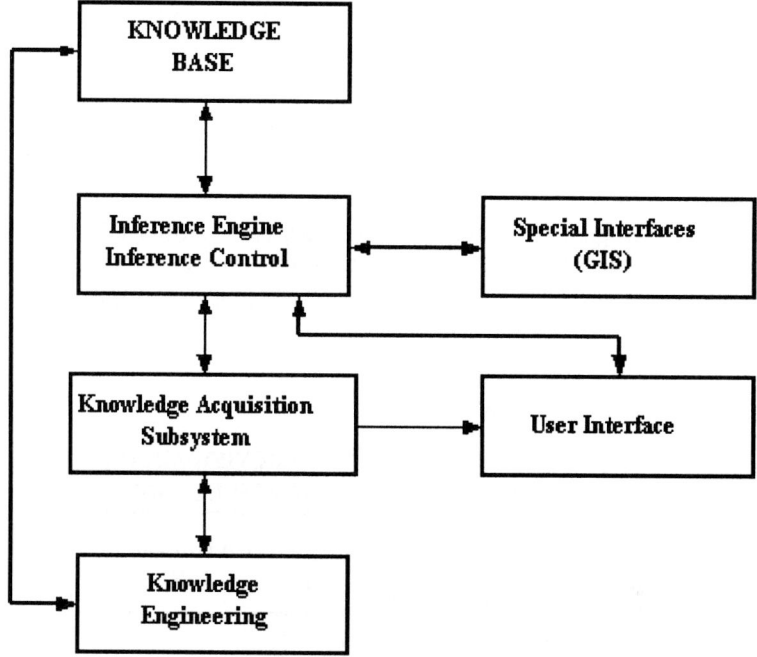

Fig. 1. The General Paradigm

These predicate-calculus format rules can be used either sequentially or simultaneously, depending on the nature of the knowledge to be implemented. For example, a rule concerning rain cancellation of activities could trigger sequential steps dealing with the need to secure areas of a site under tarpaulins, to bring certain equipment indoors, and so forth. In colder climates, the rain test would be fired simultaneously with a temperature test since the steps needed to secure a building site for freezing rain would be very different from the steps needed for non-freezing rain. The inference engine portion of the software implements these rules. The implementation can be quite complex as the rules may be interrelated both sequentially and simultaneously in unobvious ways. Obviously, the causality and direction of information flow must be carefully monitored.

The user interface portion of the expert system secures input variable information from the users of the system and delivers output variable instructions to the users. The user interface translates the information content of the output variables into useful formats such as personnel rosters, bills of materials, and route maps for drivers. Special interfaces include connections with GIS databases, real-time meteorological information, real-time traffic conditions, and real-time inventory systems with supply chain vendors. As all these databases and information streams use different protocols and conventions, exceptional challenges occur for programming the interface and translation engines for this subsystem.

The knowledge acquisition subsystem serves three distinct functions. First, this subsystem allows the knowledge engineers to formulate and enter predicate-calculus based rules. Second, this subsystem implements learning functions. The predicate-calculus based rules are not static, but are dynamically refined and modified to implement new knowledge. The knowledge engineers can use information such as statistical analysis of rule firings to eliminate inactivate rules. Similarly, problems and suggestions reported by the users can be used to modify and expand the knowledge base. Third, this subsystem serves as a front-end for the knowledge base. It arbitrates queries from the user interface, and secures and delivers information as requested by the system users. As the knowledge base deals exclusively with formal variables and is programmed using Lisp or proprietary artificial intelligence languages, it is not suitable for direct access by the users. The subsystem uses programming constructs like menus, radio buttons, and related graphical displays to make the system accessible to non-specialist users.

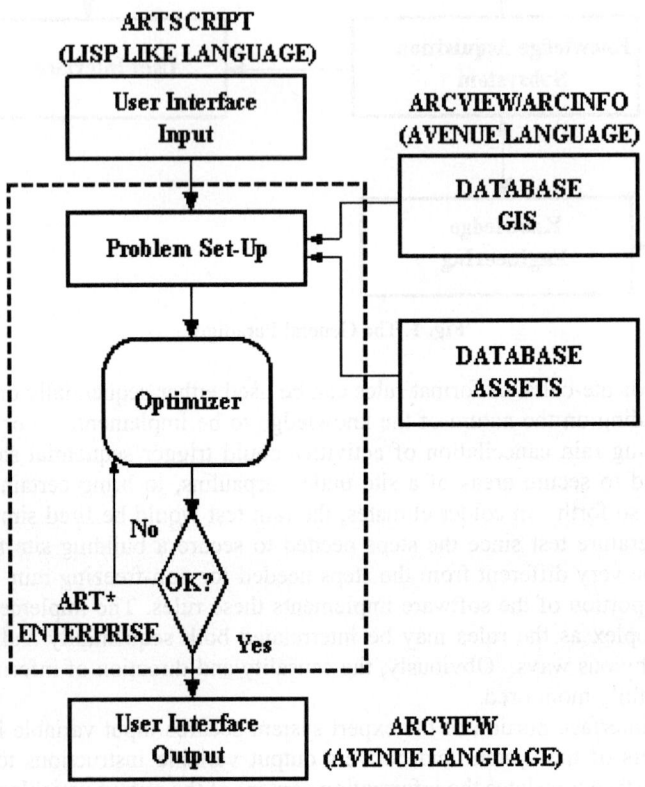

Fig. 2. Expert System Implementation

4 A Case Study: Winter Maintenance of Roads and Bridges in Iowa

Figure 2 shows the architecture of a system that was developed by the authors. The function of this system is to optimize the winter maintenance of infrastructure assets such as roads and bridges, with an emphasis on the allocation of resources for snow removal. This is a major expense for many regional and local governments. For instance, according to Black the City of New York alone spent US $100 millions on snow removal during the winter of 1995-1996 [20]. There is much interest in the development of analytical tools to support improved resource allocation, since even marginal cost efficiencies can yield very high aggregate benefits. The essential issues to be considered in snow removal are the allocation of personnel resources (drivers, mechanics, supervisors), the distribution of materials (road salt, road sand, other ice/snow removal chemicals), the deployment of equipment (snowplows, dump trucks, supervisor trucks), the observance of financial constraints (salaries, materials, repair costs), and the maximization of overall effectiveness (e.g., minimizing dead-heading or the movement of plows on roads already plowed, insuring that safety-critical and time-critical roads such as access points to hospitals, schools, and major employers are plowed quickly).

The management of these assets subsumes the larger problems of logistics and prediction. The logistics aspects include the movement of personnel, equipment, and materials. As the movement of assets becomes difficult or impossible once snow has accumulated, a great deal of planning and preparation is needed to insure that adequate stocks are in place at the right times and in the right places. Prediction focuses around several factors including meteorological data and geographic information. Knowing how much snow is expected and in what time frame is important for allocating resources and logistics planning. Geographic factors are important as basic topographical features such as road surface condition, drainage, sun and wind exposure, must be used to predict probable accumulations, thawing rates, and snow removal needs.

Both a knowledge base and a GIS database were developed to support the expert system. In order to build the knowledge base, an advisory board consisting of road maintenance engineers, supervisors, and managers was assembled. Through the knowledge elicitation process, a series of interviews was conducted with local and state winter maintenance personnel on a wide variety of issues related to their snow removal procedures. Personnel from the city, county, and state levels of government were interviewed to consider any difference in standard operating procedures by type of agency or scale of operation. Typical issues discussed during these formal interviews included route prioritization, procedures for managing drivers and vehicles, treatment of the roads with salt/sand, inter-agency and intra-agency communication and coordination, minimum snowfall thresholds to mobilize equipment, and other decision-making procedures related to the management of snow removal assets. The responses were formalized and incorporated into the expert system as outlined above.

The expert system also requires roadway data to support the allocation of routes, drivers, and other resources. For this, a GIS database for all roads in the case study area (Black Hawk County, Iowa) was obtained from the Iowa Department of Transportation (IDOT). This database consisted of an ArcView GIS shapefile and several related database files with traffic volumes, roadway inventory information, and other data. The data elements required for the snow removal management system were extracted from this database and a new ArcView shapefile was generated for incorporation into the expert system.

Because the goal of this research was to make any developed software available to small and mid-sized public works departments, relatively modest computing hardware

634 M. D. Salim et al.

and software were used. Two mid-range Pentium PCs were used to develop the expert system. Two main software development environments were used, the ART*Enterprise Expert System Shell and the ArcView GIS software. ArcView, produced by Environmental Systems Research Institute, Inc. (www.esri.com), uses a customization and application development language called Avenue. ART * Enterprise uses a Lisp like programming language called Art Script. The knowledge base, knowledge acquisition subsystem, and inference engine, were all programmed in the proprietary system language used by the ART*Enterprise Expert System Shell. In addition, ArcInfo, another GIS database package produced by the makers of ArcView, was used to sort, format, and import geographical roadway information into

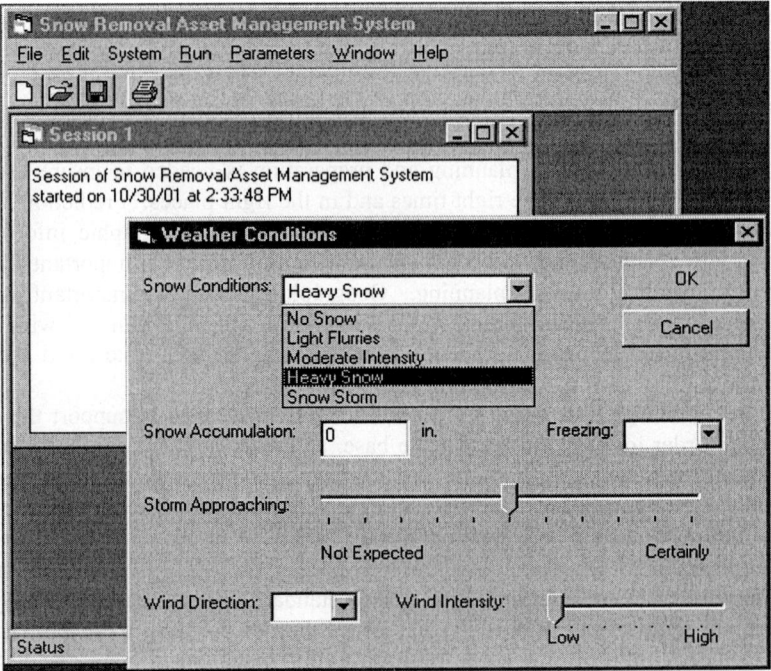

Fig. 3. User Interface of the Snow Removal Asset Management System Expert Systems Shell.

the ArcView to generate road maps and other graphical outputs for the user interface.

The knowledge base, GIS database, and programming efforts were integrated into an expert system with a user-friendly interface (Figure 3). This interface provides a series of features to guide a non-expert user in inputting the required information. For example, the interface shown in Figure 3 prompts the user to input the severity of the expected snowfall. Figure 4 shows a map generated by the system. The bold blue roads (route 4) represent second level priority (B) roads to be plowed.

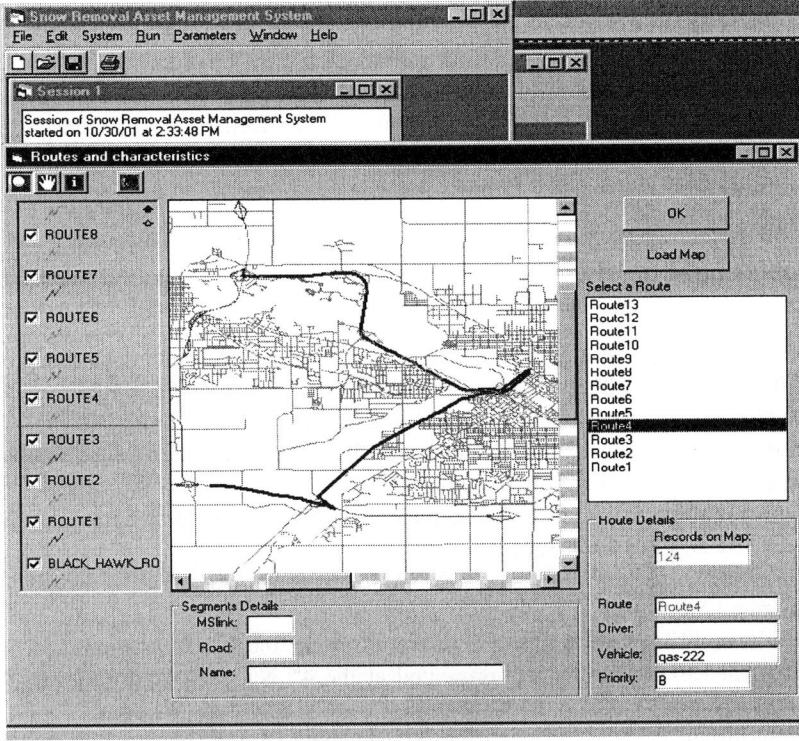

Fig. 4. Example Snow Plow Routing Map

The system is capable of importing GIS data and doing snowplow routing based on priority, and also assigning plowing equipment, as depicted in Figure 4. The assignment of the driver is also possible. An intermediate database file contains information on the allocation of assets, and other parameters such as cost, time, and kilometers (miles) of roads to be plowed.

5 Conclusions

This paper describes the development of analytical tools in a transportation asset management context through knowledge engineering methods. Specifically, a knowledge-based expert system was created to manage the allocation of snow removal assets, using a commercial expert system shell running on a mid-range computer system. Initial studies reveal that such a system could efficiently be utilized to optimize allocation of assets for plowing snow, especially in the northern part of the United States. Although the case study presented in this paper is specific to transportation infrastructure maintenance and concerns of colder climates, the methods may also be employed to other areas of management (e.g., inventory control). It is hoped that this work will be of interest to a broader audience in the civil engineering research community.

Acknowledgments

This work was funded through a grant from the Midwest Transportation Consortium of the United States using funds from the United States Department of Transportation. The authors gratefully acknowledge the contributions of their graduate students, Mr. Alavro Villavincencio and Mr. Ayhan Zora.

References

1. Federal Highway Administration, Asset Management: Advancing the State of the Art into the 21st Century Through Public-Private Dialogue, United States Department of Transportation, 1997, Publication No. FHWA-RD-97-046.
2. Federal Highway Administration, Asset Management Primer, United States Department of Transportation, December 1999.
3. Lemer, Andrew C. "Asset Management: The Newest Thing or Same-old, Same-old?" APWA Reporter, June 2000.
4. Nemmers, Charles, "Transportation Asset Management," Public Roads, July/August 1997, 49-52.
5. Markow, Mike, and Neumann, Lance, "Setting the Transportation AssetManagement Research Agenda," presented at the 4th National Transportation Asset Management Workshop, September 23-25, 2001, Madison, Wisconsin.
6. Gray-Fisher, Dena M., "Does Asset Management Deserve a Closer Look?", Public Roads, Vol. 62, No. 6, May-June, 1999, pp. 50-52.
7. Vanier, J. D., "Why Industry Needs Asset Management Tools", Journal of Computing in Civil Engineering, Vol. 15, No. 1, January, 2000, p. 35.
8. Ruben, Robert F., and Jacobs, Robert, "Batch Contraction Heuristics and Storage Assignment Strategies for Walk/Ride and Pick Systems", Management Science, Vol. 45 No. 4, April, 1999, pp. 575-577.
9. Spalding, Jan O., "Transportation Industry takes the Right-of-Way in Supply Chain", IIE Solutions, Vol. 30, No. 7, July, 1998, pp. 24-29.
10. Iraqi, Ali, Morawski, Roman Z., Barwicz, Andrzej, and Bock, Wojtek J., "Distributed Data Processing for Monitoring Civil Engineering Construction", IEEE Transactions on Instrumentation and Measurement, Vol. 48, No. 3, June, 1998, pp. 773-778.
11. Novak, Kurt, and Nimz, James, "County Creates Transportation Infrastructure Inventory," Public Works, V. 129, No. 12.
12. DeLaurentiis, John, "Building an Asset Management System - the NEIL RTA Experience," presented at the 4th National Transportation Asset Management Workshop, September 23-25, 2001, Madison, Wisconsin.
13. St. Clair, Bob, "WisDOT's Meta-Manager," presented at the 4th National Transportation Asset Management Workshop, September 23-25, 2001, Madison, Wisconsin.
14. Hung, Shih-Lin, and Jan, J. C., "Augmented IFN Learning Model", Journal of Computing in Civil Engineering, Vol. 14, No. 1, January, 2000, pp. 15-23.

15. Jia, Xudong, "Intelligis: Tool for Representing and Reasoning Spatial Knowledge", Journal of Computing in Civil Engineering, Vol. 14, No. 1, January, 2000, pp. 51-60.
16. Melhem, Hani G., Roddi, W. M. Kim, Nagaraja, Srinath, and Hess, Michael R., "Knowledge Acquisition and Engineering for Steel Bridge Fabrication", Journal of Computing in Civil Engineering, Vol. 10, No. 3, July, 1996, pp. 248-257.
17. Sayed, Tarek, and Razavi, Abdolmehdi, "Comparison of Neural and Conventional Approaches to Mode Choice Analysis", Journal of Computing in Civil Engineering, Vol. 14, No. 1, January, 2000, pp. 23-31.
18. Strandberg, Keith W., "CTA Scheduling Software for Dispatch and Maintenance", Mass Transit, Vol. 26, No. 1, January, 2000, p. 38.
19. Hart, Anthony, "Global Positioning Systems are Redefining Transit Operations in Ann Arbor", Mass Transit, Vol. 26, No. 7, November, 2000, p. 44.
20. Black, Tom, "The white stuff: it can cost plenty of green and leave budgets in the red," American City and County, Vol. 112, April, 1997, p. 32.

A Unified Approach for Spatial Object Modelling and Map Analysis Based on 2nd Order Many-Sorted Language

Oscar Luiz Monteiro de Farias and Sueli Bandeira Teixeira Mendes

Universidade do Estado do Rio de Janeiro -- Centro de Tecnologia e Ciências --
Faculdade de Engenharia, Departamento de Engenharia de Sistemas e Computação
tel.: (+55)(21)587-7442, fax.:(+55)(21)587-7374
{oscar,smendes}@eng.uerj.br

Abstract. 2nd Order Many-Sorted Language is presented here as an algebraic structure and, as such, yields an adequate tool for formalizing an extension of Peano Algebra for spatial object modeling, spatial data base information retrieval and manipulation, and also for map analysis. Here we will demonstrate that this tool provides a unified treatment for several classes of operations realized among maps, including those presented in Boolean Logic Models, Fuzzy Logic Models and Bayesian Probability Models.

Keywords: Spatial Object Modelling, Map Analysis, 2nd Order Many-Sorted Language, Algebraic Structures, Peano Algebra, Space Filling Curves, Peano-curve, Quadtreee, Geo-maps, Geo-fields

1 Introduction

A spatial object can not be represented only in an extensional way. A line, for example, has a non enumerable infinite set of points. Its representation in an extensional way it is not possible. So, spatial objects cannot be represented trough Pure Relational Algebra. Besides that, In practical terms, in computers we can not have a table with an infinite number of rows. It is well known that spatial objects need to be represented in an intensional form in databases. An intensional representation of a spatial object usually requires:

a) The storage of privileged elements of data for the object;
b) The specification of a rule generating all possible elements that belong to the spatial object, that is, a generative rule.
c) The definition of a rule for testing if a given element is a member of the object, for example, a point vis-à-vis a line or a line vis-à-vis a polygon [6].

This intensional representation needs specific algorithms to discover, for example, the topological relationships between spatial objects [7] and present some limitations when we deal with multiple maps of the same region and we need to realize special operations with these maps.

The purpose of this paper is to formalize a unified approach for modelling spatial objects and also for the operations realized on then. These operations comprise traditional ones like: set operations, geometric operations, relational operations, and also analytical operations on single maps and on multiple maps. The later ones could be: map reclassification, aggregation operations over the attributes of the objects in a map, two-map overlays, multiple-maps operations based on Boolean Logic, Fuzzy Logic or Bayesian methods.

First we present some background material related to Space Filling Curves, Peano Curves, Quadtrees, Peano Relations and Spatial Modelling. After that (in 3) , we show how we can realize several operations on spatial objects, including some algorithms. These operations are carried based on the spatial model presented in 2.2. Then we formalize a Peano Algebra Extension through a 2nd Order Many-Sorted Language (in 5). Finally (in 6) we illustrate the translation of the algorithms shown in 3, ending the paper with the traditional conclusion.

2 Modeling Spatial Objects

2.1 Space Filling Curves

Space filling curves are an attempt to represent n-dimensional systems by a one-dimensional system. The basic idea is to build a curve that travels by all points in a n-dimensional space. In the case of two-dimensions, we will need to find curves that pass for all points in a given plane. Of course, according to the Euclidian geometry, this is not theoretically possible, if we define a point as a zero-dimensional object. However, we can have a solution for this problem if we think of a point as a two-dimensional object, a square which sides tend towards zero.

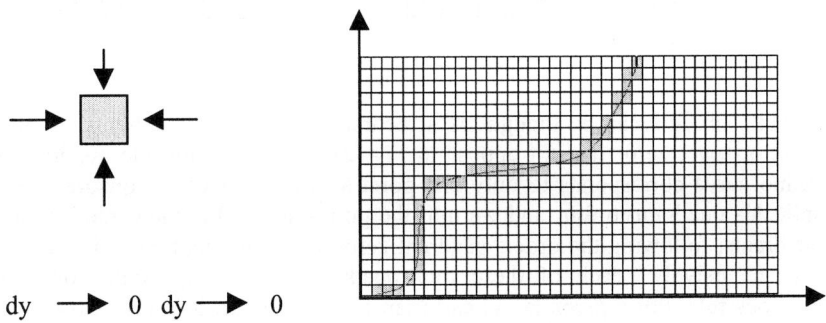

Fig. 1.

With the help of fractal geometry, it is now possible to find a curve that goes trough al these small squares and fills completely a two-dimensional space. This curve has a certain width that will also tend to zero as the sides of the squares that it traverses tend also to zero. A "good" space filling curve should observe some properties such that: i) the curve must pass only once to every point in the multi-dimensional space; ii) the curve must correspond to a bijective mapping from a one- to a multi-dimensional space, etc [6].

Two well known curves: the Peano-curve and Hilbert-curve meet most of the conditions for a "good" space filling curve. But the Hilbert curve does not provide an easy way to retrieve neighbours points in the space and is not stable when we operate through different scales. The computation of Hilbert-keys is also more difficult than the computation of Peano-keys. These keys reflect a bijective map of the n coordinates $(x_1, x_2, ..., x_n)$ on a single coordinate: the so-called Peano-key (Pk) or Hilbert-key (Hk). The initial steps for building the Peano curve are show in Fig. 2

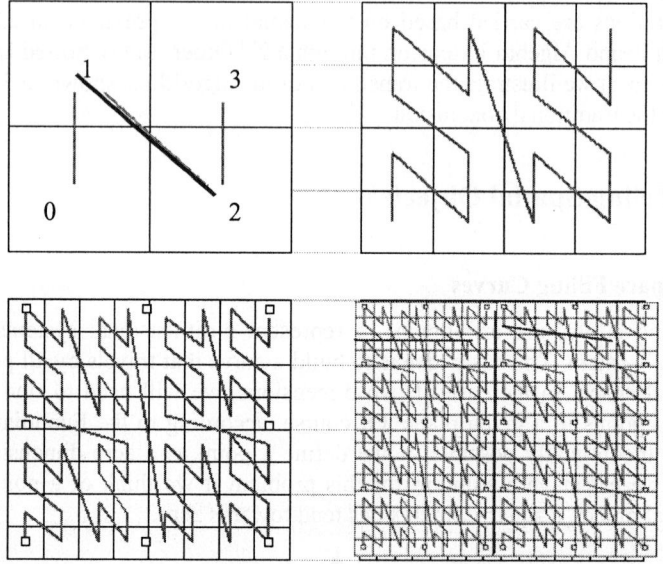

Fig. 2.

This process can be thought as a recursive division of quadrants. At the trivial case, or level zero (doesn't appear in Fig. 2), we have only one quadrant and the Peano curve corresponds to only one point at the center of the quadrant. Then, we split this quadrant in four and we unit the points situated at the center of each these quadrants, following the order: quadrant zero, quadrant one, two and three. Next we split recursively each of these four quadrants in four quadrants observing the previous rule. For a given (x, y) cartesian coordinate par, it is very easy to compute the correspondent Peano-key. We take the binary representations of coordinate x and coordinate y and build the Peano-key by the bit interleaving of the y and x coordinates. So, the position of a point in the space can be represented by its Peano-key. The Peano-curve is a kind of fractal curve. The same pattern is repeated in successive subdivisions of an n-dimensional space.

2.2 Quadtrees, Peano-Relations and Spatial Representation

We can use quadtrees to approximate an arbitrary object successively by a set of blocks at different levels, where each of these blocks is the result of a quadrant recursive subdivision. The yellow object at left in Fig. 3, for example, can be represented by blocks of different sizes belonging to three distinct levels of quadrant subdivisions. This process can be applied to any object, until a predefined resolution level is reached. This could be, for example, the resolution of a given monitor. Now, we can use the set of pairs (Peano-keys, side-length) of all the blocks that comprises the object as a spatial representation of the object. Indeed, we can represent all the objects in a two-dimensional space, for example, by the so called Peano relation PR(Object_id, PK1, side-length). The Peano relation is a kind of intensional-extensional representation for spatial objects. Its meaning is that an object is represented by the union of several tuples. Each of these tuples represents a square-region of the two-dimensional space. This region is univoquely determined by the Peano-key associated to the square and the side-length of the square. The tuples representative of the yellow object (with Object_id equal "A") at left of figure 3 would be:

PR (A, 12, 2, At_1, ..., At_t)

PR (A, 35, 1, At_1, ..., At_t)

PR (A, 36, 2, At_1, ..., At_t)

PR (A, 44, 1, At_1, ..., At_t)

PR (A, 48, 4, At_1, ..., At_t),

Where At_1, ..., At_n, is a list of attributes associated to the object.

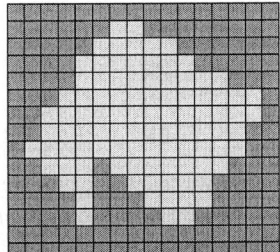

Fig. 3.

In this paper we will use the run-length-encoding representation for Peano-relations, that is: PR(Object_id, PK1, PK2, At_1, ..., At_n). In this alternative Peano relation, each object can be thought as the union of several tuples as before, but the square-region associated with a tuple means that this square occupies the region of the two-dimensional space with Peano-keys (PK) in the interval [PK1, PK2), That is PK>=PK1 and PK<PK2.

3 A Unified Approach for Modelling Spatial Objects and also for the Operations Realized on Them

Our goal in this paper is to use Peano relations and an extension of Peano-Algebra, in order to represent spatial objects and realize several different types of operations on them. These operations could be:

- Set operations (intersection, difference, union);
- Geometric operations (translation, rotation, scalling, symmetry, replication, simplification, window extraction);
- Relational operations (geometric projection, Peano Join);
- Spatial queries (query about topology);
- The traditional thematic queries (over the object attributes);
- Operations over two or more maps (binary evidence maps, index overlay with multi-class maps, Boolean models, Fuzzy Logic Models and Bayesian Models).

Due to space restrictions, we'll show here only some operations, but we hope these operations will give us some insight in the expressiveness and power of the extension to the Peano-Algebra proposed here. First we'll explain the realization of some operations based on Peano relations, and then we'll propose one particular Peano Algebra extension to generalize our approach.

First of all let us consider a Peano relation MAP (Object_Id, PK1, PK2, At1, ..., Att), that represents a set of objects spatially related, as in a two-dimensional MAP. Two objects "A" and "B" would be represented generically by the following set of "n" and "m" tuples in MAP:

MAP (A, $PK1_1$, $PK2_1$, At_{11}, ..., At_{t1})
....
MAP (A, $PK1_{1n}$, $PK2_{1n}$, At_{1n}, ..., At_{tn})
....
MAP (Object_B, $PK1_1$, $PK2_1$, At_{11}, ..., At_{t1})
....
MAP (Object_B, $PK1_m$, $PK2_m$, At_{1m}, ..., At_{tm})

A) Suppose we want to calculate the PEANO JOIN between two objects A and B. In the definition of this operation **PEANO JOIN (A, B)** we keep tuples which have the same Peano-keys both in the quadrants that constitutes B and that constitutes A, adding the others attributes of both objects. The following algorithm realize this:
// First, let us build a spatial index based in the Peano-keys:
// Classify the tuples of MAP in ascending order of PK1;
//Classify the tuples with same PK1, in ascending order of PK2.
<u>TA</u> ← first tuple pertaining to Object_A;
TB ← first tuple pertaining to Object_B;
<u>While</u> (Exist tuples in Object_A and exist tuples in Object_B) <u>do</u>;

$PKI = MAX (A.PK1, B.PK1);$
$PKF = MAX (A.PK21, B.PK2)$
if $A.PK2 >= B.PK2$
  then $TB \leftarrow$ next tuple pertaining to Object_B;
  else $TA \leftarrow$ next tuple pertaining to Object_A;
if $PKI <= PKF$
//there is an intersection between the tuples.
  then add to the PEANO_JOIN (A, B) relation tuple
    $(XX, PKI, PKF, At_1, ..., At_n, Bt_1, ..., Bt_m)$
  else; //there is no intersection between the two current tuples
end while;

B) Suppose we want to solve some the spatial query: **Is B $\subseteq$ A?** The affirmative answer to the above question implies that: *for each tuple pertaining to the B object it must exist at least one tuple of the A object such that:* $B.PK1_i >= A..PK1_k$ *and* $B.PK2_i <= A..PK2_k$, $i=1,..n$; $k=1, ..., m$. . Or, using tuple relational calculus notation: $\{i, k \mid MAP(i) \text{ and } MAP(k) \text{ and } \forall(i) \exists(k) (i..PK1 >= k.PK1 \text{ and } i.PK2 <= k.PK2\} \neq \phi$.

C) Suppose we want to solve the spatial query: **Does B intercept A?**, that is: $B \cap A = \phi$? The affirmative answer to this question implies that: *there is at least one tuple i pertaining to Object A, such that one of the two inequalities holds:* $B.PK1_k < A.PK1_i$ *and* $B.PK2_k >= A.PK1_i$.

D) The query **Is Object A adjacent to Object B?** is equivalent to: *Object B can not intercept Object A and there is at least one tuple i pertaining to object B and at least one tuple k pertaining to Object A,, for which one of the following equalities holds:* $B.PK1_i = A.Pk2_k$ *or* $B.PK2_i = A.PK1_k$.

E) Suppose we want to make a map reclassification based on attributes of the existing classes[8]. A map can be thought as *n* objects (or *n* classes) related in a two-dimensional space. We could represent the map by the Peano relation MAP (Class_Id, $PK1_1$, $PK2_1$, At_1, ..., At_t). In the general case we would have *n* classes: $Class_1$, $Class_2$, ..., $Class_n$. The reclassification operation on a map could be regarded as a function f: A $\rightarrow$ C, where A stands for the set of attributes, and C, for the set describing the new classes of the map. So the new map would be: MAP (f (At_1, ..., At_t)|Class_Id, $PK1_1$, $PK2_1$, At_1, ..., At_t), that is, f is restricted to each $Class_i$. On reality, this map would yet need to be put in the third conformance level. [6], because identical classes could be dispersed in the space.

F) Now we want to analyse two-map overlays, a powerful tool for examining the spatial patterns that arise by the interaction between two maps [8]. In this case the goal is to combine the input maps according to a set of rules (the map model) that determines for each location the class of the output map from the classes of the input maps, that is, in each MAP we have a function f_i: Class_A X Class_B $\rightarrow$ New_Class, where Class_A stands for the set of classes in MAP_A , Class_B for the set of Classes in MAP_B, and New_Class is the result of the application of the function f_i. The two maps could be described by the following Peano relations: MAP_A (Classt_Id_A, PK1, PK2, At$_1$, ..., At$_t$) and MAP_B (Classt_Id_B, PK1, PK2, At$_1$, ..., At$_t$).

The following algorithm could do the desired reclassification:

Result_Map = ϕ ;
for each Class_Id_A in MAP_A do;
    for each Class_Id in MAP_B do;
        TEMP=PEANO_JOIN (Class_Id_A, Class_Id_B);
    //now, for each location, we calculate the output class
        Result_MAP=TEMP (f(Class_Id_A,
        Class_Id_B), PK1, PK2, At$_1$, ..., At$_t$) ∪ Result_Map;
    end for;
end for;
// After the loop the Result_Map Peano relation must be put in the third
//conformity level.

We recall that the Peano Join between two Peano relations means that we keep tuples which have matching Peano keys in both Peano relations. In the above case the variable TEMP is also a PEANO relation

G) Here we are interested in map modelling with attribute tables. Given two maps:
MAP_A (Class_Id, PK1, PK2, At$_1$, ..., At$_n$) and
MAP_B (Class_Id, PK1, PK2, Bt$_1$, ..., Bt$_m$)
we want to build a new map C based in conditions f to be observed in relation to the two sets of attributes A={At$_1$, ..., At$_n$} and B = {Bt$_1$, ..., Bt$_m$}. Obviously the conditions f must be associated to the same spatial coordinates (Peano-keys) in the two input maps: MAP_A and MAP_B.
The following algorithm could do the desired reclassification:

Result_Map = ϕ;
for each Class_Id_A in MAP_A do;
    for each Class_Id in MAP_B do;
        TEMP=PEANO_JOIN(Class_Id_A,Class_Id_B);
    //now, for each location, we calculate the output class
        Result_MAP=TEMP (f(A, B), PK1, PK2, At$_1$, ..., At$_n$, Bt$_1$, ..., Bt$_m$)∪
        Result_Map;
    end for;
end for;
// After the loop the Result_Map Peano relation must be put in the third
//conformity level.

H) Boolean Logic Models, Fuzzy Logic Models and Bayesian Probability Models [4] could be used to produce an output map as the result of operations with n input maps. [8] gives an example of landfill site selection, where the site must be located, such that several conditions, over different thematic maps related to the same region, must be observed according to a given Boolean Logic Model. Boolean Logic Modelling involves the combination of several binary maps

resulting from the application of conditional operators. These binary maps could be thought as layers of evidence. These layers are combined to support some hypothesis. The n maps could be described by the following Peano relations
$MAP (i) (Class\_Id, PK1, PK2, At_{1(i)}, ..., At_{t(i)})$, $i=1,...n$.

The following algorithm could do the desired reclassification:
//n = the number of thematic maps
for i=1 to n do;
    for each Class_Id in MAP(i) do;
        $TEMP\_MAP(i) = MAP(i) (f_i (Class\_Id), PK1, PK2, At_{1(i)}, ..., At_{t(i)})$
    end for;
end for;
// f_i: Class_Id → {0, 1}, where Class_Id is the set of classes found in MAP(i).
//TEMP_MAP(i) can be thought as having only two classes, namely Class_Id_1=
//and Class_Id_2 = 0.
//Now we calculate the intersection of all /conditions applied to each MAP(i), in
//order to obtain the resulting map.

$$Result\_Map\ (\bigcap_{i=1}^{n} Class\_Id(i), PK1, PK2));$$

//Alternatively we could have:
//Result_Map=(PEANO_JOIN (Class_Id(1), .., Class_Id(n));

4   Formalizing a Peano Algebra Extension

For formalizing a Peano Algebra Extension we use a many-sorted 2nd order language, since when talking about a Peano relation, for instance PR (Object_Id, PKI, PKF, $At_1, ..., At_t$) for certain operations we can not consider the arguments of the operations as variables of the same sort. What we are trying to do here is to use the same type of language we use to formalize data structures. The need of a second order language comes from the intensional character of Peano Algebra. In pure extensionality of classical First Order Predicate Calculus the meaning of a relation $R(a_1, ..., a_n)$ is only the number of tuples that satisfies R. So, the only interpretation of sentences of Calculus is **true** (T) or **false** (F). But if we are working on an application of the Calculus, for instance, Data Base Systems, the meaning cannot be only T or F. Since we are talking about the real world we are referring to things like customers, employees and so on. The meaning of the word "employee" is not the same of the word "employer", even where the number of tuples that satisfies both relations is the same.

In the case of a Peano relation a tuple R (Object_Id, PKI, PKF, $At_1, ..., At_t$) represents a region in the two-dimensional space. Such tuple comprises a dense region in the space, with infinite number of points, and eventually can be split into n other tuples R (Object_Id, PKI_1, PKF_1, $At_1,..., At_t$), R(Object_Id,PK_2,PKF_2,$At_1, ..., At_n$), ..., R(Object_Id,PK_n,PKF_n, $At_1, ..., At_n$). Conversely, if we have m tuples with some required characteristics, then they can be aggregated to form only one tuple, after, of course applying the three conformity levels – C1, C2 and C3 [6]. In Peano relations a certain tuple can be equivalent, or have the same meaning, to several other tuples and vice-versa. This is due to the intensional aspect that is potentially enveloped in each Peano tuple. In Codd's relations this is not allowed. So, we need a language that can express some sentences of type: *"for all tuples that satisfies the conformity rules...".* Thus, if we are quantifying over tuples we are using an expression of a 2nd order language.

5   The Proposed 2nd Order Many-Sorted Language

First, we present some needed formalism and definitions:

A) 2nd Order Language L^2_τ, of type $\tau = (n_i^0, n_j^1)$, $i \in I, j \in J, I, J \subseteq N$.
B) variables $v_0, v_1, v_2, \ldots$: many-sorted individuals (type 0)
C) Variables P_0^i, P_1^i, predicates (type 1) variables of rank i, $i \in [0, n]$, $n \in N$.
D) Sentential connectives: $\neg, \vee, \wedge, \rightarrow, \leftrightarrow$.
E) Quantifiers: $\forall, \exists$.
F) Non-logical constants:
   F.1) Predicate constants: r_i, $i \in I$, of rank n_i^0.
   F.2) Operation constants: f_j, $j \in J$, of rank n_j^1.
G) Let V be the set of individual variables, P_n the set of predicate variables of rank n, $n \in N$, where N is the set of natural numbers.
H) Let Var = $V \cup (\bigcup_{j=1}^{n} P_j^i)$, be the set of all variables.
I) As in the case of first order language L_τ, the set T_{m2} of terms of L^2_τ, is the smallest set X, such that:
   I.1) $V \subseteq X \subseteq E$, where E stands for the set of all expressions;
   I.2) $e_0, e_1, \ldots e_{n-1} \in X$, $n_j^1 = n$ implies that $f_j(e_0, e_1, \ldots e_{n-1}) \in X$. That is, no type 1 variables occurs in terms.
J) Any expression of the form $r_i(t_0, \ldots, t_{n-1})$, $i \in I$, $n_i = n$, or $P_i^n(t_0, t_1, \ldots, t_n)$, $i \in N$, where $t_0, t_1, \ldots, t_{n-1}$ are terms, is called an atomic formula.
K) The AF_m be the set of atomic formulas of L^2_τ.
L) The set of formulas: F_m is the subset of X, such that:
   L.1) $AF_m \subseteq X \subseteq E$;
   L.2) If $e_0, e_1 \in X$ then $\neg e_0, e_0 \vee e_1, e_0 \wedge e_1, e_0 \rightarrow e_1, e_0 \leftrightarrow e_1, \in X$.
   L.3) If $e \in X$, $v \in Var$, then $\forall ve, \exists ve \in X$.
M) An algebraic structure for L^2_τ is just an algebraic structure of similarity type τ, exactly as in first order languages: $\Gamma = (A, R_i, F_j)$, $i \in I, j \in J$. The intention is that individuals variables $v_i \in V$, range over elements of the domain A, while predicate variables $v_i^n \in V_n$ range over n-place relations $R_i: A^n \rightarrow \{T, F\}$ on A, each $n \in N$. Thus, for instance, "$\forall x^0(\ldots)$" means *"for every truth-value (T, F)(...)"*, because a 0-place relation is essentially just a truth-value. Similarly, "$\exists y^1(\ldots)$" means *"for a subset $B \subseteq A(\ldots)$"*, because a 1-place relation $R: A^1 \rightarrow \{T, F\}$ on A is just a subset of A. Finally, "$\forall x^2(\ldots)$" means *"for every binary relation R on A (...)"*. To be precise, we define an assignment Φ over Γ to be a function

with domain Var, such that:
  i) For $v_i \in V$, $\Phi(v_i) \in A$, as in first order logic.
  ii) For $P_i^n \in V_n$, $\Phi(P_i^n)$: $A^n \to \{T, F\}$, i.e., $\Phi(P_i^n)$ is an n-place relation on A.

For $t \in T_{m2}$ we define $\Phi(t)$ as in first order language.

We introduce new satisfaction clause for atomic formulas:
  i) Φ satisfies $P_i^n\ t_0, ..., t_{n-1}$ if and only if $\Phi(P_i^n)(\Phi(t_0),...,\Phi(t_{n-1}))=T$.

We introduce also new quantifier clauses:

  ii) Φ satisfies $\forall P_i^n\ P$ if and only if for every $R_i:A^n \to \{T, F\}$, $\Phi\begin{pmatrix}P_i^n\\R\end{pmatrix}$ satisfies P

  (Here, P is a non-atomic formula, and this particular notation means that Φ satisfies P_i^n for each R).

  iii) Φ satisfies $\exists P_i^n\ P$ if and only if at least one $R_i:A^n \to \{T, F\}$, $\Phi\begin{pmatrix}P_i^n\\R\end{pmatrix}$ satisfies P.

6 Translation of Some Operations in the 2nd Order Many-Sorted Language

We have chosen two operations presented in 3 to be expressed in the proposed 2nd Order Many-Sorted Language. Here we will consider that:

  i) $=, >=, <=, \in$ are non-logical constants;

  ii) MAX(X, Y) and • are non-logical operators, where, for instance A•PK1 selects the Peano-key of object A that belongs to the Peano relation MAP (Class_Id, PK1, PK2, At$_1$, ..., At$_t$).

- Operation presented in 3.E):

  Here we introduce the notation $f_w(x_1, ..., x_n)$ to denote the function f restrict to w. Then we'll have:
  $\forall x \forall y [x = $ MAP (Class_Id, PK1, PK2, At$_1$, ..., At$_t$) $\land$ y = At$_1$, ..., At$_t$ $\to$ MAP(f_{Class_Id}(At$_1$, ..., At$_t$), PK1, PK2, At$_1$, ..., At$_t$)].

- Operation presented in 3.F):

  Let define f_i: Class_A X Class_B $\to$ New_Class as the function that has as domain the set of classes of MAP_A Cartesian Product the set of classes of MAP_B, and New_Class, as the result of operation f_w described above.
  $\forall x \forall y \forall s \forall t \forall w \forall z\ [x = $ Class_Id_A $\land$ y = MAP_A $\land$ x $\in$ y $\land$ s = Class_Id_B $\land$ t=MAP_B $\land$ s $\in$ t $\land$ w = PEANO_JOIN(Class_Id_A, Class_Id_B) $\land$ z = RESULT_MAP) $\to$ z = (f_i(Class_Id_A, Class_Id_B), PK1, PK2, At$_1$, ..., At$_t$)].

7 Conclusions

There are much research we may develop using the algebraic tool we described here. First, we want to go a step further in formalising an Algebra applied to Spatial Data Bases, similar to the approach of Relational Algebra vis-à-vis the traditional Data Bases. There, we'll give a special attention to the discussion of intensional x extensional aspects. Second, we want to investigate several operations that are done in Spatial Analysis in Geographical Information Systems and try to treat them in a unified and formal approach. Third, we think also in develop a kind of Geometric Algebra based, not in the traditional vector approach, but, instead, in the concepts related to Peano relations and quadtrees.

References:

1. Enderton, Herbert. B. A Mathematical Introduction to Logic, Academic Press, 1972..
2. Prawitz, D. Ideas and Results in Proof Theory.
3. Prawitz, D. Natural Deduction: A Proof-Theoretical Study. Almqvist & Wirksell Editors. Upsalla, 1965.
4. Pearl, J. Probabilistic Reasoning in Intelligent Systems: Networks of Plausible Inference. Morgan Kaufmann Publishers Inc., San Mateo, California, 1998.
5. Zariski, O. & Samuel, P. Commutative Algebra, vol I, 1967, D. Van Nostrand Company, Inc.
6. Robert Laurini and Derek Thompson. Fundamentals of Spatial Information Systems. Academic Press Inc, 1992.
7. Martien Molenaar. An Introduction to the Theory of Spatial Object Modelling for GIS. Taylor and Francis Ltd., 1998.
8. Graeme F. Bonham Carter. Geographic Information Systems for Geoscientists: Modelling with GIS. Pergamon Press. 1994.
9. Peter A. Burrough and Rachael A, McDonnell. Principals of Geographical Information Systems. Oxford University Press, 1998.
10. Elmasri R., Navathe, S.B. Fundamentals of Database Systems, Redwood City: Benjamin/Cummings, 1994.

Training and Application of Artificial Neural Networks with Incomplete Data

Zs. J. Viharos[1], L. Monostori[1], and T. Vincze[2]

[1]Computer and Automation Research Institute, Hungarian Academy of Sciences
POB 63, H-1518, Budapest, Hungary
{viharos,monostor}@sztaki.hu
[2]Eötvös Loránd University, Egyetem Tér 1-3., H-1056, Budapest, Hungary

Abstract. The paper describes a novel approach for learning and applying artificial neural network (ANN) models based on incomplete data. A basic novelty in this approach is not to replace the missing part of incomplete data but to train and apply ANN-based models in a way that they should be able to handle such situations. The root of the idea is inherited form the authors' earlier research for finding an appropriate input-output configuration of ANN models [16]. The introduced concept shows that it is worth purposely impairing the data used for learning to prepare the ANN model for handling incomplete data efficiently. The applicability of the proposed solution is demonstrated by the results of experimental runs with both artificial and real data. New experiments refer to the modelling and monitoring of cutting processes. Keywords: Neural Networks, Machine Learning, Applications to Manufacturing.

1 Introduction

Reliable process models are extremely important in different fields of computer integrated manufacturing [8]. Model-based solutions are efficient techniques to make difficult problems more tractable. It can contribute to elaborating new algorithms, supporting decision makers, decreasing the risk in investments, and running the systems exposed to changes and disturbances more efficiently.

Learning denotes changes in the system adaptive in the sense that learning techniques enable the system to do the same or similar task more effectively next time [9]. Artificial neural networks (ANNs) are general, multivariable, nonlinear estimators. This soft computing technique can offer viable solutions especially for problems where abilities for real-time functioning, uncertainty handling, sensor integration, and learning are essential features [9]. Successful applications in manufacturing were reported on in the literature [13].

The paper illustrates a method and a related tool able to handle and apply incomplete data arising in the field of manufacturing. After describing some real-life cases and methods for handling data sets with missing parts, the developed algorithm will be detailed. Test results based on artificial data and an application in the field of production explains the behaviour and the advantages of this novel concept. Further research issues are also enumerated.

2   Missing Data and Their Handling in Different Application Fields

The problem of missing data arises in several fields of real-life applications. This section flashes some examples from several fields, together with the applied methods for handling missing data.

Valentini *et al.* used function-based interpolation to determine a value for missing data in their study for carbon dioxide exchange of a beech forest in Central Italy [15].

Different approaches were tested to replace missing data by Pesonen *et al.* while building up a neural network to study medical data received form University Hospital of Tampere and Savonlinna Central Hospital in Finland. About 20% of the patients' data were missing, mainly the parameter of leucocyte count. They compared four methods to replace the missing data: substituting means, random values, data based on the nearest neighbour and a neural network based substitution [12].

The markets of a commercial Bank in Latin America were the field of examination where data with missing parts were arising. The researchers applied clustering methods to build different market segments and they extended the incomplete part of the individual data vectors with the most probable values of the given segment [5].

An algorithm called Expectation Maximisation was elaborated by Ghahramani and Jordan to find appropriate values for the places of missing data. Their solution was interesting in the point of view that all of the dependencies among given and missing parameters and their distribution were taken into account in the replacement procedure [4].

The problem of missing data is found in several fields of manufacturing, as pointed out by Gardner and Bicker in the case of semiconductor wafer manufacturing, as well [3].

3   A Novel Approach to Handle Missing Data by Neural Networks

Some methods were listed in the previous paragraph to solve the problem of missing data. These methods try to complete the missing part of the data vectors in several different ways. Instead of completing data vectors another aspect could be to generate neural networks, which can handle incomplete data directly. This was one of the main ideas during the research reported on here.

3.1   Earlier Results behind the Research

The authors have presented some research results to improve the applications of neural networks in the industrial field [11], [16], [17]. The introduction of a concept basically different from the assignment-oriented application of neural networks was an important milestone of the research.

The expert facing a task recognises that a model is needed to solve it, since some parameters have to be determined based on other ones. Several attempts can be found in the literature to solve various engineering assignments by using neural network based model [1], [2], [6], [7]. The authors referred to applied the same concept to select the input and output parameters of the applied neural network, namely, parameters are known in the application phase were selected as inputs and the model has to estimate the output parameters which are unknown in the assignment. Consequently, the predefined assignment determines the input-output configuration of the model.

A special requirement for modelling is to satisfy the accuracy demands acceptable in a certain situation. The requirement cannot be always met in the assignment dependent input-output configuration, i.e., it is not obvious that this setting realises the most appropriate mapping between the variables, consequently, a method finding the appropriate input-output configuration of the applied neural network model is needed. A method like this was presented by the authors in [16]. This model building process runs totally independent from the assignment, which was a new, corn idea behind. That is why the models can be considered as the general model of the system in question.

As to the application phase, a method was developed, by which various engineering tasks can be solved, using the same general model [10]. Applying the general model the method estimates the unknown parameters of the given assignment, based on the known ones. The estimation is totally independent from the input or output positions of the known or unknown parameters on the general model. This new concept ensures that several assignments (e.g. in planning, optimisation or control) can be solved by using the same general model of the system in question.

3.2 The Basic Idea in the Introduced Concept

The aim of the method reviewed here is to decide about all of these parameters whether they should act as inputs or outputs and to give also the multilayer perceptron model. The method is a modified backpropagation learning algorithm. The procedure can be illustrated in short, as follows:

repeat
{
    for all of the input-output vector pairs
    {
        forward (based on the input data vector),
        calculate the derivatives of the network weights,
        calculate the corresponding changes of the weights and sum them up
    }
    change the weights with their corresponding sum
} until a special criterion, e.g. the value of estimation error is higher than required.

Batch learning was also the basis of the convergence accelerator algorithm called SuperSab [14] applied in the algorithm described here.

A flag is used in the developed algorithm to indicate if a neuron of the ANN model is protected or not. If a neuron is in a protected state, it means neither it takes part in the forward calculation nor the derivative is calculated related to it, consequently, zero modification ratio is stored for all of the weights in the ANN structure that are directly connected to this neuron (both input and output weights). So the weights are also set into a protected state. Certainly this protected state influences the calculation of derivatives of other unprotected parts of the ANN structure. This technique combined with the sequential forward selection search method was used to find the appropriate input-output configuration of the ANN model [16]. This behaviour is shown in Fig. 1.

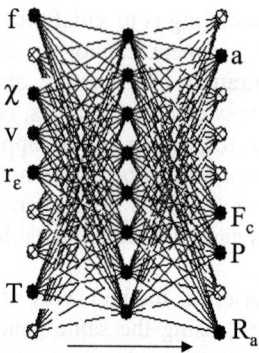

Fig. 1. Protected and unprotected states of different neurons and corresponding weights. If a graphical object is not continuous it means that this part of the ANN structure is protected and, consequently, not used

During the input-output search the protection state of a neuron is changed only after a batch is over, namely, the protection of a neuron could not be changed until all the learning data have been processed by the ANN.

3.3 Description of the Algorithm for Handling Incomplete Data

The algorithm is based on the main idea of turning the neurons corresponding to the missing part of certain data vectors into protected state and leaves the other neurons in unprotected state. Further information is needed to realise this, i.e., to describe which part of the input- and output vector is missing. A flag called validity is used for indicating whether a date in the data vector is valid or not. It shows that a so named (binary) validity vector is attached to all of the input and output vectors to describe the validity state of the data incorporated in these vectors.

The protection of the input and output layer of the ANN structure changes according to the validity vector of the data vector, in question, namely if a data is invalid in the input or output vector, the corresponding neuron is set to protected, otherwise the neuron will be unprotected. It ensures that the protections of the input and output neurons change by all the learning data vector pairs. It is necessary to explain that the changes in weights are the same as in batch learning, namely, the changes remain summed during one batch of learning. If a learning vector indicates that a weight is set to be protected in the ANN structure, the corresponding value of sum for change the weight remains the same as that of the previous learning vector. At the beginning of one batch all the weights will be changed according to the calculated sum. The algorithm can be presented as a small cycle:

repeat
{
   for all the input-output vector pairs
   {
      set the protection of the network neurons according to the validity vector of the given data vector

      forward (based on the input data vector),
      calculate the derivatives of the network weights,
      calculate the corresponding changes of the weights and sum them up

      set all the neurons unprotected
   }
   change the weights with their corresponding sum
} until a special criterion, e.g. the value of estimation error is higher than required.

This modification shows that the introduced method gives a solution for handling the missing data in the learning and application phases of the ANN model, too. The test results of this new learning and application method will be presented in the next paragraphs.

4 Experimental Tests

Some experimental tests were performed to evaluate the behaviour of the developed learning algorithm. The following situations were tested:

- based on artificial data: handling incomplete output data, handling incomplete input data, if the data are fully independent and if the data are redundant,
- based on real measurements in the field of engineering: modelling a cutting process.

In the tests the behaviour of the introduced algorithm with varying ratio of missing data was studied. As it was outlined in the previous paragraph, the so named validity vector shows which part of the input and output data vector is missing. The test stages were started always with data vectors without any missing values, indicated by validity vectors. Different ratios of missing data were simulated through various ratios of zero values in the validity vectors. The network estimation errors were used for measuring the capability of a tested method. During the test runs the difference in the estimation errors originating form the various ratios of incomplete data was eliminated.

4.1 Handling Incomplete Output Data

The first phase was to test the case where the missing data were in the output vectors. Four types of possible methods for handling missing data were compared:

1. Excluding the data vectors containing missing data, from the data set.
2. Writing a fix value into the place of missing data. Usually, data are normalised [9] before the learning stage of ANN, which transforms all the learning vector components into a predefined interval. Because this interval runs usually from 0 to 1, the value for replacing missing data was set to 0.5.
3. Writing random value into the place of missing data. The normalisation is the reason to set this value from the 0 to 1 interval.
4. Using the introduced new algorithm for handling the data with missing data.

A thousand data vectors were generated for learning and testing purposes. Two, randomly set values (x and y) gave the input vectors. Output parameters were also two dimensional, with the calculated values of x and y^2, respectively. The network estimation errors were compared in the learning and testing phases, as well. The ratio of missing output data differed in the tests. Results are shown in Fig. 2.

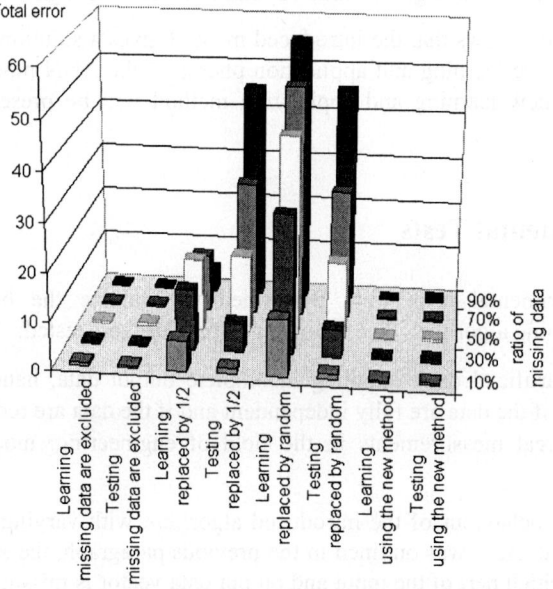

Fig. 2. Test results of handling missing output data by different methods

The results show that the introduced method can learn dependencies among parameters, in the same way as in the case where no data were missing. It shows also that it is better to use the introduced method than to replace missing data by fix or random values. The results indicate that the method is able to handle missing output data.

4.2 Handling Incomplete Input Data Containing Independent Variables

A data set having three-dimensional input and output data vectors was used in this stage. Input data vectors consisted of randomly selected independent values of x, y and z, while output vectors incorporated calculated components of x*y, x*y*z and z. A hundred data with different missing ratio were used for learning. The analysis of the learning stage shows that the errors are mainly originated from the vectors where parts of inputs were missing. This experience leads to the question whether the data vectors without a missing part were enough to learn the dependencies among parameters and whether it could be better to eliminate the data vectors form the learning data set. Comparing the estimation errors between the ANNs trained with a data set having incomplete data vectors and trained with a data set where these data vectors were eliminated. This test was repeated for different ratios of complete data vectors. The results show that it is worth eliminating the training data vectors having incomplete, fully independent input data.

4.3 Handling Incomplete but Redundant Input Data

This section analyses the introduced learning algorithm with redundant and incomplete input data. A simple task with two input parameters and one output parameter was selected. The input data are of the same value ensuring a high-level of redundancy (x, x). The output parameter is x^2. A thousand data with a third part of complete vectors, a third part were the first input, furthermore, a third part where the second input parameter was missing, were used for testing the models built up with different ratio of incomplete input vectors. Results related to estimation capabilities are shown in Fig. 4.

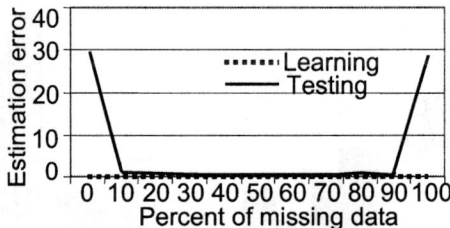

Fig. 3. Estimation errors measured on the testing data set

Fig. 4 shows an exciting result, namely, if the ANN model is built up only with complete data or only with incomplete data, the estimation error is significantly higher than using a model built up on data set having complete and incomplete input data, as well.

This fact gives the idea that if the input parameters are redundant it is worth impairing the data purposely to prepare the ANN model for handling incomplete input data successfully. This impairment means either setting some certain data as they would be missing, or extending a data set with the same but in some ratio impaired data.

5 Application of the Developed Tools for a the Cutting Process

The above paragraphs explained the test results of the introduced algorithm. These tests were based on different artificial data sets, which ensured that dependencies among modelled parameters are known. The application described in this section is based on real machining measurements, which leads to the fact that the exact dependencies among the system parameters are unknown and the analysed data set is also originated from a real process. The turning process was analysed in this experiment. The same material was cut at different speed (v), depth of cut (a) and feed per revolution (f). During the experiments the tool wear was measured and only the tools of which the flank wear (VB) had not got over a certain level, were applied, i.e. they were considered as sharp ones. 119 different experiments were performed and during the process the three components of the cutting force, the power and the cutting temperature, and after the process, the tool wear and roughness (Ra) were measured. Features [10] calculated from the measured data were used to solve a process-monitoring task, indicating that the a, f, v parameters and average and variance value of each force components and of the resultant force can be selected as input parameters. The temperature (T), average power (P) and the specific energy appropriation (u – shows the cutting energy per chip volume) give the output vectors.

A hundred data were randomly selected for learning and the remaining 19 data were used for a testing process. A special ratio of missing data was used for building up the ANN model and a data set with another ratio of missing data for testing its estimation capability. The missing data were among the input and output data, as well. The estimation errors are shown in Fig. 6.

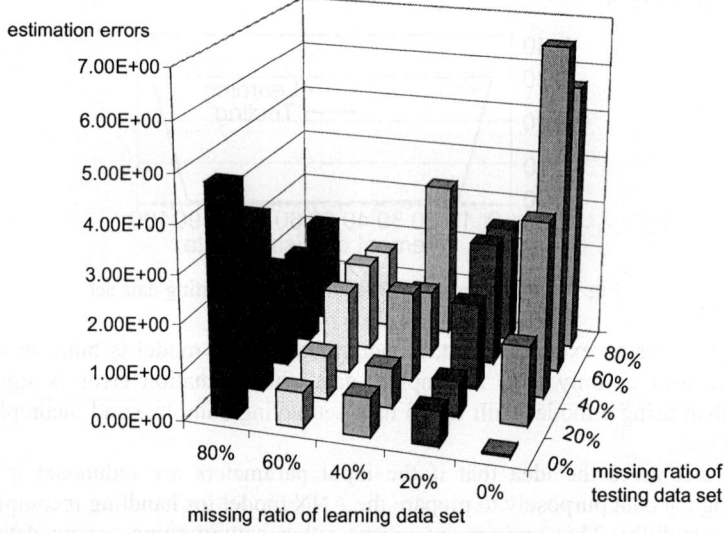

Fig. 4. Estimation errors of different ANN models for monitoring a cutting process trained on a data set having a special ratio of missing data and tested on another data set with another ratio of missing data

The figure shows several interesting outcomes:

- The best situation is to have missing data neither in the learning nor in the testing data set. No situation resulted in lower estimation errors.
- The estimation capabilities of ANN models trained on a missing ratio of 20%, 40%, 60% are very similar in their estimation capabilities.
- The estimation error of ANN models trained with no missing data is always higher than the errors of models trained with different ratio of missing data if the testing data could be incomplete. This experience conforms the idea from the previous paragraph, namely, it is worth impairing the data purposely to prepare the ANN model for handling incomplete data efficiently.
- A hundred data were used for learning and 19 for testing, which explains that Fig. 6. cannot give advice in the cases of 80% incompleteness of learning or testing data sets.

Further interesting questions are generated by the figure:

- Is there an ideal ratio for incompleteness of learning data? Does it depend on the missing ratio of testing/application data?
- What is the allowed highest level of possible incompleteness?

These and similar questions are addressed in the research work of future.

6 Conclusions

A new approach is introduced in the paper for handling incomplete data in the learning and application phases of ANN based modelling. The basic ideas behind the research were not to extend the missing data with special values but to develop a learning algorithm for the applied neural network model to be able to handle data in a situation like this. The algorithm adapts the network structure to the individual data with missing components, it selects the so-called protection state of network neurons according to the missing/existing parts of data vectors. The developed algorithm was compared with three data-extending-methods and resulted in a model with superior estimation capabilities. The algorithm was tested through artificial data and it was found that it was completely able to handle missing output data. Tests showed that it is worth eliminating learning vectors with incomplete input vectors if input parameters are totally independent. Interesting result arose in the case of incomplete input data if input parameters are redundant: *it is worth impairing the data purposely to prepare the ANN model for handling incomplete data efficiently.* Tests for monitoring a turning process were performed and reinforced the previous conclusions in a real application field, as well.

Acknowledgements

The research was partially supported by Bolyai János Research Fellowship of Dr. Viharos Zsolt János and by the "Digital enterprises, production networks" project in the frame of National Research and Development Programme of the Ministry of Education (proj. No. 2/040/2001). A part of the work was covered by the *National Research Foundation,* Hungary, Grant No. T034632.

References

1. Choi, G. H.; Lee, K. D.; Chang, N.; Optimisation of process parameters of injection modeling with neural network application in a process simulation environment, Annals of the CIRP, Vol. 43., No. 1., 1994, pp. 449-452
2. Dini, G.; A neural approach to the automated selection of tools in turning, Proc. of 2nd AITEM Conf, Padova, Sept. 18-20, 1995, pp. 1-10
3. Gardner, R.; Bicker, J.; Using machine learning to solve tough manufacturing problems, International Journal of Industrial Engineering-Theory Applications and Practice, Vol. 7., No. 4., 2000, pp. 359-364
4. Ghahramani, Z.; Jordan, M. I.; Learning form incomplete data; Report of the Massachusetts Institute of Technology, Artificial Intelligence Laboratory and Center for Biological and Computational Learning Department of Brain and Cognitive Sciences, C. B. C. L. Paper No. 108, A. I. Memo No. 1509, 1994
5. Kamakura, WA.; Lenartowicz, T.; Ratchford, BT.; Prouctivity assesment of multiple retail outlets, Journal of retailing, Vol. 72, No. 4, 1996, pp. 333-356
6. Knapp, G. M.; Wang, Hsu-Pin; Acquiring, storing and utilizing process planning knowledge using neural networks, J. of Intelligent Manufacturing, Vol. 3, 1992, pp. 333-344
7. Liao, T. W.; Chen, L. J.; A neural network approach for grinding processes: modeling and optimization, Int. J. Mach. Tools Manufact., , Vol. 34, No. 7, 1994, pp. 919-937
8. Merchant, M. E.; An interpretive look at 20th century research on modelling of machining, Inaugural Address, Proc. of the CIRP International Workshop on Modelling of Machining Operations, Atlanta, Georgia, USA, May 19, 1998, pp. 27-31
9. Monostori, L., Márkus, A., Van Brussel, H., Westkämper, E.; Machine learning approaches to manufacturing, Annals of the CIRP, Vol. 45., No. 2., 1996, pp. 675-712
10. Monostori, L.; Viharos Zs. J.; Markos, S.; Satisfying various requirements in different levels and stages of machining using one general ANN-based process model; Journal of Materials Processing Technology, Elsevier Science SA, Lausanne, 107: (1-3). NOV 22, 2000, pp. 228-235
11. Monostori, L.; Viharos, Zs.J.; Hybrid, AI- and simulation-supported optimisation of process chains and production plants, Annals of the CIRP, Vol. 50., No. 1., 2001, pp. 353-356
12. Pesonen, E.; Eskelinen, M.; Juhola, M.; Treatment of missing data values in a neural network based decision support system for acute abdominal pain, Artificial Intelligence in Medicine, Vol. 13, 1998, pp. 139-146
13. Rangwala, S. S., Dornfeld, D. A.; Learning and optimisation of machining operations using computing abilities of neural networks, IEEE Trans. on SMC, 19/2, March/April, 1989, pp. 299-314

14. Tollenare, T.; SuperSAB: fast adaptive backpropagation with good scaling properties, Neural Networks Vol. 3., 1990, pp. 561-573
15. Valentini, R.; DeAngelis, P.; Matteucci, G.; Monaco, R.; Dore, S.; Mugnozza, GES.; Seasonal net carbon dioxide exchange of a beech forest with the atmosphere, Global Change Biology, Vol. 2., No. 3., 1996, pp. 199-207
16. Viharos, Zs. J.; Monostori, L; Automatic input-output configuration of ANN-based process models and its application in machining, Book: Lecture Notes of Artificial Intelligence - Multiple Approaches to Intelligent Systems, Conference, Cairo, Egypt, May 31-June 3, Springer Computer Science Book, Springer-Verlag Heidelberg, 1999, pp. 659-668
17. Viharos, Zs.J.; Monostori, L.; Optimisation of process chains and production plants using hybrid, AI- and simulation based general process models; The Fourteenth International Conference on Industrial & Engineering Applications of Artificial Intelligence & Expert Systems, Book: Lecture Notes of Artificial Intelligence, Springer Computer Science Book, Springer-Verlag Heidelberg; Budapest, Hungary, 4-7 June, 2001, pp. 827-835

Message Analysis for the Recommendation of Contact Persons within Defined Subject Fields

Frank Heeren and Wilfried Sihn

Fraunhofer Institute for Manufacturing Engineering and Automation (IPA)
Nobelstrasse 12, 70569 Stuttgart, Germany
{feh,whs}@ipa.fhg.de

Abstract. Today, a firm's employees embody a significant source of knowledge, not only by documenting knowledge, but also by assisting other colleagues with problem solving. Due to decentralising or business networking aiming at cooperation among companies, the transparency within an enterprise as to which employees are experts in what field diminishes. The purpose of IT-systems for expert recommendation is to endow employees with easy access to experts within certain subject fields. This paper illustrates the Xpertfinder method developed at the Fraunhofer IPA that analyses explicit knowledge forms such as E-Mail- or Newsgroup messages of logged-in users for the preparation of expert profiles. Contrary to common systems Xpertfinder only uses those parts of a message entirely created by the sender. The Latent Semantic Indexing methodology is used in order to determine the subject of each message. With the aid of Bayesian Belief Networks analysis results are combined to evocative expert characteristics for anonymous display. Measures for the protection of personal data as well as future research fields are addressed.

1 Introduction

A company's employees represent an important knowledge resource and harbour a large proportion of its available knowledge pool. Personal contacts and interpersonal communication are therefore essential for knowledge transparency and -exchange within a company, as not all knowledge can be documented [1].

Organisational restructurings such as decentralisation or the set-up of internal networks for the purpose of cooperation render, however, the development and cultivation of interpersonal contacts difficult and thus lead to a lack of transparency with respect to the company-wide availability of experts and their knowledge. Duplication of work – e.g. in product- and process development – is a consequence [2].

Modern information and communication technologies are becoming established in all areas of economy and science and can alter cooperativity. Technical information assistants, for example, are gradually assuming the tasks of personal assistants [3]. In this context, E-Mail as an off-line peer-to-peer communication instrument is the most frequently used Internet service [4] and nowadays represents an important channel of communication.

Expert recommender systems support the search for persons with knowledge in certain subject fields and engender contact networks between co-workers both within the organisation as well as in ventures. The use of E-Mail- or Newsgroup messages as an indicator of knowledge in particular areas is already a component of various protocols (cf. Section 2). The presented Xpertfinder method, however, goes beyond the technological level for two reasons. Indeed, a methodology for the support of expert search is outlined that firstly carries out an automated communication analysis. The hereby-extracted contribution to the communication – the text sequence of an E-Mail or Newsgroup message personally created by the sender – is subsequently classified according to a predefined subject field and analysed. Furthermore the Latent Semantic Indexing method [5] is used in order to determine the subject of each communication contribution. With the aid of Bayesian Belief Networks analysis results are combined to evocative expert characteristics for anonymous display.

Section 2 describes methods that use E-Mail communication data to conduct expert searching. Set-up and functionality of the "Xpertfinder" prototype created at the Fraunhofer IPA are dealt with in the third section where general methodology process, communication analysis for the authentication of message authorship, similarity operations between message and subject field and combination of analysis results are illustrated. An outlook on future research areas concludes the exposé.

2 Procedure for the Support of Expert Searching with Automatic Data Acquisition

Expert recommender systems based on automatic data acquisition derive the information needed to create a knowledge profile from explicit knowledge forms. Four categories can be distinguished among these knowledge forms:

- Communication data (e.g. E-Mail- or Newsgroup-messages)
- Factual knowledge (e.g. files, documents or web sites)
- Attributed knowledge (e.g. quotations)
- Behavioural records (e.g. utilisation of resources or programmes)

The following listing describes methods and concepts that partly or exclusively utilise communication data such as E-Mail messages as a knowledge indicator.

Schwartz and Wood describe a method which analyses exclusively the sender/receiver information of E-Mail messages. Coupled with manually created graphs of persons with similar interests in specific subject fields, further persons are highlighted as a result of the search, who, very likely, are equally interested in the specified topics [6].

Procedures which consider both message text as well as address fields of E-Mails are for instance "ContactFinder" [7] and "KnowNet MailTack" [8]. ContactFinder is an agent-based method that relies on keywords and key sentences to extract the themes of messages that stand out from the remaining text by virtue of their formatting. Analogously, ContactFinder scans the messages for contained questions and automatically suggests persons who should be helpful with the posed questions. KnowNet-MailTack allows the connection and investigation of various E-Mail-based discussions through links and argument chains.

Besides the described E-Mail-message based procedures there are methods that take further files into account besides E-Mail messages for the data acquisition ("Agent Amplified Communication" [9], "Expert Finder" [10], "Yenta" [11]). Common to all is the creation of knowledge profiles via the frequency of encountered keywords. Kautz et al. enhance this with a profile of co-worker contacts, while Mattox et al. further employ the spatial proximity between name and keyword as a further indicator for knowledge.

While considering the above mentioned systems, which refer to a concrete implementation of expert recommender systems, Yimam and McDonald each present a flexible system architecture that is supposed to allow consideration of diverse data sources and data processing methods for implementing expert recommender systems. ("DEMOIR" [12]; "Expertise Recommender" [13]). Furthermore, a variety of systems for the support of expert searching are commercially available, among which "KnowledgeMail Plus" from the company Tacit Knowledge Systems is representative.

None of the mentioned systems which use the message text of an E-Mail message as indicator of knowledge conduct any tests, however, as to whether the message text has been created by the sender in its entirety. In the case of a 'forwarded' E-Mail message, for instance, the forwarded text could distort the knowledge profile of the sender since he did not create the text himself. The Xpertfinder method detailed below therefore first reduces each message to its communicative contribution and thus improves the quality of the subsequent creation of the knowledge profile.

3 The Xpertfinder Method for Recommendation of Experts

3.1 Description of the Xpertfinder Method

The basis of the Xpertfinder method is a model made up of subject fields (subject field tree). Each topic has to be described and defined. Within each subject field Xpertfinder allows the anonymous highlighting of persons who are frequently communicating. Moreover, by highlighting communications networks within or outside the subject field, Xpertfinder supports the selection of experts for the subsequent querying. Should Xpertfinder users require support with solving a problem, they may call up the relevant experts and their communications network within the corresponding subject field and select those anonymous experts that seem appropriate to provide the needed support. The user submits his/her query via E-Mail to Xpertfinder, which in turn completes the selected recipients' addresses and forwards the E-Mail. It is up to the queried experts to reply to or ignore the question.

Experts are defined in the first instance as contact persons who stand out both through a high communication intensity as well as communication contacts in specific subject fields. Automatically forwarded copies of E-Mail- or Newsgroup messages serve Xpertfinder in determining experts, whereby it performs the following four steps (cf. Fig. 1):

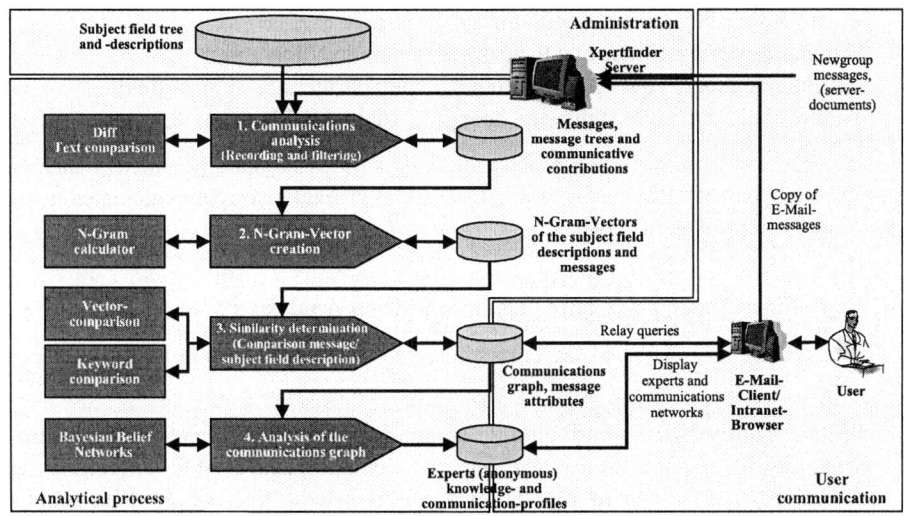

Fig. 1. Full model of the Xpertfinder process

1. Communication analysis:
   Comparison of the E-Mail- or Newsgroup messages without consideration of attachments with all sent messages of the previous weeks, in order to identify and delete identical text sequences. Erroneous evaluations caused by text sequences whose sender is not identical with their author (e.g. reply or forwarded components) can thus be eliminated.
2. Creation of N-Gram-Vectors:
   Dissection of messages and subject field descriptions in N-Grams (text sequences of length N overlapping one character with each nearest neighbour N-Gram) and calculation of their relative occurrence.
3. Determination of similarity:
   For the determination of similarity two methods are analogously employed: Firstly the similarity between vectors of E-Mail message and subject fields are calculated using Latent Semantic Indexing method [5]. Similarity values from the comparison of the message with each subject field lying above a heuristically determined significance threshold point to the topic of the communication process and hence to the sender's knowledge. On the other hand, parallel to the above mentioned method, a similarity test is performed between keywords from the message and keywords/subject indices from the subject field description. The additional keyword similarity test provides the opportunity to apply different weighing to characteristic key words.
4. Analysis of communication graphs:

The use of analysis results for expert selection requires an easily comprehensible representation of as few nominal parameters as possible. Among other the following information from input data and message analysis are available for graph analysis:

- Correlation of the message with the subject fields
- Communication partners and their major communication topics
- Communication frequency per communication partner and subject field
- Subjectively definable similarity among the subject fields

Drawing on this information probabilities for various nominal parameters such as expert status, nature of communication or personal characteristics are calculated using Bayesian Belief Networks [14].

3.2 Detailed Description of the Communication Analysis

Two objectives are related to the communication analysis (step 1 in the Xpertfinder method):

- Determination of the contribution to communication of each new incoming message, i.e. its reduction by text sequences of previously sent E-Mail- or Newsgroup messages, in order to correlate the sender to knowledge only within those subject fields where he himself authors message texts. Moreover, manipulation of knowledge profiles through repeated sending of identical messages is ruled out.
- Correlation of the communicational contributions with communication trees which are later recruited to the detailed description of communicational networks.

The implementation of communications analysis followed according to the "Basic Diff" algorithm [15]. The "Basic Diff" algorithm determines the minimum effort modification description between two symbol chains. A variation of this algorithm may be applied to E-Mail messages as follows (cf. Fig. 2).

Let there be an E-Mail message A, which is sent via Xpertfinder. Message A corresponds to the communicational contribution of A if the text sequence does not possess any overlap of at least one line of text with all messages previously submitted via Xpertfinder. The recipient of message A may send a message B which includes a portion of or the complete message A, such as "forward", "reply", or manually cut and pasted portions of A. The minimum effort modification description between the E-Mail messages A and B is composed of symbols contained in B but not in A. This modification includes browser-specific symbols to mark forwarded or replied-to components of the message – hence the need to replace these, through the use of a manually maintained stop-symbol list, with spaces – and which are thus ignored. The result of this syncopated alteration description, devoid of stop-symbols, is the communicative contribution of message B, which flows into the comparison process with the subject fields. In addition to reducing a message to its communicative contribution, the Diff-comparison informs whether a message refers to a previously sent one, as this is usually the reason for a "reply" or "forward" message. Based on this information the E-Mail message B is correlated with a message tree which already contains the previously sent message A. Each message for which there are no overlapping text sequences forms the 'root' of a new message tree.

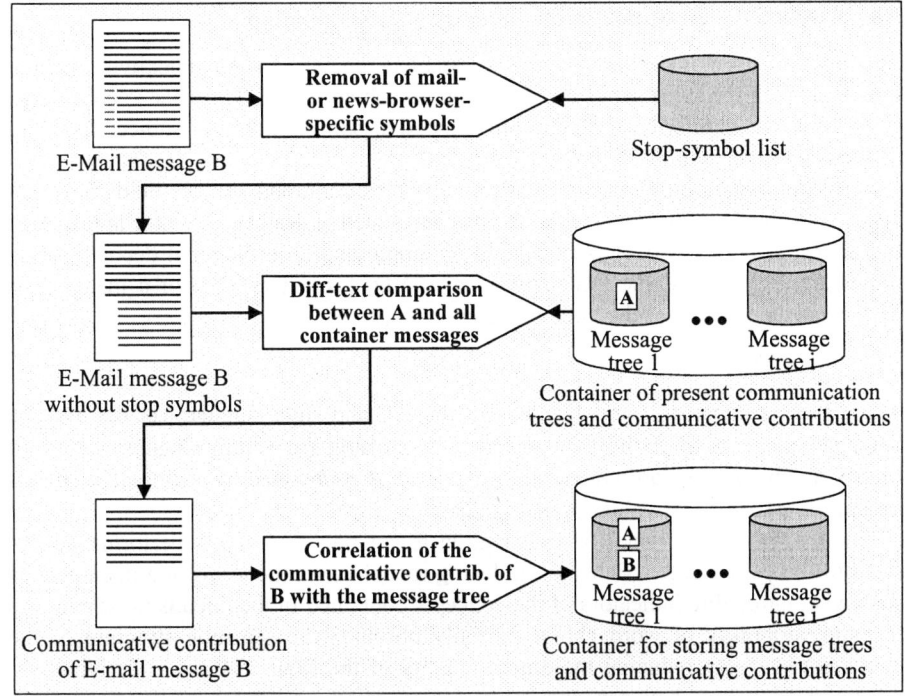

Fig. 2. Detailed model of communications analysis

3.3 Detailed Description of the Vector Creation

In preparation to the similarity comparison the subject field descriptions must be translated into vector form. The reason for this vector creation is to ensure a performing processing ability by the machine.

For this purpose the texts are dissected into N-Grams. The vector dimension corresponds to the number of the different occurring N-Grams. The weighted frequencies of each N-Gram are used as vector components. A weighting function is employed which is composed locally from a logarithmic function and globally from an entropy function [5]. The target is towards a high weighting of N-Grams which appear with high frequency within the document but rarely in all documents.

In other words, the subject field vectors can be represented as an $m \times n$ matrix. This matrix contains a line for each of m different N-Grams and a column for each of n subject fields. The components with a_{ij}

$$A = [a_{ij}] \tag{1}$$

represent the weighted N-Gram occurrences within the documents.

3.4 Detailed Description of the Similarity Determination

The determination of similarity serves to establish whether a communicative contribution deals with the same subject as one or more subject fields. This case is considered as given if the similarity between communicative contribution and subject field lies above a given threshold.

The comparison between communicative contribution and subject fields does not occur with the vectors of the initial subject field matrix. Rather, a rank-k approximation A_k ($k<<\min(m,n)$) of A is calculated by employing an orthogonal decomposition procedure 'Singular Value Decomposition' (SVD) [16]. The SVD of a matrix A is defined as the product of the three matrices

$$A = U\Sigma V^T \tag{2}$$

where the columns of U and V represent the left and right singular vectors and the diagonal elements of Σ the monotonously decreasing (in value) singular values of Matrix A. The rank-k approximation A_k of matrix A is obtained by setting all elements except the k large unity values in matrix Σ to zero.

The reason for the rank-k approximation is not only computing performance. Instead the so called Latent Semantic Indexing model views the terms in a document as somewhat unreliable indicators of the concepts contained in the document. It assumes that the variability of word choice partially obscures the semantic structure of the document [16]. By reducing the dimensionality of the Matrix A, the underlying, semantic relationships between documents are revealed, and much of the noise from differences in word usage etc. is eliminated. For a large value spectrum of k the obtainable results from the similarity comparison with matrix A_k are therefore significantly better than with the initial matrix A [5].

For the determination of similarity among subject fields, each vector has to be compared with each vector in matrix A_k. Suitable hereto is the cosine distance measure [17] which calculates the angle encompassed by two vectors.

The comparison between communicative contribution and the subject fields proceeds as follows: let the vector q be the N-Gram dissection of the communicative contribution (cf. section 3.3). This vector q is mapped in the subsequent step onto the rank-k approximated vector space A_k:

$$q' = q^T U_k \Sigma_k^{-1} \tag{3}$$

After this mapping the similarity of the mapped vector q' with any subject field can be determined through the use of the cosine distance measure.

3.5 Determination of the Expert Status

Known methods utilise either just text or just relations of communication as sources for deriving a recommendation. Therefore, a method is proposed below which calculates an expert status based on any desired variety of sources. The more activity a contact person initiates within a subject field, the more this person is considered as an expert by the system. The expert status in this case represents a statement cumulated from all sources and is maximal when all sources reach the absolutely highest value in the system. Bayesian Belief Networks are applied for this task. Contrasting with traditional statistics, which is restricted to the probability of random occurrences resulting from random experiments, Bayesian statistics goes further. It widens the concept of probability by defining the probability of statements [14]. In this usage the calculated probability reflects the suitability as contact person. Fundamental to Bayesian Belief Networks is the Bayes Theorem in the form:

$$P(B|CD) = \frac{P(B|D) * P(C|BD)}{P(C|D)} \qquad (4)$$

In the common usage of the Bayesian Theorem *B* represents the statement regarding an unknown phenomenon. *C* represents statements that contain Information pertaining to the unknown phenomenon and *D* a statement about background knowledge. *P(B|D)* is known as the Priori-probability, *P(B|CD)* as the Posteriori-probability and *P(C|BD)* as Likelihood. *P(C|D)* may be interpreted as a norming constant [14].

For each potential contact person it is thus possible to implement a Bayesian Belief Network of nodes and edges which assimilates various employee-related source data. Fig. 3 shows an exemplary network containing four such different source types.

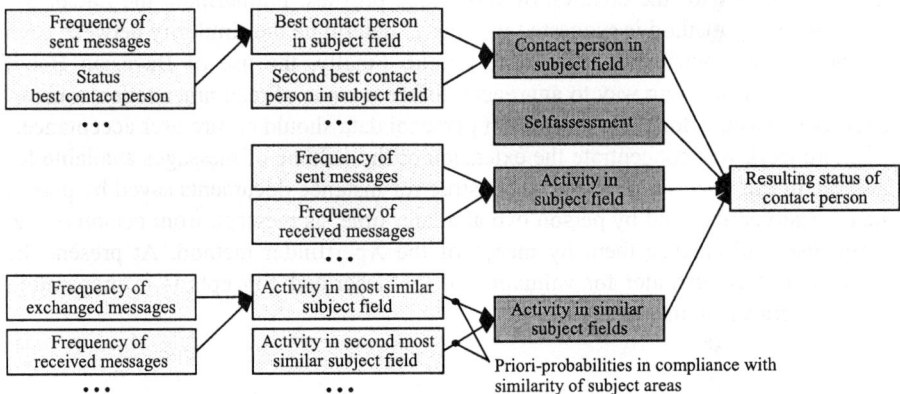

Fig. 3. Section from a Bayesian network for a contact person in a subject field

For all leafs, i. e. the outermost nodes of the network, data from the continuously running analysis is available, such as the frequency of sent messages in one subject field. These data represent the Likelihood. The significance of each page for the next following node relative to its sister nodes may be viewed as the Priori-probability, for instance when a node has, in relation to its sister node, twice as much influence on the following one. Thus it is possible to calculate the Posteriori-probability. Each ensuing node aggregates in this way the information of its predecessors, until the expert status of a person within the subject field is established.

3.6 Application of the Xpertfinder-Method

Results of prototypical deployment of the procedure show good performance for communications analysis, which thus remains applicable even to large settings. The similarity determination based on Latent Semantic Indexing and the cosine distance measure is in principle methodology-depending computing- and time intensive but still applicable in large settings, too.

Due to published experiences pertaining to a rejection by employees of expert recommender systems that offer insufficient protection of personal information [9], a considerable emphasis is put on the Xpertfinder personal privacy and data protection. The implementation of Xpertfinder prevents the scrutiny of co-workers by encoding all data for storage and representing it anonymously.

4 Summary and Outlook

The present paper contains an overview of expert recommendation procedures under the application of communication data such as E-Mail- or Newsgroup messages. The presented Xpertfinder procedure and its installation as a prototype software includes for the first time communication analysis. This prevents the use of message texts having been part of previously sent messages and thus able to falsify the knowledge profile of the sender for the creation of knowledge profiles. Furthermore the Latent Semantic Indexing method is suggested in order to determine the similarity between each communication contribution and subject field. Finally, the use of Bayesian Belief Networks is a promising way to aggregate different types of recommendation sources. Extensive measures for the protection of personal data should ensure user acceptance.

Future work will concentrate the extension of the volume of messages available for analysis. In this way it is possible to construe for instance documents saved by person one on a server and read by person two at a later time, as message from person one to person two and analyse them by means of the Xpertfinder method. At present the application of Xpertfinder for validation of the described concepts is in preparation with an aircraft manufacturer.

References

1. Polanyi, Michael: Personal Knowledge Towards A Post-Critical Philosophy London (Routledge & Kegan Paul Ltd) 1958.
2. Davenport, Thomas H.; Prusak, Laurence: Wenn ihr Unternehmen wuesste, was es alles weiß Das Praxisbuch zum Wissensmanagement Landsberg, Lech (Verlag Moderne Industrie) 1998.
3. Kuhlen, Rainer: Die Konsequenzen von Informationsassistenten Frankfurt a. M. (Suhrkamp) 1999.
4. Esser, Michael H: E-Mail im betrieblichen Einsatz Frechen (Data-Kontext-Fachverlag) 1998.
5. Dumais, Susan: Improving the retrieval of information from external sources In: Behavior Research Methods, Instruments, & Computers, Vol. 23 Issue 2 (1991) 229-236.

6. Schwartz, Michael F.; Wood, David C.: Discovering Shared Interests Among People using Graph Analysis of Global Electronic Mail Traffic In: Communications of the Association for Computing Machinery (ACM), Vol. 36, Issue 8 (1986) 78-89.
7. Krulwich, Bruce; Burkey, Chad: The Contactfinder Agent: Answering Bulletin Board Questions with Referrals In: Proceedings of the Thirteenth National Conference on Artificial Intelligence and Eighth Innovative Applications of Artificial Intelligence Conference, AAAI 96, IAAI 96, Portland, Oregon, USA (MIT Press) 1996.
8. Bettoni, Marco C.: Personal Knowledge Portfolio Tool In: Tagung Technopark Zuerich: Soft[Net]-Club "Knowledge Management" Zuerich 1999.
9. Kautz, Henry; Selman, Bart; Milewski, A.: Agent Amplified Communication In: Proceedings of the AAAI Spring Symposium on Intelligent Agents in Cyberspace Stanford, CA, USA (AAAI) 1996 3-9.
10. Mattox, Dave; Maybury, Mark; Morey, Daryl: Enterprise Expert and Knowledge Discovery, http://www.Mitre.Org/Support/Papers/Tech_Papers99_00/ Maybury_Enterprise/Index.Shtml 6.11.2000.
11. Foner, Leonard N.: Political Artifacts and Personal Privacy: The Yenta Multi-Agent Distributed Matchmaking System Massachusetts, Thesis 1999.
12. Yimam, Dawit: Expert Finding Systems for Organisations: Domain Analysis and the Demoir Approach In: Beyond Knowledge Management: Managing Expertise Escw 99 Workshop http://www.Informatik.uni-bonn.de/~prosec/ ecscw-xmws/ 1999.
13. McDonald, David W.: Supporting Nuance in Groupware Design: Moving from Naturalistic Expertise Location to Expertise Recommendation Irvine, Thesis 2000.
14. Koch, Karl-Rudolf: Einfuehrung in die Bayes-Statistik Berlin, Heidelberg New York u. a. (Springer) 2000.
15. Ukkonen, Esko: Algorithms for Approximate String Matching In: Information and Control, Vol. 64 (1985) 100-118.
16. Berry, Michael. W.; Drmac, Zlatko.; Jessu, Elisabeth. R.: Matrices, Vector Spaces, and Information Retrieval In: SIAM Review, Vol. 41 Issue 2 (1999) 335-362.
17. Salton, Gerard.; McGill, Michael J.: Information Retrieval - Grundlegendes für Informationswissenschaftler Hamburg, New York u. a. (McGraw-Hill) 1989.

An Intelligent Knowledge Processing System on Hydrodynamics and Water Quality Modeling

K. W. Chau[1], Chuntian Cheng[2], Y. S. Li[1], C. W. Li[1], and O. Wai[1]

[1]Department of Civil and Structural Engineering
Hong Kong Polytechnic University
Hung Hom, Kowloon, Hong Kong
cekwchau@polyu.edu.hk
[2]Department of Civil Engineering, Dalian University of Technology
Dalian, 116024, China

Abstract. In order to aid novice users in the proper selection and application of myriad ever-complicated algorithmic models on coastal processes, needs arise on the incorporation of the recent artificial intelligence technology into them. This paper delineates an intelligent knowledge processing system on hydrodynamics and water quality modeling to emulate expert heuristic reasoning during the problem-solving process by integration of the pertinent descriptive, procedural, and reasoning knowledge. This prototype system is implemented using a hybrid expert system shell, Visual Rule Studio, which acts as an ActiveX Designer under the Microsoft Visual Basic programming environment. The architecture, solution strategies and development techniques of the system are also presented. The domain knowledge is represented in object-oriented programming and production rules, depending on its nature. Solution can be generated automatically through its robust inference mechanism. By custom-built interactive graphical user interfaces, it is capable to assist model users by furnishing with much needed expertise.

1 Introduction

Numerical modeling of hydrodynamics and water quality is a highly specialized task that entails expert knowledge. Nowadays, the technology is quite mature with a diversity of mathematical schemes being available [1-4]. The basis of the numerical technique can be finite difference method, finite element method, boundary element method or Eulerian-Lagrangian method. The time-stepping algorithm can be implicit, explicit or characteristic-based. The shape function in the numerical analysis can be of first order, second order or higher order. The covered spatial dimensions can be one-dimensional, two-dimensional depth-averaged, two-dimensional layered, three-dimensional, etc.

Heuristics, empirical knowledge and previous experience are often entailed for any justifiable simplifications as well as the selection of the appropriate modeling technique during the formulation of numerical models on coastal hydraulics and water quality problems by the specialists. Yet, the precision and accuracy of the numerical computation depend largely on the accurate representation of the actual open boundary conditions, the adopted numerical scheme as well as model parameters [5]. The emphasis has always been focussed on the algorithmic procedures in solving some specific coastal problems. These conventional numerical models, being not user-friendly enough, lack knowledge transfers in model interpretation and effective developers/users communication. It results in significant constraints in model uses and a large gap between model developers and application practitioners.

As such, previous attempts have been made to integrate artificial intelligence technology (AI) with some one-dimensional mathematical models in order to aid users to select and apply them at ease [6-7]. These prototype expert systems employed the commercial expert system shell VP-Expert, which run on microcomputer and DOS operating environment. They are useful for practitioners to select a suitable numerical model and facilitate easy representation of heuristic reasoning. Nowadays, with the popular use of the interactive Windows platform, they are no longer adequate.

During the last decade, there has been a widespread interest in intelligent knowledge management and processing systems, which can simulate human expertise in narrowly defined domain during the problem-solving process [8]. They are able to couple together descriptive knowledge, procedural knowledge as well as reasoning knowledge. The recent advent in AI techniques renders it possible to develop these intelligent systems by employing shells of hybrid knowledge representation with established development programming environments such as Visual Basic, C++, etc. In this paper, the architecture, development and implementation of a prototype intelligent knowledge processing system on flow and water quality modeling, employing the shell Visual Rule Studio, are delineated. The primary aim of the work is to couple descriptive knowledge, procedural knowledge and reasoning knowledge for this domain problem together into an integrated system. In particular, highlight is made on the establishment of knowledge base and the interactive visual aids during the problem-solving process.

2   Acquisition and Elicitation of Domain Knowledge

The domain knowledge in hydrodynamics and water quality modeling primarily encompasses matching applicable conditions and model selection. The major task for a knowledge engineer becomes thus to represent them into an intelligent knowledge processing system. Prior to set up the prototype system, it is imperative to abstract the characteristics and applicable conditions of these diverse hydrodynamics and water quality models. The domain knowledge entailed in the development of this intelligent knowledge processing system has been encoded mainly on the basis of interviews with experienced numerical modelers and literature review. Table 1 lists the choice of some principal model parameters of diverse numerical models in common use. There are in total fifteen principal model parameters and each numerical model can be identified by a combination of some of these choices. As such, depending upon the nature and requisite tasks of the project, the most appropriate model can be selected with respect to the criteria of accuracy and computational efficiency. The knowledge base of this intelligent knowledge processing system on hydrodynamics and water quality is developed in compliance with Table 1.

Table 1. Choices of some principal model parameters in hydrodynamics and water quality modeling

Parameter	Selection 1	Selection 2	Selection 3	Selection 4	Selection 5
Numerical method	finite element	finite difference	boundary element	Eulerian-Lagrangian method	
Dimensions	1-d	2-d vertical	2-d horizontal	3-d layered	3-d fully
Co-ordinate system	rectangular	curvilinear	polar		
Scheme	explicit	implicit			
Time-stepping algorithm	single step	alternating velocity and elevation split step	alternating direction split step		
Grid	uniform	not uniform			
Stability	unconditional stable	conditional stable			
Turbulence model	mixing length	k-ε model	dispersion coefficient		
Error of scheme	first-order	second-order	higher-order		
Equation	momentum	continuity	state	density	pressure
Forcing	tide	river discharge	wind	density difference	
Initial condition	zero	non zero			
Boundary condition	zero value	first order zero	second order zero	in-out-bc	
Vertical co-ordinate	normal	sigma	refined near surface	refined near bottom	refined near specified area
Equation term	advection	Coriolis force	horizontal diffusion	decay	sediment interaction

3 The Intelligent Knowledge Processing System

The most difficult task for the novice users is to select an appropriate model and the associated model parameters. With the recent advancement of information technology, it is feasible to furnish more assistance on the selection and manipulation of the ever-complicated models on hydrodynamics and water quality that emerge constantly. When compared with the traditional algorithmic tools, the major distinct feature of an intelligent knowledge processing system is the provision of the decision support with visual window interfaces during the problem solving process.

3.1 System Architecture

Figure 1 shows the architecture of the knowledge processing system. The representation of knowledge in production rules and object-oriented programming is at the core of the process. The knowledge rules comprise two groups, namely, Rule Sets I and Rule Sets II. Based on the responses made by the user on the problem specifications, Rule Sets I generates the conditions of model selection. Rule Sets II then recommends the most appropriate model for a project with a specific task.

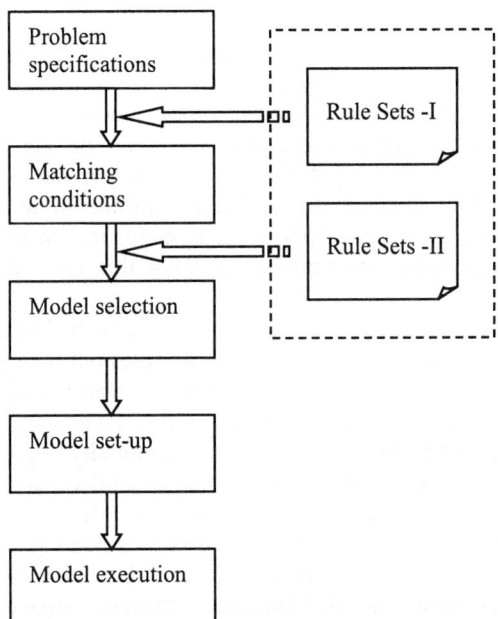

Fig. 1. Architecture of the intelligent knowledge processing system

3.2 Visual Rule Studio

Visual Rule Studio acts as an ActiveX Designer under the Microsoft Visual Basic programming environment [9]. The Production Rule Language, used by Visual Rule Studio, is a high level grammar for problem representation and abstraction designed specifically for the specification and processing of rules. Its grammar employs an object-oriented notation that is common with other programming environments, such as, C++, Java, Visual Basic, etc. It not only facilitates ease of learning, but also furnishes for the easy mapping of client objects to the objects of Visual Rule Studio.

By isolating rules as component objects, separate from objects and application logic, this shell allows developers to leverage the proven productivity. Thus rule development becomes a natural part of the component architecture development process. As such, it is no longer acceptable to require that an entire application be developed wholly within a proprietary development tool in order to realize the benefits of rules programming and knowledge processing applications. They are no longer the limitations of traditional expert system development environments.

The structure of a Visual Rule Studio object basically comprises name, properties, and attributes. The attributes are composed of name, type, facets, method, rules and demons. Visual Rule Studio possesses a robust inference engine and supports three types of inference strategies, namely, backward-chaining, forward-chaining and hybrid-chaining. These inference strategies model the reasoning processes an expert employs when solving a problem.

3.3 User Interface

Practicing hydraulic and environmental engineers are often familiar with the projects but not quite with the ever-complicated numerical models. It is a difficult task for users to select an appropriate model. Instead, well-planned interactive questionnaires, comprising the requisite problem specifications, can be easily understood and responded by users. Depending upon the replies made by the users, a well-defined user interface can duly abstract the requisite information to infer the intrinsic conditions of selection of numerical model. The major task at this stage is to glean and lay out all the pertinent determining variables that have significant effect on simulating computation.

If the nature of the project is known, the scopes of applicable mathematical models can be narrowed since every project type certainly rules out the use of some models. For instances, if reservoir routing is considered, the water level in such a storage facility may be considered horizontal. This simplifies the analysis significantly since dynamic effects are neglected and only the continuity equation needs to be considered. A finite difference approximation can be utilized to simulate the change of storage in the de Saint-Venant continuity equation. Besides, data availability as well as tasks on the hydrodynamics and water quality modeling have also significant effects on the selection of numerical model. In cases data are not sufficient, selection can merely limited to some simplified numerical schemes. Some complicated models are entailed for some special tasks to be performed, for examples, on unsteady simulation of phytoplankton growth in a coastal water system and simulation of flooding and drying of tidal waves.

A window type tabular interface is designed based on the knowledge and previous experience about numerical models. Each tab control is designed to assist the user to locate different groups of questionnaires. The purpose tab is mainly furnished for selection of either real-time or planning condition. The project tab is employed to define the nature of a project from 11 optional buttons, namely, river flood forecast, flood plain, tidal dynamics, estuarine hydrodynamics, salt water intrusion, wave propagation, eutrophication, temperature/density distribution, water pollution, wind storm propagation and outfall.

3.4 Model Selection

Once the users have entered their responses to the questionnaires, the conditions for selecting models can be generated automatically. Figure 2 shows the hierarchy of matching conditions for model selection. Problem specifications are at the first hierarchical level, which covers the key parameters that have significant effect on model selection. The answer sets, which consist of the answers for each problem specifications, constitute the second hierarchical level. The hidden condition sets of intrinsic constraints on model that are acquired through Rule Sets I compose the third hierarchical level. The fourth hierarchical level is condition sets after some repeated conditions are filtered, through which the Rule Sets II can be fired.

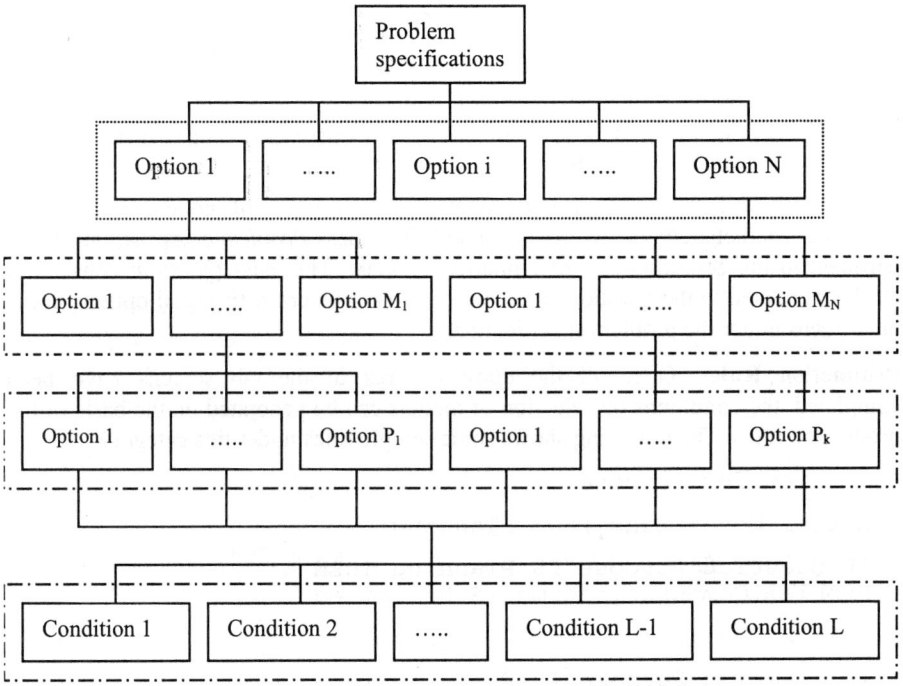

Fig. 2. Hierarchy of matching conditions for model selection

3.5 Knowledge Base

In this prototype system, the knowledge is represented in both object-oriented programming and production rules, depending on the nature of the knowledge.

Object-Oriented Programming. Under the system, three classes are employed, namely, Section, Problem and Question. For instances, part of the structure of the class Problem is shown as follows:

```
CLASS Problem

WITH ComputationTime COMPOUND Limited, Not strict,
Unlimited

WITH Accuracy COMPOUND Significant, Less Significant

WITH Current COMPOUND Large vertical current, Vertical
variation of current, Depth_averaged form, Omitted

WITH StratificationOfWater COMPOUND Significant, Less
Significant, Omitted

WITH DifferenceOfSalinity COMPOUND Significant, Less
Significant, Omitted

WITH DifferenceOfTemperature COMPOUND Significant, Less
Significant, Omitted
```

The class *Section* and its attributes related to the features of models are based on the model characteristics as shown in Table 1. The class *Problem* comprises attributes pertinent to the condition sets of numerical models. The class *Question* consists of attributes related to the questionnaires, which are consistent with visual options in the tab screens under the problem specifications.

Production Rules. Once all the questionnaires in the tab screens have been completed, the conditions on selection of models can be generated on the basis of the production rules. The following shows some sample rules under this category.

```
! Rules for Purpose

RULE 1 Conditions from Planning

IF Question.Purpose IS Planning THEN
Problem.ComputationTime IS Unlimited

RULE 2 Conditions from Planning

IF Question.Purpose IS Planning THEN
Problem.ComputationTime IS Not strict

RULE Conditions from Real_time

IF Question.Purpose IS Real_time THEN
Problem.ComputationTime IS Limited
```

Under this system, the conclusions from Rule Sets I will become the premises of Rule Sets II. The model selection depends mainly upon the model characteristics. Once these questionnaires have been responded and completed, the model characteristics can be inferred through the production rules. The following shows some sample rules under this category.

Rules for Scheme

RULE Explicit method

IF Problem.Accuracy IS Significant AND
Problem.ComputationTime IS Unlimited THEN
Section.Scheme IS Explicit

RULE Implicit method

IF Problem.Accuracy IS Less Significant AND
Problem.ComputationTime IS Limited THEN Section.Scheme
IS Implicit

RULE Semi_implicit method

IF Problem.Accuracy IS Less Significant AND
Problem.ComputationTime IS Not strict THEN
Section.Scheme IS Semi_implicit

3.6   Inference Mechanism

The inference mechanism controls the strategies that determine how, from where, and in what order a knowledge base draws its conclusions. The class *Question* comprises attributes with answer sets acquired from the questionnaires through the interactive user interfaces in the tab screens. Attributes under *Problem* are used in the conclusions drawn from the premise using attributes under *Question*, but are also used in the premise about intermediate conclusions related to attributes under *Section*. The logical relationship is thus *Question* → *Problem* → *Section*. Once the problem description in the tab screens have been entered, the process of model selection can be generated automatically, according to the Rule Sets I and Rule Sets II. This system starts the backward inference process with the attribute *Scheme*, which is one of the goals under the class *Section*. The search returns to a rule with the conclusion related to attribute *Scheme* under the class *Section*. If the premise is related to the attribute *ComputationTime* under the class *Problem* in the rule and its value remains unknown, through rule chaining, it then returns to another rule with conclusion related to the attribute *ComputationTime*. If the premise of the rule is met, the corresponding production rules will be fired. Otherwise, the search finds other rules with conclusions related to the attribute *ComputationTime*. The above processes will be continued until all the goals have been exhausted. The rules for the attribute *Purpose, Project* and *Tasks* will be searched in sequence.

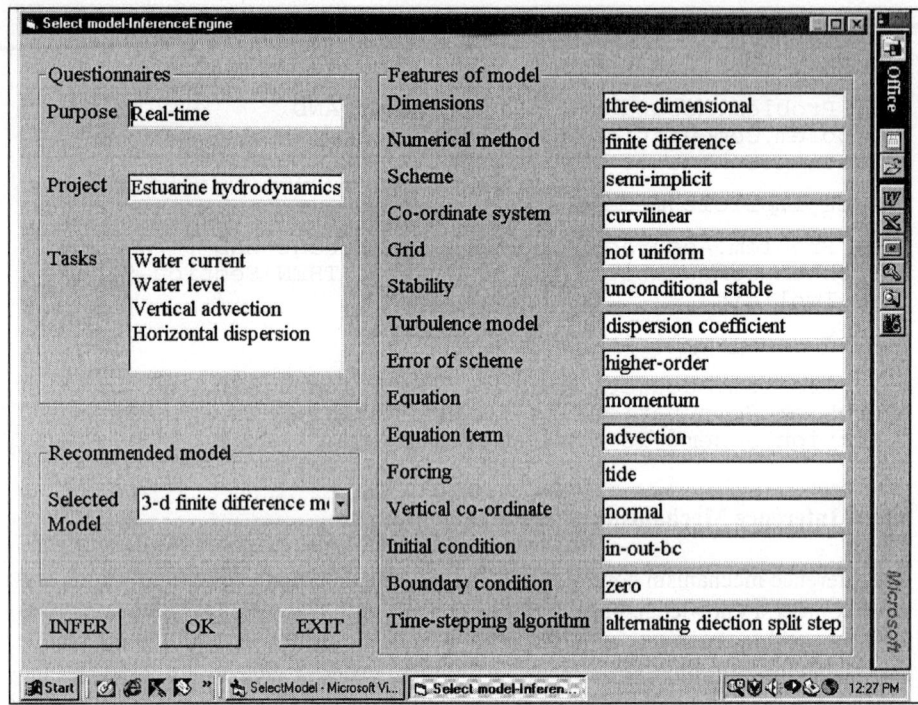

Fig. 3. Screen displaying inference results for Pearl Estuary

4 Results from Prototype System

In order to demonstrate the use of this prototype system, it is applied to Pearl Estuary in the vicinity of Hong Kong. Details of the numerical modeling of the estuary can be found in [2]. On the basis of the available information and the local characteristics for Pearl Estuary, the questionnaires are duly completed through the friendly user-interface. After the input data have been entered and a command button is clicked, a summary of the input requirements and the inference results are shown in the left and right frames of the screen in Figure 3, respectively, which are found to be consistent with the expert opinions.

5 Conclusions

A prototype intelligent knowledge processing system on hydrodynamics and water quality modeling is successfully developed and implemented. It is demonstrated that the hybrid knowledge representation approach combining object-oriented programming technique and production rule system is appropriate for this domain problem. The focus has been concentrated on the establishment of knowledge base and the interactive visual aids during the problem solving process. A user-friendly interface is capable to bridge significantly the gap between the numerical modelers and the users. By using custom-built interactive graphical user interfaces, it is able to assist designers by furnishing with much needed expertise and cognitive support in the numerical modeling of hydrodynamics and water quality.

Acknowledgement

This research study was supported by the University Research Grants of Hong Kong Polytechnic University (G-YC39) and the Natural Science Foundation of China (No. 60073037).

References

1. Blumberg, A. F., Khan, L. A., St. John, J. P.: Three-dimensional Hydrodynamic Model of New York Harbor Region. Journal of Hydraulic Engineering **125(8)** (1999) 799-816
2. Chau, K. W., Jiang, Y. W.: 3D Numerical Model for Pearl River Estuary. Journal of Hydraulic Engineering ASCE **127(1)** (2001) 72-82
3. Chau, K. W., Jin, H. S., Sin, Y. S.: A Finite Difference Model of Two-Dimensional Tidal Flow in Tolo Harbor, Hong Kong. Applied Mathematical Modelling **20(4)** (1996) 321-328
4. Chau, K. W., Jin, H. S.: Eutrophication Model for a Coastal Bay in Hong Kong. Journal of Environmental Engineering ASCE **124(7)** (1998) 628-638
5. Martin, J. L., McCutcheon, S. C., Schottman, R. W.: Hydrodynamics and Transport for Water Quality Modeling. Lewis Publishers, Boca Raton (1999)
6. Chau, K. W., Yang, W. W.: A Knowledge-Based Expert System for Unsteady Open Channel Flow. Engineering Applications of Artificial Intelligence **5(5)** (1992) 425-430
7. Chau, K. W., Zhang, X. Z.: An Expert System for Flow Routing in a River Network. Advances in Engineering Software **22(3)** (1995) 139-146
8. Chau, K. W., Chen, W.: A Fifth Generation Numerical Modelling System in Coastal Zone. Applied Mathematical Modelling **25(10)** (2001) 887-900
9. Rule Machines Corporation: Developer's Guide for Visual Rule Studio. Rule Machines Corporation, Indialantic (1998)

Uncertainty Management and Informational Relevance

M. Chachoua and D. Pacholczyk

LERIA, University of Angers
2, Boulevard Lavoisier F-49045 Angers Cedex 01, France
{chachoua,pacho}@info.univ-angers.fr

Abstract. This paper is devoted to the informational relevance notion in qualitative reasoning under uncertainty. We study the uncertainty and the relevance notions and we present a symbolic approach to deal with uncertainty. This approach enables us to represent the uncertainty in ignorance form, as in common-sense reasoning, by using linguistic expressions.

1 Introduction

Usually, the uncertainty encountered in common-sense reasoning is expressed subjectively by using linguistic expressions as *"Certain"*, *"Almost-certain"*... According several studies [8,12,1,3], the uncertainty can be expressed in two main forms: (1) *ignorance form* when the one express the information deficiency or (2) *belief form* when one refers to the information available. We have showed in our previous studies that it is more interesting to represent and to manage uncertainty in ignorance form rather in belief form. We can quote at last two reasons [1,2]: (a) *it exists some situations when uncertainty can be expressed only in ignorance form*, (b) *all uncertainty expressed in belief form can be translated in ignorance form. The reverse is not always possible.*

The first approach who allows us to evaluate the ignorance is the entropy theory of Shannon [12]. This theory is built on the substrate of the classical probability theory. By this fact, it can process only the uncertainty of probabilistic nature. To palliate this limit, two others theories have been presented. The first one, using the fuzzy set [14], has been proposed by De Luca and Termini [6]. The second theory, presented by Klir and Folger [9], use the evidence theory [7,11]. However, all these theories are of *quantitative* nature. So, in order to manage *qualitative ignorance*, we have presented recently a symbolic entropy theory devoted especially to the qualitative uncertainty [1,2,3,4].

In common-sense reasoning the uncertainty processing is based on the *relevance* of information [13,1,4]. The relevance notion allows us to extract the more relevance information to treat a problem in uncertain situation. This notion is not taking into account by the qualitative approaches of uncertainty. So, the main purpose of our study in this paper is the integration of the relevance notion in the symbolic entropy theory that we have presented.

In section 2, we explore the notions of uncertainty and informational relevance. In section 3 we present a substract of the symbolic many-valued logic that we use. In section 4, we discuss our method of representation of ignorance and informational relevance. This method consists of the definition in the language of a special many-valued predicates called *Uncert* and *Pert*. The predicate *Uncert* satisfies a set of axioms which governs the ignorance concept, that we present in section 5. In section 6 we present the theorems of ignorance management. Finally, in section 7, we show with an example how ignorance can be managed qualitatively by using informational relevance notion.

2 Uncertainty and Relevance

In this section we show the difference between ignorance and belief concepts and the relation between the notions of ignorance and informational relevance.

2.1 Uncertainty Notion

Generally, in the universe of discourse, a knowledge is considered to be *certain* if its truth value is known. In other words, this knowledge is *either true or false*. Otherwise, it is considered *uncertain*. In this sense, the uncertainty expresses the indecision on the truth value of knowledge. So, according to several studies [8,3,4], the uncertainty can be expressed in two main forms: (1) the *ignorance form*, when one expresses the *information deficiency*, (2) the *belief form*, when one refers to the *available information*. In the numerical area, the Shannon entropy theory shows the relationship between these forms [12]:

$$H(e_1, e_2, ..., e_n) = -\sum_{i=1}^{n} p_i \log(p_i) \quad (1)$$

Where p_i is the (objective) probability or `belief degree` of the event e_i and H is an entropy measure or an `ignorance measure` of the n events. The expression $-p_i \log(p_i)$ is the ignorance degree of e_i. To illustrate the relationship between the notions of ignorance and information, let us consider the case of two events A and B. According to the Shannon's theory, the realization of A produces an *information quantity* on B obtained by the following equation:

$$QI_B(A) = H(B) - H(B|A) \quad (2)$$

Where $H(B)$ is the ignorance degree of B and $H(B|A)$ designates the conditional ignorance of B knowing A obtained by:

$$H(B|A) = -\sum_{i=1}^{n} p_{(i,j)} \log(p_i(j)) \quad (3)$$

i and j denotes respectively, the valuation of the events A and B. $p_i(j)$ designates the conditional probability of j knowing i. To process conditional ignorance of two events, it is necessary to know their independence relation. This notion is defined as follow:

Definition 1. *Two events A and B (or associated formulae) are said independent, if and only if, one has:* $H(A|B) = H(A)$ *and* $H(B|A) = H(B)$.

In Shannon's theory, with Equation (1) and Equation (3) one has:

$$H(B|A) = H(A \cap B) - H(A) \qquad (4)$$

So, one can points out the following cases: (i) If A and B are equivalent events, then the *ignorance degree* of B is equal to the *information quantity* produced by the realization of A (i.e. $H_A(B) = 0$). It is the same situation when A and B are two contradictory events. (ii) One can also note the situation where A and B are independent events. Thus, in this case, one has $H(B) = H_A(B)$ and $H(A) = H_B(A)$. Consequently, $QI_B(A) = QI_A(B) = 0$. In other words, in this situation, A does not produce any piece of information on B. Thus, for these reasons, one considers that *the ignorance degree* expresses *the information deficiency degree*.

The Shannon entropy theory is usable only when the uncertainty is of probabilistic nature. To palliate this limit, two other approaches have been proposed. The first one, proposed by Luca and Termini [6], uses the fuzzy set theory. The second one, used with the evidence theory, has been presented by Klir and Folger [9]. All these theories have a *quantitative* nature. In this sense, they are not enough adapted to the *symbolic* uncertainty, like the uncertainty encountered in common-sense reasoning. Thus, we propose below a symbolic entropy theory, reproducing the fundamental properties of the preceding approaches, in the context of a symbolic many-valued logic.

2.2 Relevance Notion

There are at least three main cases where the acquired information doesn't influence on the knowledge state [13,4]: (1) *case where there is no relationship between the acquired information and the knowledge object*, (2) *case where the acquired information is redundant*, and (3) *case where the acquired information contradicts the knowledge, but its relevance (of this information) is very weak compared to the relevance of information acquired previously*[1]. In other words, in these cases the acquired information is not *relevant* for the considered knowledge object. To illustrate intuitively the informational relevance notion, let us consider the following example [4].

Exemple 21 *Let's suppose after that a road accident, the offending driver takes the flight. A policeman collects the following testimonies:*
α_1: *The offending driver car is Ford mark and red color,*
α_2: *The offending driver ran away in Paris direction,*
α_3: *The registration number of his car is 9405 WH 59,*

---

[1] This case corresponds to Axiom 4 of the symbolic entropy theory (see Section 5).

α_4: The offending driver listened music at the time of the accident,
α_5: The offending driver carried sunglasses at the time of the accident,
α_6: It snowed in Washington at the time of the accident.

All these testimonies are considered true (thus certain). The aim of the policeman's investigation is the knowledge of the offending driver identity, which is *completely unknown* (total ignorance). Intuitively, the informational relevance of these testimonies are not the same for the object of this investigation. Indeed, to recover the offending driver identity, it is more interesting to consider the testimony α_3 since it permits to recover the driver identity[2]. Testimonies that seem relatively less relevant to the investigation object (knowledge object) are α_4, α_5 and α_6. Thus, the testimonies can be classified intuitively, by increasing order of informational relevance, as follows:

$$\alpha_6 \leq \alpha_5 \leq \alpha_4 \leq \alpha_2 \leq \alpha_1 \leq \alpha_3$$

In this sense, information relevance is *gradual*. In other words, information can be relatively more or less relevant to the knowledge object. The testimony α_3 is *the most relevant*, because it is most susceptible to annihilate *all the ignorance* on the knowledge of the offending driver identity. α_6 is the *least relevant* testimony because it lets *unaltered the ignorance* on the knowledge of the offending driver identity. So, we can define the relevance as follows:

Definition 2. *An information φ is relevant for a knowledge object K, if and only if, the acquirement of φ reduces the ignorance of K. So, all acquirement information φ allowing to annihilate all ignorance of K is considered as quite relevant for K. All acquirement information φ which lets unaltered the ignorance of K is considered as not at all relevant for K. Between these extreme cases, the relevance degree of φ corresponds to the reduction degree of the ignorance of K.*

3 Many-Valued Logic Framework

The many-valued logic that we use here have been presented in [10].

3.1 Chains of De Morgan

Let $M \geq 2$ be an integer. Let us designate by $\mathcal{W}$ the interval $[1, M]$ completely ordered by the relation "$\leq$", and by "n" the application such that $n(\beta) = M + 1 - \beta$. In these conditions $\{\mathcal{W}, \wedge, \vee, n\}$ is a lattice of De Morgan with: $\alpha \wedge \beta = min(\alpha, \beta)$ and $\alpha \vee \beta = max(\alpha, \beta)$. Let $\mathcal{L}_M = \{\tau_1, ..., \tau_M\}$ be a set where: $\alpha \leq \beta \Leftrightarrow \tau_\alpha \leq \tau_\beta$. Thus, $\{\mathcal{L}_M, \leq\}$ is a chain in which the least element is τ_1 and the greatest element is τ_M. We define in $\mathcal{L}_M$, two operators "$\wedge$" and "$\vee$" and a decreasing involution "$\sim$" by the relations: $\tau_\alpha \wedge \tau_\beta = \tau_{min(\alpha,\beta)}$, $\tau_\alpha \vee \tau_\beta = \tau_{max(\alpha,\beta)}$, and $\sim \tau_\alpha = \tau_{n(\alpha)}$.

---

[2] We suppose that there exists a file of census of all cars as well as their owners at the police station.

Likewise, operators v_γ ($\gamma \in \mathcal{W}$) are defined as $\mathcal{L}_M$ in the following way: If $a = b$ then $v_a \tau_b = \tau_M$ else $v_a \tau_b = \tau_1$. In the context of the predicate many-valued logic used here, the qualitative values v_α and τ_α are associated by:

"x is v_α A" $\Leftrightarrow$ x is v_α A $\Leftrightarrow$ "x is A" is $\tau_\alpha - true$

where x and A designates respectively an object and a concept. Note that these equivalences can be viewed as generalizations of Tarski criteria [10].

In the following, the chain $\{\mathcal{L}_M, \leq, \wedge, \vee, \sim\}$ will be used as the support of the representation of truth degrees.

3.2 Interpretation of Formulae

The formal system of many-valued logic used here can be found in [10].

Definition 3. We call an interpretation structure $\mathcal{A}$ of the language $\mathcal{L}$, the pair $<\mathcal{D}, \mathcal{R}_n>$ for $n \in N$, where $\mathcal{D}$ is a non-empty set called domain of $\mathcal{A}$ and $\mathcal{R}_n$ a multi-set[3] in $\mathcal{A}$.

Definition 4. Let $\mathcal{V} = \{z_1, z_2, ..., z_n, ...\}$ be the infinite countable set of individual variables of the formal system. We call a *valuation of variables*, a sequence denoted $x = <x_0, ..., x_n, ...>$ where $\forall i, x_i \in \mathcal{D}$. So, if x is a valuation of variable, then one has: $x(n/a) = <x_0, ..., x_{n-1}, a, x_{n+1}, ...>$.

Definition 5. Let Φ be a formula of $\mathcal{SF}^4$ and x a valuation. The relation "x satisfies Φ to a degree v_α in $\mathcal{A}$" denoted $\mathcal{A} \models_\alpha^x \Phi (\alpha \in \mathcal{W})$ is defined by:
$\mathcal{A} \models_\alpha^x \mathcal{P}_n(z_{i_1}, ..., z_{i_k}) \Leftrightarrow <z_{i_1}, ..., z_{i_k}> \in_\alpha \mathcal{R}_n$
$\mathcal{A} \models_\alpha^x \neg \Phi \qquad \Leftrightarrow \mathcal{A} \models_\beta^x \Phi, \tau_\beta =\sim \tau_\alpha$
$\mathcal{A} \models_M^x V_\alpha \Phi \qquad \Leftrightarrow \mathcal{A} \models_\alpha^x \Phi$

Definition 6. Let ϕ be a formula. We say that ϕ is v_α-*true* in $\mathcal{A}$ if and only if we have a valuation x such that x v_α-*satisfies* ϕ in $\mathcal{A}$.

4 Representation of Ignorance and Relevance

Let $\mathcal{L}$ be language, $\mathcal{A}$ an interpretation structure of $\mathcal{L}$ and Ω a set of formulae of $\mathcal{SF}$ such that: $\Omega = \{\phi \in \mathcal{SF}, \mathcal{A} \models_M^x \phi \text{ or } \mathcal{A} \models_1^x \phi\}$

In other words, for a valuation x, all formulae of Ω are either *satisfied* or *not satisfied* in $\mathcal{A}$ (i.e. either true or false in $\mathcal{A}$). In the situation of information deficiency, it is impossible to known the truth value of a formula ϕ. In this case ϕ is uncertain. In our approach, the ignorance degree of ϕ is represented thanks to a particular many-valued predicate called *Uncert* which has been added in the many-valued logic.

---

[3] The notion of multi-set was introduced by M. De Glas [5]. In this theory $x \in_\alpha \mathcal{R}_n \iff x$ belongs to degree τ_α to the multi-set $\mathcal{R}_n$. Note that this multi-set theory is an axiomatics approach to the fuzzy sets theory of Zadeh [14]. In this theory $x \in_\alpha A$ is the formal representation of $\mu_A(x) = \alpha$.

[4] $\mathcal{SF}$ designates the formulae set for the system of many-valued predicates calculus.

Definition 7. Uncert *is defined as follows:*

- *Uncert is a many-valued predicate such that:* $\forall \varphi \in \Omega$ *Uncert* $(\varphi) \in \mathcal{SF}$.
- *Given an interpretation* $\mathcal{A}$ *of* $\mathcal{L}$, *for all formula of* Ω *one has:*
$\mathcal{A} \models_\gamma^{x(0/\phi)}$ *Uncert*$(V_\chi \varphi) \iff$ *Uncert*$(V_\chi \phi)$ *is* $v_\gamma -$ *true in* $\mathcal{A}$
$\iff$ "ϕ *is* $t_\chi -$ *true*" *is* u_γ-*uncertain in* $\mathcal{A}$

with $\tau_\gamma \in [\tau_1, \tau_M]$ *and* $\tau_\chi \in \{\tau_1, \tau_M\}$.

In other words, the degree τ_γ designates the *truth degree* of the formula Uncert $(V_\chi \phi)$ and the *ignorance degree* of the formula $V_\chi \phi$.

Remark 1. If there is no confusion, in the following, we will use $\mathcal{A} \models_\gamma$ Uncert $(V_\chi \phi)$ instead of $\mathcal{A} \models_\gamma^{x(0/\phi)}$ Uncert $(V_\chi \varphi)$ and $\mathcal{A} \models_\gamma$ Uncert (ϕ) instead of $\mathcal{A} \models_\gamma$ Uncert $(V_M \phi)$. Especially in the examples that we will present in the following sections, we will use $Uncert(\phi) \equiv u_\gamma -$ *uncertain* instead of $\mathcal{A} \models_\gamma Uncert(\phi)$.

According Equation (4), the symbolic conditional ignorance is defined as follow:

Definition 8. *The ignorance degree of* ψ *knowing* ϕ ($\psi|\phi$) *is obtained by:*

$$\left\{ \begin{array}{l} \mathcal{A} \models_\gamma Uncert(\phi) \\ \mathcal{A} \models_\beta Uncert(\psi \cap \phi) \end{array} \right\} \implies \mathcal{A} \models_\alpha Uncert(\psi|\phi)$$

with $\tau_\alpha = D(\tau_\beta, \tau_\gamma)$. D *is a qualitative difference operator defined bellow.*

Definition 9. D *is a qualitative difference operator satisfying the fundamental properties of the classical difference operator. So,* $\forall (\tau_\alpha, \tau_\beta, \tau_\gamma) \in \mathcal{W}^3$:
(1) $D(\tau_\alpha, \tau_1) = \tau_\alpha$, (2) $D(\tau_\alpha, \tau_\beta) = \tau_1 \Leftrightarrow \tau_\alpha = \tau_\beta$,
(3) $D(\tau_\alpha, D(\tau_\beta, \tau_\gamma)) = D(D(\tau_\alpha, \tau_\beta), \tau_\gamma)$,
(4) $\tau_\alpha \leq \tau_\beta$ *and* $\tau_\beta \leq \tau_\gamma \Rightarrow D(\tau_\gamma, \tau_\beta) \leq D(\tau_\gamma, \tau_\alpha)$.

Remark 2. In an uncertain situation, the value of τ_χ is unknown. Nevertheless, one can choose one reference truth value (τ_1 *or* τ_M). So, if one chooses the value $\tau_\chi = \tau_1$ then one will say that "ϕ is *false*" is u_γ-uncertain in $\mathcal{A}$. Otherwise, if $\tau_\chi = \tau_M$ then one will say that "ϕ is *true*" is u_γ-uncertain in $\mathcal{A}$. Usually, in common-sense reasoning, one uses the value "true" (*i.e.* $\tau_\chi = \tau_M$) as a reference truth value. To express the falseness of a statement, the human subject rather uses the **negation form** *[10,1]*. Thereafter we will use this method in our approach.

Example 1. For $M = 5$, one could introduce the following ordered set of linguistic degrees.
D_{t_5}: { *not-at-all-true, little-true, true-enough, very-true, totally-true* } which corresponds to $v_\alpha -$ *true* with $\alpha = 1, 2, ..., 5$.
D_{u_5}: { *not-at-all-uncertain, little-uncertain, uncertain-enough, very-uncertain, totally-uncertain* } which corresponds to $u_\alpha -$ *uncertain* with $\alpha = 1, 2, ..., 5$.

| Certain | Almost-certain | Uncertain-enough | Very-uncertain | Totally-uncertain |

Fig. 1. Uncertainty Scale

In the common-sense reasoning, one normally uses "*certain*" and "*almost certain*" rather than "*not at all uncertain*" and "*little-uncertain*". So, with these terms we obtain the uncertainty scale[5] represented by figure 1.

According to Definition 2, the relevance degree of an information, relatively to a knowledge object[6], corresponds to the degree of ignorance reduction produced by this information at its acquirement. As in the ignorance representation case, to represent the relevance of an information in the logical context, we introduce a special predicate named *Pert* in the language. The **truth degree** of the predicate $Pert(\varphi|\psi)$ expresses the **relevance degree of** φ **relatively to** ψ.

Definition 10. *Pert is defined as follows:*

⋄ *Pert is a many-valued predicate such that:* $\forall \varphi, \psi \in \Omega, Pert(\varphi|\psi) \in \mathcal{SF}$
⋄ *For an interpretation* $\mathcal{A}$ *of* $\mathcal{L}$, $\forall \phi, \psi \in \Omega$, *one has:*
$\mathcal{A} \models_\alpha^{x(0|\phi)} Pert(\varphi|\psi) \Leftrightarrow Pert(\phi|\psi)$ *is* v_α-*true in* $\mathcal{A}$
$\Leftrightarrow \phi$ *is* p_α-*relevant for* ψ *in* $\mathcal{A}$, *with* $\tau_\alpha \in [\tau_1, \tau_M]$.

In other words, the formula ψ represents the knowledge object (*"the offending driver identity"* in Example 21). However the formula ϕ corresponds with acquired information (testimonies in Example 21).

Remark 3. If there is no confusion, in the following, we will use:
$\mathcal{A} \models_\gamma Pert(\phi|\psi)$ instead of $\mathcal{A} \models_\gamma^{x(0|\phi)} Pert(\varphi|\psi)$ and especially in the examples $Pert(\phi|\psi) \equiv p_\gamma - relevant$ instead of $\mathcal{A} \models_\gamma Pert(\phi|\psi)$.

According to the previous study in section 2.2, we can define a relationship between the predicates *Uncert* and *Pert* (i.e. the relation between ignorance and informational relevance) as follows:

Definition 11. *The relation between* Uncert *and* Pert *is defined by:*

$$\left\{ \begin{array}{l} \mathcal{A} \models_\gamma Uncert(\psi) \\ \mathcal{A} \models_\beta Uncert(\psi|\phi) \end{array} \right\} \Longrightarrow \mathcal{A} \models_\alpha Pert(\phi|\psi)$$

with $\tau_\alpha = D(\tau_\gamma, \tau_\beta)$ *and* $\tau_\beta \leq \tau_\gamma$. *D is a qualitative difference operator.*

Example 2. For M=5, we can choose the operator D of Table 1.

The values of table 1 are interpreted as follows. If $(\tau_\gamma \neq \tau_1)$ and $(\tau_\beta = \tau_1)$, then according to Definition 2, ϕ is quite relevant for ψ (i.e. $D(\tau_\gamma, \tau_\beta) = \tau_M = \tau_5$). Indeed, in this case the acquirement of ϕ annihilated all ignorance on ψ.

---

[5] Note that this scale and the linguistic terms are chosen subjectively by a human expert.
[6] In Example 21, the knowledge object is *the offending driver identity*.

Table 1. Table of relevance calculus

$\tau_\beta \downarrow \backslash \tau_\gamma \rightarrow$	τ_1	τ_2	τ_3	τ_4	τ_5
τ_1	τ_1	τ_5	τ_5	τ_5	τ_5
τ_2	—	τ_1	τ_2	τ_3	τ_4
τ_3	—	—	τ_1	τ_2	τ_3
τ_4	—	—	—	τ_1	τ_2
τ_5	—	—	—	—	τ_1

Otherwise, if $\tau_\gamma = \tau_\beta$ then ϕ is less relevant for ψ (i.e. $D(\tau_\gamma, \tau_\beta) = \tau_1$). In the intermediate cases, according to Definition 2, the relevance degree is equal to the reduction degree of ignorance (i.e. qualitative difference between the ignorance degree of ψ and the ignorance degree of $\psi|\phi$):

$$D(\tau_\gamma, \tau_\beta) = \tau_{(\gamma-\beta+1)} \text{ with } \tau_\gamma \geq \tau_\beta$$

Remark 4. In the following, and especially in the examples, we will use as qualitative difference operator, the operator D defined in Example 2.

Exemple 41 *For M=5, one could introduce the following ordered set of relevance degrees:*
$D_{p_5} = \{$*Not at all relevant, Little relevant, Enough relevant, Very relevant, Quite relevant*$\}$ *corresponding to p_α-relevant, with $\alpha \in [1,5]$.*

Figure 2 represent the corresponding relevance scale. This scale will be used in the example that will be presented by the following.

| Not at all relevant | Little relevant | Enough relevant | Very relevant | Quite relevant |

Fig. 2. Relevance scale

The informational relevance notion can be used to process several problems in qualitative reasoning under uncertain knowledge. So, we can quote, (1) *the problem of information inconsistency in knowledge base*, and (2) *the problem of extraction of the best information to treat some problems under uncertainty*. In these cases, the informational relevance notion can be used as an objective basis criteria to establish preference between some informations or actions.

Besides, to manage rationally the uncertainty, it is necessary that the predicate *"Uncert"* satisfy an axioms set which governs this concept.

5 Axioms of *Uncert*

According to the sense that we have given in the previous section, a formula is *certain* if its *truth value*[7] is known. So, a formula ϕ true in $\mathcal{A}$ is certain in $\mathcal{A}$. From this, we derive the first axiom:

Axiom 1 $\forall \phi \in \Omega, \mathcal{A} \models_M \phi \Rightarrow \mathcal{A} \models_1 \text{Uncert}(V_M \phi)$.

Also, a formula ϕ false in $\mathcal{A}$ is certain in $\mathcal{A}$. Thus the second axiom is:

Axiom 2 $\forall \phi \in \Omega, \mathcal{A} \models_1 \phi \Longrightarrow \mathcal{A} \models_1 \text{Uncert}(V_1 \phi)$.

Besides, two equivalent formulae have the same truth value. By this fact, their ignorance degree is the same. So, the third axiom is:

Axiom 3 $\forall \phi, \varphi \in \Omega, \mathcal{A} \models_M (\phi \equiv \varphi) \Longrightarrow \{\mathcal{A} \models_\alpha \text{Uncert}(\phi) \Leftrightarrow \mathcal{A} \models_\alpha \text{Uncert}(\varphi)\}$.

The formulae ϕ and $\neg \phi$ are mutually exclusive. However, if the formulae ϕ and $\neg \phi$ are uncertain, then considering the available information, one will choose the most relevant formula. Thus, the fourth axiom is:

Axiom 4 $\forall \phi \in \Omega$, if $\{\mathcal{A} \models_\alpha \text{Uncert}(\phi) \text{ and } \mathcal{A} \models_\beta \text{Uncert}(\neg \phi)\}$ then $\mathcal{A} \models_{\alpha \wedge \beta} \text{Uncert}(V_\chi \phi)$ with $\{\text{if } \tau_\alpha \leq \tau_\beta \text{ then } \tau_\chi = \tau_M, \text{ else } \tau_\chi = \tau_1\}$

Let us consider now two formulae Ψ and Φ. Knowing the ignorance degrees of Φ and of $\Psi|Phi$, what will be the ignorance on their conjunction $\Phi \cap \Psi$?
According to Shannon's theory [12], the ignorance degree of $(\Psi \cap \Phi)$ is the *sum* of individual ignorance degrees of Φ and $\Psi|\Phi$. In the symbolic area, we use this same property. So, instead of the numerical operator "+", we use a symbolic additive operator S. This operator S satisfies the fundamental properties of classical addition. So, the last axiom is:

Axiom 5 $\forall \Phi \in \Omega, \forall \Psi \in \Omega$ if $\mathcal{A} \models_\alpha \text{Uncert}(\Phi)$ and $\mathcal{A} \models_\beta \text{Uncert}(\Psi|\Phi)$ then $\mathcal{A} \models_\gamma \text{Uncert}(\Psi \cap \Phi)$ with $\tau_\gamma = S(\tau_\alpha, \tau_\beta)$.

By the following, we will use as additive operator, the operator defined by: $S(\tau_\alpha, \tau_\beta) = max(\tau_\alpha, \tau_\beta)$.

6 Management of Ignorance

In this section we present the fundamental theorems regarding the symbolic management of the ignorance.

Theorem 1. *Conjunction of formulae.*
$\forall \Phi \in \Omega, \forall \Psi \in \Omega$, if Ψ is u_α-uncertain and Φ is u_β-uncertain, then $(\Psi \cap \Phi)$ is u_γ-uncertain, with $\tau_\gamma \leq S(\tau_\alpha, \tau_\beta)$.

---
[7] The truth values *"true"* and *"false"* correspond respectively to τ_M and τ_1

Proof. $\forall \Phi \in \Omega, \forall \Psi \in \Omega$, one has: $\mathcal{A} \models_\lambda \text{Uncert}(\Psi)$ and $\mathcal{A} \models_\beta \text{Uncert}(\Psi|\Phi)$ with $\tau_\beta \leq \tau_\lambda$. So, by changing $\Psi|\Phi$ by Ψ in Axiom 5, one obtains: If $\mathcal{A} \models_\alpha \text{Uncert}(\Psi)$ and $\mathcal{A} \models_\beta \text{Uncert}(\Phi)$ then $\mathcal{A} \models_\gamma \text{Uncert}(\Phi \cap \Psi)$ with $\tau_\gamma \leq S(\tau_\alpha, \tau_\beta)$.

Theorem 2. *Deduction rule.*
$\forall \phi \in \Omega, \forall \varphi \in \Omega$, **if we have:** φ **is** u_α**-uncertain** *and* $(\phi|\varphi)$ **is** u_β**-uncertain, then** ϕ **is** u_λ**-uncertain with** $\tau_\lambda \leq S(\tau_\alpha, \tau_\beta)$.

Proof. $\forall \phi \in \Omega, \forall \varphi \in \Omega$, if one has: $\mathcal{A} \models_\lambda Uncert(\phi|\varphi)$ and $\mathcal{A} \models_\gamma Uncert(\varphi)$ then, according to Axiom 5 one can writes: $\mathcal{A} \models_\eta Uncert(\phi \cap \varphi)$ with $\tau_\eta = S(\tau_\gamma, \tau_\lambda)$. Besides, if one has $\mathcal{A} \models_\chi Uncert(\phi)$ then $\tau_\chi \leq \tau_\eta$. So, $\tau_\chi \leq S(\tau_\gamma, \tau_\lambda)$.

Theorem 3. *Combination of uncertainties*
$\forall \Phi \in \Omega, \forall \phi \in \Omega, \forall \varphi \in \Omega$, **if we have:** ϕ **is** u_α**-uncertain, and** $(\Phi|\phi)$ **is** u_β**-uncertain** φ **is** u_γ**-uncertain and** $(\Phi|\varphi)$ **is** u_σ**-uncertain then** Φ **is** u_μ**-uncertain with** $\tau_\mu \leq S(\tau_\alpha, \tau_\beta) \wedge S(\tau_\gamma, \tau_\sigma)$.

Proof. According to Theorem 2 the equations (1) and (2) give respectively:
$\mathcal{A} \models_\mu \text{Uncert}(\Phi)$ with $\tau_\mu \leq S(\tau_\alpha, \tau_\beta)$ and $\tau_\mu \leq S(\tau_\gamma, \tau_\sigma)$.
So, $\tau_\mu \leq S(\tau_\alpha, \tau_\beta) \wedge S(\tau_\gamma, \tau_\sigma)$.

7 Example

In this example, we will use the uncertainty and relevance scales represented by Figures 1 and 2. Let us consider the following knowledge:

A_1. Uncert(Put on summer clothes to go out—Sun) $\equiv$ almost certain.
A_2. Uncert(Put on raincoat to go out—Rain) $\equiv$ almost certain.
A_3. Uncert(Put on winter clothes to go out—Snow) $\equiv$ certain.
A_4. Uncert(Remain in the house—Tornado) $\equiv$ almost certain.

Suppose that the meteorology services announce the following estimations:

I_1 Uncert(Rain) $\equiv$ uncertain enough,
I_2 Uncert(Snow) $\equiv$ almost certain,
I_3 Uncert(Tornado) $\equiv$ very uncertain,
I_4 Uncert(Sun) $\equiv$ totally uncertain.

Considering these estimations, what action will we do tomorrow?

The possible actions that we can do are: {"*Put on summer clothes to go out*","*Put on raincoat to go out*","*Put on winter clothes to go out*","*Remain in the house*"}. We suppose that these actions are inconsistent and we denote them respectively by A_1, A_2, A_3, A_4. We suppose that the action that we will do is completely unknown (i.e. totally uncertain). For each action A_i, we calculate the informational relevance degree of the available information I_j. Table 2 summarizes these relevance degrees.

According to Table 2, we can say that the available information I_2 is more relevant for the action A_3. So, we can write that $\text{Pert}((I_1 \cap I_2 \cap I_3 \cap I_4)|A_3) = \tau_4$. Then we can extract from Table 2 the most relevant action A_3. Thus, if one opts

Table 2. Table of informational relevance Pert($I_j|A_i$)

Action A_i ↓ \ $Information$ I_j →	I_1	I_2	I_3	I_4
A_1	τ_1	τ_1	τ_1	τ_1
A_2	τ_3	τ_1	τ_1	τ_1
A_3	τ_1	τ_4	τ_1	τ_1
A_4	τ_1	τ_1	τ_2	τ_1

for relevant action, then one can conclude that it is more preferable to put on winter clothes to go out.

8 Conclusion

In this paper, we have presented a first symbolic approach of uncertain reasoning which integrating the informational relevance notion. This approach offers the possibility to represent explicitly the ignorance, as it expressed subjectively in common-sense reasoning, by using linguistic values. The graduation scale used allows us to handle several special cases of uncertainty, including the situation of total ignorance. The informational relevance used allows to process several problems as, for example, the management of information inconsistency and the extraction of the best information in uncertain reasoning.

References

1. M. Chachoua. *Une Théorie Symbolique de l'Entropie pour le Traitement des Informations Incertaines*. PhD thesis, Université d'Angers, 1998.
2. M. Chachoua and D. Pacholczyk. Qualitative reasoning under uncertainty. In *11th International FLAIRS Conference, Special Track on Uncertain Reasoning*, pages 415–419, Florida,USA, 1998.
3. M. Chachoua and D. Pacholczyk. A symbolic approach to uncertainty management. *Applied Intelligence: International Journal of Artificial Intelligence, Neural Networks, and Complex Problem-Solving Technologie*, 13(3):265–283, 2000.
4. M. Chachoua and D. Pacholczyk. Qualitative reasoning under ignorance and information relevant extraction. *Knowledge and Information Systems: An International Journal*, page (to appear), 2002.
5. M. De-Glas. Knowledge representation in fuzzy setting. Rapport interne 89/48, LAFORIA, Paris, 1989.
6. A. De-Luca and S. Termini. A definition of non-probabilistic entropy in the setting of fuzzy sets theory. *Information and control*, (20):301–312, 1972.
7. A. P. Dempster. Upper and lower probabilities induced by a multivalued mapping. *Annals of mathematical statistics*, 38:325–339, 1967.
8. E. Kant. *Logique*. Librairie Philosophique J. Vrin,, Paris, 1966.
9. G. J. Klir and T. A. Folger. *Fuzzy sets, uncertainty and information*. Prentice-Hall international Edition, 1988.

10. D. Pacholczyk. A logico-symbolic probability theory for the management of uncertainty. *CCAI*, 11(4):417–484, 1994.
11. G. Shafer. *A mathematical theory of evidence.* Princeton University Press, 1976.
12. C. E. Shannon. A mathematical theory of communication. *Bell System Technical*, 27:379–423, 1948.
13. D. Sperber and D. Wilson. *La pertinence: Communication et cognition.* Les Editions de Minuit, 1989.
14. L. A. Zadeh. Fuzzy sets. *Information and control*, 8:338–353, 1965.

Potential Governing Relationship and a Korean Grammar Checker Using Partial Parsing*

Mi-young Kang[1,2], Su-ho Park[1], Ae-sun Yoon[2], and Hyuk-chul Kwon[1]

[1] AI Lab. Dept. of Computer Science, Pusan National University
San 30, Jang-geon Dong, 609-735, Busan, Korea
{kmyoung,suhopark,hckwon}@pusan.ac.kr
[2] Dept. of French, Pusan National University
San 30, Jang-geon Dong, 609-735, Busan, Korea
asyoon@pusan.ac.kr

Abstract. This paper deals with a better method to treat the various linguistic errors and ambiguities that we encounter when analyzing Korean text automatically. A natural language understanding system and the full-sentence analysis would provide a better way to resolve such problems. But the practical application of natural language understanding is still far from being achieved and full sentence analysis in the current state is not only difficult to implement but also time consuming. For those reasons a Korean Grammar Checker using the partial parsing method and the conception of potential governing relationship is implemented. The paper improves the knowledge base of disambiguation rules while trying to reduce them with the result of the linguistic analysis. The extended lexical disambiguation rules and the parsing method based on the asymmetric relation that we propose thus guarantee the accuracy and efficiency of the Grammar Checker.

1 Introduction

The performance of a grammar checker depends largely on how accurately and rapidly it can deal with various types of linguistic ambiguities and errors. The accuracy and the speed that determine the main factors of the performance are, however, supposed to be two different goals to achieve simultaneously. For example, a natural language understanding system and a full-sentence parsing method would increase the accuracy in the disambiguation and the error correcting processes. But the practical operation of a natural language understanding system is still far from being achieved, and a full-sentence parsing method is not only difficult to implement but also needs time-consuming processes in its current state. The goals of this paper are (1) to introduce

---

* This work has been supported by Institute of Information Technology Assessment. (Contract Number : AA-2000-A4-0037-0001)

the notion of the 'potential governing relationship' and the 'partial parsing method' to improve the current version of the Korean Grammar Checker that AI Lab. of Pusan National University has been developing since 1994, and (2) to demonstrate how the results obtained by linguistic analysis can ameliorate the knowledge base of disambiguation rules, thus reduce the frequent ambiguities. To this end, this paper discusses mainly a linguistic approach based on the asymmetric relation, which allows us to refine the potential governing relationship and to determine the parsing directionality and parsing scope.

The paper is composed of 5 sections. In section 2, we present the different linguistic errors and ambiguities by examining different kinds of context where we might find them. Section 3 has three sub-sections. In the first two sections, we try to define better analysis for disambiguation. We focus on a linguistic approach based on the asymmetric relation. The approach lets us refine the potential governing relationship and the directionality and the scope of parsing. Knowledge base of disambiguation rule will be described in the third sub-section. Section 4 then gives an overview of the Korean grammar checker implemented with partial parsing process. In section 5, we give our conclusions and suggestions for future work.

2 Linguistic Errors and Ambiguities

This section presents different errors and ambiguities that we meet without contextual analysis. When we parse Korean texts, we encounter several errors such as spelling errors, spacing errors, misuse of the postposition or the case marker,[1] misuse of a word or the concatenation of several words, etc.. These errors touch around syntactic, semantic and stylistic dimensions producing ambiguous situations.

First, we have the ambiguity of inherent lexical categories. In Korean text, we meet several ambiguous syntactic structures caused by homographic unities. For example, we have ambiguous structure between the noun and the adverb with the form *mos*:

1. **mos**-eul /bag-ala.[2,3] "Drive a nail!" nail-acc /to drive-imp
2. na-neun /**mos** /cham-gess-da "I can't stand it!" I-th /not /to stand-fut-end

We cannot define the category of this word before we analyze its syntactic function in a sentence.

---

[1] We call 'postpositions' all nominal suffixes except case markers like *i/ga* (nominative), *eul /leul* (accusative), *ege* or *bogo*, *hante* (dative), etc.. Those postpositions are assumed to have the same functions with the prepositions in Indo-European.

[2] Throughout this paper, we adopt the Korean standard for romanizing Korean script.

[3] The symbols and abbreviations used in this paper are as follows.
   *****: unacceptable sentence /**acc**: accusative case marker /**adv**: adverb /**caseM**: case marker /**cond**: condition ending /**conj**: conjunction /**dat**: dative case marker /**det**: deixis or numeral /**end**: ending /**fut**: future tense marker /**imp**: imperative ending /**vi**: intransitive verb /**N**: noun /**nom**: nominative case marker /**past**: past tense marker /**post**: postposition /**pres**: present tense marker /**rel**: relative ending /**vt**: transitive verb /**th**: thematic marker /**V**: verb

Let us see now the ambiguity yielded during morphological analysis. We can fall into the situation of ambiguity with inflected forms. A verb *sa-* 'to buy' is inflected with the verbal suffix of coordination in 4; and it forms an ambiguous situation with its homograph shown in example 3.

3. gyotong /**sago** 'a traffic accident' traffic /accident
4. nan-eun /sagwa-leul /**sa-go** "I'll buy an apple and" I-th /apple-acc /to buy-conj

The form *sago* can be a noun and an inflected verbal form. Starting just with the output *sago* we cannot define the category of this word.

In many cases, Korean spacing can influence deciding semantic and syntactic scope. For example, in the sentences shown below, the form *ba* changes its lexical category according to how it's spaced.[4]

5. hwaginha-**n** /**ba** /eobs-da "We haven't verified it" to verify-rel /thing /to not be-end
6. hwaginha-**nba** /sasili-da "We have verified it and it is true" to verify-cond /to be true-end

The form *ba* in sentence 5 can occupy the same syntactic position with other nouns. In other words a case marker can attach it. Diachronically the complex ending *-nba* in sentence 6 is derived from a form that is composed with the relative suffix *–n* and the dependent noun *ba*. There is no space between two morphemes belonging to two different categories as before.

Nowadays, the majority of Korean speakers distinguish hardly between /e/ and /ae/.[5] There are many errors which are related to it. The form *se-* 'to count' is a transitive verb, while the form *sae-* 'to leak' is an intransitive verb.

7. naljja-leul /**se**-eola "Count the days!" day-acc /to count-imp
8. * naljja-leul /**sae**-eola | day-acc /to leak-imp

Sentence 8 is treated as unacceptable. Because, a word with accusative case marker cannot be placed just before intransitive verbs. It touches the syntactic violation. In this pattern, we can include semantic errors caused by spelling error or the misuse of vocabulary words as well.

We have examined in this section different ambiguities. A full understanding on those ambiguity patterns according to semantic and morpho-syntactic functions of lexical entries will enable us to extend the knowledge base of the Korean grammar checker. Our morphological analyzer partly can resolve those ambiguities. If a morpheme is in multiple categories, our morphological analyzer disambiguates it mostly with adjacency conditions, which stands on the compatibility or the incompatibility of two morphemes. The scope of these adjacency conditions is intra-word. After being processed by the morphological analyzer, if a word still has ambiguities, our system starts to check the compatibility of this morpheme with the preceding or following

---

[4] In the examples there is no noun assuming a subject function. We can very easily omit the grammatical subject in Korean. We can restore an elapsed grammatical subject by analyzing the context. Here we interpret it like 'we' for the commodity.

[5] Vowel /e/ and /ae/ are front vowel. Vowel /ae/ is articulatorily lower then /e/.

words. [6] Checking grammatical errors is proceeded in the same way. In what follows, we will discuss with which kind of analysis, the grammar checker may obtain the best result in disambiguating while parsing Korean sentences.

3   For an Analysis Based on the Asymmetric Relation

To analyze a language like Korean, in which free word order tendency predominates, the dependency grammar is widely applied. It describes dependency between components in a sentence structure in distinguishing a head/dependent. According to the dependency grammar, every phrase has a head except coordinated structures. The constituency and the sub-categorization are crucial notions in dependency grammar. The first is related with the fact that it always allows the relation between two parts of a larger whole, never between the whole and one of its parts. The second is concerned with the relation between a word (i.e. governor) and the dependents (i.e. governed) that it selects. The analysis of Korean belongs to the dependency grammar and assumes, in general, right headed governing relation. For example, we understand a verb governs an adverb, or its complement in natural language analysis and the former precedes the latter. Our system deals with the knowledge based on the dependency grammar, in other words, it uses the relationship between two parts of a larger whole and the relation between a word and its dependents. But it is rather far from the classical dependency grammar. Whereas, the latter describes order and restrictions in the same formula, our system tolerates different descriptions for each case. And words and phrases are not conceived in a linguistic sense.[6] It is determined as a spacing unit and the phrase is conceived as two or more consecutive units not caring about their independency or whether or not they assume a linguistic function. With these characteristics, the governor is not always coinciding with the governor in the linguistic sense. We will discuss in what follows the kind of governors and the relation that this governor holds with its eventual dependents.

3.1   Potential Governing Relationship and Directionality of Parsing

The orientation of the parsing to detect and correct grammatical and semantic errors in our system is constructed with respect to a potential governor. As mentioned above, the governor in our system is not conceived as a linguistic sense but as a potentially incorrect word that the system identifies when checking a sentence. Those words with high possibility to be misused are collected based on corpus and heuristics and listed. The partial parsing module is triggered when this governor is detected. The partial parsing is proceeded from a selected governor to its dependents. The dependents, as described here, cannot only be a word that appear with the governor in question, but also a word that cannot appear with that governor in a sentence. The relation maintained with the first one is called 'collocation relation', and the relation maintained

---

[6] Words are linguistically considered atomic elements: they are the indivisible building blocks of syntax. And a phrase is a constituent of an expression and it is any part of the expression that, linguistically, functions as a unit.

with the last one is called 'anti-collocation relation'. If searching a word in the collocation relation with a potential governor is too time consuming, it would be better to set the system to find out a word in the anti-collocation relation with that governor. Then the partial parsing is proceeded from a selected governor until a word that forms a collocation or anti-collocation with this governor or no more possible dependents are found. Leaving the question of the role of collocation or anti-collocation in the milieu of partial parsing for the next section, we deal here with the kind of governors and the governing relation that we could have.

In dependency grammar, Korean is taken in general as a language in which the verb selects adverb or its complement. But for the current grammar checker, there are important parameters to consider. Those are time saving as well as obtaining accuracy. So the checker would count which is more time saving with accurate results. We have four possible checking directions a posteriori: right hand head, left hand head, conditional head and head with undetermined governing direction. In what follows, we will examine these closely. Consider first the candidates for governor that have left hand headed relation with their dependents:

9. gachi 'value' N /gati 'together' adv
- geugeos-eun /gachi /iss-da "It is worthy" it-th /value /to have-end
- geudeul-eun /gati /nol-n-da "They play together" they-th /together /to play-pres-end
- *geu-neun /gachi /meog-neun-da | he-th /value /to eat-pres-end

For the above example, it would be better to say that an adverb or a complement selects the verb. As a complement or an adverb is rare compared to a verb, selecting them, as a governor the system will be less solicited. All the time, the system has to have only a few rules to be considered to avoid affecting the efficiency of the system. Let us return to our question. In casual language we have many errors mixing up *gachi* and *gati*.[7] So, those words are incorporated into our knowledge base as the potential governor. And the grammar checker points out that the fourth sentence is unacceptable, because the potential governor doesn't find its fit dependent but just an element that forms an anti-collocation with it.

Let's see now right hand headed relation:

10. gage 'store' N / gagye 'household economy' N
- hwajangpum /gage 'a cosmetics store' cosmetic /store
- *hwajangpum /gagye | cosmetic /household economy

The construction of Korean compound nouns is endocentric, with the possible exception of coordinated structures. It means that they have a head. The head is on the right hand side in a construction. This linguistic aspect is reflected as it is in the parsing direction of our system. The reason why the second sentence is unacceptable is that the right hand noun cannot govern the left one. Our system controls the semantic and lexical scope of each noun belonging to compound nouns to check if the governing relation is established or not.

---

[7] Their acoustic outputs are the same: [gachi].

And we have a 'conditional governor'. We call a conditional governor a syntactic entity that can govern its dependent regarding the semantic or morpho-syntactic state of the element situated in its other wing. Many Korean speakers mix up the noun *juche* 'main body' and the noun *juchoe* 'auspice'. These two differ just by the final vowel.

11. juche 'main body' N /juchoe 'auspice' N
- hagsaeng-i /juche-ga /doe-eoss-da "It is made up mainly of students" student-nom /main body-nom /to become-past-end
- hagsaengdanche-ga /juchoe-ga /doe-eoss-da "It was sponsored by the student's organization" student's organization-nom /auspice-nom /to become-past-end

Here, the potential governor *juche* can govern its dependent *doe-* in condition of the presence of the right hand noun like *hagsaeng* while the potential governor *juchoe* can govern its dependent *doe-* in condition of the presence of the right hand noun like *hagsaengdanche*. Two nouns in a question differ in semantic features. The noun *juche* is related to a 'person' while the noun *juchoe* is to a 'thing'.

Finally, we have a head with undetermined governing direction: it is about a governor for whom the position of its dependent is not fixed. The complement of the verb can sit on the right and the left hand side of the verb. If the verb is inflected in a relative form, the parsing direction will be left-to-right. Otherwise the direction will be right-to-left. The majority of the relation between a verb and its complement is included in this class.

12. eob 'to carry on the back' vt /eobs- 'to do not exist' vi
- geu-neun /agi-leul /eob-go "he carries a baby on his back and" he-th /baby-acc /to carry on the back-conj
- geu-ga /eob-eun /agi-neun 'the baby who he is carrying on his back' he-nom /to carry on the back-rel /baby-th
- *geu-neun /agi-leul /eobs-go | he-th /baby-acc /to do not have-conj

The pattern that our system uses here coincides with the information obtained from sub-categorization of the verb. The potential governor checks until he can find his eventual dependent (i.e. an element selected by sub-categorization of the verb as a potential governor) on the left hand side. If this fails, the searching proceeds to the right hand side. The intransitive verb *eobs-* cannot govern the noun inflected with the accusative case marker. In the third sentence of example 12, the governing relation between the verb *eobs-* and the noun *agi-* with the accusative case marker has failed and the sentence to which they belong is judged as unacceptable. But here one must note the governing relation discussed enters into conflict with the case of the left-to-right governing relation quoted above. To resolve this problem the grammar checker provides rule ordering. The rules belonged to the knowledge base have a hierarchy. We will reconsider this question later.

3.2 Collocation Relation and Parsing Scope

In the previous section, we observed the possibility that the potential governor would offer to the checker the starting point for parsing and the parsing direction. The grammar check system manages this governing relation between the collocation and the anti-collocation rules. These rules are constructed in terms of the dependency grammar and define whether the current component is in collocation or not with its eventual dependent. These rules are integrated into the structure of the knowledge base shown in Fig. 1.

Entry (potential governor)	
Entry information	
Parsing direction	
Parsing scope	
Collocation et al.	
Dependent information	
Substitution	spacing; substitute
Help	number

Fig. 1. Structure of Knowledge Base

A potential governor provides the information of the checking direction and the checking scope. This checking scope is determined according to the grammatical state of the governor and the relationship that it maintains with its dependent. The governor has the information of the collocation and the anti-collocation relations with the lexical and morph-syntactic information of the eventual dependent. It also has a certain number of the replacements for when the checker finds out the relationship between the two components is in anti-collocation. And we have help to explain where the error originated from.

So far we have discussed, triggering and managing efficiently that partial parsing is done regarding the lexical category and the position of the governor and maximal scope of the checking and the items that can be ignored by the grammar checker. The following table summarizes these classifications.

Table 1. L: left hand head; R: right hand head; C: conditional head; U: head with undetermined governing direction; C.M.S.: lexical category and maximal checking scope; N.I.: negligible information; Lft: category on the left hand side: Rgt: category on the right hand side.[8]

	Governor	Dependent	C.M.S.	N.I.
L	N-post	V-end	Rgt – 2	Adv
	Adv	V-end	Rgt – 2	Adv, N
R	N	N (-caseM)	Lft –1	none
C	N-caseM	V-end	Lft – 1/ Rgt – 1	none
U	V-end	N (-caseM)	Lft – 3	Adv
	V-rel	N (-caseM)	Rgt – 2	Det, V-rel

The maximal scope of the checking is given indirectly by the means of lexical items that can be ignored by the grammar checker. Thus, we diminish the burden for the system. Let us take examples:

---

[8] Also refer to footnote 3.

13. geu-neun /daechelo /jaju /nol-n-da "He doesn't work usually" he-th /usually /often /to do not work-pres-end
14. geu-neun /mila-wa /gati /eumag-eul /deud-neun-da "He listens to music with Mila" he-th /Mila-post /together /music-acc /to listen-pres-end

With these sentences we obtain two patterns respectively:

13	noun-th	**adverb$_1$**	adverb$_2$	verb	

Parsing direction →

14	noun$_1$-th	noun$_2$-post	**adverb**	noun$_3$-acc	verb

In both sentences, the checker selects an adverb as a potential governor. The system checks if there is an eventual dependent in collocation relation or anti-collocation with it. But the system doesn't check all the items in the sentence; it can neglect *adverb$_2$* or *noun$_3$*, which are between the governor and the eventual dependent.

3.3 Disambiguation Rules and Its Application

This section presents the knowledge base of disambiguation rules in which heuristics are provided. We can make a distinction between the general disambiguation rule and the lexical disambiguation rule. The general disambiguation rule is based on morpho-syntactical analysis. To analyze a word, this rule uses the morpho-syntactic information of the word analyzed in advance as follows.

Table 2. Morpho-syntactic Conditions

a.	A dependent noun should follow a modifier (noun, deixis, verb of relative form, numeral).[9]
b.	A verb of relative form can follow another kind of modifier like deixis, numeral.
c.	A modifier like deixis and numeral can be placed between a verb relative form and a noun.
d.	A noun with an accusative marker may not follow an intransitive verb.
e.	A transitive verb may follow a noun with an accusative marker.
f.	The copula must precede a noun without any case marker.
g.	Each sentence ending mark (declarative, exclamation, interrogative, etc.) is decided regarding the verbal terminations (declarative, exclamation, interrogative ending, etc.).

The above morpho syntactical conditions *a*, *b* and *c* are related to a noun phrase while the conditions *d*, *e* and *f* concern with a verbal phrase. Here one of the heuristic rules is related to the syntactic condition of a noun phrase.

V + rel	+	Dependent noun (-caseM /-post)

---

[9] The noun, deixis, a verb of relative form and numeral can occupy the syntactic place of modifier. We call modifiers all the linguistic items, which can modify a noun. A noun can assume perfectly a function of modifier. In Korean, they are always before a noun. Let's note here there is no category of adjective in the sense of European linguistics in Korean.

With this rule we can treat the ambiguities like those in the examples, 5, 6. If the checked word and its dependent satisfy this morpho-syntactical condition, we may space two words. And one of the disambiguation rules that corresponds to the second case is shown below:

$$\frac{\text{N + no case marker}}{\text{N + nom}} + \boxed{\text{Intransitive verb}}$$

The lexical disambiguation rules are basically based on the collocation and anti-collocation rule. For example, we can constitute the rule to disambiguate two nouns *juchoe, juche* in the example 11:

$$\boxed{\text{N + nom \{- person\}}} + \boxed{juchoe \text{ (-caseM) + V}}$$

$$\boxed{\text{N + nom \{+ person\}}} + \boxed{juche \text{ (-caseM) + V}}$$

When there are many candidate rules for treating a word, they are applied regarding the priority. The Priority Conventions are established in terms of markedness (or 'naturalness')

Table 3. Rules Application Priority Conventions

A marked rule applies first.	
a.	Checking in a direction applies first.
b.	A rule with specific words applies first.

The rule finding a dependent in two orientations is more general than the rule permitting unidirectional checks. So we apply first the latter and then the former. When the rule for which the morpho-syntactic information (for example, case information) is needed and a rule for which a special word is a target, the latter is applied, as the latter is more marked than the former. But all the time the system may allow only a few rules to be considered.

4 Prime Modules of the Grammar Checker

The current system checks the syntactic, semantic and stylistic errors with the rule checker, the rule pre-interpretation module, the rule interpretation module and the error-correcting module. While using the rule interpretation module, the system activates the module for partial parsing, when there are several words between the words under checking and the anti-collocation words. The disambiguation rules are used during the activation of the rule pre-interpretation module and the module for partial parsing. Let's look closely at the structure of the grammar checker.

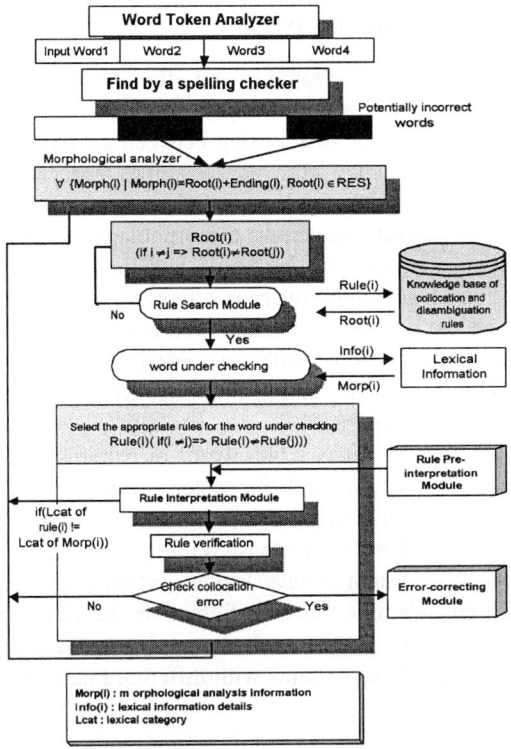

Fig. 2. Korean Grammar Checker Architecture (version 2.0)

As Fig. 2 shows it, the system has, firstly, the rules search module in Knowledge base. When the checker finds a potentially incorrect word, the rules search module responds. The rules search module is triggered and selects a certain number of rules closely related to the potentially incorrect words from the Knowledge base.

Secondly, it has the rule pre-interpretation module. This module can minimize the burden to the system by making it easy for checking and correcting errors. For this, the system replaces the rules for treating the errors by structure variables, and verifies if the lexical information of the word selected by a rule coincides with the one of the words under checking. At this stage, if the system finds an ambiguous word, the routine of disambiguation is called to disambiguate the word.

Thirdly, the rule interpretation module is inserted in the system. At this stage, with the value returned by the rule pre-interpretation module, the system examines whether the checked word can collocate with another one and whether it is in a collocation violation.

Finally, the system has the error-correcting module. This correction routine substitutes an incorrect word or an incorrect phrase with a suitable one. This routine selects help, as well, to display the errors to the end-user. The different modules discussed interplay between them when parsing the checked word and the word forming with it a collocation.

5 Conclusions and Future Work

Processing a natural language with rapidity and efficiency is fundamental for a grammar checker. Our system resolves the ambiguity problems well by using the notion of dependency implicitly. It uses the relationship between a governor and its dependents. Though the classical dependency grammar describes order and restrictions in the unique modus operandi, our system tolerates different descriptions for each case. We proposed two kinds of potential governor: a morphologically ambiguous word and a checked word heuristically selected due to its high possibility to cause errors. We proved that the notion of potential governor offers the grammar checker the starting point for parsing and the parsing orientation. We also considered this as one of the ways to obtain a result efficiently. With asymmetric relation obtained, we tried to improve the knowledge base of disambiguation rules while trying to reduce them with the result of the linguistic analysis.

Currently there are approximately 6,000 words or patterns, which have a high possibility of misuse, in our system. And the system provides about 100 rules for the disambiguation. Recently, a test was carried out using several unmodified data, and there were only 3 percent of miscorrections. While perhaps it is good enough to disambiguate the input, our system always leaves certain points to be supplemented. There remains still a considerable number of errors in the various texts, which cannot be corrected by the system. For future work, we are interested in extending our knowledge base while working on experiments with different kinds of data and making the system more efficient.

References

1. Allen, J.: Natural Language Understanding. second edition. The Benjamin/ Cummings Publishing Company, INC (1995)
2. Comrie, B.: Language Universals and Linguistic Typology. Blackwell (1989)
3. Johanna, N.: Head-marking and Dependent-Marking Grammar. In Language 62 (1986) 56-119
4. Kang, M., Park, S., Kwon, H.: A Pattern Analysis of Lexical Ambiguities and the Korean Grammar Checker with Partial Parsing. Proceeding CKITOP, Daejeon, Korea (2001) 45-52
5. Kim, M., Kwon, H., Yoon, A.: Rule-based Approach to Korean Morphological Disambiguation Supported by Statistical Method. Proceedings of 11th Pacific Asia Conference on Language. Information and Computation (1996) 237-246
6. Kim, S., Nam, H., Kwon, H.: Correction Methods of Spacing Words for Improving the Korean Spelling and Grammar Checkers. Proceedings of 5th Natural Language Processing Pacific Rim Symposium (1999) 415-419
7. Mel'čuk, I. A.: Dependency Syntax: Theory and Practice. Published by State University of New York Press, Albany (1988)
8. Nam, K.: Standard Korean Grammar. Top Publisher, Seoul (1986)
9. Spencer, A.: Morphological Theory. Blackwell (1991)
10. Tesnière, L.: Élément de Syntaxe Structurale (1959). second edition revised and corrected. Klincksieck, Paris (1999)

On-Line Handwriting Character Recognition Using Stroke Information

Jungpil Shin

Department of Computer Software, University of Aizu
Aizu-Wakamatsu City, Fukushima, 965-8580, Japan
voice: [+81](242)37-2704; fax: [+81](242)37-2731
jpshin@u-aizu.ac.jp

Abstract. For the purpose of improvement in computational time and recognition accuracy in the framework of stroke order- and number-free on-line handwriting character recognition, we performed structural analysis of the style of stroke order and stroke connection. From the real handwritten characters, chosen from among 2965 Chinese characters, we investigated the information on stroke order and connection, using the automatic stroke correspondence system. It was proved that the majority of real characters are written in fixed stroke order, and stroke order is predominantly in the standard stroke order; about 98.1 % of characters were located nearly in the standard order. Almost all stroke connections occur in the standard order (92.8 %), whereas 2 stroke connections occurred often, and stroke connections in nonstandard order occurred very rarely. In a comparison of our findings with the expected stroke connections, very few connections were found to actually occur. Moreover, we show the methods for incorporating the information on the completely stroke order- and number-free framework. The large improvement on both computational time and recognition accuracy are demonstrated by experiments.

1 Introduction

On-line recognition of cursive characters written by hand is a key issue in state-of-the-art character recognition research [6]. Much research have been conducted to mitigate the stroke order variation, stroke number variation, and stroke deformation [1,2,3,4,5,6,7,8,9,10]. On-line recognition, in contrast with off-line recognition, has the important advantage of being able to use stroke order- and connection-information, because the character pattern is expressed by ordered time sequences. Two approaches based on the stroke order and stroke connection information are considered to realize online character recognition. The first approach actively uses the stroke order-and connection-information; the second takes into account the inevitable changes in stroke order and connection from person to person, and hence realizes the "*free stroke order*" and "*free stroke number*". Our approach basically coincides with the latter, and the framework of character recognition is stroke order free and number free. Based on the framework of stroke order free and number free recognition, the information on stroke order and connection is locally available [9,10].

To build up the system for stroke order and number free recognition, the algorithm requires correct performance of stroke correspondence between input pattern and reference pattern for higher recognition performance [2,7,9,10]. In these algorithms, the enormous percentage of stroke correspondences that do not actually occur are also carried out. To realize high quality recognition on a small microprocessor with built-in memory, such as a Personal Digital Assistant (PDA), etc., the recognition time and required computational resources must be reduced. For this purpose, it is expected that reasonable level stroke-correspondence searching can be realized by using information on stroke-order and stroke-connection as they actually occur. Development of a recognition framework that incorporates information on stroke-order and stroke-connection is expected to achieve reduction of computation time and improve recognition accuracy by neglecting unreal stroke-correspondences.

In this paper, based on the assumption that the recognition is performed on the stroke order and number free framework, the useful information about stroke order and connection are investigated. That is, this paper focuses on the style analysis of stroke order variation and connection between strokes. The large improvement on both computational time and recognition accuracy are demonstrated by experiments.

2 Stroke Correspondence Search

An on-line input character is expressed as an ordered series of writing strokes, i.e.,

$$A = A_1 A_2 \cdots A_k \cdots A_N, \qquad (1)$$

where the k-th stroke A_k is the time sequence representation of local feature a_{ik} of a character, e.g., x-y coordinates or stroke direction, being expressed as

$$A_k = a_{1k} a_{2k} \cdots a_{ik} \cdots a_{Ik}, \qquad I = I(k). \qquad (2)$$

The reference pattern is similarly expressed as

$$B = B_1 B_2 \cdots B_l \cdots B_M \qquad (3)$$

$$B_l = b_{1l} b_{2l} \cdots b_{jl} \cdots b_{Jl}, \qquad J = J(l). \qquad (4)$$

Finally, N is the stroke number of the input pattern and M is the stroke number of the reference pattern, with N being equal to M for correct stroke number recognition.

The measure of dissimilarity between input pattern stroke A_k and reference pattern stroke B_l is calculated using stroke information on the shape and position; it is denoted as $\delta(k; l)$ and called the stroke distance. One-to-one stroke

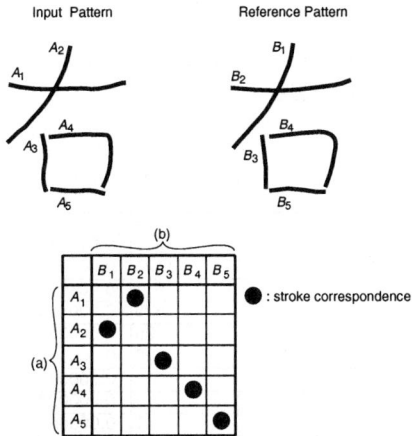

Fig. 1. Correct stroke-correspondence between (a) input pattern and (b) reference pattern

correspondence is defined by bijection $\{l(k)\}$ to stroke number l of the reference pattern from stroke number k of the input pattern as shown in Fig.1. As an evaluation standard of optimum correspondence, the sum of stroke distance $\delta(k;l)$ is used. Namely, based on stroke information, the solution of the following minimization problem is considered to give optimal stroke correspondence in which the minimum-value $D(A,B)$ is chosen as the measure of matching (i.e., the dissimilarity) [7].

$$D(A,B) = \min_{\{l(k)\}} \left[\sum_{k=1}^{N} \delta(k;l(k)) \right] \quad (5)$$

Solving this stroke correspondence determination problem provides a structural analysis of the pattern; hence, the dissimilarity $D(A,B)$ and stroke correspondence $\{l(k)\}$ between patterns are obtained as results.

3 Preparation for Analysis

The investigation data consists of 2965 categories of Chinese characters, i.e., specified characters in the first level of the Japanese Industry Standard (JIS) code set, written by 90 university students (total = 258603 characters). Students were directed to write cursively in a normal manner. The data was taken using a stylus pen on a Liquid Crystal Display (LCD) tablet. The investigation results of the stroke number change are shown in table 1. This result is mostly in agreement with [2,7,10].

Table 1. Occurrence of stroke-number changes

Stroke-number change	Occurrence number	Ratio (%)
2	13	0.000050
1	662	0.002560
0	172428	0.666767
-1	54100	0.209201
-2	18933	0.073213
-3	7268	0.028105
-4	3202	0.012382
-5	1105	0.004273
-6	517	0.001999
-7	226	0.000874
-8	99	0.000383
-9	29	0.000112
-10	11	0.000043
-11	7	0.000027
-12	1	0.000004
-13	2	0.000008
Total	258603	

Reference patterns were generated from the training data by storing average values of the loci of feature points from non-connected strokes that were extracted by rearranging the strokes according to correct stroke order. One reference pattern for each category was made.

The input character is transformed into a 256x256 mesh plane by preprocessing the steps of redundant elimination, smoothing, size normalization, and feature point extraction. As feature information, the x-y coordinates and movement directional vector between one point and the next are extracted from character data. Feature information of the input and reference patterns is placed into a_{ik} and b_{jl}, respectively.

The automatic searching for correct stroke-correspondence between input pattern and reference pattern is not an easy work, since the analysis of stroke order and stroke connection should be performed simultaneously with high accuracy.

First, using the method of Cube Search [9,10], correct stroke correspondence between input pattern and reference pattern is automatically searched for by backtracking. This algorithm has an novel advantage to efficiently search for optimal stroke correspondence in spite of (i) stroke order variation, (ii) stroke number variation due to stroke-connection, and (iii) exceptional user-generated stroke deformations. Some of the wrong stroke-correspondence is manually converted into the appropriate stroke-correspondence by observation of these characters. These errors are due to strokes written on largely different position.

4 Stroke-Order and Stroke-Connection Information

In the algorithm to realize the completely free condition for stroke-order, however, a large percentage of *unreal stroke-correspondences* is also carried out.

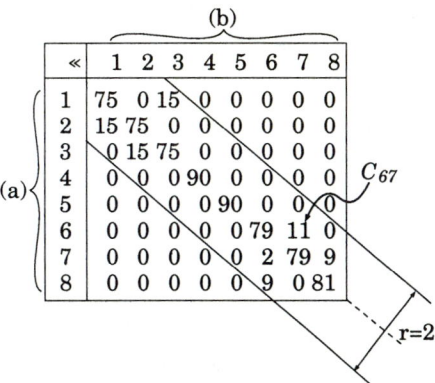

Fig. 2. Occurrence C_{kl} of correct stroke-correspondence between (a) input pattern stroke k and (b) reference pattern stroke l

$-f$	1	2	3	4	5
1	18	72	0	0	0
2	72	18	0	0	0
3	0	0	90	0	0
4	0	0	0	90	0
5	0	0	0	0	90

r=1

^	1	2	3	4	5	6
1	90	0	0	0	0	0
2	0	90	0	0	0	0
3	0	0	90	0	0	0
4	0	0	0	90	0	0
5	0	0	0	0	90	0
6	0	0	0	0	0	90

r=0

&∅	1	2	3	4	5	6	7
1	90	0	0	0	0	0	0
2	0	88	2	0	0	0	0
3	0	2	43	21	24	0	0
4	0	0	40	50	0	0	0
5	0	0	5	19	66	0	0
6	0	0	0	0	0	69	21
7	0	0	0	0	0	21	69

r=2

Fig. 3. Examples of occurring correct stroke correspondence

It is expected that a reasonable level of stroke-correspondence searching is realized by using the information of stroke-order and stroke-connection that actually occur. Further, there is no hindrance for framework of the stroke-order and number free recognition by use of statistically stable information. Development of a recognition framework that incorporates the information of stroke-order and stroke-connection is expected to achieve reduction of computation time and improve the recognition accuracy by neglecting unreal stroke-correspondences.

Table 2. Character examples having each gap

Gap	Examples
0	兄 古 功 欠 旧 幻 犬 弘 元 公 今 勾 孔 午
1	右 左 月 五 号 玄 去 甲 互 止 手 少 火 士
2	花 初 尾 秀 汝 岩 空 居 庁 舌 清 完 升 綱
3	辻 何 悲 洲 笹 豪 我 両 加 柿 報 履 霧 鑼

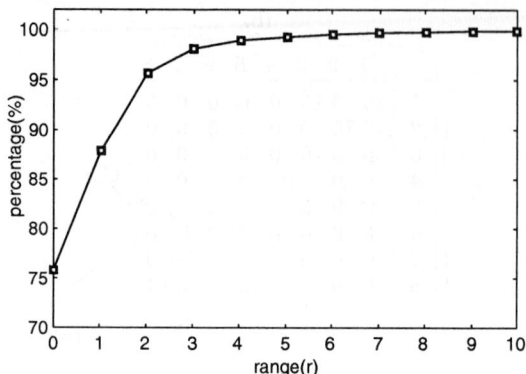

Fig. 4. Stroke correspondence distribution

4.1 Stroke-Order Analysis

Stroke-order variation among writers is mainly caused by peculiar writing style of individuals. The following are considered for stroke-order variation.

1. How much range does the fluctuation of the stroke order show in comparison with the standard stroke order, or
2. does the stroke-order change occur completely at random?

For purposes of using the stroke-order information under the framework of the stroke-order free recognition, the stroke-order change in real data is investigated from these viewpoints.

An analyzed result is shown by using the frequency of an example included in Fig. 2. C_{kl} means the number of input pattern in which the l-th stroke of the reference pattern is written as the k-th stroke of the input pattern. Note that the connected stroke is broken down into strokes corresponding to the reference pattern strokes, based on the fact that the stroke-order change of the characters with connected stroke have the same tendency as that of the characters with only single strokes.

In the example of " 姓 " based on the fact that 15 persons among 90 persons wrote the 3rd stroke to the 1st stroke incorrectly, the deviations of 15 writers are shown for the distribution of the 1st, 2nd, and 3rd strokes. In the result of stroke-order analysis, all characters of 76 % are written in accordance with standard stroke-order. Further, the distribution of the remaining characters is concentrated on the diagonal line.

In the occurrence table of Fig. 2, if the maximum distance having a non-zero element from the diagonal line is r, it is called r-gap. Figure 3 shows examples with the r-gap. Table 2 shows character examples having each gap. Figure 4 shows the percentage of the characters having the r-gap. The characters of 95.7% and 98.1% are included in the ranges of $r = 2$ and $r = 3$, respectively.

Based on the above results of stroke correspondence, the following conditions arise.

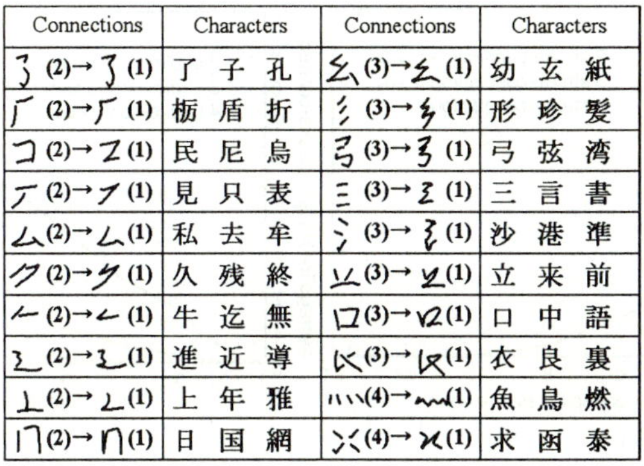

Connections	Characters	Connections	Characters
⁀(2)→⁀(1)	了 子 孔	ㄥ(3)→ㄥ(1)	幼 玄 紙
厂(2)→厂(1)	栃 盾 折	⸲(3)→⸲(1)	形 珍 髪
コ(2)→Z(1)	民 尼 烏	弓(3)→弓(1)	弓 弦 湾
ㄱ(2)→ㄱ(1)	見 只 表	三(3)→乙(1)	三 言 書
ㄥ(2)→ㄥ(1)	私 去 牟	⸲(3)→⸲(1)	沙 港 準
⁄(2)→⁄(1)	久 残 終	ㄴ(3)→ㄴ(1)	立 来 前
ㄑ(2)→ㄑ(1)	牛 迄 無	口(3)→レ(1)	口 中 語
⸲(2)→⸲(1)	進 近 導	区(3)→区(1)	衣 良 裏
上(2)→∠(1)	上 年 雅	灬(4)→灬(1)	魚 鳥 燃
冂(2)→ⁿ(1)	日 国 網	⸲(4)→⸲(1)	求 函 泰

Fig. 5. Typical stroke-connections and character examples

Table 3. Occurrence of consecutive stroke connection with (a) standard and (b) nonstandard order

type of consecutive stroke connection	(a)			
	number	ratio(%)	kinds	ratio(%)
2-strokes	94970	78.88	12779	65.78
3-strokes	15483	12.86	3695	19.02
4-strokes	1093	0.91	751	3.87
5- or more strokes	251	0.21	238	1.23
(b)				
2-strokes	8036	6.67	1513	7.79
3-strokes	417	0.35	318	1.64
4-strokes	140	0.12	117	0.60
5- or more strokes	15	0.01	15	0.08
total	120405	100	19426	100
single stroke	2580086		32717	

1. If there is no case in which the l-th stroke of the reference pattern is written as the k-th stroke of the input pattern, i.e., $C_{kl} = 0$, the (k, l) correspondence is excluded.
2. Based on the tendency that the stroke correspondences are concentrated on the diagonal line, if $|k - l| > r$, the (k, l) correspondence is excluded.

In the example of " 姓 ", only 15 pairs of $C_{kl} \neq 0$ or 34 pairs within $r = 2$ are considered as objects among with to search for correspondences.

4.2 Stroke-Connection Analysis

The stroke connection occurs mainly in fast or cursive handwriting. As a result, the number of strokes in a character is decreased. Figure 5 shows examples of characters with occurrence of stroke-connections, where some fixed shapes of stroke-connections occur. The following are considered for stroke-connection.

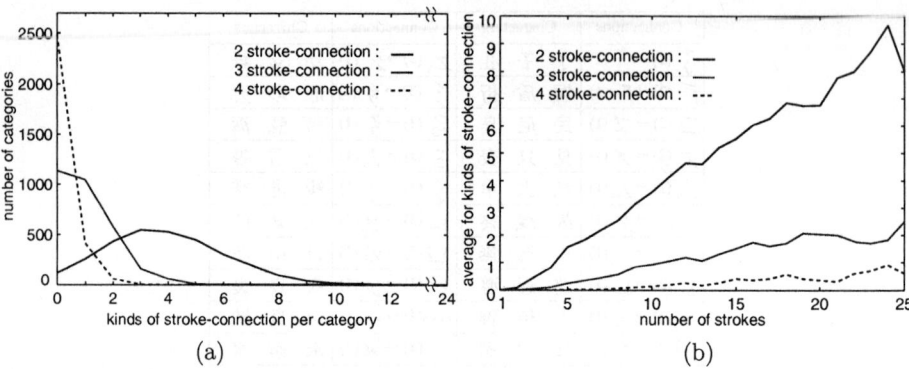

Fig. 6. (a) Number of categories having stroke-connection, (b) average for various types of stroke-connection

1. How many strokes are in one consecutive continuous-writing?
2. In one character, how many places does the independent continuous-writing occur ?
3. Are these occurrences of continuous-writing followed by the standard stroke order, or by nonstandard stroke order ?

The actual condition is complicated. The connection consisting of k strokes is represented as the k-stroke connection.

Table 3 shows the occurrences of consecutive stroke connection, which are divided into standard and nonstandard stroke order. Note that the total number of single strokes is 2580086, including 32717 kinds. Almost all stroke connections occur in the standard order (92.85 %), where 2 stroke connection occurs frequently, and stroke connections in the nonstandard order occur rarely.

The relation is examined between connection type and the number of categories, as shown in Fig. 6 (a). From these graphs, 4.82, 1.35, and 0.29 connection types for each character category can be calculated as 2-, 3-, and 4-stroke connections, respectively. As a result, the 3- and 4-stroke connection types are 29% and 6% of that of 2-stroke connection. Therefore, it could be said that the few-strokes connection occurs much frequently than the the many-strokes one.

Given a character composed of N strokes, $C(N, k)$ (the combination of n things taken r at a time) connection types of k strokes can be expected. For example, in the case of a character with ten strokes, $C(10, 2) = 45$ 2-stroke connections can possibly occur. However, the characters have a common stroke-order, and most writers write the characters according to the standard stroke pattern. Figure 6 (b) shows the average stroke-connections for each character category. As shown in figure 6 (b) and table 4, there are 4.03 2-stroke connections, and 8.95% of the expected number in the character that consisted of 10 strokes. Therefore, very fewer stroke-connections than expected occur in a character.

Based on the above results of stroke connection, the following conditions arise.

Table 4. Expected and actual occurrences of each stroke-connection type

stroke-number	expected	actual	actu./exp.(%)
	2-stroke connection		
5	$C(5,2) = 10$	1.71	17.10
10	$C(10,2) = 45$	4.03	8.95
15	$C(15,2) = 105$	5.88	5.60
20	$C(20,2) = 190$	8.04	4.23
	3-stroke connection		
5	$C(5,3) = 10$	0.32	3.20
10	$C(10,3) = 120$	1.18	0.98
15	$C(15,3) = 455$	1.74	0.38
20	$C(20,3) = 1140$	2.68	0.235
	4-stroke connection		
5	$C(5,4) = 5$	0.025	0.500
10	$C(10,4) = 210$	0.210	0.001
15	$C(15,4) = 1365$	0.456	0.0003
20	$C(20,4) = 1995$	0.440	0.0002

1. Using the connection-information shown in table 3, only the stroke connection type that occurs really is permitted.
2. In addition to the above condition, 2-stroke connection with standard order is permitted for every possible pair, based on the finding that this type occurs frequently, or
3. stroke connection with standard order is permitted for every possible pair, and stroke connection with non-standard order is permitted using the connection information.

One of the above conditions can be selected according to the quality of the writer(s), the character(s), and the recognition system. By using the stroke connection information, large improvement is expected in terms of recognition accuracy and time.

5 Experiments

The usefulness of the presented method was demonstrated by PC-performed recognition experiments (Pentium III processor, 700 MHz). Training data consisted of 2965 Chinese character categories, used as the investigation data in section 3. Reference patterns were generated from training data by storing average values of the loci of feature points from strokes extracted by rearranging according to correct stroke order. Test data was similarly provided by another 20 writers, consisting of the same characters as those of training data (total=57667 characters). Twenty categories selected at random were used for the input patterns in groups in which the stroke numbers were 4, 8, 12, 16, and 20, respectively.

The input character is transformed into a 128x128 mesh plane by preprocessing steps of redundant elimination, smoothing, size normalization, and feature point extraction. As feature information, the x-y coordinates and movement directional vector between one point and the next are extracted from character

Table 5. Recognition experiment results

	Experiment 1	Experiment 2
Search time(sec)	0.76	0.24
Recognition rate(%)	98.39	98.92

data. Feature information of the input and reference patterns is placed into a_{ik} and b_{jl}, respectively.

Using DP matching, the stroke distance $\delta(k;l)$ is calculated by the weighted sum of (1) the distance of x-y coordinate sequences and (2) the distance of directional vector sequences.

The evaluation value ρ of the difference between inter-stroke information is determined using the distance

$$\rho(k,l;p,q) = \frac{1}{4}[R(d_{ss}(A_p, A_k), d_{ss}(B_q, B_l)) + R(d_{se}(A_p, A_k), d_{se}(B_q, B_l))$$
$$+ R(d_{es}(A_p, A_k), d_{es}(B_q, B_l)) + R(d_{ee}(A_p, A_k), d_{ee}(B_q, B_l))]. \quad (6)$$

where l denotes the stroke number, B_l the l-th stroke of the reference pattern, $d_{ss}(A_p, A_k)$ the vector from the start point of A_p to the start point of A_k, $d_{se}(\cdot,\cdot)$ the vector from start point to end point, $d_{es}(\cdot,\cdot)$ the vector from end point to start point, and $d_{ee}(\cdot,\cdot)$ the vector from end point to end point. $R(\cdot,\cdot)$ is the weighted sum of the directional difference and longitudinal difference between vectors.

Recognition results were obtained using a force decision that selects the candidate with minimum dissimilarity $D(A,B)$. Optimal weighting factors were determined for each experiment.

The following experiments were performed to demonstrate the effectiveness of the proposed method.

(1) Experiment 1: Recognition experiment performing the stroke correspondence search without using stroke-order and connection information.
(2) Experiment 2: Recognition experiment performing the stroke correspondence search using stroke-order and connection information.

Table 5 summarizes the recognition experiment results. The large improvement of recognition rate is achieved by using stroke order and connection information on stroke correspondence seacht. The stroke correspondence search time by using these information can be reduced about 1/3 times. Recognition time is the sum total of about 0.65 seconds which is total of calculation time of δ and ρ, and these search times. The reasons of remaining misrecognition are (1) existence of closely similar characters and (2) heavy deformation of character.

6 Conclusion

In this paper, based on the assumption that the recognition is performed on the stroke order and number-free framework, useful information about stroke order and connection have been investigated. Using the automatic stroke correspondence system, we could observe the correct stroke order and connection in real handwritten characters from 2965 categories of Chinese characters. It was proved that the real stroke order of a character is concentrated on the standard order, in which character categories of 95.7 % and 98.1 % are included within the 2-gap. 92.85% of all stroke-connections are written in standard stroke-order. In comparison with the expected stroke connections, very few connections actually occur. We described the methods for incorporating the information in a framework that is completely stroke order-free and number-free. Both or only one side of informations of stroke order and stroke connection can be applied to the framework or the condition of that can be mitigated according to the quality. By introducing this knowledge-based technique into the stroke order- and number-free system, we have achieved reduction of computation time and improved recognition accuracy prominently.

Future work includes development of a technique for efficiently extracting and registering exceptional user-generated stroke deformations is also being considered. Further, based on this information, the reference pattern dictionary can be substantially compressed.

References

1. K. Yoshida and H. Sakoe, "Online Handwritten Character Recognition for a Personal Computer System," *IEEE Trans. Consumer Electonics*, vol. 28, no. 3, pp. 202-208, 1982.
2. T. Wakahara, K. Odaka, and M. Umeda, "Stroke Number and Order Free On-Line Character Recognition by Selective Stroke Linkage Method," *IECE Trans.*, Japan, vol. J66-D, no. 5, pp. 593-600, May 1983 (in Japanese).
3. K. Odaka, T. Wakahara and I. Masuda, "Stroke-Order-Independent On-Line Character Recognition Algorithm and Its Application," *Rev. Electrical Comm. Laboratories*, vol. 34, no.1, pp. 79-85, 1986.
4. Y. Sato and H. Adachi, "Online Recognition of Cursive Writings," *IECE Trans.*, Japan, vol. J68-D, no. 12, pp. 2116-2122, Dec. 1985 (in Japanese).
5. K. Ishigaki and T. Morishita, "A Top-Down Online Handwritten Character Recognition Method via the Denotation of Variation," *Proc. 1988 Int'l Conf. on Computer Processing on Chinese and Oriental Languages*, pp. 141-145, Aug. 1988.
6. C. C. Tappert, C. Y. Suen, and T. Wakahara, "The State of the Art in On-Line Handwriting Recognition," *IEEE Trans. Pattern Anal. Machine Intell.*, vol. 12, no. 8, pp. 787–808, Aug. 1990.
7. T. Wakahara, A. Suzuki, N. Nakajima, S. Miyahara, and K. Odaka, "Stroke-Number and Stroke-Order Free On-Line Kanji Character Recognition as One-to-One Stroke Correspondence Problem," *IEICE Trans. Inf. & Syst.*, vol. E79-D, no. 5, pp. 529–534, May 1996.

8. T. Uchiyama, N. Sonehara, and Y. Tokunaga, "On-Line Handwritten Character Recognition Based on Non-Euclidean Distance," *IEICE Trans. Inf. & Syst.*, vol. J80-D-II, No. 10, pp. 2705–2712, Oct. 1997 (in Japanese).
9. J. Shin, M. M. Ali, Y. Katayama, and H. Sakoe, "Stroke Order Free On-Line Character Recognition Algorithm Using Inter-Stroke Information," *IEICE Trans. Inf. & Syst.*, vol. J82-D-II, No. 3, pp. 382–389, Mar. 1999 (in Japanese).
10. J. Shin and H. Sakoe, "Stroke Correspondence Search Method for Stroke-Order and Stroke-Number Free On-Line Character Recognition — Multilayer Cube Search," *IEICE Trans. Inf. & Syst.*, vol. J82-D-II, No. 2, pp. 230–239, Feb. 1999 (in Japanese).

Face Detection by Integrating Multiresolution-Based Watersheds and a Skin-Color Model

Jong-Bae Kim[1], Su-Woong Jung[2], and Hang-Joon Kim[1]

[1]Dept. of Computer Engineering, Kyungpook National University
1370, Sangyuk-dong, Pook-gu, Dea-gu, 702-701, Korea
{kjblove,hjkim}@ailab.knu.ac.kr
[2]Dept. of Computer Information, Kimcheon Science College
swjung@pubnet.kcs.ac.kr

Abstract. In this paper, we propose a method to automatically segment out a human's face from a given image that consists of head-and shoulder views of humans against complex backgrounds in videoconference video sequences. The proposed method consists of two steps: region segmentation and facial region detection. In the region segmentation, the input image is segmented using multiresolution-based watershed algorithms segmenting the image into an appropriate set of arbitrary regions. Then, to merge the regions forming an object, we use spatial similarity between two regions since the regions forming an object share some common spatial characteristics. In the facial region detection, the facial regions are identified from the results of region segmentation using a skin-color model. The results of the multiresolution-based watersheds image segmentation and facial region detection are integrated to provide facial regions with accurate and closed boundaries. In our experiments, the proposed algorithm detected 87-94% of the faces, including frames from videoconference images and new video. The average run time ranged from 0.23-0.34 sec per frame. This method has been successfully assessed using several test video sequences from MPEG-4 as well as MPEG-7 videoconferences.

1 Introduction

Automatic detection and segmentation of facial regions make some major applications possible, to include model-based coding of video sequences, video indexing, videoconference and recognition or identification of faces. Currently, there is increasing interest in this area due to activities carried out in the MPEG-4 and MPEG-7 standardization processes. In the MPEG-4 and 7 context, human faces can be used to index and search images and videos, classify video scenes and segment human objects from the background [1, 2]. In addition, they help to develop tools that enable a user to access databases. Above all, the segmentation of a facial region provides a content-based representation of the image where it can be used for encoding, manipulation, enhancement, indexing, modeling, pattern-recognition and object-tracking purposes. The main objective of this research was to design a system that can segment a human face from a given videoconference image. To this end, the proposed method can generate facial regions with accurate and closed boundaries.

Until now, various techniques and algorithms have been proposed to detect and segment the human face. The model-based approach was proposed by Govindaraju *et al.* [3] where the face was defined as interconnected arcs that represent chins and hairlines. With this approach, the arcs are extracted using low-level computer vision algorithms and are then grouped based on cost minimization to detect candidate facial regions. Ten images were tested without failure. However, false alarms were often generated. The knowledge-based approach to detect human faces in complex backgrounds was proposed by Yang and Huang [4]. This algorithm consists of three levels. The higher two use mosaic images of different resolutions. The third extracts the edges of facial components. Domain knowledge and rules are applied at each level. The texture-based approach detects faces by sub-sampling different regions of the image to a standard-sized sub-image and then passes it through a neural network filter [5]. Other approaches include temple-based [6] and example-based [7], which have been studied by many researchers. These methods, however, are all computationally expensive and some can only deal with frontal views, thus allowing for little variation in size and viewpoint. Although research on face detection and segmentation has been pursued at a feverish pace, there are still many problems yet to be fully and convincingly solved. This is because the level of difficulty of the problem depends highly on the complexity level of the image content and its applications. To solve these problems, color-based face detection has recently become a new direction that has exhibited better performance [8, 9, 10, 14]. Currently, one approach makes use of skin-color to directly identify the human face, while another employs color as a feature to partition an image into a set of homogeneous regions. However, using skin-color only to detect human faces is impossible to detect faces with accurate and closed boundaries since skin-colors occur not only in faces, but also in backgrounds.

In this paper, we propose a face detection method with accurate and closed boundaries that integrates the results of region segmentation and a skin-color model. This method consists of two steps: region segmentation and facial region detection. In the region segmentation, the input image is segmented into several initial regions using a multiresolution-based watershed algorithm. The regions are then merged into a region according to spatial similarity because the regions forming an object share some common spatial characteristics. The initial region can be an object itself or part of on object that constitutes an object region. Accordingly, to segment the input image into object regions, it is necessary to merge initial regions using spatial similarity. In the facial region detection, the facial regions are identified from the results of the region segmentation stage using a skin-color model.

2 Outline of the Proposed Method

A general outline of the proposed face detection procedure is presented in Fig. 1. The procedure consists of two steps: region segmentation and facial region detection. In the region segmentation, pyramid representation creates multiresolution images using a wavelet transformation. The image at different layers of the pyramid represents various image resolutions. Images are first segmented into a number of regions at the lowest layer of the pyramid by a watershed transformation. The third-order moment values of the wavelet coefficient are then applied to merge the segmented regions. To recover the full-resolution image (original image), an inverse wavelet transformation is applied. In the facial region detection, a skin-color model is defined and facial regions are identified from the results of region segmentation using that model. In the following section on region segmentation, pyramid representation and region segmentation will be described first, followed by the wavelet and watershed transformations. The results of this segmentation will present a bias toward an over-segmentation of some homogeneous regions that will be described in further region-merging strategies. To recover the full-resolution image, we use inverse wavelet transformation. In the face region detection section, skin-colors are extracted and the facial region detection step will be described in detail.

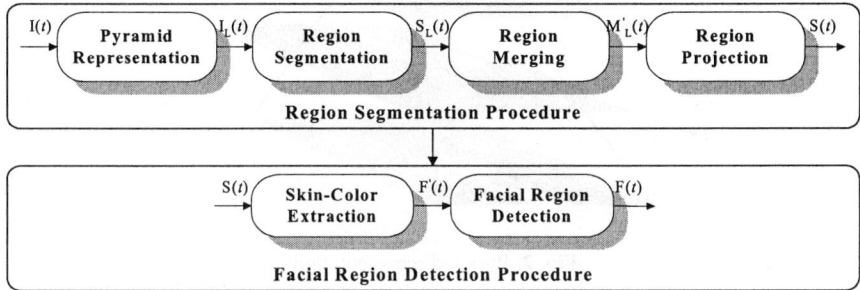

Fig. 1. Block diagram of the proposed face detection approach. I(t) is the sequence image at time t to be segmented; $I_L(t)$ is the corresponding lowest resolution image of the multiresolution image; $S_L(t)$ is the segmented image; $M'_L(t)$ is the merged image; S(t) is the recovered full-resolution image; and F(t) is the detected face region

3 Region Segmentation

3.1 Pyramid Representation

The proposed method is based on the multiresolution application of a watershed transformation for region segmentation on a low-resolution image. Multiresolution methods attempt to obtain a global view of an image by examining it at various resolution levels [11]. To represent the pyramid images, we used a wavelet transformation. Essentially, we used the Haar basis of a wavelet transformation to reduce the resolution of each frame in the sequence. This process reduces the total computation cost. In addition, a great advantage of this multiresolution method is the possibility to determine the dimensions of the regions to be segmented. Thus, over-segmentation of the watershed transformation and possibly noise in the image capturing process can be reduced. The application of Haar decomposition to the frames of a sequence will provide images of smaller resolution which can be quickly processed as well as more simply implemented by standard PC. By using Haar wavelets, the original image is first passed through low-pass filters to generate low-low (LL), low-high (LH), high-low (HL) and high-high (HH) sub-images. The decompositions are repeated on the low-low sub-image to obtain the next four sub-images.

The original image to be segmented is identified as $I(t)$, where t indicates its time reference within the sequence. Stemming from this original image, also identified as $I_0(t)$, where 0 represents a full-resolution image ($I(t) = I_0(t)$), a set of low frequency sub-sampled images of the wavelet transform $\{I_1(t), I_2(t), ..., I_L(t)\}$ at time t is generated. Fig. 2 represents the hierarchical structure where $I_L(t)$ corresponds to the sub-sampled image at level (l) of the pyramid and $I_L(t)$, the lowest available resolution image.

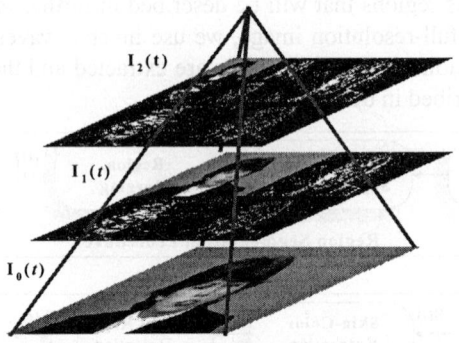

Fig. 2. Pyramid representation

3.2 Region Segmentation

After pyramid representation by wavelet transform, the lowest resolution image $I_L(t)$ is segmented through the application of a watershed transformation based on the gradient image. It is basically a region-growing algorithm, starting from markers, successively joining pixels from an uncertain area to the nearest similar region. Watersheds are traditionally defined in terms of the drainage patterns of rainfall. Regions of terrain that drain to the same points are defined to be part of the same watershed. This same description can be applied to images by viewing intensity as height. In this case, the image gradient is used to predict the direction of the drainage. By following the image gradient downhill from each point in the image, the set of points which drains to each local intensity minimum can be identified. These disjointed regions are called the watersheds of the image. Similarly, the gradients can be followed uphill to local intensity maximum in the image, defining the inverse watersheds of the image [12]. The watershed algorithm consists of initialization and flooding. Initialization puts the location of all pixels in a queue corresponding to the interior of a region in the marker. These pixels have the highest priority because they belong to their respective regions. Next, flooding assigns pixels to regions following a region growing procedure [12]. Fig. 3 shows the region segmentation of a watershed transform. Fig. 3 (a) is the original image of 'Claire', (b) is the gradient image using the Canny operator, (c) is a result of region segmentation by watershed transformation, (d) is the simplified image that has average color values within segmented regions and (e) shows a 3D image of the gradient image. Flat regions larger than 100 pixels are extracted as intensity makers.

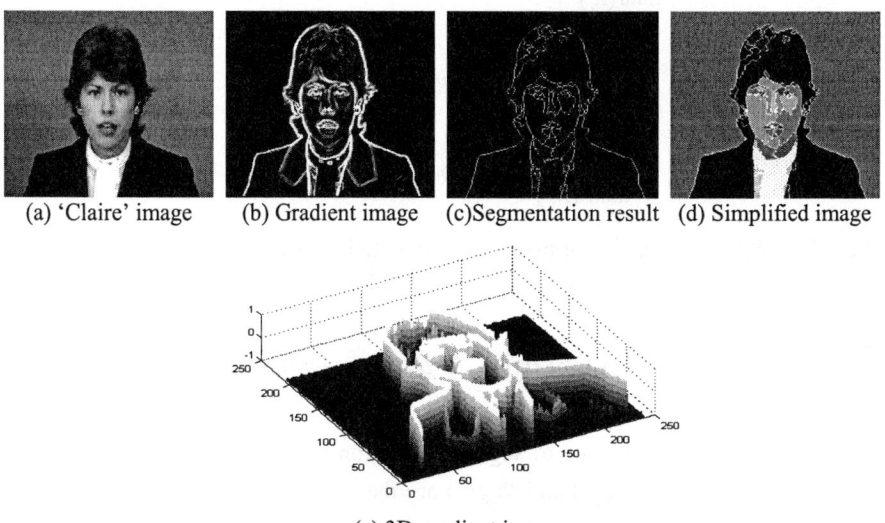

Fig. 3. Region segmentation of the watershed transformation

In the region segmentation phase, ambiguity of the region segmentation is reduced and a partition $S_L(t)$ from the image $I_L(t)$ is generated. The result of this partition $S_L(t)$ will present a bias toward over-segmentation of homogeneous regions that will make further region-merging strategies necessary.

3.3 Region Merging and Projection

In the merging phase, regions that have been over-segmented will be grouped. Generally, the above-mentioned algorithms, when followed by watershed transformations, produce meaningful region segmentations. However, when an image is degraded by noise, it will be over-segmented. Therefore, over-segmented images may require further merging of some regions. For region merging, criteria based on similarities have been proposed, each of which has a specific application [13]. Our decision on which regions to merge is determined through homogeneity and similarity criteria based on wavelet coefficients. Each of the segmented regions will have mean, second and third-order central moment values of the wavelet coefficients calculated. All the features are computed on a LL decomposed subband of a wavelet transform. For each region R_i of the segmented image $S_L(k)$ of the region segmentation phase, we calculate the mean (M), the second (μ_2)- and third-order (μ_3) central moments of the region which can be written as [13]

$$M = \frac{1}{num(R_i)}\sum\sum R_i(i,j) \quad \forall i,j \in R_i(i,j)$$

$$\mu_2 = \frac{1}{num(R_i)}\sum\sum (R_i(i,j) - M)^2 \tag{1}$$

$$\mu_3 = \frac{1}{num(R_i)}\sum\sum (R_i(i,j) - M)^3$$

where *num(R$_i$)* is the number of pixels of segmented region *i*. To merge the segmented regions using similarity criteria, we can use the following equation:

$$S_i = \frac{1}{N}(R(M_i) + R(\mu_{2i}) + R(\mu_{3i})) \quad i = 1, ..., N \tag{2}$$

where S_i is the similarity value of segmented region (*i*) and *N* is the number of segmented regions. $R(M)$, $R(\mu_2)$ and $R(\mu_3)$ are the mean, second- and third-order moment values of the segmented region *i*. If the S_i values of adjacent regions satisfy a specified value, two adjacent regions will merge. Once the merged image $M`_L(t)$ has been generated at the image partition $S_L(t)$, it must be projected down in order to recover the full-resolution image $S(t)$. The direct projection of the segmentation, based only on pixel duplication in both vertical and horizontal directions, offers very poor results. To overcome this problem, we use an inverse wavelet transformation because it generates an easily recovered and efficiently projected full-resolution image.

4 Facial Region Detection

Facial regions are identified from the segmentation results of a stochastic skin-color model, which characterizes the skin colors of human faces. Fig. 4 shows the color distribution of human faces, obtained from 200 test images in chromatic color space. As shown in Fig. 5, the color distribution of human faces is clustered in a small area of chromatic color space and can be approximated by 2D-Gaussian distribution [10, 14]. This is a three-dimensional plot with *r* and *g* forming a chrominance plane. The occurrence of each *r* and *g* in the image gives the values along the vertical axis. The histogram of the skin-color is illustrated in Fig. 4(a). Fig. 4(b) shows the mean and covariance matrix of the skin-color model obtained from 200 sample images. A skin-color model is represented by a 2D-Gaussain model with a mean vector **m** and covariance matrix Σ, where $\mathbf{m} = (\vec{r}, \vec{g})$ with

$$\vec{r} = \frac{1}{N}\sum_{i=1}^{N} r_i, \quad \vec{g} = \frac{1}{N}\sum_{i=1}^{N} g_i, \quad \Sigma = \begin{bmatrix} \sigma_r^2 & p_{x,y}\sigma_g\sigma_r \\ p_{x,y}\sigma_r\sigma_g & \sigma_g^2 \end{bmatrix} \tag{3}$$

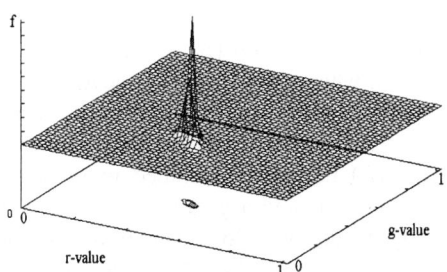

Parameters	Values
μ_r	117.588
μ_g	79.064
σ_r^2	24.132
$\rho_{X,Y}\sigma_g\sigma_r$	-10.085
$\rho_{X,Y}\sigma_r\sigma_g$	-10.085

(a) Color distribution of human faces (b) actual 2D-Gaussian parameters

Fig. 4. Color distribution of human faces and actual 2D-Gaussian parameters

5 Experimental Results

To evaluate the performance of the proposed face detection method, simulation was carried out on the frames of 'Miss America', 'Claire', 'Akiyo', 'Foreman' and 'Carphone' videoconference sequences. The pyramid image generated by the 2-scale Haar wavelet transformation and the makers for the watershed transformation were extracted from a low-resolution image. Here, flat regions larger then 100 pixels were extracted as markers. The experiments were performed on a Pentium 1.4Ghz PC with an algorithm that was implemented using Matlab 5.3. The image segmentation processing of each image took 0.29 sec on average and detection rate (the probability of correctly classifying the pixels of a facial region) was 87-94%.

This proves the proposed method is robust against noise of degraded images, tilting and rotation of the head as well as the presence of complex backgrounds. Based on both real videoconference and our conference room video sequences, the segmentation and detection of facial region aspects were studied. Here, we only present the results of 'Claire', 'Akiyo' and our conference room sequences to demonstrate the process of this algorithm. We also evaluated the results of the region segmentation step of our proposed method. To do so, we used four common measurements: the number of segmented regions, PSNR, Goodness and computation time. The Goodness function F is defined as [11]

$$F(I) = \sqrt{M} \times \sum_{i=1}^{M} \frac{e_i^2}{\sqrt{A_i}} \quad (4)$$

where I is the image to be segmented, M is the number of regions in the segmented image, A_i is the area, or the number of pixels of the ith region, and e_i is the sum of the Euclidean distance of the color vectors between the original image and the segmented image of each pixel in the region. Eq. (4) is composed two terms: the first term $\sqrt{M}$ penalizes segmentation that forms too many regions; the second penalizes segmentations having non-homogeneous regions. A larger value of e_i means that the feature of the region was not well segmented during the image segmentation process.

Table 1 shows the segmentation results of a number of segmented regions, Goodness (F), PSNR and computation times using full-resolution (I_0) and low-resolution images (I_1, I_2). The highest PSNR was observed when using direct application of the watershed algorithm for the full-resolution image. However, it also had the largest number of regions and highest computation times. For each set of segmentation results shown in Table 1, the full-resolution image (I_0) normally contained more details. However, there were also a larger number of regions and a greater value of F. The low-resolution images are the opposite: many details were lost. Therefore, the total number of regions was less and the value of PSNR was smaller. Although there were lower PSNR values, the value of F was also smaller. According to the segmentation evaluation criteria, these evaluations represent trade–off between suppressing noise and preserving details. For each set of results, the low-resolution image always had the smallest F, which means that it is the best.

In the proposed method, the use of low-resolution images is highly preferable since it gives the best objective quality when the number of regions and F are considered simultaneously. Over-segmentation and computation time after the region segmentation phase can be reduced when operating on low-resolution images. Fig. 5 and 6 show the results of region segmentation and facial region detection using the proposed approach. These images demonstrate that our approach can generate satisfactory results.

Table 1. Segmentation results of the proposed region segmentation method. I_0 is the full-resolution image (original image), I_1 and I_2 are the 1, 2-scale wavelet decomposed low-low subband images, respectively, and M is the merged image of the segmented I_2 image

Test images	Scale Levels	Number of Regions	Goodness	PSNR [dB]	Time (sec)
Claire #136	I_0	86	283.2	31.58	0.38
	I_1	36	111.7	30.79	0.24
	I_2	30	109.3	30.40	0.20
	M	7	110.5	29.98	0.02
Akiyo #2	I_0	542	302.6	31.96	0.47
	I_1	258	254.0	29.91	0.25
	I_2	65	185.2	29.23	0.22
	M	10	186.2	28.81	0.02

6   Conclusions

An efficient method for face detection using multiresolution-based watershed algorithms and a skin-color model are described in this paper. The procedure for complete face detection consists of two steps: region segmentation and facial region detection. In the region segmentation, images are segmented into arbitrary regions using multiresolution-based watershed algorithms. In the facial region detection, facial regions are identified from the results of region segmentation step using a skin-color model. The results of the region segmentation and facial region detection are integrated to provide more accurate segmentation of the facial regions. As shown in the simulation results, the proposed method generates accurate and closed boundaries of human face detection and segmentation results, which makes region- or object-based coding or videoconference systems more efficient. It can also significantly reduce the computational cost of face detection while improving detection accuracy. In the future, we plan to develop a multi-human face detection and recognition system.

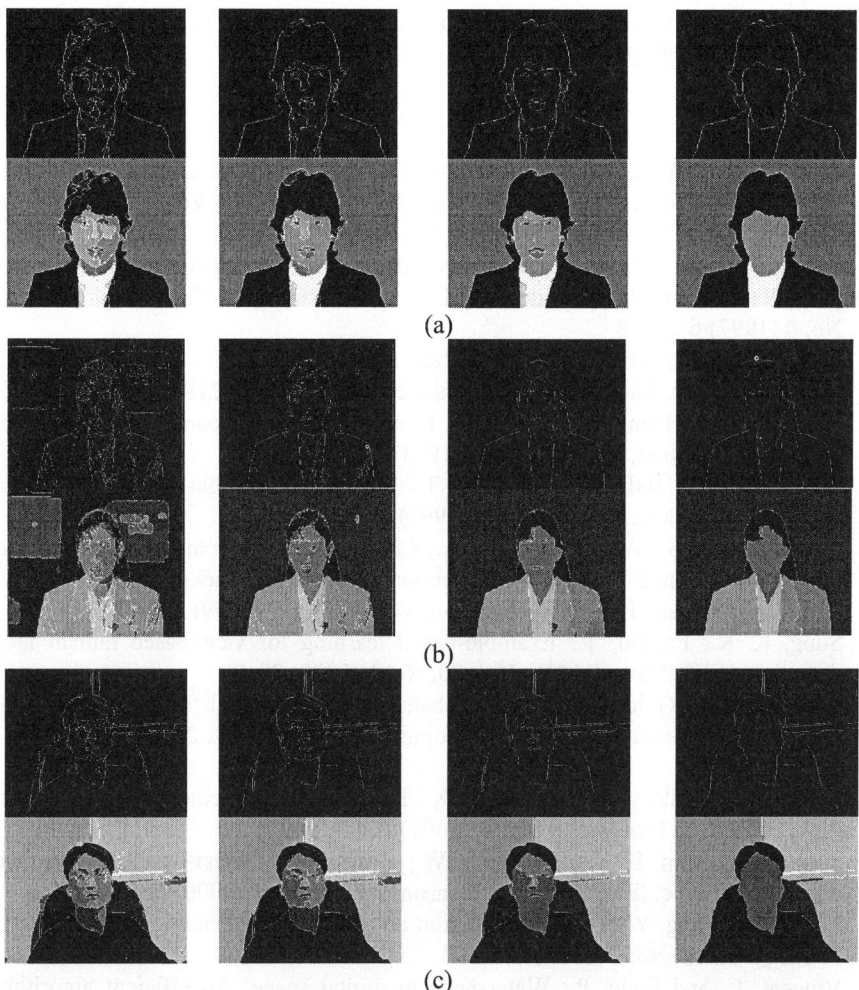

Fig. 5. Region segmentation results of the proposed method. (a), (b) and (c) are the images of 'Claire', 'Akiyo' and our conference room. The first, second and third columns show the results of segmentation of full-resolution (I_0), 1- (I_1), 2-scale wavelet decomposition images (I_2) of the pyramid. The last column shows the results of merging the third column images

 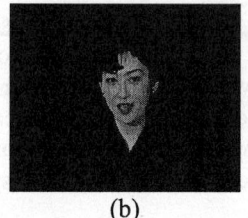 

(a) (b) (c)

Fig. 6. Results of the facial region detection. (a), (b), and (c) are the facial region images of 'Claire', 'Akiyo' and our conference room

References

1. Chai, D., and Ngan, K. N.: Face Segmentation Using Skin-Color Map in Videophone Applications, IEEE Trans. Circuit Syst. Video Tech., Vol. 9, No. 4 (1999) 551-564
2. Wang, H. and Chang, S. F.: A Highly Efficient System for Automatic Face Region Detection in MPEG Video, IEEE Trans. Circuit Syst. Video Tech., Vol. 7, No. 4 (1997) 615-628
3. Govindaraju, V., Srihari, R. K. and Sher, D. B.: A computational model for face location, in Proc. Third int. Conf. Computer Vision, (1990) 718-721
4. Yang, G. and Huang, T. S.: Human face detection in a complex background, Pattern Recognition, Vol. 27, No. 1, (1994) 53-63
5. Rowley, H. A., Baluja, S., Kanade, T.: Neural network-based face detection, IEEE Trans. PAMI., Vol. 20, No. 1 (1998) 23-38
6. Miao, J., Yin, B., Wang, K., Shen, L., Chen, X.: A hierarchical multiscale and multiangle system for human face detection in a complex background using gravity-center template, Pattern Recognition, Vol. 32, No. 7 (1999) 127-1248
7. Sung, K. K., Poggio, T.: Example-based learning for view-based human face detection, IEEE Trans. PAMI., Vol. 20, No. 1 (1998) 39-51
8. Greensapn, H., Goldberger, J. and Eshet, I.: Mixture model for face-color modeling and segmentation, Pattern Recognition Letters, Vol. 22, No. 14 (2001) 1525-1536
9. Hsu, R. L., Mottalev, M. A., Jain, A. K.: Face detection in color images, in proc. Int. Conf. Image Processing, Vol. 1 (2001) 1046-1049
10. Kim, H. S., Kim, E. Y., Hwang, S. W., Kim, H. J.: Object-based human face detection, in Proc. IEEE Int. Conf. Consumer Electronics (2000) 354-357
11. Liu, J. and Yang, Y. H.: Multiresolution color image segmentation, IEEE Trans. PAMI., Vol. 16, No. 7 (1994) 674-693
12. Vincent, L. and Soile, P.: Watersheds in digital space: An efficient algorithm based on immersion simulation, IEEE Trans. PAMI., Vol. 13, No. 6, (1998) 583-598
13. Kim, J. B., Lee, C. W., Lee, K. M., Yun, T. S., Kim, H. J.: Wavelet-based vehicle tracking for Automatic Traffic Surveillance, in Proc. IEEE Tencon'01, Vol. 3 (2001) 313-316
14. Yang, J. and Waibel, A.: A Real-Time Face Tracker, in Proc. IEEE Int. Conf. Acoustics, Speech, and Signal Processing, Vol. 3 (1996) 142-147

Social Interaction of Humanoid Robot Based on Audio-Visual Tracking

Hiroshi G. Okuno[1,2], Kazuhiro Nakadai[2], and Hiroaki Kitano[2,3]

[1] Graduate School of Informatics, Kyoto University
Kyoto 606-8501, Japan
okuno@nue.org
http://winnie.kuis.kyoto-u.ac.jp/~okuno/
[2] Kitano Symbiotic Systems Project, ERATO, Japan Science and Technolog Corp.
Mansion 31 Suite 6A, 6-31-15 Jingumae, Shibuya, Tokyo 150-0001 Japan
nakadai@symbio.jst.go.jp
http://www.symbio.jst.go.jp/~nakadai/
[3] Sony Computer Science Laboratories, Inc.
Shinagawa, Tokyo 141-0022
kitano@csl.sony.co.jp
http://www.csl.sony.co.jp/~kitano/

Abstract. Social interaction is essential in improving robot human interface. Such behaviors for social interaction may include paying attention to a new sound source, moving toward it, or keeping face to face with a moving speaker. Some sound-centered behaviors may be difficult to attain, because the mixture of sounds is not well treated or auditory processing is too slow for real-time applications. Recently, Nakadai *et al* have developed real-time auditory and visual multiple-talker tracking technology by associating auditory and visual streams. The system is implemented on an upper-torso humanoid and the real-time talker tracking is attained with 200 msec of delay by distributed processing on four PCs connected by Gigabit Ethernet. Focus-of-attention is programmable and allows a variety of behaviors. The system demonstrates non-verbal social interaction by realizing a receptionist robot by focusing on an associated stream, while a companion robot on an auditory stream.

1 Introduction

Social interaction is essential for humanoid robots in daily life, because such robots are getting more common in social and home environments, such as a pet robot at living room, a service robot at office, or a robot serving people at a party [3]. Social skills of such robots require robust complex perceptual abilities, for example, it identifies people in the room, pays attention to their voice and looks at them to identify visually, and associates voice and visual images. Intelligent behavior of social interaction should emerge from rich channels of input sensors; vision, audition, tactile, and others.

Perception of various kinds of sensory inputs should be active in the sense that we hear and see things and events that are important to us as individuals, not sound waves or light rays. In other words, selective attention of sensors represented as looking versus seeing or listening versus hearing plays an important role in social interaction. Other important factors in social interaction are recognition and synthesis of emotion in face expression and voice tones.

In this paper, we focus on audition, or sound input in localizing and tacking talkers, and report audio-visual interactions between human and humanoid. Sound has been recently recognized as essential in order to enhance visual experience and human computer interaction, and thus several contributions have been done by academia and industries [1,3,9,11]. One of social intelligent behaviors is the *cocktail party effect*, that is, a robot can attend one conversation at a crowded party and then switch to another one. Although the most important human communication means is language, non-verbal sensori-motor based behavior is nonetheless important.

Some robots realize social interaction, in particular, in visual and dialogue processing. Ono et al. use the robot named *Robovie* to make common attention between human and robot by using gestures [13]. Breazeal incorporates the capabilities of recognition and synthesis of emotion in face expression and voice tones into the robot named *Kismet* [1,2]. Waldherr et al. makes the robot named *AMELLA* that can recognize pose and motion gestures [14]. Matsusaka et al. built the robot named *Hadaly* that can localize the talker as well as recognize speeches by speech-recognition system so that it can interact with multiple people [7]. Nakadai et al developed *real-time* auditory and visual multiple-tracking system for the upper-torso humanoid named *SIG* [10].

The rest of the paper is organized as follows: Section 2 describes the real-time multiple-talker tracking system. Section 3 demonstrates the system behavior of social interaction. Section 4 discusses the observations of the experiments and future work, and Section 5 concludes the paper.

Fig. 1. *SIG* the Humanoid plays as a companion robot

2 Real-time Multiple-Talker Tracking System

2.1 SIG the Humanoid

As a test bed of integration of perceptual information to control motor of high degree of freedom (DOF), we designed a humanoid robot (hereafter, referred as *SIG*) with the following components:

- 4 DOFs of body driven by 4 DC motors — Each DC motor has a potentiometer to measure the direction.
- A pair of CCD cameras of Sony EVI-G20 for visual stereo input
- Two pairs of omni-directional microphones (Sony ECM-77S). One pair of microphones are installed at the ear position of the head to collect sounds from the external world. Each microphone is shielded by the cover to prevent from capturing internal noises. The other pair of microphones is to collect sounds within a cover.
- A cover of the body (Figure 1) reduces sounds to be emitted to external environments, which is expected to reduce the complexity of sound processing. This cover, made of FRP, is designed by our professional designer for making human robot interaction smoother as well [9].

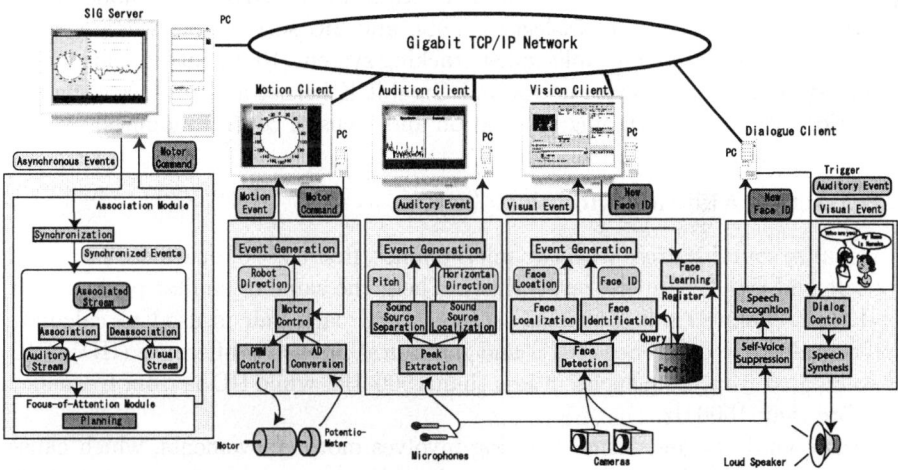

Fig. 2. Hierarchical architecture of real-time audio and visual tracking system

2.2 Architecture of Real-Time Audio and Visual Tracking System

The system is designed based on the client/server model (Fig. 2). Each server or client executes the following logical modules:

1. Audition client extracts auditory events by pitch extraction, sound source separation and localization, and sends those events to Association.
2. Vision client uses a pair of cameras, extracts visual events by face extraction, identification and localization, and then sends visual events to Association.
3. Motor client generates PWM (Pulse Width Modulation) signals to DC motors and sends motor events to Association.
4. Association module groups various events into a stream and maintains association and dissociation between streams.
5. Focus-of-Attention module selects some stream on which it should focus its attention and makes a plan of motor control.
6. Dialog client communicates with people according to its attention by speech synthesis and speech recognition. We use "Julian" automatic speech recognition system [6].

The status of each module is displayed on each node. SIG server displays the radar chart of objects and the stream chart. Motion client displays the radar chart of the body direction. Audition client displays the spectrogram of input sound, and pitch (frequency) versus sound source direction chart. Vision client displays the image of the camera and the status of face identification and tracking.

Since the system should run in real-time, the above modules are physically distributed to five Linux nodes connected by TCP/IP over Gigabit Ethernet TCP/IP network and run asynchronously. The system is implemented by distributed processing of five nodes with Pentium-IV 1.8 GHz. Each node serves Vision, Audition, Motion and Dialogue clients, and SIG server. The whole system upgrades the real-time multiple-talker tracking system [10] by introducing stereo vision systems, adding more nodes and Gigabit Ethernet and realizes social interaction system by designing association and focus-of control modules.

2.3 Active Audition Module

To localize sound sources with two microphones, first a set of peaks are extracted for left and right channels, respectively. Then, the same or similar peaks of left and right channels are identified as a pair and each pair is used to calculate interaural phase difference (IPD) and interaural intensity difference (IID). IPD is calculated from frequencies of less than 1500 Hz, while IID is from frequency of more than 1500 Hz.

Since auditory and visual tracking involves motor movements, which cause motor and mechanical noises, audition should suppress or at least reduce such noises. In human robot interaction, when a robot is talking, it should suppress its own speeches. Nakadai *et al* presented the *active audition* for humanoids to improve sound source tracking by integrating audition, vision, and motor controls [8]. We also use their heuristics to reduce internal burst noises caused by motor movements.

From IPD and IID, the epipolar geometry is used to obtain the direction of sound source [8]. The key ideas of their real-time active audition system are twofold; one is to exploit the property of the harmonic structure (fundamental frequency, $F0$, and its overtones) to find a more accurate pair of peaks in left and right channels. The other is to search the sound source direction by combining the belief factors of IPD and IID based on Dempster-Shafer theory.

Finally, audition module sends an auditory event consisting of pitch ($F0$) and a list of 20-best direction (θ) with reliability for each harmonics.

2.4 Face Recognition and Identification Module

Vision extracts lengthwise objects such as persons from a disparity map to localize them by using a pair of cameras. First a disparity map is generated by an intensity based area-correlation technique. This is processed in real-time on a PC by a recursive correlation technique and optimization peculiar to Intel architecture [5].

In addition, left and right images are calibrated by affine transformation in advance. An object is extracted from a 2-D disparity map by assuming that a human body is lengthwise. A 2-D disparity map is defined by

$$DM_{2D} = \{D(i,j)|i = 1, 2, \cdots W, j = 1, 2, \cdots H\} \quad (1)$$

where W and H are width and height, respectively and D is a disparity value.

As a first step to extract lengthwise objects, the median of DM_{2D} along the direction of height shown as Eq. (2) is extracted.

$$D_l(i) = Median(D(i,j)). \quad (2)$$

A 1-D disparity map DM_{1D} as a sequence of $D_l(i)$ is created.

$$DM_{1D} = \{D_l(i)|i = 1, 2, \cdots W\} \quad (3)$$

Next, a lengthwise object such as a human body is extracted by segmentation of a region with similar disparity in DM_{1D}. This achieves robust body extraction so that only the torso can be extracted when the human extends his arm. Then, for object localization, epipolar geometry is applied to the center of gravity of the extracted region. Finally, Vision creates stereo vision events that consist of distance, azimuth and observation time.

Finally, vision module sends a visual event consisting of a list of 5-best Face ID (Name) with its reliability and position (distance r, azimuth θ and elevation ϕ) for each face.

2.5 Stream Formation and Association

Association synchronizes the results (events) given by other modules. It forms an auditory, visual or associated stream by their proximity. Events are stored in the short-term memory only for 2 seconds. Synchronization process runs with the delay of 200 msec, which is the largest delay of the system, that is, vision module.

An auditory event is connected to the nearest auditory stream within ±10° and with common or harmonic pitch. A visual event is connected to the nearest visual stream within 40 cm and with common face ID. In either case, if there are plural candidates, the most reliable one is selected. If any appropriate stream is found, such an event becomes a new stream. In case that no event is connected to an existing stream, such a stream remains alive for up to 500 msec. After 500 msec of keep-alive state, the stream terminates.

An auditory and a visual streams are associated if their direction difference is within $\pm 10°$ and this situation continues for more than 50% of the 1 sec period. If either auditory or visual event has not been found for more than 3 sec, such an associated stream is dissociated and only existing auditory or visual stream remains. If the auditory and visual direction difference has been more than 30° for 3 sec, such an associated stream is dissociated to two separate streams.

2.6 Focus-of-Attention and Dialog Control

Focus-of-Attention control is programmable based on continuity and triggering. By continuity, the system tries to keep the same status, while by triggering, the system tries to track the most interesting object. Focus-of-Attention control is a mixed architecture based on bottom-up and top-down control. By bottom-up, the most plausible stream means the one of the highest belief factors. By top-down, the plausibility is defined by the applications. For a receptionist robot, the continuity of the current focus-of-attention has the highest priority. For a companion robot, on the contrary, the stream associated the most recently is focused. Since the detailed design of each algorithm depends on applications, the focus-of-attention control algorithm for a receptionist and companion robot is described in the next section.

Dialog control is used by a receptionist robot, which is described later. A companion robot, however, does not use it, because we want to observe the users' behaviors invoked by non-verbal actions of *SIG*.

3 Design and Experiments of Some Social Interactions

For evaluation of the behavior of *SIG*, one scenario for the receptionist robot and one for the companion robot are designed and executed. The first scenario examines whether an auditory stream triggers Focus-of-Attention to make a plan for *SIG* to turn toward a speaker, and whether *SIG* can ignore the sound it generates by itself. The second scenario examines how many people *SIG* can discriminate by integrating auditory and visual streams.

Experiments were performed in a small room of a normal residential apartment. The width, length and height of the room of experiment are about 3 m, 3 m, and 2 m, respectively. The room has 6 down-lights embedded on the ceiling.

3.1 *SIG* as a Receptionist Robot

The precedence of streams selected by focus-of-attention control as a receptionist robot is specified from higher to lower as follows:

Social Interaction of Humanoid Robot Based on Audio-Visual Tracking   731

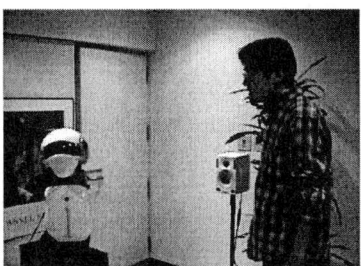 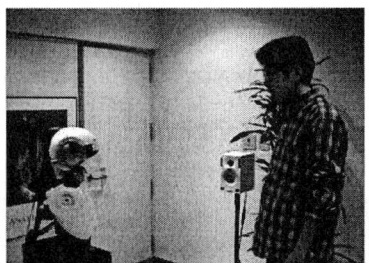

a) When a participant comes and says "Hello", *SIG* turns toward him

b) *SIG* asks his name and he introduces himself to it

Fig. 3. Temporal sequence of snapshots of *SIG*'s interaction as a receptionist robot

associated stream ≻ *auditory stream* ≻ *visual stream*

One scenario to evaluate the above control is specified as follows: (1) A known participant comes to the receptionist robot. His face has been registered in the face database. (2) He says Hello to *SIG*. (3) *SIG* replies "Hello. You are XXX-san, aren't you?" (4) He says "yes". (5) *SIG* says "XXX-san, Welcome to the party. Please enter the room.".

Fig. 3 illustrates four snapshots of this scenario. Fig. 3 a) shows the initial state. The speaker on the stand is the mouth of *SIG*'s. Fig. 3 b) shows when a participant comes to the receptionist, but *SIG* has not noticed him yet, because he is out of *SIG*'s sight. When he speaks to *SIG*, Audition generates an auditory event with sound source direction, and sends it to Association, which creates an auditory stream. This stream triggers Focus-of-Attention to make a plan that *SIG* should turn to him. Fig. 3 c) shows the result of the turning. In addition, Audition gives the input to Speech Recognition, which gives the result of speech recognition to Dialog control. It generates a synthesized speech. Although Audition notices that it hears the sound, *SIG* will not change the attention, because association of his face and speech keeps *SIG*'s attention on him. Finally, he enters the room while *SIG* tracks his walking.

This scenario shows that *SIG* takes two interesting behaviors. One is voice-triggered tracking shown in Fig. 3 c). The other is that *SIG* does not pay attention to its own speech. This is attained naturally by the current association algorithm, because this algorithm is designed by taking into account the fact that conversation is invoked by alternate initiatives.

The variant of this scenario is also used to check whether the system works well. (1') A participant comes to the receptionist robot, whose face has not been registered in the face database. In this case, *SIG* asks his name and registers his face and name in the face database.

As a receptionist robot, once an association is established, *SIG* keeps its face fixed to the direction of the speaker of the associated stream. Therefore, even when *SIG* utters via a loud speaker on the left, *SIG* does not pay an attention to the sound source, that is, its own speech. As a result, *SIG* automatically suppresses self-generated sounds. Of course, this kind of suppression is observed by another benchmark that contains the situation that *SIG* and the human speaker utter at the same time.

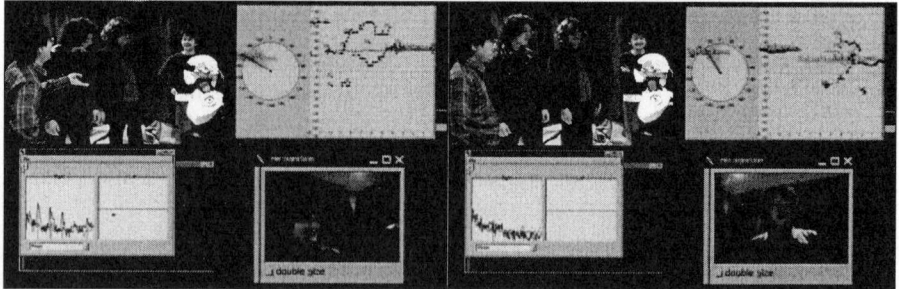

a) The leftmost man says "Hello" and SIG is tracking him

b) The second left man says "Hello" and SIG turns toward him

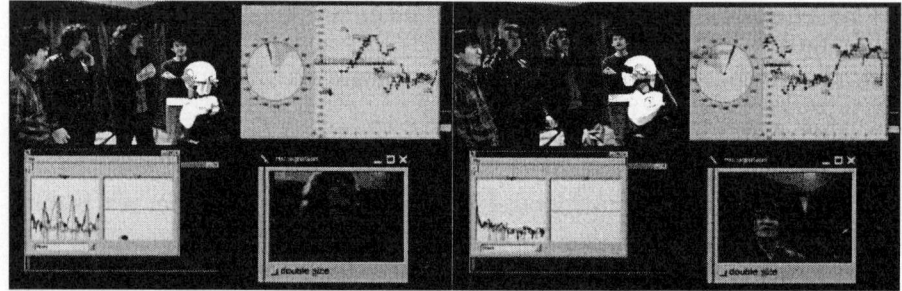

c) The second right man says "Hello" and SIG turns toward him

d) The leftmost man says "Hello" and SIG turns toward him

Fig. 4. Temporal sequence of snapshots for a companion robot: scene (upper-left), radar and sequence chart (upper-right), spectrogram and pitch-vs-direction chart (lower-left), and face-tracking chart (lower-right)

3.2 *SIG* as a Companion Robot

The precedence of streams selected by focus-of-attention control as a companion robot is as follows:

$$auditory\ stream \succ associated\ stream \succ visual\ stream$$

There is no explicit scenario for evaluating the above control. Four speakers actually talks spontaneously in attendance of *SIG*. Then *SIG* tracks some speaker and then changes focus-of-attention to others. The observed behavior is evaluating by consulting the internal states of *SIG*, that is, auditory and visual localization shown in the radar chart, auditory, visual, and associated streams shown in the stream chart, and peak extraction as shown in Figure 4 a)~d).

The top-right image consists of the radar chart (left) and the stream chart (right) updated in real-time. The former shows the environment recognized by *SIG* at the moment of the snapshot. A pink sector indicates a visual field of *SIG*. Because of using the absolute coordinate, the pink sector rotates as *SIG* turns. A green point with a label is the direction and the face ID of a visual stream. A blue sector is the direction of an auditory stream. Green, blue and red lines indicate the direction of visual, auditory and associated stream, respectively. Blue and green *thin* lines indicate auditory and visual streams, respectively. Blue, green and red *thick* lines indicate associated streams with only auditory, only visual, and both information, respectively.

The bottom-left image shows the auditory viewer consisting of the power spectrum and auditory event viewer. The latter shows an auditory event as a filled circle with its pitch in X-axis and its direction in Y-axis.

The bottom-right image shows the visual viewer captured by the *SIG*'s left eye. A detected face is displayed with a red rectangle. The top-left image in each snapshot shows the scene of this experiment recorded by a video camera.

The temporal sequence of *SIG*'s recognition and actions shows that the design of companion robot works well and pays its attention to a new talker. The current system has attained a passive companion. To design and develop an active companion may be important future work.

3.3 Observation

Horvitz and Paek developed an automated conversation system named the *Bayesian Receptionist* [4], which simulates dialog of receptionist at the front desks of buildings on the Microsoft corporate campus. It employs a hierarchy of Bayesian user models to interpret the goals of speakers by analyzing their utterances with natural language processing. Beyond linguistic features, it also takes into consideration contextual evidence, including visual findings; for example, appearance, behavior, spatial configuration, and properties of the participant.

The principles of conversational adopted by the Bayesian Receptionist may be extended to apply to *SIG*, since its current dialog system is quite naive. In this application, auditory findings as well as visual findings should be included, which is one of important future work.

In visual search, saliency map [15] is often used to determine the target of the visual attention. Usually, saliency map is built by accumulating various sensory information. In this paper, we focus on continuity of a stream that acquires the attention or emergence of a new stream, instead, to realize the sound-activated attention mechanism. Needless to say, a sophisticated focus-of-attention control mechanism based on saliency map should be exploited, which is one of future work.

4 Conclusion

In this paper, we demonstrate that auditory and visual multiple-talker tracking subsystem can improve social aspects of human robot interaction. Although a simple scheme of behaviors is implemented, human robot interaction is drastically improved by real-time multiple-talker tracking system. In fact, people pleasantly spend an hour with *SIG* as a companion robot even if its behavior is quite passive. *SIG* respond in ways that invite the interactor to participate in interaction with it and explore the principles of its functioning.

Since the application of auditory and visual multiple-talker tracking is not restricted to robots or humanoids, auditory capability can be transferred to software agents or systems. As discussed in the introduction section, auditory information should not be ignored in computer graphics or human computer interaction. By integrating audition and vision, more cross-modal perception can be attained. One of important future work is automatic acquisition of social interaction patterns by supervised or unsupervised learning. This capability is quite important to provide a rich collection of social behaviors. Other future work includes applications such as "listening to several things simultaneously" [12], "cocktail party computer", integration of auditory and visual tracking with pose and gesture recognition, and other novel areas.

Acknowledgments

We thank our current and former colleagues of Symbiotic Intelligence Group, Kitano Symbiotic Systems Project, in particular, Tatsuya Matsui and Dr. Tino Lourens, for their discussions. We also thank Prof. Tatsuya Kawahara of Kyoto University for allowing us to use "Julian" automatic speech recognition system.

References

1. BREAZEAL, C., AND SCASSELLATI, B. A context-dependent attention system for a social robot. *Proceedints of the Sixteenth International Joint Conf. on Atificial Intelligence (IJCAI-99)*, 1146–1151.
2. BREAZEAL, C. Emotive qualities in robot speech. *Proc. of IEEE/RSJ International Conf. on Intelligent Robots and Systems (IROS-2001)*, 1389–1394.
3. BROOKS, R. A., BREAZEAL, C., IRIE, R., KEMP, C. C., MARJANOVIC, M., SCASSELLATI, B., AND WILLIAMSON, M. M. Alternative essences of intelligence. *Proc. of 15th National Conf. on Artificial Intelligence (AAAI-98)*, 961–968.
4. HORVITZ, E., AND PAEK, T. A computational architecture for conversation. *Proc. of Seventh International Conf. on User Modeling* (1999), Springer, 201–210.
5. KAGAMI, S., OKADA, K., INABA, M., AND INOUE, H. Real-time 3d optical flow generation system. *Proc. of International Conf. on Multisensor Fusion and Integration for Intelligent Systems (MFI'99)*, 237–242.
6. KAWAHARA, T., LEE, A., KOBAYASHI, T., TAKEDA, K., MINEMATSU, N., ITOU, K., ITO, A., YAMAMOTO, M., YAMADA, A., UTSURO, T., AND SHIKANO, K. Japanese dictation toolkit – 1997 version –. *Journal of Acoustic Society Japan (E) 20*, 3 (1999), 233–239.

7. MATSUSAKA, Y., TOJO, T., KUOTA, S., FURUKAWA, K., TAMIYA, D., HAYATA, K., NAKANO, Y., AND KOBAYASHI, T. Multi-person conversation via multi-modal interface — a robot who communicates with multi-user. *Proc. of 6th European Conf. on Speech Communication Technology (EUROSPEECH-99)*, ESCA, 1723–1726.
8. NAKADAI, K., LOURENS, T., OKUNO, H. G., AND KITANO, H. Active audition for humanoid. *Proc. of 17th National Conf. on Artificial Intelligence (AAAI-2000)*, 832–839.
9. NAKADAI, K., MATSUI, T., OKUNO, H. G., AND KITANO, H. Active audition system and humanoid exterior design. *Proc. of IEEE/RAS International Conf. on Intelligent Robots and Systems (IROS-2000)*, 1453–1461.
10. NAKADAI, K. HIDAI, K., MIZOGUCHI, H., OKUNO, H. G., AND KITANO, H. Real-time auditory and visual multiple-object tracking for robots. *Proc. of the Seventeenth International Joint Conf. on Artificial Intelligence (IJCAI-01)*, 1425–1432.
11. OKUNO, H., NAKADAI, K., LOURENS, T., AND KITANO, H. Sound and visual tracking for humanoid robot. *Proc. of Seventeenth International Conf. on Industrial and Engineering Applications of Artificial Intelligence and Expert Systems (IEA/AIE-2001)* (Jun. 2001), LNAI 2070, Springer-Verlag, 640–650.
12. OKUNO, H. G., NAKATANI, T., AND KAWABATA, T. Listening to two simultaneous speeches. *Speech Communication 27*, 3-4 (1999), 281–298.
13. ONO, T., IMAI, M., AND ISHIGURO, H. A model of embodied communications with gestures between humans and robots. *Proc. of Twenty-third Annual Meeting of the Cognitive Science Society (CogSci2001)*, AAAI, 732–737.
14. WALDHERR, S., THRUN, S., ROMERO, R., AND MARGARITIS, D. Template-based recoginition of pose and motion gestures on a mobile robot. *Proc. of 15th National Conf. on Artificial Intelligence (AAAI-98)*, 977–982.
15. WOLFE, J., CAVE, K. R., AND FRANZEL, S. Guided search: An alternative to the feature integration model for visual search. *Journal of Experimental Psychology: Human Perception and Performance 15*, 3 (1989), 419–433.

Hybrid Confidence Measure for Domain-Specific Keyword Spotting

Jinyoung Kim[1], Joohun Lee[2], and Seungho Choi[3]

[1] Deptartment of Electronics Engineering, Chonnam National University, Yongbong-dong 300, Kwangjoo, South Korea
kimjin@dsp.chonnam.ac.kr
[2] School of Science and Engineering, Waseda University, 3-4-1 Okubo, Shinjuku, Tokyo 169-8555, Japan
vincelee64@freechal.com
[3] Departtment of Information and Communication, Dongshin University, Najoo Daeho-dong 252, Chollanam-Do, South South Korea
shchoi@white.dongshinu.ac.kr

Abstract. In this paper, we introduce three acoustic confidence measures (CM) for domain-specific keyword spotting system. The first one is a statistically normalized version of well-known CM (NCM). And the second one is a new CM based on anti-filler concept. And, finally, we propose a hybrid CM (HCM) combining the above two CMs. HCM is a linear combination of two CMs with weighting parameters. To evaluate the proposed CMs, we constructed directory service system, which is a kind of keyword spotting system. We applied our CMs to this system and compared the performance results of the proposed CM with that of the conventional CM (RLH-CM). In our experiments, NCM and HCM show superior ROC performances to the conventional CM. Especially, with HCM, the enhancement of 40% FAR reduction was achieved.

1 Introduction

In many circumstances, a reliable measure of the confidence of a speech recognizer's output is very useful and important. Basically, confidence measure (CM) is of use to reject hypotheses that are likely to be erroneous in a hypothesis test. For example, CM is a useful measure to detect an out-of-vocabulary (OOV) in continuous speech recognition. Also, CM shows its usefulness in rejecting misrecognized word in isolated-word recognition and keyword spotting system. Additionally, a reliable confidence measure may be of a practical use in recognition search (confidence estimates may be useful for ordering partial decoding hypothesis)[1-5]. Therefore, CM takes a very important role in the real services based on speech recognition.

Recently, we implemented domain-specific keyword spotting system, which is a directory service system (DSS) of our university. In this system, we applied Rahim, Lee and Huang's confidence measure (RLH-CM) [2] to reject the false-alarmed keywords and happened to know that the CM statistics are severely fluctuated depending on each word. This severe fluctuation may result in constant rejection of some words by the given CM threshold. Thus, we deduced a conclusion that confidence measure should be normalized for each word with some tricks.

On the other hand, in our keyword spotting system, the most of false alarms occurred in non-keyword utterances. Thus another way to reject unwanted keywords is to check if a detected keyword belongs to non-keywords or not. That is, we can regard the detected keyword as a non-keyword utterance. Then we can consider an anti-non-keyword (anti-filler) based confidence measure. Using this concept, we developed a new confidence measure based on anti-filler model.

Starting from the concepts mentioned above, in this paper, we propose three confidence measures: normalized CM (NCM), anti-filler based CM (AFCM) and their hybrid CM (HCM). For evaluating our proposed three CMs, we apply them to our directory service system and compare their performances with the conventional RLH-CM. The simulation results show that the proposed NCM and HCM outperform the conventional RLH-CM and HCM is most superior to the others.

This paper is organized as follows. In section 2, we explain the system setup and the data used in our experiments. In section3, we describe the performance results of RLH-CM, the conventional CM applied to our system, and its shortcomings. In section 4, we describe three new confidence measures of normalized CM, anti-filler based CM and their hybrid CM, respectively. In section 5, we present the simulation results of those CMs and compare their performances with that of the conventional RLP-CM. Finally, we conclude by summarizing our major find-outs and the outline of our future work.

2 Experimental Setup and Baseline System

We have a keyword spotting system for directory service system. The keywords set is comprised of the names of Korean universities, their colleges and the belonging departments. To establish our keyword spotting system, we collected common sentences with the help of the telephone-call-center of our university and, finally, selected 2000 sentences among them for construction of speech database. The specifications of the speech database and the recognizer are as follows.

1. Speech database: *speaker-independent (100 talkers, ~1,000 utterances per speaker)*
2. Number of words in vocabulary: *~700 words (500 keywords and 200 non-keywords)*
3. Mono-phone model based filler model *(44 Korean mono-phones, 3 states, 9 mixtures)*
4. *~4,000 tri-phones form keyword model (3 states, 3 mixtures)*
5. Number of keywords per utterance: *1~3 keywords*
6. Speech features: *8kHz sampling, 12 mel-cepstrum, log energy and their delta parameters*
7. Used test set: *10 talkers, ~ 200 utterances per speaker*

Figure 1 shows brief block diagram of our keyword spotting system. As shown in the figure, our keyword spotting system uses the token-passing based Viterbi-beam searching algorithm. Also, the recognized hypotheses are evaluated in the post-processing module, which calculates confidence measure to reject false-alarmed keywords. As briefly summarized above, our keyword spotting system is based on filler model. And our system uses mono-phone based filler. By the way, there happen

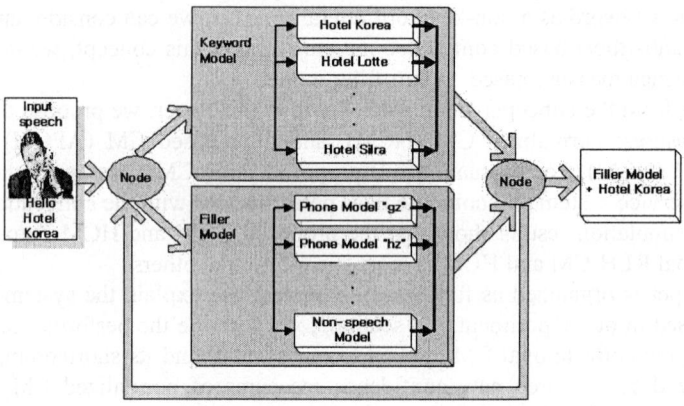

Figure 1. Keyword spotting system

to be many insertion errors of mono-phones when we permit free-transfer between mono-phones. Then, most keyword utterances are possibly replaced by series of mono-phones and this inevitably increases the possibility of misdetection rate. To solve this problem, we applied tri-gram of mono-phones to our system. Therefore, in filler model, the following restriction is used.

$$p(P_i | P_j, P_k) = \frac{N(P_i, P_j, P_k)}{N(P_j, P_k)}$$

where P_i is the i-th phoneme. To evaluate our system, we use two kinds of measures, detection ratio (DR) and false-alarm ratio (FR) defined as

$$DR = \frac{N_{CKw}}{N_{Kw}}$$

$$FR = \frac{N_{FA}}{N_{Kw} \cdot H}$$

where N_{Kw}, N_{CKw} and N_{FA} are the numbers of keywords in test DB, correctly recognized keywords and false alarmed keywords, respectively. And H is the total size of test DB in hour. In our baseline system without confidence measure, the performances are as follows.

$$DR = 93.7\% \text{ and } FR = 1.063 \text{ FA/Kw·Hr}$$

3 Conventional Confidence Measure and Its Problem

3.1 Conventional Confidence Measure and It Performance

Generally, confidence measure is used for rejecting false-alarmed keywords. In this paper, we adopted the commonly used RLH-CM [2] as a conventional CM. In our system, a keyword is a sequence of sub-word units (tri-phones) and its recognition is performed based on phoneme-units. Therefore, CM, confidence measure, is calculated as the following steps.

1) Calculate CMs of each phonemes of keyword with it's anti-phone.
2) Transform phoneme-based CMs to word CM.

In our system, we use all the mono-phones except those recognized phonemes as the anti-phone model. Assume that Θ_q is the recognized phoneme. Then, the anti-phone model Θ_p^a is represented as

$$\Theta_q^a = \bigcup_{\substack{i=1 \\ i \neq q}}^{44} \Theta_i,$$

where Θ_i is i-th phoneme, one of total 44 phonemes.

Then, CM of each phoneme is calculated by likelihood ratio of phoneme and its anti-phone model. Therefore, phoneme-level CM is given as

$$Lr_q(O_q;\Theta) = \frac{g_q(O_q) - G_q(O_q)}{|g_q(O_q)|} \quad (1)$$

where $g_q(O_q) = \log p(O_q | \Theta_q)$, $G_q(O_q) = \log p(O_q | \Theta_q^a)$ and O_q is an observation sequence corresponding to the phoneme Θ_q. Finally, word CM is calculated as

$$s_i(O;\Theta) = \log \left[\frac{1}{N(i)} \sum_{q=1}^{N(i)} \exp\{f \cdot Lr_{i(q)}(O_q;\Theta)\} \right]^{\frac{1}{f}} \quad (2)$$

where $N(i)$ is the number of phonemes of i-th hypothesis. f is a negative constant to make the calculated results positive.

Using this conventional confidence measure (RLH-CM), we can reduce FA ratio to 0.8 FA/ KW/HR with maintaining the detection ratio in our work. The ROC curve of this conventional confidence measure will be shown in Figure 4 when we compare the conventional confidence measure with our proposed confidence measures.

3.2 The Problems of Conventional Confidence Measure

The conventional confidence measure is calculated from phoneme-based CMs. Therefore, if one of the phonemes of keyword has low confidence, the word CM is also subject to be not high since it is largely influenced by the low phoneme-level CMs as shown in eq. (2). The phoneme with low CM means definitely that the corresponding phoneme is not confidential. By the way, there are two assumptions for calculation CM of eq. (2).

1) *Statistical consistency(A1)*: Every phoneme has the same statistics for its own CM. Practically, phoneme's CM is tri-phone's confidence measure.
2) *Labeling consistency(A2)*: Spoken utterance is well segmented into phoneme sequence. Although the recognized word is not correct, phoneme-level CM implies that phoneme labeling is not wrong.

However, our experimental results of confidence measure show that the statistics of each phoneme's CM is not consistent. This makes word CMs inconsistent with one another. Figure 2 shows the CM statistics for some words (For a convenient comparison-look, word index is restricted to 100). In this figure, the white box line represents the average CMs of each word and the black diamond line represents the standard deviations of word CM. From the figure, we can observe that the average CM has very different value depending on each word. Additionally, some keywords seem to be always rejected under the threshold of –0.01. This means that some words can hardly be recognized when FA ratio is kept being under reasonable value. This result reveals that the assumption A1, the statistical consistency, is not real in the practical system. On the other hand, the statistics of word CM (standard deviations) is a little stable compared with average statistics. However, the standard deviations of word CM are also different from one another. From this observation, we can conclude that, for the performance improvement of confidence measure, word-level CM should be normalized to have similar statistics.

Considering the assumption A2, as stated above, phoneme-level CM based on anti-phone model supposes that the phoneme segmentation by recognizer should be correct. This assumption is correct when the hypothesis is right since, when the hypothesis of keyword is correctly recognized, the phoneme labeling is acceptably good. However, in case of false alarm, the wrong phoneme segmentation results can be obtained. Figure 3 shows how the wrong labeling results of false alarmed utterance segment occur. In this figure, the false alarmed keyword is "yu-sz-uu-dq-ai" (yusoo-dae) while the correct content is "aa-gf-dd-xx"(aktteu) and this segment apparently belongs to garbage. From the figure, the wrong detection of keyword obviously makes bad phoneme segmentation. Thus the assumption A2 is inadequate in case of that false alarm occurs. Therefore, we need a new CM that is fit for the assumption 2 in case of false detection.

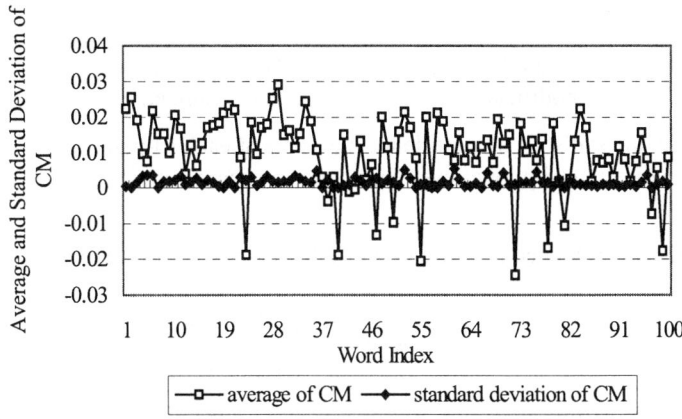

Figure 2. Statistics of word CMs

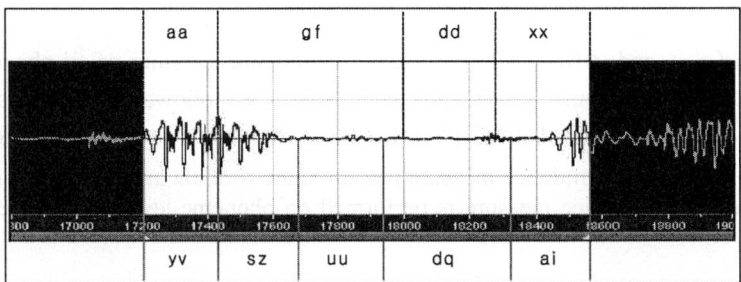

Figure 3. Example of false alarmed utterance segment

4 Proposed Confidence Measures

In this section, we describe the proposed three different confidence measures; the normalized CM, the anti-filler based CM and the hybrid CM. These CMs are proposed based on the observations explained in section 3. Therefore, we propose new CMs considering the assumptions of A1 and A2.

4.1 Normalized Confidence Measure

As explained in section 3, the statistics of word-level confidence measure are much different from one another depending on word. The reason is basically that the statistics of phoneme-level confidence measures are not consistent. Thus it is necessary to adjust phoneme-level CM for normalizing of word-level confidence measures. Of course, we may use different threshold for each word according to its CM statistics. However, this approach is inadequate to word-independent speech recognition system since it needs high cost to calculate the CM statistics and the threshold whenever the keyword list is changed.

In our keyword spotting system, the keyword HMM models are based on sub-word HMM (tri-phone) models. So, phoneme sequence is recognized with tri-phone units. Actually, the calculated CM is a confidence of the recognized tri-phone. Based on this fact, we propose a modification of the confidence measure eq. (2) as follows.

1) Obtain the statistics of confidence measures of each tri-phone
2) Normalize the tri-phone-level confidence measure with its average and standard deviation.

Therefore, the confidence measure of eq. (1-2) can be modified as

$$NLr_q(O_q;\Theta) = \frac{Lr_q(O_q;\Theta) - \mu(q_{tri})}{\sigma(q_{tri})} \quad (3)$$

$$s_i^{Nor}(O;\Theta) = \log\left[\frac{1}{N(i)}\sum_{q=1}^{N(i)}\exp\{f \cdot NLr_{i(q)}(O_q;\Theta)\}\right]^{\frac{1}{f}} \quad (4)$$

where $\mu(q_{tri})$ and $\sigma(q_{tri})$ are the average confidence measure of tri-phone composed by the phoneme q, its preceding and the following phonemes. And $NLr_q(O_q;\Theta)$ is the normalized phoneme-level confidence measure. According to eq. (4), word-level confidence measure has the same formula as in eq. (2). The normalization of confidence measure is performed on phoneme-level confidence measure.

4.2 Anti-filler Based Confidence Measure

Generally, false alarms occur in non-keyword segment without considering out-of-vocabulary (OOV). As shown figure 4, mis-detected segment has possibility to be segmented into wrong phoneme sequence. The reason is that non-keyword segment is detected as a keyword. How can we cope with this problem? One method is to assume that the detected hypothesis is not a keyword but non-keyword. With this assumption, we can verify if the recognized keyword belongs to non-keyword segment. To verify this proposition, it is necessary to obtain the correct phoneme sequence of wrongly detected segment. This can be achieved by re-searching the segment with filler-only-model. In this process, phoneme-level trigram is used to prevent insertion errors. After the proper phoneme sequence and boundaries are obtained, we suppose that this re-searched phoneme sequence is right. Of course, this assumption is not true when the hypothesis is a true keyword. Anyway, under the assumption that the detected hypothesis is not correct, we can regard the recognized keyword as anti-model of the correct filler segment. So, one approach to verify keyword segment is to check if this segment is filler segment. In this paper, we call this method anti-filler based confidence measure.

Now, we assume that we have correct phoneme labeling (phoneme sequence, boundary information and log probability of each phoneme) as

$$\{\Theta_q^C, O_q^C, G_q^C(O_q^C)\} \text{ and } q = 1,2\cdots, N_C$$

where Θ_q^C is the correct phoneme sequence, O_q^C is the observation sequence corresponding to Θ_q^C, $G_q^C(O_q^C)$ is the log probability of each phoneme segment and N_C is the number of correct phonemes. Then, the anti-filler based confidence measure (AFCM) is calculated as follows using eq. (5) and eq. (6).

1) Calculate log probability of each segment with detected hypothesis based on correct searched labeling information. That is, $g_q(U_q : V)$; optimum path and log probabilities of each node, where V is the Viterbi search results.
2) Apply confidence measure formula similar with eq. (1-2).

$$Lr_q^C(O_q^C;\Theta) = \frac{g_q^C(O_q^C) - G_q^C(O_q^C)}{|g_q^C(O_q^C)|} \quad (5)$$

$$s_i^{AF}(O;\Theta) = \log\left[\frac{1}{N_C(i)} \sum_{q=1}^{N_C(i)} \exp\{f \cdot Lr_{i(q)}^C(O_q^C;\Theta)\}\right]^{\frac{1}{f}} \quad (6)$$

We expect that AFCM works well for false alarmed segment of non-keyword part.

4.3 Hybrid Confidence Measure

In the above sections, we proposed the normalized and the anti-filler based confidence measures. We can easily guess that AFCM may not work well in the case of that the detected segment is a real keyword. In that case, the newly searched phoneme sequence may not be correct. Therefore, we can consider AFCM as an auxiliary confidence measure of other confidence measure. Thus, we propose another efficient confidence measure based on hybrid concept. The hybrid confidence measure (HCM) is defined as a linear combination of NCM and AFCM as

$$s_i^{Hybrid}(O;\Theta) = \alpha \cdot s_i^{Nor}(O;\Theta) + (1-\alpha) \cdot s_i^{AF}(O;\Theta) \quad (7)$$

where α is a weighting factor. By combining these two confidence measures, HCM has merits both in misrecognized case and false alarmed one.

5 Experimental Results and Discussion

To evaluate the proposed confidence measures, we use 10 hours utterances spoken by 10 talkers. Because our system is gender-dependent system, all the test talkers are males. Figure 5 shows the simulation results. As shown in the figure, we compared the original CM (RLH-CM), the normalized CM, the anti-filler based CM and the hybrid CM.

From the figure 4, we can observe several interesting results as follows.

1) Anti-filler based CM is not superior to the original RLH-CM. So, it is not desirable to use AFCM only.
2) Our normalization approach is reasonable. That is, CM normalization fairly enhances the performance of keyword spotting system.
3) Hybrid CM outperforms the rest of RLH-CM, AFCM and NCM. It seems that AFCM properly works in non-keyword segment.

Figure 5 shows the false alarm rates when detection rates are 7% and 10%, respectively. From the figure, it is observed that we can achieve 40% false alarm rate reduction at 7% detection ratio when using the HCM.

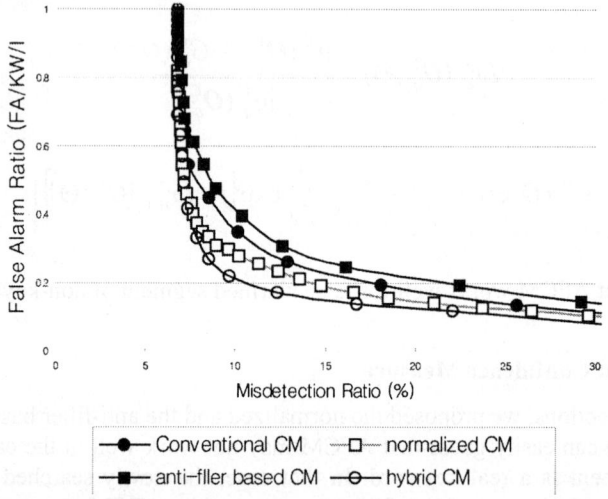

Figure 4. Comparisons of the conventional CM and the proposed CMs

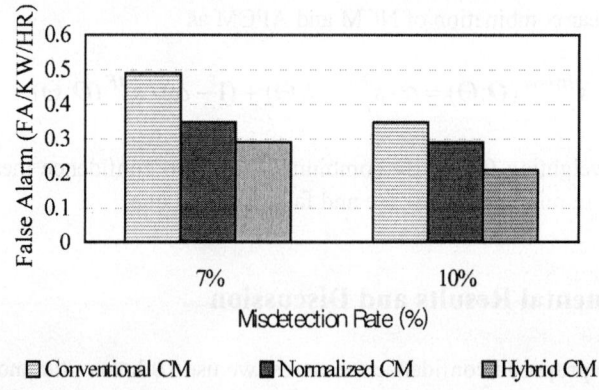

Figure 5. Comparison of the conventional RLH-CM, NCM and HCM

6 Concluding Remarks and Future Works

In this paper, we proposed the three confidence measures of NCM, AFCM and HCM and evaluated their performances compared with the conventional CM, RLH-CM, applied to the practical directory service system. Those confidence measures are devised under the facts that the assumptions of the conventional RLH-CM (statistical consistency and labeling consistency) are found not to be proper in our real system. To cope with the statistical inconsistency, we introduced a normalization approach into the conventional RLH-CM. Also we developed the anti-filler based confidence measure to overcome the labeling inconsistency problem. However, AFCM has defect in case of the correct recognition. Therefore, as an alternative confidence measure, we finally combined the two proposed NCM and AFCM to devise an efficient confidence measure, HCM. With keyword experiments, we evaluated our proposed confidence measures. HCM showed superior performance to the rest of the proposed NCM and AFCM as well as to the conventional RLH-CM. HCM achieved 40% of false alarm reduction compared with the conventional confidence measure.

However, AFCM and HCM can be applied only to domain specific keyword spotting system. This is because phoneme sequence of filler model can be searched correctly just in case of being with small number of garbage words. Thus AFCM and HCM is a kind of domain-specific confidence measure. However, since there are many applications of domain-specific keyword spotting systems, our proposed confidence measures are fairly valuable and important. On the other hand, since NCM is not domain-specific, NCM can be used for enhancing rejecting performances in every application areas. In our case, NCM shows about 30% false alarm reduction.

In the future, we will develop more robust confidence measure considering the word-level CM's inconsistency and labeling inconsistency. We are under developing a variable threshold method without CM statistics of word CMs, which is word dependent.

Reference

1. Li Jiang and Xuedong Huang, "Vocabulary-Independent Word Confidence Measure Using Subword Featuresm," Proceedings of ICSLP'98, pp.3245-3258, 1998.
2. Mazin G. Rahim, Chin-Hui Lee, Biing-Hwang Juang and Wo Chou, "Discriminative Utterance Verification Using Minimum String Verification Error (MSVE) Training," Proceedings of ICASSP'96, pp.3585-3588, 1996.
3. M. Rahim, C. H. Lee, and B. H. Huang, "Discriminative Utterance Verification for Cnnected Digits Recognition," Proceedings of EuroSpeech95, 1995.
4. Hoirin Kim, Sionghum Yi and Hangseop Lee, "Out-of-Vocabulary Rejection Using Phone Filler Model in Variable Vocabulary Word Recognition," Proceedings of ICSP Seoul, pp.337-339, 1999.
5. E. Tsiporkova, F. Vanpoucke and H. Van Hamme, "Evaluation of Various Confidence-based Strategies for Isolated Word Rejection," Proceedings of ICSLP 2000, pp.803-806, 2000.

Model-Based Debugging or
How to Diagnose Programs Automatically*

Franz Wotawa[1], Markus Stumptner[2], and Wolfgang Mayer[2],**

[1] Graz University of Technology, Institute for Software Technology
Inffeldgasse 16b/11, A-8010 Graz, Austria
wotawa@ist.tu-graz.ac.at

[2] University of South Australia, Advanced Computing Research Centre
5095 Mawson Lakes, Adelaide, SA, Australia
{mst,mayer}@cs.unisa.edu.au

Abstract. We describe the extension of the well-known model-based diagnosis approach to the location of errors in imperative programs (exhibited on a subset of the Java language). The source program is automatically converted to a logical representation (called model). Given this model and a particular test case or set of test cases, a program-independent search algorithm determines a the minimal sets of statements whose incorrectness can explain incorrect outcomes when the program is executed on the test cases, and which can then be indicated to the developer by the system. We analyze example cases and discuss empirical results from a Java debugger implementation incorporating our approach. The use of AI techniques is more flexible than traditional debugging techniques such as algorithmic debugging and program slicing.

1 Introduction

Debugging, i.e., the task of detecting, locating, and correcting faults in programs, is generally considered a very difficult and time consuming task that has, at the same time, remained crucial to software development [1]. Whereas most research deals with techniques for fault *detection*, e.g., test case generation and verification or *avoidance* (most software design methodologies fit in here), relatively little has been done in the domain of locating and correcting bugs. The main lines of research in bug location date back to the seminal papers on program slicing [2] and algorithmic software debugging [3]. In this paper we present a logic approach to this problem that is based on model-based diagnosis [4]. In this approach a logical model of the program together with the expected behavior of the program are used directly to locate and (sometimes) correct errors in the program. The logical model is automatically extracted from the program without further user interaction and without requiring a separate formal specification. The expected behavior is assumed to be given in the test cases, i.e., basically input-output vectors of used variables.

For demonstration purposes, in this paper we make use of a (non-object-oriented) subset of Java. However, it should be noted that the implementation used in the debugger

---
* Work was partially supported by Austrian Science Fund projects P12344-INF and N Z29-INF.
** Authors are listed in reverse alphabetical order.

```
program ::= id '{' stmnts '}'
stmnts ::= stmnt stmnts | ε
stmnt ::= assignment | conditional | while
assignment ::= id '=' expr ';'
conditional ::= if expr '{' stmnts '}' [ else '{' stmnts '}' ]
while ::= while expr '{' stmnts '}'
expr ::= '(' expr ')' | expr op expr | id | const
```

Fig. 1. Syntax of $\mathcal{L}$

actually covers the object-oriented parts of the language. The syntax of this subset is depicted in Fig. 1. Consider the following program:

```
1.  test {
2.      if (X=1) {
3.          Y = 1;
4.      } else {
5.          Y = 0;
6.  } }
```

and the test case $(X_2 = 0, Y_6 = 1)$ where $V_i = v$ mean that variable V has value v before executing the statement in line i. It is obvious that program *test* computes the wrong value for variable Y at line 6. This can be proven by sequentially executing statement by statement according to the control flow of the program. In this case variable Y is assigned the value 0 which contradicts the given test case.

In this situation the programmer would first search for the last statement defining variable Y – in this case, statement 5. Without further inspecting the code, it could be concluded that statement 5 should be changed to $Y = 1$. However, the addition of another test case, e.g., $(X_2 = 1, Y_6 = 0)$, may lead to a situation where changing statement 5 is not the best solution. When considering both tests the programmer would most likely then go back on the trace and finally visit statement 2. Changing the condition to $X = 0$ will lead to a program that passes both test cases.

The search for a possible bug location as described above is influenced by both the control flow of the program and the dependencies between variables that are given by the semantics of the language. Statements that are executed but do not lead to a wrong value of a different variable can be excluded from the list of bug candidates.

In our approach we consider dependencies, the control flow and the whole semantics of the program. We make use of *correctness assumptions* about the behavior of statements during the search for bug candidates. The use of assumptions is quite natural. Consider the above example where a programmer assumes that some statements are correct and others are not. During debugging the assumptions are changed and adapted until a consistent state is reached, i.e., the test cases do not contradict the assumptions. Hence, debugging can be seen as the problem of assigning correctness and incorrectness assumptions to all statements and checking the consistency of these assumptions and given test cases in the standard Model-Based Diagnosis [4,5] approach.

The paper formally characterizes the debugging process and introduces a specific logical model of the assignment language $\mathcal{L}$. While-statements are converted to nested if-statements while preserving result equivalence for the given test cases. This results in a simpler model that can be restricted to loop-free variants of programs, in contrast to our previous work [6,7,8], resulting in more effective pruning of the search space and better discrimination between diagnosis candidates. Our empirical results confirm that the approach reduces the search space for debugging and thus helps the user to focus his or her attention on relevant parts of the program.

2 Model-Based Diagnosis

In Model-Based Diagnosis, a logical sentence that represents the behavior of a system is used to determine a fault. The logical sentence is called a system description SD. In software debugging SD has to specify the behavior (or at least the interesting parts) of the program. A fault of the system, i.e., the bug in the program, is a set of system components, i.e., statements or expressions, that are responsible for a detected misbehavior, i.e., a difference between the actual and the specified behavior. Before formally defining diagnosis we first show how system descriptions are directly derived from programs.

To illustrate the resulting description, Fig. 2 shows the graphical representation of the model of program test. The structural part of the model is the system description SD which describes the connection between the components we want to diagnose, in this case Java statements, for the set of statements $STMNTS = \{S_1, S_2, S_3\}$. Each component has a set of input ports on which values produced earlier in the program execution are propagated, and an output port, on which values computed and possibly assigned during execution of the component are propagated further.

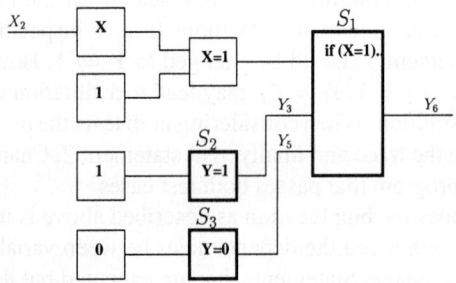

Fig. 2. Graphical representation of model test

This is complemented by the behavior model which captures the semantics of statements and expressions in a generic manner. The semantic descriptions of the various statement types are based on explicitly representing the abovementioned correctness or incorrectness assumptions in terms of predicates assigning appropriate modes to the statements. In the model described here, expressions are not assigned a mode since we only consider diagnosis at statement level. Assignment statements have the modes $Corr$ and $Incorr$ referring to the assumption of correct and incorrect behavior, respectively, with correct behavior being the default mode. Conditional statements have modes $Then$ and $Else$, corresponding to execution of the corresponding branch of the statement. For them there is no general default mode; the default depends on the branch chosen by a particular execution. The approach can be generalized to more than two modes per statement but this requires a change in the formalization [9] which is why we do not include it here.

The generic behavior model is given by the following logical sentences which must be added to the system description.

Expressions: There are 3 different types of expressions which must be taken into account for the logical model.

  Constants: The correct behavior of a constant is to evaluate to itself. In component-oriented terminology, the constant itself is propagated to the output port out. $const_k(C) \Rightarrow out(C) = k$

  Variable accesses: If a variable access is correct, the input port and the output port must have the same values. $vara(C) \Rightarrow out(C) = in(C)$

  Operators: For each operator op the following rule must be added to the system description. It says that the output value is determined by the associated to the operator, and by the input values. $fun_{op}(C) \Rightarrow out(C) = in_1(C) \; op \; in_2(C)$

  Note that for some operators not all input values need be specified in order to determine the output value. E.g., the output value of an and-operator is $\mathcal{F}$ if at least one input value is known to be $\mathcal{F}$.

Assignments: The behavior of assignments is to determine the new value of the target variable. The target variable in our model is represented by a connection with the output port. The new value itself is determined by the value of the evaluated expression. $assign(C) \Rightarrow (Corr(C) \rightarrow out(C) = in(C))$

Conditionals: A conditional either requires the then-block or the else-block to be executed depending on the value of the condition. We model this behavior by introducing two modes $Then$ and $Else$. If a conditional component C is in $Then$-mode, the value of each variable that is defined in one of the statement blocks is determined by the then-block and the condition must be true. Otherwise, C must be in $Else$ mode, the behavior is determined by the else-block, and the condition must be false.

$$if(C) \Rightarrow (default(Then(C)) \land Then(C) \rightarrow out(C,V) = in_T(C,V) \land cond(C) = \mathcal{T})$$
$$if(C) \Rightarrow (default(Else(C)) \land Else(C) \rightarrow out(C,V) = in_E(C,V) \land cond(C) = \mathcal{F})$$
$$if(C) \Rightarrow (\neg default(Then(C)) \land Then(C) \rightarrow out(C,V) = in_T(C,V))$$
$$if(C) \Rightarrow (\neg default(Else(C)) \land Else(C) \rightarrow out(C,V) = in_E(C,V))$$

The conditional statement's behavior depends on the current default mode. For example, assume that the else-branch is executed but it really should be the then-branch which is executed. In this case, the condition should evaluate to true and not to false. If the component is in mode $Then$ during diagnosis the value true would be propagated to the expression. Since an expression is not a component and cannot have an incorrect mode, this propagation may lead to unexpected results. The last assignment changing the value of a variable that is used in the expression also becomes part of the diagnosis which is not intended.

Note that there exist programs for which the default mode of a conditional cannot be determined using information obtained during the execution of a test case. Consider a conditional expression that is part of the then-branch of another conditional and suppose that the else-branch of the latter conditional is executed. In this case, the statements forming the then-branch are not executed and therefore no value is derived for the embedded conditional. As default mode for the statement part of such branches we choose the mode that is derived when the mode of the containing conditional is selected such that the previously ignored branch is executed.

The advantage of this conditional model is that the branch to be executed is determined by the mode of C and not by the connected expression components. Hence, in cases where no value of the conditional is derivable, the conditional component still has a well-defined behavior and, moreover, the output of the condition is given. This helps to restrict the number of diagnosis candidates.

In the previous section we stated that the diagnosis process is characterized by assumptions of the correctness or incorrectness of components. These assumptions are asserted and retracted during the process. The goal is to find a set of assumptions that is consistent with the given test cases:

Definition 1 (Mode assignment). *A mode assignment for statements* $\{C_1, \ldots, C_n\} \subseteq STMNTS$, *each having an assigned set of modes* ms *and a default mode* default *such that* default $\in ms(C_i)$ *for each* $C_i \in STMNTS$, *is a set of predicates* $\{m_1(C_1), \ldots, m_n(C_n)\}$ *where* $m_i \in ms(C_i)$ *and* $m_i \neq$ default(C_i).

E.g., $\{Then(S_1)\}$ and $\{Corr(S_2)\}$ are mode assignments for program test. In the default case, assignment statements are assumed to be correct. Conditional statements are assumed to be executed according to the actual test case. If the test case causes the then-block to be executed, the default behavior of the conditional is the $Then$ mode. Since we are only interested in deviations from the default behavior — which is the actual behavior of the program given a test case — mode assignments only collect the deviations, i.e., statement modes that are not equal to the default behavior.

Based on mode assignments we now define the concept of diagnosis with respect to a given test case.

Definition 2 (TC-Diagnosis). *A test case diagnosis (tc-diagnosis)* Δ *for a diagnosis problem* $(SD, STMNTS, TC)$ *is a mode assignment that is consistent with the given test case, i.e., the logical sentence* $SD \cup TC \cup \{m(C)|m(C) \in \Delta\} \cup \{d(C)|C \in STMNTS \wedge m(C) \notin \Delta \wedge d = default(C)\}$ *must be consistent.*

Consider the test case $\{X_2 = 0, Y_6 = 1\}$ of program test. The mode assignments $\{Then(S_1)\}$ and $\{\neg Corr(S_3)\}$ are both tc-diagnoses. For the test case $\{X_2 = 1, Y_6 = 0\}$ the tc-diagnoses are $\{Else(S_1)\}$ and $\{\neg Corr(S_2)\}$.

From tc-diagnoses we can generalize to diagnoses for a given set of test cases. However, since default modes of components and therefore tc-diagnoses depend on test cases, the definition of a diagnosis for all test cases must abstract from modes. For example, consider program test and the two test cases. There are two diagnoses that are associated with the conditional statement, $\{Then(S_1)\}$ and $\{Else(S_1)\}$, but $\{Then(S_1), Else(S_1)\}$ cannot be a diagnosis.

Definition 3 (Diagnosis). *A set* $\Delta \subseteq STMNTS$ *is a diagnosis for a diagnosis problem* $(SD, STMNTS, SPEC)$ *iff there exists a tc-diagnosis* Δ_{TC} *for at least one test case* $TC \in SPEC$ *were* $\Delta \cap \{C|m(C) \in TC\}$ *is not empty.*

We further say that a diagnosis (tc-diagnosis) is minimal if no proper subset of it is a diagnosis (tc-diagnosis). We compute diagnoses using Reiter's hitting-set algorithm[4]. Using this approach the notion of a conflict set characterizes a set of statements that cannot be all correct. In our case we relax this definition to statements that cannot be all in the default mode.

Definition 4 (TC-Conflict). *A set* $\{C_1, \ldots, C_n\} \subseteq STMNTS$ *is a tc-conflict iff* $SD \cup TC \cup \{d_i(C_1) | d_i = default(C_i), i = \{1, \ldots, n\}\}$ *is inconsistent.*

Diagnoses are the hitting sets of the set of conflicts. Therefore, the minimal tc-conflicts can be directly used to compute the tc-diagnoses of the same test case. We only have to add the non-default modes of the statements to the hitting sets. For example, the only tc-conflict of test and the test case $\{X_2 = 0, Y_6 = 1\}$ is $\{S_1, S_3\}$. The hitting sets are $\{S_1\}$ and $\{S_3\}$. If we consider the non-default modes of the statements we finally obtain $\{Then(S_1)\}$ and $\{\neg Corr(S_3)\}$ as tc-diagnoses. Moreover, we can compute all diagnoses directly from all tc-conflicts as stated in the following theorem.

Theorem 1. *Let T be the set of all tc-conflicts of all test cases $SPEC$. A diagnosis for $(SD, STMNTS, SPEC)$ is a hitting set of T.*

Proof. (Sketch): Every tc-diagnosis is a hitting set of that subset of T that corresponds to the same test case. Such a tc-diagnoses or a superset of it must be a tc-diagnosis of all other test cases. This follows from the definition of tc-diagnosis. If the tc-diagnosis directly is a diagnosis, then other tc-conflicts do no have an influence on this diagnosis and the diagnosis must clearly be a hitting set of T. If a superset is required to be a tc-diagnosis of all test cases, then there are other tc-conflicts that have an influence. In this case the superset must be also a diagnosis which would be detected if using T directly. Hence, all diagnoses are hitting sets of T.

The tc-conflict of test and the other test case is $\{S_1, S_2\}$. If we use both tc-conflicts to compute the hitting set, we receive $\{S_1\}$ and $\{S_2, S_3\}$ as diagnoses.

3 Loop-Free Programs

Programs with loop-statements are not easy to debug. A reason is the lack of a good model that allows to reason backwards in the program. Therefore some statements that cannot be responsible for the faulty behavior are still part of the set of computed diagnoses. This problem does not occur in programs that are comprised only of assignments and conditionals because there are good models available for both statement types. Since interesting programs contain loop-statements we have to show how to compile them to their loop-free variant without changing their semantics (as far as diagnosis is concerned).

One way of representing loops is to map them to an infinite nested sequence of conditionals. For practical purposes we can restrict the nesting depth to the maximum number of loop iterations to be considered. For our purpose, i.e., locating bugs in programs, this is always possible because we only have to consider the specified test cases. We can run the program on all test cases and measure the maximum number of iterations. The only requirement for the approach is that the resulting program must be (semantically) equivalent to the original program with respect to the given test cases. This type of equivalence is softer than the usual definition of program equivalence. In our case the variant need not be equivalent for all possible inputs.

Definition 5 (S-Equivalent). *Let V be a set of input environments. Two programs $\Pi_1, \Pi_2 \in \mathcal{L}$ are soft equivalent (s-equivalent) iff they halt on every input environment $I \in V$ with the same output environment, i.e., $\forall I \in V : eval(\Pi_1, I) = eval(\Pi_2, I)$.*

The s-equivalence of a program and its loop-free variant is an important requirement. Without s-equivalence the computed diagnoses of the variant are not guaranteed to be diagnoses of the original program. Because of our assumptions and the restricted semantics of our Java subset, the following corollary holds (given without proof).

Corollary 1. *The program Π is s-equivalent to its loop-free variant $\Gamma(\Pi)$.*

Beside s-equivalence the size of the variant and its expected runtime are important factors that influence the practicability of the approach. The size of the variant is given by the number of while-statements in the original program. In the worst case the size is determined by the maximum number of nested while-statements.

Note that in practical terms, this size increase is not a problem. The subroutines (Java methods) we are considering for debugging tend to be rather small and therefore the nesting depth of while-statements is generally quite low. Moreover, the size has (almost) no influence on the expected runtime because only those statements are executed that would be executed by the original program when using the same input environment. For example, if a loop is executed twice, then only two of the corresponding then-branches of the nested if-statements are executed. Hence, the runtime of both versions should be approximately the same.

4 Diagnosis Complexity, Remarks, and Results

Note that the approach is correct but not complete. All diagnosis results do explain the faulty behavior, but there are some cases where either the correct diagnosis is not part of the result or no diagnosis can be computed. The latter case is due to the limitations of the mapping Γ that restricts the number of possible iterations. If a multiple fault diagnosis requires more iterations than expected, it can be the case that no diagnosis is delivered back as result. Moreover, the used model is mainly for locating bugs that are specific to the wrong use of operators. A faulty variable access maybe found but this is not guaranteed. Handling this kind of error requires ways of changing the program's structure automatically during diagnosis.

Although our approach is not complete, the empirical results given in the next section why it is nonetheless useful.

4.1 Debugging Examples

This section illustrates the debugging process with the model based on a small example program shown below. The intended behavior of the program is to read a sequence of values from a file, until a value different from its predecessor is encountered. The reading is performed by an auxiliary function read_from_file, which is assumed to be a primitive of the language. Note that the program is not correct, as the condition of the loop doesn't test for equality, but is true for any value that is less than or equal to the previous value.

It is easily observed that the behavior of the program is incorrect for the test sequence $(3, 3, 2, 2, 5)$. As the model requires the program to be loop-free, the program must be expanded into a sequence of if statements as described in the previous sections. The number of iterations the loop executes is derived from the execution of the faulty program, given the test data. Using the sequence from above, the loop is executed (and is thus expanded) four times, resulting in the loop-free variant of the program that is shown in Fig. 3.

```
1  skip_equal {                      1   skip_equal_loopfree {
2     p = read_from_file ();         2      p = read_from_file ();
3     c = p;                         3      c = p;
4     while(c <= p) {                4      if (c <= p) {
5        p = c;                      5         p = c;
6        c = read_from_file ();      6         c = read_from_file ();
7  }}                                7         if (c <= p) {
                                     8            p = c;
                                     9            c = read_from_file ();
                                     10           if (c <= p) {
                                     11              p = c;
                                     12              c = read_from_file ();
                                     13              if (c <= p) {
                                     14                 p = c;
                                     15                 c = read_from_file (); }}}}}}
```

Fig. 3. Example Program

The loop-free program is transformed into a representation based on components and connections between them, according to the description in the previous sections. For this program, the model consists of 36 different components, each of them representing a statement or an expression in the program. As the diagnosis process is applied at the statement level, the components that correspond to expressions of the program are not considered diagnosis candidates, reducing the number of components that are subject to mode assignments to 14.

If the diagnosis engine is given the expected output of the program ($p = 3$ and $c = 2$) as observations, it derives 8 tc-diagnoses[1]:

$\{Else([if]_{10})\}$, $\{Ab([c =]_3), Else([if]_4)\}$, $\{Ab([c =]_6), Else([if]_7)\}$,
$\{Ab([c =]_9), Else([if]_{13})\}$, $\{Ab([p =]_{11}), Else([if]_{13})\}$, $\{Ab([p =]_{14}), Ab([c =]_{15})\}$,
$\{Ab([c =]_9), Ab([c =]_{12}) Ab([c =]_{15})\}$ and $\{Ab([p =]_{11}), Ab([c =]_{12}), Ab([c =]_{15})\}$.

Each of the tc-diagnoses represents a subset of the program's statements that can possibly be responsible for the deviations between the expected and the observed behavior of the program. The number of components assigned the *Then*- or *Else*-mode can be used as an estimate for the 'distance' between the behavior of a corrected version of the program and the behavior of the faulty program. Under the assumption that small deviations (with few statements involved) are more likely to occur, the tc-diagnosis $\{Else([if]_{10})\}$ can be considered a preferred diagnosis and presented to the user for inspection. In this example, the preferred diagnosis identifies the statement that contains the fault and, in addition, information is given about the iteration in which the fault occurs. The diagnosis indicates that the loop should exit after the second iteration (the mode of the if-statement representing the test after the second iteration is set to *Else*, which corresponds to the termination of the loop). Further inspection of the con-

---

[1] The components are represented by their corresponding statement, together with its line number.

ditional expression (using the observation that the conditional should evaluate to $\mathcal{F}$ and that the values of c and p are correct), the operator is identified as the single location of the fault. Based on the expected input- and output-values of the operator, the program can be corrected by computing a replacement for the faulty operator.

To evaluate the effectiveness of the model, first experiments with a set of example programs of varying size and structure have been performed, using a debugger prototype incorporating the diagnosis algorithm. The table below contains a brief description of the example programs, their models and the computed results. All programs implement basic algorithms (e.g. binary search, sorting, finding the maximum or sum of a sequence of numbers, computing power series and permutations, gauss-elimination, etc.) and contain a single faulty statement. The column 'Statements' contains the number of statements of the original program and the number of statements after expanding all loops. 'Components' contains the number of components the model of the program consists of and the number of components representing statements (i.e. those are considered when searching for diagnosis candidates). 'Diagnoses' lists the number of diagnoses with minimal cardinality[2] that are obtained when specifying the expected values at the end of the program. 'Hits' indicates the number of diagnoses that actually refer to a faulty statement and 'Code %' lists the fraction of the original program that has to be examined in the worst case until the fault is detected.

Name	Statements	Components	Diagnoses	Hits	Code %
sum	5 (11)	34 (11)	2	1	40
find_pair	5 (11)	34 (11)	5	2	80
skip_equal	5 (14)	36 (14)	1	1	20
bin_search	26 (59)	253 (63)	1	1	8
library	24 (39)	161 (56)	3	0	34
permutation	24 (32)	118 (32)	4	2	13
sum_powers	21 (61)	200 (61)	2	1	10
bubblesort	15 (51)	235 (54)	1	1	7
matrix	71 (191)	970 (199)	67	3	46

The results show that the model is able to locate faults in programs relatively precisely. In most cases, the number of statements to be examined is reduced significantly, even when the results are translated back to the original (unexpanded) programs. On the set of example programs, the fraction of a program that has to be examined to locate the fault is reduced to 29% on the average.

For some examples, considering only the diagnoses with minimal cardinality does not lead to the fault and diagnoses with larger cardinality have to be examined. In the library example, the inaccuracy can be explained by looking at the algorithm the program implements: the maximum value of a sequence is computed. Hence, changing the initialization of the algorithm (restricting the sequence to be considered to the element that contains the maximum value) corrects the observed behavior for the given test

---

[2] In most cases the diagnoses with minimal cardinality consist of one component, but for some examples (e.g. bin_search) the cardinality can be larger. Note that the set of diagnoses with minimal cardinality generally doesn't cover all minimal diagnoses, as some of them may be of larger cardinality.

case. The diagnoses representing these changes are smaller than the diagnoses that contain the true fault inside the loop. This is caused by the expansion of the loop, when the faulty statement is copied for each iteration. Therefore diagnoses with larger cardinality are required to locate the fault. For the example program it is sufficient to consider the 15 minimal diagnoses with cardinality 2.

Spurious diagnoses as in the library example can be eliminated by performing an iterative diagnosis process and querying the user about the expected values of variables during the execution of the program (the results were obtained by using only the output values of the program), or by using multiple test cases concurrently. A further, more powerful solution is to apply the diagnosis process on a meta-level. For example, the size of a fault indicated by a diagnosis is not measured on the loop-free program. Instead, the diagnoses can be rated based on their size after mapping them back to the unexpanded program. This approach enables the model to effectively locate faults that could not be detected when considering only diagnoses with minimal cardinality.

5 Related Research

Slicing [2] is a well known technique that is not only used for debugging but also for other applications, i.e., program analysis, software maintenance, testing, and compiler tuning. Many different slicing definitions and algorithms exist (see [10] for a survey). If we use slicing for debugging, then each slice corresponds to all statements that potentially determine the (wrong) variable value at a given point in the program. Therefore, we can view slices as conflicts. A conflict is a set of statements that if assumed to work correctly contradicts a given test case. This is exactly the case for a slice. But a slice is not required to be a minimal conflict. Moreover, there are conflicts with no corresponding slice. Consider for example the following program:

```
1.      test2 {
2.         R = D / 2;
3.         A = 3.14 * R * R;
4.         C = 3.14 * R };
```

and the test case $\{D_1 = 2, A_5 = 3.14, C_5 = 6.28\}$. A slice for test2 and variable C at location 5 is:

```
1.      test2 {
2.         R = D / 2;
4.         C = 3.14 * R }
```

With model-based diagnosis we obtain 2 conflicts. The conflict $\{St_2, St_4\}$ is equal to the slice. For the second conflict $\{St_3, St_4\}$ there is no corresponding slice. As a result, debugging using slices as conflicts may lead to the computation of too many single bug candidates which is not the case for model-based debugging. Hence, slicing is weaker for diagnosis than our approach. A similar result can be obtained for other dependency-based techniques, e.g., [11,12].

Another well-known technique for debugging is algorithmic or declarative software debugging [3,13]. Similar to the model-based approach, these also use the semantics of the language and a given test case, but do not clearly separate the knowledge about behavior from the knowledge of diagnosis. Moreover, Console et al. [14] has shown that model-based diagnosis techniques can outperform algorithmic debugging of logic programs due to the required number of user interactions before identifying a single bug.

6 Conclusion

In this paper, we have described an approach that uses Model-Based Diagnosis for the location of erroneous statements in imperative programs. We assume faults are detected due to discrepancies found in program execution on a set of test cases. The diagnosis system automatically transforms a program into a logical model by analyzing its source code. This model is then used together with a set of test cases showing desired input-output behavior to locate statements that are eligible to be the source for that fault. The language used is a subset of Java (use of a larger subset is described in [6]). Compared to earlier work, the model used is based on a transformation of the original program into a loop free form, which provides for effective search and diagnosis discrimination. Diagnosis candidates are sets of program statements which are then mapped back to locations in the source code for programmer interaction.

We have shown an empirical evaluation of different example programs using a debugger augmented with our diagnosis system, resulting in quick and direct focusing on the potentially faulty locations in the code.

Note that our approach, in contrast to verification techniques, aims at locating faults based on test cases instead of formally proving certain program properties.

Compared to traditional debugging (error location) techniques like Algorithmic Debugging and Program Slicing, our approach provides higher flexibility, the use of a generic and efficient problem solving algorithm, the ability to incorporate different models, and the ability (inherent in the problem solving algorithm) of diagnosing multiple faults.

References

1. Lieberman, H.: The debugging scandal and what to do about it. Communications of the ACM **40** (1997)
2. Weiser, M.: Programmers use slices when debugging. Communications of the ACM **25** (1982) 446–452
3. Shapiro, E.: Algorithmic Program Debugging. MIT Press, Cambridge, Massachusetts (1983)
4. Reiter, R.: A theory of diagnosis from first principles. Artificial Intelligence **32** (1987) 57–95
5. de Kleer, J., Williams, B. C.: Diagnosing multiple faults. Artificial Intelligence **32** (1987) 97–130
6. Mateis, C., Stumptner, M., Wotawa, F.: Modeling Java Programs for Diagnosis. In: Proceedings of the European Conference on Artificial Intelligence (ECAI), Berlin, Germany (2000)
7. Stumptner, M., Wotawa, F.: Debugging Functional Programs. In: Proceedings 16th International Joint Conf. on Artificial Intelligence, Stockholm, Sweden (1999) 1074–1079
8. Friedrich, G., Stumptner, M., Wotawa, F.: Model-based diagnosis of hardware designs. Artificial Intelligence **111** (1999) 3–39
9. de Kleer, J., Mackworth, A. K., Reiter, R.: Characterizing diagnosis and systems. Artificial Intelligence **56** (1992)

10. Tip, F.: A Survey of Program Slicing Techniques. Journal of Programming Languages **3** (1995) 121–189
11. Murray, W. R.: Automatic Program Debugging for Intelligent Tutoring Systems. Pitman Publishing (1988)
12. Kuper, R. I.: Dependency-directed localization of software bugs. Technical Report AI-TR 1053, MIT AI Lab (1989)
13. Lloyd, J. W.: Declarative Error Diagnosis. New Generation Computing **5** (1987) 133–154
14. Console, L., Friedrich, G., Dupré, D. T.: Model-based diagnosis meets error diagnosis in logic programs. In: Proceedings 13th International Joint Conf. on Artificial Intelligence, Chambery (1993) 1494–1499

On a Model-Based Design Verification for Combinatorial Boolean Networks

Satoshi Hiratsuka and Akira Fusaoka

Department of Computer Science, Ritsumeikan University
Nojihigashi, kusatsu-city, 525-8577, Japan

Abstract. In this paper, we propose a method to detect and correct design faults in a combinational boolean network, based on the model-based inference. We focus on the design verification for the network with multiple inverter errors. The complexity of this problem is NP-hard and it is harder than the usual verification to find a tractable algorithm. We present an effective algorithm which consists of the generation of the logical formula and its comparison to the specification for each cone in gate implementation. In this algorithm, the heuristic search method is incorporated to avoid the unnecessary backtracking based on the property that a part of the logical formula of each cone must be subformulas of functional specifications if the gate implementation is correct and irredundant.

1 Introduction

In this paper, we propose a model-based method for the design verification and diagnosis for a combinatorial boolean network. The combinatorial boolean network is an interconnection of logical gates (hereafter **CBN**) with the primary inputs and the primary outputs. The detection of design error and its automatic correction for **CBN** is well known as one of tremendously hard problems in the VLSI design due to its computational complexity. There have been the effective algorithm for this problem under the single fault assumption, namely there is at most only one design error in the circuit [1][4]. In this case, the problem is considerably simplified because the methods of classical diagnosis can be used, although it is still NP-complete. However, the single fault assumption is known to be practically not valid in the usual VLSI design. Therefore, the recent works focus on the effective diagnosis for multiple design faults of **CBN** [6][7][9]. These works use the methods to try to prove that the exclusive-or of the outputs of the gate implementation and the specification is always 0, and to diagnose the error points from the error vectors when it is disproved.

We present here another approach based on the model-based diagnosis to detect multiple design errors and correct them. The model-based diagnosis is a general method, which allows determination of the set of fault components from the set of descriptions for the normal behavior of components (SD) in the gate implementation and the observation of the incorrect behavior (OBS) [3]. We extend this method at two points in order to deal with the multiple design faults. The first is that we use the specification rather than OBS because the complete validation for the design is impossible if we use only the result of test vectors. Therefore, we incorporate the verification into the model-based diagnosis. Namely, we give a specification ($SPEC$) of **CBN** rather than OBS in addition to the gate implementation SD. The $SPEC$ is the designer's intention which is given by a logical formula of the relation between the primary inputs and outputs and it must be equivalent to the SD if the actual gate implementation is correct. By comparing the $SPEC$ with the SD, we can infer the position and type of the all design errors in the SD (verification and diagnosis) and correct them (automatic correction).

The second is that we introduce a fault model in addition to the normal behavior of the components in order to simplify the problem. The fault model is a set of descriptions for the incorrect behavior of the components [2]. It is known that there are two types of typical design error: the inverter error and line error (two types cover 98% of design errors)[1]. The inverter error means that the gate implementation contains extra inverters or missing inverters. The gate-type error such as the replacement of AND gate with OR gate is reduced to the inverter error. On the other hand, the line error means a fault of wiring. In this paper, we focus on the inverter error and present a method of verification and diagnosis working on the fault model of the inverter error. Therefore, we deal with a method to modify the design by adding or deleting the inverters so as to meet the specification. We call it the inverter problem. Since there are 2^n possible modifications for the number of gates n because we can add or delete the inverter for each line, the inverter problem is intuitively equivalent to repeat the verification (correctness check) 2^n times in the worst case. In order to reduce the search space, we introduce the heuristics called the sub-clause condition, which is based on the property that an output of any part of circuit must be contained in the specification if the part is correct.

The algorithm proposed in this paper has been implemented in **C** and currently evaluated for the simple **CBN**'s with dozens of gates.

2 Problem Description

2.1 System Description

CBN consists of interconnections of gates. We denote by x, y the set of primary input and primary output variables, respectively. To represent the interconnection of components, we also use the set of the intermediate variables u. We introduce abnormal literal for each possible error to localize malfunctioning component. The set of the abnormal literals is represented by Ab.

Description of gate implementation: The system description SD is a set of rules which describe the normal and illegal behavior of components. For example, the normal behavior of 2 fan-in NANDgate is described by the following set of clauses;

$$\{\{\bar{u}_1 \vee \bar{u}_2 \vee \bar{v}\}, \{u_1 \vee v\}, \{u_2 \vee v\}\}$$
where u_1, u_2 are input variables and v is an output variable of the gate.

On the other hand, we treat an inverter error as a rule set of irregular behaviors of the gate of which input line is incorrect. We assign the abnormality literals $A_1^i, A_2^i \cdots \in \boldsymbol{Ab}$ for each fan-in line of gate i. By using these literals, the fault model for NAND gate is given by;

$$A_1^i \supset \{\{u_1 \vee \bar{u}_2 \vee \bar{v}\}, \{\bar{u}_1 \vee v\}, \{u_2 \vee v\}\}$$
$$A_2^i \supset \{\{\bar{u}_1 \vee u_2 \vee \bar{v}\}, \{u_1 \vee v\}, \{\bar{u}_2 \vee v\}\}$$

The system description SD is a set of these gate descriptions for all components. We denote SD by $\Sigma(\boldsymbol{x}, \boldsymbol{y}, \boldsymbol{u}, \boldsymbol{Ab})$.

Specification: The Specification *Spec* consists of a set of clauses which represents the relation between the primary inputs and outputs. We represent *Spec* by
$$S(\boldsymbol{y}) \equiv \boldsymbol{y} \supset \Psi(\boldsymbol{x}), S(\bar{\boldsymbol{y}}) \equiv \bar{\boldsymbol{y}} \supset \neg\Psi(\boldsymbol{x})$$

2.2 Inverter Problem

Let assume that $SD : \Sigma$ and $Spec : \Psi$ are given. The verification and diagnosis of the **CBN** is formulated as the satisfiability check for the following formula:

$$(\exists \boldsymbol{Ab})(\forall \boldsymbol{x})(\exists \boldsymbol{y})[(\exists \boldsymbol{u})\Sigma \wedge (\boldsymbol{y} \equiv \Psi)]$$

Namely, the problem is reduced to decide the values of the all variables in $\boldsymbol{Ab}$ so as to satisfy the formula $\Sigma \wedge (\boldsymbol{y} \equiv \Psi)$ for all $\boldsymbol{x}$. If all elements of $\boldsymbol{Ab}$ are 0, the correctness is verified for the gate implementation. When some elements of $\boldsymbol{Ab}$ are 1, we can correct the design by adding or deleting the inverter for the corresponding line. The complexity of this procedure is NP-hard (it belongs to Σ_3^p). The naive search algorithm usually requires the exponential times for the number of gates so that some heuristic methods are necessary.

3 An Algorithm for CBN Verification

3.1 Outlines of the Algorithm

The part of the **CBN** which forms the value of the intermediate line i is called the $cone(i)$ for each $i \in \boldsymbol{u}$. Namely, the $cone(i)$ is the tree of gates with the root i. We denote the logical formula and its negation of the line i by $c(i)$ and $c(\bar{i})$, which are called the cone predicates. The path from i to the primary output y is denoted by $path(i, y)$. We define a function called $polarity(i, y)$ from the given gate implementation such that

$polarity(i, y) = 1$ if $path(i, y)$ contains zero or even number of inverters
$polarity(i, y) = -1$ if the odds number of inverters is contained in $path(i, y)$
$polarity(i, y) = 0$ if there are two or more paths from i to y and some of them have polarity 1 and others -1.

Note that $c(i)$ appears in $c(y)$ if $polarity(i,y) = 1$, and $c(\bar{i})$ appears in $c(y)$ if $polarity(i,y) = -1$ when the $cone(i)$ is a part of the $cone(y)$.

Roughly speaking, the core part of the proposed algorithm is constituted from the repetition of a cone processing, which consists of (1) forming the cone predicates $c(i)$ and $c(\bar{i})$, (2) simplification of cone predicates by a resolution based simplifier and (3) checking for the sub-clause condition for them. The sub-clause condition is such a condition that the current cone predicates are incorrect if the test for the condition is falsified. For the case of failure, the new $c(i)$ and $c(\bar{i})$ are formed by using another rule. When the available rules for the current gate are exhausted, the previous cone is re-examined by the backtracking mechanism. When the sub-clause condition is satisfied, the next cone is processed.

For the primary output y and the $cone(y)$, another condition is tested. Namely, the cone formula $c(y)$ and $c(\bar{y})$ are compared to the specification to check the equivalence: $c(y) \supset \Psi(y)$ and $c(\bar{y}) \supset \neg\Psi(y)$.

The outermost loop is a repetition of this core process for each primary output. If the process for the current primary output fails, the previous primary output is backtracked.

We give the general flow of the core process in Figure 1.

3.2 Sub-clause Condition

We assume that the designer intends to design the optimal circuit, so that the gate implementation is irredundant if it is correct. This means that one of the cone predicates $c(i)$ or $c(\bar{i})$ should be contained as a subformula in the specification $s(y)$, if the cone contains no error. Namely, if any clause of both $c(i)$ and $c(\bar{i})$ doesn't appear as a sub-clause of some clause in $s(y)$, the $cone(i)$ must contain errors. Moreover, if $c(i)$ exists in $s(y)$ and $polarity(i,y) = -1$, we can conclude that there must be odd times of inverter errors in $path(i,y)$. We call this property the sub-clause condition. By using this property, we can eliminate the unnecessary backtrack.

Let define a function $patherror(i,y)$ to be
$patherror(i,y) = 1$ if $path(i,y)$ contains zero or even number of the inverter errors.
$patherror(i,y) = -1$ if $path(i,y)$ contains odd number of the inverter errors.
$patherror(i,y) = 0$ if undefined.
The $patherror$ function is often useful to localize the design error. For example, assume that $patherror(i,y) = -1, patherror(j,y) = 1$ and $patherror(k,y) = 1$ for the 3 fan-in gate n in Figure 2. We can infer two cases :

(1) $A_1^n = 1, A_2^n = 0, A_3^n = 0$ and $patherror(h,y) = 1$
(2) $A_1^n = 0, A_2^n = 1, A_3^n = 1$ and $patherror(h,y) = -1$

However, the second case is very rare and can be neglected because the design errors are a few and usually scattered in the circuit. In the actual implementation, we assume that the design error is at most one for each gate. But we allow the case such that $patherror(i,y) = -1, patherror(j,y) = -1$ and $patherror(k,y) =$

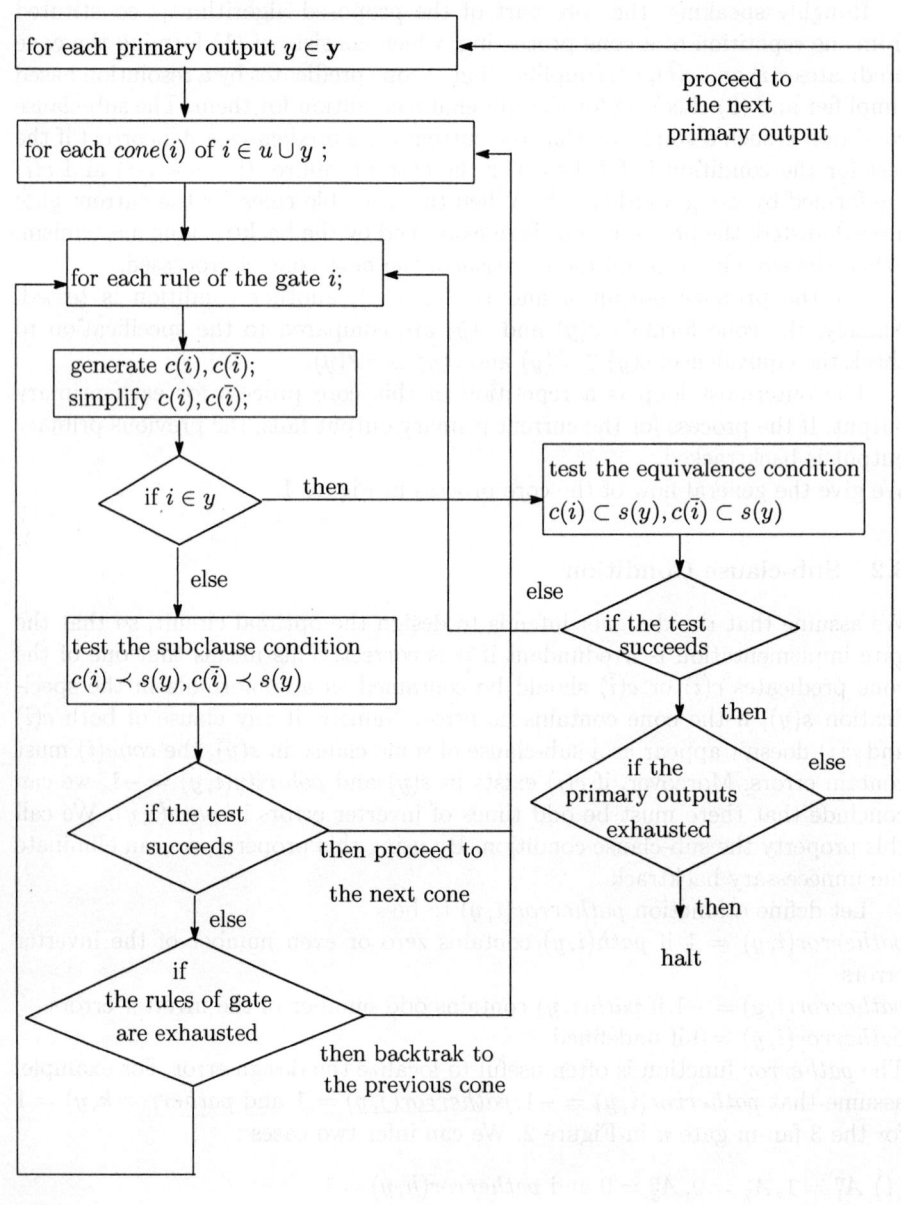

Fig. 1. Flowchart

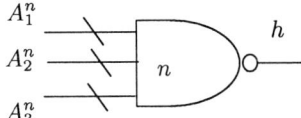

Fig. 2. 3 Fan-in Gate

-1, because it can be regarded as a gate type error. So that the AND gate in Figure 2 must be replaced with the OR gate. Namely, $A_1^n = 1, A_2^n = 1, A_3^n = 1$ and $patherror(h, y) = -1$.

Also we introduce a relation $X \prec Y$ for the sets of clauses $X = P_1 P_2 \cdots P_m$, $Y = Q_1 Q_2 \cdots Q_m$ by

$$X \prec Y \equiv \exists P_k \in X, Q_l \in Y[P_k \subseteq Q_l \vee Q_l \subseteq P_k]$$

Let assume that $\{a + b\} \in c(i)$ and $\{a + b + c\} \in s(y)$. Clearly, $c(i) \prec s(y)$. We can infer that the $cone(i)$ is possibly correct because the clause $\{a + b\}$ may be used to generate the clause $\{a + b + c\}$ by ORing c at the upper gate. Similarly, if $\{a + b + c + d\} \in cone(i)$, it may be used to generate $\{a + b + c\}$ by ANDing $\neg d$ at the upper gate. Therefore, the $cone(i)$ seems to be possibly correct at the current stage of processing. It is possible for $\{a + b + c + d\}$ to be ORed with $\{\neg d\}$ at the upper gate. In this case, the $\{a + b + c + d\} \in cone(i)$ disappears at the upper gate. Therefore, every clause of the correct $cone(i)$ does not satisfy the condition.

In the terms of these functions, the sub-clause condition is given by

For each $cone(i)$;
(1) if $\neg c(i) \prec s(y) \wedge \neg c(\bar{i}) \prec s(y)$ then $c(i)$ is incorrect.

(2) if $polarity(i, y) = 1$
if $c(i) \prec s(y)$ and $c(\bar{i}) \not\prec s(y)$ then $patherror(i, y) = 1$
if $c(i) \not\prec s(y)$ and $c(\bar{i}) \prec s(y)$ then $patherror(i, y) = -1$
if $c(i) \prec s(y)$ and $c(\bar{i}) \prec s(y)$ then $patherror(i, y) = 0$

(3) if $polarity(i, y) = -1$
if $c(i) \prec s(y)$ and $c(\bar{i}) \not\prec s(y)$ then $patherror(i, y) = -1$
if $c(i) \not\prec s(y)$ and $c(\bar{i}) \prec s(y)$ then $patherror(i, y) = 1$
if $c(i) \prec s(y)$ and $c(\bar{i}) \prec s(y)$ then $patherror(i, y) = 0$

(4) If $polarity(i, y) = 0$ then $patherror(i, y) = 0$ for the all cases.

3.3 Example

Assume that a designer who intends to design an exclusive-or function gives a schematic gate implementation of Figure 3. It contains two errors: the inverter on the line 3 is unnecessary and the inverter of the 2nd input of the gate 2 is missing.

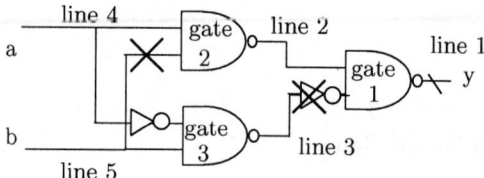

Fig. 3. Gate Implementation

Problem Description

The system description SD and the specification $Spec$ are:

Spec: $y = a\bar{b} + \bar{a}b$, $S(y) \equiv y \supset (a+b)(\bar{a}+\bar{b})$, $S(\bar{y}) \equiv \bar{y} \supset (\bar{a}+b)(a+\bar{b})$
SD:
Gate 1 $\bar{A}_1^1 \wedge \bar{A}_2^1 \supset \{1+2\}\{1+\bar{3}\}\{\bar{1}+\bar{2}+3\}$
$A_1^1 \supset \{1+\bar{2}\}\{1+\bar{3}\}\{\bar{1}+\bar{2}+\bar{3}\}$, $A_2^1 \supset \{1+\bar{2}\}\{1+\bar{3}\}\{\bar{1}+\bar{2}+\bar{3}\}$
Gate 2 $\bar{A}_1^2 \wedge \bar{A}_2^2 \supset \{2+4\}\{2+5\}\{\bar{2}+\bar{4}+\bar{5}\}$
$A_1^2 \supset \{2+\bar{4}\}\{2+5\}\{\bar{2}+4+\bar{5}\}$
$A_2^2 \supset \{2+4\}\{2+\bar{5}\}\{\bar{2}+\bar{4}+5\}$
Gate 3 $\bar{A}_1^3 \wedge \bar{A}_2^3 \supset \{3+\bar{4}\}\{3+5\}\{\bar{3}+4+\bar{5}\}$
$A_1^3 \supset \{3+4\}\{3+5\}\{\bar{3}+\bar{4}+\bar{5}\}$, $A_2^3 \supset \{3+\bar{4}\}\{3+\bar{5}\}\{\bar{3}+4+5\}$

Cone Processing

(1) We start with the $cone(3)$. Note that $polarity(3,1) = 1$.
By using the first rule for gate 3, the cone predicates are
$c(3) = \{\{a+\bar{b}\}\}, c(\bar{3}) = \{\{\bar{a}\},\{b\}\}$.
From $\{\{a+\bar{b}\}\} \not\prec s(y)$ and $\{\{\bar{a}\},\{b\}\} \prec s(y)$, we get $patherror(3,1) = -1$.

(2) For the $cone(2)$, $polarity(2,1) = -1$.
By the first rule of the gate 2, $c(2) = \{\{\bar{a}+\bar{b}\}\}, c(\bar{2}) = \{\{a\},\{b\}\}$.
From $\{\{\bar{a}+\bar{b}\}\} \prec s(y)$ and $\{\{a\},\{b\}\} \prec s(y)$, we get $patherror(2) = 0$.

(3) For the $cone(y)$, we use the third rule for the gate 1 because $patherror(3,1) = -1$. The cone predicates $c(y) = \{\{b\}\}, c(\bar{y}) = \{\{\bar{b}\}\}$. Since $c(y) \not\equiv s(y)$, we must evaluate the $cone(2)$ again (backtracking).

(4) We form the cone predicates $c(2)$ and $c(\bar{2})$ by using the second rule of the gate 2. Namely, $c(2) = \{\{a+\bar{b}\}\}, c(\bar{2}) = \{\{\bar{a}\},\{b\}\}$. Also $patherror(2,1) = 1$.

(5) For the $cone(y)$, we use the third rule for the gate 1 again. The cone predicates $c(y) = \{\{a\},\{\bar{b}\}\}, c(\bar{y}) = \{\{a+\bar{b}\}\}$. Since $c(y) \not\equiv s(y)$, we must evaluate the $cone(2)$ again (backtracking).

(6) For the $cone(2)$, the third rule of the gate 2 are used this time. So that $c(2) = \{\{\bar{a}+b\}\}, c(\bar{2}) = \{\{a\},\{\bar{b}\}\}$. Also $patherror(2,1) = 1$.

(7) For the $cone(y)$, we use the third rule for the gate 1 again. The cone predicates $c(y) = \{\{a+b\},\{\bar{a}+\bar{b}\}\}, c(\bar{y}) = \{\{a+\bar{b}\},\{\bar{a}+b\}\}$. Since $c(y) \equiv s(y)$, we can conclude that A_2^2 and A_2^1. Namely, the inverter must be inserted to the second input of the gate 2 and the inverter of the line 3 must be eliminated.

4 Implementation and Experiment

4.1 Implementation

In the actual implementation, we use a clausal form rather than BDD to deal with the cone predicate and the specification. We encode each input variable x_i by a binary number 2^i and represent a clause by using a couple of binary numbers called $base$ and $mask$. If $mask(i) = 0$ then the literal x_i does not appear in the clause. Otherwise, x_i appears in the clause if $base(i) = 1$ and $\overline{x_i}$ appears in the clause if $base(i) = 0$.

It is necessary for the sub-clause condition to work correctly that the specification $s(y)$ contains all prime implicates of y. We introduce a form of clause called the ordered clause instead of the set of prime implicates because the number of the prime implicates is usually very large. An ordered clause is such a clause that its $mask$ is $2^k - 1$ for some k. Namely, a clause must contain x_i or $\overline{x_i}$ if x_j or $\overline{x_j}$ appears in it for some $j > i$. This is the similar structure to BDD. For any clause c and a set of ordered clauses Γ, the implication $\models \Gamma \supset c$ can be checked within the linear time of the size of Γ. However, the number of ordered clauses to represent a logical formula also large, so that it is important to select the appropriate order of literals to minimize its size.

4.2 Experiment

The algorithm is currently evaluated for the simple **CBN**'s with dozens of gates. Because the design errors are actually very few, we assume that the number of design errors is at most 5 for each circuit which are selected at random. We evaluate the total numbers of clauses of cone predicates and the number of clause generation for simplification (both for the space consumption), and the elapsed time.

In Table 1 and Table 2, we present the results of experiments for the 4bit carry-lookahead adder **74Xseries 74182**. In this algorithm, the existence of more than two inverter faults on a single path increases number of backtracks and it leads to larger amount of calculation, in general.

Table 3 and Table 4 show the benchmark circuit C432 of ISCAS-85 and the result of its evaluation result. The c432 circuit is the 27(3×9)-channel interrupt controller. However, for simplicity, we reduced the channels to $\frac{1}{3}$ (3×3).

4.3 Discussions

1. We use a principle that a part of the cone predicate or its negation must appear in the specification as a subformula, if the design is correct. Namely, the cone predicate and the specification share some subformula. Roughly speaking, the size of the shared formula is considered to represent the certainty for the correctness of the cone. In this paper, we measure the size of shared subformula by the number of the clauses that have the sub-clause relationship with the specification and use it to manage the backtracking efficiently.

Table 1. 74Xseries 74182

gates	2-fanin AND: 8, 3-fanin AND: 3, 4-fanin AND: 3, 5-fanin AND: 2, 2-fanin XOR: 8, 2-fanin OR: 1, 3-fanin OR: 1, 4-fanin OR: 1, 5-fanin OR: 1, total: 25
lines	primary inputs: 7, primary outputs: 5, intermediate lines: 23, total: 35
gate rules	81 (388 clauses)
specification clauses	1784 (ordered clause)

Table 2. Results (CPU: Athlon 1GHz)

errors	number of clauses for cone predicates	number of backtracks	CPU time (sec.)
5	11131	88	0.256
4	50196	36	0.410
3	19728	59	0.251
2	2230	5	0.172
1	1821	1	0.147
0	1813	1	0.176

Table 3. C432 (simplified)

gates	2-fanin AND: 2, 2-fanin OR: 2, 2-fanin NAND:15, 2-fanin NOR: 6, 3-fanin NAND:6, 4-fanin NAND:3, 2-fanin XOR: 6, total: 40
lines	primary inputs: 12, primary outputs: 5, intermediate lines: 35, total: 52
gate rules	126 (444 clauses)
specification clauses	397 (prime implicate)

Table 4. Results (CPU: Pentium4 2GHz)

errors	number of clauses for cone predicates	number of backtracks	CPU time (sec.)	clauses used for simplification
5	935008	967	10.078	870996
5	10514745	12631	114.188	10286623
4	24975	3	0.172	18384
4	25808	45	0.203	24781
3	24972	3	0.188	18329
3	66181	42	0.656	55294
2	3860983	9204	21.937	5598009
2	13394	0	0.094	9924
1	13394	0	0.094	9924
1	291257	330	2.875	265371
0	13394	0	0.094	9924

2. Since we do not use any heuristics except the sub-clause condition, it may fall to the exhaustive search when the cone predicate satisfies the sub-clause condition in spite that it is wrong. We show some cases in Table 4 (the cases of huge CPU-time). To avoid unnecessary backtracks, we plan to introduce another heuristics using the sub-clause conditions of the cone predicates for the exclusive-OR of the specifications and the outputs, since if a cone predicate has not the shared subformula with the exclusive-OR, the cone is estimated to be correct probably.
3. We use the clausal form to represent the logical formula because it is convenient for the evaluation of the sub-clause relation. However, the clausal representations needs more memory space than BDD representation. Therefore, a linked memory structure for clausal form ,like BDD's one, should be introduced to handle larger circuits.
4. The method described in this paper aims at to detect the design faults of components, like inverter errors or gate-type errors, under the assumption that the wiring structure is correct. Therefore, it is difficult to diagnose the wiring faults, especially the missing wire faults, although the incorrectness of the design is proved . For these types of design errors, another search methods are required.

5 Concluding Remarks

In this paper, we propose a method of model-based diagnosis and verification for design errors of a combinatorial boolean network. Especially, we give a heuristic search method for the design errors to reduce the generation of the unnecessary clause. It works efficiently for the simple circuits in 74LS series. The benchmark test for circuits of the 1000 gates class is left for the future work.

References

1. Chung, P., Wang, Y., and Hajj, I. N. 1994. Logic Design Error Diagnosis and Correction, *IEEE Trans. on VLSI systems,*, 2.3:320-331.
2. Dressler, O., and Struss, P. 1996. The Consistency-based Approach to Automated Diagnosis of Devices, in *Principles of Knowledge Representation edited by G. Brewka* 267-311,*CSLI Publications*.
3. Reiter, R. 1987. A Theory of Diagnosis From First Principle, *Artif. Intell.*, 32:57-95.
4. Tomita, M., Jiang, H., Yamamoto,T., and Hayashi, Y. 1990. An Algorithm For Locating Logic Design Errors, *Proc. of IEEE/ACM int. conf. CAD* , 468-470.
5. Tison, P. 1967. Generation of Consensus theory and application to the minimization of Boolean function, *IEEE Trans. Electronic Computer* , 446-456.
6. Gupta, A., and Ashar, P. 2000. Fast Error Diagnosis for Combinational Verification, *13th International Conference on VLSI Design* , 442-448.
7. Gupta, A., and Ashar, P. 1998. Integrating a Boolean satisfiability checker and BDDs for combinational verification, *Proceedings of VLSI Design 98* , 222-225.
8. Wotawa, F. 2001. Using Multiple Models for Debugging VHDL Designs, *IEA/AIE 2001* , 125-134.
9. Jain, J., Mukherjee, B., and Fujita, M. 1995. Advanced Verification Techniques Based on Learning, *Proceedings of 32nd Design Automation Conference* 24.3:420-426.

Optimal Adaptive Pattern Matching

Nadia Nedjah and Luiza de Macedo Mourelle

Department of Systems Engineering and Computation, Faculty of Engineering
State University of Rio de Janeiro, Rio de Janeiro, Brazil
{nadia,ldmm}@eng.uerj.br
http://www.eng.uerj.br/~ldmm

Abstract. We propose a practical technique to compile pattern matching for prioritised overlapping patterns in equational languages into a minimal, deterministic, adaptive, matching automaton. Compared with left-to-right matching automata, adaptive ones have a smaller size and allow shorter matching time. They may improve termination properties as well. Here, space requirements are further reduced by using directed acyclic graphs (*dags*) automata that shares all the isomorphic subautomata. We design an efficient method to identify such subautomata and hence avoid duplicating their construction while generating the minimised dag automaton.

1 Introduction

Pattern matching of terms is performed according to a prescribed traversal order. The pattern matching order in the lazy reduction strategy may affect the size of the matching automaton, the matching time and in the worst case, the termination properties of term evaluations. Recall that the lazy strategy is the top-down left-to-right lazy strategy used in most lazy functional languages [1], [3], [4], [7], [13], [14]. It selects the leftmost-outermost redex but may force the reduction of a subterm if the root symbol of that subterm fails to match a function symbol in the patterns (for more details see [9], [5]). So, the order of such reductions coincides with that of pattern matching. Using left-to-right pattern matching, a subject term evaluation may fail to terminate only because of forcing reductions of subterms when it is unnecessary before declaring a match. For the left-to-right traversal order, such unnecessary reductions are required to ensure that no backtracking is needed when matching fails [2], [5], [6], [8].

With appropriate data structures, pattern matching can be performed in any given order. So, one way to improve efficiency i.e., matching times, which may improve termination too, consists of adapting the traversal order to suit the input patterns. We identify such an adaptive traversal order for a given pattern set and construct the corresponding automaton.

We first introduce from [12] a method that for any given traversal order constructs the corresponding adaptive matching automaton. *Indexes* are positions whose inspection is necessary to declare a match. Inspecting them first for patterns which are not essentially strongly sequential [9] allows us to engineer adaptive traversal orders that should improve space usage and matching times as shown in [12], [14]. When no index can be found for a pattern set, we showed [9] that a position, which is an index for a maximal number of high priority patterns can always be identified. Selecting such a position attempts to improve matching times for terms that match patterns of high priority. A *good* traversal order, i.e. one that improves space, time and termination properties [12], inspects positions that are indexes/partial indexes for a pattern set.

In this paper, space requirements of matching automata are further reduced by using a directed acyclic graph (*dag*) automaton that shares all the isomorphic subautomata, which are duplicated in the tree automaton. We design an efficient method to identify such subautomata and avoid duplicating their construction while generating the dag automaton. We generalise some results of [11] to be able to construct adaptive dag automata directly. Then, we compare left-to-right matching automata to obtained adaptive automata.

2 Preliminaries

In the rest of the paper, we will use some notation and concepts defined as follows: symbols in a *term* are either function or variable symbols; the non-empty set of function symbols that appear in the patterns $F = \{a, b, f, g, h, ...\}$ is *ranked* i.e., every function symbol f in F has an *arity* which is the number of its arguments and is denoted $\#f$; a term is either a constant, a variable or has the form $ft_1 t_2...t_{\#f}$ where each t_i, $1 \le i \le \#f$, is itself a term. Terms are represented by their corresponding abstract tree. We abbreviate terms by removing the usual parentheses and commas. This is unambiguous in our example since the function arities will be kept unchanged throughout, namely $\#f = 3$, $\#g = 1$, $\#h = 1$, $\#a = \#b = 0$. Variable occurrences are replaced by ω, a meta-symbol that is used since the actual symbols are irrelevant here. A term containing no variables is said to be a *ground* term. We generally assume that patterns are linear terms, i.e. each variable symbol can occur at most once in them. Pattern sets will be denoted by L and patterns by $\pi_1, \pi_2, ...$, or simply by π. A term t is said to be an *instance* of a (linear) pattern π if t can be obtained from π by replacing the variables of π by corresponding subterms of t. If term t is an instance of pattern π then we denote this by $t \triangleleft \pi$.

Here, we assume that we are free to choose any order for pattern matching terms and their evaluation proceeds using the adaptive strategy. In particular, if $f(t_1, ..., t_n)$ is the term being matched at the root, the argument t_n may be reduced before the argument $t_j, j \le i$ if the pattern-matcher visits the position at which t_n is rooted before that at which t_j is rooted.

Definition 1. A term t *matches* a pattern $\pi \in L$ if, and only if, t is an instance of π, i.e. $t \triangleleft \pi$ and t is not an instance of any other pattern in L, i.e. higher priority than π.

Definition 2. A *position* in a term is a path specification, which identifies a node in the parse tree of the term. Position is specified here using a list of positive integers.

The empty list Λ denotes the position of the root of the parse tree and the position $p.k$ ($k \geq 1$) denotes the root of the kth. argument of the function symbol at position p.

Definition 3. A *matching item* is a pattern in which all the symbols already matched are now *ticked* i.e., they have the *check-mark* ✓. Moreover, it contains the hollow *matching dot* • that only designates the *matching symbol* i.e., the symbol to be accepted next. The position of the matching symbol is called the *matching position*. A *final* matching item, namely one of the form π•, has the *final* matching position, which we write ∞. Final matching item may contain unchecked positions. These positions are irrelevant for announcing a match and so must be labelled with the symbol ω. Matching items are associated with a rule name.

The term obtained from a given item by replacing all the terms with an *unticked* root symbol by the placeholder _ is called the *context* of the items. For instance, the context of the item $f^\checkmark\, a\text{•}g\,\omega a\,a^\checkmark$ is the term $f(\_, \_, a)$ where the arities of f, g and a are as usual 3, 2 and 0. In fact, no symbol will be checked until all its parents are all checked. So, the positions of the placeholders in the context of an item are the positions of the subterms that have not been checked yet. The set of such positions for an item i is denoted by $up(i)$ (short for unchecked positions).

Definition 4. A *matching set* is a set of matching items that have the same context and a common matching position. The *initial* matching set contains items of the form •π because we recognise the root symbol (which occurs first) first whereas, *final* matching sets contain items of the form π•, where π is a pattern. For initial matching sets, no symbol is ticked. A final matching set must contain a final matching item i.e., in which all the unticked symbols are ωs. Furthermore, the rule associated with that item must be of highest priority amongst the items in the matching set.

Definition 5. Suppose $L \cup \{\pi\}$ is a prioritised pattern set. Then π is said to be *relevant for L* if there is a term that matches π in $L \cup \{\pi\}$. Otherwise, π is *irrelevant for L*. Similarly, an item π is *relevant for (the matching set) M* if there is a term that *deterministically* matches π in $M \cup \{\pi\}$.

Since the items in a matching set M have a common context, they all share a common list of unchecked positions and we can safely write $up(M)$. The only unchecked position for an initial matching set is clearly the empty position Λ.

3 Adaptive Tree Matching Automata

States of adaptive tree automata are labelled by matching sets. Since here the traversal order is not fixed a priori (i.e., it will be computed during the automaton construction procedure), the symbols in the patterns may be accepted in any order. When the adaptive order coincides with the left-to-right order, matching items and matching sets coincide with matching items and matching set respectively [9].

We describe adaptive automata by a 4-tuple $\langle S_0, S, Q, \delta \rangle$. S is the state set, $S_0 \in S$ is the initial state, $Q \subseteq S$ is the set of final states and δ is the state transition function. The states are labelled by matching sets, which consist of original patterns together with extra instances of the patterns which are added to avoid backtracking in reading the input. In particular, the matching set for S_0 contains the initial matching items formed from the original patterns and labelled by the rules associated with them. The transition function δ of an adaptive automaton is defined using three functions namely, *accept*, *choose*, and *close*: $\delta(M, s) = close(accept(M, s, choose(M, s)))$. The function *accept* and *close* are similar to those of the same name in [12], and *choose* picks the next matching position. For each of these functions, we give an informal description followed by a formal definition, except for *choose* for which we present an informal description. Its formal definition will be discussed in detail in the next section.

- *accept*: For a matching set and an unchecked position, this function accepts and ticks the symbol immediately after the matching dot in the items of the matching set and inserts the matching dot immediately before the symbol at the given unchecked position. Let t be a term in which some symbols are ticked. We denote by $t_{\bullet p}$ the matching item which is t with the matching dot inserted immediately before the symbol $t[p]$. The definition is $accept(M, s, p) = \{(\alpha s^{\checkmark} \beta)_{\bullet p} \mid \alpha \bullet s \beta \in M\}$.

- *choose:* This function selects a position among those that are unchecked for the matching set obtained after accepting the symbol given. For a matching set M and a symbol s, the *choose* function selects the position which should be inspected next. The function may also return the final position ∞. In general, the set of such positions is denoted by $up(M,s)$. It represents the unchecked positions of $\delta(M, s)$, and consists of $up(M)$ with the position of the symbol s removed. Moreover, if the arity of s is positive then there are $\#s$ additional unchecked terms in the items of the set $\delta(M, s)$, assuming that ωs are of arity 0. Therefore, the positions of these terms are added (see [12] for details).

- *close*: Given a matching set, this function computes its closure in the same way as for the left-to-right matching automaton. As it is shown in the following, the function adds an item $\alpha \bullet f \omega^{\#} \beta$ to the given matching set M whenever an item $\alpha \bullet \omega \beta$ is in M together with at least one item of the form $\alpha \bullet f \beta'$. The definition of function *close* allows us to avoid introducing irrelevant items [12].

For a matching set M and a symbol $s \in F \cup \{\omega\}$, the transition function for an adaptive automaton can now be formally defined by the composition of the three functions accept, choose and close is $\delta(M, s) = close\ (\ accept\ (M, s, choose\ (M, s))\ )$.

Example 1. Consider the set $L=\{1{:}hgga\omega\omega a,\ 2{:}hgfa\omega\omega aa,\ 3{:}h\omega b\}$. An adaptive automaton for L using matching sets is shown in Fig. 1. The choice of the positions to inspect will be explained later. Transitions corresponding to failures are omitted, and the ω-transition is only taken when there is no other available transition, which accepts the current symbol. Notice that for each of the matching set in the automaton, the items have a common set of symbols checked. Accepting the function symbol h from the initial matching set (i.e., that which labels state 0) yields the set $\{1{:}h^{\checkmark}\bullet gga\,\omega\omega a,\ 2{:}h^{\checkmark}\bullet gfa\,\omega\omega aa,\ 3{:}h^{\checkmark}\bullet \omega b,\ 3{:}h^{\checkmark}\bullet g\omega\omega b\}$ if position 1 is to be chosen next. Then, the closure function adds the item $h^{\checkmark}\bullet g\omega\omega b$. Provided with a matching set M, the *choose* function selects positions that are labelled with a function symbol in at least one item of M. If more than one position is available then the leftmost is selected. In short, the traversal order used avoids positions labelled with ω in every item of M.

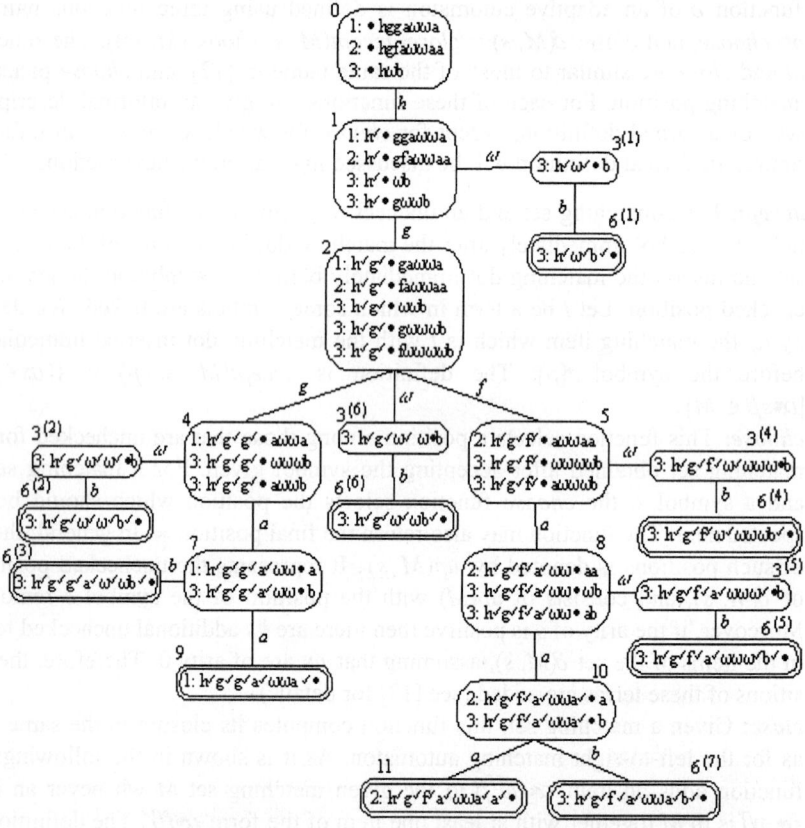

Fig. 1. An adaptive tree automaton for $\{1:hgga\omega\omega a, 2:hgfa\omega\omega aa, 3:h\omega b\}$

4 Optimal Adaptive Matching Automata

The tree automaton described above is time efficient during operation because it avoids symbol re-examination. However, it achieves this at the cost of increased space requirements. The unexpanded automaton corresponding to the pattern set of Fig. 1, and to which no patterns are added, is given in Fig. 2. In that automaton, states are also labelled with the matching position. For instance, state 0 investigates position Λ and states 4 and 5 both scans position 1.1.1. Furthermore, final states are also labelled with the number of matched rule. The non-deterministic automaton is much smaller. For instance, $hg\omega b$ is only recognised by backtracking from state 2 to state 1 and then taking the branch through state 3 instead. But in Fig. 1 a branch recognising $hg\omega\omega b$ has been added to avoid backtracking, thereby duplicating the existing sub-branch which recognises the b in hab. We can see similar duplication in several other branches of Fig. 1; those identified by sharing the same main state numbers.

By sharing duplicated branches, tree automata can be converted into an equivalent but smaller directed acyclic graph (*dag*) automata. States, which recognise the same inputs and assign the same rule numbers to them are functionally equivalent, and can be identified. For instance, the dag automaton corresponding to the automaton of Fig. 1 is given in Fig. 3. The state number is thereby reduced from 22 to 12.

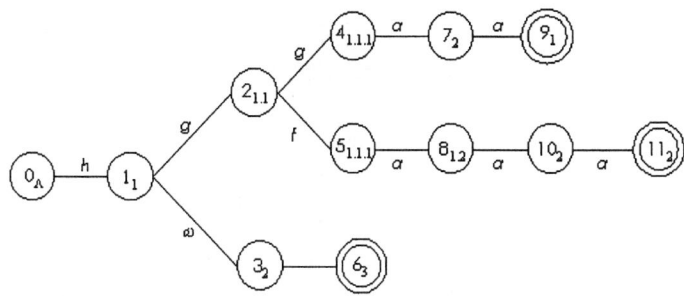

Fig. 2. Unexpanded adaptive tree automaton for {1:hggaωωa, 2:hgfaωωaa, 3:hωb}

In tree-based adaptive matching automata, functionally identical (or isomorphic) subautomata may be duplicated. Using directed acyclic graphs instead of trees can then improve the size of adaptive automata. In this section, we define the notion of matching item equivalence so that a .i.automaton:dag;-based adaptive automaton can be generated.

The illustrative example hides the complexity of recognising duplication where a number of suffixes are being recognised, not just one. The required dag automaton can be generated using finite state automaton minimisation techniques but this may require a lot of memory and time. The obvious alternative approach consists of using the matching sets to check new states for equality with existing ones while generating the automaton. In the case of equality, the new state is discarded and the existing one is shared. However, comparison of matching sets may be prohibitively expensive and it may well require bookkeeping for all previously generated matching sets. A major aim of this paper is to show how to avoid much of this work. First, we must characterise states that would generate isomorphic subautomata.

Definition 6. Let $i_1 = r_1:\alpha_1 \bullet \beta_1$ and $i_2 = r_2:\alpha_2 \bullet \beta_2$ be two matching items and p a position in $up(i_1) \cup up(i_2)$. i_1 and i_2 are *equivalent* if, and only if, $r_1 = r_2$ and the symbols labelling unchecked positions in i_1 and i_2 are as follows, otherwise, i_1 and i_2 are *inequivalent*.

$$\begin{cases} \alpha_1\beta_1[p] = \alpha_2\beta_{21}[p] & \text{if } p \in up(i_1) \cap up(i_2) \\ \alpha_1\beta_1[p] = \omega & \text{if } p \in up(i_1) \setminus up(i_2) \\ \alpha_2\beta_2[p] = \omega & \text{if } p \in up(i_2) \setminus up(i_1) \end{cases}$$

Definition 7. Two matching sets M_1 and M_2 are *equivalent* if, and only if, to every item i in $M_1 \cup M_2$ there correspond items $i_1 \in M_1$ and $i_2 \in M_2$ which are equivalent to i. Otherwise, the sets are *inequivalent*.

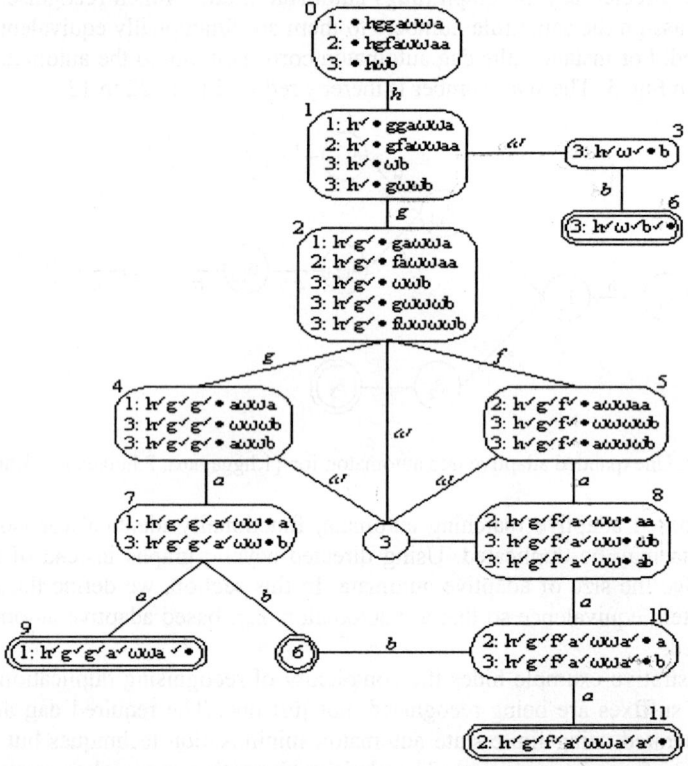

Fig. 3. An adaptive dag automaton for {1:hggaωωa, 2:hgfaωωaa, 3:hωb}

For instance, in the adaptive matching automaton of Fig. 1, the matching sets labelling the states $3^{(1)}$ and $3^{(2)}$ are equivalent. So, if two matching sets are equivalent, then their items differ only in those unchecked positions that occur in either of the matching sets. Moreover, for every item in which such positions occur, they are labelled with the symbol ω and so irrelevant for declaring a match. Therefore, equivalence as formalised in Definitions 6 and 7 is the right criterion for coalescing nodes of the adaptive tree automaton to obtain the equivalent adaptive dag automaton.

Lemma 1. Two matching sets generate identical automata if they are equivalent.

For equivalent sets, using the same traversal order strategy, function *choose* will certainly select the same positions. They must occur in the intersection of their unchecked position sets. So equivalent matching sets generate equivalent adaptive automata. We believe this equivalence is actually necessary as well as sufficient to combine corresponding states in the automaton. Equivalent matching sets may have different contexts, as can be seen in Fig. 1.

5 Adaptive Dag Automata Construction

In this section, we describe how to build the minimised dag automaton efficiently without constructing the tree automaton first. This requires the construction of a list of matching sets in a suitable order to ensure that every possible state is obtained, and a means of identifying potentially equivalent states.

The items in matching sets all share a common context. Hence, the matching position of any item is an invariant of the whole matching set. The states of the tree automaton can therefore be ordered using the left-to-right total ordering on the common matching positions of their matching sets. The dag automaton is constructed as indicated in the following algorithm.

```
algorithm DagAutomaton(pattern set L);
    A ← ∅; l ← {⟨M₀ ← (∀π∈L, •π), S₀⟩};current ← 0;
    do
        for each s∈F∪{•} do
            compute •(M, s);
            if ∄ ⟨M', S'⟩ ∈l | ••(M,s)≡M' then
                create state S' labelled with •(M,s);
                add transition S ——s——→ S' to A;
                add pair ⟨•(M,s), S'⟩ to l;
            else  add transition S ——s——→ S' to A;
        current ← current + 1;
    while current ≠ null;
end algorithm.
```

We iteratively construct the machine using a list l of matching sets in which the sets are ordered according to the matching position of the set. So the initial matching set is first and final matching sets come last. Each set in l is paired with its state in the automaton and a pointer is kept to the current position in l.

The list l represents the equivalence classes of matching states where the assigned matching position is that of the representative which is generated first. In each new set which is generated, the position of the current matching set is incremented at least one place to the right. So new members of l are always inserted to the right of the current position. This ensures that all necessary transitions will eventually be generated without moving the pointer backwards in l. It is easy to see from the definition of the *close* function that added patterns cannot contain positions that were not in one of the original patterns. So l only contains sets with matching positions from a finite collection and, as each set can only generate a finite number of next states all of which are to the right, the total length of l is bounded and the algorithm must terminate.

The list l represents the equivalence classes of matching states where the assigned matching position is that of the representative which is generated first. New members of l are always inserted to the right of the current position. This ensures that all necessary transitions will eventually be generated without moving the pointer backwards in l. It is easy to see from the definition of the *close* function that added patterns cannot contain positions that were not in one of the original patterns. So l only contains sets with matching positions from a finite collection and, as each set can only generate a finite number of next states, the total length of l is bounded and the algorithm must terminate. Furthermore, the tree and dag automata clearly accept the same language and the automaton is minimal, in the sense that, by construction, none of the matching sets labelling the states in the automaton are equivalent.

6 Automaton State Equivalence

In this section, we show how matching sets can frequently be discriminated easily so that the cost of checking for equivalence is minimised. Comparison of suffixes in the matching sets is completely avoided. We look at two properties to help achieve this. One is the set of rules represented by patterns in the matching set and the other is the matching position.

In Fig. 3, all inequivalent matching sets are distinguished by the use of the rule set. Thus the criterion is useful in practice. However, it will clearly not be sufficient in general. In the opposite direction, it is also sometimes easy to establish equivalence. Combining it with the matching position, we have the following very useful result, which enables the direct checking of equivalence to be avoided entirely in Example 1.

Theorem 1. Matching sets that share a common matching position and rule set are equivalent.

Proof: It suffices to show the kernels of the matching sets are equivalent, since then the function *close* will add equivalent items to both sets. Let M_1 and M_2 be two matching sets that share the same rule set and common matching position p. Let $i_1 = r{:}\alpha_1{\cdot}\beta_1$ and $i_2 = r{:}\alpha_2{\cdot}\beta_2$ be any two items associated with the same rule in their respectively kernels. The definitions of *accept* and *close* guarantee that the suffixes consist of a suffix of the original pattern π_r of the rule preceded by a number of copies of ω. To identify this suffix, let p' be the maximal prefix of the position p corresponding to a symbol in π_r. This is either the whole of p or is the position of a variable symbol ω, either ways, substitutions made by *close* for variables before p' in either i_1 or i_2 have already been fully passed in the prefix, and no substitution has yet been made for any variable further on in π_r. So if β is the suffix of π_r that starts at p' then the items i_1 and i_2 must have the form $\alpha_1{\cdot}\omega^{n_1}\beta$ and $\alpha_2{\cdot}\omega^{n_2}\beta$ for some $n_1, n_2 \geq 0$. It is clear that those items are equivalent in the sense of Definition 6. We conclude that M_1 and M_2 are equivalent. ∎

Although matching sets that share these two properties are equivalent, the matching positions of equivalent matching sets are not necessarily identical. An example can be found in [9].

7 Automaton Complexity

In this last main section, we evaluate the space complexity of the adaptive dag automaton by giving an upper bound for its size in terms of the number of patterns and symbols in the original pattern set. The bound established considerably improves left-to-right dag automata bound [5], [10]. The size of the left-to-right dag automaton for a pattern set L is bounded above by: $1+|L|+(2^{|\pi|}-1)(\Sigma_{\pi \in L}(|\pi|-1))$ (see [10] for details).

Theorem 2. The size of the adaptive dag automaton for a pattern set L is bounded above by $2^{|L|} * |F|$.

Proof: Consider the left-to-right dag automaton for pattern set L. Let M_1 and M_2 be two matching sets labelling two sets S_1 and S_2 in the dag automaton. Assume that M_1 and M_2 have the same rule set. Now, consider the contexts of their relevant items. Since symbols are scanned in the left-to-right order, one of the contexts c_1 must be an instance of the other context c_2. Hence, there exists a total order among theses prefixes. This argument can be generalised to all the states of the dag automaton that share the same rule set with M_1 and M_2. Consequently, the number of of such prefixes, and hence the number of states, is bounded above by the size of the largest context, which is in turn bounded above $|F|$. Furthermore, as there are $|L|$ distinct patterns, there are at most $2^{|F|}$ different rule sets. This yields the upper bound $2^{|L|}*|F|$ for the size of the adaptive dag automaton. ∎

Although the adaptive tree automata that inspect indexes first are smaller (or the same) than the tree left-to-right automaton for any given pattern set, it is not necessarily true that the equivalent adaptive dag automata are smaller than the left-to-right dag automaton. For instance, consider the pattern set L = {1:$fa\omega a\omega$, 2:$fa\omega b\omega$, 3:$fa\omega c\omega$, 4:$fa\omega d\omega$, 5:$fa\omega e\omega$, 6:$f\omega bbb$, 7:$f\omega a\omega \omega$} where f has an arity of 4 and a, b, c, d and e are constants. Assuming a textual priority rule and using the traversal order, which is described in Section 3, the adaptive dag automaton obtained for L (where most of inspected positions are indexes) is shown in Fig. 4. It has 17 states whereas the left-to-right dag automaton for L, given in Fig. 5, includes 15 states only.

The matching times, however, using the adaptive dag automaton remain better than that using the left-to-right dag automaton as sharing equivalent subautomata does not affect the matching times. In particular, with the adaptive dag automaton of Fig. 4, the matching times for terms that match either of the patterns of all the rules except 6 and 7 are optimal. However, using the left-to-right dag automaton of Fig. 5, every term needs at least three position inspections. Also, from the contrived nature of the example, this situation of the left-to-right dag automata having lesser states than adaptive ones seems to occur only for rare examples.

8 Conclusion

First, we described a practical method that compiles a set of prioritised overlapping patterns into an equivalent deterministic adaptive automaton. With ambiguous patterns a subject term may be an instance of more than one pattern. To select the pattern to use, a priority rule is usually engaged. The matching automaton can be used to drive the pattern matching process with any rewriting strategy [11].

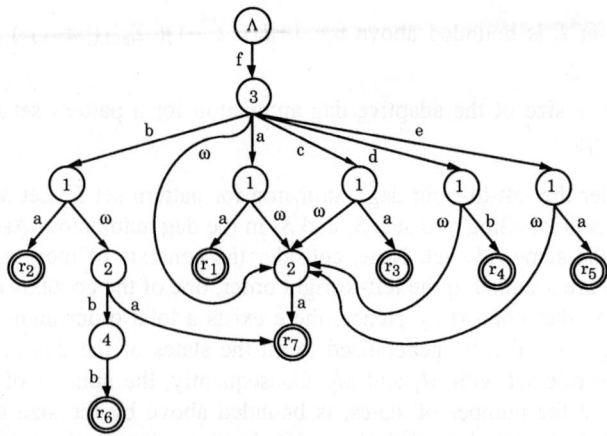

Fig. 4. An adaptive dag automaton for L

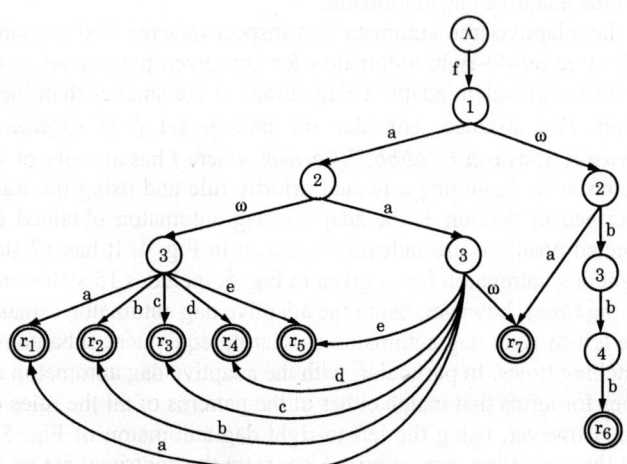

Fig. 5. The left-to-right dag automaton for L

In the main body of the paper, we described a method to generate an equivalent minimised dag adaptive matching automaton very efficiently without constructing the tree automaton first. We directly built the dag-based automaton by identifying the states of the tree-based automaton that would generate identical subautomata. By using the dag-based automata we can obtain adaptive pattern-matchers that avoid symbol re-examination without much increase in the space requirements. A theorem which guarantees equivalence in terms of several simple criteria was then applied to establish improved upper bounds on the size of the dag automaton in terms of just the number of patterns and symbols in the original pattern set.

References

1. A. Augustsson, *A Compiler for Lazy ML*, Proceedings ACM Conference on Lisp and Functional Programming, ACM, pp. 218-227, 1984.
2. J. Christian, *Flatterms, Discrimination Nets and Fast Term Rewriting*, Journal of Automated Reasoning, vol. 10, pp. 95-113, 1993.
3. N. Dershowitz and J. P. Jouannaud, *Rewrite Systems*, Handbook of Theoretical Computer Science, vol. 2, chap. 6, Elsevier Science Publishers, 1990.
4. A. J. Field and P. G. Harrison, *Functional Programming*, International Computer Science Series, 1988.
5. A. Gräf, *Left-to-Right Tree Pattern matching*, Proceedings Conference on Rewriting Techniques and Applications, Lecture Notes in Computer Science, vol. 488, pp. 323-334, 1991.
6. C. M. Hoffman and M. J. O'Donnell, *Pattern matching in Trees*, Journal of ACM, **29**(1):68-95, 1982.
7. P. Hudak and al., Report on the Programming Language Haskell: a Non-Strict, Purely Functional Language, Sigplan Notices, Section S, May 1992.
8. A. Laville, *Comparison of Priority Rules in Pattern Matching and Term Rewriting*, Journal of Symbolic Computation, 11:321-347, 1991.
9. N. Nedjah, *Pattern matching Automata for Efficient Evaluation in Equational Programming*, Ph.D. Thesis, University of Manchester-Institute of Science and Technology, Manchester, UK, (Abstract and Contents in the Bulletin of the European Association of Computer Science, vol. 60, November 1997.)
10. N. Nedjah, C. D. Walter and S. E. Eldridge, *Optimal Left-to-Right Pattern matching Automata*, Proceedings of the Sixth International Conference on Algebraic and Logic Programming, Southampton, UK, Lecture Notes in Computer Science, M. Hanus, J. Heering and K. Meinke Editors, Springer-Verlag, vol. 1298, pp. 273-285, 1997.
11. N. Nedjah, C. D. Walter and S. E. Eldridge, *Efficient Automata-Driven Pattern matching for Equational programs*, Software-Practice and Experience, **29**(9):793-813, John Wiley, 1999.
12. N. Nedjah and L. M. Mourelle, *Improving Time, Space and Termination in Term Rewriting-Based Programming*, Proc. International Conference on Industrial & Engineering Applications of Artificial Intelligence & Expert Systems, Budapest, Hungary, Lecture Notes in Computer Science, Springer-Verlag, vol. 2070, pp. 880-890, June 2001.
13. M. J. O'Donnell, Equational Logic as Programming Language, MIT Press, 1985.
14. R. C. Sekar, R. Ramesh and I. V. Ramakrishnan, *Adaptive Pattern matching*, SIAM Journal, **24**(6):1207-1234, 1995.

Analysis of Affective Characteristics and Evaluation of Harmonious Feeling of Image Based on 1/f Fluctuation Theory*

Mao Xia, Chen Bin, Zhu Gang, and Muta Itsuya

(P.O.Box 206, Beijing University of Aeronautics and Astronautics
Beijing 100083, China)
{moukyou,bin_chen_buaa}@263.net

Abstract. Affective information processing is an advanced research direction in the AI world. Affective Information of image is taken as the objective of research in this paper. The influence of color vision properties' histograms of image on human emotions is analyzed. Then based on $1/f$ fluctuation theory, a model of two-dimensional $1/f$ fluctuation is established, on which analysis is made on the fluctuation characteristics of image, resulting in a new algorithm proposed to objectively evaluate the harmonious feeling of image. After that, psychological testing method of SD is applied to verify the uniformity of the objective and subjective evaluations. At last, a conclusion is drawn that the image with $1/f$ fluctuation is harmonious and beautiful.

1 Introduction

In recent years, with the rapid development of signal processing technology, to process image and voice signal freely has been achieved. Beyond that, it is also expected to be able to process affective signal with computer. In early days, research on affections was restricted within psychological area. But now it has been connected with electronics through technologies of signal processing, image processing, computer graphics, audio signal processing, and artificial neural network, etc. Nowadays, many countries, such as the USA and Japan, have begun their research on affection science by utilizing these methods. [1-5]

This paper aims to analyze the affective characteristics of different images. Firstly, color vision histograms of image and its basic statistical properties are analyzed, and a two-dimensional fluctuation model is set up, on which a new method of analyzing fluctuation of images based on $1/f$ fluctuation theory is proposed [6]. Then, SD (Semantic Difference) method is adopted to evaluate images of different fluctuation features. As a result, the algorithm of analyzing fluctuation characteristics of images proposed in this paper is proven feasible, and the images with $1/f$ fluctuation characteristics can render to people a feeling of harmony and beauty.

---

* Supported by the National Nature Science Foundation of China (No.60072005).

2 HSV Histogram of Image and Basic Statistical Analysis of Image's Affective Characteristics

Psychological investigation is conducted to 50 images respectively on the three affective characteristics of monotonous, harmony and muss. From them, thirty images are selected and divided into three groups according to different affective features.

The color vision properties of image, namely H, S, V, have a direct influence on the harmonious effect of images. This paper takes the H, S, V histograms and their basic statistical properties as a means to judge the affective characteristics of images. Fig.1~3 displays a monotonous image, a mussy image and a harmonious image together with their respective histogram.

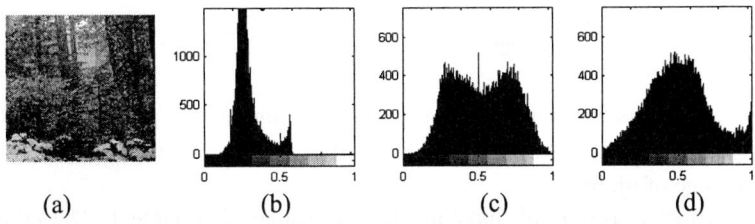

Fig. 1. Harmonious image and its histograms of color vision properties. (a) is original image; (b) is histogram of H; (c) is histogram of S; (d) is histogram of V

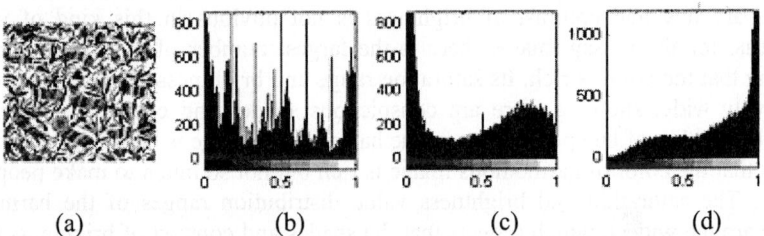

Fig. 2. Mussy image and its histograms of color vision properties. (a) is original image; (b) is histogram of H; (c) is histogram of S; (d) is histogram of V

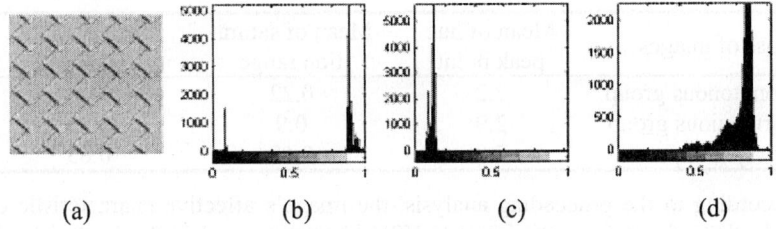

Fig. 3. Monotonous image and its histograms of color vision properties. (a) is original image; (b) is histogram of H; (c) is histogram of S; (d) is histogram of V

The hue histogram of Fig.1.(a) (harmonious image) demonstrates that there are two peak points of hue in this image, namely, this image is mainly composed of two kinds of colors. Meanwhile, the saturation histogram shows that the distribution range is wide, reflecting that the shades of this image are rich; and the value histogram shows that the image's brightness value is mainly distributed in the central area and transits smoothly from high brightness to low brightness. From Fig.2 (the mussy image), we can see there are many hue peak points that are distributed in the whole area while the saturation and brightness value fill up the whole area as well. Fig.3 (monotonous image), however, shows that there are few hues and its hue, saturation and brightness value are all distributed in a relatively narrow range. For each group, $\bar{H}$, mean value of hue peak point number, $\bar{S}$, mean value of saturation distribution range, $\bar{V}$, mean of brightness value distribution range are defined as the statistical variables of image's affective characteristics.

$$\bar{H} = \frac{1}{N}\sum_{i=1}^{N} H_i \quad \bar{S} = \frac{1}{N}\sum_{i=1}^{N} S_i \quad \bar{V} = \frac{1}{N}\sum_{i=1}^{N} V_i \quad . \tag{1}$$

Here, N=10, is the number of every group's samples. H_i, S_i, V_i are respectively the number of hue peak points, saturation distribution range and brightness value distribution range of ith image.

The statistical results of thirty images are displayed in table 1. The table 1 illustrates that, among the three kinds of images, the monotonous image displays the least number of hue peak points and its saturation distribution range and brightness value distribution range are the narrowest as well. This tells us that the color is not rich, shades are few and contrast of brightness is not obvious in this kind of images. Whereas, for the mussy image, there is the largest number of hue peak points, reflecting that the color is rich. Its saturation range and brightness value range are comparatively wide, showing there are conspicuous shades and contrast of brightness. That the number of hue peak points of the harmonious image is between the other two shows that the color of harmonious image is rich but not so much to make people feel mussy. The saturation and brightness value distribution ranges of the harmonious image are the widest, which reflects that the shades and contrast of brightness of harmonious images are the richest.

Table 1. the statistical results of image's HSV properties

Class of images	Mean of hue peak point	Mean of saturation range	Mean of value range
Monotonous group	2.2	0.22	0.4
Harmonious group	2.9	0.9	0.83
Mussy group	3.8	0.67	0.63

According to the proceeding analysis, the image's affective characteristic can be roughly figured out from the image's HSV histograms and its basic statistical variables. But these characteristics are all "static". In order to tell the dynamic properties of image's affective characteristics, the power density spectrum of image's HSV properties should be further observed.

3 Analyses of Image's Affective Characteristics Based on 1/f Fluctuation Theory

There is no absolutely static thing in nature. Everything must vary with temporal or special change. The random change of a physical variable can be called fluctuation. All fluctuations can be classified in accordance with the relation between its power spectra density and frequency. Reference [6] proposed that there exists three typical fluctuations in nature: One is completely irregular white fluctuation (white noise), its power spectra density is a constant in all frequency range, so it always is called $1/f^0$ fluctuation. This type of fluctuation made people upset. The second is strong correlative Brownian fluctuation (Brownian motion). Its power spectra density is inverse proportional to square of frequency f. So it can be called $1/f^2$ fluctuation. This type of fluctuation makes people feel monotonic. Between these two fluctuations is $1/f^1$ fluctuation whose power spectra density is proportional to $1/f^1$. The $1/f^1$ fluctuation can make people feel harmonious. This is because this kind of fluctuation is in accordance with the fluctuations of people's α brain wave and heart-throb when in pacific and comfort state. Based on this theory, the following will propose a 2-D $1/f$ fluctuation model and utilize it to analyze the harmonious affective properties of image.

3.1 The Definition of 2-D 1/f Fluctuation

According to the reference [6], the model of 1-D $1/f$ fluctuation is

$$S(f) = Kf^{-1} . \tag{2}$$

Here, S is the power spectral density; f is frequency.

When rotate the $1/f$ curve which is in X-Z plane around the Z axial, we can get the $1/f$ curved plane as the Fig.4 shows.

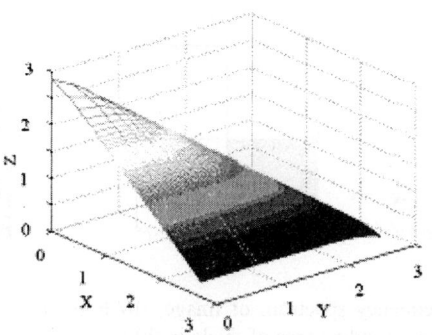

Fig.4. $1/f$ fluctuation curved plane

The model of $1/f$ curved plane is

$$S(r) = K \times r^{-1} \tag{3}$$

Here, $r = \sqrt{x^2 + y^2}$. Then in cylindrical coordinates, any plane of $\theta = \theta_0$ will intersect with this $1/f$ curved plane by a $1/f$ curve.

$$\begin{cases} S(r) = K \times r^{-1} \\ \theta = \theta_0 \end{cases} \quad (4)$$

Eq. (5) enables 2-D $1/f$ fluctuation and 1-D $1/f$ fluctuation to tally well. From this, the model of 2-D $1/f$ fluctuation can be defined as follows:

$$S(r) = K \times r^{-1} \quad (5)$$

3.2 Samples of the Power Spectrum Information of Image

First, 2-D Fourier transforms of H and S and V of image are made to obtain relevant information on frequency spectrum.

$x(m,n)$ is defined as the value of one color vision properties of image pels, and $X(k,l)$ denotes corresponding frequency spectrum. Then

$$X(k,l) = \sum_{m=0}^{N-1}\sum_{n=0}^{N-1} x(m,n) W_N^{km} W_N^{ln} \quad (6)$$

Here, $W_N^{km} = e^{-j2\pi\frac{km}{N}}, W_N^{ln} = e^{-j2\pi\frac{ln}{N}}, (k,l,m,n = 0,1,2,\cdots N-1)$, and its corresponding power spectral density is

$$S(m,n) = \frac{1}{N^2} | X(m,n) |^2 \quad (7)$$

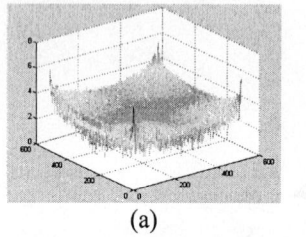
(a)

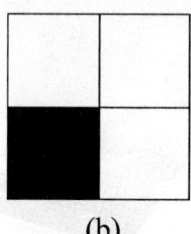
(b)

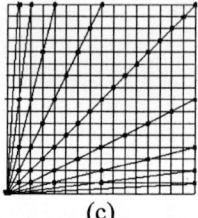
(c)

Fig. 5. (a) Amplitude frequency spectrum of image, (b) Four parts of amplitude frequency spectrum of image, (c) corresponding area of shade in (b)

The frequency spectrum of image is shown in Fig.5. In order to determine the fluctuation property of image, the frequency spectrum is sampled from different angles based on the definition of 2-D fluctuation and Eq. (5). In this way, the 2-D fluctuation is converted to several 1-D fluctuations.

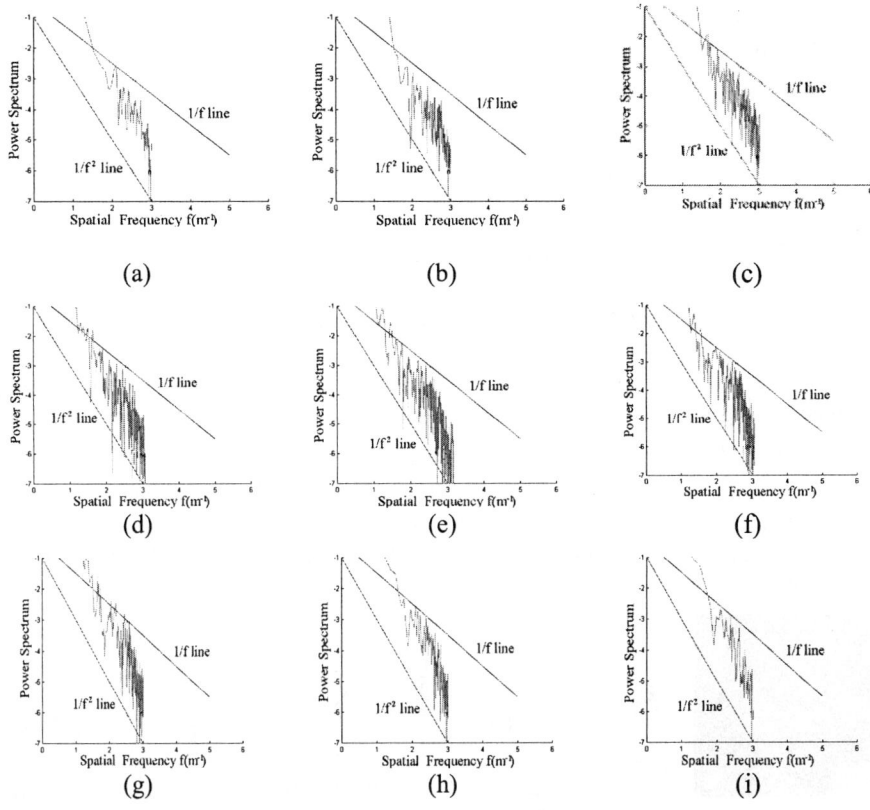

Fig. 6. (a)~(i) are nine samples of power spectrum of H

Because the amplitude frequency spectrum is axially symmetrical, data from 9 curves of a quarter area in Fig.6 are used to exemplify the sampling method. Here $X_{s,k}$ is the data of the kth curve, and X is the value of amplitude frequency spectrum of image. Then when $0 \leq n \leq m < N/2, n = 0,1,2,3\cdots, m = \frac{32}{2^k}n, k = 1,2,3,4,5$

$$X_{s,k}(n) = X(m,n) \tag{8}$$

When $0 \leq n \leq m < N/2, n = 0,1,2,3\cdots, m = \frac{2^k}{32}n, k = 6,7,8,9$

$$X_{s,k}(n) = X(n,m) \tag{9}$$

Using this method, the power spectral densities of the nine curves (of hue H) can be calculated.

The Fig.6 shows that all the nine curves of power spectral densities tend to be nearly the same as "$1/f$ line". Therefore, we can know that the image conforms to the 1/f fluctuation law. In the research, a great many lines are sampled. Their trends are all the same as these nine curves, thus enabling us to conclude that these nine curves can reflect the fluctuation law of the whole image.

3.3 Analysis of Affective Characteristics

Analysis of fluctuation properties of image can be carried out according to the following steps:

(1) Using Eq.(7) and Eq.(8) to calculate respective power spectral density of H,S,V of image;
(2) Using Eq.(9) and Eq.(10) to sample the power spectral density and obtain nine curves;
(3) Using the least mean-square method to fit each curve respectively to one line and then averaging the slopes of these lines. We can regard these averages as the fluctuation eigenvalue.

Fig.7 (a) is an image of natural scenery. The power spectral density and eigenvalue of this image are computed according to the above-mentioned steps. Fig.7 (b), (c), and (d) are the graph of the fifth curve in power spectrums of H, S and V of the image respectively.

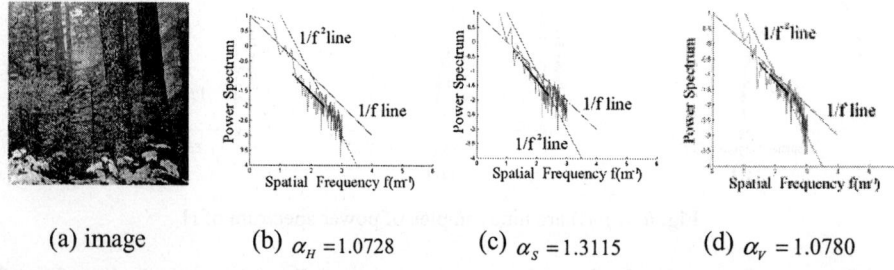

(a) image (b) $\alpha_H = 1.0728$ (c) $\alpha_S = 1.3115$ (d) $\alpha_V = 1.0780$

Fig. 7. Picture 1 $\alpha = 1.1541$. (a) is original image; (b), (c), and (d) are the graph of the fifth curve in power spectrums of H, S and V of the image respectively

From this graph we can see that all the sampled curves of H,S,V are close to the $1/f$ trend and the eigenvalue of this image $\alpha = 1.1541$. So we can say that Fig.7 (a) is an image nearly corresponding to 1/f law, which, as "1/f fluctuation theory" has pointed out, can give people a feeling of harmony and comfort.

The graphs of the fifth line in power density spectrum of H, S and V of the image Fig.8 (a) are shown in Fig.8 (b), (c) and (d). It can be seen that the trend of low frequency part of sampling data is near horizontal line. The eigenvalue of this image is $\alpha = 0.2432$, so this image gives people feeling of mussy.

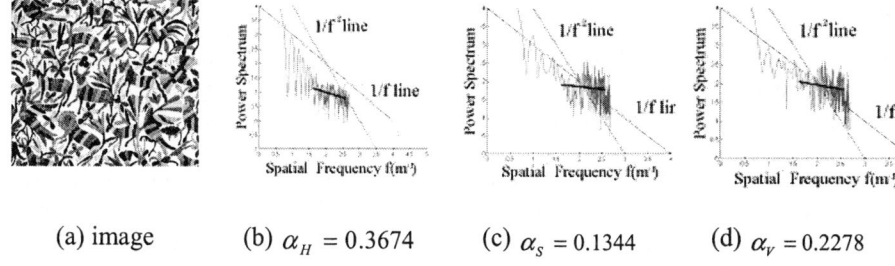

(a) image   (b) $\alpha_H = 0.3674$ (c) $\alpha_S = 0.1344$ (d) $\alpha_V = 0.2278$

Fig. 8. Picture 2 $\alpha = 0.2432$. (a) is original image; (b), (c), and (d) are the graph of the fifth curve in power spectrums of H, S and V of the image respectively

The graphs of the fifth line in power density spectrum of H, S and V of the image Fig.9 (a) are shown in Fig.9 (b), (c) and (d).

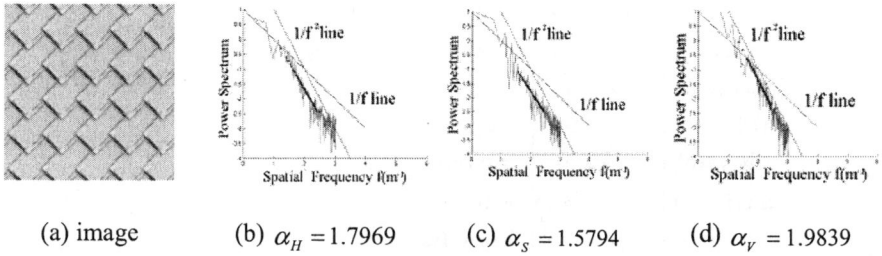

(a) image   (b) $\alpha_H = 1.7969$ (c) $\alpha_S = 1.5794$ (d) $\alpha_V = 1.9839$

Fig. 9. Picture3 $\alpha = 1.7867$. (a) is original image; (b), (c), and (d) are the fifth spectra curves of H, S and V respectively

It can be seen that the trend of low frequency part of sampling data is between the $1/f^2$ line and $1/f$ line. The eigenvalue of this image $\alpha = 1.7867$, so these images give people feeling of monotonous as what "$1/f$ fluctuation theory" has said.

4 The Evaluation on Harmony of Image

The proceeding analysis is an objective evaluation of images. However, before we can decide whether the objective evaluation really reflect the feelings of people, the subjective analysis must be carried out as well. Different people have different characters and live in different environments, which would result in the differences of evaluations on the same thing. Hence, the feeling of harmony is subjective. Even for the same person, his or her feeling will change in accordance with different emotions. It is very difficult to decide a universal criterion for feeling of harmony. Thus, it is quite necessary to seek a method of evaluation that can take into account differences of various people.

4.1 SD (Semantic Difference) Method[7]

We selected 42 male and female undergraduates between 20 and 26 years old to evaluate images when the temperature is about 20 degrees centigrade. The relative evaluation method is adopted to evaluate images that are close to $1/f^0$, $1/f$, and $1/f^2$ images.

The evaluation on one thing is made after considering various factors. We investigate these factors respectively with SD scale method. We choose the following seven factors that can be described with pairs of antonym: monotonous-mussy, noisy-quiet, depressive-joyful, worldliness-elegant, decadent-inspiring, artificial-natural, ugly-beautiful. Each factor is divided to 5 levels (-2, -1, 0, +1, +2). By means of investigated results of these factors from the 42 undergraduates, we can assess feelings of images. Fig.8 shows the results of the three images in Fig.1.

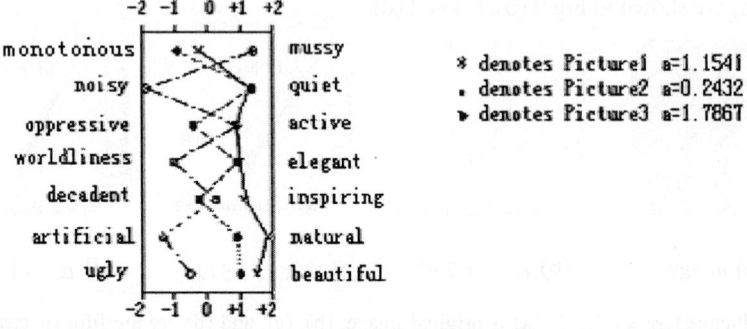

Fig. 10. The mean values of seven measured facts

In Fig.10, it can be seen obvious that Picture 2 close to $1/f^0$ fluctuation is mussy, noisy and unnatural, so that this picture will make people upset; Picture 3 close to $1/f^2$ fluctuation has properties of monotonous, so that it will give people feeling of being dull; Picture 1 between monotonous and mussy has the characters of being natural and beautiful, thus rendering to people a feeling of harmony and making people feel comfortable.

5 Conclusion

The proposed model of 2-D fluctuation in this paper offers an objective criterion for evaluating 2-D fluctuation properties. Moreover, the algorithm analyzing fluctuation properties of images testifies that the image conforming to $1/f^0$ fluctuation law will make people feel upset, the image to $1/f^2$ fluctuation law will make people feel dull and the image to $1/f$ fluctuation law will give people a feeling of harmony and make people feel comfort. Finally, the result of utilizing SD method to conduct psychological tests on different kinds of images proves the consistency between subjective evaluation and objective evaluation.

References

1. Shuji Hashimoto: Introduction to Kansei Information Processing [J]. Journal of Japan Society for Image information Media, 1998.52(1): 41-45.
2. Saburo Tsuji: Kansei Information Processing [J]. Journal of Japan Society for Robot, 1999, 17(7): 916-919. 9(6): 861-869.
3. Sloman Aaron: Review of Affective Computing [J]. AI Magazine, 1999,20(1), Spring.
4. Healey J. A.: Wearable and Automotive Systems for Affect Recognition from Physiology [D], Dept. of Electrical Engineering and Computer, May 2000 Science, MIT.
5. Mao Xia: Kansei Information Processing[J].Journal of Telemetry, Tracking, and Command, 2000,21[6]:58-62.
6. MushaT: $1/f$ Fluctuation [J], Journal of Japan Society for Physical Application, 1977, 46(12) : 1144 – 1155.
7. Yang Buomin: The Methods of Psychological Statistics [M], Guangming Daily Press, 1989.

Collective Intelligence and Priority Routing in Networks

Tony White, Bernard Pagurek, and Dwight Deugo

Carleton University, School of Computer Science
Ottawa, Canada K1S 5B6
(arpwhite,bernie,deugo@scs.carleton.ca)

Abstract: This paper describes how biologically-inspired agents can be used to solve complex routing problems incorporating prioritized information flow. These agents, inspired by the foraging behavior of ants, exhibit the desirable characteristics of simplicity of action and interaction. The collection of agents, or swarm system, deals only with local knowledge and exhibits a form of distributed control with agent communication effected through the environment. While ant-like agents have been applied to the routing problem, previous work has ignored the problems of agent adaptation, multi-path and priority-based routing. These are discussed here.

1 Introduction

Networks today have a wide range of applications running on them. Many would assume that making best use of network capacity implies load balancing; however, this is a simplistic assumption and ultimately it is user perception of the quality of service offered by the network that is important. Anyone who has used streaming audio or IP telephony services on the Internet will certainly appreciate this.

Circuit planning in a large network is a hard problem. An off-line or planning solution is possible. In this approach, the set of connections to be created is known in advance and routes for them computed to optimize a fitness function. Typically, the fitness function seeks to balance load across nodes and links in the network and may take account of constraints of the devices themselves and user routing preferences.

On-line approaches are also possible. In traditional networks, routing protocols are often used that attempt to maintain a global view of the network. Several agent-oriented approaches have recently been proposed that appeal to principles drawn from Swarm Intelligence [5], [2], [7] and others. In an on-line approach, agents compute routes for connections in order to optimize their connection routing cost, where cost may represent an aggregate statistic of delay, utilization, reliability and other factors. In these approaches, a global view of the network is not maintained, and we deal only with information that can be measured locally. Swarm approaches are robust with respect to the loss of individual routing agents. Beyond the routing domain, the appeal of swarms of biologically-inspired agents for industrial problem solving has recently been appreciated [4]. Research into the problems and potential of multiple, interacting swarms of agents is just beginning [8]. This paper builds on prior work by proposing routing solutions for creation of multi-cast routes, in an environment that supports traffic prioritization. It appeals to the SynthECA agent architecture recently proposed [8].

This paper consists of 3 further sections. The next section introduces elements of the SynthECA architecture pertinent to this paper. The following section describes the algorithms used to solve routing problems, and the results of applying them. The paper then summarizes its key messages.

2 SynthECA Agents

Agents in the SynthECA system can be described by the tuple, $A = (E,R,C,MDF,m)$. The important components pertinent to this paper are the ideas of a chemical (C) and a Migration Decision Function (MDF). A detailed description of the architecture can be found in [8].

The chemical concept is used in order to provide communication between agents and to create dissipative fields within the environment. The chemical concept is used to provide communication with, and sensing of, the environment and provides the driving force for agent mobility. A chemical consists of two components, an encoding and a concentration. When the encoding uses the alphabet $\{1, 0, \#\}$ in a string of length m, we say that we are using a Binary Array Chemistry of order m.

The MDF is a function or rule set that is used to determine where an agent should visit next. The MDF typically uses chemical and link cost information in order to determine the next hop in its journey through the network or may simply follow a hard-coded route through the network. An important consideration in designing an MDF is that it should take advantage of gradients in chemicals that are present in the network. In doing so, agents may take advantage of the actions of other agents. Particular agents may want to move up a gradient (attraction) or down a gradient (repulsion). *The MDF may take advantage of the pattern matching properties of the language used for the chemistry of the system.* Consider a chemical encoding consisting of 2 bits. We might include a term in the MDF consisting of the chemical 1#, where the # symbol matches either a 0 or a 1.

Consider a scenario where an agent has the choice of two links. Link 1 has concentrations Ch(10, 0.1) and Ch(11, 0.7). Link 2 has concentrations Ch(10, 0.5) and Ch(11, 0.6). Thus, an agent moving up a gradient indicated by the 1# pattern would follow link 2 because Ch(1#, 1.1) is sensed for that link. Similarly, an agent moving up the gradient indicated by the 11 pattern would follow link 1. This example is crucially important for the priority discussion later in the paper.

3 Swarm Routing

The swarm algorithm solution to this routing problem relies on the movements of artificial agents on the associated graph designed to make the global shortest path emerge. The communications network is represented in this paper as a weighted graph where the vertices correspond to switching nodes and the edges represent the physical links. When a connection request is made, a colony of agents is created and a Connection Creation Monitoring Agent (CCMA) is created on the source node. The functions of the CCMA are to decide when a path has emerged and when the current path is no longer the shortest path and that path re-planning should occur.

There are three classes of routing-related agent. *Explorer* agents search for a path from a source to a destination. *Allocator* agents allocate resources on the links used in a path. *Deallocator* agents deallocate resources on the links used in a path.

In establishing multiple point-to-point connections, the problem becomes more constrained since the connections consume bandwidth and that after a while, some links might run out of available bandwidth. The extension is quite straightforward, the graph edges have an associated available bandwidth and a condition is added for the agents to use a given edge: it should have enough bandwidth. Every time a path has emerged and a connection has been established the amount of available bandwidth is decreased on every edge of the path thereby adding additional bandwidth constraints to the graph. Three routing problems have been solved. These are described in the next 3 sections.

3.1 Point to Point Routing

For this case, the algorithm is quite straightforward. Explorer agents are created by the CCMA and leave the node in order to explore the network following their local rules.

The explorer agent has two modes of behaviour. If travelling towards its destination, it finds links at each node which the agent has not yet traversed, and which have enough bandwidth available for this connection. It selects a link from this set based on the probability function $p_{ijk}(t)$. Having selected a link, the selected link is added to the tabu list. The cost of the journey so far is updated and then the link to the next node is traversed. An agent whose path cost exceeds a given threshold dies.

At the destination, the explorer agent switches to trail-laying mode. When travelling back to the source node the agent pops the tabu list and moves over the link just popped dropping pheromone at a constant rate proportional to the cost of the route found.

The CCMA at the source node maintains a set of statistics relating to the set of routes that have been found so far in both point-to-point and point to multi-point connections. This is achieved by querying the returning agents (and agents which are sent from the destination node to this node) about the path that they took. This information is maintained by their tabu list. The node records the frequency of agents following a particular route over a given time period (a moving window). It also records details such as the total cost of the route. A good route is one for which a proportion of agents in the current time window exceeds a specified limit; e.g., 95%. When the limit is exceeded, the node sends out an allocator agent that creates the connection by allocating resources in the network.

Once an allocator agent is dispatched, if it does not succeed in establishing the connection because, for example, another connection used all the available bandwidth, it simply backtracks. In the meantime, explorer agents continue to explore the problem space. The CCMA at the source node may send out an allocator agent again when the path emergence criteria are satisfied, or may choose to delay sending the allocator out again until the problem space settles to a steady state. Chemicals laid down on a link evaporate over time. This is controlled by a constant evaporation rate, r.

3.2 Point to Multi-point Routing

Point to multi-point connections can be regarded as multiple point to point connections starting from the same source node. The only modification to the previous point to point algorithm concerns the allocator. Rather than sending a different allocator for each destination, identical allocators are sent from the source toward the destinations. Only the first allocator passing on the link will allocate the bandwidth, and fan out points are created on bifurcation nodes.

3.3 Cycle (or Multi-path) Routing

Cycle or multi-path routing can be regarded as two node and link disjoint paths (excluding source and destination nodes) that connect a source node to a destination node. The only modification to the original algorithm for the explorer agent is that upon reaching the destination it turns around and finds a path back to the source node that does not use any of the nodes or links used in the outward journey. Paths of this type are frequently constructed for the purpose of fault tolerance. SONET networks are constructed using cyclical paths.

3.4 The Route Allocation Algorithm

There are two phases to the movement of an explorer agent: an outward exploring mode and a backward trail-laying mode. The algorithm used by an explorer agent for a *single* connection is shown below.

```
do:
  Set t:= 0
  For every edge (i,j), set S_ij(t) := 0,
    cr_k := 0
  Place m agents on source node.
  Explorer agents created at frequency
    e_f
end
Set i := 1 {tabu list index}
For k := 1 to m do
  Place starting node, s, of the k^th agent
    in Tabu_k[i].
Repeat until destination reached:
  Set i := i + 1
  For k := 1 to m do
    Choose next node, p_ijk(t) {eqtn 1}
    Move the k^th agent to node j.
    cr_k = cr_k + C_ij(u)
    If cr_k > cr_max then
      kill the k^th explorer agent.
    Insert node j in Tabu_k[i].
    At destination go to {**}.
end

{**}
While i > 1
  Move to node Tabu_k[i].
  S_ij(t) := S_ij(t) + ph(cr_k)
  i := i - 1
end
{At the source node}
do:
  If the path in Tabu_k is the same as
    b% of
    paths in PathBuffer then create and
    send an allocator agent
  if t > T_max then
    create and send an allocator agent
    for shortest path found
end
```

In the algorithm above, the following symbols are used:

$S_{ij}(t)$ is the quantity of pheromone present on the link between the ith and j^{th} nodes,
$C_{ij}(u)$ is the cost associated with the link between the ith and j^{th} nodes at utilization, u.
cr_k is the cost of the route for the kth explorer agent.
$Tabu_k$ is the list of edges traversed.
T_{max} is the maximum time that is allowed for a path to emerge.
PathBuffer is the array of paths obtained by the (up to m) explorer agents.
cr_{max} is the maximum allowed cost of a route.
$ph(crk)$ is the quantity of pheromone laid by the kth explorer agent.
$p_{ijk}(t)$ is the probability that the kth agent will choose the edge from the ith to the jth node as its next hop given that it is currently located on the ith node.

More generally, for multiple simultaneous connection finding, with one chemical used for each connection to be computed, the probability, $p_{ijk}(t)$, with which the kth agent chooses to migrate from its current location, the ith node, to the jth node at some time, t, is given by:

$$p_{ijk}(t) = F_{ijk}(t) / N_{ik}(t), R < R^*$$
$$= H_{ij}(t) \text{ otherwise} \tag{1}$$

$$N_{ik}(t) = \Sigma_{l \text{ in } A(i)} F_{ilk}(t) \tag{2}$$

$$F_{ijk}(t) = \Pi_r [S_{ijr}(t)]^{\alpha}_{kr} [C_{ij}(u)]^{-\beta} \tag{3}$$

$$F_{ijk}(t) = \Pi_r [S_{ijr}(t)]^{\alpha}_{kr} [C_{ij}(u)]^{-\beta}, j = j^{max}$$
$$= 0 \text{ otherwise} \tag{4}$$

where:

α_{kr}, β are control parameters for the kth agent and r^{th} chemicals for which the kth agent has receptors, $\alpha_{kr} = 0$ if the agent does not have a receptor for the rth chemical,
$N_{ik}(t)$ is a normalization term,
A(i) is the set of available outgoing links for node i,
$C_{ij}(u)$ is the cost of the link between nodes i and j at a link utilization of u,
$S_{ijr}(t)$ is the concentration at time t of the rth chemical on the link between nodes i and j,
R is a random number drawn from a uniform distribution (0,1],
R* is a number in the range (0,1],
$H_{ij}(t)$ is a function that returns 1 for a single value of j, j*, and 0 for all others at some time t, where j* is sampled randomly from a uniform distribution drawn from A(i),
$F_{ijk}(t)$ is the migration function for the kth agent at time t at node i for migration to node j,
j^{max} is the link with the highest value of the product: $\Pi_r [S_{ijr}(t)]^{\alpha}_{kr} [C_{ij}(u)]^{-\beta}$.

3.5 The Agent Creation Algorithm

Due to the pheromone sensitivity of the agents, it is possible for premature convergence of the route-finding algorithm to occur. Sometimes, the agents can be attracted by that trail in such a way that they will not explore other links any more. To avoid being locked into a local optimum, the pheromone sensitivity has to be modified. With lower pheromone sensitivity, the agents are more likely to explore other links. The problem is when to lower the sensitivity and what new pheromone sensitivity to give to the agents.

Each explorer agent encodes its α and β sensitivity values that are used in the calculation of $p_{ijk}(t)$. Initially, these values are selected randomly from a given range. Hence, the population of m agents initially sent out into the network has a range of sensitivity values. When these agents return to the source node, having found a route to a given destination, the route cost, cr_k, is used to update the fitness value, $f(\alpha,\beta,k)$, associated with the (α,β) pair. The equation used to update $f(\alpha,\beta,k)$ is given by:

$$f_{new}(\alpha,\beta,k) := f_{old}(\alpha,\beta,k) + \gamma(crk - f_{old}(\alpha,\beta,k)), \ 0 < \gamma < 1.$$

An agent returning with a lower route cost than the current $f(\alpha,\beta,k)$ will cause $f(\alpha,\beta,k)$ to decrease. However, several agents must return with the same crk value before $f(\alpha,\beta,k)$ approaches cr_k. A discrete space for (α,β) was chosen in order to ensure that updates to $f(\alpha,\beta,k)$ would occur. A discounted feedback mechanism as shown above is required in this system because of the stochastic nature of the search. The same (α,β) encoding may result in several different crk values and $f(\alpha,\beta,k)$ represents the average over all possible routes found in the network. Clearly, with γ set to zero it is possible to ignore previous searches with a given encoding.

As noted earlier, the source node retains a path buffer that contains m paths. The source node also retains m (α,β) pairs and their associated fitness values. When new agents need to be created and sent out to explore the network, the fitness values are used to create new (α,β) pairs. First, parent (α,β) encodings are selected based upon their $f(\alpha,\beta,k)$ values. The lower the value of $f(\alpha,\beta,k)$, the more likely the (α,β) encoding is to be chosen.

The way we have achieved this is with a Genetic Algorithm-like (GA) process. As stated above, each agent has its own cost and pheromone sensitivity. At the very beginning, all the agents have random sets of parameters that are defined within a given range.

When an agent returns, its set of parameters is stored along with the cost of the route found. Its parameters are linked to the cost of the path found. This cost has the same role as the fitness function of a GA. When creating a new agent, the sets of all of the last returning agents are considered. An intermediate population of parameters is created: each set has a probability of being chosen proportional to its fitness. Some random parameters are automatically added to the population. Given that the encoding is a bit string, the spectrum of parameter values is discrete, a property which is essential for the use of the updating equation for $f(\alpha,\beta,k)$. The negative values allow agents to flee the main trail and therefore to explore new links. If these values are useful they will be stored for future 'breeding', otherwise they will be forgotten. Then the genetic operators such as mutation and crossover are carried out. Finally, agents with the corresponding set of parameters are created and sent out to explore the graph.

This approach differs from a conventional GA in that, in this algorithm, we are trying to *avoid* the convergence of the population because it tends to lead to local optima. This is perhaps closer to work on co-evolving populations because the environment of the agents (the network and its representation, the graph) is modified by their actions. There is considerable inter-play between the pheromone laying activities of one agent with the cost of a path found by another agent and, therefore, the fitness associated with the (α,β) encoding of pheromone and cost sensitivity values.

3.6   Experimental Setup

Two graphs were used during experimental investigation of the adaptive system for the point to point, point to multi-point and multi-path problems. These are shown in Figure 1 and Figure 2. The numbers associated with the edges in these networks represent the costs of the edges at zero edge utilization. Each edge is considered to have a capacity of 63 units. For problem one, the point to point path finding scenario, ten randomly generated traffic profiles were created for all source-destination pairs with bandwidth requirements sampled uniformly from the set {0, 2, 4, 6, 8, 10} bandwidth units. A bandwidth requirement of zero units was taken to mean that no path need be calculated for the source-destination pair. Paths were calculated such that the utilization of the network increasing by the bandwidth requirements of the traffic as paths emerged. All paths were computed in parallel. Initial network edge utilizations of 0, 30, and 50% were considered in order to test the effects of four different cost functions. For problem three, the same randomly generated traffic profiles were used for experimentation. For problem two, ten randomly generated traffic profiles were created with 2, 3 or 4 destinations. Bandwidth requirements for the point to multi-point requests were identical to problem one.

A population size of 50 was used with path emergence considered to have occurred when 90% of the population follows a given path. A maximum of 100 cycles of the path finding algorithm was allowed before path calculations were stopped and 20 agents per cycle were sent out into the network for path finding. The value of α was allowed to vary in the range -0.25 to 3 and the value of β was allowed to vary in the range -0.125 to 1.5. A total of 16 bits was allowed for the encoding of α and also for β. When adaptive search was contrasted with its non-adaptive counterpart, with constant α and β, values of 2 and 1 respectively were used. These constant values were found to be a reasonable compromise for path finding. A value of 10 was chosen for the constant of proportionality for the quantity of chemical to be laid. An indirect representation was used with mapping of bit strings into floating point values in the above ranges in such a way as to cover the ranges uniformly. Values of 0.8 and 0.01 were used for the probabilities of crossover and mutation respectively. Single point crossover was used as the crossover operator.

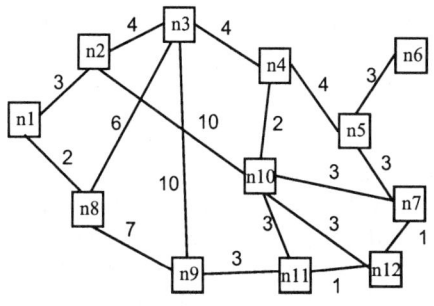
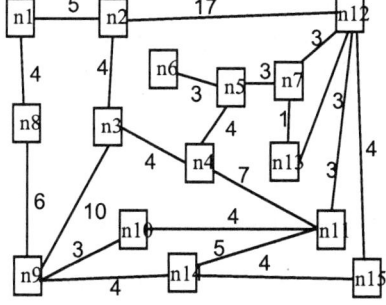

Fig. 1. Experimental Graph   **Fig. 2.** Experimental Graph

3.7 Results

Comprehensive results are reported elsewhere [8]. However, comparing the results of using static routing to the adaptive, swarm-based routing described above, the swarm system routed 18% more traffic in the network. We have observed that the algorithm with adaptive chemical sensitivity values is able to adapt to a new situation *much* more quickly when using these adaptive parameters. The time needed to discover the new path was typically 30% lower (and often much, much lower) when compared to an algorithm using static parameters. When individual components were caused to fail, traffic was quickly re-routed to remaining network paths.

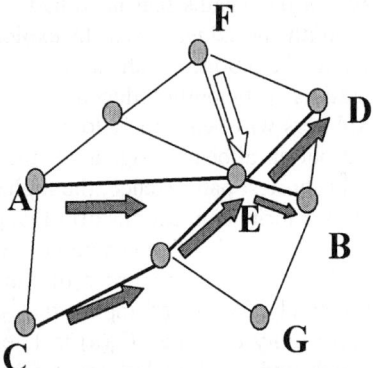

Fig. 3. Priority Routing

3.8 Multi-priority Routing

In many networks we would like to route application traffic with varying priorities. In the model proposed here, the idea is to move existing traffic to longer routes when a higher priority connection request enters the system. Consider, for example, the scenario as represented in Figure 3. Here, two routes have already been computed: AB and CD. These are indicated with the dark arrows. A new connection request now enters the system (FG) as shown by the white arrow. The idea, then, is to have the AB and CD explorer agents (which continue to explore the network even after route allocation) to recognize that the environment has changed and to deallocate the existing route and find one that avoids the higher priority traffic.

In adding multi-priority routing, the explorer algorithm proceeds as before, with one modification. The MDF of the explorer agent does not sense the lower priority connection pheromones but has two receptors instead of one used in the basic routing system. In fact, we use a two class pheromone system. The first class of pheromones has the same meaning as in the earlier sections of this paper, i.e., it is a member of the Binary Array Chemistry of order N. The second class of pheromones is a different length chemical, i.e., it is a member of the Binary Array Chemistry of order N^p, where $N^p \neq N$. In this way, the utilization (and therefore the cost) of a link appears lower to a higher priority connection when compared to a lower priority connection. This requires that chemicals have a well-defined and standardized encoding. As before, all explorer agents find the shortest path using the algorithms described in the previous sections and, upon path emergence, an allocator is sent out to allocate resources in the network. The allocator differs here from the previous design. In a multi-priority system, the allocator deposits pheromone on its way back from the destination having allocated resources for the connection. The reason for the allocator depositing pheromone on its way back from the destination is that we want to ensure that resources have been allocated before communicating the existence of the new connection to other connections in the network. The allocation pheromone, or a-chemical, *is* sensed by lower priority connection explorer agents. Upon return of the allocation agent, the lower priority explorer agents, or lp explorers as we shall refer to them, begin to experience higher costs for links on their paths that have had resources allocated for the higher priority connection. With the added costs, lp explorers will deposit smaller quantities of their connection pheromones, indicating a reduced confidence in the path. If confidence in the path is sufficiently reduced, i.e., it is no longer the shortest path, the majority of lp explorers will begin to follow the new shortest path. At this point, the CCMA will send out an allocation agent for the new shortest path and a deallocation agent for the old shortest path. Hence, the higher priority connection has forced the movement of lower priority connections off of its path.

The above is probably best illustrated with an example. Consider again the network and connections shown in Figure 3. Let the order of the connection and priority chemistries be 8 and 4 respectively. The latter implies that 4 priority levels exist. Let all edges in the network have unity cost, i.e., $C_{ij}(u) = 1$, except edges FD and GH which have $C_{ij}(u) = 2$. The paths indicated by dark arrows for connections AB and CD are then the shortest paths and will be found by explorer agents. Now, consider the introduction of the high priority connection FG. The shortest path for this connection is FE-EB-BG. Assuming that the allocator for this connection drops 3 units of the a-chemical, lp explorer agents for the AB connection will begin to experience higher costs for the AE-EB path, now 4, which makes the path longer than the shortest path, now AE-ED-DB with cost 3. After some time, the CCMA agent for the AB connection will observe that the majority of its explorer agents follow the new shortest path and will send deallocation and allocation agents out into the network in order to move the connection to the new shortest path.

Obviously, the above example has been constructed to demonstrate the path moving algorithm. However, it is sensitive to the quantity of a-chemical deposited in the network by the allocator agent for the higher priority connection. If the concentration of a-chemical is too low, the lower priority connections sharing links with the high priority connection will not be forced to move. This limitation can be overcome by having the CCMA monitor the quality of service associated with the connection. If the measured quality of service falls below the required service level agreement, an agent is sent out into the network that deposits higher concentrations of the a-chemical, thereby increasing the effective cost of the links as seen by the lp explorer agents.

Returning to the above example, assume that the high priority connection allocation agent deposited only 1 unit of the a-chemical. In this case, the cost of the path for the AB connection does not appear to change, and no re-routing occurs. The CCMA for the high priority question, seeing its quality of service to be below the agreed value, sends an agent out into the network to deposit a further quantity of the a-chemical, say 2 units. The rate of increase of a-chemical concentration achieved by sending out multiple agents that deposit a-chemical we call the *priority momentum factor*. This is described in [7,8]

At this point, the quantity of a-chemical on the high priority connection links is 3, as before, and re-routing consequently takes place. With a lower priority momentum factor the required number of agents depositing a-chemical to achieve re-routing will be higher, but re-routing will inevitably occur when the route is made "expensive enough" as seen by the lp explorer agents.

4 Conclusions

This paper has shown that multi-agent swarm techniques can solve complex routing problems on networks. The main strengths of the algorithm are its robustness, the simple nature of the agents, and that it continues searching for new solutions even if a very good one was found. The integration of Genetic Algorithms with the basic Swarm Algorithm has improved the speed of convergence of the routing algorithm. The generality of the SynthECA architecture was validated when adding multiple priority levels to the algorithm proved straightforward. The algorithms described here work best on networks where connections are long lived [3]. For the algorithm to operate on a real network, several things are required. First, agent mobility must be supported. Second, chemical concentrations storage is needed. Finally, on the source node, the CCMA must be present in order to determine whether a route has emerged. All of the above points can be addressed with the application of Java, a mobile agent framework and the use of many of the Virtual Managed Component (VMC) concepts found in [6]. Besides these requirements, consideration must be given to the amount of bandwidth taken up by routing agents. Research addressing resource usage has been forthcoming [1].

References

1. Boyer, J., Pagurek, B., White, T. 1999, Methodologies for PVC Configuration in Heterogeneous ATM Environments Using Intelligent Mobile Agents. In Proceedings of the 1st Workshop on Mobile Agents and Telecommunications Applications (MATA '99).
2. Di Caro G. and Dorigo M. 1998, AntNet: Distributed Stigmergetic Control for Communications Networks. *Journal of Artificial Intelligence Research (JAIR),* 9:317-365.
3. Pagurek B., Li Y., Bieszczad A., and Susilo G. 1998, Configuration Management In Heterogeneous ATM Environments using Mobile Agents, Proceedings of the Second International Workshop on Intelligent Agents in Telecommunications Applications (IATA '98).
4. Parunak H. Van Dyke 1998, Go to the Ant: Engineering Principles from Naturally Multi-Agent Systems, In *Annals of Operations Research.* Available as Center for Electronic Commerce report CEC-03.
5. Schoonderwoerd R., Holland O. and Bruten J. 1997, Ant-like Agents for Load Balancing in Telecommunications Networks. Proceedings of Agents '97, Marina del Rey, CA, ACM Press pp. 209-216.
6. Susilo, G., Bieszczad, A. and Pagurek, B. 1998, Infrastructure for Advanced Network Management based on Mobile Code, Proceedings IEEE/IFIP Network Operations and Management Symposium NOMS'98, New Orleans, Louisiana.
7. White T., Pagurek B. and Oppacher F. 1998, Connection Management using Adaptive Mobile Agents, Proceedings of the International Conference on Parallel and Distributed Processing Techniques and Applications (PDPTA'98).
8. White T. 2000, SynthECA: A Society of Synthetic Chemical Agents, Ph.D. diss., Carleton University.

An Agent-Based Approach to Monitoring and Control of District Heating Systems

Fredrik Wernstedt and Paul Davidsson

Department of Software Engineering and Computer Science
Blekinge Institue of Technology
Soft Center, 372 25 Ronneby, Sweden
{fredrik.wernstedt,paul.davidsson}@bth.se

Abstract. The aim is to improve the monitoring and control of district heating systems through the use of agent technology. In order to increase the knowledge about the current and future state in a district heating system at the producer side, each substation is equipped with an agent that makes predictions of future consumption and monitors current consumption. The contributions to the consumers, will be higher quality of service, e.g., better ways to deal with major shortages of heat water, which is facilitated by the introduction of redistribution agents, and lower costs since less energy is needed for the heat production. Current substations are purely reactive devices and have no communication capabilities. Thus, they are restricted to making local decisions without taking into account the global situation. However, a new type of "open" substation has been developed which makes the suggested agent-based approach possible.

1 Introduction

Agent technology is currently a very active area of research and is widely applied in research labs all around the world. However, few industrial applications exist, in particular where the problem domain is truly distributed and heterogeneous, i.e., the type of domains in which agent technology is supposed to excel. We will here present some initial results from a project concerned with such a domain, namely district heating (which is very similar to district cooling).

The control of district heating systems can be seen as a just-in-time [9] production and distribution problem where there is a considerable delay between the production and the consumption of resources. The reason for the delay may be either long production time or, as in the case of district heating, long distribution time. Another characteristic of this class of problems is that resources need to be consumed relatively quickly after they have arrived to the consumer. In order to cope with these problems it is essential to plan the production and distribution so that the right amount of resources is produced at the right time.

District heating systems are inherently distributed both spatially and with respect to control. A customer, or more commonly, a set of customers, is represented by a substation embedded within the district heating network. Currently, the substation instantaneously tries to satisfy the demands of its customers without considering the amount of available resources or the demands of other substations. Each substation can be viewed as a "black-box" without communication capabilities, making local decisions without taking into account the global situation. Thus, today a district heating network is basically a collection of autonomous entities, which may result in behaviour that is only locally optimal. For instance, during a shortage in the network, resource allocation is unfair since consumers close to the production source will have sufficient amount of heat, while those distantly located will suffer.

Another consequence of the way that current substations work is that the producers only have very limited information concerning the current state of the district heating system. This together with the considerable distribution time, results in that the amount of heat to produce and deliver to the substations is typically based on uninformed estimates of the future heat demand, made by the control engineer at the production plant. In order to ensure sufficient heat supply, the tendency has been to produce more heat than necessary and hence an important waste of energy [1, 3].

The ABSINTHE (Agent-based monitoring and control of district heating systems) project is an effort aimed at dealing with these problems. It is a collaboration with Cetetherm AB, one of the world-leading producers of substations (heat exchanger systems). ABSINTHE is part of a larger venture "Intelligent District Heating", which aims at revolutionising the district heating industry, mainly by developing a new type of "open" substation, which is based on a communication and computation platform developed by Siemens. The main goal of ABSINTHE is to develop a decision support system for the district heating system operators that makes it possible to reduce the surplus production by increasing the knowledge about the current and future state of the system. To implement this we equip each substation with an agent that continually makes predictions of future consumption and monitors current consumption. However, in order to deal with situations where there is a shortage of heat in the network, there is also a fully automated part of the system supporting cooperation between substations. To do this we introduce redistribution agents that are able to impose minor, for the customer unnoticeable, restrictions on a set of substations. Another goal of the project is to more fairly deal with situations where there is a global shortage of heat by using different modes when issuing restrictions.

We begin by briefly describing the district heating domain and the involved hardware technology. This is followed by a description of the multi-agent system that has been developed within the project in order to solve the problems discussed above. Finally, we provide conclusions and pointers to future work.

2 District Heating Systems

The basic idea behind district heating is to use cheap local heat production plants to produce hot water (in some countries steam is used instead of water). The water is then distributed by using pumps at approximately 1-3 m/s through pipes to the customers where it may be used for heating both tap water and the radiator water. The cooled water then returns to the production plant forming a closed system (see Fig. 1).

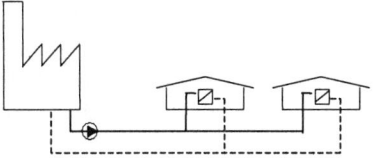

Fig. 1. A simple district heating network containing one heat producer and two consumers

At the customer side, there is a *substation* (see Fig. 2). It is normally composed of two or three heat exchangers and a control unit, which receives hot water from the district heating network. The substation heats both cold tap water and the water in the radiator circuit by exchanging the required heat indirectly from the primary flow of the distribution network. The hot network water is returned to the network at a somewhat lower temperature. Both the temperature of the returning water and the flow rate in the network are dependent on the consumption of substations. When the water, returned by substations, arrives at the heat production plant it is heated and again pumped into the distribution network.

Several different energy sources may be used for heating, e.g., waste energy, byproduct from industrial processes, geothermal reservoirs, otherwise combustion of fuels as oil, natural gas etc. is used. If the demand from the customers is high several heat producing units must be used. A district heating system in a large city can be very complex, containing thousands of substations and hundreds of kilometers of distribution pipes. In addition, they are dynamical as new substations may be added or old substations may be replaced by new ones with different characteristics.

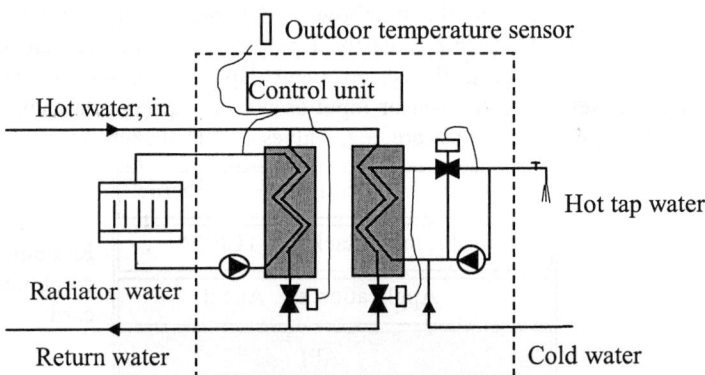

Fig. 2. A substation consisting of heat exchangers (the shaded boxes), control valves, pumps and a control unit. The radiator system (household heating) is controlled by the control unit, using information about actual outdoor temperature

Most district heating control systems of today are strictly reactive, i.e., they only consider the current state and do not predict what is likely to happen in the future. As the distribution time from the heat production plant to the customers is large, the decision on how much heat to produce becomes complicated. Ideally, the control engineer knows the amount to produce several hours ahead of consumption. Load prediction is difficult since many factors are unknown or uncertain which force the operators to make coarse estimations of future consumption based mainly on experience and simple rules-of-thumb [13]. As a consequence, and in order to be sure to satisfy the consumers, district heating systems are typically run with large margins producing more heat than necessary [1, 3]. Furthermore, operators are usually busy with keeping the heating plants running and the time available for making production decisions is therefore limited [1].

Consumption in a district heating network is mainly composed of two parts [14]:

- The heating of buildings, which mainly is a linear function of the outdoor temperature.
- The consumption of tap water, which mainly is dependent on consumption patterns, e.g., social factors.

The tap water consumption of a substation is very "bursty" even in large buildings, and therefore very difficult to predict, whereas the radiator water consumption is "smoother" and therefore relatively easy to predict assuming that reliable weather predictions are available.

Due to the rising demand of automation of building services (heating, ventilation, and air-conditioning etc.) Siemens have developed the Saphir, an extendable I/O platform with an expansion slot for a communication card, suitable for equipment control. Easy and quick access to sensor data is provided by a Rainbow communication card in the expansion slot (see Fig. 3). It has previously not been possible to develop, or make commercially available, such an advanced platform due to high costs.

The Saphir containes a database that continuously is updated with sensor data from the I/O channels by a small real-time operating system, which is directly accessible from the Rainbow card. On the Rainbow card a small computational platform (a handheld PC) makes it possible to easily deploy software and by that providing the possibility to host an agent. Hence, an agent deployed on such a platform could potentially read all connected sensor input as well as send commands over the I/O channel to actuators on the hardware, e.g., valves on a heat exchanger.

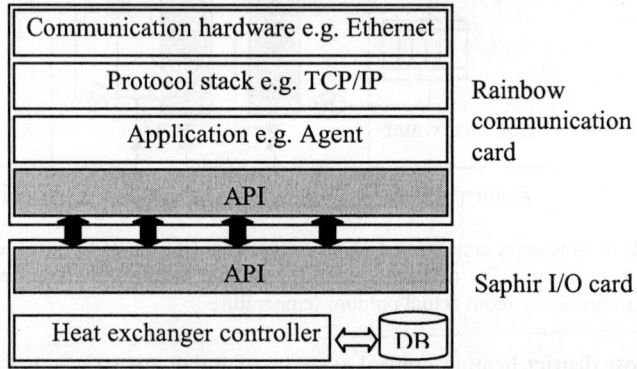

Fig. 3. The Rainbow communication and computation card is here shown on top of the Saphir hardware interface card

3 Software System Architecture

In this chapter we will first discuss the advantages and disadvantages of a distibuted compared to a centralized solution to the problem of district heating monitoring and control. We will then argue for an agent-based approach, and eventually suggest a multi-agent architecture.

3.1 Distributed versus Centralized Approaches

District heating systems are by their very nature physically distributed. Thus, if we aim at a monitoring and control approach based on knowledge about the current state of the system, at least sensor data must be collected via a distributed system. The question is whether also computation and control should, or, need to, be distributed.

In principle, it is possible to continuously collect all sensor data at each substation, do all computations necessary for the control of the system at a single central computer, and then send control signals to each of the substations. In some problem domains, it is possible to increase the utilization of resources if a global picture of the system state is available. In the district heating domain, this is certainly true for the part of the problem related to the production, which we implement as a decision support system (basically a monitoring system) for the network and production plant operators. However, we argue that one subproblem, the construction of the local consumption predictions, should be performed locally. The reason is that these computations may involve substantial amounts of sensor data, which otherwise need to be communicated to a central computer. By doing the consumption predictions locally, less communication is necessary without reducing the quality of the predictions. Also the computations involved in making the forecasts can be computationally resource demanding, and performing these in a centralized fashion would hardly be feasible. One could also argue that some of the sensor data used should be constrained for local usage only, due to its potentially sensitive nature.

Furthermore, to assume that it is easy to collect and use the sensor information from each entity in the network in a centralized fashion is somewhat wrong. Since substations, pumps and valves, etc., often are manufactured by different organisations, it would be a complex task to keep track of all these aspects centrally. Developing local monitoring and control software adopted for each type of substation, but with the same inteface to the rest of the software system, seems as a much more natural approach.

A more general argument against centralized approaches for problems as complex as the management of district heating systems (where a large number of parameters and constraints should be taken into account), is that when the problems are too extensive to be analysed as a whole, solutions based on local approaches often allow them to be solved more quickly [12].

Regarding, the part of the problem that concerns how to fairly deal with shortages of heat in the system, also a semi-distributed approach, were the control is distributed to clusters of substations, has been considered. A completely centralized approach may result in severe communication problems without achieving greater fairness. It would use the same number of messages as the semi-distributed approach, but with a possible communication bottleneck at the central computer. Also, each message would need to travel a longer route which would increase the total network load. A completely distributed approach, on the other hand, would result in a larger number of messages being sent than the semi-distributed approach without achieving greater fairness.

3.2 Why an Agent-Based Approach?

A general advice on when to consider agent technology is to be requirement-driven rather than technology-driven. Thus, we should investigate whether the characteristics of the target domain match the characteristics of the domains for which agent-based systems has been found useful. Parunak [10] argues that agents are appropriate for applications that are modular, decentralised, changeable, ill-structured, and complex. We argue that district heating systems has all these characteristics:

- *Modular:* Each entity of a district heating system, i.e., substations, heat production plants, pumps etc., can be described using a well-defined set of state variables that is distinct from those of its environment. Also, the interface to the environment can be clearly identified for each entity.
- *Descentralised:* The entities of a district heating system can be decomposed into stand-alone geographically distributed autonomous nodes capable of performing useful tasks without continuous direction from some other entity .
- *Changeable:* The structure of a district heating system may change as new entities are added or old enitites are replaced. In addtion, there are short-term changes in the system when individual substation or parts of the network are malfunctioning.
- *Ill-structured:* All information about a district heating system is not available when the monitoring and control system is being designed.
- *Complex:* District heating systems are considered to be very complex systems [3]. The entities of a district heating system exhibits a large number of different behaviours which may interact in sophisticated ways. In addition, the number of entities in a district heating system can be very large, up to a couple of thousands.

There are also more general arguments for choosing an agent-based approach [4]. From a methodological perspective the concept of agents introduces a new level of abstraction that provides an easier and more natural conceptualisation of the problem domain. Other advantages are increased, e.g., *robustness*, the distribution of control to a number of agents often implies no single point of failure, *efficiency*, less complex comuptations and communication are necessary if control is distributed, *flexibility*, the use of agent communications languages that support complex interaction between entities provides a flexibility that is difficult to achieve using traditional communication protocols, *openness*, by having a common communication language, agents implemented by different developers are still able to interact with each other, *scalability*, it is easy to add new agents to a multi-agent system, and finally, *economy*, since agent technology provides a natural way to incorporate existing software.

3.3 Multi-Agent System Architecture

The MAS architecture we suggest is composed of one type of agents associated with the producers, responsible for the interaction between the heat production plant and the other agents of the MAS, and another type of agents associated with the consumers, responsible for the interaction between the substation and the other agents of the MAS.

In a domain with limited resources, such as district heating, agents must coordinate their activities with each other to satisfy group goals [6]. To identfy group goals is complicated by two types of conflicts between individual goals. First, there is a conflict between consumer agents, who wants to maximize the comfort of the consumers by taking as much heat from the network as the consumer asks for, and the producer agents, who want to produce as little heat as possible to reduce costs. To deal with this conflict we define the following group goal: produce as little heat as possible while maintaining sufficient level of customer satisfaction. The second type of conflict is between consumer agents when there is a shortage of heat in a part of the network. In this situation each of the consumer agents in that part of the network wants to satify their consumer's demand, which is impossible. To deal with this conflict we define the following group goal: when there is a shortage, the available heat should be shared fairly between the consumers. Satisfaction of these group goals could be achieved by either :

- competition, where each agent competes for resources, i.e., a self-interested approach (This approach is similar to the current situation, where a substation only consider the demands of their customer.), or
- cooperation, by letting the overall goal of primarily ensuring consumption of tap water, in favour of radiator consumption, affect all consumers, i.e., where one consumer could reduce radiator consumption to benefit tap water consumption for another consumer.

We implement cooperation to achieve redistribution of resources. This is practically possible since pressure propagates by the speed of sound in water and that district heating systems are, at least in principle, parallel coupled, i.e., resources not consumed by one consumer is available for other consumers in the vicinity. Also, we need the capability to restrict consumers if they try to consume more in total than predicted. For these reasons we introduce another type of agent, redistribution agents, which are responsible for a cluster of consumer agents and has both a mediator and decision maker role [5]. The mediator role includes receiving the consumptions and predictions from the consumers in the cluster, summarize them, and distribute the result to the producer agent (see Fig. 4). The decision maker role concerns what actions to take, i.e., impose restrictions upon consumer agents, to maintain an overall acceptable consumption rate (which is defined by the predictions made earlier).

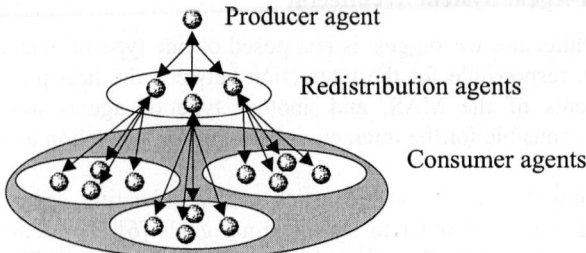

Fig. 4. Each redistribution agent manages a cluster of consumer agents

The *consumer agents* are responsible for the interaction with the substations (see Fig. 5). It monitors the actual heat consumption by reading the substation's sensors and decides which data to send to the redistribution agent. The agent continually evaluates previously made predictions using historical data and creates new predictions of future consumptions and sends these to the redistribution agent. The consumption is sent rather often, e.g., each minute, whereas the predictions are sent at larger interval, e.g., for each 10 minutes period.

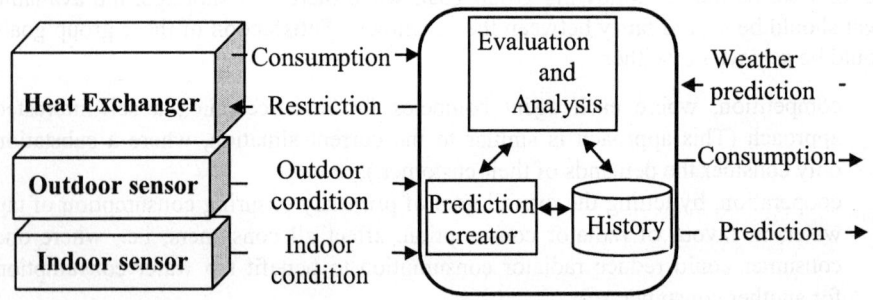

Fig. 5. Consumer agent architecture and its interaction with the environment

The *redistribution agents* are responsible for collecting predictions and monitoring the total consumption of a cluster of consumer agents. If the redistribution agent notices that the cluster is trying to consume more heat than predicted it invoke restrictions of consumption to the cluster. Restrictions can be invoked in two modes, one where each consumer agent in the cluster is to reduce its consumption with the same amount and one where different priority values are used, e.g., there might be reason that a hospital has a higher priority than a university.

A restriction will enforce consumers to not use any radiator water during the next consumption interval. If this is not enough to compensate for the excessive consumption, the redistribution agent will also impose restrictions on tap water consumption. This redistribution strategy (and all other strategies based on radiator water restrictions) will lead to a radiator water deficit that needs to be compensated (otherwise the temperature in the households, eventually, will fall). In order to do this, each substation will individually compensate by using more radiator water than predicted when the tap water consumption is less than predicted. Compensation for tap water deficits work in a similar way, but on the cluster level (see Fig. 6).

Another problem that has to be solved by the redistribution agent is how to cope with the "bursty" consumption of tap water without commanding unnecessary restrictions. The approach used here is to let the cluster use more tap water than predicted (measured in minute averages) in the beginning of a consumption period and then gradually lower the allowed average consumption towards the predicted average consumption.

The *producer agent* receives predictions of consumption from the redistribution agents and is responsible for the interaction with the control system of the heat production plant (possibly including human operators). The producer agent is also responsible for monitoring the actual consumption of consumer agents. The figures of consumption may be used to calculate the returning temperature, i.e., the producer would know in advance the temperature of the water to heat. The producer agent is

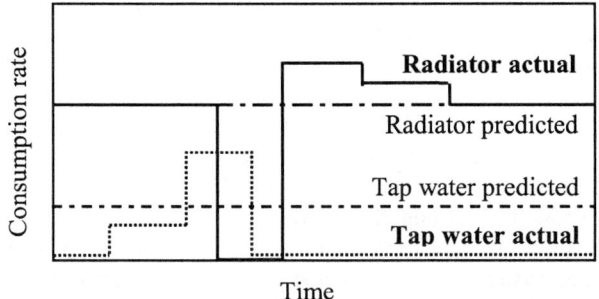

Fig. 6. A schematic view of the consequences of a restriction

also capable of imposing proactive restrictions for consumers to reduce the cost of producing heat to cover for short temporary heating needs. This will especially be the case if the predicted consumption marginally exceeds the capacity of the primary heating plant or only exceeds the capacity for a short time period (see Fig. 7).

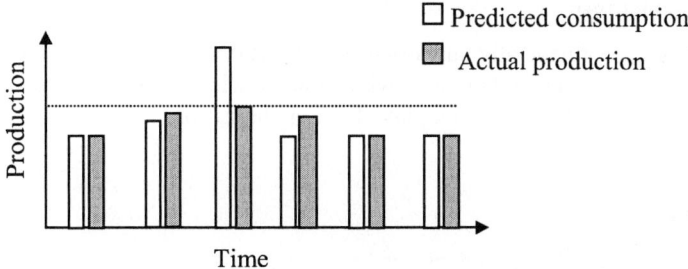

Fig. 7. A predicted consumption for a limited time period above the capacity of the primary production source (the dashed line) can be covered by a temporary imposing restrictions of consumption

The system is dependent on the truthfulness of agents, e.g., individual consumer agents should not lie to gain benefits. However, we believe that this will not pose a problem since the system can be viewed as closed with respect to development and deployment of the agents, for instance by introducing certification procedures.

4 Conclusions and Work in Progress

In the short term the suggested project will adapt and introduce agent technology in an industrial application were the use of leading edge information technology currently is very low. To our knowledge, agent technology has never been applied to monitoring and control of district heating systems. It will provide a novel combination and integration of existing technologies, which will open up new possibilities.

The contributions to the final users, i.e., the operators and the consumers, will be higher quality of service, e.g., better ways to deal with major shortages of heat water, and lower costs, i.e., less energy is needed to produce the heat water. Since the heating of water often is associated with burning fuel that pollutes the air in one way or another, the project obviously contributes to increase the quality of life for the inhabitants. We also believe that the introduction of advanced information and communication technology will enhance the work situation for the network operator staff, e.g., through the new possibility for remote diagnosis of heat exchanger systems.

We have implemented a simplified version of the MAS described in this paper using Jade [2] and applied the Gaia methodology [15] for the design. Initial simulation ex-periments have shown that it is possible to reduce the production by at least 4 percent while maintaining the overall consumer satisfaction, or quality of service. However, there are indications that this figure can be improved considerably.

The implemented system is not a finished product, however, it could at this stage be considered an emulated system. To reach the next maturity level [11], prototype, we are in progress of testing the approach on real domain hardware, but under laboratory conditions. Furthermore, the prediction mechanism of the consumer agents needs to be improved and we are looking at, e.g., neural network-based approaches [3, 7].

Acknowledgements

This work has been financially supported by VINNOVA (The Swedish Agency for Innovation Systems). The authors also wish to acknowledge the valuable input from the ABSINTHE project members employed at Cetetherm AB.

References

1. Arvastsson, L.: Stochastic Modeling and Operational Optimization in District Heating Sys-tems, Doctoral Thesis, Lund Institute of Technology, Sweden (2001).
2. Bellifemine, F., Poggi, A., and Rimassa, G.: Developing multi-agent systems with a FIPA-compliant agent framework. Software: Practice and Experience, Vol. 31(2), John Wiley & Sons, Ltd, New York (2001) 103-128.
3. Canu, S., Duran, M., and Ding, X.: District Heating Forecast using Artificial Neural Net-works, International Journal of Engineering, Vol. 2(4) (1994).

4. Davidsson, P. and Wernstedt, F.: Software Agents for Process Monitoring and Control, Journal of Chemical Technology and Biotechnology, 2002 (To appear).
5. Ferber, J.: Multi-Agent System, An introduction to Distributed Artificial Intelligence, Addison Wesley (1999).
6. Huhns, M., Stephens, L.: Multiagent Systems and Societies of Agents. In Weiss, G. (ed.) Multiagent Systems, MIT Press (1999).
7. Malmström, B., Ernfors, P., Nilsson, D., and Vallgren, H.: Short-term forecasts of district heating load and outdoor temperature by use of on-line-connected computers. Värmeforsk, 1996:589, ISSN 0282-3772 (in Swedish).
8. Nwana, H., Lee, L., and Jennings, N.R.: Coordination in Software Agent Systems, The British Telecom Technical Journal, 14 (4) (1996) 79-88.
9. Ohno, T.: Toyota Production System. Productivity P, US (1988).
10. Paranak, H.: Industrial and Practical Applications of DAI. In Weiss, G. (ed.) Multiagent Systems, MIT Press (1999).
11. Parunak, H.: Agents in Overalls: Experiences and Issues in the Deployment of Industrial Agent-Based Systems. International Journal of Cooperative Information Systems Vol. 9(3) (2000) 209-227.
12. Rinaldo, J. and Ungar, L.: Auction-Driven Coordination for Plantwide Optimization. Foundations of Computer-Aided Process Operation FOCAPO (1998).
13. Weinspach, P.M.: Advanced energy transmission fluids. International Energy Agency, NOVEM, Annex IV, Sittard, Netherlands (1996).
14. Werner, S.: Dynamic heat loads from fictive heat demands. Fjärrvärmeföreningen, FOU 1997:10, ISSN 1402-5191 (in Swedish).
15. Wooldridge, M., Jennings, N.J., and Kinny, D.: The Gaia Methodology for Agent-Oriented Analysis and Design. Journal of Autonomous Agents and Multi-Agent Systems, Vol. 3(3) (2000) 285-312.

Using Machine Learning to Understand Operator's Skill

Ivan Bratko and Dorian Šuc

Faculty of Computer and Information Sc., University of Ljubljana
Tržaška 25, 1000 Ljubljana, Slovenia
{ivan.bratko,dorian.suc}@fri.uni-lj.si

Abstract. Controlling complex dynamic systems requires skills that operators often cannot completely describe, but can demonstrate. This paper describes research into the understanding of such tacit control skills. Understanding tacit skills has practical motivation in respect of communicating skill to other operators, operator training, and also mechanising and optimising human skill. This paper is concerned with approaches whereby, using techniques of machine learning, controllers that emulate the human operators are generated from examples of control traces. This process is also called "behavioural cloning". The paper gives a review of ML-based approaches to behavioural cloning, representative experiments, and an assessment of the results. Some recent work is presented with particular emphasis on understanding human tacit skill, and generating explanation of how it works. This includes the extraction of the operator's subconscious sub-goals and the use of qualitative control strategies. We argue for qualitative problem representations and decomposition of the machine learning problem involved.

1 Introduction

Controlling a complex dynamic system, such as an aircraft or a crane, requires operator's skill acquired through experience. In this paper we are interested in the question of understanding of tacit human skills, and designing automatic controllers by transfer of operators' skill into a controller. One approach would be to attempt to extract the skill from the operator in a dialogue fashion whereby the operator would be expected to describe his or her skill. This description would then be appropriately formalised and built into an automatic controller.

The problem with this approach is that the skill is sub-cognitive and the operator is usually only capable of describing it incompletely and approximately. Such descriptions can only be used as basic guidelines for constructing automatic controllers, because as discussed for example in [18,4], the operator's descriptions are not operational in the sense of being directly translatable into an automatic controller.

Given the difficulties of skill transfer through introspection, an alternative approach to skill reconstruction is to start from the *manifestation* of the skill.

Although an operational description of the skill is not available, the manifestation of the skill *is* available in the form of traces of the operator's actions. One idea is to use these traces as examples and extract operational models of the skill by machine learning (ML) techniques.

Extracting models of a real-time skill from operators' behaviour traces by machine learning is also known as *behavioural cloning* [9,10]. In general there are two goals of behavioural cloning:

- To generate *good performance* clones, that is those that can reliably carry out the control task.
- To generate *meaningful* clones, that is those that would help to understand the operator's skill by making the skill symbolically explicit.

The second goal is important for several reasons:

- Operationalising human operator's instructions for controlling a system. Operator's instructions are a useful source of information, but are normally too incomplete and imprecise to be directly translatable into a control program.
- Flexibly modifying and optimising the induced clones to prevent undesired patterns in their behaviour.
- Understanding what exactly a human operator is doing and why. This is of practical importance regarding the capture of exceptional operators' skill and its transfer to less gifted operators. The operator's control strategy would ideally be understood in terms of goals, sub-goals, plans, feedback loops, causal relations between actions and state conditions etc. These conditions are to be stated in terms of information that is easily accessible to the operator, e.g. visually. It can be argued that such information should be largely qualitative, as opposed to the prevailing numerical information.

Behavioural cloning has been studied in various dynamic domains, including: pole balancing [10], flying a Cessna aircraft [13,1], flying the F16 [11,5], operating cranes [18,12], electrical discharge machining [6]. Fig. 1 illustrates the crane and the "acrobot" systems that have been, in addition to pole balancing and air piloting, often used in such studies.

In this paper we focus on the question of understanding operators' control strategies, and generating explanation of how and why these strategies work.

2 Inducing "Direct Controllers"

The following is a simple and often used procedure of applying machine learning (ML) to recovering control strategy from example execution traces. A continuous trace is normally sampled so that we have a sequence of pairs ($State_i$, $Action_i$) ordered according to time. $State_i$ is the state of the system at time i, and $Action_i$ is the operator's action performed at time i. In this simple approach, the sequence of these state-action pairs is viewed as a set of examples, thereby ignoring the time order. This simplification can be justified by the formal argument that the control action is fully determined by the state of the controlled system. Many

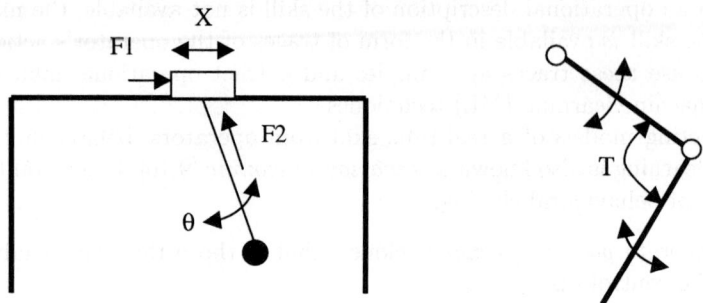

Fig. 1. Two frequently used dynamic systems in experiments in behavioural cloning. The arrows indicate degrees of freedom and possible control actions. Top: container crane with horizontal and vertical control forces; task is to move the load to a goal position while controlling the swinging of the rope at the goal. Bottom: the "acrobot" (double-linked pendulum), hinged at the top, with control torque at the middle joint; task is to keep increasing oscillation until a full cycle is made over the top bar

machine learning programs can be viewed as reconstructors of functions $y = f(x1, x2, ...)$ from sets of given pairs of the form:

$$(x1_i, x2_i, ...), y_i)$$

The arguments $x1$, $x2$, ... are usually called attributes, and function value y is called the *class* value. So the usual formulation of behavioural cloning as a ML task is as follows: the system's state variables $x1$, $x2$ etc. are considered as attributes, and the action as the class variable. In behavioural cloning, the most frequently used ML techniques have been decision tree learning and its variation regression tree learning. In fact, any technique that reconstructs a function from examples can be used, including neural networks. However, most studies in behavioural cloning are concerned with the use of symbolic learning techniques, thus excluding neural networks. There is an important reason for this limitation. Namely, the goal of skill reconstruction is, in addition to good performance, also *understanding* the operator's control strategy: What is the operator's control strategy, why is he doing this or that?

The result of machine learning, using the formulation of the learning problem above, is a controller in the form of a function from system states to actions:

$$Action = f(State) = f((x1, x2, ...))$$

This controller maps the system's current state into an action *directly*, without any intermediate, auxiliary result. Therefore such controllers will be called *direct* controllers, to be distinguished from "indirect" controllers discussed later.

It has been shown empirically that direct controllers, although they appear to be the most natural, suffer from serious limitations. Part of the reason of direct controllers' inferiority can be explained by the following. The representational decision of treating sequences as sets is debatable: operator's decisions almost always depend on the history and not only on the current state of the system. In controlling the system, the operator pursues certain goals over a time period and then switches to other goals, etc. So both the state *and* the current goal determine the action, although the goal is not part of the system's state but only exists in the operator's mind.

In spite of these reservations, most of the known studies in behavioural cloning treat example traces as sets. The conclusions from these studies can be summarized as follows. On the positive side, successful clones have been induced using standard ML techniques in all the domains. Also, the so-called clean-up effect, whereby the clone surpasses its original, has been occasionally observed in all the domains. However, the "direct" approaches to behavioural cloning typically suffer from the following problems:

- They lack robustness in the sense that they do not provide any guarantee of inducing with high probability a successful clone from given data.
- Typically, the induced clones are not sufficiently robust with respect to changes in the control task.
- Although the clones do provide some insight into the control strategy, they in general lack conceptual structure and representation that would clearly capture the structure of the operator's control strategy.

In particular the last deficiency is indicative of slow progress towards really understanding the operator' skill. The reasons for mixed success of direct controllers usually lie in the representation used in skill reconstruction. The usual representation is inherited from traditional control theory, and is entirely numerical and unstructured. More appropriate representations are largely qualitative and involve goals and history (not just the current state of the system) and qualitative trends.

3 Inducing Indirect Controllers

We say that a controller is "indirect" if it does not compute the next action directly from the current system's state, but uses in addition to the state some other, intermediate information. A typical such additional information is a subgoal to be attained before attaining the final goal.

Subgoals often feature in operator's control strategies. Also, their skill is often best explained in terms of subgoals. Examples of operators' own explanations of their strategies are studied in [18]. The first idea to exploit this fact is to automatically detect, from operator's traces, such subgoals. This was studied in [14] with an approach based on the theory of linear quadratic controllers. The problem of subgoal identification was treated as the inverse of the usual problem of controller design. The usual problem is: given a system's model and a goal, find the actions that achieve the goal. The problem of goal identification is: given the actions (in operator's trace), find the goal that these actions achieve.

More precisely, find the goal so that the actions performed are QR-optimal with respect to this goal.

This approach was applied successfully to behavioural cloning in the crane domain [14]. However, the limitation of the approach is that it only works well for the cases in which there are just a few subgoals. A counter example is the acrobot domain. The task of completing a full cycle requires a sequence of oscillations, each of them having an increased amplitude. Such a trajectory consists of dense subgoals (illustrated in Fig. 2). For such cases a more appropriate idea is to generalise the operator's trajectory. Such a generalised trajectory can be viewed as defining a continuously changing subgoal (Fig. 2).

Subgoals and generalised trajectories are not sufficient to define a controller. A model of the system's dynamics is also required. Therefore, in addition to inducing subgoals or a generalised trajectory, this approach also requires the learning of approximate system's dynamics, that is a model of the controlled system. The next action is then computed "indirectly" roughly as follows: (1) compute the desired next state (e.g. next subgoal), and (2) determine an action that brings the system to the desired next state. Fig. 3 illustrates this approach and contrasts it to the induction of direct controllers. The next action is computed so as to minimise the difference between the predicted next state of the system and the generalised trajectory. This can be done in various ways, and the "difference" between by dynamics model predicted next state and the trajectory is defined accordingly.

By way of indirect controllers, the problem of behavioural cloning is decomposed into two learning problems: (1) learning trajectory, and (2) learning system's dynamics. It has been shown experimentally that this decomposition leads to much better performance than the induction of direct controllers [15,16]. The overall controller's performance of the induced clones is less sensitive to the classification accuracy achieved in the two learning tasks than in the case of direct controllers. This is demonstrated for several domains (including crane, acrobot and bicycle control) in [12,17]. In that work, the learning tasks were accomplished by the program Goldhorn [7] for inducing generalised trajectory, and locally weighted regression for learning approximate system's dynamics. Ideas of using subgoals in behavioural cloning in aeroplane flying has also been discussed by Bain and Sammut [1].

4 Using Qualitative Representations

An important representational decision in behavioural cloning is concerned with the choice between a numerical representation, or a qualitative representation of induced descriptions. For example, whether to use system state variables as attributes for machine learning, or some other, qualitative attributes of the given traces.

An operator, executing his skill on an actual physical system, or on a graphical simulator, hardly uses the numerical state variables. First, some of these variables values are hard to observe precisely, for example the angular velocity

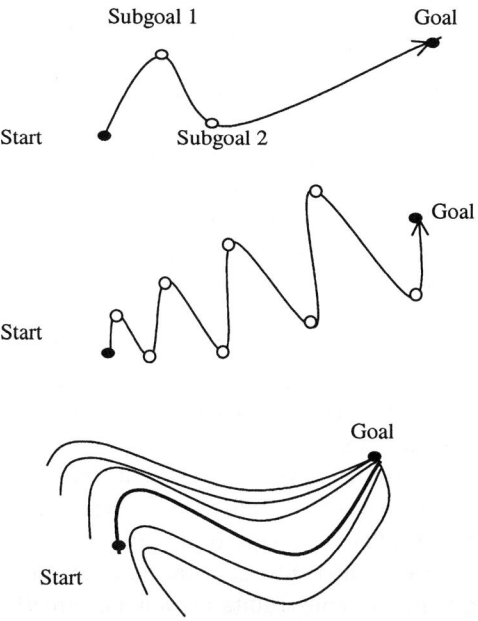

Fig. 2. Top: A trajectory and two subgoals. Middle: A trajectory with dense subgoals. Bottom: A generalised trajectory; the thicker line corresponds to the original operator's example trace; the thinner lines generalise this original trajectory to the rest of the problem space

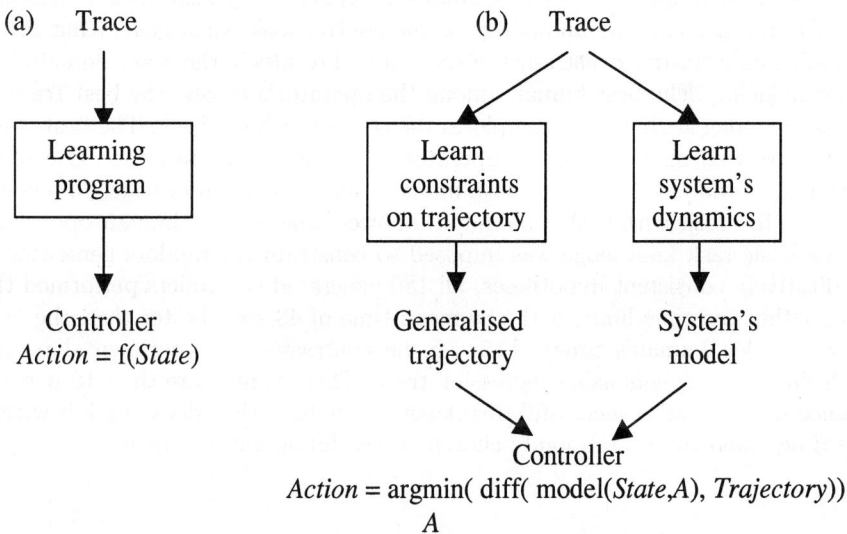

Fig. 3. (a) Induction of direct controllers. (b) Induction of indirect controllers

of the rope in crane control. Also, the operator surely does not online evaluate an arithmetic formula to decide on his next action. Instead, he has to base his control decisions on other, easily recognisable patterns in terms of qualitative features of the current state and recent behaviour of the system. Such qualitative features include the following ones, well known in qualitative physics: the sign of a variable (for example, in pole balancing, the pole leaning left or right), the direction of change (pole rotating left), variable crossing a landmark (pole upright), variable just reached local extreme.

There is strong experimental evidence that in behavioural cloning qualitative representations are more suitable than numerical representations. The use of qualitative attributes in the induction of direct controllers for the pole-and-cart task was studied in [3]. It was shown that the qualitative approach resulted in clones whose style of control was much closer to the original human operator's control. Also, the resulting qualitative controllers provided a much better explanation of the operator's strategy.

One way of using qualitative representation in indirect controllers was explored by Šuc [12]. The idea was to simplify the generalised trajectory, induced by Goldhorn in the form of differential equations. By means of qualitative abstraction, these equations are turned into qualitative constraints on the operator's trajectories. Such qualitative constraints cannot be directly used for the computation of the next action. First, a quantitative constraint consistent with the qualitative constraint is needed, and this quantitative constraint is used in the concrete numerical calculation. There are many possible quantitative constraints consistent with the qualitative constraint and any of them corresponds to a candidate controller that is qualitatively equivalent to the operator's strategy. So this approach defines an optimisation space over which an optimal controller can be sought. Since qualitative constraints are explicit, they also make it possible to take into account the knowledge of the control task when generating corresponding quantitative constraints. Experimental results in the crane domain [12] are convincing. The best human among the operator's traces (the best trace of the best among six humans) completes the control task in 51 sec. The best clone qualitatively consistent with this operator's strategy does it in 38 sec. Out of 90 controllers randomly generated with minimal background knowledge, 83 percent perform the task within the maximum allowed time set for human operators. When basic task knowledge was imposed to constrain the random generator of qualitatively consistent hypotheses, all 180 generated controllers performed the task within the time limit, with the mean time of 48 sec. (better than the best ever recorded human's time). This can be contrasted to experimental results with direct controllers using regression trees. There is no more than 10 percent chance of inducing a successful direct controller (one that does the job within the time limit) from a randomly chosen successful operator's trace.

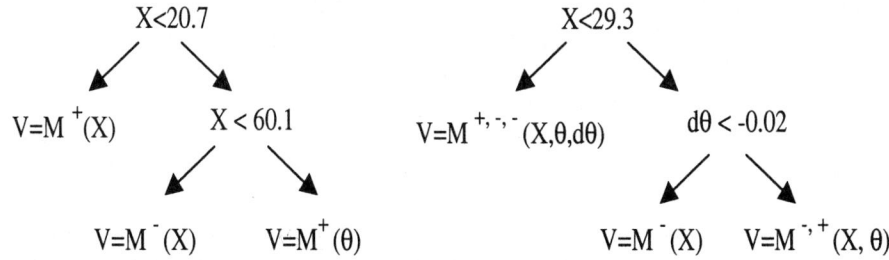

Fig. 4. Left: qualitative tree induced from a control trace of operator S. Right: qualitative tree from a trace of operator L. The left branches in the trees always correspond to "yes", and the right branches to "no". X and V are the position and the desired velocity of the cart; Φ and $d\Phi$ are the angle and the angular velocity of the rope

5 Induction of Qualitative Control Strategies

In the previous section, qualitative control strategies were induced from operator's control traces indirectly. First, a differential equations model was induced with the GoldHorn program, and then a qualitative strategy was obtained from these differential equations by means of qualitative abstraction.

Here we discuss a more direct way of inducing qualitative constraints directly from numerical data. For this purpose, the system QUIN (QUalitative INduction; [12]) was developed that induces qualitative constraints, in the form of *qualitative trees*, from numerical data. Qualitative trees are similar to decision trees as used in learning of decision trees. So internal nodes of a qualitative tree correspond to tests on variable values, just as in decision trees. In contrast to decision trees, the leaves are assigned qualitative constraints among the variables. Qualitative constraints used in QUIN are a generalisation of the monotonically increasing/decreasing constraints, written as M^+ or M^-, often used in qualitative reasoning (e.g. [8]). For example, $y = M^+(x)$ means: y is a monotonically increasing function of x; whenever x increases, y also increases. QUIN generalises these monotonicity constraints to multiple arguments. For example, $P = M^{+,-}(T,V)$ means: when temperature T increases and volume V stays unchanged then pressure P also increases, and when V increases and T stays unchanged then P decreases.

Fig. 4 shows two qualitative trees induced from control traces of two operators controlling a simulated crane [12]. They define the qualitative relation between the cart's desired velocity and the other variables in the system, when the goal of control is to reach a goal state near $X=60$, $V=\Phi=d\Phi=0$. The controller applies a force that makes the cart's actual velocity similar to the desired velocity. The two trees nicely illustrate the qualitative differences in the control styles of the two operators. Initially, when far from goal, operator S just accelerates the cart ($V=M^+(X)$). Then, when getting closer to the goal, he decelerates. Only when very close to the goal, he pays attention to the angle and in a quite sensible way finally controls the swing ($V=M^+(\Phi)$). L's qualitative strategy is considerably more sophisticated. Obviously, he pays attention to the load swing already at the initial stages of the task and makes active effort at reducing the swing. Due to this, in actual (quantitative) execution traces, operator L is more successful and achieves shorter times to complete the task. He can afford greater accelerations and occasionally larger rope angle since he has the skill of reducing the angle at any time.

It should be noted that, to execute qualitative control strategies such as those in Fig. 4, they have to be converted to quantitative controllers as described in Section 4.

Fig. 4 illustrates that qualitative induction gives good insight into tacit skill. Qualitative trees provide some explanation of how such skill works. However, qualitative trees give static relations only, and a good intuition is still required from the user about the dynamics of the system to understand the mechanics of how a qualitative strategy achieves the goal. One remaining question is: Is it possible to generate from a qualitative control strategy explanation that involves behaviour in time? We look now at one approach to answering this.

The idea is to use a qualitative model of the dynamics of the controlled system and a qualitative control strategy, together with a qualitative simulator, to generate explanation of how the strategy works in time. We have explored this idea by using a Prolog implementation [2] of the qualitative simulation algorithm QSIM [8] with the qualitative strategies of Fig. 4. We used the following (approximate) qualitative model, in terms of qualitative constraints used in QSIM, of the crane:

$$deriv(X,V), \ deriv(V,A), \ A = M^+(F + M_0^+(\Phi)),$$
$$deriv(\Phi, d\Phi), \ deriv(d\Phi, dd\Phi), \ dd\Phi = M_0^-(F + M_0^+(\Phi))$$

The constraint deriv(X,Y) means Y is the time derivative of X. The notation $Y=M_0^+(X)$ means Y is monotonically increasing function of X, where $Y(0)=0$.

As an example, consider the swing control with S's qualitative strategy near the goal position expressed as $V=M_0^+(\Phi)$. Let qualitatively the current state of the crane be such that the cart velocity is zero, the angle is positive and angular velocity zero, and let the goal be to reach a steady state with all these three variables equal zero. The initial state does not satisfy S's constraint $V=M_0^+(\Phi)$ ($V=0 \neq \Phi=$pos). Therefore, using the qualitative simulator, the crane's state space was searched for a qualitative control trace that brings the crane to a state that satisfies S's constraint. From then on, S's qualitative strategy was executed by respecting this qualitative constraint. This is done by simply adding S's constraint to the qualitative model of the crane, and with this the simulator then was constrained to search within S's strategy, all the time correspondingly adjusting control force F. Notice that due to the branching in qualitative simulation (also including some spurious behaviours) several control traces were thus generated. One of them is shown in Fig. 5. It represents one way of explaining how at least one aspect of S's qualitative control strategy works.

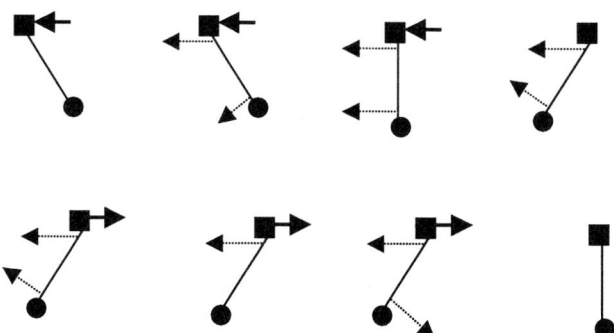

Fig. 5. A qualitative control trace for the crane that explains how S's constraint $V = M_0^+(\Phi)$ steadies the crane. This trace consists of eight qualitative states, shown by eight diagrams above. The time order is from left to right. The dotted arrows indicate the velocity of the cart and the angular velocity of the rope. The solid arrows indicate the control force. No arrow means that the corresponding variable is zero. For example, the first diagram illustrates the qualitative state where the cart's velocity is zero, the rope's angle is positive, the angular velocity is zero, and the control force is negative (pushing to left)

6 Conclusions

The problem of behavioural cloning of operators' skill has been discussed in the paper, together with several recently developed approaches. In particular, this paper focuses on approaches that facilitate the understanding of tacit human skills. When applying machine learning to this problem, two kinds of approaches were distinguished: direct and indirect controllers. Although much more widely used, the induction of direct controllers clearly has serious difficulties. In contrast, the induction of indirect controllers enables a decomposition of the cloning task, which gives much more robust results and better controllers. This decomposition involves the concept of generalised operator's trajectory. Advantages are:

- The resulting controllers are much more robust than direct controllers.
- Qualitative abstraction of induced trajectories enables better explanation of the human's control strategy.
- Qualitative induction facilitates the detection of qualitative differences in control styles of different operators.
- Qualitative control strategies, induced from operators' control traces, together with qualitative simulation, enable automated generation of explanation of how an operator's tacit skill works.

Acknowledgements

The work reported in this paper was partially supported by the European Fifth Framework project Clockwork and the Slovenian Ministry of Education, Science and Sport.

References

1. Bain, M., Sammut, C.: A framework for behavioural cloning. In: Furukawa, K., Michie., D, Muggleton, S. (eds.): *Machine Intelligence 15*. Oxford University Press (1999).
2. Bratko, I.: *Prolog Programming for Artificial Intelligence, 3rd edition*, Chapter 20. Addison-Wesley (2001).
3. Bratko, I.: Qualitative reconstruction of control skill. In: Ironi, L. (ed.): *Qualitative reasoning : 11th International Workshop*, 41–52. Cortona, Italy (1997).
4. Bratko, I., Urbančič, T.: Control skill, machine learning and hand-crafting in controller design. In: Furukawa, K., Michie., D, Muggleton, S. (eds.): *Machine Intelligence 15*. Oxford University Press (1999).
5. Camacho, R.: Inducing models of human control skills. In *Proc. 10th European Conf. Machine Learning*. Chemnitz, Germany (1998).
6. Karalič, A., Bratko, I.: First Order Regression. *Machine Learning*, (1997), 26:147–176.
7. Križman, V.: *Automated Discovery of the Structure of Dynamic System Models*. Ph.D. Dissertation, Faculty of Computer and Information Sc., University of Ljubljana (in Slovenian) (1998).
8. Kuipers, B.: *Qualitative Reasoning: Modeling and Simulation with Incomplete Knowledge*. MIT Press (1994).
9. Michie, D.: Knowledge, learning and machine intelligence. In: Sterling, L. (ed.): *Intelligent Systems*, 2–19. New York: Plenum Press (1993).
10. Michie, D., Bain, M., Hayes-Michie, J.: Cognitive models from subcognitive skills. In: Grimble, M., McGhee, J., Mowforth, P. (eds.): *Knowledge-Based Systems in Industrial Control*. Stevenage: Peter Peregrinus (1990).
11. Michie, D., Camacho, R.: Building symbolic representations of intuitive real-time skills from performance data. In: Furukawa, K., Muggleton, S. (eds.): *Machine Intelligence and Inductive Learning*. Oxford: Oxford University Press (1994).
12. Šuc, D: *Machine reconstruction of human control strategies*. Ph.D. Dissertation, Faculty of Computer and Information Sc., University of Ljubljana, Slovenia (2001).
13. Sammut, C., Hurst, S., Kedzier, D., Michie, D.: Learning to fly. In: *Proc. 9th International Workshop on Machine Learning*, 385–393. Morgan Kaufmann (1992).
14. Šuc, D., Bratko, I.: Skill reconstruction as induction of LQ controllers with subgoals. In: Mellish, C. (ed.): *Proc. 15th International Joint Conference on Artificial Intelligence*, vol. 2, 914–920 (1997).
15. Šuc, D., Bratko, I.: Skill modelling through symbolic reconstruction of operator's trajectories. In: Brandt, D. (ed.): *Automated systems based on human skill, 6th IFAC Symposium*, 35–38. Kranjska gora, Slovenia (1997).
16. Šuc, D., and Bratko, I.: Problem decomposition for behavioural cloning. In: Mántaras, R., Plaza, E. (eds.): *Proc. 11th European Conference on Machine Learning*, 382–391. Springer (2000).

17. Šuc, D., Bratko, I.: Skill modelling through symbolic reconstruction of operator's trajectories. *IEEE Transaction on Systems, Man and Cybernetics, Part A* (2000), 30(06):617–624.
18. Urbančič, T., Bratko, I.: Reconstructing human skill with machine learning. In Cohn, A., ed., *Proceedings of the 11th European Conference on Artificial Intelligence*, 498–502. John Wiley & Sons, Ltd (1994).

Reactive Load Control of Parallel Transformer Operations Using Neural Networks

Fakhrul Islam, Baikunth Nath, and Joarder Kamruzzaman

Gippsland School of Computing & Information Technology
Monash University, Churchill, Australia 3842
{Fakhrul.Islam,Baikunth.Nath,
Joarder.Kamruzzaman}@mail1.monash.edu.au

Abstract. Artificial Neural Network (ANN) is used in various fields including control and analysis of power systems. ANN in its learning process establishes the relationship between input variables by means of its weights updating, and provides a good response to another nonidentical but similar input. This paper proposes the use of neural network to control the on-load tap changer of parallel operation of two transformers supplying power to a local area. For simplicity, only two transformers are considered although operation of multiple transformers can be dealt with in a similar manner. A synthetic data set relating to tap changer operation sequence was used for training a backpropagation network to decide automatically on transformer's on-load tap changer whether to raise, lower or hold the same desired position. Preliminary results show that a trained neural network can be successfully used for on load tap changing operation of transformers.

Keywords: Parallel transformers, neural networks, reactive load control, tap changer

1 Introduction

Most of the power transmission and distribution lines and transformers operate in parallel to supply electricity. Power sharing and supply voltage levels are usually maintained by controlling the automatic tap changer control system of the transformers. A typical existing system consists of a control panel with relays and complex circuits to control the tap changing of transformers using master and follower principle. While running in parallel, one of the transformers is selected as master and the remaining as followers. The followers always change their tap positions as done by the master. Transformers connected in parallel must have their tapings interlock system that should be active only in parallel operation. The interlock prevents different tap settings on the parallel transformers from giving rise to an excessive reactive current, which could damage the transformers. At present, the system is not fully automated as an operator attendance is needed for any tripping of master transformer or when its termination from master operation becomes necessary for any reason.

The nature of the above problem suggests that an Artificial Neural Network (ANN) model can be developed to control the power sharing and supply voltage of transformers running in parallel. Measuring quantities from the instrument transformers and transducers can be processed and fed into the neural network. The neural network can be trained to adjust the voltage level and tapings to minimize reactive current of paralleled transformers. In this paper, we present a concise description of the key ideas and the problem. We also present some experimental results supporting our ideas.

2   Reactive Load Sharing of Parallel Transformers

Transformers parallel operation with different voltages between taps results a circulating current accrued to unequal sharing of reactive power [6]. The circulating currents are dangerous for the transformer. The conventional way to keep circulating current minimum in transformers parallel operation imposes some restrictions that ensures i) exactly the same input bus voltage ii) all the transformers and its tap changers should be identical and operate at the same tap positions. Transformer tap changer control using reactive power sharing technique can relax these limitations. Moreover, it will simplify the system. Single line diagram of two transformers operating in parallel is shown in Fig.1.

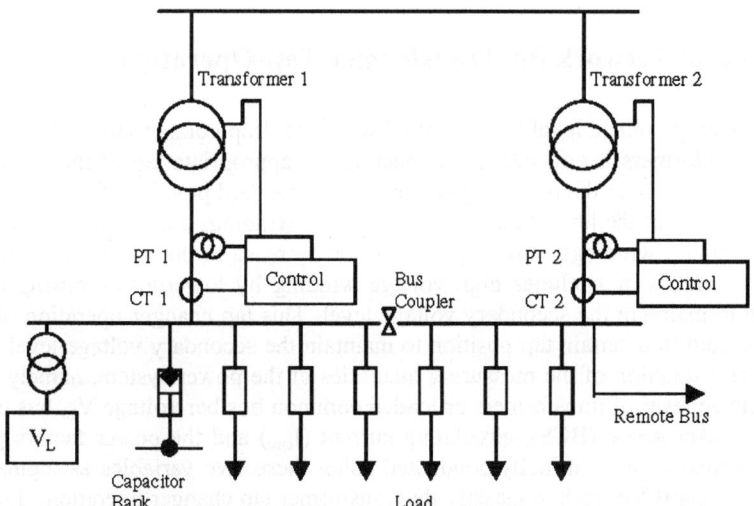

Fig. 1. Single line diagram of two transformers operating in parallel

Figure 2 shows the load current (I_{load}), the secondary currents (I_{ct1} and I_{ct2}) measured by the current transformer CT1 of transformer 1, current transformer CT2 of transformer 2 respectively. The circulating current (I_{circ}) shown flowing in the two transformers are the results of unequal sharing of reactive load by the transformers.

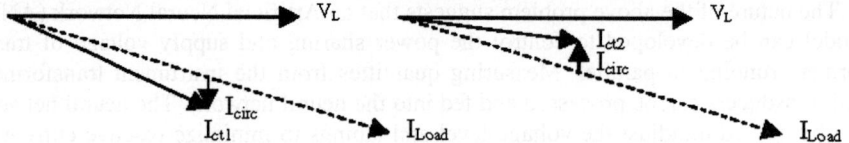

Fig. 2. Current sharing diagram of parallel Transformers

Power factors at which the two transformers are operating can be measured from the individual output currents of the transformers and the bus voltage. Network power factor (pf_n) can be obtained from the summation of all the output currents of the feeders and bus voltage or alternately from the total MVA and MW output.

$$pf_n = \frac{MW}{MVA} \tag{1}$$

Current I_{ct1} lags I_{load} indicating that it takes more reactive power whereas current I_{ct2} leads I_{load} indicating that it takes less reactive power. Therefore, transformer tap changer can be controlled comparing the individual transformer's output power factor (pf_t) with the network power factor (pf_n) at the desired voltage level (V_L). When each transformer's pf_t matches with the network pf_n the circulation current will be minimized and the transformers will operate safely.

3 Neural Network for Transformer Tap-Operation

The secondary voltage level (V_L) justified with line drop compensation (LDC) of the power transformers is adjusted by connecting the appropriate tap of the transformer winding according to the incoming voltage level. Standard practice of the tap-position numbering is that the lowest number connects full windings and highest number connects minimum windings. So, any increase or decrease in the incoming voltage tap changer includes or excludes high voltage winding by lowering or raising the tap position to maintain the secondary voltage level. This tap changer operation of raise, lower or hold in a certain tap position to maintain the secondary voltage level can be treated as a function of the measuring quantities of the power system, namely secondary voltage level of transformers or loaded common bus bar voltage V_L, bus coupler circuit breaker status (BCS), circulating current (I_{circ}) and the power factors pf_t and pf_n. Experiments were initially conducted using these five variables as inputs to an Artificial Neural Network to classify the transformer tap changer operation. Later the same experiments were conducted with four variables where pf_t and pf_n were excluded and the ratio of them (pf_r) defined as $pf_r = pf_t/pf_n$ is used as the fourth variable.

A transformer tap changer operation described above is categorized as a classification problem. In neural net literature, many models have been proposed for classifying inputs [1-3]. Performance of the models depends on many different factors including complexity and size of the problem, architecture and learning algorithm [4].

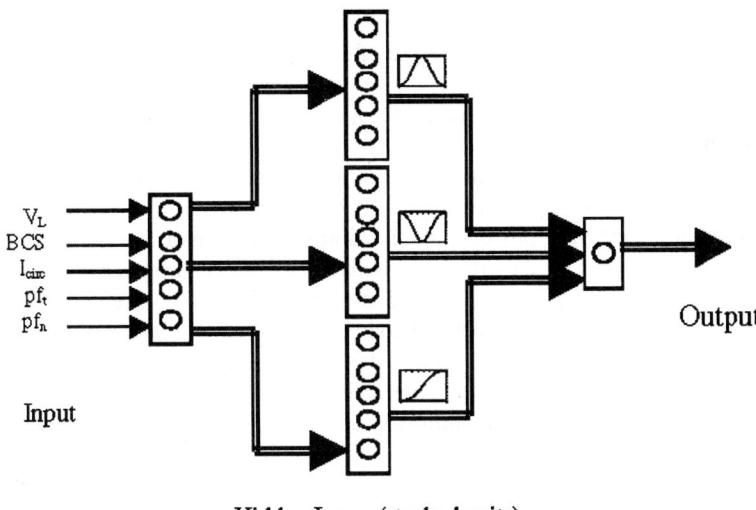

Fig. 3. ANN architecture for transformer tap-operation classifications

The network architecture used for this problem consists of 15 hidden neurons in three groups. The network is trained by backpropagation learning algorithm [2, 5]. Each of the groups is applied different activation functions (Gaussian, Gaussian complement and hyperbolic tangent) as shown in Fig. 3. Different activation functions are applied to the hidden units to allow them to detect different features in a data pattern processed through the network. Experiments with this particular problem show that use of these three activation functions with the architecture depicted in Fig. 3 produces better results instead of using the same activation function for all hidden units. Input and output units are activated with linear and logistic activations respectively.

4 Input Data for Transformer Operation

Sets of data related to sequence of tap changer operation of the transformers are taken from a range of real and reactive power variations of a substation supplying a local area. The load pattern varies tremendously between summer and winter. Power factor, in summer, is poorer than that in winter as summer load takes air-conditioning and industrial load that has more reactive components. The first input (V_L) is taken from the substation operating through out the year at the desired operating voltage level. Voltage ranges above and below the desired voltage levels are also included in that data set. These two voltage ranges contribute to the undesired operating voltage levels and hence included in the data set for training and testing purposes. Power factors of transformers for various dissimilar tap-changers positions are calculated using the sample data from the substation taking above described load pattern.

Input variable component BCS is a discrete variable (supervisory status), which has only two values 1 or 0 (close or open). Like BCS circulating current (I_{circ}) can be considered as a variable of supervisory status i.e.

$$I_{circ} = 0 \text{ absence of circulating current}$$
$$= 1 \text{ presence of circulating current}$$

The classification levels or the sequence of tap-changer operations are defined according to the input vectors.

The total data set is divided into two main groups, coarse distributed and fine distributed. Each of these groups covers the full problem domain of the two transformer parallel operation supplying power in the local area. Coarse and fine distributed input vectors are distinguished by the difference between transformer power factor (pf_t) and network power factor (pf_n). This difference is much higher for coarse distribution and very low for fine distribution. A reduced coarse distributed data set is picked up from the main data set ensuring that it covers the problem domain. The objective is to check the generalization capability of the trained net using the reduced coarse distributed data and testing with the remaining data set.

5 Network Performance

Figure 4 shows the artificial neural network output values against the target values (-1, 0, 1) for the five input variables, where (a) displays responses for the smaller coarse training data and (b) displays results for fine training data. Target values -1, 0, 1 represents the transformer tap-operation such as tap raises, on hold or lowers respectively. The ANN output consists of a range of values against each of the transformer tap-operation classes. These ranges of the ANN output values related to tap-operation classes overlap between each other. This overlapping contributes to incorrect tap-operation and is higher in the case of coarsely distributed training data set and lower for randomly chosen training data set. The ANN output values are interpreted as follows:

$$\text{If network output value } > 0.5 \quad \text{tap position lowers}$$
$$< -0.5 \quad \text{tap position rises,}$$
$$\text{otherwise} \quad \text{tap on hold}$$

This interpretation, in fact, provides a partition between the correct and erroneous responses of the ANN outputs for each of the transformer tap-operation classes.

Figure 5 shows the ANN output values against target values when the same experiment was conducted with four input variables. In this case the ANN output ranges related to tap-operation classes produces shorter ranges and smaller overlapping between classes, especially when randomly chosen training data set is used. It gives more correct responses, for similar interpretation, than the previous case of using five variables. Incorrect responses for differently chosen training data are summarized in Tables 1 and 2.

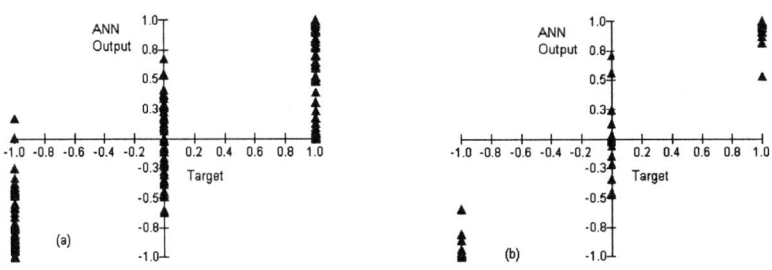

Fig. 4. Scatter graph for five variable training data (a) reduced coarse, (b) random

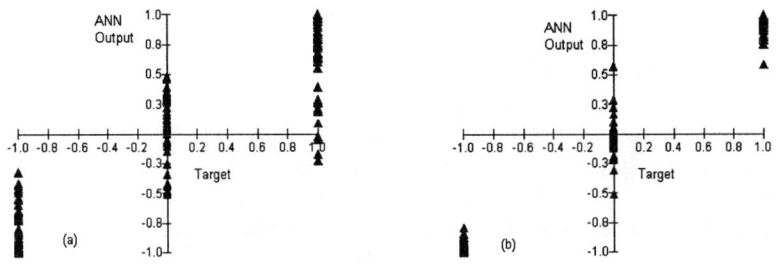

Fig. 5. Scatter graph for four variable training data (a) reduced coarse (b) random

Table 1. Neural network responses with five variables

Data distribution Training			Data distribution Testing			False responses			Correlation Coefficient
Coarse	Fine	Random	Coarse	Fine	Random	Train data	Test data	Total	
63			105	105		4	34	38	0.9216
168				105		3	29	32	0.9466
	168		105			0	21	21	0.9642
		168			105	1	1	2	0.9945

Table 2. Neural network responses with four variables

Data distribution Training			Data distribution Testing			False responses			Correlation Coefficient
Coarse	Fine	Random	Coarse	Fine	Random	Train data	Test data	Total	
63			105	105		2	21	23	0.9397
168				105		0	23	23	0.9749
	168		105			0	6	6	0.9725
		168			105	2	0	2	0.9936

6 Conclusion

This paper explores the possibility of using artificial neural network to control parallel operation of two power transformers based on the principle of minimizing circulating currents between the transformers. A neural net model has been developed using backpropagation learning rule. The input variables to this model that give a better generalization ability of the network and hence higher accuracy in controlling the operation have been identified. Initial experiments show promising results.

7 Future Research Direction

A friendly use of the neural network for working people will be to train the network with specified coarse distributed input values as they set the related conventional relays. Development of a robust network that could produce correct decision even with this limited data set would be of very much practical use and is one of the foremost goals of this research. The network would also have the capability to co-operate with other voltage regulating agents used in the system [7]. Other alternative approaches like Fuzzy rule base controller and neuro-fuzzy controller will be explored. Future research will also be aimed to analyze its constructive and practicable features along with verifications to identify any tractable mathematical properties.

References

1. S. Fahlman and C. Lebiere, The Cascade-Correlation Learning Architecture. Advances in Neural Information Processing Systems (D. Touretzky, Ed.), Vol 2, Morgan Kaufman, San Mateo, CA (1990) 524-532.
2. L. Fauset, Fundamentals of Neural Networks. Prentice Hall, Englewood Cliffs, NJ, (1994).
3. M. Lehtokangas, Modelling with constructive backpropagation, Neural Networks, Vol. 12 (1999) 707-716.
4. D. Popovic, D. Kukolj and F. Kulic, Monitoring and assessment of voltage stability margins using artificial neural networks with a reduced input set. IEE Proc., General Transmission and Distribution, Vol. 145 (4) (July 1998) 355-362.
5. D. E. Rumelhart et al, Parallel Distributed Processing, Vol 1. MIT Press (1986).
6. V. P. Thornley and N. J. Hiscock, Improved Voltage Quality through Advances In Voltage Control Technique. VA Tech Reyrolle ACP Ltd U K, Fundamental Limited UK.
7. H. F. Wang, Multi-agent co-ordination for the secondary voltage control in power-system contingencies. IEE Proc., General Transmission and Distribution, Vol. 148 (1) (January 2001) 61-66.

Author Index

Ahmad, H. 435
Albermani, F. 119
Angus, D. 618
Anson, M. 393
Ashfield, B. 146

Back, B. 557
Bayes, T. 232
Beetstra, J. 129
Belward, J. 514
Bi, J. D. 557
Bin, C. 780
Bond, W. P. 567
Braendler, D. 190
Bratko, I. 812
Briscoe, G. 314
Burrage, K. 514

Chachoua, M. 680
Chalidabhongse, J. 356
Chang, E. 303
Chau, K. W. 119, 393, 670
Chen, P. 514
Chen, S.-F. 325
Cheng, C. 670
Chien, S.-I. 17
Chinnan, W. 356
Choi, B. 450
Choi, S. 736
Clay, P. 232
Crispin, A. 232

Daoudi, K. 253
Darwen, P. J. 525
Davidsson, P. 801
Debenham, J. 335
Debnath, R. 293
Deugo, D. 146, 790
Dillon, T. 303
Dong, H. 284
Dunn, S. 220

Emch, M. 628

Farias, O. L. M. de 638
Feng, Y. 284
Ferrari, A. B. 77
Finzi, A. E. 596
Fusaoka, A. 758

Galitsky, B. 482
Gang, Z. 780
Goh, A. 346

Ha, N. D. H. 47
Hackney, R. 232
Han, F. 284
Hanan, J. 514
Hannessen, D. 129
Heeren, F. 660
Hendtlass, T. 190, 211, 241, 502, 618
Henry, D. S. 264
Hesselink, H. 129
Hewahi, N. M. 435
Hiratsuka, S. 758
Hirayama, J. 136
Hu, C.-K. 1

Irrgang, R. 607
Islam, F. 824
Ito, T. 460, 547
Itsuya, M. 780

Jiang, Y. 325
Jung, S.-W. 493, 715

Kamruzzaman, J. 824
Kang, M.-y. 692
Kiem, H. 47
Kim, H.-J. 67, 715
Kim, I.-C. 17
Kim, J. 736
Kim, J.-B. 67, 715
Kim, K.-H, ... 57
Kitano, H. 725

Author Index

Kitano, H. 725
Kitterer, H. 200
Kiwprasopsak, S. 536
Kravis, S. 607
Ku, S. 363
Kwon, D. 586
Kwon, H.-c. 493, 692

Laumanns, M. 200
Laumanns, N. 200
Lavagna, M. 596
Lee, B.-H. 57
Lee, B. 363
Lee, D. 363, 403
Lee, J. 403, 736
Lee, K. H. 403
Lee, S. 586
Lee, S. 586
Levinski, O. 9
Li, C. W. 670
Li, Y. S. 670
Lim, S. H. 470
Ling, T. C. 470
Loh, G. H. 264

Matsuo, T. 460
Mayer, W. 746
Mendes, S. B. T. 638
Michalopoulos, D. 1
Mitra, D. 567
Mitrpanont, J. L. 425, 536
Monostori, L. 157, 649
Mouhoub, M. 575
Mourelle, L. de M. 88, 768
Murray, G. 211
Murtagh, N. 275

Nakadai, K. 725
Nam, J. 57
Natarajan, V. 346
Nath, B. 824
Nedjah, N. 88, 768
Nguyen, A. H. 47
Nyakoe, G. N. 179

Ohki, M. 179
Ohkita, M. 179

Okuno, H. G. 725
Omlin, C. 36
Oppacher, F. 146

Pacholczyk, D. 680
Pagurek, B. 790
Park, S.-h. 692
Park, T.-W. 493
Peucker, S. 220
Phang, K. K. 470
Phoha, V. V. 450
Plengpung, T. 425

Quan, V. H. 47

Ramos, C. 414
Randall, M. 168
Rauber, T. W. 383
Reedman, D. 232
Room, P. 514
Rossetto, S. 383
Runqiang, B. 514

Saba, W. S. 373
Salami, M. 241, 502
Salim, M. D. 628
Santos, J. 414
Shin, J. 703
Sihn, W. 660
Skliarova, I. 77
Sklyarov, V. 108
Strauss, T. 628
Stumptner, M. 746
Šuc, D. 812
Sung, K.-H. 493

Tabuchi, S. 179
Takahashi, H. 293
Takeda, F. 136
Tantisirithanakorn, A. 356
Taylor, G. 232
Thawonmas, R. 136
Tian, W. 450
Tin, L. T. 47
Trung, P. N. 47

Vale, Z. 414
Varejão, F. M. 383

Viharos, Zs. J. 649
Vincze, T. 649
Vollebregt, A. 129

Wai, O. 670
Walsh, D. 36
Wang, D. 303
Wechasaethnon, P. 356
Wernstedt, F. 801
White, T. 146, 790
Won, Y. 57
Wotawa, F. 746

Xia, M. 780
Yaacob, M. Hj. 470
Yim, J. 25
Yin, X.-R. 325
Yip, J. 575
Yoon, A.-s. 692
Yu, X. 284

Zhang, H. 557
Zhou, Z.-H. 325
Zitar, R. A. 99
Zudor, E. I. 157

Lecture Notes in Artificial Intelligence (LNAI)

Vol. 2155: H. Bunt, R.-J. Beun (Eds.), Cooperative Multimodal Communication. Proceedings, 1998. VIII, 251 pages. 2001.

Vol. 2157: C. Rouveirol, M. Sebag (Eds.), Inductive Logic Programming. Proceedings, 2001. X, 261 pages. 2001.

Vol. 2159: J. Kelemen, P. Sosík (Eds.), Advances in Artificial Life. Proceedings, 2001. XIX, 724 pages. 2001.

Vol. 2160: R. Lu, S. Zhang, Automatic Generation of Computer Animation. XI, 380 pages. 2002.

Vol. 2166: V. Matoušek, P. Mautner, R. Mouček, K. Taušer (Eds.), Text, Speech and Dialogue. Proceedings, 2001. XIII, 542 pages. 2001.

Vol. 2167: L. De Raedt, P. Flach (Eds.), Machine Learning: ECML 2001. Proceedings, 2001. XVII, 618 pages. 2001.

Vol. 2168: L. De Raedt, A. Siebes (Eds.), Principles of Data Mining and Knowledge Discovery. Proceedings, 2001. XVII, 510 pages. 2001.

Vol. 2173: T. Eiter, W. Faber, M. Truszczynski (Eds.), Logic Programming and Nonmonotonic Reasoning. Proceedings, 2001. XI, 444 pages. 2001.

Vol. 2174: F. Baader, G. Brewka, T. Eiter (Eds.), KI 2001: Advances in Artificial Intelligence. Proceedings, 2001. XIII, 471 pages. 2001.

Vol. 2175: F. Esposito (Ed.), AI*IA 2001: Advances in Artificial Intelligence. Proceedings, 2001. XII, 396 pages. 2001.

Vol. 2182: M. Klusch, F. Zambonelli (Eds.), Cooperative Information Agents V. Proceedings, 2001. XII, 288 pages. 2001.

Vol. 2190: A. de Antonio, R. Aylett, D. Ballin (Eds.), Intelligent Virtual Agents. Proceedings, 2001. VIII, 245 pages. 2001.

Vol. 2198: N. Zhong, Y. Yao, J. Liu, S. Ohsuga (Eds.), Web Intelligence: Research and Development. Proceedings, 2001. XVI, 615 pages. 2001.

Vol. 2203: A. Omicini, P. Petta, R. Tolksdorf (Eds.), Engineering Societies in the Agents World II. Proceedings, 2001. XI, 195 pages. 2001.

Vol. 2225: N. Abe, R. Khardon, T. Zeugmann (Eds.), Algorithmic Learning Theory. Proceedings, 2001. XI, 379 pages. 2001.

Vol. 2226: K.P. Jantke, A. Shinohara (Eds.), Discovery Science. Proceedings, 2001. XII, 494 pages. 2001.

Vol. 2246: R. Falcone, M. Singh, Y.-H. Tan (Eds.), Trust in Cyber-societies. VIII, 195 pages. 2001.

Vol. 2250: R. Nieuwenhuis, A. Voronkov (Eds.), Logic for Programming, Artificial Intelligence, and Reasoning. Proceedings, 2001. XV, 738 pages. 2001.

Vol. 2253: T. Terano, T. Nishida, A. Namatame, S. Tsumoto, Y. Ohsawa, T. Washio (Eds.), New Frontiers in Artificial Intelligence. Proceedings, 2001. XXVII, 553 pages. 2001.

Vol. 2256: M. Stumptner, D. Corbett, M. Brooks (Eds.), AI 2001: Advances in Artificial Intelligence. Proceedings, 2001. XII, 666 pages. 2001.

Vol. 2258: P. Brazdil, A. Jorge (Eds.), Progress in Artificial Intelligence. Proceedings, 2001. XII, 418 pages. 2001.

Vol. 2275: N.R. Pal, M. Sugeno (Eds.), Advances in Soft Computing – AFSS 2002. Proceedings, 2002. XVI, 536 pages. 2002.

Vol. 2281: S. Arikawa, A. Shinohara (Eds.), Progress in Discovery Science. XIV, 684 pages. 2002.

Vol. 2293: J. Renz, Qualitative Spatial Reasoning with Topological Information. XVI, 207 pages. 2002.

Vol. 2296: B. Dunin-Kęplicz, E. Nawarecki (Eds.), From Theory to Practice in Multi-Agent Systems. Proceedings, 2001. IX, 341 pages. 2002.

Vol. 2298: I. Wachsmuth, T. Sowa (Eds.), Gesture and Language in Human-Computer Interaction. Proceedings, 2001. XI, 323 pages.

Vol. 2302: C. Schulte, Programming Constraint Services. XII, 176 pages. 2002.

Vol. 2307: C. Zhang, S. Zhang, Association Rule Mining. XII, 238 pages. 2002.

Vol. 2308: I.P. Vlahavas, C.D. Spyropoulos (Eds.), Methods and Applications of Artificial Intelligence. Proceedings, 2002. XIV, 514 pages. 2002.

Vol. 2309: A. Armando (Ed.), Frontiers of Combining Systems. Proceedings, 2002. VIII, 255 pages. 2002.

Vol. 2313: C.A. Coello Coello, A. de Albornoz, L.E. Sucar, O.Cairó Battistutti (Eds.), MICAI 2002: Advances in Artificial Intelligence. Proceedings, 2002. XIII, 548 pages. 2002.

Vol. 2317: M. Hegarty, B. Meyer, N. Hari Narayanan (Eds.), Diagrammatic Representation and Inference. Proceedings, 2002. XIV, 362 pages. 2002.

Vol. 2322: V. Mařík, O. Štěpánková, H. Krautwurmová, M. Luck (Eds.), Multi-Agent Systems and Applications II. Proceedings, 2001. XII, 377 pages. 2002.

Vol. 2336: M.-S. Chen, P.S. Yu, B. Liu (Eds.), Advances in Knowledge Discovery and Data Mining. Proceedings, 2002. XIII, 568 pages. 2002.

Vol. 2338: R. Cohen, B. Spencer (Eds.), Advances in Artificial Intelligence. Proceedings, 2002. X, 197 pages. 2002.

Vol. 2358: T. Hendtlass, M. Ali (Eds.), Developments in Applied Artificial Intelligence. Proceedings, 2002 XIII, 833 pages. 2002.

Lecture Notes in Computer Science

Vol. 2318: D. Bošnački, S. Leue (Eds.), Model Checking Software. Proceedings, 2002. X, 259 pages. 2002.

Vol. 2319: C. Gacek (Ed.), Software Reuse: Methods, Techniques, and Tools. Proceedings, 2002. XI, 353 pages. 2002.

Vol.2320: T. Sander (Ed.), Security and Privacy in Digital Rights Management. Proceedings, 2001. X, 245 pages. 2002.

Vol. 2322: V. Mařík, O. Stěpánková, H. Krautwurmová, M. Luck (Eds.), Multi-Agent Systems and Applications II. Proceedings, 2001. XII, 377 pages. 2002. (Subseries LNAI).

Vol. 2323: À. Frohner (Ed.), Object-Oriented Technology. Proceedings, 2001. IX, 225 pages. 2002.

Vol. 2324: T. Field, P.G. Harrison, J. Bradley, U. Harder (Eds.), Computer Performance Evaluation. Proceedings, 2002. XI, 349 pages. 2002.

Vol 2326: D. Grigoras, A. Nicolau, B. Toursel, B. Folliot (Eds.), Advanced Environments, Tools, and Applications for Cluster Computing. Proceedings, 2001. XIII, 321 pages. 2002.

Vol. 2327: H.P. Zima, K. Joe, M. Sato, Y. Seo, M. Shimasaki (Eds.), High Performance Computing. Proceedings, 2002. XV, 564 pages. 2002.

Vol. 2329: P.M.A. Sloot, C.J.K. Tan, J.J. Dongarra, A.G. Hoekstra (Eds.), Computational Science – ICCS 2002. Proceedings, Part I. XLI, 1095 pages. 2002.

Vol. 2330: P.M.A. Sloot, C.J.K. Tan, J.J. Dongarra, A.G. Hoekstra (Eds.), Computational Science – ICCS 2002. Proceedings, Part II. XLI, 1115 pages. 2002.

Vol. 2331: P.M.A. Sloot, C.J.K. Tan, J.J. Dongarra, A.G. Hoekstra (Eds.), Computational Science – ICCS 2002. Proceedings, Part III. XLI, 1227 pages. 2002.

Vol. 2332: L. Knudsen (Ed.), Advances in Cryptology – EUROCRYPT 2002. Proceedings, 2002. XII, 547 pages. 2002.

Vol. 2334: G. Carle, M. Zitterbart (Eds.), Protocols for High Speed Networks. Proceedings, 2002. X, 267 pages. 2002.

Vol. 2335: M. Butler, L. Petre, K. Sere (Eds.), Integrated Formal Methods. Proceedings, 2002. X, 401 pages. 2002.

Vol. 2336: M.-S. Chen, P.S. Yu, B. Liu (Eds.), Advances in Knowledge Discovery and Data Mining. Proceedings, 2002. XIII, 568 pages. 2002. (Subseries LNAI).

Vol. 2337: W.J. Cook, A.S. Schulz (Eds.), Integer Programming and Combinatorial Optimization. Proceedings, 2002. XI, 487 pages. 2002.

Vol. 2338: R. Cohen, B. Spencer (Eds.), Advances in Artificial Intelligence. Proceedings, 2002. X, 197 pages. 2002. (Subseries LNAI).

Vol. 2340: N. Jonoska, N.C. Seeman (Eds.), DNA Computing. Proceedings, 2001. XI, 392 pages. 2002.

Vol. 2342: I. Horrocks, J. Hendler (Eds.), The Semantic Web – ISCW 2002. Proceedings, 2002. XVI, 476 pages. 2002.

Vol. 2345: E. Gregori, M. Conti, A.T. Campbell, G. Omidyar, M. Zukerman (Eds.), NETWORKING 2002. Proceedings, 2002. XXVI, 1256 pages. 2002.

Vol. 2346: H. Unger, T., Böhme, A. Mikler (Eds.), Innovative Internet Computing Systems. Proceedings, 2002. VIII, 251 pages. 2002.

Vol. 2347: P. De Bra, P. Brusilovsky, R. Conejo (Eds.), Adaptive Hypermedia and Adaptive Web-Based Systems. Proceedings, 2002. XV, 615 pages. 2002.

Vol. 2348: A. Banks Pidduck, J. Mylopoulos, C.C. Woo, M. Tamer Ozsu (Eds.), Advanced Information Systems Engineering. Proceedings, 2002. XIV, 799 pages. 2002.

Vol. 2349: J. Kontio, R. Conradi (Eds.), Software Quality – ECSQ 2002. Proceedings, 2002. XIV, 363 pages. 2002.

Vol. 2350: A. Heyden, G. Sparr, M. Nielsen, P. Johansen (Eds.), Computer Vision – ECCV 2002. Proceedings, Part I. XXVIII, 817 pages. 2002.

Vol. 2351: A. Heyden, G. Sparr, M. Nielsen, P. Johansen (Eds.), Computer Vision – ECCV 2002. Proceedings, Part II. XXVIII, 903 pages. 2002.

Vol. 2352: A. Heyden, G. Sparr, M. Nielsen, P. Johansen (Eds.), Computer Vision – ECCV 2002. Proceedings, Part III. XXVIII, 919 pages. 2002.

Vol. 2353: A. Heyden, G. Sparr, M. Nielsen, P. Johansen (Eds.), Computer Vision – ECCV 2002. Proceedings, Part IV. XXVIII, 841 pages. 2002.

Vol. 2358: T. Hendtlass, M. Ali (Eds.), Developments in Applied Artificial Intelligence. Proceedings, 2002 XIII, 833 pages. 2002. (Subseries LNAI).

Vol. 2359: M. Tistarelli, J. Bigun, A.K. Jain (Eds.), Biometric Authentication. Proceedings, 2002. XII, 373 pages. 2002.

Vol. 2361: J. Blieberger, A. Strohmeier (Eds.), Reliable Software Technologies – Ada-Europe 2002. Proceedings, 2002 XIII, 367 pages. 2002.

Vol. 2363: S.A. Cerri, G. Gouardères, F. Paraguaçu (Eds.), Intelligent Tutoring Systems. Proceedings, 2002. XXVIII, 1016 pages. 2002.

Vol. 2367: J. Fagerholm, J. Haataja, J. Järvinen, M. Lyly. P. Råback, V. Savolainen (Eds.), Applied Parallel Computing. Proceedings, 2002. XIV, 612 pages. 2002.

Vol. 2374: B. Magnusson (Ed.), ECOOP 2002 – Object-Oriented Programming. XI, 637 pages. 2002.